U0291353

国家自然科学基金资助项目

内蒙古藏传佛教建筑形态演变研究 (50768007)

The Research on the Evolution of the Form of Inner Mongolia Tibetan Buddhism Architecture (50768007)

漠南蒙古地域藏传佛教召庙建筑的比较及其探源研究 (51168032)

The Research on the Comparison and Origin of Tibetan Buddhism Temple Architecture in Monan Mongolia Region (51168032)

漠北蒙古地区藏传佛教建筑形态演变研究 (51768050)

Research of the Evolution of Tibetan Buddhism Architectural Style in Mobei Mongolia (51768050)

基于整体保护的内蒙古藏传佛教建筑遗产价值体系构建 (51668049)

Based on the integrated protection conservation of Tibetan Buddhism Architectural Heritage Value System Construction in Inner Mongolia (51668049)

内蒙古
召庙建筑

上册

INNER MONGOLIA TEMPLE ARCHITECTURE (VOL.1)

张鹏举 著

ZHANG PENGJU

中国建筑工业出版社

CHINA ARCHITECTURE & BUILDING PRESS

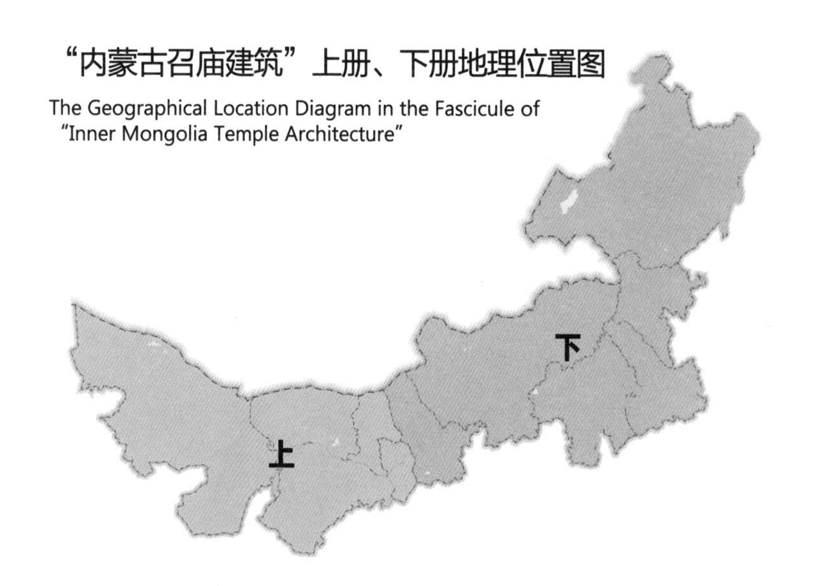

"内蒙古召庙建筑"上册、下册地理位置图

The Geographical Location Diagram in the Fascicule of "Inner Mongolia Temple Architecture"

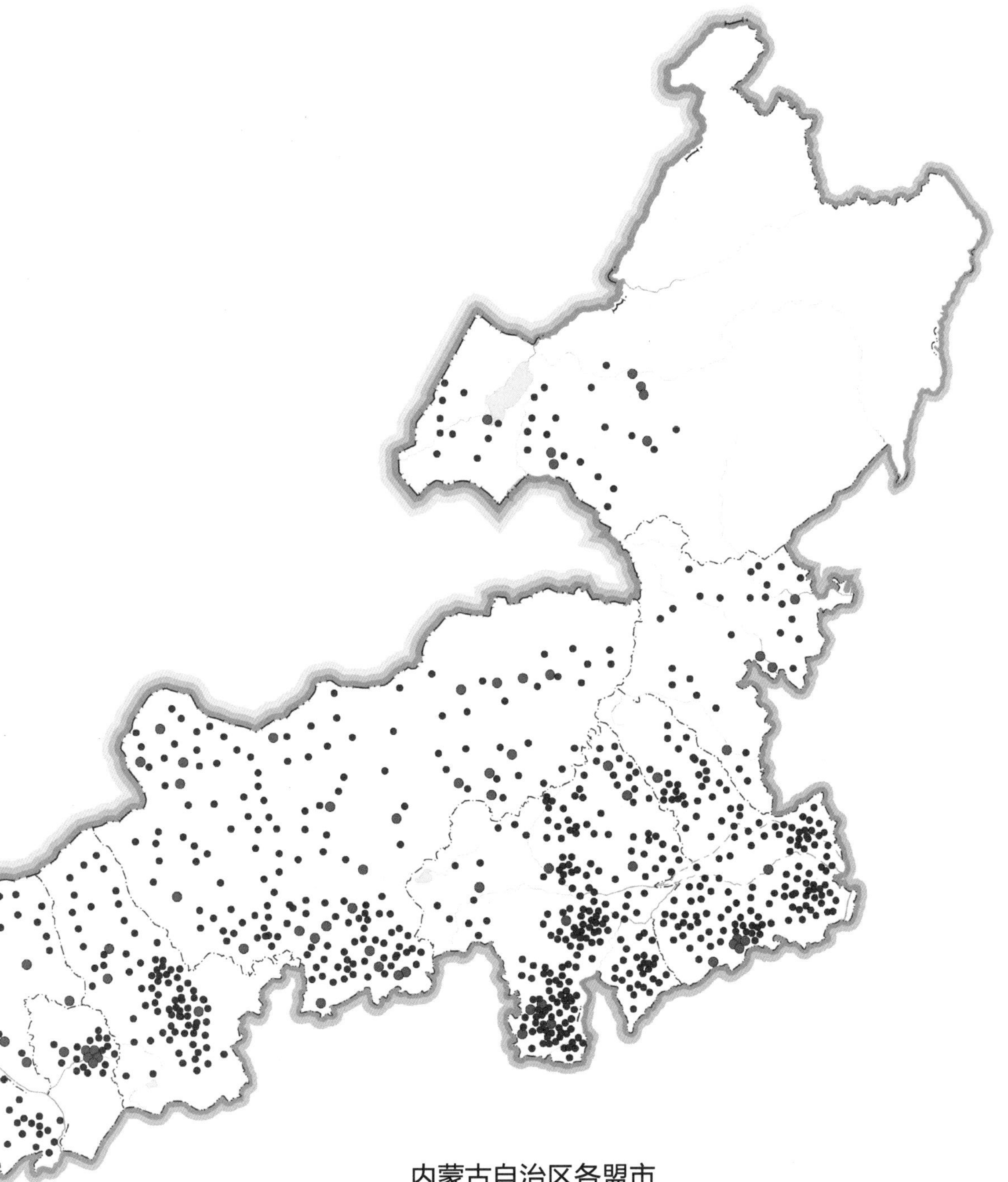

内蒙古自治区各盟市
中华人民共和国成立初期寺庙地理位置分布示意图

Prefectures and Cities of Inner Mongolia
Diagram of Temple Location and Distribution in the establishment of the People's Republic of China

图例：
Legend：

内蒙古地区现存召庙
The Existing Temples in Inner Mongolia Region

内蒙古地区曾有召庙
The Temples Ever Existed in Inner Mongolia Region

底图来源：内蒙古自治区自然资源厅官网 内蒙古地图
审图号：蒙S（2017）028号

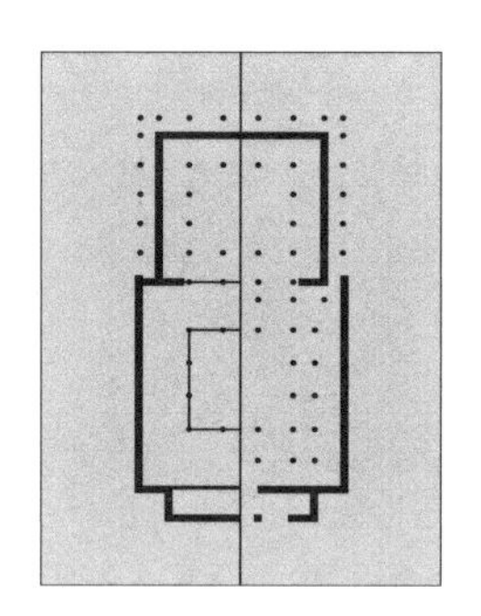

自明末藏传佛教以格鲁派再次传入漠南蒙古之后，作为其召庙的核心建筑，经堂和佛殿普遍采用一经堂一佛殿、前经堂后佛殿的布局形式，且重要召庙基本都采用二者相接的空间模式，形制为门廊→经堂→佛殿。常用于召庙内措钦大殿以及各个扎仓大殿。

Since Gelug of Tibetan Buddhism was introduced into the Monan Mongolia to late Ming dynasty again, the ceremony hall and the Buddha hall, the main motif of architecture in Tibetan Buddhism Gelug temples, conformed to the universal layout mode of one ceremony hall and one Buddha hall, namely, ceremony hall in the front and Buddha hall in the rear. The spatial mode of combination ceremony hall and Buddha hall were generally adopted in the important temples, which formed with portico, ceremony hall and Buddha hall, and it was commonly applied in Buddha hall of Oratory Palace and different Gyuto Palaces.

本书摘要

本书是关于内蒙古召庙建筑的学术专论和资料汇集，分上册、下册。

内容分为三部分：第一部分是综述，系统论述了内蒙古地域召庙建筑形态的影响因素、发展的历史分期以及一般的共性特征等，是本书的阅读背景；第二部分是召庙建筑的档案资料，对全区范围内重要历史遗存的召庙及其建筑进行了逻辑整理和系统归档，主要内容包括召庙简介、历史沿革、保存状况、建筑做法、技术档案、测绘图纸及现状照片等，是本书的主体内容；第三部分为相关附录，内容包括现存其他召庙的档案简表、不同历史时期召庙数量列表以及召庙不同名称的汉、蒙古、藏文对照表，是本书的补充内容。

本书对内蒙古地域召庙建筑的静态保护与动态发展具有参考价值，对发掘地域建筑文化和相关学术研究方面具有指导价值，同时，本书成果将成为此类项目后续研究的基础素材，适合相关专业科研人员、高校教师和学生使用。

Abstract

The book is an academic monograph and material collection of Inner Mongolia Temple architecture,divided into two volumes.

The book is composed of three parts. The first part is the summary which systematically discusses the factors affecting the form, historical stages and general features. It is the reading background of the book. The second part contains the file documents of the temples, which systematically and logically filed the important remaining historical temples in the region including temples introduction, history, preservation condition, construction practice, technological files, mapping and current photos. It is the main body of the book. The third part is related appendix including the file table of other extant temples, the numbers of temples in different historical stages and table of comparison of temple names. It is the supplement of the book.

The book has the reference value on the static protection and dynamic development of Inner Mongolia Tibetan Buddhist architecture and has the instructive value on exploring the regional architectural culture and the related academic research at the same time. Meanwhile the result of this book will be the basic materials of the following study of such project, which is applicable to the relevant professional research staff, college teachers and students.

序一

本书作者张鹏举同志曾是我的博士生，扎根内蒙古从事创作与研究已经三十多年了。其中，研究内蒙古的传统建筑是他长期坚持的一项重要工作，为编此书，张鹏举和他的团队花了大量的时间，投入了大量的精力，吃了不少苦。内蒙古这么大，从东到西走一遍都很不容易，更何况一座庙一座庙地去调研、测绘，有些地方，如沙漠、山区和草原深处更是道路不通，人迹罕至，因此，把全区现存的召庙做一次详细的调研整理，这是一件难能可贵的事情，他们的工作做得很扎实。这套书最有价值的应是召庙资料的档案集成。在我看来，这套书至少有以下三个方面的意义：

一、全书的内容不是简单的普查结果，而是基于文献和现状调研两个方面的资料归纳。全书把召庙建筑的相关资料信息按照一种逻辑进行了程式化的归档整理，为后人在这方面的继续研究提供了方便。

二、这个资料集成实际上是抢救了历史信息。在我国不少地区，尤其是少数民族地区，当地的人们一度缺乏对文物建筑的保护意识。近年，遇到经济大发展，建设性的破坏就自然十分严重，许多有价值的历史信息已经不可再生了，很让人痛心。从书中看到，内蒙古的情况也是如此，因而，建立这套召庙的建筑档案就相当于及时抢救了这些信息，这为日后的建筑文物保护工作奠定了很好的基础。

三、近年来，地域性的建筑创作少见优秀作品，其中的一个主要原因是创作者缺少文化根基和历史视野。这使得类似的大多数作品仍处于语言和符号的形式表层，他们对于前人的建造智慧和哲学思想等没有去认真挖掘和总结，尤其是一些年轻的地方建筑师更是表现得有些浮躁，只顾生产，不去积淀。从这一点上看，这套书的努力也是很有现实意义的。

内蒙古具有广阔的空间，2011年，我曾去内蒙古工业大学，初步领略到了张鹏举和他团队的精神面貌，祝愿他们在民族建筑研究和地域性建筑创作方面走出一条路来，取得好成绩。

彭一刚

彭一刚

中国科学院院士

Foreword I

The author of this book, Zhang Pengju, a former PhD student of mine, has been rooted in Inner Mongolia for more than thirty years in creation and research. He and his team have spent a lot of time, invested a lot of energy and suffered a lot to compile this book. Inner Mongolia is vast in territory, and some places such as deserts, mountains and prairies are impassable and uninhabited. Actually, it is uneasy to go from the east to the west. But they have dedicated their efforts to investigating and mapping temple by temple. Therefore, it is very commendable to do exhaustive survey and systematical list about the existing temples in the whole region. Obviously, their achievements are hard-won. The most valuable part of the book should be the archival integration of temple materials. In my opinion, this collection is significant in at least three ways:

First, the content of the book is not a simple survey result, but the material summary based on the two aspects of documents and the survey of current status. The books file the relevant information of the temples in logical stylization, which provides convenience for later research of future generations.

Second, these integrated materials actually save historical information. In some areas of China, especially the ethnic minority regions, the local people once were lack of the protection consciousness of the cultural architecture relics. In recent years, it is quite distressing that constructive destructions are suffered seriously by the economic development, and many valuable historical relics have not been reproduced. According to the books' Introduction, the similar situation is also happened in Inner Mongolia. Therefore, establishing a set of files of the temple buildings is equivalent timely to rescue these information, which has laid a good foundation for future protection work of the architectural relics.

Third, in recent years, the outstanding works in regional architectural creation are produced rarely. One of the main reasons is the creators are lack of cultural accumulation and historical perspective, which the most of similar works are still in the form surface of the architectural languages and symbols. They do not study and summarize carefully about the predecessors' constructive wisdom and philosophic methods. Especially some young local architects are somewhat impetuous, and they would rather produce than accumulate. From this point of view,t he efforts of creating the set of books are also practically significant.

Inner Mongolia has a vast space. In 2011, I visited the Inner Mongolia University of Technology and got a first glimpse of the spirit of Zhang Pengju and his team, wishing them well in their research on ethnic architecture and the creation of regional architecture.

Peng Yigang

Academician of the Chinese Academy of Sciences

序二

内蒙古召庙建筑是内蒙古地域深厚文化结构的一个组成部分，其历史遗存具有重要的社会学、人类学和建筑学意义。《内蒙古召庙建筑》是一套难得的完整收录内蒙古地区该类建筑相关历史文献和现状资料的专业书籍。此书翔实介绍、总结了内蒙古地区召庙建筑的历史分期、影响因素及其共性特征，同时对现存100余座召庙进行了系统立档：包括召庙简介、历史沿革、保存状况、建筑做法、测绘图纸和现状照片等。这套书所涵括的基础资料对此类建筑的静态保护和动态发展都具有重要现实意义。

据了解，目前学界对内蒙古召庙建筑的关注度并不高，如今，这些召庙建筑不少面临着数量递减，历史信息缺失，建设性破坏的危机，其历史遗存部分亟待整体性保护,其复原新建部分则迫切需要专业性研究，而这些都需要完备的基础研究工作作为支撑。基于此，此书的出版可以成为当前各项针对内蒙古召庙保护工作的现实基础。事实上，早有此认识的张鹏举及其团队，申请完成了多项国家自然科学基金项目，调研测绘、斟查文献，历时十多年方汇成此书，可谓工作扎实，且难能可贵。

身处当地的张鹏举，既是杰出的地域建筑师，又是地区建筑教育的领头人，同时致力于推动学科建设和发展。他和他的团队经过多年的努力，已经建构了内蒙古地域建筑学理论和实践框架。研究内蒙古地域传统建筑正是这个框架中文化历史部分的内容，而编撰《内蒙古召庙建筑》是这项系统工作的部分成果。尽管本书涉及交叉学科，但从成果看出，张鹏举及其团队具备了担当此项任务的学术视野和研究能力！

孟建民

孟建民

中国工程院院士

Foreword II

Inner Mongolian temple architecture is an integral part of the deep cultural fabric of the Inner Mongolian region, and its historical remains are of great sociological, anthropological and architectural significance. Inner Mongolia Temple Architecture is a rare and complete collection of historical documents and current information on this type of architecture in Inner Mongolia. The book provides a detailed introduction and summary of the historical phasing, influencing factors and common features of temple architecture in Inner Mongolia, as well as a systematic documentation of over 100 existing temples, including an introduction to the temples, their history, state of preservation, architectural practices, mapping drawings and photographs of their current state. The basic information contained in this collection is of great relevance to the static conservation and dynamic development of such buildings.

It is understood that the academic community has not paid much attention to the architecture of Inner Mongolia's temples. Today, many of these temples are facing a crisis of decreasing numbers, lack of historical information and constructive damage, and their historical remains need to be protected as a whole, while their restored and newly built parts are in urgent need of professional research, which needs to be supported by complete basic research work. Based on this, the publication of this book can serve as a practical basis for the current efforts to preserve the temple in Inner Mongolia. In fact, Zhang Pengju and his team, who have been aware of this for a long time, have applied for and completed a number of National Natural Science Foundation of China projects, researched and mapped, and studied the literature for more than ten years before turning this book into a solid and invaluable work.

Locally based, Zhang Pengju is both a distinguished regional architect and a leader in regional architectural education, as well as dedicated to promoting the building and development of the discipline. He and his team have, over the years, constructed a framework for the theory and practice of Inner Mongolian regional architecture. The study of traditional Inner Mongolian regional architecture is part of the cultural-historical component of this framework, and the compilation of Inner Mongolian Temple Architecture is part of the outcome of this systematic work. Despite the interdisciplinary nature of the book, the results show that Zhang Pengju and his team have the academic vision and research capacity to undertake this task!

Meng Jianmin
Academician of the Chinese Academy of Engineering

前言

内蒙古召庙建筑自16世纪以来伴随着藏传佛教在内蒙古地域的传播与发展逐渐形成了鲜明的地域特色，积淀为一种独有的历史文化遗产，成为内蒙古地域深层文化结构的重要组成部分。它的广泛创立与发展曾促进了草原游牧建筑文化向定居方向的转化。

纵观这段历史，学者们对于内蒙古地域藏传佛教文化的研究十分丰富，且语种繁多，但对此类建筑文化方面的探究却寥寥可数，且多集中于近现代，并以个体研究或图片展示为主要特点，如（清）葛尔丹旺楚克多尔济著、巴·孟和校注的《梅日更召创建史》；张驭寰、林北钟著的《内蒙古古建筑》；金峰整理注释的《呼和浩特召庙》；乔吉编著的《内蒙古寺庙》；（日本）长尾雅人著、白音朝鲁译的《蒙古学问寺》等。这些著作及文献均对本书的成稿具有重要的参考意义。

本书是国家自然科学基金资助项目《内蒙古藏传佛教建筑形态演变研究》（项目编号：50768007）、《漠南蒙古地域藏传佛教召庙建筑的比较及其探源研究》（项目编号：51168032）和《漠北蒙古地区藏传佛教建筑形态演变研究 》(项目编号：51768050)、《基于整体保护的内蒙古藏传佛教建筑遗产价值体系构建》(项目编号：51668049)的研究成果之一。

上述前两个项目是在内蒙古召庙建筑面临数量递减、历史信息缺失和建设性破坏等全面危机的情况下展开的。课题首先对内蒙古自治区现行区划范围内的研究对象进行了全面的调研和测绘，并对内蒙古周边地区的同类召庙进行了初步调研，在此基础上，完成了资料的系统归档，同时进行了建筑形态方面的相关研究。这项工作历时6年，在内蒙古自治区境内，调研了全部具有一定规模及研究价值的遗存召庙110座，并测绘了其中极具历史价值的召庙24座；后两个项目进一步扩大调研范围并向保护研究层面延伸。

本书即是在上述成果的基础上编著而成。为了便于阅读，全书按地理区域分为上、下两册：上册由内蒙古中西部地区（阿拉善盟、巴彦淖尔市、鄂尔多斯市、包头市、呼和浩特市）的23座重点召庙和28座其他召庙构成；下册由内蒙古中东部地区（锡林郭勒盟、乌兰察布市、赤峰市、通辽市、呼伦贝尔市、兴安盟）的33座重点召庙和26座其他召庙构成。由于内蒙古地域辽阔，召庙的数量大、分布广、路线长，加之重要文献多为蒙古文、藏文版本，增加了调查研究的难度。在课题历时的6年多时间里，虽竭尽全力，但仍感力不从心，且越深入越认识到其具有广阔的研究空间，因而本书只是阶段性的成果，谬纰之处，一定不少，权当引玉之砖，更望同仁共同参与。

感谢如下人员：

全文审稿的乔吉、乌云，民族学角度审稿的何生海，英文翻译的刘卓媛、杨丹宇，参与课题的韩瑛、杜娟、白丽燕、高旭、白雪、韩秀华、李国保、额尔德木图。

Preface

Since 16 century, Inner Mongolia Temple architecture gradually formed a distinct regional characteristics with the spread and development of Tibetan Buddhism in Inner Mongolia area , and accumulated as a unique historical and cultural heritage, which became an important part of Inner Mongolia cultural structure . The creation and development had promoted the transformation of grassland construction from nomadic style to settlement.

Throughout this period of history, scholars dedicated greatly to the research on Tibetan Buddhist Culture in Inner Mongolia Region in a wide variety of languages. Their products focused on the main characteristics of the individual studies and the picture shows in modern times with little exploration to architectural cultures, such as *"The Foundation History of Meirigeng Temple"* wrote by Gehr Dan Wangchuck Doll (from Qing Dynasty) and collated and annotated by Ba Menghe; *"Inner Mongolia Ancient Architecture"* wrote by Zhang Yuhuan and Ling Beizhong; *"Huhhot Temples"* collated and annotated by Jing Feng; *"Inner Mongolia Temples"* edited by Qiao Ji; *"Mongolia Study on Temples"* wrote by Nagao Masahito (from Japan) and translated by Baiyinchaolu etc. These works and documents are of great reference significance for this book.

The book is one of the research result of NSFC project "The Research on the Evolution of the Form of Inner Mongolia Tibetan Buddhism Architecture" (project number: 50768007) , "The Research on the Comparison and Origin of Tibetan Buddhism Temple Architecture in Monan Mongolia Region" (project number: 51168032),"Research of the Evolution of Tibetan Buddhism Architectural Style in Mobei Mongolia"(project number:51768050) and "Based on the integrated protection conservation of Tibetan Buddhism Architectural Heritage Value System Construction in Inner Mongolia"(project number:51668049)

The above two projects were conducted under the situation of overall crisis such as decreasing numbers, historical information deficiency , constructive destruction in Tibetan Buddhism buildings in Inner Mongolia . The project firstly conducted a comprehensive survey and mapping on the research objects and the temples in surrounding areas in current Inner Mongolia Autonomous Region and completed the file system of the materials on this basis , meanwhile , undertook the relevant research in architecture form . This work lasted for 6 years, and surveyed the extant 110 temples with certain scale and research value in Inner Mongolia Autonomous Region and mapped 24 temples of them with historical value; The latter two projects further expand the research scope and extend to the protection research level.

The book is edited on the basis of above results and it is divided into two volumes according to the geographical areas: the first volume contains 23 key temples and 28 other temples in western Inner Mongolia area (Alxa Prefecture, Bayannaoer City, Erdos City, Baotou City, Hohhot City); The second volume contains 33 key temples and 26 other temples in Eastern Inner Mongolia area (Xilingol Prefecture, Ulanqab City, Chifeng City, Tongliao City, Hulunbeir City, Hinggan Prefecture) .Because Inner Mongolia has vast territory, large number of temples with wide distribution and long route, and the important documents are mostly written in Mongolian and Tibetan, it is very difficult to conduct the survey and the research. In 6 years, although we try our best, but still feel unsatisfied, and aware of the wide research space of this field. So the book is only a phased achievement, hoping colPrefectures to participate in the further study.

My sincere thanks to:

Qiaoji and Wuyun for full text review, He Shenghai from the perspective of Ethnology, Liu Zhuoyuan and Yang Danyu for English translation, Han Ying, Dujuan, Bai Liyan, Gao Xu, Bai Xue, Han Xiuhua, Li Guobao, Erdemt for participating the project.

下面是具体参加调研的成员：

◆ 阿拉善盟地区：

张鹏举 高旭 李国保 宝山 付瑞峰 王志强 韩瑛 额尔德木图 白丽燕

◆ 巴彦淖尔地区：

韩瑛 李国保 宝山 韩秀华 付瑞峰 王辉胜 汤湛 弓志光 乔恩懋 秦格乐

◆ 包头地区：

张鹏举 白雪 韩瑛 白丽燕 高旭 韩秀华 李国保 杜娟 宝山 张宇 孟一军 王辉胜 庄苗 郝慧敏 董雪峰 汤湛 贺伟 田琳 托亚 薛剑 房宏伟 卢文娟 王小伟 艾力夫 卢小亮 冯旭 刘丁 马辰 王新 李诺 陈伟业 吴洁 杨彬 呼木吉乐 赵扬 赵宣 贾凌云 傅森 董萧迪 吕昱达 马辰 付瑞峰 董鹏 布音敖其尔 乔恩懋 庄健 庞磊 王志强 李源河 王璞 姬煜 何鑫

◆ 鄂尔多斯地区：

张鹏举 杜娟 宝山 苍雁飞 高亚涛 李国保 白丽燕 韩瑛 白雪 薛剑 王辉胜 托亚 青格乐 汤湛 刘磊 贺伟 庄苗 郝慧敏 董铁鑫 董雪峰 潘瑞 武月华 王红丽 程日启 石宝宝 贾金华 梅永发 王强 王兴家 马涛 杨润宾 王新 李永正 李伟 李欣楠 李源河 杨耀强 高维小 崔永在

◆ 呼和浩特地区：

张鹏举 白丽燕 萨日朗 李国保 宝山 韩秀华 韩瑛 杜娟 房宏伟 汤湛 贺伟 王辉胜 刘磊 白雪 田琳 薛剑 托亚 庄苗 董雪峰 郝慧敏 刘旭 武月华 王志强 潘瑞 赵扬 赵宣 王新 艾力夫 卢小亮 董萧迪 冯旭 刘丁 李诺 陈伟业 吴洁 杨彬 马辰 呼木吉乐 贾凌云 傅森 王红丽 王强 石宝宝 贾金华 梅永发 程日启 王兴家 杨润宾 马涛

◆ 乌兰察布地区：

张鹏举 韩秀华 额尔德木图 张宇 李国保 高旭 布音敖其尔 吕昱达 马辰 付瑞峰 董鹏 乔恩懋 庄健 呼木吉乐 庞磊 王志强 李源河 王璞 姬煜

◆ 锡林郭勒盟地区：

张鹏举 额尔德木图 韩瑛 白雪 贺龙 李国保 栗建元

◆ 赤峰地区：

张鹏举 白丽燕 李国保 宝山 白雪 杜娟 托亚 高旭 付瑞峰 乔恩懋 贺伟 房宏伟 苍雁飞 薛剑 张宇 卢小亮 刘洋 卢文娟 栗建元 艾力夫 何鑫 姬煜 姜楠 王英旭 韩傲 布音 乔恩懋 王志强 杨力吉 马辰 呼木吉乐

The specific research members are as follows:

◆ Alxa Prefecture:

Zhang Pengju,Gao Xu, Li Guobao, Bao Shan, Fu Ruifeng, Wang Zhiqiang, Han Ying, Erdemt, Bai Liyan

◆ Bayannur area:

Han Ying, Li Guobao, Bao Shan, Han Xiuhua, Fu Ruifeng, Wang Huisheng, Tang Zhan, Gong Zhiguang, Qiao Enmao, Qin Gele

◆ Baotou area:

Zhang Pengju, Bai Xue, Han Ying, Bai Liyan, Gao Xu, Han Xiuhua, Li Guobao, Du Juan, Bao Shan, Zhang Yu, Meng Yijun, Wang Huisheng, Zhuang Miao, Hao Huimin, Dong Xuefeng, Tang Zhan, He Wei, Tian Lin, Toyah, Xue Jian, Fang Hongwei, Lu Wenjuan, Wang Xiaowei, Alev, Lu Xiaoliang, Feng Xu, Liu Ding, Ma Chen, Wang Xin, Li Nuo, Chen Weiye, Wu Jie, Yang Bin, Humu Jill, Zhao Yang, Zhao Xuan, Jia Lingyun, Fu Sen, Dong Xiaodi, Lü Yuda, Ma Chen, Fu Ruifeng, Dong Peng, Buinoch, Qiao Enmao, Zhuang Jian, Pang Lei, Wang Zhiqiang, Li Yuanhe, Wang Pu, Ji Yu, He Xin

◆ Erdos area:

Zhang Pengju,Du Juan, Bao Shan, Cang Yanfei, Gao Yatao, Li Guobao, Bai Liyan, Han Ying, Bai Xue, Xue Jian, Wang Huisheng, Toyah, Qing Gele, Tang Zhan, Liu Lei, He Wei, Zhuang Miao, Hao Huimin, Dong Tiexin, Dong Xuefeng, Pan Rui, Wu Yuehua, Wang Hongli, Cheng Riqi, Shi Baobao, Jia Jinhua, Mei Yongfa, Wang Qiang, Wang Xingjia, Ma Tao, Yang Runbin,Wang Xin, Li Yongzheng, Li Wei,Li Xinnan, Li Yuanhe, Yang Yaoqiang, Gao Weixiao, Cui Yongzai

◆ Hohhot area:

Zhang Pengju, Bai Liyan, Sagiran,Li Guobao, Bao Shan, Han Xiuhua, Han Ying, Du Juan, Fang Hongwei, Tang Zhan, He Wei, Wang Huisheng, Liu Lei, Bai Xue, Tian Lin, Xue Jian, Toyah, Zhuang Miao, Dong Xuefeng, Hao Huimin, Liu Xu, Wu Yuehua, Wang Zhiqiang, Pan Rui, Zhao Yang, Zhao Xuan, Wang Xin, Alev, Lu Xiaoliang, Dong Xiaodi, Feng Xu, Liu Ding, Li Nuo, Chen Weiye, Wu Jie, Yang Bin, Ma Chen, Humu Jill, Jia Lingyun, Fu Sen, Wang Hongli, Wang Qiang, Shi Baobao, Jia Jinhua, Mei Yongfa, Cheng Riqi, Wang Xingjia, Yang Runbin, Ma Tao

◆ Ulaqab area:

Zhang Pengju, Han Xiuhua, Erdemt, Zhang Yu, Li Guobao, Gao Xu, Buinoch, Lü Yuda, Ma Chen, Fu Ruifeng, Dong Peng, Qiao Enmao, Zhuang Jian, Humu Jill, Pang Lei, Wang Zhiqiang, Li Yuanhe, Wang Pu, Ji Yu

◆ Xilingol Prefecture:

Zhang Pengju, Erdemt, Han Ying, Baixue, He Long, Li Guobao, Li Jianyuan

◆ Chifeng area:

◆ 通辽地区：

张鹏举 李国保 宝山 高旭 付瑞峰 乔恩懋 白丽燕 白雪 杜娟 托亚

◆ 兴安盟地区：

张鹏举 房宏伟 栗建元 宝山 额尔德木图 李国保 白丽燕

◆ 呼伦贝尔地区：

张鹏举 宝山 额尔德木图 白丽燕 房宏伟 栗建元 李国保

张鹏举

2019年10月28日于内蒙古工业大学

Zhang Pengju,Bai Liyan, Li Guobao, Bao Shan, Bai Xue, Du Juan,Toyah, Gao Xu, Fu Ruifeng, Qiao Enmao ,He Wei, Fang Hongwei, Cang Yanfei, Xue Jian, Zhang Yu, Lu Xiaoliang, Liu Yang, Lu Wenjuan, Li Jianyuan,Alev, He Xin, Ji Yu, Jiang Nan, Wang Yingxu, Han Ao, Bu Yin, Qiao Enmao, Wang Zhiqiang, Yang Liji, Ma Chen,Humu Jill

◆ Tongliao area:

Zhang Pengju, Li Guobao, Bao Shan, Gao Xu, Fu Ruifeng, Qiao Enmao, Bai Liyan, Bai Xue, Du Juan, Toyah

◆ Higgan Prefecture:

Zhang Pengju, Fang Hongwei, Li Jianyuan, Bao Shan, Erdemt, Li Guobao, Bai Liyan

◆ Hunlunbeir area:

Zhang Pengju, Bao Shan, Erdemt, Bai Liyan, Fang Hongwei, Li Jianyuan, Li Guobao

Zhang Pengju

Inner Mongolia University of Technology

October 28th, 2019

目录

第一部分

综述

Part One Summary

第一章 内蒙古地域现存召庙建筑概况

内蒙古自治区是我国召庙建筑分布最为广泛的地区之一。作为漠南蒙古历史地域的主体区域[1]，内蒙古自治区境内的召庙在建筑风格、创建模式等方面充分体现了其不同于漠北、漠西蒙古历史地域的特点。课题组[2]对今内蒙古自治区境内110座召庙[3]进行了普查调研、实地测绘，从而整理出自治区境内现存召庙的建筑信息，并尝试归纳出其地域性特征与空间分布规律，为后继科研工作打下了基础。

内蒙古召庙大多于明清时期建造，虽有元以及在此之前兴建的少量寺庙，但均于清朝时期改建或改宗为藏传佛教召庙。因而，历经数百年风雨留存下来的召庙古建筑多以明、清两朝古建筑为主，其中清朝遗迹居多数。自改革开放以来，尤其是近20年，随着中国社会环境的巨大变化，内蒙古地域的藏传佛教进入一个新的时期，召庙的建设也开始逐步活跃起来，各地对已拆毁或残存[4]的召庙进行了整体重建、部分新建以及修缮维新，也有少量召庙得以新建。

本章将在引用清朝、民国及中华人民共和国成立后编撰的各类文献资料的基础上，将已调研的内蒙古地域110座召庙按照西部、中部、东部地区进行归纳并将其名称、具体地点及现存主要建筑以表格的形式加以描述。

第一节 西部地区

一、阿拉善盟地区

阿拉善盟位于内蒙古自治区最西部。巴丹吉林、腾格里、乌兰布和三大沙漠横贯整个阿拉善地域，沙漠地貌占到全盟总面积的三分之一。由于地处亚洲大陆腹地，为内陆高原，远离海洋，周围群山环抱，这一地区为典型的大陆性气候，干旱少雨，风大沙多。

阿拉善盟地区由康熙年间设立的阿拉善厄鲁特旗与额济纳土尔扈特旗两旗组成，定牧于河套以西，不设盟，通常称作套西二旗[5]，1980年成立阿拉善盟。在清代，套西二旗共有召庙40座。[6]

阿拉善盟境内的召庙普遍呈现传播模式独特、体系完整清晰、古建筑保护较好等特点。阿拉善地区在内蒙古地域藏传佛教传播史上具有独特的地位。据学者考证，六世达赖喇嘛仓央嘉措于1716—1746年在阿拉善地区弘法建寺，并圆寂于门吉林庙。[7]南寺现仍有专供六世达赖喇嘛灵塔的殿宇。一些召庙的创建史亦与六世达赖有密切联系。[8]阿拉善和硕特旗寺庙由三大寺庙系统[9]、八大寺庙[10]及三大寺庙之属庙[11]组成。额济纳土尔扈特旗有3座寺庙，其中一座系民国年间由蒙古国移民主持新建。阿拉善盟召庙分布广泛，且多坐落于人烟稀少、交通不便的偏远地带，有的甚至位

1 漠南，或幕南一词最早出现于“汉书”，后历代沿用，至清代正式形成地理区域概念。历史上的漠南蒙古，除包括今内蒙古自治区行政辖区外另包括黑龙江省、吉林省、辽宁省、河北省、山西省等周边省份的部分地区。课题组在国家自然科学基金资助项目——《漠南蒙古地域藏传佛教召庙建筑的比较及其探源研究》（项目编号为51168032）中将对漠南蒙古地域内的藏传佛教召庙予以系统研究。

2 课题组，为完成国家自然科学基金资助项目——《内蒙古地区藏传佛教建筑形态演变的研究》（项目编号为50768007）而成立的以张鹏举教授带头的专门调查研究小组。

3 关于藏传佛教寺院名称，有寺院、寺庙、召庙等多种统称，蒙古语有苏莫sum、交（即召）jao、黑得hiid、呼热hure等不同称谓。课题组在调研过程中总结出，内蒙古西部地区多称“召”（如鄂尔多斯与呼和浩特地区），东部地区则多称“庙”sum（如锡林郭勒盟与东部各盟市），故选择“召庙”，以便反映出这一约定俗成的文化现象。课题组所调研的召庙可依据法事活动的延续现状分为有正常法事活动的召庙及无法事活动的召庙两大类，依据现有建筑形态可分为复建召庙（再可细分为原地复建召庙与异地复建召庙两类）、新建召庙及被遗弃召庙三大类。实地调研的召庙包括部分被遗弃的召庙遗存，课题组视现仍有部分残破殿宇的召庙为召庙遗存，其范围不包括已无殿堂残留的召庙遗址。

4 内蒙古地域内藏传佛教召庙的损毁主要由几次历史事件所至。如同治年间的回民起义（1869年）、丑牛之乱（1913年）、日本占领期及国民党统治时期的小规模战乱、苏蒙红军入境（1945年）以及中华人民共和国成立后的土改运动及“文化大革命”运动（1966年）等。也因一些自然灾害或人为原因，部分召庙或个别殿宇被烧毁。

5 参见：周清澍主编.内蒙古历史地理［M］.呼和浩特：1991：176—177．

6 具体数据为，阿拉善厄鲁特旗37座召庙，额济纳土尔扈特旗3座召庙。关于阿拉善地域召庙数量统计依据，请参见:贺·却木布拉,图布吉日嘎拉编著.阿拉善宗教史录.巴音森布尔,总70—71期：9;额济纳旗文史资料（专辑二）.额济纳旗政协文史资料研究委员会出版,1986，8:85—87.民国36年（1947年），在“西北论衡”1卷第1期刊登的“阿拉善概况”一文中，统计了阿拉善21座召庙及各庙僧数。参见:阿拉善盟史志资料选编（第一辑）.贺希格都仁翻译.阿拉善盟地方志办公室编印,1989.

7 该说主要依据为南寺第一世迭斯尔德呼图克图阿格旺道尔吉著“仓央嘉措传”，在阿拉善至今广泛传播着仓央嘉措弘法的传说。参见：贾拉森.再现辉煌的广宗寺1757—2007［M］.阿拉善广宗寺印，2007.

8 六世达赖喇嘛仓央嘉措参与寺庙选址或启示新建召庙的历史信息多以口传文学形式留存至今，如仓央嘉措与朝克图库伦庙的兴建等。

9 阿拉善三大寺庙系统指衙门庙、南寺、北寺三大寺庙。参见：贾拉森.缘起南寺［M］.呼和浩特：内蒙古大学出版社，2003：8.

10 阿拉善八大寺庙指清廷或民国政府赐予寺匾的八座寺庙，即衙门庙（延福寺）、南寺（广宗寺）、北寺（福因寺）、门吉林庙（承庆寺）、朝克图库伦庙（昭化寺）、图库木庙（妙华寺）、阿贵庙（宗乘寺）、希贵庙（方等寺）。参见：松儒布.阿拉善北寺史（蒙文）［M］.北京：民族出版社，2003：8．

11 阿拉善厄鲁特旗37座寺庙中有28座庙为三大寺庙之属庙，另有6座由旗扎萨克衙门或地方行政组织管辖的寺庙。参见：贺·却木布拉，图布吉日嘎拉编著：阿拉善宗教史录．载 巴音森布尔．总70—71期：9．

Chapter One
Survey of the Extant Remaining Temple Architectures within Inner Mongolia

As the main area of former Monan Mongolia, Inner Mongolia is one of the regions, which has the widest distribution of architectures. After the investigation and mapping of 110 temples in Inner Mongolia, the research group of Inner Mongolia University of Technology, which studies on Tibetan Buddhism, sorted out the architecture information of almost all the Tibetan Buddhism temples within Inner Mongolia, and made an attempt to sum up their regional characters and regulation of distribution, which made a solid foundation for the coming researches.

Tibetan Buddhism temples in Inner Mongolia were mostly built in Qing and Ming Dynasty. Although there are a few temples built in Yuan Dynasty or even earlier, they were all rebuilt or changed into Tibetan Buddhism temples in Qing Dynasty. As a result, the ancient Tibetan Buddhism temples, which remained after several hundred years of ups and downs, were mostly Ming and Qing architectures in which Qing buildings were in the majority. Besides, after the reform and opening up of China, especially in the recent two decades, with the great change of Chinese social environment, Tibetan Buddhism of Inner Mongolia entered a new era. The construction of Tibetan Buddhism temples increased over time. Those demolished and survival temples were respectively rebuilt, renewed and repaired, a few temples were newly built.

This chapter will start with another angle, geospatial dimension, to analyze and expound Inner Mongolia Tibetan Buddhism Temples systematically. On the basic of quoting documents and literatures that were compiled or published in Qing, the Republic of China and after the establishment of People's Republic of China, we divided the 110 temples in Inner Mongolia region into the western region, the central part and the eastern region, each of them will be explained.

I.The Western Region

1. Alxa Prefecture

Alxa Prefecture locates in the westernmost end of Inner Mongolia, 1/3 of the land is desert. Three deserts, Badanjilin, Tenger and Ulanbu go across the whole Alxa area; the great majority of the Prefecture does not have a convenient transportation. Due to the location of Central Asian Continent and the land form of interior plateau which is far from the ocean and encircled by mountains, the climate here is the typical continental one. It rains a little and strong wind blows with sand.

The region of Alxa was consisted of Elute Banner and Ejin Tuerhute Banner which were established in the period of emperor Kangxi. People settled to the west of Hetao. Established without any Prefecture, this area was generally called the two Western Banners. In 1980, Alxa Prefecture was established. In the Qing dynasty, there were 40 temples in the two Western Banners.

As its features, Tibetan Buddhism temples within Alxa have a unique developing mode, complete and clear system; its ancient architectures were well-preserved. Alxa has an assured place on the list of Tibetan Buddhism development history within Inner Mongolia. According to the research of scholars, Sangs-Ryyas Rgyamtsho, the 6th Dalai Lama who built temples and spread Buddhism in Alxa between 1716 and 1746, died in Menjilinmiao Temple. Stupa of Sangs-Ryyas Rgyamtsho was still enshrined in a palace of Nansi Temple. The 6th Dalai Lama was closely related to the building history of some temples. Temples in Elute Banner consisted of three main temples system, eight major temples and subordinated temples of the three main temples.There are three temples in Ejin Tuerhute Banner; one of them was built by Mongolia immigrants in the period of the Republic of China. Temples in Alxa distributed widely.

于戈壁、大漠深处。其中，位于巴彦浩特的衙门庙与位于大漠深处的巴丹吉林庙目前保存较为完好，而其他召庙均在“文化大革命”时期受到不同程度的破坏。需要说明的是，额济纳旗喀尔喀庙及新西庙大殿为以往学界很少关注或拍摄过的召庙。

课题组对阿拉善盟境内的15座[1]藏传佛教召庙进行了调研普查工作,具体情况如下表所示。

阿拉善盟地区调研的现存召庙

召庙名称	所在位置	基本信息	
		调研情况	现存主要建筑
南寺（广宗寺）	阿拉善左旗巴润别立镇	普查	大雄宝殿、多吉帕母庙、弥勒殿、财神殿、药师殿、转经阁、白塔、活佛府
衙门庙（延福寺）	阿拉善左旗巴彦浩特镇王府街	普查、测绘	山门、鼓楼、钟楼、天王殿、大雄宝殿、观音殿、吉祥天女殿、阿拉善大殿、三世佛殿
巴丹吉林庙	阿拉善右旗巴丹吉林沙漠腹地	普查	山门、大雄宝殿、白塔、活佛府
北寺（福因寺）	阿拉善左旗贺兰山北段	普查	大雄宝殿、白塔
达力克庙	阿拉善左旗豪斯布尔都苏木陶力嘎查	普查	大雄宝殿、活佛府
库日木图庙	阿拉善右旗雅布赖苏木境内	普查	大雄宝殿、时轮殿、八白塔、阿贵山洞
朝克图库伦庙	阿拉善右旗朝格图呼热苏木驻地	普查	大雄宝殿、白哈五王殿、六世达赖都贡、观音殿、寺庙管理所
图库木庙（妙华寺）	阿拉善左旗图克木苏木境内	普查	大雄宝殿、千佛殿、佛塔、活佛府、观音殿
沙日扎庙	阿拉善左旗乌力吉苏木境内	普查	大雄宝殿、护法殿、转经阁
夏日格庙	阿拉善右旗阿拉坦敖包苏木境内	普查	大经堂、护法殿、喇嘛僧舍、高僧修行山洞、寺庙管理用房、白塔
红塔庙	阿拉善左旗敖伦布拉格镇境内	普查	胜乐金刚殿
阿拉腾特布西庙	阿拉善右旗朝格苏木	普查	大雄宝殿、密宗殿、八白塔、活佛府、转经阁
额济纳西庙	额济纳旗达来呼布镇驻地	普查	大雄宝殿、喇嘛食堂、寺庙管理用房、白塔
额济纳新西庙	额济纳旗东风城宝日乌拉嘎查境内	普查	大雄宝殿、千佛殿、佛殿、活佛府、观音殿
喀尔喀庙	额济纳旗达来呼布镇苏泊淖尔苏木驻地	普查	传经殿堂

二、巴彦淖尔市地区

巴彦淖尔市位于河套平原和乌拉特草原上，该地区属中温带大陆性季风气候，光照充足，热量丰富，降水量少，气候干燥，风大沙多，温差较大，四季分明。

巴彦淖尔市现辖区由清时乌兰察布盟乌拉特三旗[2]、阿拉善厄鲁特旗、伊克召盟部分地区组成。1956年建巴彦淖尔盟，2004年撤盟设为巴彦淖尔市。因此一些召庙原分属不同盟旗，如阿贵庙为阿拉善八大寺庙之一。该地域召庙以乌拉特

1 据课题组收集到的信息，阿拉善盟境内现存19座召庙，由于交通条件及时间所限，实际调查16座召庙，其中包括一处召庙遗址。参见：高旭等.内蒙古阿拉善盟藏传佛教寺庙调查报告，2010，8.

2 乌拉特三旗，也称乌拉特三公旗，原游牧于今呼伦贝尔一带，顺治五年（1648年）编为三旗，中旗及前旗扎萨克为镇国公，后旗扎萨克为辅国公，故称三公旗，分别称东公旗、中公旗、西公旗。参见：周清澍主编.内蒙古历史地理［M］.呼和浩特：内蒙古大学出版社1991：169 .

Most of them locate in the remote region with difficult transportation, a few even in the distant desert. Among them, Yamenmiao Temple of Bayanhot Town and Badanjilinmiao Temple in the desert were well-preserved. Other temples were destroyed in different degrees in the "Culture Revolution". Worthy of mention is the Karkamiao Temple of Ejin Banner and the main hall of Xinximiao Temple were seldom concerned or shot before.

The research group investigated 15 temples within Alxa Prefecture; details are presented in the form below.

Extant Tibetan Buddhism Temples Researched within Alxa Prefecture

Name	Location	Basic Information	
		Research Method	Extant Architecture Status
Baronhiid Temple (Guangzongsi Temple)	Barunbieli Town in the south of Helan Mountain, Left Alxa County	General survey	Mahavira Hall, Dorjipam Temple, Maitreya Temple, Mammon Temple, Pharmacist Hall,Circumambulation Terrace,The Pagoda, Buddha Hall
Yamen Temple (Yanfusi Temple)	Northern side of Wangfujie Street, Bayanhaote Town, Left Alxa County	General survey, mapping	Monastery Gate, Drum Tower,Bell Tower, The Heaven King Hall, Mahavira Hall,Avalokiteshvara Hall, Lakshmi Temple, Alxa Big Hall, III-Buddha Temple
Badaijiren Temple	Right Alxa County, hinterland of Badanjilin Desert	General survey	Monastery Gate, Mahavira Hall,The Pagoda,Buddha Hall
Drunhiid Temple (Fuyinsi Temple)	Northern Segment of Helan Mountain, Left County	General survey	Mahavira Hall,The Pagoda
Darhe Temple	Taoli Gacha, Haosibuerdu Sumu, Left Alxa County	General survey	Mahavira Hall, Buddha Hall
Hurimut Temple	Within Yabulai Sumu, Right Alxa County	General survey	Mahavira Hall, Calachakra Temple,Babai Pagoda,Agu Cave
Chaoketuhure Temple	In the station of Zhaogetuhure Sumu, Right Alxa County	General survey	Mahavira Hall, Pehar Gyalpo Hall, Dalai-IV Temple,Avalokiteshvara Hall, The Temple Management
Tuhum Temple (Miaohuasi Temple)	Within Tukemu Sumu, Left Alxa County	General survey	Mahavira Hall, Thousand Buddha Temple, Pagoda,Buddha Hall, Avalokiteshvara Hall
Sharaja Temple	Within Wuliji Sumu, Left Alxa County	General survey	Mahavira Hall, Dharmapalas Hall, Circumambulation Terrace
Xiarige Temple	within Alatanaobao Sumu, Right Alxa County	General survey	Main Assembly Hall,Dharmapalas Hall, Lama House,Hierarch Practice Cave, The Temple Administration,The Pagoda
Olansubrega Temple	Within Aolunbulage Town, Left Alxa County	General survey	Chakrasamvara Temple
Alatentebxi Temple	Chaoge Sumu, Right Alxa County	General survey	Mahavira Hall, Vajrayana Hall, Babai Pagoda, Buddha Hall, Circumambulation Terrace
West Ejina Temple	Station of Dalaihubu Town, Ejin County	General survey	Mahavira Hall, Lama Dining Hall, The Temple Administration, The Pagoda
New West Ejina Temple	Within Baoriwula Gacha, Dongfeng Town, Ejina County	General survey	Mahavira Hall, Thousand Buddha Temple, Buddha Hall, Buddha Hall, Avalokiteshvara Hall
Halehe Temple	Station of Suponaoer Sumu, Dlaihubu Town, Ejina County	General survey	Circumambulation Hall

2. Bayannaoer City

Located in Hetao Plain and Ulat Prairie, Bayannaoer has a middle temperate continental monsoon climate which means sufficient sunshine and rich heat energy. Small amount of rain brought it droughty weather, huge winds with sand, big temperature difference and clear distinction between four seasons.

The current precinct of Bayannaoer City consists of the 3 Wulate Counties of Ulanqab Prefecture, Alxa Elute County and part of Yikezhao Prefecture in Qing Dynasty. In 1956, Bayannaoer Prefecture was established and then replaced by Bayannaoer City in 2004. Therefore, some of the temples were belonging to different Prefectures and counties, for example, Aguimiao Temple was one of the eight major

三公旗召庙为主，三公旗原有召庙80余座[1]，因行政区划的变革，其主要召庙现已划归包头市。

巴彦淖尔市召庙普遍呈现特性鲜明、建筑风格以藏式为主等特点。阿贵庙为内蒙古地域现存唯一一座按照宁玛派仪轨活动的藏传佛教红教派召庙。现存召庙多为原中公旗巴音善岱庙、西公旗梅日更庙的属庙及佐庙。与西公旗多数召庙一样，点布斯格庙曾以蒙古语诵经而著称。巴彦淖尔市召庙多以藏式建筑为主，且主要殿宇保存完好。

课题组对巴彦淖尔市境内的5座召庙进行了调研普查工作，具体情况如下表所示。

巴彦淖尔市地区调研的现存召庙			
召庙名称	所在位置	基本信息	
		调研情况	现存主要建筑
点布斯格庙	乌拉特前旗白音华镇境内	普查	大雄宝殿、喇嘛僧舍
阿贵庙	磴口县沙金套海苏木境内	普查	时轮金刚殿、大雄宝殿、护法殿、财神殿、金刚亥母殿
善岱古庙	乌拉特后旗巴音镇	普查	山门、大雄宝殿、密宗殿、活佛府、五座白塔
东升庙	乌拉特后旗巴音宝力格镇	普查	大雄宝殿
哈日朝鲁庙	乌拉特后旗潮格温都尔苏木境内	普查	大雄宝殿

三、鄂尔多斯市地区

鄂尔多斯市位于内蒙古自治区西南部，地处鄂尔多斯高原腹地，该地区属北温带半干旱大陆性气候区，冬夏寒暑变化大，全年多盛行西风及北偏西风。

鄂尔多斯市前身伊克昭盟辖地由清顺治年间设立的鄂尔多斯6旗，即左翼中旗（郡王旗）、左翼前旗（准噶尔旗）、左翼后旗（达拉特旗）、右翼中旗（鄂托克旗）、右翼前旗（乌审旗）、右翼后旗（杭锦旗）及乾隆年间增设的右翼前末旗（扎萨克旗）7旗[2]组成，七旗会盟于王爱召[3]，故称伊克昭盟。鄂尔多斯七旗共有召庙300座[4]，召庙分布密度很高，仅以达拉特旗为例，旗境内共有召庙72座。其数远超出日本学者桥本光宝所估算的内蒙古每旗平均有20座寺庙的数量。[5]

鄂尔多斯市召庙普遍呈现发源较早、体系完整、大型召庙留存较为普遍等特点。鄂尔多斯草原为藏传佛教最早传入的地区之一。“藏传佛教格鲁派与蒙古之间关系的引进者”——呼图黑台·彻辰·洪台吉[6]建议土默特部阿拉坦汗迎请三世达赖索南嘉措，为藏传佛教的传播起到极大的作用。索南嘉措于第二次前来蒙古途中，在其驻地停留三个多月，之后到博硕克图吉农驻地后专为他指出了修建三世佛寺的地点，据称，此三世佛寺之地为王爱召。[7]一些召庙体系均得以较完整地保存下来，显示出特定历史时期内的召庙网[8]及主次关系。鄂尔多斯七旗旗庙或各旗境内最大的召庙[9]多数已恢复法事，如准格尔召、乌审召、鄂托克召、展旦召等。

课题组对鄂尔多斯市境内的18座藏传佛教召庙进行了普查调研及测绘，具体情况如下表所示。

四、包头市地区

1 参见：莫德力图.乌兰察布史略（第十一辑）[M].政协乌兰察布盟委员会文史资料研究委员会编，1997：347。另依据民国时期编纂的地方志记载，乌拉特三公旗召庙却不及80座。民国26年（1937年）编《绥远通志稿》卷54记载，乌拉特三公旗共有召庙67座。文献以各旗旗治为中心，详细记录了召庙所处方位及僧数。嘉庆十六年（1811年）编《乌拉特寺庙名录》详细记载了各大召庙的初建与清廷赐匾时间。参见：内蒙古师范大学图书馆藏古籍《乌拉特寺庙名录》（编号为02077）。

2 参见：周清澍.内蒙古历史地理[M].呼和浩特：内蒙古大学出版社，1991：170—174.

3 王爱召，即广惠寺，俗称伊克召，意为大召。始建于1607年，竣工于1613年，民国末年被烧毁。参见：萨·那日松，特木尔巴特尔.鄂尔多斯寺院（蒙古文）[M].海拉尔：内蒙古文化出版社，2000：33—38.

4 关于鄂尔多斯七旗召庙的总数，依据民国26年（1937年）编《绥远通志稿》卷54记载，共有召庙247座。民国28年（1939年）编《伊克昭盟志》记载，共有召庙274座。参见：绥远通志馆编纂.绥远通志稿.民国26年（1937年），内蒙古人民出版社整理再版.绥远通志稿（第七册 卷50至卷60）2007，8；边疆通讯社修撰．伊盟左翼三旗调查报告书．民国28年（1939年）；内蒙古图书馆编．鄂托克富源调查记、准郡两旗旅行调查记、伊盟左翼三旗调查报告书、伊盟右翼四旗调查报告书、伊克昭盟志、伊克昭盟概况（下册）．呼和浩特：远方出版社，2007，11．

5 参见：（日本）桥本光宝著.海勒图愓·陶克敦巴雅尔译．蒙古喇嘛教(蒙古文)．呼和浩特：内蒙古人民出版社,2009,12:147.

6 呼图黑台·彻辰·洪台吉为阿拉坦汗之兄——鄂尔多斯部之始祖衮必里克墨尔根吉农之孙，《蒙古源流》的作者萨冈·彻辰之曾祖父。参见：乔吉.蒙古佛教史——北元时期1368—1634[M].呼和浩特：内蒙古人民出版社，2007：108—109.

7 见“三世达赖喇嘛传”，引自乔吉.蒙古佛教史—北元时期1368-1634[M].呼和浩特：内蒙古人民出版社，2007：67．

8 召庙网即指在同一行政区划内的召庙按其主次等级，构成主庙与属庙的隶属关系的特定现象。如乌审召为主庙，查干庙、海流图庙、陶日木庙等召庙为乌审召的属庙。属庙接受主庙的管辖，承担学部的分设、活佛的避暑修行等宗教事务。

9 这些规模宏大的召庙在其俗称上均冠以所属扎萨克旗名称，成为鄂尔多斯召庙的一大特点，如乌审召、鄂托克召、准格尔召等。

temples of Alxa.Wulatesan County, which has more than 80 temples before, is the biggest temple owner in Bayannaoer City. But its major temples were charged by Baotou City now due to the change of administrative division.

Temples in Bayannaoer City can be distinguished by its distinct features and major architectural style of Tibet. In the area of Inner Mongolia, Aguimiao Temple is the only existing Mongolian Lamaism temple which was built in the structure of holding traditional religious ritual of Ningma sect. Most of the existing temples are the subsidiary and side temples of Bayinshandaimiao Temple of the former Zhonggong County and Meirigengmiao Temple of Xigong County. Temples in Bayannaoer City were mostly built in Tibetan style and their main halls were all well-preserved.

The research group investigated five temples within Bayannaoer City; details were presented in the form below.

Extant Tibetan Buddhism Temples Researched within Bayannaoer City			
Name	Location	Basic Information	
		Research Method	Extant Architecture Status
Debseg Temple	Within Baiyinhua Town, Front Wulate County	General survey	Mahavira Hall, Lama House
Agui Temple	Within Shajintaohai Sumu,Dengkou County	General survey	Kalachakra Hall, Mahavira Hall,Dharmapalas Hall, Mammon Temple, Vajravarahi Hall
Shaletew Temple	Bayin Town of Back Wulate County	General survey	Monastery Gate, Mahavira Hall, Vajrayana Hall, Buddha Hall, The Five Pagodas
Dongxure Temple	Bayinbaolige Town of Back Wulate County	General survey	Mahavira Hall
Harichaolu Temple	Within Chaogewenduer Sumu, Back Wulate County	General survey	Mahavira Hall

3.Erdos City

Lies in the southwest of Inner Mongolia Autonomous Region, Erdos City locates in the hinterland of Erdos Plateau which has a semiarid continental climate in North Temperate Zone. Seasons change clearly, west and north-westerly wind blows all the year.

Yikezhao Prefecture, the predecessor of Erdos City, consist of seven counties including six counties of Erdos: the Left Wing Middle County (Junwang County), the Left Wing Front County (Zhunger County), the Left Wing Back County (Dalate County), the Right Wing Middle County (Etuoke County), the Right Wing Front County (Wushen County), the Right Wing Back County (Hangjin County) which were established in the period of emperor Shunzhi of Qing Dynasty and one county of the Right Wing Frontend County newly established in the period of emperor Qianlong. Seven counties met in Wang'aizhao Temple, which was the origin of the name of Yikezhao Prefecture. 300 temples are located in seven counties of Erdos. The density of temples distribution is quite higher. Just taking an example of Dalate county, there are 72 temples in it. This number of temples are far beyond an average of 20 temples of each county estimated by the Japanese scholar Hashimoto Hikaru in Inner Mongolia. Most temples in Erdos have long history and complete system, major temples were well-preserved. Erdos Prairie is one of the areas that introduced Tibetan Buddhism early. Hutuheitai Chechen Hongtaiji, the facilitator of the relationship between Gelu sect of Tibetan Buddhism and Mongolia, suggested Alatan Khan of Tumd Tribe to invite the Third Dalai Lama, Suonanjiacuo, who made a great contribution to the development of Tibetan Buddhism. On his way of coming to Mongolia for the second time, Suonanjiacuo stayed in Hutuheitai's station for three months. Later he pointed out the spot of building Three-Buddha Temple for Boshuoketujinong when he arrived at Bo's station. Someone figured out that the temple is just Wang'aizhao Temple. Some temple architectures were well-preserved which may present us a net of temples and the relation between major and side ones. Most of The Seven Temples in Erdos or biggest ones in each county such as Zhungeerzhao Temple, Wushenzhao Temple, Etuokezhao Temple, Zhandanzhao Temple resumed religious rites.

The research group investigated 11 temples within Erdos; details were presented below.

Extant Tibetan Buddhism Temples Researched within Erdos City			
Name	Location	Basic Information	
		Research Method	Extant Architecture Status
Zhungarzhao Temple	Zhungeer Sumu, Zhungeer County	General survey and mapping	Mahavira Hall, Maitreya Hall, The Five Paths Lam Lnga Temple,Lama House, Buddha Hall, Avalokiteshvara Hall, Padma Sambhava Hall,Arira Hall, Thousand Buddha Hall
Wushenzhao Temple	Wushenzhao Town, Wushen County	General survey and mapping	Mahavira Hall, Longevity Buddha hall, Fawang Hall, Maitreya Hall, Tara Hall,Pharmacist Hall, Jaronkasiar Pagoda, Kalachakra Hall, The Upper Hall
Hailiut Temple	Bayinchaida Sumu, Wushen County	General survey	Vajrabhairava Hall, Guhyasamaja Hall ,Dharmapalas Hall,Sakyamuni Buddha Hall, Monastery Gate, Avalokiteshvara Hall, Pharmacist Hall,Five-Buddhas Hall, The Heaven King Hall

鄂尔多斯市地区调研的现存召庙			
召庙名称	所在位置	基本信息	
		调研情况	现存主要建筑
准格尔召	准格尔旗准格尔苏木	普查、测绘	大雄宝殿、弥勒殿、五道庙、大常署、佛殿、观音殿、莲花生殿、舍利殿、千佛殿
乌审召	乌审旗乌审召镇	普查、测绘	大雄宝殿、长寿佛殿、法王殿、弥勒殿、度母殿、药师佛殿、扎荣噶沙尔塔、时轮金刚殿、德都殿
海流图庙	乌审旗巴音柴达木苏木	普查	大威德金刚殿、密集金刚殿、护法殿、弥勒佛殿、山门、观音殿、药师佛殿、五神殿、天王殿
陶亥召	伊金霍洛旗纳林陶亥召镇新庙村	普查	山门、天王殿、西转经阁、东转经阁、大雄宝殿、时轮金刚殿、法相殿、白塔
特布德庙	前旗上海庙镇特布德嘎查	普查	大雄宝殿、佛爷墒、山门、古塔遗址
鄂托克召	鄂托克旗阿尔巴斯镇脑高岱嘎查	普查	大雄宝殿、佛爷墒、大常署、山门、时轮金刚殿、白塔
展旦召	达拉特旗展旦召镇展旦召嘎查	普查	天王殿、大雄宝殿、舍利殿、大常署、药神殿
公尼召	伊金霍洛旗公尼召苏木喇嘛敖包嘎查	普查	大雄宝殿、四青塔、白塔、舍利塔
乌兰木伦庙	伊金霍洛旗乌兰木伦镇	普查	大雄宝殿、圆满宝堂寺、苏力德庙、喇嘛庙、佛爷墒、大常署、白塔
乌拉庙	伊金霍洛旗扎萨克镇孟克庆嘎查	普查	宗喀巴殿、护法殿、北方神殿、大雄宝殿、山门、东转经阁、西转经阁、东配殿、西配殿、白塔
哈日根图庙	鄂托克前旗查干陶勒盖苏木哈日根图嘎查	普查	大雄宝殿
阿日赖庙	鄂托克前旗昂素镇阿日赖嘎查	普查	大雄宝殿、度母殿、护法殿
苏里格庙	鄂托克旗苏米图镇苏里格嘎查	普查	大雄宝殿、佛爷墒、转经阁、山门、偏殿、财神殿、牌楼
沙日召	杭锦旗独贵塔拉镇沙日召嘎查	普查	大雄宝殿、护法殿、白塔
沙日特莫图庙	杭锦旗伊和乌苏苏木宝日胡术嘎查	普查	天王殿、大雄宝殿、菩提白塔、弥勒殿、活佛墒、度母殿、护法殿
哈毕日格庙	杭锦旗阿日斯楞图苏木巴音宝力格嘎查	普查	大雄宝殿、五道庙、度母庙、山门、白塔
查干庙	乌审旗乌审召镇查干庙嘎查	普查	白塔、护法殿、大雄宝殿、活佛墒
陶日木庙	乌审旗陶利苏木陶日木庙嘎查	普查	天王殿、八白塔、吉祥护法殿、喇嘛墒、僧舍、菩萨庙、西殿、东殿、时轮塔

包头市地处内蒙古高原的南端，全市由中部山岳地带、北部高原草地和南部平原三部分组成。该地区属半干旱中温带大陆性季风气候。

包头市现辖区由清时乌兰察布盟乌拉特三公旗、喀尔喀右翼旗、茂明安旗及归化城土默特旗部分地区组成。1923年初置包头设治局，1938年改建包头市。[1]因此一些召庙原分属不同盟旗，如美岱召、希拉木仁庙为归化城土默特旗召庙。

包头市境内的召庙普遍呈现形制高、特性明显、历史影响深刻、古建筑保护较好等特点。如梅日更召、昆都仑召、百灵庙分别为乌拉特西公旗、乌拉特中公旗、喀尔喀右翼旗旗庙，各自掌管旗境内所有召庙的宗教事务。五当召为蒙古地区最著名的学问寺，梅日更召为现今唯一一座延续蒙古语诵经制度的召庙，美岱召为蒙古地区建造最早的召庙[2]之一，又是具有城寺合一、政教一体的独特形制的召庙等。在古建筑保存方面，五

1 参见：周清澍.内蒙古历史地理［M］.呼和浩特：内蒙古大学出版社，1991：233.

2 关于蒙古地区第一座藏传佛教格鲁派寺庙，学界有不同的表述，可以总结为，1574年建于青海湖畔的恰布恰庙（仰华寺）为蒙古人首次建造的格鲁派寺庙；1579年在归化城新建的大召为漠南蒙古地区第一座格鲁派寺庙。但据学界已有成果看，此两座召庙之前蒙古地区仍有一些召庙，如1572年始建于福化城内的阿拉坦汗家庙（即美岱召），再如博格达查干喇嘛在阿拉坦汗与三世达赖索南嘉措在蒙古地区弘法以先，在归化城北大青山南麓修建了西喇嘛洞（广化寺）旧庙。请参阅：乔吉著《蒙古佛教史——北元时期1368-1634》（2007年）、金峰整理注释《呼和浩特召庙》（1982年）、乌·那仁巴图、贾拉森等编著《蒙古佛教文化》（1997年）等书籍。

Continued Table

Extant Tibetan Buddhism Temples Researched within Erdos City			
Name	Location	Basic Information	
		Research Method	Extant Architecture Status
Taohaizhao Temple	Xinmiao Village, Nalintaohaizhao Town, Yijinhuoluo County	General survey	Monasty Gate, The Heaven King Hall, The west-circumambulaion Terrace, The East-circumambulation Terrace, Mahavira Hall, Kalachakra Hall, Exotoric Buddhism Hall, The Pagoda
Tubed Temple	Tebude Gacha, Shanghaimiao Town, Front County	General survey	Mahavira Hall,The Buddha House,Monastery Gate, The Old Pagodas Site
Otogzhao Temple	Naogaodai Gacha, Aerbasi Town, Etuoke County	General survey	Mahavira Hall, The Buddha House,Lama House,Monastery Gate, Kalachakra Hall, The Pagoda
Zandanzhao Temple	Zhandanzhao Gacha, Zhandanzhao Town, Dalate County	General survey	The Heaven King Hall, Mahavira Hall, Sarira Hall,Lama House, Pharmacist Hall
Gongnizhao Temple	Lamaaobao Gacha, Gongnizhao Sumu, Yijinhuoluo County	General survey	Mahavira Hall The Four- green Pagoda, The Pagoda, Sarira Hal
Ulanmuren Temple	Ulanmulun Town, Yijinhuoluo County	General survey	Mahavira Hall, Yuanmanbaotang Temple, Suld Temple, Lamasery, The Buddha House, Lama House,The Pagoda
Wula Temple	Mengkeqing Gacha, Zhasake Town, Yijinhuoluo County	General survey	Tsongkhapa Hall,Dharmapalas Hall, The North Holy Shrine, Mahavira Hall, Monastery Gate, The East-circumambulation Terrace, The West-circumambulation Terrace, The East-side Hall, The West-side Hall,The Padoga
Harigent Temple	Harigentu Gacha, Chagantaolegai Sumu, Front Etuoke County	General survey	Mahavira Hall
Arilai Temple	Arilai Gacha, Angsu Town, Front Etuoke County	General survey	Mahavira Hall, Tara Hall, Dharmapalas Hall
Surehi Temple	Sulige Gacha, Sumitu Town, Etuoke County	General survey	Mahavira Hall, The Buddha House, Circumambulation Terrace, Monastery Gate, The Side Hall, Mammon Hall, Memorial Arch
Sharizhao Temple	Sharizhao Gacha, Duguitala Town, Hangjin County	General survey	Mahavira Hall, Dharmapalas Hall, The Pagoda
Sharitemet Temple	Baorihushu Gacha, Yihewusu Sumu, Hangjin County	General survey	The Heaven King Hall, Mahavira Hall, The Boddhi Pagoda, Maitreya Hall, The Buddha House, Tara Hall, Dharmapalas Hall
Habirig Temple	Bayinbaolige Gacha, Arisilengtu Sumu, Hangjin County	General survey	Mahavira Hall, Five paths lam lnga Temple, Tara Hall, Monastery Gate, The Pagoda
Chagan Temple	Chaganmiao Gacha, Wushenzhao Town, Wushen County	General survey	The Pagoda, Dharmapalas Hall, Mahavira Hall, The Buddha House
Taorim Temple	Taorimumiao Gacha, Taoli Sumu, Wushen County	General survey	The Heaven King Hall, The Babai Pagoda,Dharmapalas Hall, Lama House, Monk House ,Bodhisattva Temple, The West Hall, The East Hall, The Kalachakra Stupa

4. Baotou City

Baotou City lies in the south end of Inner Mongolia Plateau, the whole city consist of mountains in the middle, plateau prairie in the north and plain in the south. It has a climate of semiarid continental monsoon in the temperate zone.

The present precinct of Baotou City consists of Wulatesangong County of Ulanqab Prefecture, Karka Right Wing County, Maomingan County and part of Tumd County of Guihua Town in Qing Dynasty. In the beginning of 1923, Baotou Administrative Institution was established and then changed into Baotou City in 1938. Therefore, some of the temples were belong to different counties and Prefectures. For example, Maidarzhao Temple and Xiaramuren Temple were under the administration of Tumd County of Guihua Town.

Temples in Baotou City were highly built with distinct features and deep historical influence; its ancient architectures were well-preserved. Meirigen Temple, Hundele Temple, and Batahalaga Temple were respectively under the administration of Wulatexigong County, Wulatezhonggong County and Karka Right Wing County. The temples were in charge of all the religious affairs within their own area. Badgar Temple is the most famous Knowledgeable Temple in Mongolia; Meirigen Temple is the only temple that still chants sutras in Mongolian. Maidarzhao Temple is one of the earliest-built temples in Mongolia and is unique for its combination with the city

当召为国家在20世纪70年代初拨款修葺的召庙，其余召庙古建筑保存也较好。该地域与最早引入藏传佛教的呼和浩特、鄂尔多斯地域相毗邻，其境内召庙有独特的历史地位。

课题组对包头市境内6座召庙[1]进行普遍调查及详细测绘工作，具体情况如下表所示。

包头市地区调研的现存召庙			
召庙名称	所在位置	基本信息	
		调研情况	现存主要建筑
美岱召	土默特右旗美岱召村	普查、测绘	大雄宝殿、乃琼庙、达赖庙、太后庙 、观音殿、西万佛殿、琉璃殿、八角庙、泰合门城门楼、罗汉殿、角楼
五当召	固阳县吉忽伦图乡	普查、测绘	大雄宝殿、菩提道学殿、甘珠尔活佛府、显宗殿、斋戒殿、洞阔尔活佛府、时轮殿、苏卜盖殿、章嘉活佛府、护法殿、阿会殿、根丕庙
梅日更召	九原区	普查、测绘	天王殿、护法殿、佛殿、活佛仓、舍利殿、活佛府
昆都仑召	昆都仑区	普查、测绘	大雄宝殿、小黄庙、天王殿、度母殿、玛尼殿、东活佛府、西活佛府、哈斯尔殿、王爷府、白塔
百灵庙	达尔罕茂明安联合旗百灵庙镇	普查、测绘	大雄宝殿、显宗殿、丹珠尔殿、甘珠尔殿、天王殿、新东配殿、新西配殿、白塔
希拉木仁庙	达尔罕茂明安联合旗希拉慕仁镇	普查、测绘	大雄宝殿、佛爷府、六世活佛供殿、天王殿、护法殿

第二节 中部地区

一、呼和浩特市地区

作为内蒙古自治区首府的呼和浩特市位于华北西北部、内蒙古中部的土默川平原，属中温带大陆性季风气候，四季气候变化明显，年平均气温由北向南递增。其境内地貌主要为以北部大青山和东南部蛮汉山为主的山地地形以及南部、西南部土默川平原为主的平原地形。

呼和浩特市由清时归化、绥远二城，天聪年间设立的土默特二都统旗及雍正至乾隆年间设置的归化城厅、绥远城厅、萨拉齐厅、清水河厅、和林格尔厅、托克托城厅等诸厅组成。[2]呼和浩特地区为藏传佛教最早传入并大量集中兴建召庙的区域，故被民间誉为具有“七大召、八小召、七十二个绵绵召”的召城。据道光十五年（1835年）成书的“宝鬘”记载，呼和浩特地区共有大小召庙近70座。[3]

呼和浩特市境内的召庙普遍呈现发源较早、体系完整、名望高、建筑风格以汉式为主等特点。1571年藏区僧人阿兴喇嘛出现于土默特部，通过讲授佛法，使阿拉坦汗皈依佛教[4]，使藏传佛教再度传入了蒙古高原。明末至清初的归化城及城北山中的岩洞成为早期佛教弘法者及苦行僧的修业道场，为藏传佛教的扩布起到核心作用。呼和浩特各召庙的等级有序，管理严谨。呼和浩特掌印扎萨克达喇嘛印务处设于大召，并管辖归化城及土默特旗境内[5]15座藏传佛教召庙的宗教事务。其中由扎萨克达喇嘛管辖的7座庙[6]为七大召，达赖喇嘛管辖的8座庙[7]为八小召，其余10座为属庙[8]。呼和浩特地区汇聚了内齐托音、席力图等地位显赫的呼图克图及多位在早期藏传佛教传

1 课题组实地调研了8座召庙，除下表所列6座庙外，有包头召（福徵寺）、夏日朝鲁庙2座召庙。见韩英等“内蒙古包头地区藏传佛教建筑调查”2010,10.

2 参见：周清澍.内蒙古历史地理［M］.呼和浩特：内蒙古大学出版社，1991：191—192．

3 参见：（德）W.海西希《宝鬘》蒙古喇嘛寺院及蒙古编年史，伊西班丹著，1835年成书，哥本哈根，于1961年影印出版：75.

4 乔吉.蒙古佛教史——北元时期1368-1634［M］.呼和浩特：内蒙古人民出版社，2007,9．

5 呼和浩特召庙多位于归化城城区及周边土默特旗境内，但召庙所处具体区位不局限于上述区域，如1733年建托里布拉克庙（仁佑寺）位于当时的"外蒙古"地区（现为蒙古国）。参见：金峰整理注释《呼和浩特召庙》（1982年）。

6 这七座召庙为：大召（无量寺）、席力图召（延寿寺）、小召（崇福寺）、朋苏克召（崇寿寺）、拉布齐召（宏庆寺）、班迪达召（尊胜寺）、乃莫齐召（隆寿寺）。见金峰整理注释《呼和浩特召庙》（1982年）。

7 这八座召庙为：西喇嘛洞（广化寺）、东喇嘛洞（崇禧寺）、西乌素图召（庆缘寺）、美岱召（寿灵寺）、太平召（宁祺寺）、慈寿寺、广福寺、乔尔吉召（延禧寺）。见金峰整理注释《呼和浩特召庙》（1982年）。

8 这十座属庙为席力图召（延寿寺）、小召（崇福寺）、朋苏克召（崇寿寺）、西乌素图召（庆缘寺）等四座召庙的属庙，其名称为：东乌素图召（广寿寺）、查干哈达召（永安寺）、希拉木仁庙（普会寺）、五塔寺（慈灯寺）、岱海庙（荟安寺）、登奴素山召（善缘寺）、吉特库召、里素召（增福寺）、乔尔吉拉让巴召（法禧寺）等九座庙再加后成为八小召的席力图召属庙乔尔吉召（延禧寺）。见金峰整理注释《呼和浩特召庙》（1982年）。

and the politics. In aspect of ancient-architecture-preserving, Badgar Temple Temple was repaired in the 1970s by using the national grant; other ancient temples were also well preserved. The city is adjacent to the cities which introduced Tibetan Buddhism very early such as Hohhot and Erdos. Its temples have unique historical values.

The research group investigated 6 temples within this area; details were presented below.

Extant Tibetan Buddhism Temples Researched within Baotou City			
Name	Location	Basic Information	
		Research Method	Extant Architecture Status
Maidarzhao Temple	Maidarzhao Village, Right Tumd County	General survey and mapping	Mahavira Hall,Arhat Hall, Dalai Temple,The Queen Temple, Avalokiteshvara Hall, The West Ten southands Temple, Glass Hall,The Octagonal Temple, Taihe Gate, Arhat Tepmple, Turret
Badgar Temple	Jihuluntu in Gu Yang County	General survey and mapping	Mahavira Hall, Lamarim Hall, Ganzhurwa House, Exotoric Buddhism Hall, Nongnai Hall, Dongkhor Buddha House, The Kalachakra Stupa, Su Bu Gai Hall,Icang-skya Khutu House, Dharmapalas Hall, Arhat Hall, Gempi Temple
Merigen Temple	Jiuyuan District	General survey and mapping	The Heaven King Hall, Dharmapalas Hall,Buddha Temple, Khutu House, Sarira Hall, The Buddha Hall
Hundele Temple	Hundele District	General survey and mapping	Mahavira Hall, Xiao Huang Temple, The Heaven King Hall, Tara Hall, Mani Hall,The Eastern Buddha Hall, The Weastern Buddha Hall, Hasier Hall, Wangye House,The Pagoda
Batahalaga Temple	Bailingmiao Town, Daerhan Maoming'an United County	General survey and mapping	Mahavira Hall, Exotoric Buddhism Hall, Tangyur Hall, Kangyur Hall,The Heaven King Hall, The New East-side Hall, The New West-side Hall, The Pagoda
Xiaramuren Temple	Xiaramuren Town of Damao County	General survey and mapping	Mahavira Hall, The Buddha House, IV-Buddha Hall, The Heaven King Hall, Dharmapalas Hall

II. The Middle Region

1. Hohhot City

Hohhot, the capital of Inner Mongolia Autonomous Region is located in Tumochuan Plain in central Inner Mongolia, northwest of Northern China, under the temperate continental monsoon climate with distinct four seasons and the annual average temperature increasing from north to south. The regional landform mainly contains the mountainous terrain composed by Mount Daqing in north and Mount Manhan in southeast and the plain terrain of Tumochuan in south and southwest.

Hohhot City is composed of Guihua City, Suiyuan City, two Tumed Dutong Counties, established in Tiancong period and several countries set up from Yongzheng to Qianlong period, such as Guihua County, Suiyuan County Salaqi County, Qingshuihe County, Helinger County, Togtoh County etc.. Hohhot is the region where Tibetan Buddhism was earliest introduced and the temples were built in large amount and concentration. So it is known as a temple city of "Seven Big Temples ,Eight Small Temples, Seventy-two Temples with No Names". According to *Bao Man* completed in the 15th year of Daoguang (1835), there are nearly 70 temples in Hohhot area.

The temples in Hohhot are mainly characterized by early origin, integrate system, prestigious fame,Han-style construction etc. In 1571 Tibetan monk AXing Lama appeared in Tumed; through teaching Dharma, Aletan Khan converted to Buddhism, the Tibetan Buddhism once again entered into Mongolia plateau. Guihua City and the caves in the northern city in late Ming Dynasty to the early Qing Dynasty became the practicing field of early Buddhist advocates and Sadhu,serving as the center of spreading Tibetan Buddhism. The temples of Hohhot are hierarchically structured with strict management. Yinwuchu (temple management department) under Zhasakeda Lama who was in power in Hohhot was in Dazhao Temple and administered the religious affairs of 15 Tibetan Buddhism Tmeples in Guihua City and Tumed region, among which 7 temples under the jurisdiction of Zhasakeda Lama are known as "seven big temples", 8 temples under Dalai Lama are "eight small temples",the rest 10 temples are subordinate temples. In Hohhot area, there gathered Neiqituoyin, Xilitu and some eminent Hutuketu and many Hutuketu reincarnation system which played an important effect in the early history of spreading Tibetan Buddhism. And the position of Zhasakeda

播史上起到重要影响的呼图克图转世体系。而呼和浩特掌印扎萨克达喇嘛一职不局限于本地，也由京城、上京、热河、多伦诺尔等地的呼图克图担任。呼和浩特召庙建筑以汉式建筑为主，只有席力图召大雄宝殿、城西郊乌素图召庆缘寺、法禧寺的大雄宝殿为汉藏结合式殿宇。

课题组较为系统、完整地对整个呼和浩特市地区内现存的，具有代表性的7座藏传佛教召庙[1]进行了调研与测绘。具体情况如下表所示。

呼和浩特市地区调研的现存召庙

召庙名称	所在位置	基本信息	
		调研情况	现存主要建筑
大召	玉泉区大召前街	普查、测绘	牌楼、山门、天王殿、钟鼓楼、菩提过殿、大雄宝殿、九间楼、乃春庙、藏经阁、大白伞盖佛殿、公中仓、菩萨殿、玉佛殿、弥勒殿
席力图召	玉泉区石头巷北端	普查、测绘	牌楼、天王殿、菩提过殿、大雄宝殿、九间楼、古佛塔、护法殿、观音殿、度母殿、活佛府、长寿塔
小召	玉泉区小召前街北端	普查	牌楼
五塔寺	玉泉区五塔寺街	普查、测绘	山门、度母殿、观音殿、三世佛殿、大日如来佛殿、金刚座舍利宝塔、不空成就佛殿、阿弥陀佛殿、阿门佛殿、宝生佛殿、白塔
乌素图召	回民区攸攸板镇西乌素图村	普查、测绘	庆缘寺、长寿寺、法禧寺、罗汉寺、药王寺等五座寺院的主要建筑
喇嘛洞	土默特左旗	普查	护法殿、佛爷府、喇嘛洞、龙王殿、广化寺、舍利塔、聚佛塔、统化塔、菩提塔
乃莫齐召	玉泉区	普查、测绘	大雄宝殿

二、乌兰察布市地区

乌兰察布市地处内蒙古自治区中部，地形自北向南由蒙古高原、乌兰察布丘陵、阴山山脉、黄土丘陵四部分组成。该地区地处中温带，属大陆性季风气候，四季特征明显。因大青山横亘该地区中部，又形成了前山地区比较温暖，雨量较多，后山地区则是多风的特殊气候。

乌兰察布盟由清时四子部落旗、喀尔喀右翼旗、茂明安旗、乌拉特三公旗6个扎萨克旗组成。因行政建置的历史沿革，现乌兰察布市境内召庙为四子王旗召庙及原察哈尔右翼四旗[2]召庙。

乌兰察布市境内的召庙普遍呈现现存数量少、建筑风格因旗而异等特点。四子王旗境内曾有24座召庙，8座诵经会。[3]现存希拉木仁庙、王府庙2座庙，西拉哈达庙仅存1座残破的殿，图库木庙仅存3座残破的僧舍。[4]原察哈尔右翼四旗仅存阿贵庙1座庙，现卓资县巴音查干庙仅存2座石狮[5]，位于商都县的原上都马群旗巴德玛图庙有1座新建殿宇。[6]建筑风格上四子王旗境内现存召庙均为藏式，而察哈尔召庙以汉式或汉藏结合式为主。

课题组实地调研了现乌兰察布市辖区内尚存的藏传佛教召庙3处[7]，具体情况如下表所示。

1 呼和浩特市现已复建的召庙还包括拉布齐召（宏庆寺）、查干哈达召（永安寺）两座召庙。因政局变动及社会变革原因，呼和浩特市及其周边的土默川地域的召庙在清末及民国年间已开始破败或被遗弃。俄国旅行家阿·马·波兹德涅耶夫（1851—1920）于光绪十九年（1893年）在归化城考察，详细记载了各大召庙的状况。据其旅行记记载，当时的朋苏克召（崇寿寺）、拉布齐召（宏庆寺）等召庙已完全倒塌或荒废。一些召庙的呼图克图转世者已无人去寻找。参见：（俄）阿马波慈德涅夫著.蒙古及蒙古人．刘汉明等译．呼和浩特：内蒙古人民出版社，1983：84—85．

2 原察哈尔右翼四旗，又称绥东四旗，即清朝察哈尔八旗中的西边正黄、正红、镶红、镶蓝四旗。民国时期先后由隶察哈尔绥远特别行政区、绥远省、巴彦塔拉盟，中华人民共和国成立后由乌兰察布盟管辖，主体区域被编为察哈尔右翼前、中、后三旗。

3 关于清时四子部落旗的召庙数量，有不同记载。内蒙古档案馆藏《四子部落旗实录》（民国）记载有28座召庙。《绥远通志稿》（卷54）亦记载28座召庙。上述两种文献详细记载了各召庙方位与僧数。《乌兰察布寺庙》记载有32座寺庙，15个诵经会。《解放前四子部概况》记载共有32座召庙与诵经会，并详细记录了各召庙的殿宇与拉布隆数量，《四子王旗简史》记载有24座召庙，并详细记载了各召庙的创建时间。课题组依据后两种文献得出以上数据。参见：《四子部落旗实录》（四子王旗文史资料第五辑）2005,10；满都麦，莫德尔图主编．乌兰察布寺院．海拉尔：内蒙古文化出版社,1996,5；李德尔道尔吉编著．解放前四子部概况．呼和浩特：内蒙古人民出版社，2010,7；丹·赛音巴雅尔著．四子王旗简史．四子王旗地方志办公室出版，1997,12．

4 课题组2009年四子王旗境内召庙现状调研资料，此次调研包括该旗境内10座寺庙遗址。

5 课题组2012年乌兰察布市寺庙补充调研资料。

6 课题组2010年锡林郭勒盟寺庙调研资料。由镶黄旗哈音海日瓦庙派遣一名喇嘛主持法事。

7 课题组调研普查的寺庙涉及乌兰察布市境内6座召庙，此3座庙为现仍延续法事活动的召庙。

Lama,who was in charge of Hohhot was not confined to the local area,but could be held by Hutuketu from places as Beijing, Rehe,Duolunnur etc. The temple constructions are mainly architecture of Han style. Only the Main Hall in Xilituzhao Temple, Qingyuan Temple in Wusutuzhao in the west suburb of the city, the Main Hall in Faxi Temple are Han-Tibetan combination style.

The research group conducted the survey and mapping of the 7 extant and representative Tibetan Buddhism Temples in Hohhot systematically and completely. Specific cases are as follows.

Extant Temples Researched within Hohhot City			
Name	Location	Basic Information	
		Research Method	Extant Architecture Status
Dazhao Temple	Front Dazhao Street, Yuquan District	General survey and mapping	Memorial Arch, Monastery Gate, The Heaven King Hall, Bell-drum Tower,Boddhi-path Hall, Mahavira Hall, Nine-spaces Building, Arhat Hall, Epositary of Buddhist Sutra,Ushnisha Sita Tapatra Hall,Gong Zhong Cang,Bodhisattva Hall, The Jade Buddha Hall, Maitreya Hall
Xiretzhao Temple	North end of Shitou Lane, Yuquan District	General survey and mapping	Memorial Arch, The Heaven King Hall,Boddhi-path Hall, Mahavira Hall, Nine-spaces Building,The Old Pagoda,Dharmapalas Hall, Avalokiteshvara Hall, Tara Hall, Buddha Hall, The Pagoda of Longevity
Xiaozhao Temple	North end of front Xiaozhao Street, Yuquan District	General survey	Memorial Arch
Tavensobrega Temple	Wutasi Street, Yuquan District	General survey and mapping	Monastery Gate Tara Hall, Avalokiteshvara Hall, Buddhas of three-Yugas Hall, Vairocana Hall, Dorje Sarira Pagoda,Amoghasiddhi Hall, Amitabha Hall, Amon Hall, Ratnasambhava Hall,The Pagoda
Osutozhao Temple	West Wusutu Village, Left Tumd County	General survey and mapping	Qing Yuan Temple, Longevity Temple, Fasi Temple, Arhat Hall, Pharmacist Temple etc.
Lama-in-agui Temple	Left Tumd County	General survey	Dharmapalas Hall, The Buddha House, Lama Cave, Longwang Hall, Guang Hua Temple, Sarira Hall, Buddhas Stupatup Buddha House, Tong Hua Pagoda, Boddhi Pogoda
Emqizhao Temple	Yuquan District	General survey and mapping	Mahavira Hall

2.Ulanqab City

Ulanqab City is located in central Inner Mongolia Autonomous Region. The landform is composed from north to south by Mongolia plateau, Ulanqab hills,Yinshan Mountains and loess hills. The Area is situated in mid temperate zone and belongs to continental monsoon climate with distinct features of four seasons. Because Mount Daqing sits in the middle of the area, the special climate is formed, warmer and rainy in the area of front mountain and windy in back mountain.

Ulanqab Prefecture is composed of six Zhasake Counties in Qing Dynasty as Sizibuluo County, Kerke Right-wing County, Maoming'an County,and Wulatesangong County. Because the historical evolution of administration, the temples in Ulanqab region are Siziwangqi Temple and the original Four Chahar Right-wing Counties Temple.

The temples in Ulanqab City are characterized by small numbers and different architectural styles in different Counties etc. There were 24 temples, 8 Chanting Temples in Siziwang County region with the extant temples as follows: Xiramuren Temple, 2 Wangfu Temple, and only 1 remaining wrecked hall in West Lahada Temple, only 3 remaining wrecked monk houses, only 1 Agui Temple in the original Four Chahaer Right-wing Counties, only 2 remaining stone lions in the original Bayinchagan Temple in Zhuozi county, 1 newly-built hall in the original Badematu Temple in Maqun of Shangdu in Shangdu country. The architectural style of the extant temples in Siziwang County are all Tibetan style, while only Chahar temple is mainly in Han- style and the combination of Han and Tibet.

The research group investigated the only 3 extant Tibetan Buddhism temples in Ulanqab City area on the spot; the specific circumstances are as shown in the following table.

Extant Temples Researched within Ulanqab City			
Name	Location	Basic Information	
		Research Method	Extant Architecture Status
Xiaramuren Temple	Hongor Sumu,Siziwang County	General survey and mapping	Exotoric Buddhism Hall,Dharmapalas Hall, Vajrayana Hall
Ordon Temple	Chaganbaolige Sumu, Siziwang County	General survey	Mahavira Hall, Rajang Hall, Stupa
Agui Temple	Hayanhudong Sumu,Right-wing Back Chahar County	General survey	Mahavira Hall, Argu Cave, The Rear Hall, Buddha Hall

乌兰察布市地区调研的现存召庙			
召庙名称	所在位置	基本信息	
		调研情况	现存主要建筑
希拉木仁庙	四子王旗红格尔苏木	普查、测绘	显宗殿、护法殿、密宗殿
王府庙	四子王旗查干宝力格苏木	普查	大雄宝殿、扎仓殿、舍利塔
阿贵庙	察哈尔右翼后旗哈彦忽洞苏木	普查	大雄宝殿、阿贵洞、后殿、活佛府

三、锡林郭勒盟地区

锡林郭勒盟位于内蒙古自治区中部。该地区以高原为主体，兼有多种地貌，总体地势是南高北低，东、南部多低山丘陵，盆地错落其间，西、北部地形平坦，零星分布一些低山丘陵和熔岩台地。这一地区的主要气候特点是风大、干旱、寒冷，为华北最冷的地区之一，降雨量则由东南向西北递减。

锡林郭勒盟由清时乌珠穆沁、浩齐特、阿巴嘎、阿巴哈纳尔、苏尼特五部十旗组成。1958年撤销察哈尔盟建制，所辖正蓝旗、镶白旗、正白旗、镶黄旗4旗划归锡林郭勒盟。锡林郭勒盟现存召庙有32座[1]，召庙现存数量居自治区之首。

锡林郭勒盟境内的召庙普遍呈现现存数量较大、分布均匀、召庙形态多样、等级形制高、建筑风格以汉式为主、现存大雄宝殿居多等特点。该盟现存内蒙古地区已罕见的诵经会建筑若干座，如苏尼特左旗敖兰胡都格诵经会、查干陶勒盖诵经会[2]等。原锡林郭勒盟五部均有召庙留存，其中原乌珠穆沁右翼旗旗属六大寺庙[3]中的五座寺庙仍在延续法事活动。现已划归锡林郭勒盟的多伦诺尔县原为著名的喇嘛城，其两大寺庙[4]为清帝敕建庙，共有来自藏区、蒙古本土的14位活佛在此设仓，清廷四大呼图克图[5]均在此列。清廷在此设立喇嘛印务处，命章嘉呼图克图为札萨克达喇嘛，掌管蒙古喇嘛教事务，制定蒙古各旗选派僧人到该寺礼佛的制度。锡林郭勒盟召庙建筑以汉式建筑为主，仅苏尼特左右翼二旗的寺庙多为藏式建筑。寺庙布局多取沿中轴线均衡排列形式，从南至北分别有旗杆、影壁、山门、天王殿（山门与天王殿有时合为一体）、钟鼓楼、大雄宝殿、左右配殿、弥勒佛殿、敖包等建筑设置。“文化大革命”时期锡林郭勒盟寺庙多数被拆毁，但由于地域原因，一些召庙的大雄宝殿作为以召庙聚落为基础新建的基层行政单位的粮库或战备粮粮库，留存至今。

课题组对锡林郭勒盟境内的27座藏传佛教召庙展开了调研普查工作，具体情况如下表所示。

锡林郭勒盟地区调研的现存召庙			
召庙名称	所在位置	基本信息	
		调研情况	现存主要建筑
毕鲁图庙	苏尼特右旗朱日和镇毕鲁图嘎查	普查	大雄宝殿
查干敖包庙	苏尼特左旗查干敖包苏木阿如宝拉格嘎查	普查、测绘	西拉布隆活佛府
巴音乌素诵经会	苏尼特左旗洪格尔苏木希如昌图嘎查	普查	遗址（仅存已无顶的朱都巴殿、天王殿与院墙）
杨都庙	阿巴嘎旗洪格尔高勒镇镇政府所在地	普查	大雄宝殿、显宗殿、菩提道学殿

1 因道路险阻等客观原因，未能实地调研4座庙，即西乌珠穆沁旗彦吉嘎庙（已恢复法事活动）、乌兰乌台庙（复建召庙，位于旅游区）、苏尼特左旗宝日汗喇嘛庙（召庙遗存）、正镶白旗会苏庙（仅存残破的大殿，被牧户用作棚圈）及阿巴嘎旗兀良哈庙（2011年年末由纳·布和哈达先生电话告知，仅存一座小殿）。

2 诵经会，蒙古语称“呼日拉”，为一种独特的召庙形态，通常由民众自行组织，建造小尺度殿宇或搭建蒙古包，邀请一名或数名喇嘛按时诵经礼佛，满足僧众信仰生活需求的召庙雏形。当诵经会在僧侣数量、殿宇规模等方面达到一定程度后，经旗衙、大召庙等的准许，升级为正式召庙。内蒙古各盟旗均有诵经会，其中以锡林郭勒盟为最多。这与草原广阔的地理空间、游牧生活方式及分散而居的社会生活形态有密切联系。如在阿巴嘎左右翼二旗、阿巴哈纳尔右翼旗有50余座诵经会。而在苏尼特左翼旗境内曾有85座诵经会。参见：那·布和哈达主编．锡林郭勒寺院（蒙古文）.海拉尔：内蒙古文化出版社，1999,4；达·查干编写．苏尼特左旗寺庙（蒙古文）（内部资料）．苏尼特左旗政协文史办公室出版,2000,12．

3 原乌珠穆沁右翼旗旗属六大寺庙或称六大寺庙分别为：浩勒图庙（施恩寺）、乌兰哈拉嘎庙（宝成寺）、喇嘛库伦庙（集慧寺）、新庙（密宗广普寺）、彦吉嘎庙（施缘寺）、敖包图葛根庙。除乌珠穆沁右翼旗旗庙——敖包图葛根庙外，其余五座已恢复法事活动。

4 多伦诺尔两大寺庙分别为汇宗寺与善因寺。前者由康熙帝敕建，后者由雍正帝敕建。参见：任月海.多伦汇宗寺［M］.北京：民族出版社，2005,6．

5 清廷四大呼图克图指章嘉、葛尔丹席力图、济隆、敏朱尔四位呼图克图转世体系。

3.Xilingol Prefecture

Xilingol Prefecture is located in the middle of Inner Mongolia Autonomous Region. It is a district with a variety of landforms and plateau as the main body. The overall terrain is high in south and low in north, more low hills in east and south, with basons among them, and flat terrain in West and North dotted with some low hills and lava tablelands. The major climate characteristics are strong wind, drought and cold and the region is one of the coldest in North China with rainfall descending from southeast to northwest.

Xilingol Prefecture is composed by five sections and ten Counties in Qing Dynasty of Ujimqin, Haoqite, Abag, Abagahanaer, Sonid. Chahar Prefecture was revoked in 1958, and the four Counties—Zhenglan County, Xiangbai County, Zhengbai County, and Xiangbai County under the jurisdiction were turned to put under Xilingol Prefecture. There are 32 temples in Xilingol Prefecture ,ranking the first in the number of extant temples in the Autonomous Region .

The temples in Xilingol Prefecture are characterized by large quantities, uniform distribution, various forms, high grade and Han architectural style, more extant Main Halls etc. There are several rare extant Chanting architectures in this region, such as Olonhudug Chanting Temple in Left Sonid County, Chagantaolegai Chanting Temple etc. There are extant temples in the original Five Sections of Xilingol Prefecture, among which 5 of the 6 big temples of the original Right Ujimqin County still continue the religious activities. Duolun Nur county, now under Xilingol Prefecture, was the famous Lama City. The two big temples in it were imperial temples built in Qing Dynasty. Altogether 14 living Buddhas from Tibet and local Mongolia area settled here including the Four Hutuketu of Qing. Qing government established Yinwuchu here and ordered Zhangjia Hutuketu as Zhasakeda Lama handling Lama affairs in Mongolia, formulating the system of selecting monks to worship Buddha from each County Mongolia. The temples in Xilingol Prefecture were built mainly in Han style while only the temples of Left Sonid County and Right Sonid County are mostly Tibetan architecture. The layout of the temples mostly adopted balanced arrangement along the axis, respectively with flagpole, screen wall, temple gate, The Heaven King Hall (sometimes the two are combined together), Bell and Drum Tower, Main Hall, side hall, temple hall, Obo and some other buildings from the south to the north. During the "Culture Revolution" ,most temples in Xilingol Prefecture were demolished, but due to the geographical reasons, some Main Halls of the temples are preserved until now as the grain depots of the basic administrative units newly organized on the basis of temple settlement .

The research group surveyed and investigated 27 Tibetan Buddhism Temples in Xilingol Prefecture; the specific results are shown in the following table.

Extant Temples Researched within Xilingol Prefecture

Name	Location	Basic Information	
		Research Method	Extant Architecture Status
Bilut Temple	Bilutu Gacha,Zhurihe County,Right Sonid County	General survey	Mahavira Hall
Chagan Obo Temple	Arubaolage Gacha,Chagan Obo Sumu ,Left Sonid County	General survey, mapping	Silabulong Buddha House
Bayinwusu Chanting Temple	Xiruchangtu Gacha,Honger Sumu,Left Sonid County	General survey	Remains (only Zhuduba Hall, The Heaven King Hall without roofs, and yard walls)
Yangdu Temple	The location of Hongergaole , Town Government,Abaga County	General survey	Mahavira Hall, Exotoric Buddhism Hall, Lamrim Hall
Beis Temple	Xilinhaote City	General survey, mapping	Vajrayana Hall, Exotoric Buddhism Hall,Mahavira Hall, Ming Gan Hall
Xin Temple	Daotenur Town,East Ujimqin County	General survey	Calachakra Hall, Monastery Gate
Wanggai Temple	Balagergaole Town ,Ujimqin County	General survey	Calachakra Hall
Hoqit Temple	Jirenguole Sumu ,West Ujimqin County	General survey	Mahavira Hall, Khutu House, Stupa, The West-side Hall, The East-side Hall,The Heaven King Hall
Hoh Temple	Duolunnur County	General survey, mapping	Lai Jian Temple Gate, The Heaven King Hall,The Rear Hall, The Courtyard of Icang-skya Khutu
Xiara Temple	Duolunnur County	General survey, mapping	Monastery Gate, The Heaven King Hall, Bell Tower, Drum Tower
Boritologai Temple	Yihenaoer Sumu,Zhengxiangbai County	General survey	Mahavira Hall, The East-side Hall, The Weat-side Hall, Stupa
Burd Temple	Burd Sumu ,Zhengxiangbai County	General survey and mapping	Mahavira Hall
Chagantaolegai Chanting Temple	Daerhanwula Sumu, Left Sonid County	General survey	The Small Scripture Hall, Cang

续表

锡林郭勒盟地区调研的现存召庙			
召庙名称	所在位置	基本信息	
		调研情况	现存主要建筑
贝子庙	锡林浩特市	普查、测绘	密宗殿、显宗殿、大雄宝殿、明干殿
新庙	东乌珠穆沁旗道特淖尔镇	普查	时轮殿、山门
王盖庙	乌珠穆沁旗巴拉噶尔高勒镇	普查	时轮殿
浩齐特庙	西乌珠穆沁旗吉仁郭勒苏木	普查	大雄宝殿、活佛拉布隆、佛塔、西配殿、东配殿、天王殿
汇宗寺	多伦诺尔县	普查、测绘	敕建庙山门、天王殿、后殿、较完整的章嘉活佛仓院落
善因寺	多伦诺尔县	普查、测绘	山门、天王殿、钟楼、鼓楼
宝日陶勒盖庙	正镶白旗伊和淖尔苏木	普查	大雄宝殿、东配殿、西配殿、佛塔
布日都庙	正镶白旗布日都苏木	普查、测绘	大雄宝殿
查干陶勒盖诵经会	苏尼特左旗达尔汗乌拉苏木	普查	小经堂、庙仓
敖兰胡都格诵经会	苏尼特左旗赛汗戈壁苏木达尔汗乌拉区	普查	大经堂、僧舍
吉日嘎朗图庙	阿巴嘎旗那仁宝拉格苏木吉日嘎朗图嘎查	普查	后七丈大殿
汉贝庙	阿巴嘎旗别力古台镇	普查	僧房
浩勒图庙	西乌珠穆沁旗浩勒图音高勒镇镇政府所在地	普查	山门、耳房、配殿、大雄宝殿、佛塔、八座舍利塔、大白塔
乌兰哈拉嘎庙	西乌珠穆沁旗乌兰哈拉嘎苏木政府所在地	普查	舍利殿、活佛拉布隆、大经堂
喇嘛库伦庙	东乌珠穆沁旗乌力雅苏台镇	普查	大雄宝殿、配殿、山门、钟楼、鼓楼、白塔
宝拉格庙	东乌珠穆沁旗宝拉格苏木	普查	大雄宝殿、舍利殿、山门、佛塔
嘎黑拉庙	东乌珠穆沁旗道特淖尔镇嘎黑拉区	普查	大经堂、山门、配殿
哈音海日瓦庙	镶黄旗新宝拉格镇	普查	大雄宝殿、山门、东西配殿、藏式小殿、钟楼、鼓楼
明如拉葛根庙	正蓝旗五一牧场	普查	15座僧舍及庙仓
玛拉日图庙	正蓝旗上都镇	普查	天王殿、钟楼、鼓楼、大雄宝殿、东西配殿、佛塔
玛拉盖庙	太仆寺旗贡宝拉格苏木	普查	大殿、东配殿、西配殿
善达庙	正镶白旗伊和淖尔苏木善达区	普查	大雄宝殿、安居殿
扎嘎苏台庙	正蓝旗扎嘎苏台苏木政府所在地	普查	后殿、僧舍

第三节 东部地区

一、赤峰市地区

内蒙古第一人口大市赤峰市位于自治区东南部。该地区呈三面环山，西高东低，多山多丘陵的地貌特征，大体可被分为四个地形区，即北部山地丘陵区、南部山地丘陵区、西部高平原区、东部平原区。赤峰市地区冬季漫长而寒冷，春季干旱多大风，夏季短促炎热、雨水集中，秋季短促、气温下降快、霜冻降临早，属中温带半干旱大陆性季风气候区。

赤峰市前身为昭乌达盟，1983年撤盟设地级赤峰市。昭乌达盟由清时阿鲁科尔沁旗、巴林左右翼2旗、克什克腾旗、翁牛特左右翼2旗、敖汉左右翼2旗、奈曼旗、喀尔喀左翼旗、扎鲁特左右翼2旗组成。现辖区也包括原卓素图盟喀喇沁旗部分地区。

赤峰市境内的召庙普遍呈现建筑历史悠久、创建历程独特、建筑风格因旗而异等特点。赤峰市召庙及其建筑的历史久远，个别召庙前身能够追溯至辽代，如巴林左旗格里布尔召原为辽代佛寺。一些召庙的创建史本身折射出地域文化变迁史，记录了藏传佛教替代地域本土宗教形态或汉传佛教的独特历程。如喀喇沁旗龙泉寺始建于辽代，兴盛于元、明、清三代，并于清朝年间被改为藏传佛教召庙。法轮寺建于辽、金、元时期的灵隆寺废墟之上。赤峰市一些召庙的创建者为清朝下嫁蒙古王公的宗室格格及其驸马。如康熙帝第三女固伦荣宪公主下嫁巴林右翼旗扎萨克郡王乌日衮，新建东瓦房庙。[1]至锡林郭勒盟东部地区

1 除固伦荣宪公主（1673—1728）于康熙四十五年（1706年）所建东瓦房庙（荟福寺）外，在巴林右翼旗另有清太宗皇太极之五女固伦淑慧长公主（1632—1700）于康熙六年（1667年）所建西瓦房庙（圆会寺）。此两座庙与阿贵庙（阐化寺）、古日古勒台庙（嘉佑寺）被称为巴林四大寺庙。参见：刘冰，顾亚丽编著．草原姻盟——下嫁赤峰的清公主．呼和浩特：远方出版社，2007，4；嘎拉增、呼格吉乐图等编．昭乌达寺院（蒙古文）．海拉尔：内蒙古文化出版社，1994，10；中共中央内蒙古分局宗教问题委员会编．内蒙古喇嘛教（上册）．1951，3．

Continued Table

Extant Temples Researched within Xilingol Prefecture			
Name	Location	Basic Information	
		Research Method	Extant Architecture Status
Olonhudug Chanting Temple	Daerhanwula district,Saihan Gobi Sumu,Left Sonid County	General survey	The Big Scripture Hall, Monk House
Jarigalangt Temple	Jirigalangt Gacha,Narenbaolage Sumu,Abag County	General survey	The Rear Seven-spaces Hall
Hamb Temple	Bieligutai Town , Abag County	General survey	Monk House
Holot Temple	The location of Haoletuyingaole Town Government ,West Ujimqin County	General survey	Monastery Gate, Aisle, The Side Hall, Mahavira Hall, Stupa, The Eight Stupas,The Pagoda
Ulanhalag Temple	The location of Wulanhalaga Sumu government,West Ujimqin County	General survey	Sarira Hall,Rabulong Buddha,The Big Scripture Hall
Lamiinhure Temple	Wuliyasutai Town,East Ujimqin County	General survey	Mahavira Hall, The Side Hall, Monastery Gate, Bell Tower, Drum Tower, The Pagoda
Bulag Temple	Baolage Sumu ,West Ujimqin County	General survey	Mahavira Hall, Stupa, Monastery Gate, Stupa
Gahile Temple	Heila District ,Daotenaoer Town,East Ujimqin County	General survey	The Big Scripture Hall, Monastery Gate, The Side Hall
Hayinhairiwa Temple	Xinbaolage Town , Xianghuang County	General survey	Mahavira Hall, Monastery Gate, The East-side Hall, The West-side Hall, The Small Tibetan Hall, Bell Tower, Drum Tower
Mingrulagegen Temple	Wuyi Pasture,Zhenglan County	General survey	The Fifteen Monk Houses and Congs
Malarite Temple	Shangdu Town, Zhenglan County	General survey	The Heaven King Hall, Bell Tower,Drum Tower, Mahavira Hall, The Side Halls, Stupa
Malagai Temple	Gongbaolage Sumu, Taipusi County	General survey	The Main Hall, The East-side Hall,The West-side Hall
Shangde Temple	Shanda District,Yihenaoer Sumu,Zhengxiangbai County	General survey	Mahavira Hall,Living Hall
Jagestai Temple	The location of Jagestai Sumu Government,Zhenglan County	General survey	The Rear Hall, Monk House

Ⅲ The Eastern Region

1. Chifeng City

Chifeng, the largest city in Inner Mongolia in terms of population, is located in the southeast of Inner Mongolia. Embraced on three sides by green hills, it is high in the west and low in the east and rather mountainous and hilly. Generally, this area can be divided into four topographic zones, i.e. the northern mountainous and hilly zone, the southern mountainous and hilly zone, the western high plain zone and the eastern plain zone. Chifeng City has long and cold winter, arid and windy spring, short, torrid and rainwater-concentrated summer, and short autumn during which temperature drops rapidly and frost comes early, so it belongs to temperate semi-arid continental monsoon climate zone.

Chifeng City is a Prefecture-level city founded in 1983 and formerly was Zhaowuda Prefecture, which comprised Ar Khorchin County Baarin Left County and Baarin Right County, Hexigten County, Ongniud Left County and Ongniud Right County, Aohan Left County and Aohan Right County, Naiman County, Khalkha Left County, Jarud Left County and Jarud Right County in Qing Dynasty. In addition, now the jurisdiction of this city also covers the former Harqin County of Zhuosutu Prefecture.

Temples in Chifeng are generally featured by long history, unique development course and county-oriented architectural style. Some of them can be traced to Liao Dynasty, for example, Gelibuer Temple in Baarin Left County is a Buddhism temple from Liao Dynasty. The establishment history of some temples reflects regional cultural changes and presents the unique course that Tibetan Buddhism temples replaced native temples or Chinese Buddhism. For instance, Longquan Temple in Harqin County was built in Liao Dynasty, flourished in Yuan Dynasty, Ming Dynasty and Qing Dynasty, and was changed into Tibetan Buddhism temple in Qing Dynasty; Falun Temple was built on the ruins of Linglong Temple of Liao Dynasty, Jin Dynasty and Yuan Dynasty. Some temples in Chifeng were founded by the emperor's imperial clan princesses who married mongolian princes and dukes and their husbands. For example, Hesuorongxian Princess, the third daughter of Kangxi Emperor, married Wurigun, Prefectual Governer of Jasak of Baarin Right County and built East Wafang Temple. Tibetan type temple that was rare in the eastern region of Xilingol Prefecture reemerged in Chifeng. For example,

已很罕见的藏式召庙建筑在赤峰市再度出现，如阿鲁科尔沁旗罕庙、根坯庙、巴拉奇如德庙等召庙以藏式建筑为主，而原喀喇沁三旗[1]的召庙均为汉式建筑。显现了内蒙古地域召庙建筑风格之带状分布的规律性。

课题组对赤峰市境内11座藏传佛教召庙展开了普查调研工作，具体情况如下表所示。

赤峰市地区调研的现存召庙

召庙名称	所在位置	基本信息	
		调研情况	现存主要建筑
东瓦房庙（荟福寺）	巴林右旗大板镇	普查、测绘	山门、密宗殿、护法殿、天王殿、大雄宝殿、战神殿、钟楼、鼓楼、普觉寺、司命神殿
格里布尔召（真寂寺）	巴林左旗查干哈达乡	普查	大雄宝殿、四大天王殿、喇嘛僧舍
查干布热庙（梵宗寺）	翁牛特旗乌丹镇北4公里处	普查、测绘	天王殿、护法殿、长寿佛殿、钟楼、鼓楼、罗汉堂、藏经阁、关公庙、大雄宝殿、时轮金刚殿、转经阁、弥勒殿、度母殿
罕庙	阿鲁科尔沁旗罕苏木境内	普查	山门、天王殿、大经堂、大雄宝殿、护法殿、罗汉殿、喇嘛僧舍、寺庙管理用房、查干活佛府、哈木尔活佛府
巴拉奇如德庙	阿鲁科尔沁旗巴拉奇如德苏木境内	普查	萨布腾拉哈木殿、大雄宝殿、葛根活佛府、喇嘛僧舍
根坯庙	阿鲁科尔沁旗罕苏木境内	普查	四大天王殿、玛尼殿、关公殿、钟鼓楼、大雄宝殿、护法殿、喇嘛僧舍
龙泉寺	喀喇沁旗锦山镇境内	普查	大雄宝殿、山门、僧舍
福会寺	喀喇沁旗锦山镇境内	普查	山门、天王殿、钟楼、鼓楼、长寿佛殿、大雄宝殿、释迦牟尼佛殿、弥勒佛殿、观音殿、莲花生大师殿、护法殿、东西配殿
马日图庙（法轮寺）	宁城县大城子镇	普查、测绘	大雄宝殿、护法殿、菩萨殿、韦驮殿、地藏殿、药师殿、长寿殿、钟鼓楼、关公殿、旃檀殿、天王殿、灵隆寺
灵悦寺	喀喇沁旗锦山镇	普查	天王殿、钟楼、鼓楼、前殿、东西厢房、东西配殿、大雄宝殿、大经堂、藏经阁、转经阁
毕如庙	克什克腾旗经棚镇	普查	大雄宝殿、四大天王殿、海尔汗庙、布格德殿、东西配殿、召佛殿、影壁

二、通辽市地区

通辽市位于内蒙古自治区东部，总体地势南部和北部高，中部低平，呈马鞍形，其北部为大兴安岭南麓余脉的石质山地丘陵，南部为辽西山地边缘的浅山、黄土丘陵区，中部为西辽河流域沙质冲积平原。该地区属典型的半干旱大陆性季风气候，春季干旱多风，夏季炎热，雨热同季。

通辽市现辖区由清时哲里木盟、卓索图盟、昭乌达盟部分地区组成。清时的哲里木盟辖科尔沁左右翼6旗、郭尔罗斯左右翼2旗、杜尔伯特旗、扎赉特旗。1999年撤哲里木盟，设地级通辽市。

通辽市境内的召庙普遍呈现创建模式独特、个别区域召庙集中分布等特点。漠南蒙古地域唯一一座喇嘛旗[2]，即在旗境内施行政教合一体制的席力图库伦[3]扎萨克喇嘛旗位于该地域，故在蒙古地域藏传佛教传播史上又增添一种独特的召庙创建模式。今库伦旗三大寺[4]与吉祥天女庙均由该旗历代札萨克达喇嘛主持兴建，并由其掌管宗教事务。召庙分布较集中，个别召庙，如象教寺成为席力图库伦札萨克达喇嘛的执政中心及全旗政教合一的掌印机构，兼具寺庙与衙门的双重职能。

1 喀喇沁三旗，即原卓索图盟天聪年间所设喀喇沁左旗（乌公旗）、右旗（喀喇沁王旗）、中旗（马公旗或喀喇沁贝子旗）。1912年隶热河省，中华人民共和国成立后喀喇沁三旗辖地分隶内蒙古自治区赤峰市喀喇沁旗、宁城县等旗县及辽宁省喀左旗等旗县。参见：政协喀喇沁旗文史资料委员会．喀喇沁旗文史资料（第二辑）．1985,12:13；周清澍主编.内蒙古历史地理［M］.呼和浩特：内蒙古大学出版社，1991：224．

2 喇嘛旗为清朝在蒙古地区施行的盟旗制度之一种。喇嘛旗在旗境内施行政教合一的制度。有清一代，在蒙古地区设七个喇嘛旗，施行政教合一体制，统领政教。7个喇嘛旗中5个在漠北蒙古（即喀尔喀蒙古），内蒙古与青海各有一个喇嘛旗。席力图库伦又称小库伦，明万历年间察哈尔部曾驻牧于此，始建寺庙。

3 席力图库伦又称小库伦，明万历年间察哈尔部曾驻牧于此，并始建寺庙。满殊希里呼图克图阿兴喇嘛来此后以“满殊希里库伦”著称，后其弟被清廷封为“席力图达尔罕乔尔吉”，席力图库伦之称由此而来。参见：呼日勒沙编．哲里木寺院（蒙古文）．海拉尔：内蒙古文化出版社，1993：109—116．

4 库伦三大寺为兴源寺、福缘寺、象教寺。

Han Temple, Genpi Temple and Ar Balaqirude Temple in Khorchin County are Tibetan type while the temples in the former three Harqin County are Chinese type, which reflects the regularity of zonal distribution of temple styles in Inner Mongolia region.

The research team surveyed 11 Tibetan Buddhism temples in Chifeng. The table below presents the details.

Extant Temples Researched within Chifeng City			
Name	Location	Basic Information	
		Research Method	Extant Architecture Status
East Huhger Temple (Huifu Temple)	Daban Town of Baarin Right County	General survey and mapping	Monastery Gate, Ajrayana Hall, Dharmapalas Hall, The Heaven King Hall, Mahavira Hall, Zhan Shen Hall, Drum Tower, Bell Tower,Pujue Temple , Siming (Life-control) Hall
Geliver temple (Zhenji Temple)	Chaganhada Township of Baarin Left County	General survey	Mahavira Hall, The Four Heavenly Kings Hall,Lama House
Chaganbure Temple (Fanzong Temple)	4km north of Wudan Town of Ongniud County	General survey and mapping	The Heaven King Hall,Dharmapalas Hall, Longevity Hall, Bell Tower, Drum Tower,Arhat Hall, Depositary of Buddhist Sutra,Guan Gong Temple, Mahavira Hall, Kalachakra Hall,Circumambulation Terrace, Maitreya Hall, Tara Hall
Han Temple	Han Township of Arkhorchin County	General survey	Monastery Gate, The Heaven King Hall, Main Assembly Hall, Mahavira Hall, Dharmapalas Hall, Arhat Hall, Lama Housc, Thc Tcmplc Administration, Chagan Buddha Hall, Hamar Buddha Hall
Balaqirude Temple	Balaqirude Township of Arkhorchin County	General survey	Samtanlahan Hall, Mahavira Hall, The Buddha Hall, Lama House
Gempi Temple	Han Township of Arkhorchin County	General survey	The Heaven King Hall, Mani Hall, Guan Gong Hall, Drum-bell Tower, Mahavira Hall, Dharmapalas Hall, Lama House
Longquan Temple	Jinshan Town of Harqin County	General survey	Mahavira Hall, Monastery Gate, Monk House
Fuhui Temple	Jinshan Town of Harqin County	General survey	Monastery Gate, The Heaven King Hall, Bell Tower, Drum Tower, Longevity Hall,Mahavira Hall, Sakyamuni Buddha Hall,Maitreya Hall, Avalokiteshvara Hall, Padma Sambhavav Hall,Dharmapalas Hall, The East-side Hall, and The West-side Hall
Maritu Temple (Falun Temple)	Dachengzi Town of Ningcheng County	General survey and mapping	Mahavira Hall, Dharmapalas Hall, Bodhisattva Hall,ayanningb Hall, Pharmacist Hall, Longevity Hall, Bell-drum Tower, Guan Gong Hall, Zandan Hall, The Heaven King Hall, Ling long Temple
Lingyue Temple	Jinshan Town of Harqin County	General survey	The Heaven King Hall, Bell Tower, Drum Tower, The Front Hall, The East-side House, The West-side House, The East-side Hall, The West-side Hall, Mahavira Hall, Main Assembly Hall, Depositary of Buddhist Sutra, Circumambulation Terrace
Biru Temple	Jingpeng Town of Hexigten County	General survey	Mahavira Hall, The Four Heaven Kings Hall, Hair Han Hall, Bugd Hall, The East-side Hall and The West-side Hall, Sakyamuni Buddha Hall, Screen Wall

2. Tongliao City

Tongliao City is located in the east of Inner Mongolia Autonomous Region. Its relief is saddle-shaped, namely, high in the south and north and low in the middle. In its north lies the rocky hilly area at the foot of the south of Great Khingan, in its south lie the shallow mountain area and loess hilly region on the edge of western Liaoning mountainous region, and in its middle lies the sandy fluvial plain in West Liaohe Basin. This area belongs to the typical semi-arid continental monsoon climate, arid and windy in spring and torrid in summer, and the rainy and hot weather is in the same season.

Now Tongliao City is composed of some areas of Jirem Prefecture, Josota Prefecture and Jouda Prefecture of Qing Dynasty. Jirem Prefecture, in Qing Dynasty had jurisdiction over 6 Khorchin left and right counties, 2 Gorlos left and right counties, Dorbod County and Jalaid County. And in 1999, Jirem Prefecture, was renamed Tongliao, a Prefecture-level city.

Generally temples in Tongliao City have a unique establishment mode and some temples are concentrated individual areas. The only Lama county in Monan Mongolian region, namely, Xilitu Kulun Jasak Lama county that enforces the system of unification of the government and the temple is located in this area, so a unique temple establishment mode was created in the communication history of Tibetan Buddhism in Mongolia region. Now three major temples of Kulun County and Siddhi Lakshmi Temple were built under the auspices of Jasak Lama of this county in all ages and the religious affairs were also administrated by them. The temples are relatively concentrated, individual temples like Xiangjiao Temple become the administration center of Xilitu Kulun Jasak Lama and the theocratic power-holding institution of the whole county, and they served as temple and bureaucracy.

There is another establishment mode of temples in Tongliao, that is, the temples were built by Buddhism promotion person in the early communication history

通辽市召庙的另一种创建模式为由藏传佛教早期传播史上的弘法者从呼和浩特地区东游至靠近清朝政治中心的科尔沁部新建召庙。如受土默特阿勒坦汗之邀，来蒙古地区弘法的满殊希里呼图克图阿兴喇嘛为第一代席力图喇嘛。达赖喇嘛派往蒙古地区掌教的第四任代表——迈达日呼图克图来到蒙古地区后先为呼和浩特美岱召主持开光仪式，后东行至科尔沁部创建迈达日葛根庙。

课题组对通辽市境内7座藏传佛教召庙进行了调研普查工作，具体情况如下表所示。

通辽市地区调研的现存召庙

召庙名称	所在位置	基本信息	
		调研情况	现存主要建筑
兴源寺	库伦旗库伦镇内	普查	山门、鼓楼、钟楼、天王殿、护法殿、罗汉殿、大雄宝殿、玛尼殿
象教寺	库伦旗库伦镇内	普查	山门、弥勒殿、莲花生殿、药师佛殿、长寿佛殿、玉柱堂、度母殿
福缘寺	库伦旗库伦镇内	普查	山门、东西厢房、三世佛殿、五间楼
吉祥天女庙	库伦旗库伦镇内	普查	女神殿、罗汉堂、护法殿、白塔
迈达日葛根庙（格尔林庙）	库伦旗格乐林苏木所在地	普查	山门、大雄宝殿
希拉木仁庙（吉祥密乘大乐林寺）	市区西拉木伦公园北侧	普查	天王殿、大雄宝殿、观音殿、四座白塔、护法殿、财神殿、弥勒殿
板子庙	扎鲁特旗格日朝鲁苏木巴彦宝力格嘎查境内	普查	大经堂、山门、活佛府、喇嘛僧舍、寺庙管理用房

三、兴安盟地区

兴安盟位于内蒙古自治区的东北部，地处大兴安岭向松嫩平原过渡带的兴安盟由西北向东南分为四个地貌类型，即中山地带、低山地带、丘陵地带和平原地带，全年四季分明，属温带半干旱季风气候。

兴安盟建立于1946年，由哲里木盟分出。现辖区由清时的哲里木盟科尔沁右翼三旗及扎莱特旗部分地区组成。

兴安盟境内的召庙普遍呈现创建模式独特、召庙遗存较少、召庙原建筑以藏式为主等特点。兴安盟召庙亦呈现与通辽市召庙相同的一种创建模式，即藏传佛教早期传播史上的弘法者从呼和浩特地区东游至靠近清朝政治中心的科尔沁部新建召庙。如呼和浩特小召第一世内齐托音呼图克图于天聪至顺治初年间在科尔沁部十旗王公的资助下兴建巴音和硕庙。陶赖图葛根庙、王爷庙等召庙的建筑均为藏式建筑，显现了内蒙古地域召庙建筑风格之带状分布的规律性。而与其接壤的通辽市召庙以汉式建筑为主。与内蒙古地域内众多召庙相同，召庙聚落对该地域的城镇化发展奠定了空间设施基础，如乌兰浩特市市区在某种程度上以王爷庙作为聚落化发展的始点，召庙名称成为该市俗称。

课题组对兴安盟境内的4座召庙展开了调研普查工作，具体情况如下表所示。

兴安盟地区调研的现存召庙

召庙名称	所在位置	基本信息	
		调研情况	现存主要建筑
巴音和硕庙（遐福寺）	科右中旗巴彦胡硕镇	普查	西热喇嘛仓、博格达仓、堪布喇嘛仓
陶赖图葛根庙（葛根庙）	乌兰浩特市葛根庙镇	普查	大雄宝殿、山门、天王殿、舍利殿、金刚殿、梵通寺、菩提济渡寺、观音殿、护法殿、观音殿（在建）
王爷庙	乌兰浩特市普惠街	普查	山门、天王殿、大雄宝殿、五小庙、密宗殿
昂格日庙	扎赉特旗巴彦扎拉嘎	普查	大雄宝殿、山门、僧舍

of Tibetan Buddhism when they travelled from Hohhot eastwards to Khorchin that was close to political center of Qing Dynasty. For example, Manshu Xili Khutukhtu Axing Lama invited by Alatan Khan of Tumed to promote Buddhism in Mongolia region was the first Xilitu Lama and his younger brother was called "Xilitu Darhan Joel", there out the name of Xilitu Kulun. After coming to Mongonlia region, the fourth representative Maidari Khutukhtu assigned by Dalai Lama to take charge of religious affairs presided over the consecration ceremony of Meidai Temple in Hohhot and then travelled eastwards to create Maidari-gegen Temple in Khorchin.

The research team surveyed 7 Tibetan Buddhism temples in Tongliao City. The table below presents the details.

Extant Temples Researched within Tongliao City

Name	Location	Basic Information	
		Research Method	Extant Architecture Status
Xingyuan Temple	Kulun Town of Kulun County	General survey	Monastery Gate, Drum tower, Bell tower, The Heaven King Hall, Dharmapalas Hall, Arhat Hall, Mahavira Hall, Mani Hall
Xiangjiao Temple	Kulun Town of Kulun County	General survey	Monastery Gate, Maitreya Hall, Padma Sambhava Hall, Pharmacist Hall, Longevity Hall, Yu Zhu Hall, Tara Hall
Fuyuan Temple	Kulun Town of Kulun County	General survey	Monastery Gate, The Sides Houses, Buddhas of Three-yugas Hall, The Five-spaces Building
Auhen Ohin-engri Temple	Kulun Town of Kulun County	General survey	Nv Shen Hall(Godness Hall), Arhat Hall, Dharmapalas Hall, The Pagoda
Maidari-gegen Temple (Gerlin Temple)	Gerlin Township of Kulun County	General survey	Monastery Gate,Mahavira Hall
Xarauren Temple (Jixiang Micheng Dale Temple)	North side of Xilamulun Park in urban area	General survey	The Heaven King Hall,Mahavira Hall, Avalokiteshvara Hall, The Four Pagodas, Dharmapalas Hall, Mammon Hall, Maitreya Hall
Banzan Temple	Bayan Baolige Gacha of Gerichalu Township of Jarud County	General survey	Main Assembly Hall, Monastery Gate,The Buddha Hall,Lama House,The Temple Administration

3. Hinggan Prefecture

Hinggan Prefecture is located in the northeast of the Inner Mongolia Autonomous Region and in the transition zone of Great Khingan and Songnen Plain. It has four geomorphic types from the northwest to the southeast, that is, middle mountain area, low mountain area, hilly area and plain area. And this Prefecture with four distinct seasons belongs to temperate semi-arid monsoon climate.

Hinggan Prefecture was set up in 1946 and was separated from Jirem Prefecture. Now it has jurisdiction over three Khorchin right counties and some areas of Jalaid Prefecture.

Temples in Hinggan Prefecture are featured by unique establishment mode, remaining of fewer temples and Tibetan type-oriented original temples. Meanwhile, they also present the establishment mode the same as that of temples in Tongliao City, that is, the temples were built by Buddhism promotion person in the early communication history of Tibetan Buddhism when they travelled from Hohhot eastwards to Khorchin that was close to political center of Qing Dynasty. For example, the first Neiqituoyin Khutukhtu of Hohhot temple built Bayan Hesuo Temple under the auspices of princes and dukes of ten counties of Khorchin from Tiancong to early years of Shunzhi. Taulayitu-gegen Temple, Wangye Temple and other temples are Tibetan type, which reflects the regularity of zonal distribution of temple styles in Inner Mongolia region. However, temples in neighboring Tongliao City are mainly Chinese type. Same as many temples in Inner Mongolia region, temples settlement laid spatial facilities foundation for urbanization development of this area. For example, Ulanhot City takes Wangye Temple as its starting point of settlement development to some extent, so names of temples may be the common name of this city.

The research team surveyed 4 temples in Hinggan Prefecture. The table below presents the details.

Extant Temples Researched within Hinggan Prefecture

Name	Location	Basic Information	
		Research Method	Extant Architecture Status
Bayin-Huxo Temple (Xiafu Temple)	Bayan Hushuo Town of Khorchin right and middle Counties	General survey	Xire Lama Cang, Bogd Cang, Hamb Lama Cang
Taulait-gegen Temple (Gegen Temple)	Gegen Temple Town of Ulanhot City	General survey	Mahavira Hall, Monastery Gate, The Heaven King Hall, Sarira Hall, Dorje Hall, Fan Tong Temple, Namoroljangqobling Temple, Avalokiteshvara Hall, Dharmapalas Hall, Avalokiteshvara Hall (building)
Wang Temple	Puhui Street of Ulanhot City	General survey	Monastery Gate, The Heaven King Hall, Mahavira Hall, Wu Xiao Temple, Vajrayana Hall
Anggeri Temple	Bayan Zhalaga of Jalaid County	General survey	Mahavira Hall, Monastery Gate,Monk House

四、呼伦贝尔市地区

呼伦贝尔市是中国面积最大的一个地级市。从地形地貌特点来看，属于高原型地貌，处于亚洲中部蒙古高原的东北缘，形成了大兴安岭山地、呼伦贝尔高原、河谷平原低地三个地形单元，西部位于内蒙古高原东北部，北部与南部被大兴安岭南北向直贯境内，东部为大兴安岭东麓，东北部平原则是松嫩平原的边缘，地形总体特点为是西高东低。这一地区气候分布特点是以大兴安岭为分界线，岭东区为季风气候区，为半湿润性气候，岭西区为大陆气候区，为半干旱性气候，全年气候总特征为冬季寒冷干燥，夏季炎热多雨，年温差、日温差大。

现辖地由清时新巴尔虎、陈巴尔虎、厄鲁特、索伦、达斡尔、鄂伦春等部编成的呼伦贝尔总管八旗及布特哈总管八旗辖地构成。1948年设呼伦贝尔盟，1945年设纳文慕仁盟，1949年合并两盟，简称呼纳盟，1953年撤销此建制。[1]1954年改称呼伦贝尔盟，2001年撤呼伦贝尔盟，设地级呼伦贝尔市。内蒙古各盟旗中，位居呼伦贝尔地区的巴尔虎、布里亚特、索伦等旗召庙数量最少。[2]

呼伦贝尔市境内的召庙普遍呈现创建模式独特、召庙遗存较少等特点。呼伦贝尔市召庙的一种创建模式为随着部族大规模的迁移，在迁入地新建召庙，以此延续在迁出地时的宗教传统，该模式实为一种召庙迁建行为。如从喀尔喀蒙古及俄罗斯境内迁至呼伦贝尔草原的巴尔虎、布里亚特等部，延续佛法传统，在迁入地创建召庙。巴尔虎部于1734年从喀尔喀车臣汗部迁至呼伦贝尔时已有90余名僧人。[3]因此，新建召庙满足僧众需求已是理所当然的事。布里亚特部迁至呼伦贝尔后创建了锡尼河庙，寺庙与北部布里亚特[4]各召庙常有交往。[5]呼伦贝尔境内召庙遗存十分罕见，仅有甘珠尔庙、新巴尔虎西庙等寺庙有少量古建筑遗存。

课题组对呼伦贝尔市境内7座藏传佛教召庙展开了调研普查工作，具体情况如下表所示。

呼伦贝尔市地区调研的现存召庙

召庙名称	所在位置	基本信息	
		调研情况	现存主要建筑
甘珠尔庙	新巴尔虎左旗阿木古朗宝丽格苏木	普查	大雄宝殿、护法殿、药师殿、天王殿、桑吉德莫洛姆殿、龙王殿、罗汉殿、显宗殿
新巴尔虎西庙（达西朋斯克庙）	新巴尔虎右旗阿尔山苏木	普查	大殿、东配殿、西配殿、山门
阿尔山庙	新巴尔虎左旗	普查	大殿、配殿、山门、舍利塔
巴音库仁庙	陈旗巴音库仁镇	普查	山门、大雄宝殿、护法殿
呼和庙	鄂温克旗巴音镇	普查	山门、大雄宝殿
锡尼河庙	鄂温克旗	普查	大殿、显宗殿、东配殿、西配殿、三座护法殿
达尔吉林庙	海拉尔区	普查	天王殿、三学日光殿、兜率尊胜殿、佛学院、山门、钟楼、鼓楼、活佛府

1 周清澍主编.内蒙古历史地理［M］.呼和浩特：内蒙古大学出版社，1991：178,253.
2 据伪满洲国蒙政部1936年的调查，兴安北省蒙旗（即呼伦贝尔）有寺庙42座。但寺庙具体名称不详。据1951年的调查，呼伦贝尔蒙旗寺庙有新巴尔虎左右翼两旗甘珠尔庙、老楞庙、陈巴尔虎旗巴彦库仁庙、索伦旗南屯庙、锡尼河庙、厄鲁特庙等6座。参见：中共中央内蒙古分局宗教问题委员会编．内蒙古喇嘛教（上册）．1951：72,84．
3 参见：米希格道尔吉编著．达西朋苏格庙（蒙古文）.（内部资料）．新巴尔虎右旗文化中心，2000.11：2．
4 北部布里亚特指留守俄罗斯境内故土的布里亚特人。1922年，最初有160户布里亚特人逃避俄罗斯境内的战乱，经呼伦贝尔副都统衙门的准许，迁入呼伦贝尔境内，驻牧于锡尼河草原。从1751年至1830年，俄罗斯境内布里亚特蒙古人中已有30余座规模较大的藏传佛教召庙。参见：宝敦古德·阿·毕德编，华赛·都古尔扎布校订．布里亚特蒙古简史（蒙古文）［M］.海拉尔：内蒙古文化出版社，1983：65,28．
5 布里亚特人迁至呼伦贝尔后，锡尼河庙的僧侣与留居故土的布里亚特各召庙间有较为密切的往来。[编者注，参见：都·呼·勒格其得编著．锡尼河庙回忆录（蒙古文）．（内部资料）.2008：49．]

4. Hulunbeir City

Hulunbeir City is a Prefecture-level city with the largest area in China. In terms of geographic and geomorphic characteristics, Hulunbeir belongs to plateau landscape and lies at the northeast margin of Mongolian Plateau in Central Asia, and three landform units including Great Khingan mountainous region, Hulunbeir Prairie and Valley Plain lowland generated. Its west is in the northeast of Inner Mongolian Plateau, its north and south are run through by Great Khingan in north-south direction, its east is the east foot of Great Khingan and its northeast is the margin of Songnen Plain; generally it is high in the west and low in the east. This area is featured by the climate distribution that takes Great Khingan as the boundary, with monsoon climate zone with sub-humid climate in the east of Great Khingan and continental climate zone with semi-arid climate in the west of Great Khingan. And for the general climatic characteristics all the year around, this area is cold and arid in winter and hot and rainy in summer, and both annual temperature difference and daily temperature difference are large.

Now this area comprises 8 counties under jurisdiction of Hulunbeir and 8 counties under jurisdiction of Butaha, including New Barag, Old Barag, E Lute, Sauron, Daghur, Oroqen, etc. Hulunbeir Prefecture was set up in 1948, Nawenburen Prefecture was set up in 1945 and these two Prefectures were combined and called Hu-Na Prefecture in 1949 and this structure was withdrawn in 1953. In 1954, it was named Hulunbeir Prefecture which was abolished in 2001and changed to Hulunbeir City. Among the Prefectures and counties of Inner Mongolia, Barhu, Buliyate, and Suolun in Hulunbeir area own the least number of temples.

Temples in Hulunbeir City are featured by unique establishment mode and remaining of fewer temples. Some temples in Hulunbeir City were built with large-scale migration of the tribe, and the migrants continued the religious tradition of out-migrating places on this account. Such a mod is actually a kind of temple relocation. For example, Barga, Buryat and other tribes migrated from Khalkha Mongolia and Russia to Hulunbeir Prairie continued Buddhism tradition and built temples in ingoing places. Barag tribe had more than 90 monks when migrating from Khalkha Chen Khan group to Hulunbeir. So naturally it is necessary to build new temples to meet the demand of monks. Buryat tribe built Xini River Temple after migrating to Hulunbeir and this temple kept frequent contact with temples in the northern Buryat. Remains of temples in Hulunbeir are very rare, and only Kanjur Temple, Western New Barag Temple and so on have fewer ancient relics.

The research team surveyed 7 Tibetan Buddhism temples in Hulunbeir. The table below presents the details.

Extant Temples Researched within Hulunbeir City

Name	Location	Basic Information	
		Research Method	Extant Architecture Status
Ganjur Temple	Amu Gulang Baolige Township of New Barag Left County	General survey	Mahavira Hall, Dharmapalas Hall, Pharmacist Hall,The Heaven King Hall, Sanjay DeMo los mu Hall,Long Wang Hall, Arhat Hall, Exotoric Buddhism Hall
Western New Barag Temple (Daxipengsuge Temple)	Arxan Township of New Barag Right County	General survey	The Main Hall, The East-side Hall, The West-side Hall, Monastery Gate
Arxan Temple	New Barag Left County	General survey	The Main Hall, The Side Hall, Monastery Gate,Stupa
Bayin Kuren Temple	Bayan Kuren Town of Old Barag County	General survey	Monastery Gate, Mahavira Hall,Dharmapalas Hall
Huhe Temple	Bayan Townshi of Ewenki County	General survey	Monastery Gate, Mahavira Hall
Xinehen Temple	Ewenki County	General survey	The Main Hall, Exotoric Buddhism Hall, The East-side Hall, The West-side Hall, Three Dharmapalas Halls
Darjilin Temple	Hailar District	General survey	The Heaven King Hall, Mahavira Hall,Tsongkhapa Hall,Buddha Institute, Monastery Gate, Bell Tower, Drum Tower, Buddha Hall

第二章 内蒙古地域召庙建筑形态的影响因素

在藏传佛教输入蒙古地域并逐步传播、发展、兴盛的漫长过程中，召庙建筑在这一地域的入驻、生长、成形均受到了当时当地社会中诸多因素的影响，其中，主要有政治、文化以及自然三个方面的因素。

第一节 政治因素的影响

政治因素是藏传佛教能够进入蒙古地域并得到自上而下传播、发展的首要因素，藏传佛教传入之后在本地域形成的政教合一制度使得该地域召庙及殿堂的规制、形态及材料的表达均带有鲜明的政治色彩。

一、广建召庙

自藏传佛教传入蒙古地域以来，基于明确政治考虑的各朝各代统治者尊崇、扶植蒙古地域藏传佛教的重要手段之一便是兴建召庙。

（一）各时期的兴建情况

1. 元朝时期

这一时期，基于元朝统治者政治大局下的宗教政策，召庙大多位于大都、上都等政治中心以及内地的政治战略重心城镇，乃至全国较有影响力的名山大川。以元大都和五台山为例。

（1）此时期在元大都建造、修缮的较有规模且极具代表性的召庙就有：大护国仁王寺、大圣寿万安寺（白塔寺）、大兴教寺、崇国北寺、大承华普庆寺、大天寿万宁寺、大崇恩福元寺（南镇国寺）、大永福寺（青塔寺）、寿安山卧佛寺、大承天护圣寺等。[1]

（2）此时期在五台山也兴建、重建了十余座召庙，例如：万圣佑国寺、圆照寺、普恩寺、铁瓦寺、寿宁寺、殊祥寺等。[2]

2. 明朝时期

明朝时期，在阿勒坦汗的影响下，以土默特地区、鄂尔多斯地区等漠南蒙古地区为首率先掀起了兴建召庙的热潮。随着此时期内蒙古地域第一座召庙——大召的建立，蒙古地域的其他地区也纷纷效仿，广建召庙、普宣佛教。在短短几十年内，仅呼和浩特地区就建造了数十座典型的召庙，其中个别召庙至今依然是内蒙古地域召庙之典范，如大召（无量寺）、席力图召（延寿寺）、小召（崇福寺）、乌素图召（庆缘寺）等，甚至到了明末清初，连地名也被冠以“召城”的称谓。

3. 清朝时期

在极力推行、贯彻“兴黄教以安蒙”的对蒙宗教政策的清朝时期，蒙古地域的藏传佛教发展到极盛。清朝政府全力提倡在蒙古地方兴建召庙，仅内蒙古地域的召庙就可达千座之多，且深入分布于内蒙古的各旗，平均每旗就有二十多座召庙。特别是康熙、雍正、乾隆时期，召庙普及整个内蒙古草原，兴建运动达到了狂热程度，乾隆年间为最高峰。这一时期兴建的诸多庙宇甚至由清政府直接出资、出人兴建，如多伦诺尔（今锡林郭勒盟多伦县）的汇宗寺、善因寺等。

（二）政治参与召庙建设的情况

根据调研结果显示，内蒙古地域现存的召庙大多初建于明清时期，并且这一时间段内建造的召庙，从多个方面表现出明显的特点：

1. 参政于选址

虽然，明清时期的统治者极力推行藏传佛教，但多数召庙并不是想建就建，想建哪里就建哪里，更多的时候，这些召庙是应政事之需而生的，因此，召庙兴建的位置具有重大的政治意义。例如，乌兰察布市的百灵庙是因其所处区域地理位置的重要性，出于明显的政治军事目的而建。百灵庙建庙之后就成为当时内蒙古北部地区的政治、经济、文化中心，也是漠南通往漠北、漠西、俄罗斯等地的交通要道之一。

2. 融政于庙堂

明清时期一些较有规模的重要召庙，均为政教合一的产物，召庙建筑的功能、形态也受到了影响，其不仅仅是信众礼佛、拜佛的场所，同时也成为无上皇权的象征，其中，以呼和浩特市的大召大雄宝殿最为典型。清崇德五年（1640年），清

1 参考：乔吉.蒙古族全史（宗教卷）．[M].呼和浩特：内蒙古大学出版社，2011：79—88.

2 参考：乔吉.蒙古族全史（宗教卷）．[M].呼和浩特：内蒙古大学出版社，2011：92—96.

Chapter Two
The Factors Influencing the Architectural Form of Temple Architectural in Inner Mongolia Region

In the long process of Tibetan Buddhism's entering into Mongolia region and gradually spreading, developing, the settling, growing and forming of Tibetan Buddhism temple architecture were all influenced by many factors in the society there and then, among which the politics, culture and nature were the major aspects.

I. The Influence of Political Factors

Politics was the main factor to make Tibetan Buddhism enter into Mongolia region and spread from upper class and develop. The regional theocratic system formed after the entering of Tibetan Buddhism made the expression of the regulations, shapes and materials of the temples and halls carry a distinct political colour.

1. Widely Building Temples

Since the entering of Tibetan Buddhism into Mongolia region, based on the clear political consideration, one of the important means used by the rulers of every dynasty to worship and foster Tibetan Buddhism in Mongolia region is to build temples.

(1) Constructions in Various Periods

A. Yuan Dynasty

During this period, based on the religious policy of Yuan rulers, Tibetan Buddhism temples were mostly located in political centers such as Dadu, Shangdu etc. and some mainland political strategic cities, even famous mountains and rivers with great influence in the country. Yuan Dadu and Mount Wutai are two examples.

a. The representative temples built and renovated in Yuan Dadu in this period were: Dahuguorenwang Temple, Dashengshouwan'an Temple (White Pagoda Temple), Daxingjiao Temple, North Chonguo Temple, Dachenghuapuqing Temple, Datianshouwanning Temple, Dachongenfuyuan Temple (South Zhenguo Temple), Dayongfu Temple (Green Pagoda Temple), Wofo Temple in Mount Shouan, Dachengtianhusheng Temple etc.

b. In this period, more than ten temples were built and rebuilt on Mount Wutai, such as Wanshengyouguo Temple, Yuanzhao Temple, Puen Temple, Tiewa Temple, Shouning Temple, Shuxiang Temple etc.

B. Ming Dynasty

During Ming Dynasty, under the influence of Altan Khan , Tumd Area, Erdos Region and other Mongolia regions in the south of the desert took the led to boom the construction of Tibetan Buddhism temples. With the establishment of the first Tibetan Buddhism temple in Mongolia region—Dazhao (Big Temple), other regions followed one after another to construct the temples and spread Buddhism widely. In the short span of a few years, only in Hohhot area ,dozens of typical Tibetan Buddhism temples were built, among which some are still models of Inner Mongolia Temples, such as Dazhao , Xilituzhao (Yanshou Temple), Xiaozhao (Chongfu Temple), Wusutuzhao etc.. Even at the end of the Ming Dynasty, and the early years of Qing Dynasty, the region was called "city of temples" .

C. Qing Dynasty

Tibetan Buddhism was developed to the climax under the policy of "spreading Buddhism to settle Mongolia", which was greatly promoted and carried out in Qing Dynasty. The Qing government advocated to build temples in Mongolia. Only in Inner Mongolia area, Tibetan Buddhism temples amounted to more than 1000, distributed in each division, county of Inner Mongolia with the average of more than 20 temples in each County. Especially in the four periods of Kangxi, Yongzheng, Qianlong led by Beijing, Tibetan Buddhism temples were built in the country, with Chengde as the center, while in Inner Mongolia region with Duolunnur and Hohhot as the center. Tiberan Buddhism temples were popular all over Inner Mongolia grasslands and the feverish construction activities reached the peak in Qianlong period. Many temples built in this period were even invested and constructed directly by Qing government, such as Huizong Temple and Shanyin Temple etc.in Duolunnur (now Xilingol Prefecture Duolun county).

(2) Political Participation in the Construction of Temples

According to the survey results, most extant Tibetan Buddhism temples in Inner Mongolia region now were built in the early Ming and Qing period, showing obvious characteristics intervention from multiple aspects:

A. Participating in Siting

Although the rulers of Ming and Qing Dynasties made an utmost effort to promote Tibetan Buddhism temples, most temples were not built randomly .More often they were built to meet the political needs and the sites possessed great political significance.For example, Bailing Temple (Lark Temple) in Ulanqab City was built with obvious political and military purposes due to its regional importance of geographic position . Bailing Temple then became a political, economic, cultural center in northern Inner Mongolia and one of the main vital communication lines from Monan (south of desert) to Mobei (north of desert) and Russia.

B. Melting Politics with Temples

Some large and important temples in Ming and Qing Dynasties were mostly products of caesaropapism which affected the function and form of the temples construction. The temples were not only the places Buddhist believers respected and worshiped Buddha, but also became the symbol of supreme imperial power. The Main Hall in Dazhao in Hohhot was the most typical one among them.

太宗命令蒙古土默特部重修大召，并亲赐寺名为无量寺；清康熙三十六年（1697年），大召扩建，大召大雄宝殿的屋顶改覆为象征皇权的黄琉璃瓦，将殿内皇太极坐过的座位设置为“皇帝宝座”，并将黄金铸造的“万岁金牌”供奉在银佛前，因此大召被尊为“帝庙”或“皇庙”。[1]

3. 推政于建造

藏传佛教，在元朝时期是统治者安抚地区的利器，明朝时期是统治者争权夺利的权杖，至清朝时期则成为统治者“驭藩”的有力武器，清王朝已经不满足于为召庙选址、将召庙皇家化等较为简单形式的政治控制，而是直接参与蒙古藏传佛教召庙的建设当中，将宗教载体完全政治化，锡林郭勒盟多伦汇宗寺的建造开创了朝廷直接在蒙古地区建造召庙的先例。

（三）召庙兴建的方式

在内蒙古地域，召庙采取属从方式兴建并如巨大的织网遍及整个草原，例如，阿拉善盟地区原有召庙28座，其中8座拥有清帝或民国大总统授予的寺名匾额，因此有阿盟八大寺之说。事实上，5座有匾额的召庙和其他无匾额的召庙均分别属于南寺、北寺（福因寺）、衙门庙（延福寺）三大召庙系统，是该三大寺的属庙。[2]

二、植入类型

综观内蒙古地域召庙的建筑，在丰富的形态背后，其基本的类型并不多。据此，在调研归纳和文献研究的基础上，总体考察明清时期由政治力量自上而下进行的形态植入主要分为两类。

（一）间接植入的类型

由于明朝在其整个统治时期并没有全程推奉某一宗教。因此，间接植入自然成为明朝时期蒙古地域修建召庙的主要方式。需要说明的是，这种间接植入的方式在清朝时期也有一定的流行。

召庙被间接地植入本地域的情况大体可分为两种：

1. 明清时期的蒙古上层贵族出于一定的目的，如笼络部族民心、树立政治地位、讨好当朝统治者等，通过独自出资或与其统领的民众共同出资的方式在蒙古地方修建召庙，并待合适的机会奏报当朝统治者为召庙恭请庙名或殿名。

2. 明清时期的高僧喇嘛为弘扬佛法、传道修行，在贵族或众多信众的资助下修建召庙，召庙上报朝廷登记在册后，当朝统治者赐予寺名。

这两种情况的共同点在于，国家政权没有直接参与召庙的筹划与修建，但在政策方面给予了极大支持，肯定了召庙在蒙古地方的政治作用和宗教统治地位，使得召庙成为当朝统治蒙古地方的重要工具。二者的不同点在于，由于召庙建造主持者的不同，产生了本地域两种不同性质的召庙，即崇祀寺[3]与学问寺[4]。

这一时期，间接植入内蒙古地域的召庙以汉藏合璧为特征的大召为始，在本地域推广繁殖开来。此后，在广阔的蒙古地域大量修建的召庙多采用汉藏相结合的形式，且该形式至清朝中期完善至顶峰，其主要特征是：在汉式或汉藏相结合的建筑布局下，建筑单体多采用汉、藏地区的建造方法，兼具汉、藏建筑风格。

（二）直接植入的类型

出于“驭藩”的目的，善于汲取历史经验的清廷在其统治的整个时期，比前朝更为积极地奉行全力支持藏传佛教的国策，不仅“因俗而治”取得了成功，巩固了北部边疆的安全与稳定，也加快、加深了藏传佛教在蒙古地域的推广，使得该地域藏传佛教的发展达到前所未有的顶峰。因此，与明朝的间接植入方式不同的是，清朝蒙古地域出现了召庙被直接植入的情形，即清廷直接参与召庙的筹划与修建，并给予相关的政策、经济支持。众多的敕建庙在清王朝的恩惠下如雨后春笋般林立于蒙古地域乃至全国，其中，最为典型的是锡林郭勒盟的汇宗寺与善因寺。

锡林郭勒盟的多伦诺尔距离北京只有七百里，是内地通往东、西、北部蒙古地区的交通要冲。为融洽蒙古与清廷中央的关系，加强、巩固朝廷对蒙古地区的统治，康熙皇帝亲赴多伦诺尔与蒙古各部会盟，后“从诸部所请，即其地建庙”，即在会盟地点敕建汇宗寺以巩固、纪念多伦诺尔会盟的政治成果。从决定建庙到立碑为

---

1 参考：蒙古学百科全书·宗教卷（蒙古文版）．［M］.呼和浩特：内蒙古人民出版社，2007：323.

2 参考：贾拉森. 缘起南寺［M］.呼和浩特：内蒙古大学出版社，2003：1.

3 崇祀寺，此类召庙的主要任务是为帝王家祈福、诵经，政治色彩浓厚。

4 学问寺，此类召庙的主要任务是研究、学习藏传佛教教理，建此类召庙的主要任务是研究、学习藏传佛教教理、教义，从宗教之本源推行、弘扬佛法精神及理念。通常，召内设有藏传佛教的学部，在大的召庙，五部俱全。（参考：长尾雅人著，蒙古学问寺［M］. 白音朝鲁译.呼和浩特：内蒙古人民出版社，2004，8）.

In the year 1640 (Chongde 5th year, Qing),Qing Taizong ordered Mongolia Tumd Tribe to renovate Dazhao and gave the name "Wuliang Temple" in 1697 (Kangxi 36th year, Qing),Dazhao was expanded with the roof of the main hall changed to be covered by yellow glazed tiles symbolizing imperial power,and the seat sat by Huangtaiji was set as "the emperor's throne ", and the "long live golden medal" cast by gold was enshrined and worshiped in front of the silver Buddha.So Dazhao was honored as "the king's temple" or "the temple of the emperor".

C. Establishing Politics in Construction

Tibetan Buddhism was an edged tool used by the rulers of Yuan Dynasty to placate territories and truncheon in Ming Dynasty to contend for rights and benefits and the powerful weapon to "control the vassal states" in Qing Dynasty. Qing government was not content with the simple political control of siting the temples and making temples imperialized, it is involved into the construction of Mongolia Tibetan Buddhism Temples to make religious carrier politicalized. Huizong Temple in Duolun, Xilingol Prefecture served as an precedent of the Tibetan Buddhism Temples directly built by the government in Mongolia region.

(3) Methods of Constructing Temples

In Inner Mongolia, temples are built in subordinating way and covered the prairie as a huge net. For example, there are 28 temples in Alxa Prefecture, 8 of which were awarded the name- plaques by Qing emperors or the leaders of old China. Thus appeared the expression —"8 big temples in Alxa". In fact, 5 temples with the plaques and other temples without plaques respectively belong to the three big temple system of the South Temple, the North Temple (Fuyin Temple) and Yanfu Temple and the subordinate temples of them.

2. Embedded Types

Viewing the construction of Tibetan Buddhism Temples in Inner Mongolia region, behind the rich forms, there are only a few basic types. Accordingly, on the basis of investigation summary and documents research, the overall inspection suggests the types that the political power of Ming and Qing Dynasty embedded into architectural forms from top to bottom are divided into two categories.

(1) The Indirect Embedding Type

Since Ming government hadn't pushed to believe in certain religion throughout the Dynasty. Thus ,Embedding indirectly naturally became the main way to build Temples in Mongolia region in Ming Dynasty.What should be stated is the indirect embedding way was also kind of popular in Qing Dynasty.

The indirect embedding of Tibetan Buddhism Temples into the region generally fell into two catergories:

A. In Ming and Qing Dynasties, the nobles, for certain purposes such as, pleasing people, setting up political positions, flattering the current government and so on, built temples in Mongolia region through investment alone or joint investment, and requested the current rulers to give the names to the temples or halls on suitable occasions.

B. To promote Buddhism and doctrine practice, the eminent monks and Lamas built temples with the financial aid of the nobles and believers. After the temples were reported and registered, the current rulers gave the names to the Temple or the Hall.

The common points in these two cases are: the state power was not directly involved in the planning and construction of temples, but gave great support in policy and affirmed the political function and religious dominance of the temples in Mongolia region and made the temples become the weapons of the current government to control Mongolia region. The different points lie in the appearance of two kinds of temples with different nature due to different construction presiders—Chongsi Temple and Xuewen Temple.

During this period, the temples embedded indirectly into Inner Mongolia region with the beginning of Dazhao Temple which featured with combination of Han and Tibet were spred and developed in the region. Since then ,many temples built in the vast Mongolia region adopted Han-Tibetan style which was improved and reached the peak in mid of Qing Dynasty. The main feature was: in the construction layout of Han or Han-Tibetan, single building adopted the construction methods of Han and Tibet region, with both Han and Tibetan architectural style .

(2) The Direct Embedding Type

For the purpose of "controlling vassal states", Manchu Qing court, good at absorbing historical experiences, pursued the policy of supporting Tibetan Buddhism more actively than previous dynasties. It not only succeeded in "following the custom to rule", and also accelerated and deepened the promotion of Tibetan Buddhism in Mongolia region. Thus the regional development of Tibetan Buddhism reached an unprecedented peak. So, the different embedding way with Ming Dynasty was that direct embedding of temples appeared in Mongolia region in Qing Dynasty ,that is the direct involvement of Qing government in planning and constructing of the temples and giving relevant policy and economic support. Many of the imperial temples sprouted like mushrooms in Mongolia region and even the country under the support of Qing Dynasty, among which the most typical were Huizong Temple and Shanyin Temple in Xilingol Prefecture .

Only 700 Chinese miles from Beijing, Duolunnur region in Xilingol Prefecture is the gateway of mainland

记，康熙皇帝曾多次到多伦诺尔视察，甚至召庙的施工建设、布局陈设以及召庙喇嘛、主持的选派都要亲自操持。根据建设的后续工程的历史记录来看，召庙建设的开支主要是由清政府的内务府和工部负责，特别是召庙中心轴线上各个殿宇内的佛像和供奉都是由内务府和工部主持造办的，而建设召庙的工匠则基本上是由清政府从京城和直隶地区调集来的。此外，召庙建设中所需的一些重要建筑构件，如琉璃瓦、鸱吻等均取自京城。[1]

相较于明朝而言，清朝虽在蒙古地域极力扶植藏传佛教，但自身仍然十分推崇汉地儒家文化并深受其影响。由于汇宗寺显著的敕建性质，其召庙建筑群从整体布局到单体建筑在建筑形态和建造技艺方面都直接植入了汉式风格及做法。因此，以多伦汇宗寺为代表的蒙古地域敕建庙多采用以汉式建筑布局、汉式殿堂风格以及汉式结构做法,例如清雍正五年（1727年）紧邻汇宗寺敕建的善因寺是由宫廷样式房“样式雷”设计，由汉族工匠营建的典型的纯汉制官式召庙建筑群。

总体而言：

1. 以藏式建筑为母本和以中原官式建筑为母本成为植入的两大基本类型。

(1)以藏式寺庙建筑为母本的建筑体系。

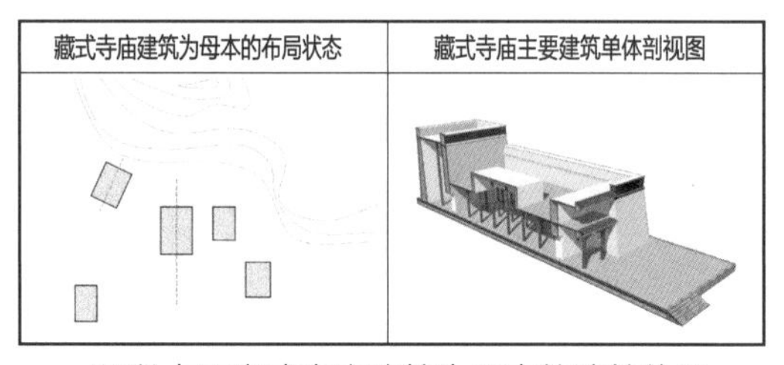

(2)以中原官式寺庙建筑为母本的建筑体系。

中原官式建筑为母本的布局状态	汉式寺庙主要建筑单体剖视图

2. 政治上的植入带有草创性质，形式上的不完备在以后的演变中逐渐成熟起来,如大召的主体建筑为汉藏结合式，其他附属建筑则为纯汉式，反映了一种在草创时期只关注政治效应的状态。

3. 宏大是政治植入建筑形式的主要特征，大召和汇宗寺的主体建筑都极力表现了气势的宏大，它让草原人民欢欣鼓舞的同时感受到了一种宗教的征服力量，这正是统治者所需要的(图1)。

图1 南寺（广宗寺）大雄宝殿

同时，基于明清两朝特定的政治背景，内蒙古地域召庙建筑的营造直接接纳和采用了部分来自藏地的建筑手法，但由于建造这些或间接或直接输入蒙古地域召庙的多是当地或周边的汉族工匠以及熟悉汉式营造技术的蒙古族匠人，故而蒙古地域召庙建筑的特点是以藏式寺庙建筑为建造蓝本，并带有鲜明的多民族性和近地域风格。

第二节 文化因素的影响

内蒙古地域召庙受到藏族地区建筑文化和汉地官式建筑文化的共同影响，同时又在发展的过程中与本地域的建筑传统相结合，形成了丰富多样的建筑形式，可以说，内蒙古地域的召庙是藏、汉、蒙古文化的结合体。

一、迎源于藏区

对内蒙古地域召庙建筑产生最直接影响的是藏文化的重要组成部分——藏传佛教文化自身。在内蒙古地域，这种影响力表现在以下几个方面：

（一）召庙选址

内蒙古地域，除了受政治因素影响较大而需统治者抉择位置外，还存在一些召庙，其庙址的选择与定位同藏地一样，是由专门学习天文地理的喇嘛来完成的。

（二）建筑布局

内蒙古地域的部分召庙通过模仿藏地组团式建筑布局标示其宗教属性及本源，因此，本地域出现了藏式建筑布局，其中，以包头市五当召最具代表性，而这种布局形式的根源则是模仿藏传佛教中代表佛教宇宙模式的曼荼罗图式的结果（图2）。

1 参考：任月海.多伦汇宗寺［M］.北京：民族出版社，2005：24—42.

transportation to the east, west and north Mongolia. In order to harmonize the relationship between Mongolia and Qing Dynasty and strengthen and consolidate the rule of Mongolia region, Emperor Kangxi went Duolunnur in person to meet Mongolian tribes for alliances. Then "for the request of the tribes, temples are to be built in the region"— imperial Huizong Temple was built in the location of meeting to consolidate and memorize the political achievements of Duolunnur Union. From the decision to build the temple to the erection of the memorial, Emperor Kangxi went to inspect Duolunnur several times, and even took personal charge of the construction of the temple building, the layout of furnishings and selection of temple's chief Lama. According to the history record of the follow-up project, the main cost of temple construction was under the charge of the Imperial Household Department and the Ministry of Works of Qing government, especially the statues of Buddha and objects of worship in various halls on the central axis were all built and made by them. The craftsmen building the temples were basically collected by Qing government from the capital and the direct subordinate regions. In addition, the building components for the temples, such as glazed tile, ornaments on roof ridge, etc.were all taken from the capital.

Compared with Ming Dynasty, though Qing Dynasty strongly fostered Tibetan Buddhism in Mongolia region, it still respected Han's Confucian culture very much and was influenced by it. Because of the significant imperial feature of Huizong Temple, the buildings directly embedded Han style and practices in the architectural form and construction skills from overall layout to single building. Therefore, represented by Huizong Temple in Duolun, the imperial temples in Mongolia region mostly adopted Han-style layout, Han-style temple halls, and Han structural practices. For example, in the year 1727 (5th years, Qing Emperor Yongzheng), the imperial Shanyin Temple which was close to Huizong Temple was the typical pure Han official temple building group built by Han craftsmen, designed with the palace style of " Lei Family".

Generally speaking,

A. The buildings based on Tibetan construction and the official construction in Central Plain became two basic types of embedding.

B. The political embedding expressed that the immature feature and the incompletion in forms grew mature in the following evolution .For example, the main building of Dazhao was the combination of Han and Tibet while the other outbuildings were pure Han style, reflecting the immature state of only concerning political effects.

C. The main feature of the building with political embedding was grandness. The main building of Dazhao and Huizong Temple were shown as great and grand, which made the grassland people feel a kind of religious force at the time of inspiration. That was just what the rulers needed.

At the same time, based on the specific political background of Ming and Qing Dynasties, the construction of Inner Mongolia Tibetan Buddhism Temples directly accepted and adopted part of building techniques from Tibet, but as the people who built the temples (which were introduced to this region directly or indirectly) mostly were local or neighboring Han craftsmen and Mongolian craftsmen familiar with techniques of Han-style construction, therefore the characteristic of Tibetan Buddhism architecture in Mongolia region was: taking Tibetan temple building as the construction blueprint, with a distinctive multi-ethnic and near-regional style.

Ⅱ. The Influence of Cultural Factors

The Tibetan Buddhism temples in Inner Mongolia region were influenced by the combined effect of the architectural culture of Tibetan area and the official architectural culture in Han area and also integrated with the local architectural tradition in the process of development at the same time. So the temples in Inner Mongolia region are the combination of Tibetan ,Han and Mongolian culture.

1.Depriving from Tibetan Area

The most direct influence to the architecture of Tibetan Buddhism Temples in Inner Mongolia region is from an important part of Tibetan culture—Tibetan Buddhist culture. In Inner Mongolia region ,the influence is manifested in the following aspects:

(1) Temple Siting

In Inner Mongolia region, besides the temples which needed siting by the rulers for important political factors, there were also some temples whose site selection and positioning, just as in Tibet, were decided by Tibetan Lamas specialized in study of astronomy and geography .

(2) Construction Layout

Some temples in Inner Mongolia region marked their

图2 五当召总平面图

另外，基于藏传佛教特殊的宗教仪式，在内蒙古地域的召庙中，与宗教仪式相匹配的空间有两类：

1. 跳神空间

跳神空间是举行宗教法舞——跳神仪式的专门场所，通常设于整个召庙前部、中部或者重要殿堂前部较为宽敞的室外空间（图3）。

图3 跳神

2. 转经空间

内蒙古地域的召庙将藏地的活动引入本地域，并加以改造利用，如在定位敖包之内与召庙建筑群之外设置转经道，沿重要殿堂内或外布置的转经道、转经空间以及贯通大殿室内外的转经空间；甚至会在召庙中专门设立一个转经空间，如玛尼亭、玛尼殿等。

除此之外，在第三章中提到的两类聚落的产生以及两类殿堂的划分均是基于藏传佛教中对于不同空间领域的理解与要求。

（三）殿堂空间

对于内蒙古地域召庙建筑而言，藏式建筑文化对其产生的影响具体表现在两个方面：

1.都纲法式及其变异形式的运用

都纲法式具体的建筑形制是：建筑一层平面纵横排列柱网，中间部位凸起方形或近方形的垂拔空间，此垂拔通高二层，并在东、南、西三个方向开窗采光。建筑二层平面呈“回”字形，中部为经堂凸起的垂拔空间，垂拔井外侧建房间一圈，这些房间大多用作管理用房或储藏室等。建筑外观为四周平顶，中部垂拔之上为坡屋顶，多采用歇山式，偶见藏式平顶（多在纯藏式殿堂中运用）。如上所述殿堂中心部分突起，设高侧窗，平面呈“回”字形的一种定型化的建筑规制即“都纲法式”。

在内蒙古地域，有一些召庙的经堂是完全按照都纲法式建造的，如包头市五当召的大雄宝殿经堂、昆都仑召大殿经堂，通辽市库伦兴源寺的大殿经堂，呼和浩特市席力图召的大殿经堂等。通常这种做法在清朝兴建的经堂中比较普遍（图4）。

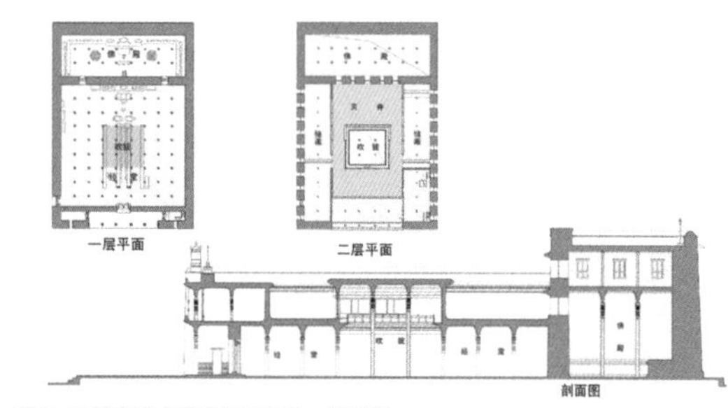

图4 昆都仑召大雄宝殿平面图、剖面图

但内蒙古地域也有很多大殿经堂，其原型虽是都纲法式，但已经发生了一些变异，如下表所示。

religious nature and origin by imitating the layout of Tibetan group buildings, therefore, Tibetan architectural layout appeared in the local region. Badgar Temple was the most representative among them and the origin of this layout was the result of modeling Mandala schema representing Buddhist universe model in Tibetan Buddhism.

Besides, based on the special religious ritual of Tibetan Buddhism, in Tibetan Buddhism temples in Inner Mongolia area, there are two spaces matching with the building space:

A. Space for Religious Dance

Dance space is the place to hold religious dance — specialized sites for religious ritual dance, usually located on the spacious outdoor space in front of or in the middle of the temples, or in the front of the important halls.

B. Praying Space

Tibetan Buddhism temples in Inner Mongolia region introduced Tibetan prayer activities into the area, and tried to transform and utilize it, such as, setting circumanbulation inside the fixed Obo and outside the temple groups, or along the inside or outside of the important halls or through the main hall, and even specially setting up space for twirling prayer wheels in the temples, such as Mani Pavilion, Mani Temple etc..

The classification of the two settlements and two kinds of halls referred in the Chapter Three, is based on the understanding and requirements of Tibetan Buddhism to different spatial domains.

(3) Hall Space

For Tibetan Buddhism temple buildings in Inner Mongolia region, the impact of Tibetan architectural culture is performed in two aspects:

A. The Application of Dugangfa Style and Its Variation

The specific architectural form is: the first floor of building has vertical and horizontal arrangement of column, with the convex vertical space in square or nearly square in the middle, which is usually two floors high with windows in the East, South, West for lighting. The building plane in the second floor was the shape of the character "Hui" (回) with the protruding vertical space in the middle of the Hall. Outside of space, rooms are built in a circle, which are mostly used as management or storage rooms. The appearance of the building is flat around, with slope roof in the middle of the central vertical space. Most are built in Saddle Roof Style, occasionally in Tibetan flat roof (used in pure Tibetan Hall).As is mentioned above, The fixed building regulation with protrusion in the Hall center, a clerestory and flat "Hui" (回) Shape is called "Dugangfa Style".

In Inner Mongolia area, the halls of some Tibetan Buddhism temples were built completely according to Dugangfa style, such as Suguqin Hall in Badgar Temple, Baotou, Main Hall of Hundele Temple, Xingyuan Temple in Kulun, Tongliao and Xilituzhao in Hohhot City etc.. Usually, this practice was quite common in the halls built in Qing Dynasty.

The prototypes of many great halls in Inner Mongolia region are Dugangfa Style with some variations, as shown in the following table:

Moreover, to illustrate specially, the pattern of some halls built in the mid of Ming Dynasty is a two-floor pavilion-style building, but the layer of the second floor is

都纲法式及其变异			
典型都纲法式剖面	都纲法式变异1	都纲法式变异2	都纲法式变异3
昆都仑召大雄宝殿	梅日更召大雄宝殿	大召大雄宝殿	五当召显宗殿
图例：斜线部位为经堂，点画线部位为附属空间或佛殿			

此外，需特别说明的是：建于明中叶的一些经堂，其格局就是一个两层的楼阁式建筑，只不过二层的楼面被取消，使一、二层连通，做成垂拔空间。与都纲法式不同的地方是，这种经堂无二层，也就是无天井及围绕天井一圈的房间；经堂外墙伸出一层屋面层，作用及形状类似女儿墙。这种做法产生于藏传佛教第二次传入之初期（明中叶），成熟、完善于清中叶，且采用这种做法的殿堂在内蒙古地域数量较多，如呼和浩特市大召大雄宝殿（图5）、乌素图召庆缘寺大殿、包头市美岱召大雄宝殿等。

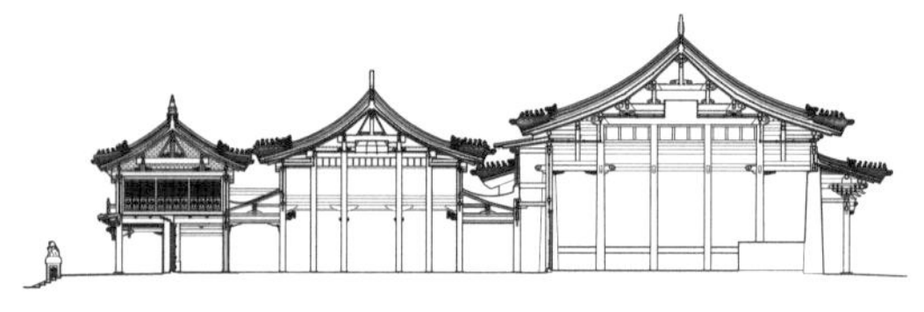

图5 呼和浩特市大召大雄宝殿剖面图

具体而言，内蒙古地域藏传佛教召庙殿堂内的垂拔有如下特点：

(1) 垂拔的大小及位置

1) 对于大雄宝殿

一般情况下，在横向上，垂拔位于明间及旁边的次间，占据柱网中间的3间或5间；在纵向上，垂拔占据柱网中间的2间或3间。

2) 对于扎仓大殿[1]

垂拔设置的位置及大小相对比较自由，有的虽有2层，但没有垂拔（如呼和浩特市乌素图召的法禧寺大殿）；有的有垂拔，却很小，只有一个柱间距大小，且位于中间部位（如鄂尔多斯市乌审召的德都庙大殿、准格尔召五道庙）；有的只有一层，而无垂拔（如包头市昆都仑召的小黄庙、呼和浩特市席力图召的长寿寺大殿）；有的是面阔三间、进深五间、高两层、后部高三层的阶梯式垂拔（如包头市五当召的显宗殿）。

(2) 垂拔的布置

垂拔内部会有若干条长条状布质经幢垂下，且四周挂满唐卡及刺绣。通常，殿堂垂拔二层背面墙体均为实墙，不开窗，绘有壁画，其余三面开窗，窗为汉式槛窗，不用藏式窗。有的垂拔周围有一圈回马廊，可以看到一层的经堂，例如：昆都仑召大雄宝殿、五当召大雄宝殿、五当召显宗殿等。

2. 藏式做法的变通性使用

内蒙古地域众多召庙建筑的普遍共性特点之一便是以藏式为基本母题进行地域化，因此，藏式做法在本地域得到了变通式的使用，具体内容详见第四章的第三节。

二、取异于汉地

内蒙古地域召庙建筑在其发展的过程中，除了受到藏传佛教发源地西藏以及甘青地区召庙建筑形态文化的影响，更有来自汉地（特别是北京、山西、陕西等地区）建筑文化对其产生的影响，并在召庙、建筑形态上呈现出明显的过渡性与多元性。汉族传统的建筑工艺与建筑风格在内蒙古地域的召庙中表现为两类：一是以庄严肃穆为性格的官式殿宇；二是以风格地域化为特点的民居式庙堂。

（一）庄严性官式殿宇

内蒙古地域不少藏传佛教召庙建筑都采用了此种模式，其特点在于：

（1）建筑群体布局以伽蓝七堂式为基本模板。

（2）建筑单体以汉地木构架宫殿式建筑为建造范式。

1 参考：第五章第三节的相关介绍。

canceled, so that the first and second floor are connected to be a vertical space. The difference with Dugangfa style is: there is no second floor in this kind of hall, that is no patio and a circle of rooms around the patio; there is a floor of room surface outside the outerwalls, with a similar function and shape of the parapet. This approach appeared in the early days (mid of Ming) of the second time of Tibetan Buddhism's incoming. Being mature and perfect in the middle of Qing Dynasty, there were a large number of halls using this pattern in Inner Mongolia area, such as the Main Hall of Dazhao Temple in Hohhot, the Hall of Qingyuan Temple in Wusutuzhao and the Main Hall of Maidarzhao Temple in Baotou etc..

Specifically speaking, the vertical space in the halls of Inner Mongolia Tibetan Buddhism temples has the following characteristics:

a.The Size and Position of the Vertical Space

(a) Tsochin Hall

In general, in horizontal, vertical space is located in the central room and the side room occupying the 3 or 5 rooms in the middle of the column; in vertical, vertical space occupies middle 2 or 3 rooms of the column.

(b) Dratsang Hall

The size and position of the Vertical space are relatively free, although some have 2 floors, but no vertical space (e.g. the Main Hall of Faxi Temple of Wusutuzhao in Hohhot City); some have vertical space but very small with only one column spacing, located in the middle part (such as Degedusumo Hall of Wushenzhao, Erdos City and Wudao Temple in Zhunger); some only have one floor without vertical space (such as Xiaohuang Temple of Hundele Temple in Baotou, the Main Hall of Changshou Temple of Xilituzhao in Hohhot City); some are three-room wide , five -room deep, two-room high ladder type vertical spaces with three floors high at the rear (such as Xianzong Temple of Badgar Temple, Baotou).

b. Layout of the Vertical Space

Inside the vertical space, several strip-shaped cloth Dhvaja are hanging, with Thangka and embroidery around. Usually, the two- floor back wall of the vertical space of the hall is solid wall without windows and with wall paintings on it; the remaining three sides have cage windows of Han style instead of Tibetan window style. Some vertical spaces have a circle of veranda around and the prayer hall, on which the first floor can be seen. For example, the Main Hall of Hundele Temple and Badgar Temple, Xianzong Hall of Badgar Temple.

B. The Flexible Use of Tibetan Practices

One of the common characteristics of many Tibetan Buddhism temples architecture in Inner Mongolia region is to be regional based on Tibetan style. Therefore, Tibetan practices have been used flexibly in local area. Details are found in the Third Part of Chapter Five.

2.Difference from Han

In the process of the development, the construction of Tibetan Buddhism temples in Inner Mongolia region was influenced not only by the culture of temple construction form in Tibet—the origin of Tibetan Buddhism, and Gan Qing District, but also by the architectural culture from Han district (especially Beijing, Shanxi, Shaanxi and other regions) and the obvious transition and cultural diversity are presented on temples and architectural forms. The traditional architectural technology and architectural style of Han in Tibetan Buddhism Temples in Inner Mongolia region fall into two categories: one is the official Halls with solemn characteristics; the other is the residential temples characterized by regional features.

(1) Solemn Official Temple

Some Tibetan Buddhism temple buildings in Inner Mongolia region use this model, which is characterized in the following:

a. The building layout is based on Qianlan Seven-hall Style.

b. The single building uses the palace architecture of wooden frame in Han area as building paradigm.

The solemn official Temples with strong Han style consist of the style of pure Han and Han-Tibetan combination, the following contents mainly introduce the Han elements in the halls with these styles:

A. Structural System

a. The Wood Frame

具有浓郁中原风格的庄严性官式殿宇包括纯汉式及汉藏结合式，以下内容主要介绍纯汉式及汉藏结合式召庙殿堂中的汉式元素：

1. 结构体系

(1) 木构架

该类殿宇结构体系采用以木构架为主、砖为辅的结构体系，即主体构架由柱、梁、枋、檩等组成，且常见的为五架梁。

(2) 柱子

此类殿堂内的柱子，其样式基本都是原样照搬汉式，除用柱头的斗栱承托梁架外，部分柱子上还刻有龙浮雕，如呼和浩特大召佛殿银佛座前有两根雕有蟠龙造型的木柱；部分柱子则在其上用沥粉贴金绘龙，如美岱召大雄宝殿内的柱子；还有些柱子柱头上雕有龙头浮雕，如呼和浩特市乃莫齐召佛殿柱头。由于内蒙古地域整体粗放的建造技艺，柱子大多没有明显的收分，直接将完整的原木刨平、上漆、立柱，柱断面也多为不规则的圆形，少数制作讲究、等级较高的殿堂内也采用其他断面，如方形、楞八楞等。

(3) 斗栱

在内蒙古地域，斗栱通常使用在等级相对较高、建筑体量相对较大的殿堂中，例如，呼和浩特市大召的大雄宝殿、鄂尔多斯市准格尔召的佛殿等。该类殿堂内外，使用斗栱的柱子通常是檐柱以及与外墙围合成转经甬道的柱子，个别地区斗栱略有变异，如准格尔召佛殿斗栱出昂极多。

2. 瓦石作

(1) 台基

内蒙古地域召庙殿堂的台基为普通台基，基本构造多为四面砖墙，里面填土，上面墁砖的台子。

(2) 墙体

墙体采用砖，偶见局部采用石的情况。由于在歇山顶下、两山之下做廊用以转经，因此，此类殿堂墙体的特点在于，殿堂山墙无墀、挑檐石等构件，而是直接将砖向上砌筑，并向内收回。

(3) 屋顶

该类殿堂多采用歇山顶，或单檐或重檐，屋顶上施筒板瓦，或琉璃或陶质。通常，正脊两端有正吻，由龙头形装饰张大口将正吻咬接，吻下山面有吻座，仙人、走兽则施于垂脊之上，等级、规制不同，个数不同。

3. 平面特征

此类殿堂平面多为矩形或近方形，面阔5—9间，进深亦为5~9间，由于建筑主体采用四柱牵制一间的原则，因而建筑物中柱的布置极为规则。

4. 立面特征

此类殿堂立面中，汉式元素出现率极高，例如，框槛和隔扇，其中，框槛是不动部分，隔扇是可动部分；在中槛与下槛之间安装门与窗，上槛与中槛之间或为墙体或为横披；窗多采用槛窗，做在槛墙之上；门大多采用板门形制，偶有使用隔扇门的情况。

明清时期，由于植入蒙古地区的召庙建筑更多的是基于该时期的政治因素，特别是清朝时期由政府直接干预建造的召庙。因此，性格庄严肃穆的官式殿宇更受到青睐，甚至此类殿堂直接由建造皇家宫廷殿阁的宫廷样式房承建。

5. 布局特点

在汉地建筑文化的影响下，内蒙古地域召庙建筑的布局出现了纯汉式及汉藏结合式。总体而言，这两种形式均以伽蓝七堂制为布置基础，或直接使用，或取其布置形式的精髓所在。此种布局方式下的召庙由山门、天王殿、钟楼、鼓楼、东西配殿和大殿组成，强调中轴线，建筑对称，并且以中轴线后部设置的汉式或汉藏结合式建筑风格的大殿为布置中心，如锡林郭勒盟的贝子庙。采用汉式建筑形态的藏传召庙多出现在内蒙古东部地区，以锡林郭勒、赤峰、通辽地区最为集中。

（二）风格化民居式庙堂

地域风格化的内蒙古藏传佛教庙堂形同汉地中原的民居，在召庙建筑的组成当中多见于活佛府（拉卜隆）、庙仓等部分，其中，以山陕风格最为突出，具体有以下两种表现方式：

1. 四合院式

通常，内蒙古地域召庙佛仓内的殿堂采用四合院式，建筑单体与四合院内的民居几无差别，如包头市梅日更召的葛根仓、阿拉善盟的丹巴达尔（图6）。

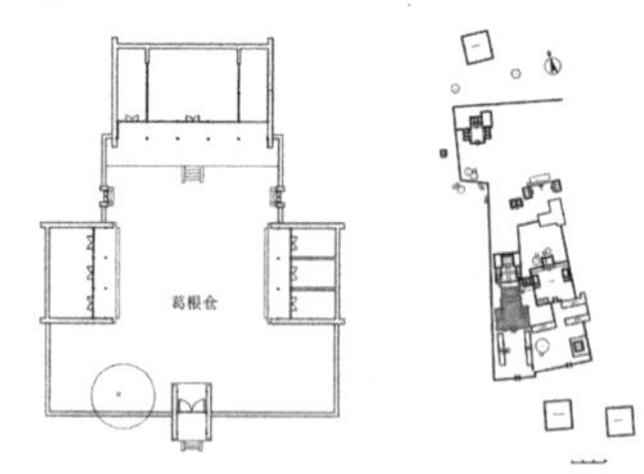

图6 梅日更召葛根仓平面图

The main structure system of these temples is wood frame as the main structure supplemented by bricks. The main framework consists of columns, beams, square-purlins, and five-frame beams are commonly used.

b. The Pillars

The style of the pillars in the halls is the complete copy of Han style. Besides supporting the frame with Dougong (the bracket system), part of the poles are carved with dragons in relief, such as the two wooden pillars with carved dragons in front of the silver Buddha in Dazhao Temple Buddha Hall in Hohhot, while some pillars are decorated by embossed paintings and gold foil paintings with dragon patterns, such as the pillars in the Main Hall of Maidarzhao Temple; also some pillars are carved with reliefs of dragon heads on chapiters, such as Naimoqizhao in Hohhot City. Because of the general extensive building techniques in Inner Mongolia region, most pillars have no clear contracture, and the whole log is planned to be painted and erect; and most column sections are irregular round shape; only a few have delicate manufacture while the higher grade halls also use other sections, such as square, a small octagonal and so on.

c.Dougong

In Inner Mongolia area, Dougong is commonly used in the halls of relatively high-grade and massive constructions, for example, the Main Hall of Dazhao in Hohhot City, Buddha Hall of Zhungerzhao in Erdos City etc. Inside and outside such halls ,the pillars used as Dougong are usually eave columns and the columns which are combined with the outwalls to be the prayer corridor. The Dougongs have slight variation in some areas, such as Buddha Hall in Zhungarzhao Temple.

B. Tile Stone

a. Stylobate

The stylobates of the temple halls in Inner Mongolia region are common ones. The basic structure is the platform with surrounding brick walls, earth filled inside ,bricks on it.

b.Walls

Walls are made with bricks, occasionally partly with stones. The walls are characterized in: as the corridors are made under Saddle Roof and Gable Roof for prayer, therefore, gable walls have no parts as steps and cornice, but with bricks directly masoned and withdrawing inward.

c. Roofs

The halls mostly adopted Saddle Roof style, or single eave or multiple eaves, with cylinder tiles of glaze or ceramic on the roof. Usually, there are Zhengwen on the two edges of the highest ridge, snapping-in Zhengwen with the mouth of dragon-head decoration, and there is the seat under the Wen. Celestial beings, animals are applied to the vertical ridge above, and the numbers are different with grades and regulations.

C. Planar Features

The planes of this kind of hall are mostly rectangular or nearly square, 5-9 rooms wide, 5-9 rooms deep. As the main body of the building adopts the principle of four-column holding one room, the layout of the columns is very regular.

D. Facade Features

The elements of Han appeared with a high rate in the facade of this kind of halls, for example, frame sill and sash, wherein the frame is the fixed part, the sash is the movable part; and doors and windows are installed between the middle frame and mud still, walls or Hengpi(parts of wall) in the middle of headsill and mudsill. Sill windows are mostly used, settled above the door sill; most of the doors use the smaller version of the door shape of Taihemen, with occasional partition board.

During Ming and Qing Dynasties, the embedding of Tibetan Buddhism temple architecture into Mongolia area was based more on the political factors of the period, especially the temples built directly under the intervention of Qing government. Therefore, solemn official Temples were more popular. This kind of halls were even built under the sample of royal palace.

E. Characteristics of Distribution

Under the influence of Han architectural culture, pure Han style and Han-Tibetan combination appeared in the layout of temple construction in Inner Mongolia area. Generally speaking, these two forms are based on Qianlan Seven-Hall System—used directly or took the essence of its arrangement form. The layout mode of the temples is made by Monastery Gate, The Heaven King Hall, Bell Tower, Drum Tower, Side Halls and the Main Hall, emphasizing the central axis, the symmetry of the construction, and the layout center is the Main Hall with Han style or Han-Tibetan combination style set at the rear part of the central axis. such as Beis Temple in Xilingol Prefecture. The Tibetan temples built in Han architectural form mostly appeared in the eastern region of Inner Mongolia, with Xilingol, Chifeng, Tongliao as the parts of concentration.

(2) Temples with Residential Style

Inner Mongolia Tibetan Buddhism temples with Regional style are like residential houses in the central Han area, especially expressed in the parts of Dastcang and Focang etc.. wherein, Shanshan style is most prominent, the following are the two specific ways :

A. The Style of Quadrangle Courtyard

Usually, the hall of Focang in Inner Mongolia Tibetan Buddhism temple adopted the style of quadrangle courtyard. The single building has no difference with the residential houses in the courtyard, such as Gegentsang in Merigenzhao, Baotou, Danba Dahl in Alxa Prefecture.

What should be noticed is whether it is in west , central, or east, the temples built with local style in the quadrangle shape in Inner Mongolia district are mostly Buddha Palace

但值得注意的是，无论是西部、中部，还是东部地区，内蒙古地域以四合院的方式塑造具有地方风格庙堂的，多是各召庙内具有礼佛功能的佛爷府、活佛住所。这些院落虽是活佛私用之处，但其正房并非生活起居之用，而是礼佛、敬佛、诵经之所，是典型的庙堂。

2. 窑洞式

内蒙古地域内最南部靠近晋、陕北部的地区，出现了具有明显窑洞元素的召庙建筑，特别是鄂尔多斯、呼和浩特地区，如鄂尔多斯市的海流图庙。

此外，召庙选址亦受到汉地佛寺建筑文化的影响。藏传佛教传入蒙古地域后，随着宗教文化的发展，许多蒙古贵族皈依佛门，并仿效中原地区“舍宅为寺”的做法，捐府邸为召庙，内蒙古地域的许多召庙均以这样的形式出现，最早且最具代表性的例子是明代阿拉坦汗时期，在今包头萨拉旗境内修建的官邸——美岱召（美岱召是一座集政治、军事、文化一体的历史名城，后逐步演变为以庙宇建筑为主体的著名“寺城”）。此外，在清代，本地域的众多召庙前身均为王爷家庙，如赤峰市的福会寺等。这些召庙或位于蒙古贵族领地中地理位置优越、水草肥美风景秀丽之地或直接置身于蒙古贵族府邸之内。

三、得风于蒙古

藏传佛教深入蒙古社会后，召庙建筑以贴近蒙古地域文化的状态通过某种特定主题的方式表现出来。

（一）蒙古民族的生活方式

逐水草而居的蒙古游牧民族没有固定的生活建筑，长期的生存方式使草原上的人们形成了一种相对自由开放的生活习性，而这种习性早已沉淀在草原游牧民族的集体无意识中，成为草原文化的一个有机组成部分。正是这种草原文化使得蒙古人在迎接藏传佛教到来的同时，保留了其自由的布局方式，也正是由于与藏族有着几近相同的生活习惯，藏式的自由布局更容易在蒙古草原上扎根。当然这种布局方法在草原深处和没有外界干预力量的情况下表现得更加突出，如包头市地区昆都仑召的建造。位于内蒙古包头市西北乌拉山与大青山分界的昆都仑河沟口右岸的昆都仑召原为乌拉特中公旗旗庙，直属清朝理藩院管辖，其历史沿革与蒙古乌拉特部迁徙定居至乌拉山的历史息息相关。1633年，成吉思汗的胞弟哈萨尔的后裔率乌拉特部归附后金。乌拉特部驻牧乌拉山、大青山的山地后，将此地域命名为“昆都仑”，昆都仑召的名称也由此而生。此外，蒙古地域存在的不拘于形式，且与召庙密切相关的聚落也深受本地域自由游牧生活的影响，甚至召庙属地范围的规定也采用跑马的形式来确定，如以一匹成年壮实的马一天（天数不确定，由蒙古贵族或者宗教上层决定）内所跑的最远距离为半径画圆，其内为召庙所辖之属地（如包头市的昆都仑召）。

（二）地域建筑文化

蒙古地域主题性的建筑语汇——蒙古包成为本地域召庙建筑的形式元素之一。

藏传佛教传入的早期，由于该教派宗教文化传播的深度及广度不够，没有实力在广大的蒙古地域范围内深度推广并营造其召庙建筑，因此，蒙古民众的蒙古包就是早期藏传佛教喇嘛弘法的主要场所，甚至在蒙古草原上出现了专门用于喇嘛居住、诵经、做法的蒙古包。

藏传佛教发展到一定程度之时，并没有完全抛弃蒙古包这一原始的蒙古地域建筑语汇，而是将蒙古包直接引入召庙建筑，例如，召庙殿堂顶部覆以蒙古包形穹顶，蒙古包作为进入大雄宝殿的前导空间等，即使是在建造艺术与技术发达的现、当代，本地域的民众也同样常常将此形态广泛应用（图7）。

图7 蒙古包形穹顶殿堂（蒙古国迈达里庙）

（三）本土宗教崇拜

在藏传佛教入主蒙古草原之前，蒙古人固有的宗教信仰是萨满教[1]，是一种融合蒙古族的自然崇拜、祖先崇拜、图腾崇拜等多种民间信仰

1 萨满一词源于满—通古斯语，意为兴奋、激动、不安或癫狂的人，为萨满教巫师的专称，因此，该类信仰被称为萨满教。萨满教是一种基于万物有灵论基础上的世界性原始宗教形态，该教传统始于史前时代并且遍布世界。这种自发而生的宗教形态，广义上是指流行于亚欧大陆北部广大地区以及北美、澳大利亚等地，包括流行于其地理范围中的各种原始巫术在内的信仰；狭义上是指以西伯利亚地区为中心的阿尔泰语系各民族所信仰的宗教；本文涉及的萨满教则主要是指流行于蒙古地域为蒙古民族所信仰的原始宗教。（参考：孙懿.从萨满教到喇嘛教——蒙古族文化的演变［M］.北京：中央民族大学出版社，1998：1—2.）

and the residence of Living Buddha with the function of respecting Buddha. The courtyard is a private place of Buddha, but its principal rooms are used as typical temples for respecting Buddha and chanting sutra. instead of living places.

B. Style of Cave Dwelling

The far south of Inner Mongolia region is near the northern area of Shanxi and shaanxi, in which the temple architecture with obvious elements of cavehouse appeared,especially in Erdos, Hohhot area, such as Hailiut Temple in Erdos City.

In addition, the temple siting is also affected by Buddhist architectural culture in Han area. After Tibetan Buddhism spred into Mongolia area, along with the development of religious culture, many nobles of Mongolia converted to Buddhism, and followed the practice of "offering houses to be temples" in Central Plains region, donating mansion for temples. Many Tibetan temples in Inner Mongolia region are in such form. The earliest and the most representative example is Maidarzhao Temple (a famous historical city combined with politics, military, and culture, and gradually evolved into a famous "Temple City" with temple construction as main body) built in Alatan Khan period in Ming Dynasty in Salaqi, Baotou. Furthermore, in Qing Dynasty, the predecessor of many Tibetan Buddhism temples were ancestral temple of princes, such as Fuhui Temple in Chifeng City etc.. These temples were located in the places of Mongolian nobles' possession, with superior geographical position and beautiful scenery of the fertile plants and water or directly in the nobles' mansion in Mongolia.

3. Learning from Mongolia

After Tibetan Buddhism went deep into Mongolia society, temple construction was built through a method with a specified subject in the manner to close the regional culture of Mongolia region.

(1) Living-style of Mongolian People

Following water and grass, Mongolia nomads had no fixed buildings to live, the long-term survival way made the grassland people form a kind of relatively open and free living habits, and this habit had deposited in the collective consciousness of prairie nomads and became an organic part of grassland culture. It was this grassland culture that made the Mongols welcome Tibetan Buddhism and retain its free layout at the same time. Because Mongolian people had almost the same life habits with Tibetan, Tibetan free layout rooted more easily in the Mongolia grassland. Of course, this layout was more prominent in the far grasslands under the condition without outside intervention, such as the construction of Hundele Temple in Baotou. Hundele Temple, located on the right bank of the mouth of Hundele River which divides Mount Wula and the Mount Daqing, was the former temple of Wulate Zhonggong County, directly under the jurisdiction of Qing Dynasty vassal state-governing department. Its history is closely related to the history of migration and settlement of Mongolian Wulate tribe. In 1633, the descendant of Genghis Khan's brother Hasar submitted Wulate tribe to Houjin. After they settled and herded in Mount Wula and Mount Daqing, the area was named "Hundele", from which the name of Hundelezhao came. In addition, the settlements existing in Mongolia region which were not stick to forms and closely related to Tibetan Buddhism temples were also influenced by the regional free nomadic life, even temples territorial scope were also determined by horse racing, drawing a circle by the maximum distance of the radius which an adult horse ran to in a day (the number of days were uncertain, decided by the Mongolia aristocracy or religious upper class) as the territory of a temple (such as Hundele Temple in Baotou City).

(2) The Regional Architectural Culture

The key architectural word in Mongolia region–yurt became one of the elements of the forms of regional Tibetan Buddhism temple building .

In the early entering of Tibetan Buddhism, because the lack of the depth and breadth of the spreading of the religious culture and having no strength to promote deeply and build the temples in the vast Mongolia region ,the yurts were the places the early Tibetan Buddhist Lama romoted Buddhism and even in Mongolia grassland appeared the yurts exclusively for lamas to live, chant, and practice .

When Tibetan Buddhism developed to a certain extent, yurt, the original Mongolian regional architectural vocabulary was not completely abandoned, instead it was introduced directly to the Tibetan Buddhism temple building, for example, the top of the temples were covered with a yurt-shaped dome, and yurt served as a leading space into the Main Hall etc.. Even in contemporary and current time with advanced construction art and technology, this form is still widely used by local people.

(3) Local Religious Worship

Before Tibetan Buddhism entered into Mongolia grassland, Mongolian religious belief was Shamanism, which was a religious pattern combined with Mongolian nature worship, ancestor worship, totemism and a variety of folk beliefs and once existed, prevailed and deep-rooted

于一体的宗教模式，曾一度长期广泛流行于蒙古各部，且根深蒂固。藏传佛教在政治因素的导入下，开始与蒙古萨满教展开了长期的斗争，斗争的结果是：在蒙古上层的绝对承认和强力庇护下，以对萨满教的禁废、镇压、改造而终结。“大多数皈依喇嘛教的蒙古人，不过是将萨满教信仰依照喇嘛教的形式进行了某些改造。喇嘛们积极搜集民间传诵的各种祷词，然后添加新的宗教内容，以旧瓶装新酒的方式促使喇嘛教在民间传布。与此同时，隐匿于民间的萨满教也接纳了喇嘛教的众神，改用喇嘛的祈祷仪式和经文举行祭祀，它蜕去了旧的躯壳，披上了喇嘛教的外衣，这正是萨满教在新的环境下借以抵制喇嘛教迫害的方式。”[1]

虽然萨满教与藏传佛教的斗争是以对萨满教的禁废而结束，但萨满教中的一支——白萨满[2]以接受藏传佛教的改造，成为藏传佛教的一部分为代价，保存了较为完整的势力，且该派萨满的教义、仪式也与藏传佛教进行了大规模的糅合再生，蒙古萨满甚至穿上了喇嘛服，改用喇嘛所用法器，同喇嘛一起参与各项喇嘛教法会、仪式等等。这种仪轨的通融在影响召庙建筑形态的演变方面主要包括三个内容：

1. 敖包[3]崇拜

古匈奴时期，游牧于北方草原的匈奴人就多以石头或沙土堆成的敖包状物作为道路或者牧场边界的地标，在广阔的大草原上用它来辨别方向、区分不同的游牧场所。后期，随着蒙古萨满教的逐步发展与深入，蒙古人将敖包演变为祭祀天神、自然神、祖先、英雄人物等的祭坛，而整套相关的敖包祭祀活动则由萨满主持。藏传佛教真正意义上深入蒙古民众内部后，曾专属于萨满教的敖包祭祀成为藏传佛教中的一项重要仪式，敖包也转为蒙古地域藏传佛教的膜拜体，并表现出鲜明的地域特征。

(1) 空间聚落化

1）敖包集中在召庙周围

传统敖包所选择的建造基址为山顶、隘口、湖畔、路旁、滩中等特殊却显而易见的地方，并且所在地域必定水草肥美，环境适宜。此外，这些敖包的大部分均以所在地域之山名或地名或水名而命名。藏传佛教将其转宗后，敖包多出现在召庙内部及周围地域，甚至用以标至召庙的庙界所在。

2）敖包聚落化

被赋予藏传佛教新意后的敖包，其排列组合形式大体分为两种，一是以中间大敖包为中心呈”一“字形排列；二是以中心大敖包为圆心，呈”十“字形或三角形排列；每组所含敖包个数不同，多使用三、七、十三、十九、二十五等藏传佛教中偏爱的数字；敖包聚落中部大敖包具有核心性，代表藏传佛教宇宙中的须弥山（图8）。

图8 锡林郭勒盟贝子庙额尔敦敖包

(2) 类型丰富化

内蒙古地域的敖包，根据其祭祀内容的不同，分为多种类型，例如：与英雄人物有关的敖包。与成吉思汗有关的敖包、与蒙古族圣物崇拜有关的敖包（苏力德敖包）、贵族敖包（旗王爷敖包或诺颜敖包）、部落或阿塔天神敖包（与萨满教有直接关系）以及演变后的喇嘛教召庙敖包、家族家庭敖包等。其中，与藏传佛教相关的敖包又分为很多种，例如，用于纪念喇嘛的敖包（如花敖包）、用于藏传佛教仪式的敖包（如甘珠尔敖包）等。转宗后的敖包不但没有丧失其最初的功用，反而在此基础上日益丰富，类型走向

1 参考：罗布桑却丹．蒙古风俗鉴．内蒙古社会科学院图书馆抄本．

2 在蒙古萨满教与喇嘛教的斗争中，萨满教为了自身的生存而发生了内部分化，从而出现了两个教派分支：白萨满教和黑萨满教，即所谓的亲佛派和排佛派。其中，以郝伯格泰博为首的妥协派或亲佛派被称为“白萨满派”；反对藏传佛教及其理念，提出“斩下僧人头，祭坛作牺牲”口号，不为佛祖诵经祈祷，不与喇嘛积聚做法，不进入召庙佛堂，不就桌椅而席地坐，顽强地沿袭着蒙古萨满教古老传统的排佛派则被称为“黑萨满派”。

3 敖包，蒙古语，意为堆子，或者译成“脑包”“鄂博”，意思是由木、石、土堆成的堆状物。这种由人工堆成的“木头堆”“土堆”或“石头堆”遍布蒙古各地，是蒙古族重要的祭祀载体。（参考：孙懿.从萨满教到喇嘛教——蒙古族文化的演变［M］.北京：中央民族大学出版社，1998：17．）

in Mongolia for a long time. Tibetan Buddhism, under the intervene of political factors, began the long struggle with Mongolian Shamanism. The result was: with the absolute recognition and strong protection of Mongolian upper class, Shamanism was banned, repressed, and transformed. "Most of the Mongols converted to Lamaism only made certain transformation of Shamanism faith in accordance with Lamaism. Lamas made efforts to collect various popular folk prayers, and then added new religious contents, just as putting new wine in old bottles, and in this way promoted Lamaism people to people. At the same time, the Shamanism hidden in the folk area also accepted the gods of Lamaism and turned to hold fete in prayers ceremony and scriptures of Lamas. It sloughed off the old dress, put on a coat of Lamaism. This was the way Shamanism resisted the persecution of Lamaism in the new environment."

Although the struggle between shamanism and Tibetan Buddhism ended with the ban of Shamanism, a branch of shamanism—White Shamanism saved relatively complete force at the cost of accepting Tibetan Buddhism transformation, becoming a part of Tibetan Buddhism. And the religious beliefs, rituals of this branch blended with Tibetan Buddhism and regenerated on a large scale. Mongolian Shamans even put on the dress of Lahma, used Lamas instruments, participated together with Lamas in the Lamaism assembly, rituals and so on. The accommodation of such rituals influenced the evolution of temple architectural forms in three contents:

A. Obo Worship

In the period of ancient Hun, the nomadic Huns in the northern grassland used Obo which was a pile of stones or sand as a road or pasture boundary landmark to find the direction, to distinguish different nomadic places in the vast grassland. Later, with the gradual development and spread of Shamanism in Mongolia, Mongolians developed Obo to the altar offering sacrifices to gods, the God of nature, ancestors and heroes, and the whole set of relevant Obo Rituals was hosted by Shamans. After Tibetan Buddhism virtually went deep into Mongolians, Obo Festival which once belonged to Shamanism had become one of the important rituals in Tibetan Buddhism while Obo was turned to be the Tibetan Buddhist worship object in Mongolia region and showed distinctive regional characteristics.

a. The Space Settlement

(a) Obos around Temples

The building site of traditional Obo was selected on mountaintops, mountain pass, lakeside, roadside, beaches and such special but visible places. The location must have plump plants and plenty of water with appropriate environment. In addition, most Obos were named after the mountains, places, or water in the region. After they were transformed by Tibetan Buddhism, most Obos were located inside the temples or around the region, which were even the marks of the temple locations.

(b) Obo Settlement

The permutations and combinations of Obos endowed with new ideas by Tibetan Buddhism can be generally divided into two types; one form is arranged in a line with the Obo in the middle as a center, or arranged in a cross-shape or triangle around the big Obo in the middle as the center of the circle; the number of Obos in each group is different, most of which is three, seven, thirteen, nineteen, twenty-fifth and other figures favored by Tibetan Buddhism; the big Obo in the middle of Obo settlement has the function of a core representing Mount Meru in the universe of Tibetan Buddhism.

b.The Various Types

Obos, in Inner Mongolia region, according to different ritual contents, are in various types, for example: those related to heroes; to Genghis Khan; and to Mongolian hierolatry (Su Lide Obo); aristocratic Obao (County Prince Obo or Nuoyan Obo); tribal or Atta gods Obo(related directly to Shamanism); as well as Lamaism temples Obo after evolution; family Obo, etc..Among them, the Obo related to Tibetan Buddhism are divided into many kinds, for example, Obo used to commemorate Lamas ; Obo used in Tibetan Buddhist rituals (such as the Kangyur Obo) etc.. After convertion , Obao didn't lose the original function, but became rich and multiple on the basis of theses types.

c.The Diversification of Forms

The original Obo was only pile of stones, sand, and

多元化。

(3) 形式多样化

原始的敖包仅为用石头、沙土、树枝垒起来的堆状物，藏传佛教传入蒙古后，敖包的形式及做法发生了变化，其基本模式为：建于平地或圆坛之上，以石头为材料环叠三层或多层，并在此基础上逐步向上聚拢叠加，越往上越小越尖，顶上树立藏传佛教的玛尼杆。这些敖包根据所处地域的不同，亦在此基础上发生演变，例如，东部地区的敖包顶部多为圆穹状，敖包东、西、北侧各树立木杆三根，其上分别刻有日、月、云图案，并用彩色绸带将其与中心敖包相连接，彩带上悬挂哈达、风马旗等装饰；而在西部鄂尔多斯地区，敖包前均设有风马旗台，敖包向阳处则设置佛龛和香烛台。

2. 成吉思汗崇拜

藏传佛教传入蒙古地域后，对待基于萨满教祖先崇拜的成吉思汗崇拜的态度是较为特殊的，不但没有将其禁查严止，反而将成吉思汗纳入喇嘛教崇奉体系内，供之于喇嘛教庙堂之上。成吉思汗在蒙古喇嘛高僧的著作中被奉为印度教的创造之神——大梵天，或奉为金刚菩萨转世。例如，漠南蒙古藏传佛教领袖一世章嘉呼图克图阿噶旺罗布桑（1642—1714）在1690年左右编写的祭祀祈愿文中，将成吉思汗奉为大梵天王（佛教产生后，被吸收成为释迦牟尼佛的护法）的化身，使之成为佛祖释迦牟尼的护法神。根据具体的需要，成吉思汗在画像中的表情或安详或威严，其塑像也直接被供奉于四大天王殿或护法殿内，如鄂尔多斯市地区沙日特莫图庙（菩提济度寺）的天王殿。此外，也有将其他著名的蒙古人佛化的情况，如将阿勒坦汗佛化，将三娘子佛化等。

3. 苏力德[1]崇拜

同样在蒙古文化中占有举足轻重地位的苏力德崇拜也被成功地引入藏传佛教，苏力德则成为蒙古地域召庙中的重要标志之一。

(1) 作为建筑装饰

召庙广布于蒙古草原后，作为蒙古人崇拜对象的苏力德以建筑装饰的角色出现在殿堂建筑中，最为普及的方式是配合四角经幢、祥麟法轮等藏式装饰出现于殿堂屋顶，以标示出蒙古地域召庙之特色。

(2) 作为召庙空间中的节点

1）殿堂空间

除作为殿堂建筑装饰外，苏力德亦是殿堂空间布局中的重要节点，它的出现往往预示着重要殿堂的出现。换句话说，在蒙古地域的召庙中，通常能够在重要殿堂前发现立有苏力德，且苏力德的位置大多居于统领建筑布局的主轴线上(图9)。

图9 锡林郭勒盟毕鲁图庙（左图）和苏力德（右图）

此外，苏力德亦会出现于整个召庙的庙门之外，也同样居于整个召庙的中心轴线之上，用以标志召庙之起始。

可见，苏力德是蒙古地域召庙空间中重要的标志性节点。

2）生活空间

在召庙的附属生活空间中，苏力德同样是聚落单元的重要组成部分，通常情况下，被布置在院落或院落中正房的正前方。以内蒙古鄂尔多斯市乌审旗为例，据当地有关部门的统计，全旗蒙古族总户数共计9291户，在其门前立有苏力德的户数是8640余户，拥有苏力德的户数达93%之多，其他7%的居户之所以未立苏力德，也是由于

1 苏力德，蒙古语，意为“长矛”“旗帜”。蒙古人认为，苏力德是长生天赐予成吉思汗的神矛，是成吉思汗征战所向披靡的标志，因而，被蒙古民族当作象征着精神力量的战旗，蒙古民族世世代代以敬献哈达、神灯、全羊、圣酒的等方式祭祀苏力德。

branches. After Tibetan Buddhism was introduced into Mongolia, Obo forms and practices had been changed, the basic pattern were: built in flat or on circular altar, stacking round to three or more floors with stones, and on this basis, gradually gathered together and stacked upward, the upper,the smaller with Tibetan Buddhist Mani rod on the top. These Obos evoluted on this basis according to different areas , for example, the top of Obo in the east was a circular dome, three wooden poles standing in the east, west, north, which were respectively engraved on the patterns of sun, moon, cloud, and used the colored ribbons hanging with hada, Wind Horse flags to connect with the center, and in western Erdos region, there were always Wind Horse Flag platform in front of Obo, and shrines ,and incense holders were in the sunshine side.

B. Worship of Genghis Khan

After Tibetan Buddhism spred in Mongolia area, the attitude towards the worship of Genghis Khan based on the worship of Shamanism ancestors was special, Genghis Khan was not banned and stopped but was brought into Lamaism faith system, and settled in Lama Temple. Genghis Khan, in Mongolia Lama Buddhist writings is regarded as the Hindu god of creation—Brahman, or as a diamond Bodhisattva reincarnation. For example, in 1690 A.D, in the ritual prayers written by Monan Mongolia Tibetan Buddhist leader, Zhang Jia I, khotokto Aga Wang Luobusang (1642–1714), Genghis Khan was worshiped as personification of Brahma (took as Buddha Dharma after the appearance of Buddhasim), and made it become Dharma protector of Buddha Sakyamuni. According to the specific needs, the expression of Genghis Khan in the portrait is quiet or severe and his statues are also directly enshrined in the SidaThe Heaven King (Four Heaven Kings) Hall or Dharma Hall, such as The Heaven King Hall in Putijidu Temple in Erdos City. In addition, some other famous Mongolians were also turned to Buddhas, such as Altan Khan, the Third Lady etc..

C. Worship of Su Lide

The worship of Su Lide, which has an important position in Mongolia culture had also been successfully introduced into Tibetan Buddhism. Su Lide had become one of the major signs of Tibetan Buddhism Temples in Mongolia area.

a. As Architectural Decoration

After Tibetan Buddhism temples were widely distributed in Mongolia grasslands, Su Lide, as an object of Mongolian worship appeared as building decorations in temple architecture. The most common way is putting on the palace roof matching with four angle Dhvaja (kind of Buddhist flag), Xianglin Dharma-cakra and other Tibetan decorations, marking the characteristics of Tibetan Buddhism Temples in Mongolia region .

b. As the Node of Temple Space

(a) Hall Space

Besides being the decoration in hall building, Su Lide is also the important node in the layout of hall space, it often indicates the appearance of the Main Hall. In other words, in Tibetan Buddhism Temples in Mongolia area , Su Lide can usually be found in front of the Main Hall, and is mostly located on the principle axis guiding the construction layout.

In addition, Su Lide also appeares outside the tempe gate, and is located on the central axis of the whole temple, used to mark the beginning of temples.

So, Su Lide is the important symbolic node of Mongolia regional temple space.

(b) Living Space

In the affiliated living space of temples, Su Lide is also an important part of the settlement unit. Usually, Su Lide is arranged in the courtyard or in front of the principal room. Take Wushen County in Erdos City of Inner Mongolia as an example, according to the statistics of local authorities, the total number of Mongolian families in the whole district reaches 9291 households, among which, 8640 families have Su Lide at the front doors, at the rate of 93%. The other 7% of the households do not erect Su Lide due to the limitation of living conditions, but their daily habits are deeply affected by the Su Lide culture.

c.The Temple Halls with Su Lide Worship as the Theme

居住环境条件的限制，但其日常的生活习惯都深受到苏力德文化的影响[1]。

(3) 以苏力德祭祀为主题的召庙殿堂

作为民族崇拜对象，关于苏力德的祭祀活动同样风靡整个蒙古部族，例如，每年阴历三月十七日，蒙古人都会举行隆重的苏力德祭祀仪式。基于蒙古地域特殊的文化传统，藏传佛教在其深入之时，将苏力德祭祀纳入本宗，而以苏力德为主题的召庙殿堂也随之出现，成为区别于西藏、青海、甘肃等藏区藏传佛教寺庙的特征之一。以鄂尔多斯地区为例：

作为两个独立的崇拜系统，苏力德及成吉思汗都由鄂尔多斯部守护、管理、供奉、祭祀，但苏力德崇拜与祖先成吉思汗崇拜紧密相连，不可分割，因此，在鄂尔多斯蒙古族中，这两种崇拜祭祀活动都得以体现。在鄂尔多斯部内，有专司祭祀的达尔扈特人，且有大达尔扈特和小达尔扈特之别，其中，大达尔扈特人是专门守护供奉成吉思汗八白室的人，而小达尔扈特人则是守护、供奉苏力德的人，这种区分是从忽必烈时代留传下来的，有着严格的规定。随着藏传佛教在蒙古地域的普及，鄂尔多斯地区不但将成吉思汗佛化为藏传佛教的护法神，还在本地区出现了专门供奉苏力德，用于为苏力德诵经的召庙殿堂，如乌兰木伦庙，由此成为内蒙古地域召庙极为显著的蒙古特点。

第三节 自然因素的影响

一、地理因素影响下的召庙选址

根据前面的章节可知，内蒙古地域地形地貌丰富，如位于内蒙古自治区中南部的阴山山脉，横贯内蒙古高原与河套平原之间，东连大兴安岭，西至北山而没入阿拉善的茫茫沙海之中，因此，内蒙古地域的召庙对庙址的选择也因具体情况而定，但是，通常情况下，地形地貌、地质以及水源是其考虑的重要因素。

1. 地形地貌

内蒙古地域召庙优先选择山地地形以及平原、高原中地理位置相对较高的场所建造召庙以利于排除雨雪积水，防止洪涝淹没房屋，便于交通往来等。例如，召庙沿阴山山脉从西到东依次分布：巴彦淖尔市的阿贵庙位于狼山支脉；包头市的梅日更召、昆都仑召等位于乌拉山支脉；包头市的美岱召、萨尔沁召以及呼和浩特市的喇嘛洞、乌素图召等位于大青山支脉；而并称为阿拉善盟的“南寺”与“北寺”的广宗寺和福因寺等广布于贺兰山山脉。此外，根据召庙名称可以直接判断其所处地理位置的地貌特点。例如，内蒙古地域的一部分召庙被称为“阿贵庙”，这是蒙古语的译音，意思“山洞庙”，根据其名称可知这类召庙均建在多山的地区，凿山建庙，以得到清修的空间，内蒙古的呼和浩特市、阿拉善盟、巴彦淖尔市、乌兰察布市等地区均有这样的召庙。

通常，内蒙古地域召庙建筑受地形地貌的影响最为直接地体现在召庙建筑布局上：建于山地的召庙多顺应地形、地势，依山而建，建筑布局灵活自由，如五当召、阿贵庙；位于平原或高原中的平缓之地的召庙，多采用较为规整的布局，工整严谨，如大召、贝子庙等。

2. 地质

除了地形、地貌之外，地质构造也是选址时需要考虑的重要内容。通常，召庙会派专人考察所选位置土质的密实性、土质等，且常有建寺人将准备建庙地方的土壤送给高僧以作勘定之用。

3. 水源

召庙多选择河流凸岸且高于常年洪水水位之上的台地上建造殿堂。

大多数依山而建的召庙由于地势的原因周围会存在季节河，为召庙的生产和生活带来了一定的方便，如包头市的梅日更召、五当召，呼和浩特市的乌素图召等。

平原地区的召庙在选址时也都考虑到了水源的重要性，因此也将建筑群体建在了靠近水源的地方，如呼和浩特市的大召、席力图召、五塔寺、乃莫齐召，包头市的昆都仑召，锡林郭勒盟的汇宗寺等。

1 参考：乌审旗民族事务局.分管民族问题和宗教问题 [Z] .

As a national worship object, ritual activities about Su Lide also swept the entire Mongolian tribes, for example, in March 17th of every lunar year, the Mongols would hold a grand ceremony for Su Lide. Based on the special traditional culture in Mongolia region, Tibetan Buddhism brought Su Lide ritual into the sect in the process of its deep entering, and the subsequent appearance of the Tibetan Buddhism temples with the theme of Su Lide became one of the features distinguishing from other Tibetan Buddhism temples in Tibetan area as Tibet, Qinghai and Gansu. Take Erdos area as an example:

As two independent worship systems, Su Lide and Genghis Khan are both guarded, managed, enshrined and worshiped by Erdos ,but Su Lide worship is closely and inseparately related to the worship of ancestor Genghis Khan. Therefore, these two kinds of worship activities can be embedded in the Mongolian nationality in Ordos. Within Erdos, there are Dahl Hu Te who specialize in ritual, with the distinction of Big Dahl Hu Te and Little Dahl Hu Te, among whom Big Dahl Hu Te specialize in guarding Babai Room enshrining Genghis Khan, while Little Dahl Hu Te are guardians and dedicated people of Su Lide. This distinction is passed down from the Kublai era with strict rules. With the popularity of Tibetan Buddhism in Mongolia region, Genghis Khan was turned to the Dharma Protector of Tibetan Buddhism in Erdos Area, and in the region appeared Tibetan Buddhism Temples specialized in enshrining and chanting for Su Lide, such as Wulan Mulun Temple, which became the obvious Mongolian characteristic of Tibetan Buddhism temples in Inner Mongolia area .

Ⅲ. The Influence of Natural Factors

1. Temple Site under the Influence of Geographic Factors

According to the preceding chapters, Inner Mongolia region is rich in terrains and landforms, for example,the Yinshan Mountains in south-central Inner Mongolia Autonomous Region crosses the Inner Mongolia plateau and Hetao Plain, east to Greater Khingan Range, west to northern Mountain, extending into the vast desert of Alxa. Therefore, the selection of Tibetan Buddhism Temples in Inner Mongolia area was subjected to the specific circumstances .But usually, topography, geology and water resource were the important factors under consideration.

A. Topography and Geomorphology

The sites of Tibetan Buddhism Temples in Inner Mongolia area were given preference to mountainous terrain as well as plain and relatively higher geographical places in plateau easy to drain off water from rain and snow and to prevent flood , to be convenient for traffic and so on. For example, temples along the Yinshan Mountains from east to west: Agui Temple in Bayannur City is in the branch of Lang Mountain, Merigen Temple, Hundele Temple in Baotou City are in the branch of Wulashan Mountain, Maidarzhao Temple in Baotou, Sarqinzhao Temple and Lamadong, Wusutuzhao Temple in Hohhot are located in the branch of Daqing Mountain, and so-called Southern Temple and Northern Temple—Guangzongsi Temple and Fuyinsi Temple and so on are widely distributed in the Helan Mountains. Moreover, according to the names of temples,people can directly determine the geomorphological features of the location. For example, a part of the Inner Mongolian temples are called “Agui Temple”, which is the transliteration of Mongolian, meaning “Cave Temple”. According to its name, people can learn these temples were built in mountainous areas in order to get quiet place to cultivate. There are such temples in Hohhot City, Alxa Prefecture, Bayannur City, Ulanqab City and other regions in Inner Mongolia.

Usually, the influence of topography and landform to Inner Mongolia Tibetan Buddhism temples is most directly reflected in the layout of the temple building: the temples built on the mountains mostly conform to the topography, terrain, and close to the mountains, with flexible building layout, such as Badgar Temple, Agui Temple. The temples located in the plains or the flat land of plateau often adopt more regular layout, neat and rigorous, such as Dazhao Temple, Beis Temple etc..

B. Geology

In addition to topography and landform, geological structure was also an important part in considering the site. Usually, specially-assigned person by the temple would investigate the soil density, soil texture and so on in the selected location, and the builders often sent the soil of the place prepared to built the temples to the eminent monks for surveying and making determination.

C. Water Source

The temples mostly chose convex bank of rivers and the platform higher than the flood level to construct temples.

Seasonal rivers can be found around the majority of hillside temples due to the terrain, which bring some convenience to the production and life in the temples, such as Merigen Temple, Badgar Temple in Baotou City, Badgar Temple in Hohhot City etc..

The importance of water was also under the consideration of the temple site on plains. So buildings were also built in the areas near water, such as Dazhao Temple, Xilituzhao Temple, Wutasi Temple, Naimoqizhao Temple in Hohhot City, Hundele Temple in Baotou, Huizong Temple in Xilingol Prefecture etc..

2. Temple Architecture under Climatic Conditions

To seek a shelter from the wind and rain and a pleasant interior climatic environment is one of the original motives to build houses. Among the natural factors affecting and deciding the architectural style of Tibetan Buddhism temples, climate condition is a most basic factor of universal meaning.

A. Building Orientation

It is cold in winter in Inner Mongolia region, in order to

二、气候条件下的召庙建筑

遮风避雨、寻求宜人的内部气候环境，是建造房屋的原始动因之一。在影响和决定召庙建筑风格的自然因素中，气候条件是一个最基本的因素，也是最具有普遍意义的因素。

1. 建筑朝向

内蒙古地域冬季寒冷，为了能够得到更好的室内热环境和光环境，殿堂建筑朝向多选为南偏东或南偏西方向。当然也有特例，目前在调研中仅发现阿拉善盟的巴丹吉林庙、乌兰察布市的王府庙、锡林郭勒盟的毕鲁图庙为东朝向。

2. 建造构造

西藏地区和内蒙古地区的气候存在一定程度的相似性，西藏多风雪较寒冷，内蒙古多风沙较干燥，因此，藏式的窄条窗，无论是藏式盲窗还是藏式明窗，均适应内蒙古的气候条件。故内蒙古地域召庙建筑多采用藏式窗，且多为长方形，较内地门窗用材小。窗上可开启面积较小，可以适应蒙古地区高寒气候特点，并有一定的防风沙能力。但是，由于藏地昼夜温差大，阳光、紫外线强烈，为了躲避寒流和烈日，其建筑外墙的洞口和数量都尽可能减小些；而内蒙古地域的气候相对于藏区来说则较为缓和，但冬季仍旧寒冷漫长，因此内蒙古地域召庙建筑的南向开窗较大，以争取更多的日照。此外，由于藏地降水稀少，建筑采用平屋顶即可；而内蒙古地域的雨水稍多，因此常常将藏地平屋顶与汉地利于排水的坡屋顶结合起来使用。

总体而言，对内蒙古地域的召庙建筑来讲，政治因素是其产生、发展、成熟的主导因素，它对召庙建筑形态的影响多体现在建筑的形制、规格、等级的取舍上；文化和自然因素则是作为召庙建筑形态丰富多样的直接原因。

get better indoor thermal environment and light environment, the building orientations of halls are mostly in south, east or south-west, with some exceptions that only Badanjili Temple in Alxa Prefecture, Wangfu Temple in Ulanqab City, Bilutu Temple in Xilingol Prefecture are found in east orientation in the survey.

B. Construction Structure

The climate in Tibet area is similar to that in Inner Mongolia area to a certain degree .It is windy and snowy and colder in Tibet while windy and sandy and drier in Inner Mongolia. Therefore, Tibetan narrow windows, hidden windows or out windows are all adapted to the climate conditions in Inner Mongolia. The Tibetan Buddhism temples in Inner Mongolia region mostly adopted Tibetan windows, with the shape of rectangular, using less materials in doors and windows than in mainland. The smaller openable parts in windows are adapted to the cold climate in Mongolia area, with a certain function of anti-sandy wind. However, because of big temperature difference in day and night, strong sunshine and ultraviolet radiation in Tibet, in order to keep shelter from cold air and blazing sunshine ,the openings and the numbers of them are minimized as possible in the exterior walls; in Mongolian area, the climate is moderate comparing with that of Tibet, but still cold in long winter, so the south windows of Inner Mongolia regional Tibetan Buddhism temples are larger in order to strive for more sunshine. In addition, since there is little rainfall in Tibet, buildings only adopt flat roof; while in Inner Mongolia regions with a little more rain, the combination of Tibetan flat roof and drainable-slope roof in Han are used.

Overall, in terms of Tibetan Buddhism temple buildings in Inner Mongolia region, the political factor was the dominant factor for its generation, development and maturity, which influenced the form of temple construction mostly in the alternatives of form style, specifications and levels. Cultural and natural factors were the direct causes of the rich variety of temple architectural forms.

第三章 内蒙古地域召庙建筑形态演变的历史分期

从蒙古地域召庙与本地域政治、文化、自然的关系及其自身规模、类型、技术水平等方面来看，可把蒙古地域召庙的建设演变分成四个历史时期：初期、发展期、繁荣期及后期。

第一节 或改或借的初期

元朝至明初是蒙古地域藏传佛教发展的初期，亦是蒙古地域召庙发展的初期。

这一时期，藏传佛教虽然得到了蒙古上层的大力提倡和扶持，但仅被蒙古贵族尊崇信仰，并未在广大普通蒙古民众中普及，随着元王朝的没落，蒙古藏传佛教与西藏佛教主流失去联系，甚至在元末明初时曾一度消失。

该时期蒙古地域藏传佛教的召庙以及殿堂有以下特点：

1. 热衷修建于重镇的召庙

蒙元时期的政治当局虽大力扶持藏传佛教，但其重点扶持、兴建及修复的召庙多集中在元朝统治的中心，例如大都、上都等地区以及内地重要城镇和名山胜地等，召庙并未普建于广阔的蒙古地域。

2. 保护性修复下的汉式召庙

元朝政治者虽极力推崇藏传佛教，但并没有大规模地建造藏传佛教召庙，而是对前几朝（如唐、宋、辽、金等朝）遗留下来的汉式佛寺加以保护，并进行修复及改造，从而达到使之转宗为藏传佛教召庙的目的。因此，这一时期的殿堂多汉式，且结构做法多为各朝遗风，风格迥异。

3. 藏式的缺失

这个时期的蒙古地域正处于藏传佛教传入的初始阶段，由于大规模的传教活动并没有展开，宗教活动仅限于传教喇嘛携藏传佛教教义、教理活动于蒙古上层社会，因而传统的藏式殿堂并没有在蒙古地域普及，更无藏族匠人将成熟的藏式结构做法引入本地域。

4. 蒙古毡包式殿堂的运用

该时期，由于藏传佛教在蒙古地域传播的特点（普及性不强，无建造大量的、成规模的召庙的条件），本地域出现了以适应游牧生活为特点的，以蒙古包为代表的毡包式殿堂。

由于政治、历史、自然等原因，初期的蒙古地域藏传佛教召庙已不复存在，或仅存遗址，但总体而言，这一时期，专门营建的藏传佛教召庙较少，且大都是在其他宗教建筑、草原民居的基础上略作发展，或直接借用，或适度改造。

第二节 汉风浓郁的发展期

发展期主要是指明中叶至清初，即以阿勒坦汗在青海仰华寺与三世达赖喇嘛索南嘉措会晤为起点至清初约100多年间。

这一时期，藏传佛教召庙或直接或间接地被植入蒙古地域，其主要建造者为蒙古、藏、汉族匠人，在殿堂营造方面还处于探索阶段。随着内蒙古地域第一座藏传佛教格鲁派召庙——大召的建立，蒙古地区兴起了广建召庙的热潮。该时期，召庙内的殿堂多以汉式为主（殿堂风格及构造做法），兼有藏式做法（局部采用），殿堂发展之势初见端倪，具体如下：

1. 藏式为源

通过消化与吸收藏族地区措钦、扎仓建筑的布局，该时期召庙内的殿堂多采用前部外廊、中部经堂、后部佛殿相依纵向布置的形式。此外，该时期的殿堂呈现出一定的特点：佛殿层数为一层，外设一圈廊道为转经道；在经堂后两侧分别设有门，使转经道与经堂得以连接，以满足殿内做法事时转经之用。此类做法是间接沿用西藏藏传佛教建筑发展初期殿堂特点的结果（图10）。

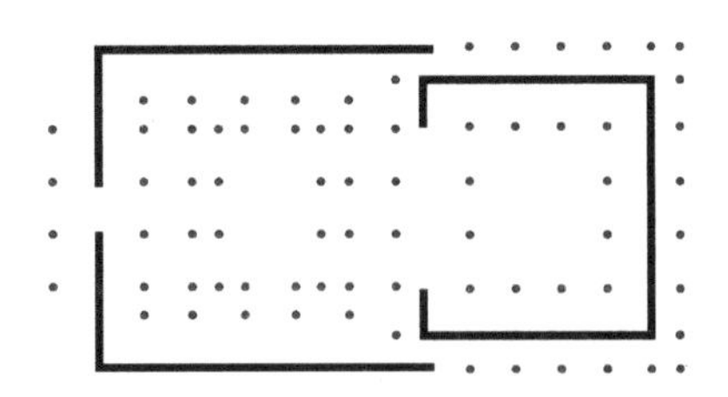

图10 发展期藏式典型平面图

此外，在这一时期，经堂多为一层，且中部有垂拔以解决建筑内部通风、采光问题，这种做法则是藏式都纲法式做法的变异形式。

Chapter Three
Historical Stages of Evolution of Architectural Form of Temples Architecture in Inner Mongolia Region

According to the introduction of the above chapters, from the relations between Tibetan Buddhism temples in Inner Mongolia region and the local politics, culture , nature and from the aspects of their own sizes , types ,technology etc. , the construction evolution of Tibetan Buddhism temples in Mongolia can be divided into four historical stages: the Initial Stage, the Developing Stage, the Prosperous Stage and the Later Stage.

I. The Initial Stage of Modifying or Borrowing

From Yuan Dynasty to early Ming Dynasty ,it was the initial developing stage of Tibetan Buddhism in Mongolia region, also the early period of the development of Tibetan Buddhism temples.

During this period, Tibetan Buddhism had been worshiped and believed by Mongolian nobles with the vigorous promotion and support by Mongolian upper classes,but still not popular in the general ordinary Mongolian people. With the decline of Yuan Dynasty, Mongolia Lamaism lost contact with the mainstream of Tibetan Buddhism, even once disappeared in late Yuan Dynasty and early Ming Dynasty.

The features of Tibetan Buddhism temples and Halls in Mongolia area in this period are as follows:

A. Building Temples in Important Cities

Although the political authorities in Mongolia Yuan period made effort to support Tibetan Buddhism, the key temples which were supported, built and renovated were mostly concentrated in the centre of Yuan Dynasty, for example, Dadu, Shangdu etc. and the important towns and famous mountains and spots in mainland. Tibetan Buddhism temples were not built everywhere in the vast region of Mongolia.

B. Han-style Temples under Protective Renovation

Though the governors of Yuan Dynasty strongly respected Tibetan Buddhism, they didn't build Tibetan Buddhism temples in large-scale, instead, they protected, renovated and reformed the Han-style Buddhism temples left by the previous dynasties (such as Tang, Song, Liao, Jin etc.) so as to achieve the purpose of converting to Tibetan Buddhism Temples. Therefore, the halls in this period were mostly Han-style with the structures and characteristics of different dynasties.

C. The Vacancy of Tibetan Style

This period was the initial stage of Tibetan Buddhism entering into Mongolia region, because the large-scale missionary activities hadn't been expanded, the religious activities were confined to the missionary Lamas with religious doctrines and teachings in Mongolia society, thus the traditional Tibetan temples were not popular in Mongolia region and no Tibetan builders brought the mature Tibetan structure into the district.

D. The Use of Mongolia Yurt Hall

In this period, as a result of spreading characteristics of Tibetan Buddhism in Mongolia area (small popularization, no conditions of building a large number temples of big scale), Mongolia yurt, as the representative of the yurt type hall appeared in the region to fit for the characteristics of the nomadic way of life.

Due to political, historical, natural and other reasons, the Tibetan Buddhism temples of the initial stage in Mongolia region no longer exists, or only remain relics. But in general, in this period, there were less specialized construction of Tibetan Buddhism temples, with most of them only slightly developing, directly borrowing or moderately transforming on the basis of other religious buildings and grassland residential houses.

II. The Developing Period Rich in Han style

The developing period mainly indicates mid Ming Dynasty to early Qing Dynasty, namely about 100 years from the meeting of Altan Khan and Dalai Lama III, Suonancuojia in Yanghua Temple in Qinghai to the beginning of Qing Dynasty.

During this period, Tibetan Buddhism temples were embedded into Mongolia region directly or indirectly. The main builder were Mongolian, Tibetan, Han artisans, and the temple construction was still under the stage of exploration. With the building of the first Tibetan Buddhism Gelu Sect temple—Dazhao Temple in Inner Mongolia region (the earliest Tibetan Buddhism temple in Mongolia in Ming and Qing period), the wide construction of temples in Mongolia boomed. In this period, the halls in the temples began to be built mainly by Han style (style and structure of main hall), combined with Tibetan style (partly).Specific as follows:

A. Tibetan Style as the Source

Through digestion and absorption of layout of Cuoqin and Dratsang construction in Tibetan areas, the hall in this period mostly adopted the setting form of vertical arrangement of attaching front veranda, central scripture hall to back Buddhist hall. And the hall showed some characteristics in this period: only one floor in the Buddhist hall with a round of outer corridor as circumanbulation; on both sides at the back of scripture hall were doors to connect circumanbulation and scripture hall to meet the needs of prayers in Dharma events in the hall. Such practices were the results of indirect application of the hall features of Tibetan Buddhist architecture in the early developing period.

Moreover, during this period, most halls are one-floor with vertical space in the middle to solve the problem of air flowing and lighting inside the building, which were the variant forms of Tibetan Dugangfa style.

2. 汉风浓郁

这一时期，从事建筑营造活动的多是汉族工匠以及掌握了汉族建造技术的当地蒙古族工匠，而藏区的建筑文化也以零散、不系统的方式陆续传入蒙古地域，因此，在发展期的本地域藏传佛教召庙中，汉式布局、汉式风格以及汉式做法占到相当大的比重。例如，建于本时期的乌素图召庆缘寺大殿、美岱召大雄宝殿、大召大雄宝殿等，这一类召庙，无论是从结构体系还是在装饰、装修上，均采用了汉式做法，仅仅在较少部分的构造做法上采用藏式做法，如外墙、边玛檐墙等。

第三节 藏式风靡的成熟期

清康熙、雍正、乾隆时期是蒙古地域召庙发展的成熟期，在此期间，召庙以燎原之势遍布蒙古草原。

这一时期，朝廷为大力扶持和发展蒙古地域的藏传佛教，不惜花费大量人力、物力、财力兴建召庙，乾隆年间达到了最高峰，大量的敕建庙宇也于此时期成批出现。相较于发展期而言，藏族建筑文化也开始不断地被引入蒙古草原，而蒙古地域的匠人则根据蒙古民族及当地主要驻民的文化接受模式，对汉地建筑文化及藏族建筑文化进行了筛选、吸收和重构。这个时期，蒙古地域召庙殿堂空间呈现如下特点：

1. 平面布局灵活自由

经堂与佛殿的平面布局较为灵活，后文总结的接合式、结合式和分离式组合形式均有使用。

2. 转经道的布置出现变体

佛殿周围转经道的布置形式也较前期更加自由。有的虽保留了原来佛殿后面转经道，但仅仅是出于建筑结构的需要，而非转经之用（如包头市昆都仑召的小黄庙）；有的殿堂干脆将转经道去掉（如包头市昆都仑召的大雄宝殿）；还有的建筑将转经道向前廊延伸，绕建筑一周（如今包头市达茂旗希拉木仁庙的大殿）。

3. 纯藏式召庙殿堂的出现

至清中叶，西藏建筑文化已经比较完整、系统、全面地介绍了蒙古地域，纯藏式的殿堂在内蒙古地域出现。

4. 都纲法式的引入与发展

很多内蒙古地域召庙殿堂直接使用了典型的都纲法式，如包头市的昆都仑召大雄宝殿、五当召大雄宝殿等，此外，亦有其他都纲法式[1]的变异形式出现，并形成定制。

5. 大面积经堂的出现与殿堂内部内容的增多

由于这一时期的召庙僧人数量大增，大面积经堂应运而生，例如，通辽市库伦三大寺中的兴源寺大殿、包头市五当召的大雄宝殿以及昆都仑召的大雄宝殿，这些殿堂均有九九八十一间。此外，殿堂内部的内容也日趋完善，如殿堂一层局部或二层设有储藏及管理用房等。

第四节 无创无新的后期

清道光年间至1949年，约100多年时间是蒙古地域召庙发展的后期。

清朝末期，清朝政府内忧外患，无暇顾及蒙古地域藏传佛教，加之战乱连年，蒙古民众困苦不堪，致使蒙古地域召庙逐年衰败。至清朝封建统治被推翻，本地域的藏传佛教势力和影响又随之进一步被削弱，召庙的建设基本处于停止状态。因此，对于后期的内蒙古地域召庙而言，营造活动相对减少；殿堂数量急剧锐减；建筑活动的重点转移至殿堂建筑的修缮工作上，而非新建；建造技术上也只是延续前期的做法，没有创新。

在此需要说明两点：

1. 内蒙古地域的很多召庙，由于多方面的原因，其建造、修缮过程可能跨越多个历史时期及朝代，因此，以上各个历史时期是以召庙内大部分重要殿堂的建造年代、风格、做法等作为划分依据的。

2. 由于政治、历史、自然等多方面的原因，内蒙古地域现存的召庙殿堂多为发展期以后的建筑，其中，明清时期的遗存召庙有较高的研究价值。

1 都刚法式："都刚"为藏语音译，意为"聚集的房屋"，是喇嘛僧众聚会、诵经、祈祷的场所。都刚法式是这类殿堂常见的一种定型化的建筑规制：殿堂平面呈回字形，中心部分突起，设高侧窗以获得光线和中心感，象征曼荼罗宇宙图式。

B. Rich in Han style

In this period, the people engaging in construction activities mostly are Han artisans and local Mongolian artisans mastering construction technology of Han, and Tibetan architectural culture was introduced into Mongolia region in scattered and unsystematic way. Therefore, in developing local Tibetan Buddhism temples, there was a large portion of layout, style and practice of Han. For example: Hall in Qingyuan Temple in Wusutuzhao Temple built in the period, the Main Hall of Maidarzhao Temple and Dazhao Temple and so on. This kind of temples all adopted the practice of Han from structural system or decoration, with only a little part using Tibetan practice in structure, such as outer walls, eaves etc.

Ⅲ. Maturity Stage with Prevailing Tibetan Style

The periods of Kangxi, Yongzheng and Qianlong in Qing Dynasty belonged to the mature stage of the development of Tibetan Buddhism Temples in Mongolia area. During this period, Tibetan Buddhism temples spred to all over the Mongolia grassland.

During this period, the royal court spent a great amount of manpower, material resources and financial power in building Tibetan Buddhism temples to support and develop Tibetan Buddhism in Mongolia area and reached the summit in Qianlong period. Many temples were built by the government appeared in the period. Compared with the developing period, Tibetan architectural culture was also constantly introduced into the grasslands of Mongolia, and the craftsmen in Mongolia region selected, absorbed and reconstructed the architectural culture of Han and Tibet according to the cultural acceptance model of Mongolian people and the local civilians. During this period, the temple space was characterized by:

A. Flexible and Free Plane Layout

In this period, the layout of the sculpture hall and Buddha hall were more flexible, the summary of the joint type, combination type and separation type in the following chapters were all used.

B. The Variant Layout of Circumanbulation

In this period , the layout around the hall was more free than before. Some retained the original circumanbulation at the back of the hall, but only for the need of building structure instead of praying (such as Xiaohuang Temple in Hundele Temple in Baotou City); some simply removed circumanbulation (such as the Main Hall in Hundele Temple in Baotou City); and some extended circumanbulation to the front corridor ,around the building (such as the Hall in Puhui Temple in Baotou City).

C. The Appearance of Pure Tibetan Buddhism Temples

To the mid of Qing Dynasty, Tibetan architectural culture had been introduced to Mongolia area completely, systematically and comprehensively. Pure Tibetan temple halls appeared in Inner Mongolia area.

D. The Introduction and Development of Dugangfa Style

During this period, many Tibetan Buddhism temples in Inner Mongolia area directly applied typical Dugangfa Style, such as the Main Hall in Hundele Temple in Baotou City and Badgar Temple etc.. Furthermore , other variant forms of Dugangfa Style appeared and formed the regulation.

E. The Appearance of Large Hall and the Increase in the Content

Since the significant increase in the numbers of monks in the temples in this period, the hall with a large area emerged at the request, for example, the Hall of Xingyuan Temple in the Three Big Temples in Kulun, Tongliao, Suguqin Hall in Badgar Temple Temple ,Baotou City and the Main Hall in Hundele Temple in Baotou City which all have eighty-one rooms .Furthermore, the internal contents of the halls were also perfecting, such as arranging the storage and management rooms on part of the first floor or second floor.

Ⅳ. The Later Stage with no Creation and Innovation

The 100 years or so from Daoguang period of Qing Dynasty to the founding of the People's Republic of China in 1949 belonged to the later stage of the development of Tibetan Buddhism temples in Mongolia region.

At the end of Qing Dynasty, Qing government was under domestic trouble and foreign invasion, having no time to support and promote Tibetan Buddhism in Mongolia area, and Tibetan Buddhism temples declined year by year with years' wars, chaos and the great hardship and bitterness suffered by Mongolian people. When Qing Dynasty was overthrown, the power and influence of the regional Tibetan Buddhism were further weakened with it and the construction of temples was ceased. Therefore, for the later Tibetan Buddhism temples in Inner Mongolia region, the construction activities were relatively decreased; the number of halls had been drastically reduced; the focus of the construction activities was shifted to the renovation of halls instead of building new ones, and the construction technology just continued the previous practice without innovation.

Two points need stating here :

A. There are many temples in Inner Mongolia area, due to various reasons, the process of construction and perfection crosses many historical periods and several dynasties. Therefore, the dividing of each historical period is based on the building age, style and practice etc. of the most important halls in the temples.

B. Due to political, historical, natural and other reasons, the extant Tibetan Buddhism temples in Inner Mongolia region are all the constructions after the developing period, among which the temples in Ming and Qing Dynasties are of great research value.

第四章 内蒙古地域召庙建筑形态的一般特征

自藏传佛教在内蒙古地域开始传播、发展，作为其植入本地域重要手段的召庙建筑文化逐步深入到蒙古民族文化中，并且逐渐积淀成为具有鲜明地域特色的历史文化遗产。

根据前述章节内容可知，自窝阔台汗之三子阔端首礼藏传佛教至今，上下约七百年间，明清时期是内蒙古地域藏传佛教发展与传播的鼎盛时期，从域外以不同方式被引入内蒙古地域的召庙建筑无论在数量上还是在质量上均达到了较高水平，且内蒙古地域召庙建筑基本上都是这一时期新建的，部分召庙建筑遗存至今。

通过对文献资料的研究以及对内蒙古地域召庙建筑的实地调研发现，召庙在内蒙古地域的扎根生长的发展过程中，既保留了藏、汉建筑形制的重要特征，又在本地域其他因素的影响下，融合了鲜明的地域特征，表现出一定的形态共性。

第一节 布局多元

内蒙古地域召庙的构成直接承袭了藏、甘、青地区藏传佛教召庙的基本内容，即主要包括经堂、佛殿、活佛府（拉卜隆）、扎仓、白塔、僧舍等内容。这些召庙构成要素的排布方式在政治、文化、自然因素的影响下，呈现出多元化的形态，如有汉地佛教寺院中传统的“伽蓝七堂式”及其变体（如呼和浩特市地区的大召、多伦汇宗寺等）,也有以主要殿堂为中心象征佛教宇宙中心的曼荼罗式布局及其变体（如包头市地区的昆都仑召、阿拉善地区的南寺等），又有依山而建的自由式布局（巴彦淖尔市地区的阿贵庙等），还有多轴线组团式布局（如呼和浩特市地区的乌素图召、锡林郭勒盟地区的贝子庙等），更有许多依当地地形而建的综合型布局。

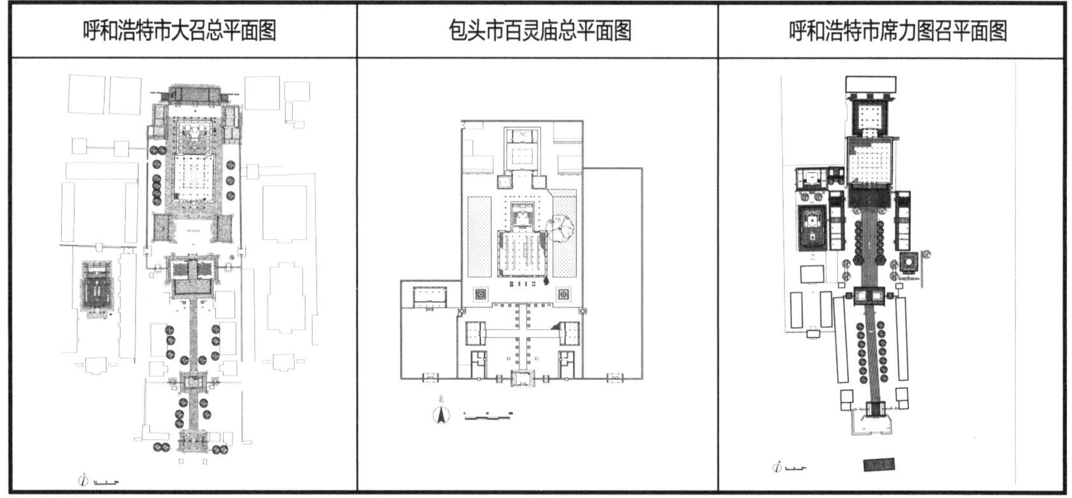
呼和浩特市大召总平面图　包头市百灵庙总平面图　呼和浩特市席力图召平面图

伽蓝七堂式

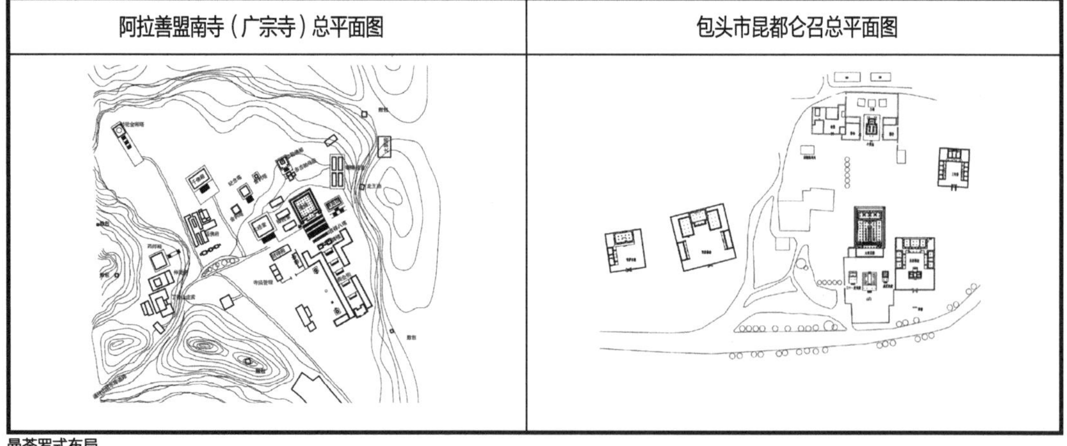
阿拉善盟南寺（广宗寺）总平面图　包头市昆都仑召总平面图

曼荼罗式布局

Chapter Four
The General Features of Temple Architecture in Inner Mongolia Region

Since the Tibetan Buddhism in Inner Mongolia region began to spread and develop, temples as the important means of this region was being embedded into Mongolian all aspects of life and production, and gradually became with distinct regional characteristics accumulation of the historical and cultural heritage.

According to the above section content that, since Godan, the third son of Ogadai Khan, was first ceremony of Tibetan Buddhism so far about seven hundred years, the Ming and Qing Dynasties were heyday of development and dissemination of Tibetan Buddhism in Inner Mongolia region. During this period, Tibetan Buddhism temple architectures in Inner Mongolia region which were introduced extraterritorially in the different ways in terms of quantity or quality had reached a high level. Furthermore, the new temple architectures of Inner Mongolia regional Tibetan Buddhism were basically set up in this period, and some of them were remained today.

Through literature study and on-the-spot investigation of Inner Mongolia region Tibetan Buddhism temples found that Tibetan Buddhism temples rooted growth in development process in Inner Mongolia region, not only retained the important features of the Tibetan and Han architectural structure, but also integrated the distinctive regional characteristics to show some form of common under the influence of the other geographical factors. Therefore, the concrete summaries are as follows.

I. Diversified Layout

The formation of Inner Mongolian Tibetan Buddhism temples was inherited the basic structures directly by Tibet, Gansun and Qinghai Tibetan Buddhism temples including mainly Ceremony Hall, Buddha Hall, Hudut (Labreng), Gyuto, the White Pagoda, Monk Houses etc.. The assignment of configuration in these temples presented diversified forms under the influence of political, cultural and natural factors, such as The Traditional Jalan Seven-hall Style and its variants in Han Buddhism temples (Dazhao in Hohhot City, and Huizong Temple in Duolun etc.), the main hall as the core being symbol of Mandara layout of Buddhism universe center and its variants (Hundele Temple in Baotou City, Guangzong Temple in Alxa Prefecture etc.) freestyle layout of the hillside (Agu Temple in Bayannur Prefecture etc.), Multi-axis group layout (Wusutuzhao in Hohhot City, and Beizi Temple in Xilingol Prefecture etc.), and many integrated layouts with the local terrain.

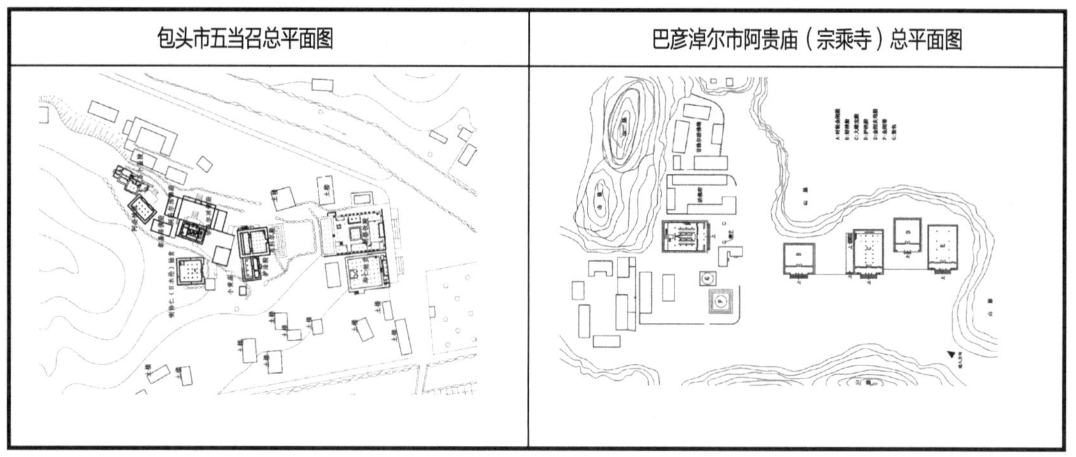

自由式布局

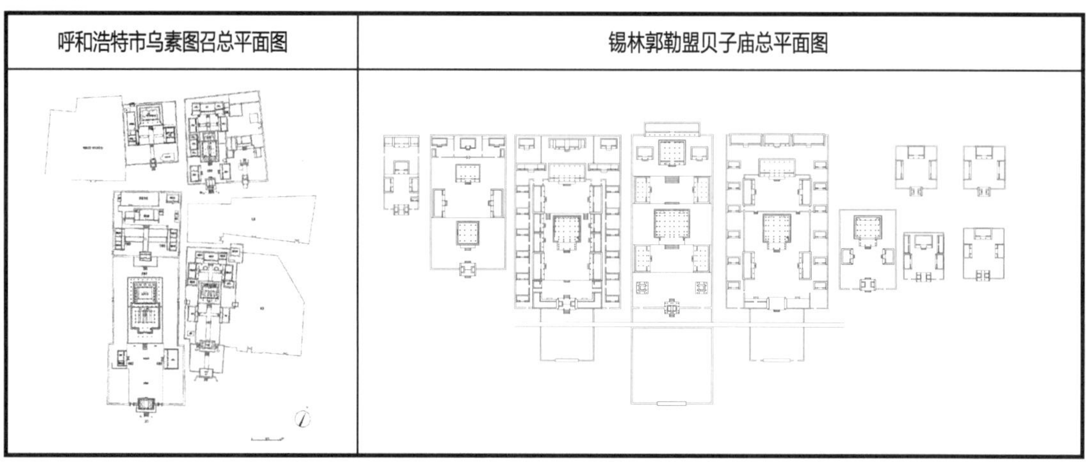

多轴线组团式布局

第二节 类型丰富

长期以来，学界普遍将内蒙古地域藏传佛教召庙大致概括为藏式、汉式和汉藏混合式（或汉藏结合式）三种，学界亦有学者认为，本地域的建筑存在蒙古式及其混合式[1]。

从总体布局来讲，藏式是指沿袭藏族传统的山地布局风格，以自由的、顺应山势的建筑布局手法来安排召庙内的建筑，单体之间没有明确的关系，仅仅是利用地形，将主要建筑置于相对较高的重要位置，与低矮的次要建筑形成对比，从而形成整个建筑群体的艺术形象（如包头市地区的五当召）；所谓汉式，是指殿堂建筑的总体布局采用伽蓝七堂制及其变体形制，即同内地佛寺相似，有着明确的轴线关系，通常轴线贯穿前后，且主要建筑均置于轴线中部或后部（如锡林郭勒盟地区的贝子庙）；而汉藏混合式较为明显的特征则是在以具有轴线关系的重要殿堂组为中心，其他殿堂进行自由式布局。

从单体建筑形态来讲，藏式是指建筑单体具有显著藏区建筑特征，如藏式结构（石木混合结构）、藏式平屋顶、藏式边玛檐墙等；汉式是指建筑单体建造风格完全采用内地传统官式建筑营造的殿堂类型，该类建筑仅在装饰方面标示其藏传佛教的属性；汉藏混合式则是融有汉、藏建造技术及艺术为一体的形式。

其实，这种通俗易通的概括仅仅是一种粗放的风格归类，其中每一类的形态又极其丰富。从基本类型上看，内蒙古地域的召庙大致可分为两类：其一是移植明清时期中原官式建筑形制和汉地民间丰富多彩的建筑风格的召庙，较为典型的实例是赤峰市马日图庙（法轮寺）、锡林郭勒盟的汇宗寺（图11）。其二是移植甘、青、藏地区藏传佛教建筑风格的召庙，其单体建筑形制的典型特征是有一层碉房和多层都纲法式，如阿拉善

1 参考：包昌德.概述蒙古族建筑文化［A］.呼和浩特，2010：43—49；包慕萍.蒙古游牧文明的城市和建筑体系的特质［A］.呼和浩特，2010：50—559；色音.试论蒙地开垦后蒙古族民居建筑的变化［A］.呼和浩特，2010：91—94；刘润民.蒙古族传统建筑文化的几点思考［A］.呼和浩特，2010：107—1129；张晓东.蒙古式建筑探析［A］.呼和浩特，2010：170—174.

II.Rich Types

Over the years, scholars generally summarized Inner Mongolia region Tibetan Buddhism temples as Tibetan type, Han type and Han-Tibetan hybrid type (or Han-Tibetan combined type). Others believed that the architecture of the local region existed in Mongolian type and its hybrid type.

In terms of overall layout, the Tibetan type is the inheritance of Tibetan traditional mountain layout style, arranging the inner constructions in the temples with conforming freely to the architectural layout of the mountain-style. there is no clear relationship between the individual buildings, but the main buildings is placed in relatively high important position in contrast with the low secondary buildings, so as to form the whole building complex artistic image (such as Badgar Temple in Baotou City). The so-called Han type refers to the whole layout of the temple architecture adopting the system of the Jalan Seven-hall Style and its variants of structure, which is similar with the mainland Buddhism temples. It has a clear axial relationship that the axis usually runs though the front and the rear of the temple. And the main constructions are built up the middle and the back of the axis (such as Beizi Temple in Xilingol Prefecture). However, the remarkable features of the Han-Tibetan hybrid is the center of the important hall with the axial relationship, the other halls are adopted the freestyle layout.

As far as the individual building form is concerned, the Tibetan type construction presents that the individual building has the significant features of the Tibetan architecture such as the Tibetan mixed structure (stone and wood structure), the Tibetan flat roof and the Tibetan eaves wall, etc.. The single building of Han type is completely introduced by the style of the inland official architecture temple, and the kind of temple is only marked in the decoration of the properties of Tibetan Buddhism. Han-Tibetan hybrid type is combined with construction technology and the arts of Han and Tibetan.

In fact, the understandable generalizations are merely the extensive style classifications, and each form of them is extremely rich. On the basic types, Tibetan Buddhism temples in Inner Mongolia region could be broadly divided into two categories: one is the transplantation of temples which originated from the central plains official construction modality in Ming and Qing Dynasties and from the folk colorful constructional style in Han. The typical temples are like Falun Temple in Chifeng City and Huizong Temple in Xilingol Prefecture. Another one is the transplantation of Tibetan Buddhism temples from Gansu, Qinghai and Tibet. The remarkable characteristic of its single building is a layer of stone house and multilayer of the Dugang style such as Badaijiren Temple. in Alxa Prefecture and Shaletew Temple in Bayannur City.

A wide variety of integrations between the two basic types of Inner Mongolia regional Tibetan Buddhism temples present abundant form, which mainly build up the space of the Dugang style with the Han's roof (such as Dazhao Temple in Hohhot City) and the veranda temple with Tibetan eaves wall (such as Xiaramuren Temple in Baotou City). In addition, being based on the differences of the various scales and grades of the single building, the same integrated construction is also emerged more or less in the diversity of the modality. And even there are the multiplicities of roof and the inconsistencies of the local design, the type of the Tibetan Buddhism temple architecture in Inner Mongolia was extremely rich.

盟巴丹吉林庙、巴彦淖尔市善岱古庙（图12）。

这两种基本类型之间多种多样的融合方式使得内蒙古地域召庙建筑呈现出极为丰富的形态，主要表现为都纲法式空间上加汉式屋顶（如呼和浩特市地区的大召）、副阶周匝平面加藏式檐墙（包头市希拉木仁庙）（图13）等。此外，基于不同建筑单体的规模、等级等各方面的差异性，同种融合方式下建筑亦呈现出形制方面或多或少的差异，加之屋顶类型的多样性、地方做法的不一致性，使得内蒙古地域召庙建筑的类型极为丰富。

图11 马日图庙大雄宝殿

汇宗寺大雄宝殿

图12 巴丹吉林庙大雄宝殿

善岱古庙大雄宝殿

图13 大召寺大雄宝殿

希拉木仁庙大雄宝殿

第三节 藏式为母

众所周知，自佛教传入吐蕃，经历了多次的起起落落，西藏地区的佛教建筑从最初的在建筑形式上模仿印度、尼泊尔以及中原佛寺逐渐演变为融合、吸收当地土生土长的建筑特征后形成的“藏式”风格建筑。此类寺院建筑风格在藏传佛教格鲁派占据藏区佛教上风时期基本定型，传入内蒙古地域后，成为该地域主要的建筑类型母题，主要表现为以下几个方面：

1. 通常情况下，殿堂由门廊、经堂、佛殿三部分构成；

2. 殿堂后部是佛殿，通常是整座建筑的最高点和重点所在；

3. 通常殿堂前廊为两层，采用多楞柱或方柱，柱头设托木，梁柱多饰以藏式彩画和浮雕；

4. 经堂空间采用都纲法式；

5. 屋顶上饰有祥麟法轮、苏力德、经幢、风马旗的藏式宗教元素；

6. 外墙多为砖石结构，且有明显收分（上小下大），外墙墙面多为白色，整体呈现稳定的梯形状，建筑立面呈“两实夹一虚”的效果；

7. 建筑外墙开窗，窗洞较小，外饰黑色梯形窗套；

8. 檐口处是整个殿堂外观装饰的重点之一，重要殿堂檐口处多做棕红色边玛檐墙条，并施以单层或多层檐板，且边玛檐墙条上饰以铜饰或鎏金饰，多为“六字真言”或“八吉祥”。

以上这些做法及元素都广泛地出现在内蒙古地域纯藏式殿堂及汉藏混合式殿堂中，特别是纯藏式殿堂，与甘、青、藏地区的藏传佛教寺庙相比，内蒙古地域的纯藏式殿堂表现出以下几个方面的不同：

1. 相对于藏区的殿堂，本地域殿堂的规模较小；

2. 建筑外墙一般用砖砌筑，纹理较西藏地区的建筑纹理更为细腻；

3. 通常以红色涂料模仿边玛檐墙以达到藏式装饰效果。

整体而言，虽然，内蒙古地域的召庙建筑受到当地工匠及当地做法的影响，在融合蒙古民族、汉民族元素之后展现出较为鲜明的地域特点。但是，由于内蒙古地域的藏传佛教源自西藏地区，加之蒙古高原地域的地形、地貌、自然气候等特征与青藏高原有一定相似性，因此，内蒙古地域此类文化建筑的形式、风格亦多承袭西藏地区的藏传佛教建筑，且这种情况较为普遍、明显，特别是内蒙古中西部地区的殿堂表现得尤为突出。更为重要的是，对于内蒙古地域由经堂和佛殿组成的殿堂来说，不论其建筑外观是哪种风格，其建筑平面和空间形式都具有一致特征，即平面多为矩形，内部空间大多为都纲法式或模仿都纲法式的“凸”字形空间。此外，除少数经堂、佛殿合一的小型殿堂外，内蒙古地域的召庙殿堂基本上都是前经堂、后佛殿的模式，并且建筑整体由门廊、经堂、佛殿三部分构成，从而成为内蒙古地域召庙建筑普遍共有的形式母题。

Ⅲ.Tibetan Type-oriented

As known to all, Buddhism witnessed multiple rises and falls since its introduction to Tubo (ancient name for Tibet). Buddhism architectures initially imitated the architectural form of Buddhism temples in India, Nepal and central plains of China, gradually integrated and absorbed local architectural features, and then formed the Tibetan type architectures. Architectural style of such temples basically took into shape when Gelug of Tibetan Buddhism dominated the Buddhism in Tibetan region, and it became the main motif of architectural form in this region, mainly involving the following aspects:

A. In general, the palace hall is composed of portico, ceremony hall and Buddha hall;

B. At the rear of the palace hall lies the Buddha hall, which is usually the zenith and key part of the entire building;

C. Generally, the front portico is of two-floor, polygon prism or square column is applied, column cap refers to bolster, and beam column is mostly decorated with Tibetan type colored drawing and relieve;

D. The ceremony hall is of Dugang style;

E. The roof is decorated with Tibetan religious elements like propitious unicorn and dharma-cakra, trident, Dhnaja and Himo flags;

F. The exterior wall is mostly of masonry structure and has obvious contracture (small at top and large at bottom) and presents a steady trapezoid-shape on the whole, the surface of exterior is mostly white, and building facade shows the effect of "one virtual in two solids";

G. Windows of small size were set on the exterior wall and externally decorated with black trapezoid window casings;

H. Cornice is the key part of decorative appearance of the entire palace hall and that of important palace hall is generally provided with reddish-brown side edge bars, single-layer or multiple-layer is applied, and side edge bars are decorated with copper objects or gold-plating objects, mostly "Six-word Memoirs" or "Eight Auspiciousnesses";

All ways and elements above were widely applied in the full Tibetan type palace halls and Han-Tibetan hybrid type palace halls in Inner Mongolia region, especially the full Tibetan type palace halls. Compared with Tibetan Buddhist temples in Gansu, Qinghai and Tibet, full Tibetan type palace halls in Inner Mongolia region have the following distinctives:

A. Local palace halls are of small scale compared with those in Tibetan region;

B. Generally exterior walls are built with bricks and their textures are more delicate as compared with those in Tibetan region;

C. Generally red paint is used to imitate the side edge wall to achieve the effect of Tibetan type decoration.

On the whole, Tibetan Buddhism temples in Inner Mongolia region presented the vivid geographical features after integrating Mongolian and Han elements due to the effect of local craftsman and local techniques. However, Tibetan Buddhism in Inner Mongolia region originated from Tibet and Mongolian Plateau share universal and obvious similarity with Qinghai-Tibet Plateau in respect of the landform, natural climate and other characteristics, especially the palace halls in the middle and west of Inner Mongolia. More importantly, the palace halls composed of ceremony hall and Buddha hall in Inner Mongolia region are identical in their architectural plane and space form regardless of the style of their architectural appearance, that is, the plane is mostly rectangular and the inner space is mostly raised space of Dugang style or that imitating Dugang style. In addition, except few small-scaled palace halls integrating ceremony halls and Buddha halls into one, Tibetan Buddhist palace halls in Inner Mongolia region are basically the mode of ceremony hall in the front and Buddha hall in the rear, and the entire palace hall is composed of portico, ceremony hall and Buddha hall, thus generating the universe motif of Tibetan Buddhist temples in Inner Mongolia region.

第四节 规制式微

与上述共性特征伴生的特点是规制式微，即极为明确、规矩的形制要求和格式在内蒙古地域召庙建筑中表现得较为微弱。

一般而言，宗教建筑都有明确的组织规制，尤其在总体布局、殿堂形制和建筑形式方面表现得较为突出。区别于一般性宗教建筑，尽管内蒙古地域召庙建筑以藏式的平面特征为核心母本，但上述共性特征中体现的形制同藏地佛教建筑以及中原官式建筑中体现的体系化的形制相比，其规制显得较为微弱，且形制呈现多元化的变通。

在布局方面，不管是藏式自由布局的藏传佛教召庙，还是中原伽蓝七堂制控制下的汉地佛寺，二者均有着十分清晰的规制模式，但这些规制在内蒙古地域却表现得十分灵活、自由。在形式方面，上述两种成熟的建筑文化也分别有着十分清楚的规制，但当政治力量将其植入蒙古草原后，这两种建筑文化在这一特殊地域的特殊发展过程却使这种清晰的规制逐渐式微化，如阿拉善盟地区的广宗寺主殿，该殿堂建筑外围增加柱廊后，格鲁派竖向三段、下实上虚的典型立面特征被改变，而其两实夹一虚的横向特征也变得模糊。此外，同上述情况一样，内蒙古地域内改变中原官式建筑规制的召庙建筑更是比比皆是。

第五节 粗放的建造技艺

在任何类型建筑的发展史中，建造技术都起着支持或制约的作用，同时，它为建筑风格的产生提供了某种可能和限制。通过对内蒙古地域召庙现状的实地调研可知，在该地域召庙的建造过程中，负责营建的匠人们并没有严格地遵循藏区藏式和中原汉式建筑的传统建造技艺，而是采用了较为粗放实用的建筑方式来灵活应对具体工程的实际情况，这也成为内蒙古地域召庙建筑的重要特征之一。

就藏式风格的殿堂而言，可将其建造大体分为土工、木工、砖石瓦工和构造装饰几个主要部分。在西藏地区，寺院建筑对营建其各部分所选用的材料都有着较为严格的限制和规定，但在内蒙古地域，因各个召庙所处的自然地理环境差异较大，因而在召庙建筑中所用建筑材料也不尽相同。例如墙体材料，在西藏地区，匠人一般用石材砌筑墙体；而在内蒙古的有些地区（如沙漠、草原地区），由于能够符合作为建筑材料条件的石料的获得极为不易，匠人通常会选用更易于取得并方便运输的砖来代替石料；在一些极为偏远的地区，匠人更是取用当地的土坯来代替石材或砖作为建筑材料。再如藏传佛教召庙建筑的重要特征之一——边玛檐墙，在青藏地区，边玛檐墙都是用捆扎晒干的红柳堆砌而成，而内蒙古大部分地区的召庙并没有完全按照藏区传统做法制作边玛檐墙，而是在保持基本形态不变的情况下将边玛檐墙的做法大大简化，通常都是在墙体砌筑完成后，在边玛檐墙的位置涂红色颜料假饰红柳材料从而与下部墙体区分开来。这种“假冒”边玛檐墙的简化做法在本地区运用地极为广泛，成为一种较为普及的地方建筑语言。

就汉式风格的召庙而言，本地域殿堂建筑的木作、石作、屋顶相对于中原汉地的都有着不同程度的简化，如斗栱、柱础等。

因此，对于内蒙古地域召庙建筑而言，不论是哪种建造方式，从整体到细部，其做工及技艺均较为粗放。

当然，也有少数建造考究的例子，如雍正五年（1727年），在多伦诺尔所建的善因寺就是由宫廷样式房“样式雷”仿故宫中宫殿设计，出自汉族宫廷匠人之手的召庙。但相较于多伦善因寺等少数由皇帝直接拨款并直接指派专人营造的大型召庙外，在广袤的内蒙古地域，更多召庙的建设采用的是较为粗放的建造技艺。

第六节 近地域性特征

内蒙古地域召庙建筑是藏地佛教建筑文化、汉地官式建筑文化和本地域蒙古文化的融合体。然而，在内蒙古地域召庙建筑发展的过程中，除了受到上述建筑形态的影响外，更有来自临近汉地（宁夏、山西、陕西、河北、北京、吉林、辽宁等地区）的地域建筑文化对其产生的直接影响，并在内蒙古地域召庙建筑形态上呈现出明显的近地域性。

首先，内蒙古地域周边汉地建筑文化在对建筑的组织方式和空间的认知方面都对本地域召庙建筑的整体及细部的表现产生了或浅或深的影响；其次，综合对内蒙古地域召庙普查调研结果的分析发现，本地域召庙建筑的形态特征由西部向东部呈现出由显著的藏地建筑风格逐步趋于汉地建筑特征明确的特点；同时，由于携带着所在

Ⅳ. Deformed Form and Style

The above common features are accompanied with deformed form and style, that is, Tibetan Buddhism temples in Inner Mongolia region have no definite and proper form requirements and style.

Generally, religious architectures have definite structure and form, especially for the overall layout, palace hall form and architectural form. Differing from common religious architectures, Tibetan Buddhist temples in Inner Mongolia region takes Tibetan type plane features as core type, but the form and style presented by the aforementioned common features are deformed and diversified compared with the systematic form and style of Buddhist temples in Tibet and official architecture temples in Central Plain.

As to the layout, Tibetan Buddhist temples in freestyle Tibetan type layout and Han Buddhism temples under the control of the Central Plain Jalan Seven-hall Style have very clear forms and styles, which are very flexible and free in Inner Mongolia region. As to the form, the aforementioned two mature architectural cultures have the definite forms and styles respectively, which are gradually deformed due to the special development process of these two cultures in this special region after they are implanted into Mongolian steppe by political force. For example, after the portico was added on the periphery of the main palace of Guangzong Temple in Alxa Prefecture, Gelug typical facade features of three sections vertically and solid at bottom and virtual at top were changed, the transverse characteristics of "one virtual in two solids" also became blurred. Moreover, similarly, the temples that change the form and style of official architecture temples in Central Plain can be found everywhere in Inner Mongolia region.

V. Extensive Building Techniques

Building techniques play the role of supporting or restricting the development of architectures of any style. Meanwhile, they provide certain possibility and restriction to the emergence of architectural style. The on-the-spot investigation of Tibetan Buddhist temples in Inner Mongolia region shows that the craftsmen did not strictly follow the traditional building techniques of Tibetan type architectures in Tibet and Chinese type architectures in Central Plain, but adopted the extensive building way to flexibly cope with actual situation in specific projects, which is an important feature of Tibetan Buddhism temples in Inner Mongolia region.

In terms of Tibetan type palace halls, the building can be generally divided into soil works, wood works, brick and tile works and structural decorations. In Tibet, the temples specified strictly restrictions and regulations on materials for building each part; however, in Inner Mongolia region, various materials were used for building temples due to large difference in natural and geographical conditions. Taking wall materials for example, generally stones were used for building walls in Tibetan region; in some regions (such as desert and grassland) of Inner Mongolia, usually the craftsmen replaced stones with bricks that can be easily accessed and conveniently transported for it was extremely uneasy to obtain the stones that met building conditions; and in some extremely remote regions, the craftsmen replaced stones or bricks with local adobes. Taking side edge wall, one of key parts of Tibetan Buddhism temples, for another example, they were built of bundled and dry rose willow in Qinghai-Tibet area while such walls were not built for temples in most areas of Inner Mongolia fully in accordance with traditional way of Tibet and the building way was greatly simplified under the condition of keeping basic shape the same. Generally after finishing the wall, red pigment was painted on side edge wall to imitate rose willow materials so as to distinguish it from wall at lower part. Such a simplified way of imitating side edge wall was widely applied in this region and became a universe local architectural language.

For Han type temples, carpenter's work, stonework and roof of local palace halls were simplified at different levels as compared with those of Central Plain Han, such as bucket arch, plinth, etc..

Therefore, regardless of building way, the workmanship and techniques of Tibetan Buddhism temples in Inner Mongolia region are generally extensive from the entire to detail.

Certainly there are some exquisite architectures. For example, Shanyin Temple built in Duolun in A.D.1727 (5th year of Emperor Yongzheng) was accomplished by imperial craftsmen upon palace-style architecture Lei style with the design imitating Hall of Central Harmony of the Imperial Palace. However, compared with Duolun Shanyin Temple and other few large-scale temples built by specific person directly designated by the emperors with the fund directly allocated by the emperors, many Tibetan Buddhist temples in the vast Inner Mongolia region adopted the extensive building techniques.

Ⅵ. Approximate Regionality

Tibetan Buddhism temples in Inner Mongolia region are the integration of Tibetan Buddhism architectural culture, Han official architecture culture and local Mongolian culture. However, in addition to the impact from the aforementioned architectural form during the development of Tibetan Buddhism temples in Inner Mongolia region, there is also direct impact from architectural cultures nearby Han (Ningxia, Shanxi, Shaanxi, Hebei, Beijing, Jilin, Liaoning, etc.), and the obvious approximate regionality is presented in the form of Tibetan Buddhism temples in Inner Mongolia.

Firstly, Han official architecture culture around Inner Mongolia region produced slight or deep impact on the entire and details of local Tibetan Buddhism temples in

地域建筑技艺的工匠自古就是建造技艺传播的主要途径之一，因而，根据调研发现，本地域周边的中原汉风对内蒙古地域召庙建筑的影响还表现在建造技术方面。内蒙古广阔地域上的原驻居民多以游牧生活为主，逐水草而居，没有较为固定的聚居地，这样的生活方式决定了他们还不具备掌握较为先进的定居性建筑建造技艺的客观条件。因此，当召庙建设兴起之时，内蒙古地域召庙建筑的建造大都是由部分藏族工匠，大量的周边汉地工匠以及部分由汉、藏族工匠培训的蒙古族工匠完成的。尽管，由于所建召庙所处自然地理位置、兴建缘起、财力支持状况以及建设周期等各方面因素的不同，本地域内的召庙建筑在建造技术方面存在着较大的差异，但都十分清晰地反映着邻近地域的影响。

terms of structural form of the architecture and spatial cognition. Secondly, the combination of analysis on result of general investigation of Tibetan Buddhism temples in Inner Mongolia region shows that the shape of local Tibetan Buddhism temples evidently and gradually has the definite Han architecture feature from original Tibetan architectural culture from the west to the east. Meanwhile, craftsmen with building techniques are one of major methods for spreading building techniques and ways, so the investigation shows that surrounding Central Plain Han style also impacted the building techniques of Tibetan Buddhism temples in Inner Mongolia region. The original residents in the vast Inner Mongolia region mostly normalized and moved from place to place in search of water and grass and no fixed settlement formed, so such a life style determines they have not possessed the objective conditions for grasping the advanced building techniques. Therefore, at the time when temples construction emerged, Tibetan Buddhism temples in Inner Mongolia region were mostly built by many surrounding Han craftsmen and a few Mongolian craftsmen trained by Han and Tibetan craftsmen. Notwithstanding, due to various factors like natural location, construction emergence, financial support and construction cycle, Tibetan Buddhism temples in this region differed in respect of building techniques but still clearly reflected the impact from neighboring region.

第五章 内蒙古地域召庙建筑形态的构成与秩序

第一节 聚落

众所周知，蒙古族自古以来就是一个逐水草而居的游牧民族，因此，作为蒙古地域最早的固定性宗教聚落的召庙群对该地域居住聚落的产生发挥了重要的作用。

大体而言，内蒙古地域召庙建筑文化影响下的聚落大体可以分为两类：

一、居民聚落

随着藏传佛教逐步深入普通蒙古民众的物质、精神生活，召庙亦成为蒙古人心之向往的神圣之地。当某一区域内有召庙建立，该区域内的牧民便在以召庙为中心的一定范围内进行贸易活动，出现了普通民众暂时落脚的居民性聚落，它们是草原城市的雏形，是现今内蒙古众多城市的前身。这种以召庙为中心的蒙古地区城市的构建和形成，除历史变迁的因素以外，基本成为内蒙古地区城市形成和发展的基础，如锡林郭勒盟的锡林浩特市[1]、呼和浩特市等城市便是这样形成的。

二、喇嘛聚落

随着藏传佛教的兴盛繁荣，召庙建筑也日益发展起来，这种发展首先表现在召庙规模上。随着召庙规模的逐步扩张，喇嘛人数的日益增加，数量有限、功能简单的喇嘛僧舍显然不能满足众多喇嘛的日常生活起居，喇嘛聚落应运而生。此类聚落逐渐与居民性聚落相连接，共同构成蒙古地域的早期城镇。因此，内蒙古地域很多地方素有“先有召，后有城”的说法，典型的有鄂尔多斯市乌审召周边的喇嘛聚落等。

基于僧俗有分的宗教思想，喇嘛聚落临近召庙以方便喇嘛日常礼佛、敬佛，一般距离约在几里地左右；而居民聚落离召庙有数十里之远，一则以示僧俗有别，二则防止召庙内的喇嘛与世俗之人有过多的接触（藏传佛教格鲁派以严格的教规著称，严禁喇嘛无故在外留宿）。

总体而言，内蒙古地域在召庙建筑文化影响下的聚落有如下形式特点：

（一）聚落围绕多级中心，逐层展开

该种形式的聚落以“点”的形态构建整个大环境的核心，其生成有着很明显的自发特征。通常这种自发特征首先以召庙大殿为一级核心点，聚落便以这一点为中心向外以同心圆或放射状或均衡或不均衡地向外拓展、延伸，从而构建整个空间体系。二级、三级中心则是召庙或聚落中的其他建筑。

（二）聚落呈单元片状分布

此类聚落以群组的形式出现，每组为一单元以“面”或“片”的形态构建边界相对模糊的空间，其领域被进一步细化，如商贸区域、居住区域等。不同聚落片区的布置有着一定的向心性，“心”为召庙或召庙内的某一座或某一组殿堂，但中心归属感不如第一种形式的聚落强烈。

（三）聚落分布充满偶然性，不拘泥于形式

这一类聚落几近自由式，其形式的形成充满偶然且完全不被限制。通常产生这种形式的原因有二：

1. 聚落的形成不是在短时期内完成的，而是经历了上百年的生长过程。因此，在整个漫长的成型过程中，聚落或因没有进行统筹规划或因不同时期的建造者随意选择可以避开树木、坑洼的适宜场地而使得整个聚落具有了随机性和偶然性。

2. 蒙古民族自古就是一个豪放的游牧民族，其特定的民族文化使得蒙古人对空间的认识有着自己独特的思维，长期在广阔草原游牧的生活使得他们不习惯将自己限定在某种有着明确界定的空间内，反映在整个聚落上便是松散、开放、自由且不拘泥于形式。

第二节 召庙

受到甘、青、藏地区建筑文化、中原内地及周边汉地建筑文化、蒙古民族文化等多方面影响的内蒙古地域藏传佛教召庙，在内蒙古独特的自然地理气候条件下形成了极具地域特色的多元化布局形式。

一、神俗有分

在藏传佛教的教义规定中，神俗两世界是必须被严格区分的，作为藏传佛教神圣领域之所在

1 参考：任月海.多伦汇宗寺［M］.北京：民族出版社，2005：151—158．

Chapter Five
Composition and Form of Temple Architecture in Inner Mongolia Region

I. Settlements

As known to all, Mongolians are a nomad moving from place to place in search of water and grass, so the earliest Tibetan Buddhism temples in the fixed religious settlements in Mongolia region played an important role in emergence of settlements in this region.

In general, settlements under the impact of architectural culture of Tibetan Buddhism temples in Inner Mongolia region can be divided into two categories:

1. Residential Settlements

With the gradual integration of Tibetan Buddhism into material and spiritual lives of ordinary Mongolian people, Tibetan Buddhism temple became a holy place aspired to by Mongols. So if a temple was built in certain region, the herdsmen in this region would nomadize in certain range centering on the temple. Up to Qing Dynasty, residential settlements basically took into shape with the development and promotion of farming technique in Inner Mongolian region. And such settlements are the perform of prairie city and predecessor of many current cities in Inner Mongolia, including Xilingol Prefecture Xilinhot City, Hohhot City, etc.

2. Lama Settlements

With the flourish and prosper of Tibetan Buddhism, temples also developed increasingly, first marked by the scale. So the quantity-limited and function-simple monk houses could not meet the demand of everyday life of the increasing Lama due to gradual expansion of temple scale, and Lama settlements emerge as the times require. Afterwards, such settlements coordinated with residential settlements and jointly formed the early towns in Mongolia region. Therefore, the saying "Temple emerged first and then the city" is popular in Mongolia region. And typical Lama settlements include those around Uxin Ju of Erdos City.

Contrarily speaking, based on religions thought that god and people differ, Lama settlements and temples are neighboring in favor of Buddha worship and respect of Buddha, and generally the distance is about several Chinese miles. However, residential settlements are dozens of Chinese miles from temples for showing the boundary between god and people and preventing Lama in temples wandering among the people (Gelug of Tibetan Buddhism is famous for precepts and Lama are strictly prohibited to accommodate outside).

In general, settlements under the impact of architectural culture of Tibetan Buddhism temples in Inner Mongolia region have the following formal features:

(1) Settlements are Laid Layer by Layer around Multi-Level Centers

For settlements in such form, the core of entire environment is constructed in the form of "Point", and its generation features obvious spontaneity. Usually such spontaneity first take temple palace as level-1 core point, correspondingly settlements, centering on this point, evenly or unevenly expand and spread outwards in the shape of concentric circle or radial pattern, thus forming the whole three-dimensional system. Level-2 and level-3 centers are other architectures in temples or settlements.

(2) Unit Slice-like Layout of Settlements

Such settlements emerged by groups, and each group formed the space with relatively fuzzy boundaries in the shape of "plane" or "slice", so the domain was further thinned into commercial area, residential area, etc.. And the layout of different settlement area is of certain centrality, and the "center" refers to the temple or certain palace hall or certain group of palace halls in temple, but its belongingness is not as strong as that of the first type.

(3) Occasionality of Settlement Layout and Unbound Form

Such settlements are approximately freestyle, and the generation of its form is of occasionality and not restricted. Generally such form emerged due to two causes:

A. The settlements took into shape upon growing several hundred years. Therefore, during long shaping period, the settlements are of randomicity and occasionality for being kept away from trees and bumpy places due to none overall planning or haphazard selection by builders.

B. The Mongolia is a bold and unconstrained nomad since ancient times, its specific culture enabled the Mongolians have its own original way of understanding the space, and long-period nomadic life in vast grassland made them unaccustomed to restrict themselves in definitely limited space, which refers to the looseness, openness, freeness and unbound form.

II. Temples

The diversified layout with regional characteristics was generated under the unique natural, geographical and climatic conditions of Inner Mongolia for Tibetan Buddhism temples in Inner Mongolia region that was affected by the architectural culture of Gansu, Qinghai and Tibet, and that of inland Central Plain as well as surrounding Han and Mongolian culture.

1. God and People Differ

The god and people are strictly distinguished in doctrines of Tibetan Buddhism. Temples, as the holy place of Tibetan Buddhism, are also clearly separated from scattered secular community outside. For example, it was strictly prohibited to build residential houses and retail shops around temples; the earthly beings cannot accommodate around temples indefinitely; both residential settlements and Lama settlements were kept distant from temples.

的召庙，与外围世俗性的外散社区也是明确分离的，例如，严禁在召庙周边建造各类民舍与商铺；世俗之人不可无限期留地宿于召庙周边；无论是居民聚落还是喇嘛聚落均需与召庙保持一定距离。

既然世俗生活与佛教圣界是有明确界限的，那么必定存在代表界限的标志物。在蒙古地域，通常使用小石堆、敖包和院墙作为圣界标志。

1. 区别于汉地佛寺，蒙古地域的多数召庙采用小石堆、敖包等蒙古化的标志极为开放地标明庙界之所在。庙界以内，各个殿堂或有规律或自由地排布开来，例如，乌兰察布市四子王旗的多数召庙均以小石堆作为圣界标志；包头市的梅日更召没有院墙，分隔世俗世界与净土世界的是召庙四周的四座敖包。

2. 通常情况下，位于城镇中的召庙以及采用汉式佛寺伽蓝七堂制布局特点显著的召庙多用院墙作为隔离以区分神俗世界，如锡林郭勒盟的贝子庙、呼和浩特市的大召等。

二、建筑布局

由第三章内容可知，一般来说，内蒙古地域召庙建筑的布局主要受藏地召庙建筑布局以及汉地佛寺建筑布局的影响。

（一）轴线式

内蒙古地域的多数召庙深受汉文化的影响，所以汉地建筑布局方式中常见的单元体、单轴布局、方阵布局、主轴方阵等布局形式在该地域都有实例，但又不是完全照搬，而是以适应地形变化的方式进行灵活处理，并进行了一定的取舍（如呼和浩特市大召寺内增加了菩提过殿）；很多召庙省去了钟鼓楼而增加了护法殿等（图14）。

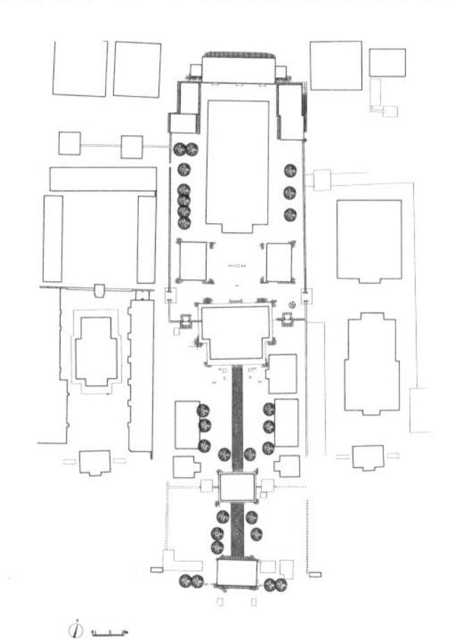

图14 大召总体布局依中轴对称

（二）中心自由式

基于曼荼罗中心与四周关系的影响，藏传佛教格鲁派的寺庙基本上都是遵循以主要殿堂为中心，其他建筑散布于其四周的自由式布局做法，内蒙古地域的部分召庙亦是如此，如内蒙古召庙成熟期的代表——五当召即模仿此种布局方式（图15）。

图15 五当召建筑群以大雄宝殿为中心自由分布

（三）自由式

此外，受到蒙古游牧文化影响的召庙则多以纯粹的自由式布局出现，没有任何规律可循，如乌兰察布市的希拉木仁庙（图16）。

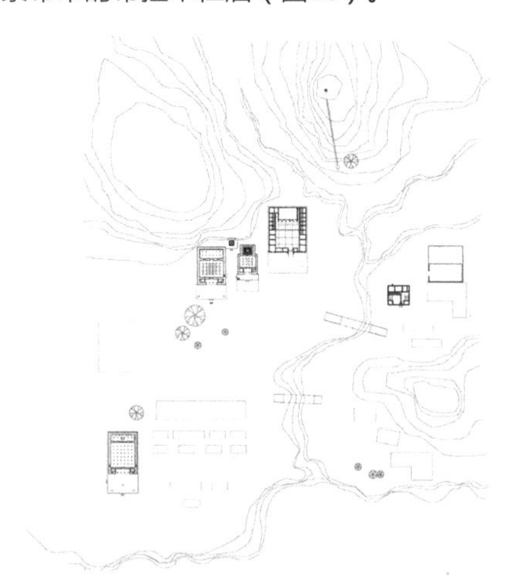

图16 希拉木仁庙呈自由式布局

很多情况下，上述三种形式相互渗透，例如，呼和浩特市的席力图召是以典型的轴线式布局，但召庙内两路轴线均以各自轴线上的重要大殿为重心，且西侧的轴线以东侧中轴线为重心。

There is a clear boundary between mundane life and holy Buddhist place, so the mark representing the boundary necessarily exists. Generally the barrans, obo and courtyard walls were taken as the boundary with the holy place in Inner Mongolia region.

A. Different from Han temples, many Tibetan Buddhism temples in Inner Mongolia region marked the boundary of temples with barrans, obo and other Mongolian signs in an extremely open manner. Within the temple, each palace hall was regularly or freely arranged. For example, most temples in Siziwang County of Ulan Qab City use barrans as boundary of holy place; Merigen Temple has no courtyard and is separated from the secular world with four obo.

B. In general, temples in cities and those featuring the layout of system of the Jalan Seven-hall Style of Han Dynasty temples mostly adopt courtyards for separating the celestial world and secular world, such as Beizi Temple in Xilingol Prefecture, Dazhao Temple in Hohhot City, etc..

2. Architectural Layout

According to those introduced in Chapter Three, generally the layout of Tibetan Buddhism temples in Inner Mongolia region is mainly affected by Tibetan temple layout and Han temple layout.

(1) Axial Line Type

Most Tibetan Buddhism temples in Inner Mongolia region were deeply affected by Han culture, so the common unit body, uniaxial layout, square matrix layout, chief axis layout and other forms of Han architectural layouts were put into practice in this region, but they were not fully imitated, but flexibly applied in a manner of adapting to topographical change upon appropriate use and reject (for example, Buddhism bodhi palace was built in Dazhao in Hohhot City; many temples have Dharmapalas Halls instead of Bell-drum Towers).

(2) Center Freestyle

Due to the impact from the relation between Mandara center and surrounding, Tibetan Buddhism Gelug temples basically conformed to the freestyle layout taking main palace as center and scattering other buildings around, so did some temples in Inner Mongolia region, for example, the representative Badgar Temple in mature period of Inner Mongolia Tibetan Buddhism temples do imitate such layout.

(3) Freestyle

In addition, temples affected by Mongolian nomadic culture emerged in the pure freestyle layout mostly without any rules, such as Xiaramuren Temple in Ulanqab City.

In most cases, the aforementioned styles interpenetrated. For instance, Xiretzhao Temple in Hohhot is in typical axial line type layout while both axial lines in the temple focus on respective major palaces and axial line in the west takes central axis of the east side as center.

第三节 建筑

根据使用功能，内蒙古地域的召庙建筑可以分为主体宗法建筑、世俗生活建筑两类。

一、宗法建筑

藏区的藏传佛教格鲁派寺庙有着严格、完善的寺庙管理体系。这些寺庙被分为措钦（寺院级）、扎仓（经学院级）、康村（地域性的僧团组织）[1]三级，且每级均设专门的委员会进行管理。内蒙古地域的召庙承袭了这种管理体制，并在这种体制下，措钦大殿、扎仓大殿、拉康成为内蒙古地域召庙内的主体宗法建筑，通常被统称为殿堂，主要是指召庙中的经堂、佛殿，亦代表召庙中各个重要建筑房屋。

措钦，藏语音译，是寺庙里最高一级的组织机构，措钦大殿则成为全寺僧众聚会、习经、举行法事活动的殿堂。因此，措钦大殿是寺院级的大殿，也是整个寺庙的宗教活动中心。在内蒙古地域，不同召庙对措钦大殿的称呼也不同，但以诸如“大雄宝殿”“某某寺大殿”这样的称谓居多。通常，在内蒙古地域召庙中的众多建筑单体中，措钦大殿建筑体量为最大，在整个召庙建筑群中占有最高地位。

扎仓，藏语音译，意为僧院、学部或学院，是寺庙中完整独立的一级组织，亦是藏传佛教僧众学习经典的学院。藏传佛教格鲁派学经制度完善，开设的学科门类也很齐全，寺庙根据不同的经学内容，设立了不同的扎仓，并以相关的学习内容或学部名称为扎仓命名，不同“专业”的喇嘛在不同的扎仓学习佛法。内蒙古地域各个召庙拥有的扎仓数目不同，大的召庙五大扎仓[2]齐备，小的召庙仅有一个扎仓或无扎仓，如位于锡林郭勒盟地区的贝子庙有五大扎仓，坐落于阿拉善盟地区的延福寺在历史上曾设有四大扎仓。一般而言，具有完整五大扎仓的召庙即为典型的学问寺，如五当召。由此可知，扎仓大殿是分属于各个学部的主要殿堂建筑。由于在一些规模较小或情况特殊的召庙内，仅设置一个学部或者未设置学部（如呼和浩特市地区的大召），所有僧侣不分“专业”地研修佛法，因此，在诸如此类的召庙中就没有扎仓大殿，措钦大殿则成为整个召庙中所有僧侣均能使用的综合建筑。

需要说明的是，通常情况下，内蒙古地域的措钦大殿和扎仓大殿都是将辅助用房（如储藏室）、僧侣聚会习经的经堂、信众礼佛膜拜的佛殿等建筑内容有机组织在一起而形成的一幢大型建筑，二者仅在等级、规模方面有所区别（一般而言，措钦大殿等级高、规模大，扎仓大殿次之）。

在藏区，除措钦大殿和扎仓大殿外，还有一类藏语称之为拉康的殿堂，即独立式佛殿。这类殿堂是专门用于供奉佛像、经书、佛塔的场所，是整个寺庙内最为神圣的地方，且根据主供物的不同，拉康可被称为释迦牟尼佛殿、弥勒佛殿、护法殿、灵塔殿等。历史上，独立式佛殿一度曾是藏传佛教寺院中最重要的建筑物，在经历了经堂与佛殿相分离的变革后，拉康仍然以其在寺院中必不可少的地位而独立发展，且功能也越来越单一化。同措钦大殿和扎仓大殿一样，这类殿堂也被引入内蒙古地域，但相较于前两者而言，数量较少、规模较小，且地位等级较低（如鄂尔多斯市地区准格尔召的六臂护法殿）。在此需要辨明的是，此处提到的独立式佛殿是指功能单一，仅用于礼佛且有独立建筑空间与体量的建筑单体，区别于措钦大殿和扎仓大殿内的佛殿。

二、生活建筑

藏传佛教召庙内用于生活的建筑主要有活佛府（活佛生活起居之处）、喇嘛住所（或称僧舍）。

活佛府，藏语称之为拉让，常作拉布隆，是活佛及其侍从的住处，亦为管理活佛私有财产的机构，通常以院落的形式出现，且根据活佛地位的高低，其规模大小也有所不同。

喇嘛住所，是一般喇嘛居住的地方，其数量最多，有院落式的（如准格尔召的喇嘛住所），也有独立式的（如五当召的喇嘛住所）。

在生活建筑中常包含一个或几个小型的殿堂，这些殿堂亦有经堂、佛殿之功能，是世俗生活建筑群的建筑重心、宗教中心，用来满足活佛及喇嘛日常礼佛诵经功课（该类宗教活动等级低于措钦大殿和扎仓大殿内举行的正式宗教活动）。甚至在个别居所中，建筑内部的某个角落被独立出来用于安放佛像、诵经祈祷、研习佛法。

基于神俗有分的宗教思想，即使是在召庙

---

1 参考：陈耀东.中国藏族建筑［M］.北京：中国建筑工业出版社，2006：200—201，383.

2 五大扎仓，藏传佛教格鲁派五大学部所在，包括却伊拉扎仓、卓德巴（密宗）扎仓、丁科尔扎仓（时轮学部）、满巴扎仓（医学部）、喇嘛日木扎仓（菩提道学部）。

Ⅲ. Architectures

According to the function, Tibetan Buddhism temples in Inner Mongolia region can be classified into main patriarchal architecture and secular living architecture.

1. Patriarchal Architecture

Tibetan Buddhism Gelug temples have strict and perfect temple management system. These temples are divided into Tsochin (temple level), Gyuto (secular school level) and Khangtsen (regional Sangha) , and special committee was established to manage each level. Tibetan Buddhism temples in Inner Mongolia region inherited such management system, under which Tsochin Hall, Dratsang Hall and Buddha's shrine became the main patriarchal architectures in Tibetan Buddhism temples in Inner Mongolia region, and generally they are called palace halls, mainly referring to ceremony hall and Buddha hall in temples and also representing all important houses in temples.

Tsochin, a Tibetan transliteration, is the highest organization in temple. Tsochin Hall is the palace hall where all Buddhist monks of the temple get together, learn scriptures and develop Buddhism activities, so it is of temple level and the center of religious activities. In Inner Mongolia region, different temples have different names of Tsochin Hall, but the titles like "Great Hall" and "Main Palace of Temple" prevailed. In general, Tsochin Hall is the largest and most important among numerous individual buildings of Tibetan Buddhism temples in Inner Mongolia region.

Gyuto, a Tibetan transliteration, means monastery, department or school, and it is a complete and independent level-1organization, and also a school where Tibetan Buddhism monks learn the scriptures. Tibetan Buddhism Gelug system has perfect learning system and complete subjects. And various Gyutos are built according to different contents of scriptures and are named as related learning content or department name, and Lama of different "majors" shall learn Buddhism doctrines in different Gyutos. The number of Gyutos in Temples in Inner Mongolia region differs, five Gyutos1in large temples and only one or none Gyuto in small ones. For example, Beizi Temple in Xilingol Prefecture has five Gyutos; Yanfu Temple located in Alxa Prefecture has had four Gyutos. Generally speaking, temples with five Gyutos are the typical learning temple, such as Badgar Temple. So it is clear that Gyuto Hall is the main palace building belonging to each department. For some small and special temples only have a or none department (such as Dazhao in Hohhot City), all monks research Buddhism doctrines without classification of "majors", so no Gyuto Hall was built in such temples and Tsochin Hall became a complex building that can be used by all monks.

It is important to note that generally Tsochin Hall and Gyuto Hall in Inner Mongolia region refer to a large-scale architecture organically integrating auxiliary houses (like storeroom), Ceremony Hall where monks get together and learn the scriptures and Buddha Hall where believers worship and bow, and they differ only in the level and scale (generally Tsochin Hall is of high level and large scale, followed by Gyuto Hall).

In addition to Tsochin Hall and Gyuto Hall, there is also a kind of palace hall named Buddha's Shrine in Tibet, that is, an independent Buddha Hall. Such a hall is specifically used for worshiping Buddha statue, scriptures and Buddhism pagoda and is the holy place of the temple. And according to the different main sacrifices, Buddha's shrine can be called Sakyamuni Buddha Hall, Maidar, Dharmapalas Hall and Tower Hall. In history, the independent Buddha hall had been the most important architecture in Tibetan Buddhism temples one time, and Buddha's shrine still develops independently as its indispensable position in temples after the separation of ceremony hall and Buddha hall and its function becomes more and more simplified. Similar to Tsochin Hall and Gyuto Hall, such palace halls were also introduced into Inner Mongolia region, but in small number, small scale and low grade (such as Six-arm Dharmapalas Hall in Zhungarzhao Temple in Erdos City) as compared with Oratory Palace and Gyuto Palace. It needs to state that independent Buddha halls mentioned here refer to the individual building with independent space and simplified function and only for Buddha worship and they are different from Buddha halls in Tsochin Hall and Gyuto Hall.

2. Living Architecture

The architectures for living in Tibetan Buddhism temples mainly include Buddha House (where Living Buddha lives) and Lama Houses (or called Monk Houses)

Buddha House, Rajang in Tibetan and commonly Labulong, is the place where Living Buddha and his attendants live and also the agency where private property of Living Buddha is managed. Generally it is courtyard and differs in scale.

Lama Houses are the places where common Lama live and of the largest number, and they are either courtyard (such as Lama Houses in Zhungarzhao Temple) and independent type (such as those in Badgar Temple).

There is also one or several small palace hall(s) with the function of ceremony hall and Buddha hall in living architectures and they are the building core and religious center of the secular living architectures and for meeting the demand of everyday Living Buddha and Lama for worshiping Buddha and chanting scriptures (such religious activities are inferior to the formal ones held in Tsochin Hall and Gyuto Hall). Even in individual houses, certain corner of the architecture is separately used for placing Buddha statue, chanting scriptures and praying and researching Buddhism doctrines.

Generally, based on religions thought that god and people differ, living architectures serving Lama in the temples are different from patriarchal architectures, for example, the grade and scale of the former are lower than those of the latter, and they need to be separated. However, there is still exception. For example, Buddha House also has

内，为喇嘛服务的生活建筑通常要与宗法建筑有所区别，如前者的等级、规模低于后者，前者与后者需相互间隔。但也有例外，例如活佛府内也有主供神像的佛殿，且活佛府的地位极高（如锡林郭勒盟的汇宗寺与珠轮寺），甚至高于宗法建筑——大殿。

在内蒙古地域的藏传佛教召庙中，主体宗法建筑的殿堂空间始终扮演着十分重要的角色。一般而言，殿是安奉佛祖、菩萨塑像以供僧人信众礼拜、祈祷之处；而堂则是供喇嘛僧徒说法、习经、行道之所。因此，殿堂是整个召庙内的核心建筑空间，召庙中的其他建筑单体均以其为重点参照进行布置，甚至定位。在内蒙古地域，殿堂又是最初整个召庙的源，有了它才会有其他功能的单体逐步出现，无论随时间会增建多少其他单体建筑，殿堂总是保持着它在整个召庙中的核心地位，甚至在内蒙古地域的某些地区，一座殿堂就是一个召庙。而内蒙古地域殿堂中最重要的空间是经堂、佛殿，二者的分合形制则尤显突出。

第四节 经堂、佛殿的分合形制

自窝阔台汗之三子阔端引入藏传佛教至清末，内蒙古地域的召庙建筑先后经历了三个封建朝代，其间共600多年，在本地域已经发展得较为成熟，并且已经形成了具有地域特点的建筑空间。根据大量的实地调研结果发现，内蒙古地域召庙的经堂与佛殿在数量上形成了一一对应的关系，即一座佛殿对应一座经堂。且二者的相对位置关系均为前后式，即一座佛殿对应一座经堂并被组织在一条轴线上，经堂在前，佛殿在后。内蒙古地域不同地区召庙的殿堂在这种空间模式下，根据自身的具体情况采取了灵活多变的组合方式，即在保证“前经堂后佛殿”这种空间形态不变的情况下，佛殿与经堂的结合方式呈多样性，概括起来大体有三种：接合式、结合式、分离式。

一、数量及相对位置关系

（一）数量对应关系

在内蒙古地域，并未发现类似西藏地区藏传佛教殿堂的“一经堂多佛殿”或“一佛殿多经堂”模式，经堂与佛殿的数量关系仅为“一经堂一佛殿”，即一座经堂仅对应一座佛殿。

（二）相对位置关系

在藏区，“一经堂一佛殿”模式又可分为并列式、自由式、前后式（见下表），但在内蒙古地域的召庙中并未发现以并列式或自由式模式出现的经堂与佛殿。内蒙古地域经堂与佛殿的相对位置关系均为前后式，即一座经堂与对应的一座佛殿被组织在一条轴线上，经堂居前，佛殿置后。

经堂与佛殿相对位置关系		
并列式	自由式	前后式

二、三种组合方式

（一）接合式殿堂

接合式是指经堂与佛殿紧密拼接在一起的一种组合方式，其空间模式是门廊→经堂→佛殿，是一种在内蒙古地域十分常见的模式，常用于大雄宝殿和各学部大殿。通常，接合式大雄宝殿与接合式学部大殿的形制一模一样，即均为前经堂后佛殿，二者的内部空间也大致相同，仅在建筑体量与规模上有所区别。根据措钦大殿或扎仓大殿后部佛殿形状的不同，可将接合式殿堂分为两类。

1. 经堂与长条形佛殿接合的殿堂

特点：佛殿与经堂在同一建筑内的不同空间，二者以一墙分隔；经堂后部用作分隔的墙两侧开门，以联通佛殿；佛殿无外绕围廊，进深较窄；佛殿的面阔小于（如包头市达茂旗的普会寺大殿）或等于（如包头市的昆都仑召大雄宝殿）经堂的面阔；建筑高度上，佛殿高于经堂。

包头市的昆都仑召大雄宝殿、达茂旗普会寺大殿以及鄂尔多斯市准格尔召的五道庙、观音庙、千佛殿等殿堂均采用此种组合方式。该类殿堂的平面简图及交通流线分析如图17所示。

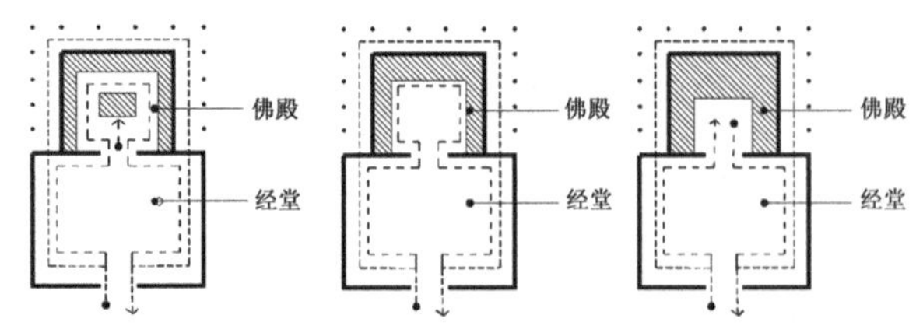

图17 经堂与长条形佛殿接合的殿堂平面简图及交通流线分析

the Buddha hall mainly for worshiping Buddha statue and it is of extremely high position (Huizong Temple and Zhulun Temple in Xilingol Prefecture) and its position is even higher than patriarchal architecture—major hall.

Among Tibetan Buddhism temples in Inner Mongolia region, palace hall space of main patriarchal architecture always plays an important role. Generally, the palace is for offering the statues of Bodhisattva Buddha for worshiping and praying by monks and believers while the hall is for expounding Buddhism doctrine, learning scriptures and spreading morality. So the palace and hall are the core space of the entire temple and all other individual buildings in the temples are arranged and even oriented with reference to them. And in Inner Mongolia region, the palace is also the initial source of temples, and the individuals with other functions gradually emerged due to the existence of palace hall. No matter how many other individual buildings emerged with time going, the palace hall always maintains its core position in the temple. Moreover, a palace hall generally is a temple in some areas of Inner Mongolia region. Therefore, the separation and combination form of ceremony halls and Buddha halls, the most important spaces in palace halls of Inner Mongolia region, becomes prominent accordingly and will be separately described in the following section.

Ⅳ. Separation and Combination Form of Ceremony Halls and Buddha Halls

Since the introduction of Tibetan Buddhism by Godan, the third son of Ogadai Khan (Wokuotaihan) to late Qing Dynasty, Tibetan Buddhism temples in Inner Mongolia region witnessed three feudal dynasties, over 600 years in total, and the temples has become mature and had architectural space with regional characteristics. According to numerous on-the-spot investigation results, ceremony halls and Buddha halls of Tibetan Buddhism temples in Inner Mongolia region have a one-to-one correspondence in their quantities, that is, a Buddha hall corresponds to a ceremony hall; and relative position of them is front-rear type, that is, a Buddha hall corresponds to a ceremony hall, they are arranged in an axial line and generally the ceremony hall is in the front and Buddha hall is in the rear. Under such spatial mode, palace halls of different temples in Inner Mongolia region adopted flexible combination ways according to respective condition. It means that, under the condition ensuring spatial form "ceremony hall in the front and Buddha hall in the rear" unchanged, Buddha hall and ceremony hall are combined in diversified ways. Generally there are three kinds: connection type, combination type and separation type.

1. Quantitative Relation and Relative Position Relation

(1) Quantitative Relation

In Inner Mongolia region, the mode of "one ceremony hall and multiple Buddha halls" or "one Buddha hall and multiple ceremony halls" that similar to Tibetan Buddhism palace in Tibet was not found, and the quantitative relation between ceremony hall and Buddha hall is only "one ceremony hall and one Buddha hall", namely, a ceremony hall only corresponds to a Buddha hall.

(2) Relative Position Relation

The mode "one ceremony hall and one Buddha hall" in Tibet can be classified into side-by-side style, freestyle and front-rear style. However, temples in Inner Mongolia region have no side-by-side style or freestyle ceremony hall and Buddha hall, and relative position relation between them is front-rear style, that is, a ceremony hall and a Buddha hall corresponded are arranged in an axial line and generally the ceremony hall is in the front and Buddha hall is in the rear.

2. Three Combination Modes

(1) Connection Type Palace Hall

Connection type is a combination mode tightly splicing ceremony hall and Buddha hall. The spatial model refers to portico → ceremony hall → Buddha hall, and it commonly applied in Inner Mongolia region and usually used in great hall and department halls. Generally, connection type great hall shares the same form with connection type department hall, namely, ceremony hall in the front and Buddha hall in the rear, inner spaces of them are also identical and they only differ in building volume and scale. According to the difference in the shape of rear Buddha hall of Tsochin Hall or Gyuto Hall, connection type palace hall can be classified into:

A. Palace Hall of the Connection of Ceremony Hall and Elongated Buddha Hall

Characteristics: Buddha hall and ceremony hall are in different places of the same temple and separated by a wall; the wall for separating at rear of ceremony hall has doors on both sides to connecting with Buddha hall; Buddha hall has no external peridrome and shallow depth; the building width of Buddha hall is smaller than (such as the main palace of Puhui Temple in Damao County of Baotou City) or equal to (such as Great Hall of Hundele Temple in Baotou City) that of ceremony hall; and Buddha hall is higher than ceremony hall.

Great Hall of Hundele Temple in Baotou City, main palace of Puhui Temple in Damao County as well as Wudao Temple, Kwun Yam Temple, Thousand Buddha Temple and other halls in zhungarzhao Temple in Ordos City adopt this combination mode. And plan sketch and traffic streamline analysis of such palace halls are as follows.

B. Palace Hall of the Connecting Ceremony Hall and Buddha Hall with External Peridrome

Characteristics: Buddha hall and ceremony hall are in different places of the same temple; the former is in a large single space, the plane is square and the roof is generally a Han type hipped-gable roof, and ceremony hall is also a large single space, the plane is square or approximately square and the roof is a Tibetan type truncated roof or Han type hipped-gable roof; inside the palace hall, the lattice

2. 经堂与外绕围廊佛殿接合的殿堂

特点：佛殿与经堂在同一建筑内的不同空间；佛殿为单一的大空间，平面呈方形，屋顶通常为汉式歇山顶，经堂亦为单一的大空间，平面呈方形或近似方形，屋顶采用藏式平屋顶或汉式歇山顶。殿堂内部，用槅扇门或墙分隔佛殿与经堂，经堂后部中间开门，以联通佛殿。在佛殿与经堂屋顶的连接方面，通常的做法是前经堂的坡屋顶与后佛殿的坡屋顶屋檐口对接，或后佛殿的坡屋顶作为前经堂平屋顶的泛水而相互连接起来。建筑高度上，佛殿高于经堂。佛殿为汉地传统木构建筑的副阶周匝形式，即佛殿外绕有围廊，围廊常常与经堂后部两侧的出口相连，联通的经堂内部与外部围廊形成了贯通内外的完整转经道。佛殿内部的交通是根据佛殿空间本身的大小及佛像的摆设来确定，小一点的佛殿采用停滞式，大一点的佛殿，采用环绕式，若佛殿中部安放佛像，则采用停滞式与环绕式并用的模式。

呼和浩特市的乌素图召庆缘寺大殿、大召大雄宝殿、席力图召古佛殿以及包头市的美岱召大雄宝殿等殿堂均采用此种组合方式。该类殿堂的平面简图及交通流线分析如图18所示。

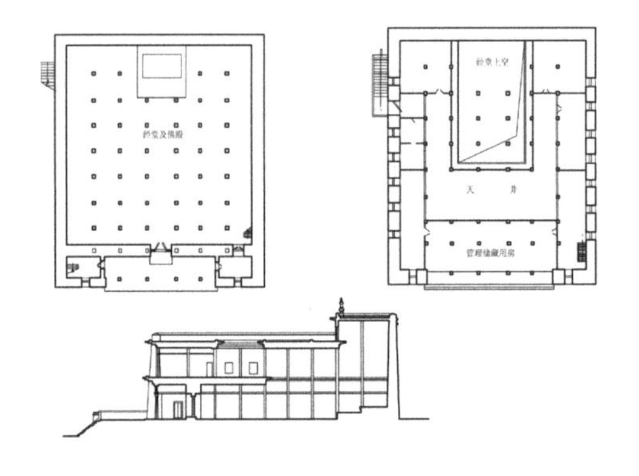

图18　经堂与外绕围廊佛殿接合的殿堂平面简图及交通流线分析

（二）结合式殿堂

结合式是指经堂与佛殿被组织在同一幢建筑中的同一个空间内，建筑兼有经堂、佛殿二者之功能的一种组合方式，其空间模式是门廊→经堂和佛殿。因此，以该种组合方式形成的殿堂空间既有经堂的特征，又有佛殿的特征。在内蒙古地域，这种模式常用于学部大殿，大雄宝殿通常不采用，如呼和浩特市的乌素图召长寿寺大殿、包头市的五当召显宗殿等。该类殿堂的平面简图及交通流线分析如图19所示。

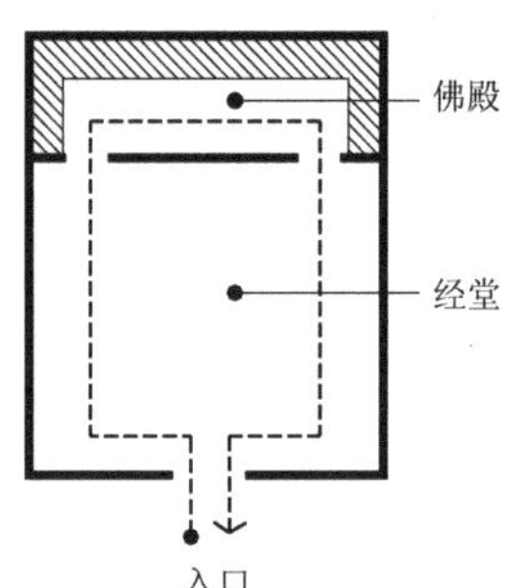

图19　殿堂平面简图及交通流线分析

（三）分离式殿堂

分离式是指佛殿与经堂分别位于不同的建筑空间内，二者均有独立的建筑形象，彼此分离不连接的一种组合方式，其空间模式是门廊→经堂→室外（庭院、过道等）→佛殿。

在内蒙古地域，产生经堂与佛殿分建情况的原因是多方面的：

1. 受到内地佛寺分建多座殿堂于中轴线上的影响（如呼和浩特市席力图召的经堂与其后的佛殿）；

2. 受到殿堂所在位置地形的影响（如通辽市库伦旗兴源寺的经堂与其后的玛尼佛殿）；

3. 由于火灾、扩建等方面的原因，后来加建殿堂的情况（如包头市梅日更召的经堂与其后的弥勒佛殿）。

虽然经堂与佛殿分建，但二者之间相互联系：在相对位置关系上，二者不仅处于同一条轴线上，还会通过加砌围墙（如包头市梅日更召的大殿）或建厢房（如鄂尔多斯市准格尔召的大殿）的方式相互围合，形成具有一定面积的庭院，使得前部经堂与后部佛殿产生空间对话。通常庭院的形状、大小并无定制，可以是方形的（如鄂尔多斯市准格尔召的大殿），也可以是矩形的（如包头市梅日更召的大殿）。该类殿堂的平面简图及交通流线分析如图20所示。

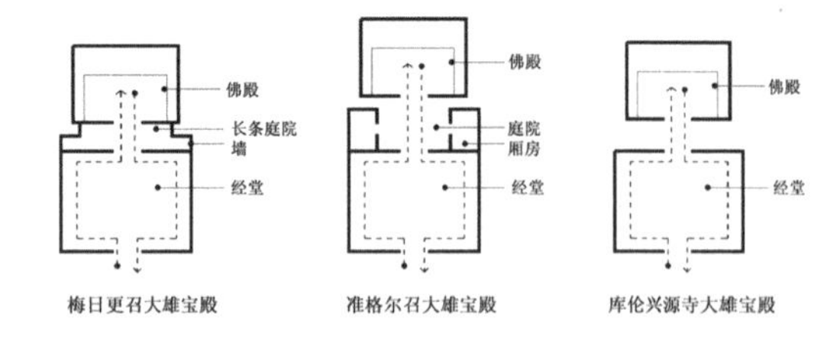

图20　分离式殿堂简图及交通流线分析

door or wall separates Buddha hall and ceremony hall, in the rear of which there is a door in the middle to connect with Buddha hall; for connection of roofs of Buddha hall and ceremony hall, generally the eaves of pitched roof of front ceremony and pitched roof of rear Buddha hall are spliced, or pitched roof of rear Buddha hall serves for flashing of front ceremony hall and then is interconnected; Buddha hall is higher than ceremony hall; Buddha hall is in the form of setting up a peridrome outside hall, such a form originated from Han traditional wooden architecture, and the peridrome is usually connected with two exits on both sides in the rear of ceremony hall, and the connected interior of ceremony hall and external peridrome formed a complete circumambulation stretching through the inside and outside; the traffic inside Buddha hall is determined according to the size of Buddha hall and layout of Buddha statues, small Buddha hall adopts detention type while large Buddha hall adopts encircling type, and if Buddha statue is placed in the middle of Buddha hall, the mode integrating detention type and encircling type shall be applied.

Main palace of Qingyuan Temple of Wusutuzhao, Great Hall of Dazhao and Ancient Buddha hall of Xilitu Temple in Hohhot City as well as Great Hall of Maidarzhao Temple in Baotou City and so on adopt such combination mode. Plane sketch and traffic streamline analysis of such palace halls are as follows.

(2) Combination Type Palace Hall

The combination type is a combination mode tightly arranging ceremony hall and Buddha hall in the same space of the same building that combines the functions of ceremony hall and Buddha hall, and the spatial model refers to portico → ceremony hall and Buddha hall. So palace hall space in such combination type has the features of ceremony hall and Buddha hall. In Inner Mongolia region, such a type is generally applied in department hall, not in great hall, such as Changshou Temple hall of Wusutuzhao in Hohhot City and Xianzong Hall of Badgar Temple in Baotou City. Plane sketch and traffic streamline analysis of such palace halls are as follows.

(3) Separation Type Palace Hall

Separation type is a combination mode arranging Buddha hall and ceremony hall in different spaces, under which the two halls have independent architectural figures and are separated. And the spatial model refers to portico → ceremony hall → outdoor (courtyard, corridor, etc.) → Buddha hall.

In Inner Mongolia region, the separation of ceremony hall and Buddha hall is caused due to many factors:

A. The impact from the situation that most inland temples were separately built in central axis of multiple palace halls (such as ceremony hall and Buddha hall behind it of Xilituzhao in Hohhot City);

B. The impact from the location of palace hall (such as ceremony hall and Marni Buddha hall behind it of Xingyuan Temple in Kulun County of Tongliao City);

C. The situation that new palace hall was built due to fire, expansion and other causes (such as ceremony hall and Maitreya hall behind it of Merigenzhao in Baotou City);

Although ceremony hall and Buddha hall were built separately, they are interrelated: in relative position relation, they are not only in the same axial line, but also shall be enclosed mutually via building an enclosure wall (such as the palace hall of Merigenzhao Temple in Baotou City) or building apprentice(such as main palace of Zhungarzhao Temple in Erdos City), thus forming a courtyard covering an area and enabling the front ceremony hall and the rear Buddha hall have a spatial link. Generally the shape and scale of courtyard can be square (such as main palace of Zhungarzhao Temple in Erdos City) or rectangular (such as main palace of Merigenzhao Temple in Baotou City). Plane sketch and traffic streamline analysis of such palace halls are as follows.

第二部分

召庙

Part Two Temples

编写说明　Illustration of Compling

本书对召庙建筑信息予以系统归档，对各召庙的历史信息予以详细的整理。编写者在通览各类召庙文献的基础上，以一种特定的程式建构了每座召庙的创建与兴衰史。并以叙述文体与基本概况表格两种预定格式，将所搜集的历史信息系统归入相关内容环节，为便于查阅，对本书所涉及的各召庙历史文本的构成、名称翻译、召庙排序、文献情况等予以说明。

一、召庙简介说明

召庙简介以五段式文本构成。编写者在预先设定若干历史信息条目后，将其整合至五个段落，再将文献所载信息依次填入相应段落与条目内。因此，文本中某一类信息的缺失完全与原文献有关，即文献中无此类信息记载。

各自然段内容设置为：

第一段：召庙原属行政区划、召庙等级（盟庙、旗庙、佐庙、属庙等）、清廷或民国政府所赐寺匾时间、名称及文字种类、所辖属庙名称与数量、召庙特色及珍藏文物、出自该庙的杰出历史人物。

第二段：召庙创建史（初建人、时间、迁址次数、定址地点及相关重要历史信息）、召庙名称由来及释义。文中若引用缺乏历史考证，且有较高学术价值的传说故事等口传文本，句前加“据传”，予以注明。

第三段：建筑风格、盛期规模（主要殿宇的名称、尺度、层数、建筑风格）、活佛府名称及数量、庙仓数量、佛塔数量、僧舍规模（以院落、间数计算）。

第四段：所设学部名称及数量、活佛系统数量、名称及转世总次数。

第五段：最后一次被毁坏时间及留存殿堂名称与数量、正式恢复法会时间、重建召庙建筑时间及当前规模。

二、基本概况说明

基本概况由10个主要信息栏组成。以下对各信息栏的内容筛选原则与方法予以说明。

1. 召庙名称：为便于一般读者易于理解，并能够反映出本土文化惯例，在目录与简介部分中，使用了寺院俗称。例如选用“席力图召”，而未选用其钦赐名“延寿寺”。关于名称的选择也有下列特殊情况。

下列两种情况下，选用了正式名称：

（1）召庙俗称已无法确定时使用正式名称，如“龙泉寺”“灵悦寺”。

（2）在内蒙古地域，常见有名称相对称的两座召庙。若使用俗称，名称所指不够准确，且易于混淆，如多伦诺尔汇宗寺俗称为“青庙”（或东大仓、旧庙），而善因寺为“黄庙”（或西大仓、新庙）。故取钦赐名予以区别。

（3）多个召庙组成召庙群时，取俗称显得更为必要。例如，乌素图召由法成广寿寺、庆缘寺、长寿寺、法禧寺等多个召庙组成，乌素图召为俗称。

（4）由于方位认知与语言表述的不同，单个召庙有不同俗称。如广宗寺的蒙古语俗称为“巴荣黑德”（即西寺），福因寺的蒙古语俗称为“忠黑德”（即东寺），因此，采纳当前惯例，取俗称“南寺”与“北寺”。

关于汉语正式名称，如有清廷御赐的若干寺匾，选择主庙寺匾作为正式名称。通常有两种情况：

（1）以殿宇为单位：召庙内主要殿宇均有钦赐寺匾时，选择大雄宝殿的寺匾。

（2）以庙宇为单位：召庙群内宗庙与子庙（或分院）均有钦赐寺匾时，选择宗庙寺匾，如“庆缘寺”为乌素图召的宗庙。

2. 所在地：现属行政区划（某旗或县—某苏木或乡—某嘎查或村），不采取召庙所处位置的地理环境及方位描述。

3. 坐标：以谷歌网络卫星地图或手持全球定位系统为主要数据来源。

4. 初建年：填写始建年或初建工程的竣工年，因为多数文献中此项信息是模糊的。

5. 盛期喇嘛数—现有喇嘛数：在数字前（或后）填写近（或余）。

6. 盛期时间：如文献中无确定记载，未填该栏。因召庙盛期通常以僧人数量、殿宇规模、学部种类、庙产数量作为主要评价尺度，编写者依据上述变量，适当予以判断，但通常选择某时间段。

7. 保护等级：以实地调研时间为准。

8. 历代活佛：各召庙的活佛（呼图克图、葛根、呼比勒汗、沙布隆）转世体系不同，故采纳了下列办法：

（1）有一位葛根，若干名沙布隆的情况下，简介部分予以简述，而在栏内只填入葛根的历代转世。

（2）有东、西两名活佛（或此外另有若干名沙布隆），将两位活佛历代转世全部填入。

（3）有多名呼图克图的情况下，将每一位呼图克图的名称及系列依次填入，而不采取选择一位呼图克图，将其历代转世填入的方法。

（4）有若干名活佛，而其中某活佛有自己的独立寺院，且其寺院已收录在本书中，则不填入表内。

（5）一些活佛的前几世转世于印度或我国西藏、安多藏区，且与召庙无关联的情况下，从建造召庙的第一世向后推延，依次填入。

因文献信息的不全，导致某信息栏的空缺。备注一栏内主要填入活佛俗称、转世的盟旗等信息，如历代活佛排序不明，则注明是几世活佛。

9. 历史沿革：以召庙建筑史为主，以时间顺序罗列了具体殿堂的新建、修缮、扩建、烧毁（或拆毁）、重建、改建、加建的时间（只记年，不记月、日），以及重要历史事件的发生年（达赖、班禅等宗教领袖的驾临时间、各学部的创设年）。

10. 现状描述由实地调研者依据召庙现状描写填入。

三、蒙古文、藏文名称的汉译标准

文本涉及的主要名称有活佛（或普通僧人、王公施主）的法名（或名称）、佛名、殿堂名称、扎仓名称四大类型。

1. 藏语人名汉译：依据蒙古语读音音译，而未使用汉语惯用翻译。

如采用“嘉木苏”，而未采用“嘉措”，采用“贾拉森”，而未采用“坚赞”。

2. 佛名：依据学界已有惯例直译。

如“阿日雅拉布”译为“观音”，“森格杜玛”译为“狮面佛母”。

3. 殿堂名称：主要采取了下列5种方法：

（1）以汉传佛教惯用名称直译相对应的殿堂名称：如将“朝克钦独贡”译为“大雄宝殿”，“玛哈仁扎音独贡”译为“天王殿”。

（2）以学部名称直译冠以扎仓名称的殿堂，如将“却日殿”译为“显宗殿”，将“洞阔尔殿”译为“时轮殿”。

（3）以蒙古语义译冠以所藏经卷名称的殿堂，如将“收藏大藏经的配殿”译为“甘珠尔殿”或“丹珠尔殿”。

（4）根据语义概括义译以特定功能命名的经堂佛殿，如将“雅日乃殿”译为“安居殿”，将“农乃殿”译为“斋戒殿”。

（5）依据佛名翻译标准直译专供某神像的佛殿，如将“德木齐格音独贡”译为“胜乐金刚殿”，“迈达日音独贡”译为“弥勒殿”。

四、召庙排序说明

本套书所选56座主要召庙及附录中所列54座召庙的排序依据及方法如下。

1. 排序依据

（1）历史时期内的召庙主次等级、规制。

（2）建造时间及社会影响。

（3）地理区位及部族分布：采取从西到东的地理空间排列，如锡林郭勒盟召庙的排列顺序为：苏尼特—阿巴嘎—阿巴哈纳尔—浩奇特—乌珠穆沁—察哈尔等。

（4）古建筑现存程度及代表性。

2. 排序方法：综合考虑上述4项依据，尽量表现出召庙在历史时期内的宗属关系、社会影响、地域及部族代表性。若主庙与属庙同被选进，将连排在一起，展现其历史关联性（虽然现已无所属关系），若有单独属庙，排在最后。

五、文本引用文献概况及引用法说明

课题组收集的文本种类繁杂，可以依据不同的标准加以分类，即依据所著时代分为：清代文稿、民国文献及中华人民共和国成立后的文献资料。依据撰写文字分为：蒙古文文献、汉文文献、藏文文献及其他外文文献（日、英、俄）。依据文本形式分为公开出版物（含内部资料）、官方文件、作者手稿。

1. 文献概况

（1）由于多种客观原因，清代编写的召庙史原文献已十分少见。目前只有呼和浩特喇嘛印务处编《呼和浩特各召庙活佛莅席年班及召庙兴建调查》、伊希巴拉丹著《宝鬘》《梅日更召创建史》、萨·那日松辑录《鄂尔多斯人历史文献集》（第四、五辑）等少量召庙史文献及《内齐托音传》《普和寺历代活佛传》等一些活佛传记。此外，在《归绥道志》（光绪三十三年，即1907年）等地方志中有零星记载。

（2）民国时期文献所涉及的召庙名称及所处位置较详细，但具体内容方面不全面。主要有绥远通志馆编纂《绥远通志稿》（民国26年，即1937年）、蒙藏委员会调查室编印《伊盟左翼三旗调查报告书》（民国30年，即1941年）等文献及各类国内外游人的游记、活佛回忆录等。

（3）中华人民共和国成立后的文献可分为“文化大革命”前及近20年公开出版的召庙志及各类地方志、召庙通志两种。“文化大革命”前的文献以中共中央内蒙古分局宗教问题委员会编的《内蒙古喇嘛教》（1951年）为主。近二十年的召庙类书籍较多，随着藏传佛教建筑的普遍重建，出现各类单本召庙志，其历史信息多有不妥之处或多为学界所质疑，但对于个别召庙的历史，是不可忽视的重要参考资料。

（4）课题组实地调研记录为了解建筑现存规模、近年修缮重建状况等信息的重要依据。

2. 具体引用法

引用原则为：

（1）以召庙志为主，地方志、召庙通志、游记为辅，如编写“毕鲁图庙简介”时“毕鲁图庙史”为首选资料，而《锡林郭勒盟寺院》等书为附带参考资料。

（2）每座召庙简介的参考资料限定在三本主要参考文献范围内，个别召庙引用四本文献。在两种情况下使用此方法：一为参考文献（单项召庙志）过多时选择主要版本。二为文献记载过少时拼贴多种文献加以整理。通常质量较好的一本召庙志足可以提供所需各种信息。

（3）以清代文献为主。单项信息若有多种记载，择取主要文献信息，其余说法列在其后。

（4）以蒙古文文献（包括召庙僧侣用藏文撰写，后再蒙译的文献）为主，汉文文献为辅。

在收集资料过程中，多位学者同仁予以指导与支持。提供资料的学者有阿拉善盟达莱、斯·苏雅拉图、苏和、铁木尔布和；鄂尔多斯市彻·哈斯毕力格图、曹纳木、其勒木格；巴彦淖尔市巴图苏和；呼和浩特市金峰、乌·那仁巴图、拉希其仁；乌兰察布市丹·赛音巴雅尔；锡林郭勒盟纳·布和哈达、达·查干、沙·东希格、吉格米德彻仁；赤峰市纳·宝音贺希格、乌恩巴雅尔；呼伦贝尔市讷黑图、宝力道巴特尔、斯仁巴图。

阿拉善盟地区

Alxa League

底图来源：内蒙古自治区自然资源厅官网 内蒙古地图
审图号：蒙S（2017）028号

阿拉善盟辖阿拉善右旗、阿拉善左旗、额济纳旗。1980年成立阿拉善盟。现辖区由清代康熙年间设立的阿拉善和硕特旗与额济纳土尔扈特旗两旗组成，定牧于河套以西，不设盟，通常称作套西二旗。盟境内曾有40座藏传佛教寺庙，现存15座已恢复重建或尚有建筑遗存的寺庙，课题组实地调研15座寺庙。

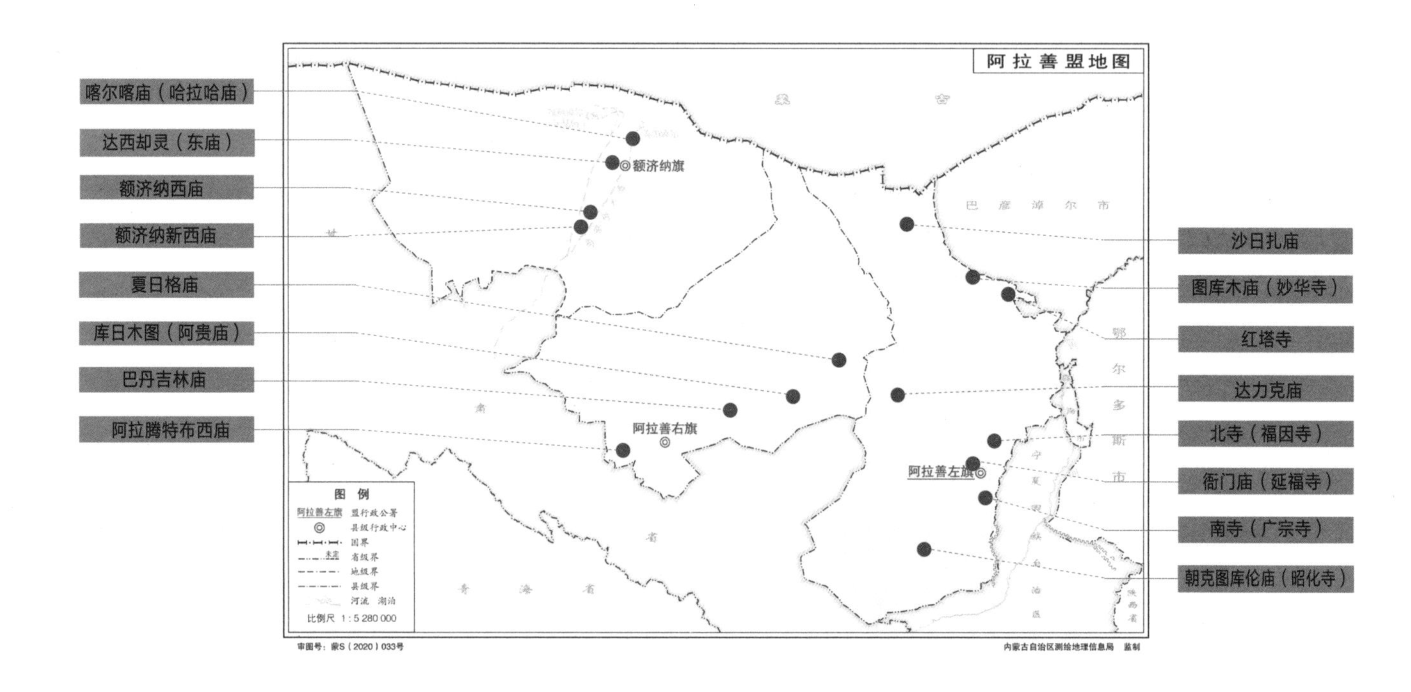
阿拉善盟地图
喀尔喀庙（哈拉哈庙）
达西却灵（东庙）
额济纳西庙
额济纳新西庙
夏日格庙
库日木图（阿贵庙）
巴丹吉林庙
阿拉腾特布西庙
沙日扎庙
图库木庙（妙华寺）
红塔寺
达力克庙
北寺（福因寺）
衙门庙（延福寺）
南寺（广宗寺）
朝克图库伦庙（昭化寺）
额济纳旗
阿拉善右旗
阿拉善左旗
巴彦淖尔市
鄂尔多斯市
图例
比例尺 1：5 280 000
审图号：蒙S（2020）033号
内蒙古自治区测绘地理信息局 监制

1 南寺(广宗寺) Baronhiid Temple

1 南寺(广宗寺)Baronhiid Temple

南寺建筑群

南寺为原阿拉善和硕特旗寺庙，位居阿拉善三大寺庙系统及八大寺庙之首。乾隆二十五年（1760年），清廷御赐蒙古、汉、满、藏四体“广宗寺”匾额。寺庙管辖门吉林庙、朝格图库伦庙、图克木庙、沙日扎庙、查干郭勒庙、达日巴照格洞、额尔德尼召、道布吉林庙、石门寺等9座属庙。南寺是六世达赖喇嘛的寺院，以供奉六世达赖仓央嘉措灵塔而著称。寺庙珍藏唐玄奘的铃杵、六世达赖喇嘛的五佛冠、八世班禅所赐银壶、光绪皇帝所赐玉如意等各类稀世文物。

康熙五十八年（1719年），六世达赖仓央嘉措在阿拉善之地弘法时初到赛音希日格河，看中此处的祥瑞地形和殊胜景色，遂决定在此建寺弘法。当时，此地有一座供奉无量寿佛和弥勒佛的小庙。1746年，六世达赖圆寂于阿拉善门吉林庙，其弟子阿格旺道尔吉遵照六世达赖之遗言，在其师所选之地，经十年准备工作，于乾隆二十一年（1756年）始建寺庙。依据寺院所处方位，习称西寺。

寺庙建筑风格为汉藏结合式建筑。1760年时南寺的庙宇僧舍共197间，而到1869年时已达2859间。在“文化大革命”前寺庙有72间黄楼庙、81间大雄宝殿、49间密宗殿、弥勒殿、金刚亥母殿、三世佛殿、显宗殿、时轮殿、医药殿、持斋庙、上护法殿、下护法殿、法轮殿、吉祥斋等大小殿堂数座，两处活佛府，大小庙仓多处及僧舍上百处。

“文化大革命”中寺庙重遭严重损毁。1981年起南寺僧众开始了重建工作，经二十余年的建设，修复并建全了黄楼庙、赞康殿、六世达赖喇嘛灵塔、三世佛殿及各学部经堂等主要建筑。

参考文献：

[1] 贾拉森.缘起南寺（汉文）.呼和浩特：内蒙古大学出版社，2003,8.
[2] 贾拉森.再现辉煌的广宗寺1757－2007（汉文）.阿拉善广宗寺印, 2007.
[3] 松儒布.阿拉善南寺史（蒙古文）.北京:民族出版社,2004,5.

<table>
<tr><td colspan="8">南寺 · 基本概况 1</td></tr>
<tr><td rowspan="3">寺院蒙古语藏语名称</td><td>蒙古语</td><td>ᠪᠠᠷᠠᠭᠤᠨ ᠬᠡᠶᠢᠳ</td><td rowspan="2">寺院汉语名称</td><td>汉语正式名称</td><td colspan="3">广宗寺</td></tr>
<tr><td>藏语</td><td>དགའ་ལྡན་བསྟན་རྒྱས་གླིང་།</td><td>俗称</td><td colspan="3">阿拉善南寺</td></tr>
<tr><td>汉语语义</td><td>西庙</td><td colspan="2">寺院汉语名称的由来</td><td colspan="3">清廷赐名</td></tr>
<tr><td>所在地</td><td colspan="4">阿拉善左旗贺兰山南巴润别立镇境内</td><td>东经 105° 48′</td><td>北纬</td><td>38° 39′</td></tr>
<tr><td>初建年</td><td colspan="2">乾隆二十二年(1757年)</td><td>保护等级</td><td colspan="4">自治区级文物保护单位</td></tr>
<tr><td>盛期时间</td><td colspan="2">清代同治年间</td><td colspan="2">盛期喇嘛僧/现有喇嘛僧</td><td colspan="3">约1500人/115人</td></tr>
<tr><td rowspan="2">历史沿革</td><td colspan="7">1756—1757年，将原有小庙扩建成9间，并新建49间大雄宝殿与庙仓。
1757年，从朝格图库伦把潘代加木草林寺全盘搬至新寺址，并把六世达赖灵塔请至南寺，专门修建一座大殿供奉在其中。
1828年，新建81间大雄宝殿与49间密宗殿。
1928—1932年，五世迭斯尔德呼图克图桑吉嘉木苏掌管阿拉善和硕特旗旗印五年。1933年，扩建并修缮黄楼庙。1934年，九世班禅驾临南寺。
1848年，重新装饰了大雄宝殿。
1869年，清同治年间回民起义中除时轮殿与金刚亥母殿之外的所有寺庙建筑遭受严重损毁。六世达赖法体临时被安置于延福寺。战乱平息后，重建81间大雄宝殿，于1913年修缮并扩建六世达赖灵塔殿，俗称黄楼庙，之后各学部大小殿宇相继落成。
1971年，南寺所有殿堂全部被拆除。
20世纪80年代，全盟境内各寺庙僧人聚集在延福寺恢复法事，原南寺的几名僧人在南寺遗址上搭建帐篷、蒙古包恢复法事，并在原活佛仓遗址上新建15间佛殿。
1989年，在原活佛拉布隆遗址上新建35间大雄宝殿。
1999年，在六世达赖灵塔殿遗址上新建六世达赖纪念塔。
2000—2001年，新建下护法殿、黄楼庙。此后，陆续新建药师殿、时轮塔等殿塔。
2001年，六世达赖喇嘛灵塔最后装藏完毕</td></tr>
<tr><td>资料来源</td><td colspan="6">［1］贾拉森.缘起南寺（汉文）.呼和浩特:内蒙古大学出版社,2003，8.
［2］贾拉森.再现辉煌的广宗寺1757—2007（汉文）.阿拉善广宗寺印,2007.
［3］松儒布.阿拉善南寺史（蒙古文）.北京: 民族出版社,2004，5.</td></tr>
<tr><td rowspan="2">现状描述</td><td colspan="4" rowspan="2">南寺现存建筑共十余座，依山顺坡势而建,庙宇全部为20世纪80年代再建，结构形式多为砖混结构，黄庙为钢筋混凝土结构。建筑装饰风格以藏式为主，汉式为辅的装饰风格。调研期间，大经堂、辩经堂处于施工阶段</td><td>描述时间</td><td colspan="2">2010/08/18</td></tr>
<tr><td>描述人</td><td colspan="2">付瑞峰</td></tr>
<tr><td>调查日期</td><td colspan="2">2010/08/18</td><td>调查人员</td><td colspan="4">高旭、宝山、李国保、付瑞峰、王志强</td></tr>
</table>

南寺基本概况表1

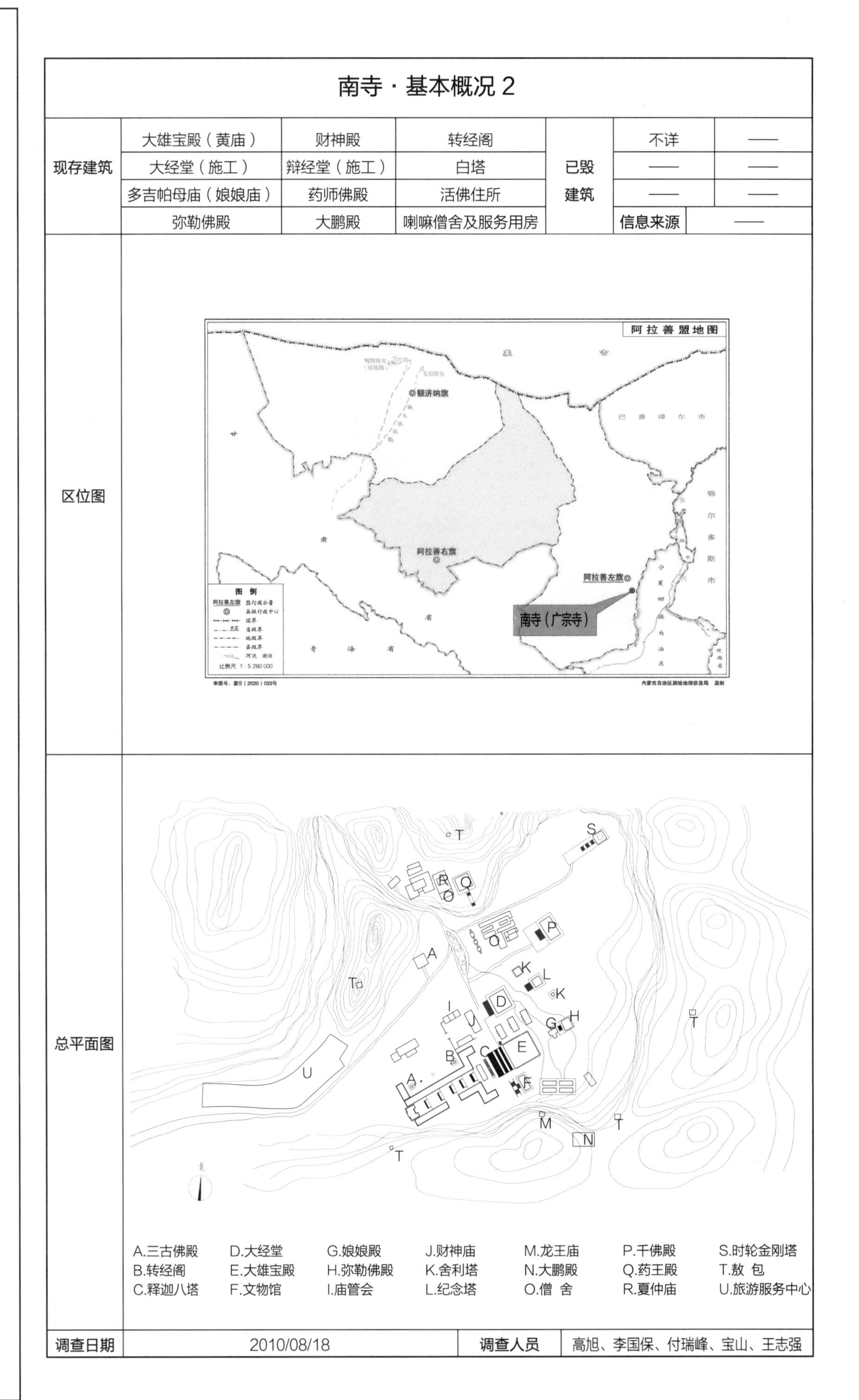

南寺 · 基本概况 2

现存建筑	大雄宝殿（黄庙）	财神殿	转经阁	已毁建筑	不详	——
	大经堂（施工）	辩经堂（施工）	白塔		——	——
	多吉帕母庙（娘娘庙）	药师佛殿	活佛住所		——	——
	弥勒佛殿	大鹏殿	喇嘛僧舍及服务用房		信息来源	——
区位图						
总平面图						
调查日期	2010/08/18		调查人员	高旭、李国保、付瑞峰、宝山、王志强		

南寺基本概况表2

A.三古佛殿 B.转经阁 C.释迦八塔 D.大经堂 E.大雄宝殿 F.文物馆 G.娘娘殿 H.弥勒佛殿 I.庙管会 J.财神庙 K.舍利塔 L.纪念塔 M.龙王庙 N.大鹏殿 O.僧 舍 P.千佛殿 Q.药王殿 R.夏仲庙 S.时轮金刚塔 T.敖 包 U.旅游服务中心

南寺总平面图

1.1 南寺·大雄宝殿

单位:毫米

建筑名称	汉语正式名称	——	俗称	黄庙							
概述	初建年	1999年（原黄庙初建于1757年，但“文化大革命”期间损毁）	建筑朝向	西	建筑层数	四					
	建筑简要描述	新黄庙建于1999年，为砖木混合结构，装饰风格为汉藏结合									
	重建重修记载	以前修缮，具体时间不详，调研期间未曾修缮									
		信息来源	寺庙喇嘛口述								
结构规模	结构形式	砖木混合结构	相连的建筑	无	室内天井	经堂为都纲法式，佛殿为天井					
	建筑平面形式	长方形	外廊形式	回廊							
	通面阔	27000	开间数	9间	明间	3900	次间	3000	梢间	3000	次梢间 3000 尽间 3000
	通进深	26100	进深数	9间	进深尺寸（前→后）	3000→3000→3000→3000→3000→3000→3000→3000→3000					
	柱子数量	36	柱子间距	横向尺寸	3900，3000	（藏式建筑结构体系填写此栏，不含廊柱）					
				纵向尺寸	3000						
	其他	——									
建筑主体（大木作）（石作）（瓦作）	屋顶	屋顶形式	藏式密肋平屋顶、汉式歇山顶、硬山卷棚顶	瓦作	黄琉璃瓦						
	外墙	主体材料	青砖	材料规格	240×120×60	饰面颜色	暗红色涂料				
		墙体收分	有	边玛檐墙	有	边玛材料	水泥、涂料				
	斗栱、梁架	斗栱	无	平身科斗口尺寸	——	梁架关系	汉藏混合				
	柱、柱式（前廊柱）	形式	藏式	柱身断面形状	圆形	断面尺寸	直径 D=550	（在没有前廊柱的情况下，填写室内柱及其特征）			
		柱身材料	木材	柱身收分	有	栌斗、托木	有	雀替	无		
		柱础	不详	柱础形状	——	柱础尺寸	——				
	台基	台基类型	须弥座	台基高度	11930	台基地面铺设材料	石材				
	其他	——									
装修（小木作）（彩画）	门(正面)	板门	门楣	有	堆经	有	门帘	无			
	窗（正面）	藏式盲窗	窗楣	有	窗套	有	窗帘	无			
	室内隔扇	隔扇	有	隔扇位置	经堂西北角						
	室内地面、楼面	地面材料及规格	方砖500×500，木板900×200	楼面材料及规格	水泥砂浆抹平						
	室内楼梯	楼梯	有	楼梯位置	正殿中央左侧	楼梯材料	木材	梯段宽度	780		
	天花、藻井	天花	有	天花类型	井口天花	藻井	无	藻井类型	——		
	彩画	柱头	有	柱身	无	梁架	有	走马板	无		
		门、窗	有	天花	有	藻井	——	其他彩画	——		
	其他	悬塑	无	佛龛	有	匾额	无				
装饰	室内	帷幔	无	幕帘彩绘	无	壁画	有	唐卡	有		
		经幡	有	经幢	有	柱毯	有	其他	——		
	室外	玛尼轮	有	苏勒德	无	宝顶	有	祥麟法轮	有		
		四角经幢	有	经幡	无	铜饰	有	石刻、砖雕	有		
		仙人走兽	1+5	壁画	有	其他	——				
陈设	室内	主佛像	仓央嘉措舍利塔	佛像基座	须弥座						
		法座 有	藏经橱 无	经床 有	诵经桌 有	法鼓 有	玛尼轮 无	坛城 有	其他 ——		
	室外	旗杆 无	苏勒德 无	狮子 有	经幡 无	玛尼轮 有	香炉 有	五供 无	其他 ——		
	其他	——									
备注	黄庙又称黄楼庙，是六世达赖的塔殿，是广宗寺最重要的建筑，它建于原葛根公馆——大乐光昭喇卜愣的旧址										
调查日期	2010/08/18	调查人员	付瑞峰	整理日期	2010/08/18	整理人员	付瑞峰、李国保				

大雄宝殿基本概况表1

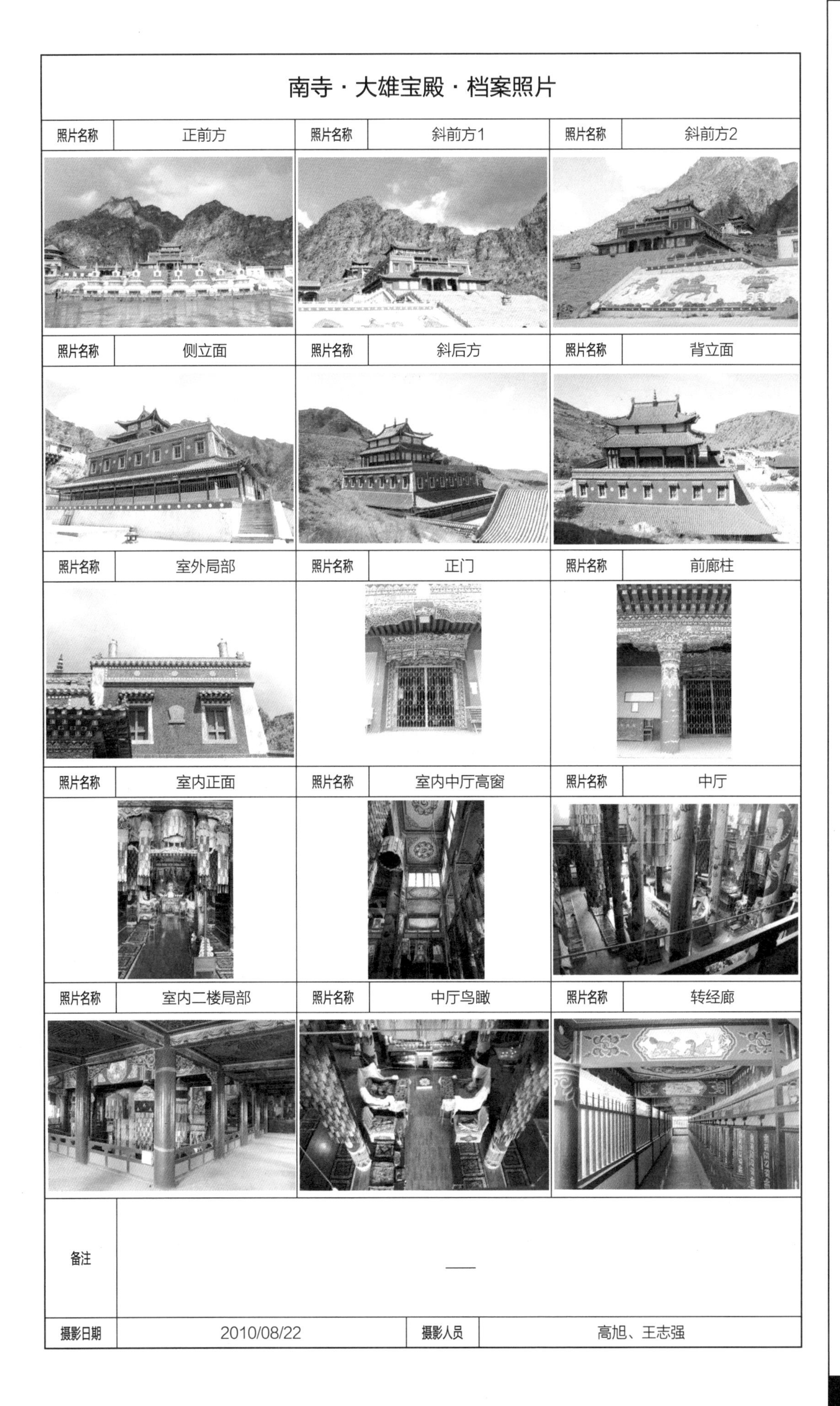

南寺·大雄宝殿·档案照片

照片名称	正前方	照片名称	斜前方1	照片名称	斜前方2
照片名称	侧立面	照片名称	斜后方	照片名称	背立面
照片名称	室外局部	照片名称	正门	照片名称	前廊柱
照片名称	室内正面	照片名称	室内中厅高窗	照片名称	中厅
照片名称	室内二楼局部	照片名称	中厅鸟瞰	照片名称	转经廊
备注	——				
摄影日期	2010/08/22	摄影人员	高旭、王志强		

大雄宝殿基本概况表2

大雄宝殿正前方

大雄宝殿斜前方（左图）
大雄宝殿顶棚（右图）

大雄宝殿侧面（左图）
大雄宝殿室内正面（右图）

1.2 南寺·多吉帕母庙

单位:毫米

建筑名称	汉语语义	——				俗称	娘娘庙										
概述	初建年	1998年（原多吉帕母庙初建于1757年，但“文化大革命”期间损毁）				建筑朝向	西	建筑层数	—								
	建筑简要描述	汉式砖木混合结构体系，装饰风格为汉藏结合															
	重建重修记载	至1998年再次修缮，其后到调研期间未曾修缮															
		信息来源	寺庙喇嘛口述														
结构规模	结构形式	砖木混合结构	相连的建筑	无	室内天井	无											
	建筑平面形式	长方形	外廊形式	前廊													
	通面阔	9000	开间数	3间	明间	3000	次间	3000	次间	——	梢间	——	尽间	——			
	通进深	10500	进深数	5间	进深尺寸（前→后）	1500→3000→3000→1500→1500											
	柱子数量	——	柱子间距	横向尺寸	——	（藏式建筑结构体系填写此栏，不含廊柱）											
				纵向尺寸	——												
	其他	——															
建筑主体（大木作）（石作）（瓦作）	屋顶	屋顶形式	硬山卷棚顶	瓦作	布瓦												
	外墙	主体材料	青砖	材料规格	240×120×60	饰面颜色	灰白色										
		墙体收分	无	边玛檐墙	无	边玛材料	——										
	斗栱、梁架	斗栱	无	平身科斗口尺寸	——	梁架关系	六檩卷棚										
	柱、柱式（前廊柱）	形式	汉式	柱身断面形状	圆形	断面尺寸	直径D=180	（在没有前廊柱的情况下，填写室内柱及其特征）									
		柱身材料	木材	柱身收分	有	栌斗、托木	无	雀替	有								
		柱础	有	柱础形状	圆形	柱础尺寸	直径D=200										
	台基	台基类型	普通台基	台基高度	1300	台基地面铺设材料	红砖										
	其他	——															
装修（小木作）（彩画）	门(正面)	隔扇	门楣	无	堆经	无	门帘	无									
	窗（正面）	槛窗	窗楣	有	窗套	无	窗帘	无									
	室内隔扇	隔扇	无	隔扇位置	——												
	室内地面、楼面	地面材料及规格	红砖240×120×60	楼面材料及规格	——												
	室内楼梯	楼梯	无	楼梯位置	——	楼梯材料	——	梯段宽度	——								
	天花、藻井	天花	有	天花类型	海漫天花	藻井	无	藻井类型	——								
	彩画	柱头	有	柱身	有	梁架	有	走马板	无								
		门、窗	有	天花	有	藻井	——	其他彩画	——								
	其他	悬塑	无	佛龛	有	匾额	无										
装饰	室内	帷幔	有	幕帘彩绘	无	壁画	有	唐卡	有								
		经幡	有	经幢	有	柱毯	无	其他	无								
	室外	玛尼轮	无	苏勒德	无	宝顶	无	祥麟法轮	无								
		四角经幢	无	经幡	无	铜饰	无	石刻、砖雕	无								
		仙人走兽	无	壁画	无	其他	——										
陈设	室内	主佛像	如来佛	佛像基座	须弥座												
		法座	无	藏经橱	无	经床	有	诵经桌	无	法鼓	无	玛尼轮	无	坛城	无	其他	——
	室外	旗杆	无	苏勒德	无	狮子	有	经幡	无	玛尼轮	无	香炉	无	五供	无	其他	——
	其他	——															
备注	——																
调查日期	2010/08/18	调查人员	付瑞峰	整理日期	2010/11/18	整理人员	付瑞峰、李国保										

多吉帕母庙基本概况表1

南寺·多吉帕母庙·档案照片

照片名称 	整体院落斜前方	照片名称 	院落门	照片名称 	斜后方1
照片名称 	斜后方2	照片名称 	北立面	照片名称 	室外局部
照片名称 	前柱廊	照片名称 	前廊柱柱头	照片名称 	前廊柱柱础
照片名称 	槛窗	照片名称 	室内正视图	照片名称 	天花
照片名称 	室内柱子	照片名称 	室内装饰	照片名称 	室内地面
备注	——				
摄影日期	2010/08/22	摄影人员	高旭、王志强		

多吉帕母庙基本概况表2

1.3 南寺 · 弥勒佛殿

单位:毫米

建筑名称	汉语正式名称	——	俗称	弥勒佛殿													
概述	初建年	2006年（原黄庙初建于1757年，但“文化大革命”期间损毁）	建筑朝向	西	建筑层数	四											
	建筑简要描述	弥勒佛殿重新建于2006年，为砖木混合结构体系，装饰风格为汉藏结合															
	重建重修记载	2006年修缮，后到调研期间未曾修缮															
		信息来源	寺庙喇嘛口述														
结构规模	结构形式	砖木混合结构	相连的建筑	无	室内天井	有											
	建筑平面形式	长方形	外廊形式	回廊													
	通面阔	12550	开间数	5间	明间	2970	次间	2960	次间	1900	梢间	——	尽间	——			
	通进深	10460	进深数	4间	进深尺寸（前→后）	950→1500→3070→3000											
	柱子数量	——	柱子间距	横向尺寸	——	（藏式建筑结构体系填写此栏，不含廊柱）											
				纵向尺寸	——												
	其他	——															
建筑主体（大木作）（石作）（瓦作）	屋顶	屋顶形式	重檐歇山屋顶	瓦作	黄琉璃瓦												
	外墙	主体材料	青砖	材料规格	不详（粉刷）	饰面颜色	红色、白色										
		墙体收分	有	边玛檐墙	无	边玛材料	——										
	斗栱、梁架	斗栱	无	平身科斗口尺寸	——	梁架关系	四柱十梁										
	柱、柱式（前廊柱）	形式	藏式	柱身断面形状	圆形	断面尺寸	直径 D=400	（在没有前廊柱的情况下，填写室内柱及其特征）									
		柱身材料	木材	柱身收分	有	栌斗、托木	有	雀替	无								
		柱础	有	柱础形状	圆形	柱础尺寸	直径 D=440										
	台基	台基类型	普通台基	台基高度	2300	台基地面铺设材料	水泥砂浆										
	其他	——															
装修（小木作）（彩画）	门(正面)	板门	门楣	有	堆经	有	门帘	无									
	窗（正面）	无	窗楣	——	窗套	——	窗帘	——									
	室内隔扇	隔扇	有	隔扇位置	二楼天井四周空间分割												
	室内地面、楼面	地面材料及规格	大理石方砖600×600	楼面材料及规格	水泥砂浆												
	室内楼梯	楼梯	无	楼梯位置	——	楼梯材料	——	梯段宽度	——								
	天花、藻井	天花	有	天花类型	海漫天花	藻井	无	藻井类型	——								
	彩画	柱头	有	柱身	无	梁架	有	走马板	无								
		门、窗	有	天花	有	藻井	——	其他彩画	须弥座								
	其他	悬塑	无	佛龛	有	匾额	无										
装饰	室内	帷幔	无	幕帘彩绘	有	壁画	无	唐卡	有								
		经幡	有	经幢	有	柱毯	有	其他	——								
	室外	玛尼轮	无	苏勒德	无	宝顶	有	祥麟法轮	有								
		四角经幢	无	经幡	无	铜饰	无	石刻、砖雕	无								
		仙人走兽	1+3	壁画	有	其他	——										
陈设	室内	主佛像	弥勒佛	佛像基座	须弥座												
		法座	无	藏经橱	无	经床	有	诵经桌	有	法鼓	无	玛尼轮	无	坛城	无	其他	——
	室外	旗杆	无	苏勒德	无	狮子	有	经幡	无	玛尼轮	无	香炉	有	五供	无	其他	——
	其他	——															
备注	重新修缮的弥勒佛殿为砖混结构仿木结构，所以梁架关系一栏中按照木构梁架结构填写																
调查日期	2010/08/18	调查人员	付瑞峰	整理日期	2010/11/18	整理人员	付瑞峰										

弥勒佛殿基本概况表1

南寺·弥勒佛殿·档案照片

照片名称 	斜前方	照片名称 	正立面	照片名称 	斜后方
照片名称 	殿背立面	照片名称 	正门	照片名称 	柱廊
照片名称 	室外局部	照片名称 	转经廊	照片名称 	室外柱头
照片名称 	室内正面	照片名称 	室内侧面	照片名称 	室内佛龛
照片名称 	室内柱子	照片名称 	室内天井	照片名称 	室内局部
备注	——				
摄影日期	2010/08/22	摄影人员	高旭、王志强		

弥勒佛殿基本概况表2

弥勒佛殿正前方

弥勒佛殿斜后方（左图）
弥勒佛殿侧廊（右图）

弥勒佛殿正前方（左图）
室内佛像（右图）

1.4 南寺·财神殿

单位:毫米

建筑名称	汉语正式名称	——	俗称	财神殿
概述	初建年	1981年	建筑朝向	西	建筑层数	一
	建筑简要描述	财神殿为砖木混合结构，装饰风格为汉藏结合
	重建重修记载	1999年修缮后，至调研期间未曾修缮
		信息来源	寺庙喇嘛口述
结构规模	结构形式	砖木混合结构	相连的建筑	无	室内天井	无
	建筑平面形式	长方形	外廊形式	回廊
	通面阔	15030	开间数	5间	明间	2950	次间	3020	次间	3110	梢间	——	尽间	——
	通进深	10980	进深数	4间	进深尺寸（前→后）	1850→3060→3010→3200
	柱子数量	——	柱子间距	横向尺寸	——	（藏式建筑结构体系填写此栏，不含廊柱）
				纵向尺寸	——	
	其他	——
建筑主体（大木作）（石作）（瓦作）	屋顶	屋顶形式	歇山屋顶	瓦作	黄色琉璃瓦
	外墙	主体材料	青砖	材料规格	240×120×60	饰面颜色	灰白色
		墙体收分	有	边玛檐墙	有	边玛材料	水泥
	斗栱、梁架	斗栱	无	平身科斗口尺寸	——	梁架关系	不详（吊顶）
	柱、柱式（前廊柱）	形式	藏式	柱身断面形状	圆形	断面尺寸	直径D=400	（在没有前廊柱的情况下，填写室内柱及其特征）
		柱身材料	木材	柱身收分	有	栌斗、托木	有	雀替	无	
		柱础	无	柱础形状	——	柱础尺寸	——	
	台基	台基类型	普通台基	台基高度	1500	台基地面铺设材料	石材
	其他	——
装修（小木作）（彩画）	门(正面)	板门	门楣	有	堆经	有	门帘	无
	窗（正面）	藏式明窗	窗楣	有	窗套	有	窗帘	无
	室内隔扇	隔扇	无	隔扇位置	——
	室内地面、楼面	地面材料及规格	水泥砂浆	楼面材料及规格	——
	室内楼梯	楼梯	无	楼梯位置	——	楼梯材料	——	梯段宽度	——
	天花、藻井	天花	无	天花类型	——	藻井	无	藻井类型	——
	彩画	柱头	有	柱身	无	梁架	——	走马板	无
		门、窗	有	天花	——	藻井	——	其他彩画	——
	其他	悬塑	无	佛龛	有	匾额	无
装饰	室内	帷幔	无	幕帘彩绘	无	壁画	无	唐卡	有
		经幡	有	经幢	有	柱毯	有	其他	——
	室外	玛尼轮	无	苏勒德	无	宝顶	无	祥麟法轮	无
		四角经幢	无	经幡	无	铜饰	有	石刻、砖雕	无
		仙人走兽	1+1	壁画	有	其他	——
陈设	室内	主佛像	财神	佛像基座	须弥座
		法座	有	藏经橱	无	经床	有	诵经桌	有	法鼓	有	玛尼轮	无	坛城	无	其他	——
	室外	旗杆	无	苏勒德	无	狮子	无	经幡	无	玛尼轮	无	香炉	有	五供	无	其他	——
	其他	——
备注	——
调查日期	2010/08/18	调查人员	付瑞峰	整理日期	2010/11/18	整理人员	付瑞峰

财神殿基本概况1

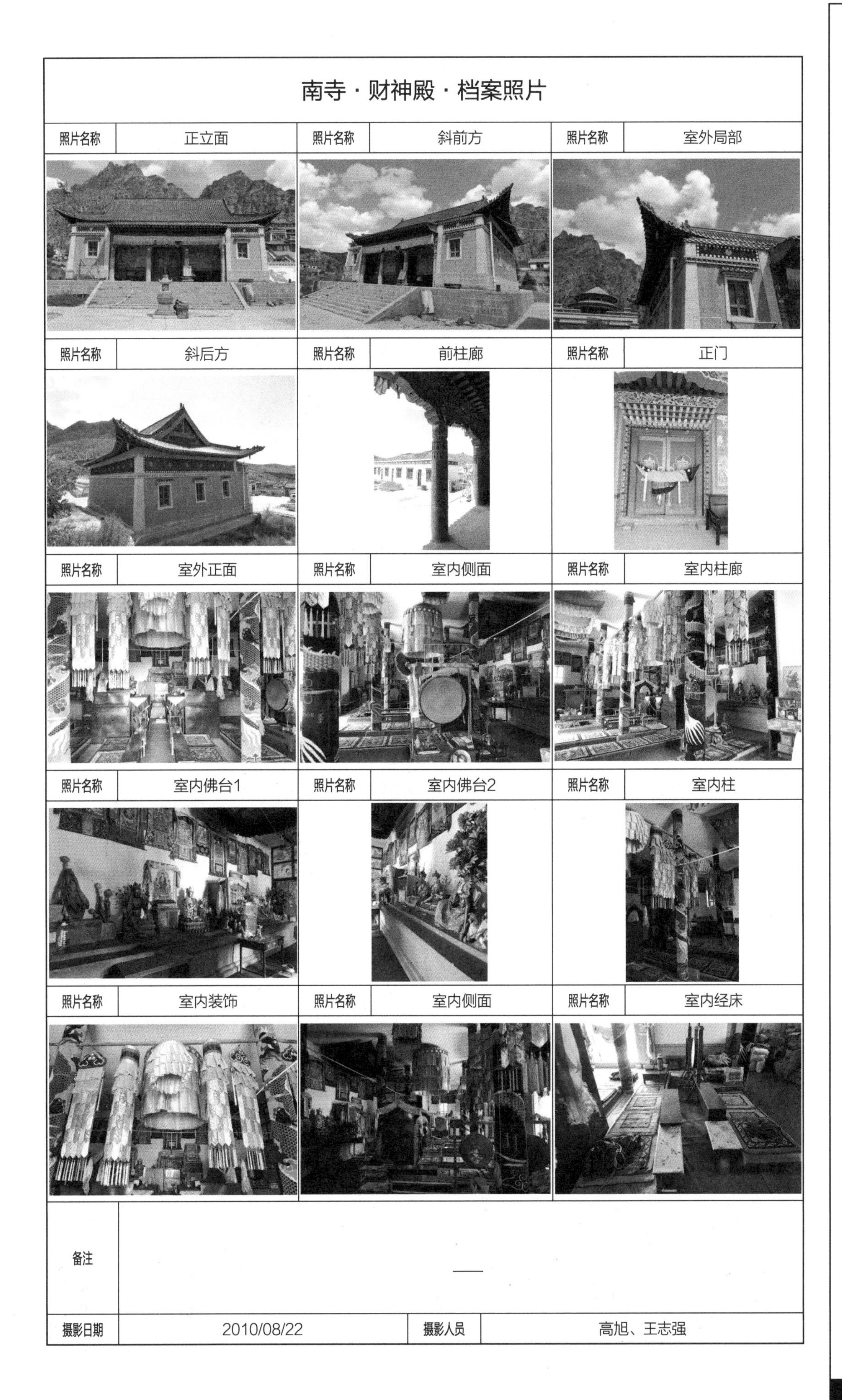

南寺·财神殿·档案照片

照片名称	正立面	照片名称	斜前方	照片名称	室外局部
照片名称	斜后方	照片名称	前柱廊	照片名称	正门
照片名称	室外正面	照片名称	室内侧面	照片名称	室内柱廊
照片名称	室内佛台1	照片名称	室内佛台2	照片名称	室内柱
照片名称	室内装饰	照片名称	室内侧面	照片名称	室内经床

备注	——		
摄影日期	2010/08/22	摄影人员	高旭、王志强

财神殿正前方

财神殿室外柱子（左图）
财神殿斜后方（右图）

财神殿室内柱（左图）
财神殿室内正面（右图）

1.5 南寺 · 大经堂

大经堂正前方(左图)
大经堂柱头（右上图）
大经堂斜后方（右下图）

1.6 南寺 · 大鹏殿

大鹏殿斜前方（左上图）
大鹏殿室内正面（左下图）
大鹏殿室外柱子（右图）

1.7 南寺·千佛殿

千佛殿正前方（左图）
千佛殿周围环境（右上图）
千佛殿斜后方（右下图）

1.8 南寺·药师佛殿

药师佛殿斜前方（左图）
药师佛殿二层斜后方（右上图）
药师佛殿二层正前方（右中图）
药师佛殿二层斜前方（右下图）

1.9 南寺·展览馆及活佛府

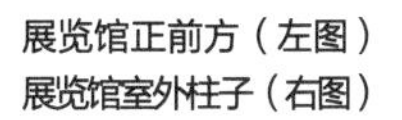

展览馆正前方（左图）
展览馆室外柱子（右图）

活佛府斜前方（左图）
展览馆斜前方（右图）

1.10 南寺·白塔及敖包

释迦八塔（左图）
石刻（右图）

白塔（左图）
敖包（右图）

2 衙门庙(延福寺) Yamen Temple

2 衙门庙(延福寺) Yamen Temple

衙门庙建筑群

衙门庙为原阿拉善和硕特旗寺庙，系阿拉善三大寺庙系统及八大寺庙之一，也是阿拉善最早建立的藏传佛教寺院。乾隆二十五年（1760年），清廷御赐满、蒙古、汉、藏四体“延福寺”匾额。寺庙管辖布尔汗乌拉庙、布日嘎苏台庙、巴丹吉林庙、额日博黑庙、夏日格庙、希日陶勒盖庙、固始庙、敖包图庙、阿贵庙、红塔庙、玛尼图庙等11座属庙。定远营城域的城隍庙、关帝庙、孔庙、财神庙等寺庙道观的住持也由衙门庙委任。

衙门庙作为府庙、王庙及内庙，其发展与阿拉善历任扎萨克的建设密不可分。康熙二十五年（1686年），阿拉善第一代扎萨克贝勒浩如来率所部从新疆迁至阿拉善后将其带来的佛像与圣物供奉在一处小佛堂中。雍正九年（1731年），清廷将定远营赐予浩如来之子阿布作王府。阿布在任期间先后修缮、改建了城中固有的寺庙建筑，竣工后命名为王府庙，成为扎萨克府邸专用寺庙。从浩如来至达尼扎纳，共十代扎萨克在任期间均不同程度地修缮、扩建了该寺。

寺庙建筑风格为汉藏结合式建筑。该寺在其最盛时占地面积6700多平方米，有48间大雄宝殿、25间医药殿、25间密宗殿、25间显宗殿、3间观音殿、15间三世佛殿、12间藏经楼、12间阿拉善神殿、12间安居殿、1间法轮殿、2间天王殿、钟楼、鼓楼等大小13座殿堂，共计326间，庙仓、甘珠尔巴上师活佛拉布隆共计185间，僧舍550间。

“文化大革命”中寺庙严重受损，一些殿堂被用作机修厂和库房，从而幸免于难。1979年起，开始修缮殿宇8座，拉布隆1处，庙仓房舍175间，正式恢复了法会。

参考文献：

［1］杨永忠.延福寺与佛教.阿拉善盟文史（第十辑）,2004，12.

［2］政协内蒙古自治区委员会文史资料委员会.内蒙古喇嘛教纪例（第四十五辑），1997，1.

衙门庙·基本概况 1

寺院蒙古语藏语名称			寺院汉语名称		
	蒙古语	[illegible]		汉语正式名称	延福寺
	藏语	དགེ་རྒྱས་གླིང་།		俗称	王爷庙
	汉语语义	王府庙	寺院汉语名称的由来		清廷赐名

所在地	阿拉善左旗巴彦浩特镇王府街北侧	东经	105° 40′	北纬	38° 50′
初建年	雍正九年（1731年）	保护等级	自治区级文物保护单位		
盛期时间	——	盛期喇嘛僧/现有喇嘛僧数	约547人/25人		

历史沿革	
	1731年起，修缮并增建定远营旧有寺庙，新建三层大雄宝殿，创建了王府寺。 1737—1739年，由仓央嘉措倡导新建“广慈菩提寺”，从衙门庙抽出100名僧人到新寺住修。 1745年，新建周围有108个法轮，双层49间大雄宝殿。 1797年，罗布桑丹毕贡布将延福寺参尼扎仓分离于原寺管辖，带领一些弟子到定远营之北新建北寺。 1805年，增建观音殿、度母殿。 1805—1840年，扩建天王殿，两边增建18间偏殿，将药师殿扩建至6间。 1922年，新建25间经论大殿和庙仓。 1932年，修缮大雄宝殿。 1937年，新建带阁楼的12间阿拉善神殿。 1941年，阿拉善神殿因失火烧为灰烬。 “文化大革命”中，寺庙部分建筑被拆毁。 1979年起，开始修缮殿宇、庙仓，正式恢复了法会
	资料来源： ［1］杨永忠.延福寺与佛教.阿拉善盟文史（第十辑）,2004，12. ［2］政协内蒙古自治区委员会文史资料委员会.内蒙古喇嘛教纪例（第四十五辑），1997，1.

现状描述	现存延福寺的十余座遗存建筑的结构体系基本完好，装饰精美。活佛府因“文化大革命”期间被毁，后又重建，现存建筑群为藏式部局，单体形式丰富，保存完好	描述时间	2010/08/17
		描述人	付瑞峰

调查日期	2010/08/17	调查人员	高旭、李国保、宝山、王志强、付瑞峰

衙门庙基本概况表1

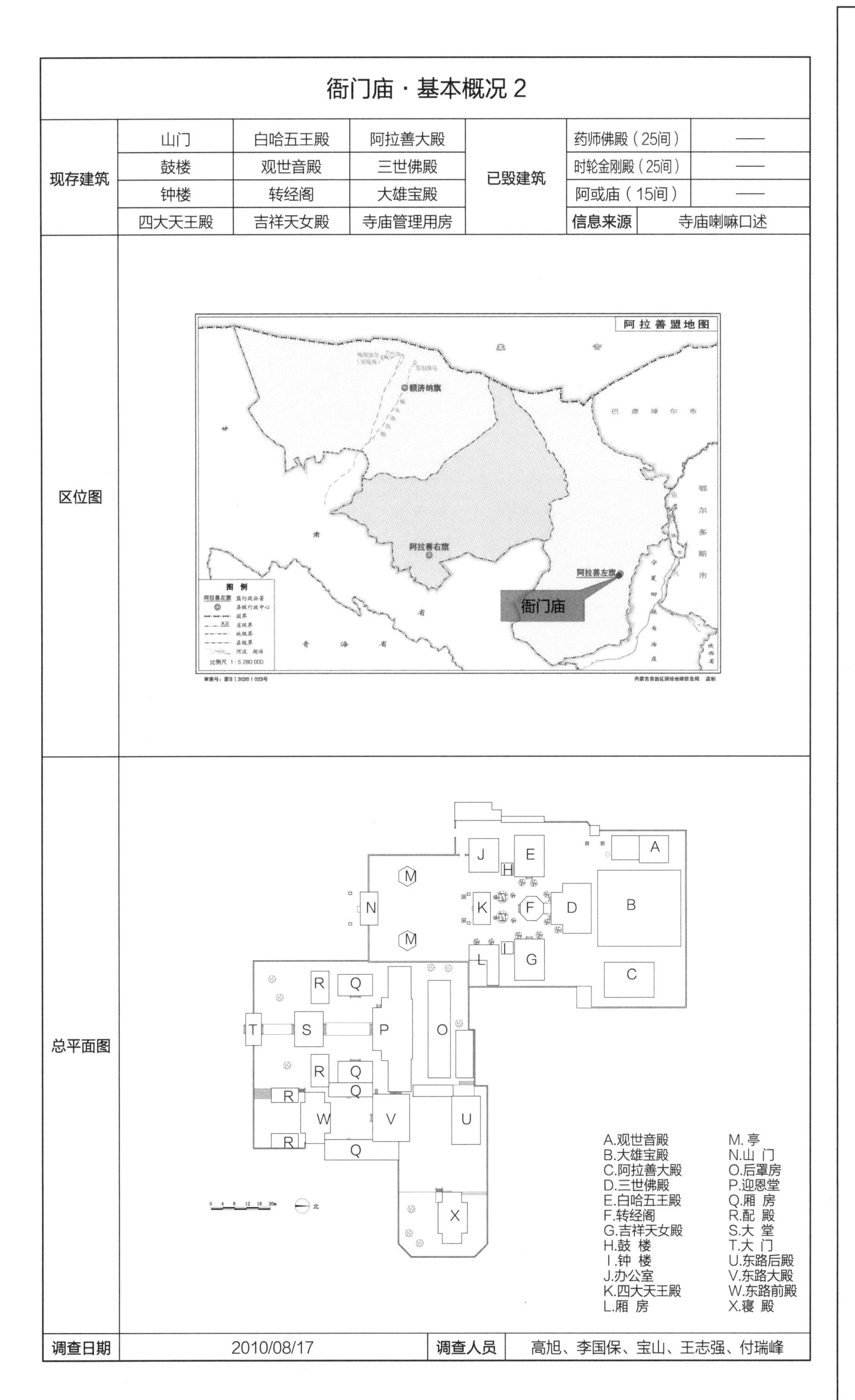

衙门庙 · 基本概况 2

现存建筑				已毁建筑		
现存建筑	山门	白哈五王殿	阿拉善大殿	已毁建筑	药师佛殿（25间）	——
	鼓楼	观世音殿	三世佛殿		时轮金刚殿（25间）	——
	钟楼	转经阁	大雄宝殿		阿或庙（15间）	——
	四大天王殿	吉祥天女殿	寺庙管理用房		信息来源	寺庙喇嘛口述
区位图						
总平面图						
调查日期	2010/08/17		调查人员	高旭、李国保、宝山、王志强、付瑞峰		

衙门庙基本概况表2

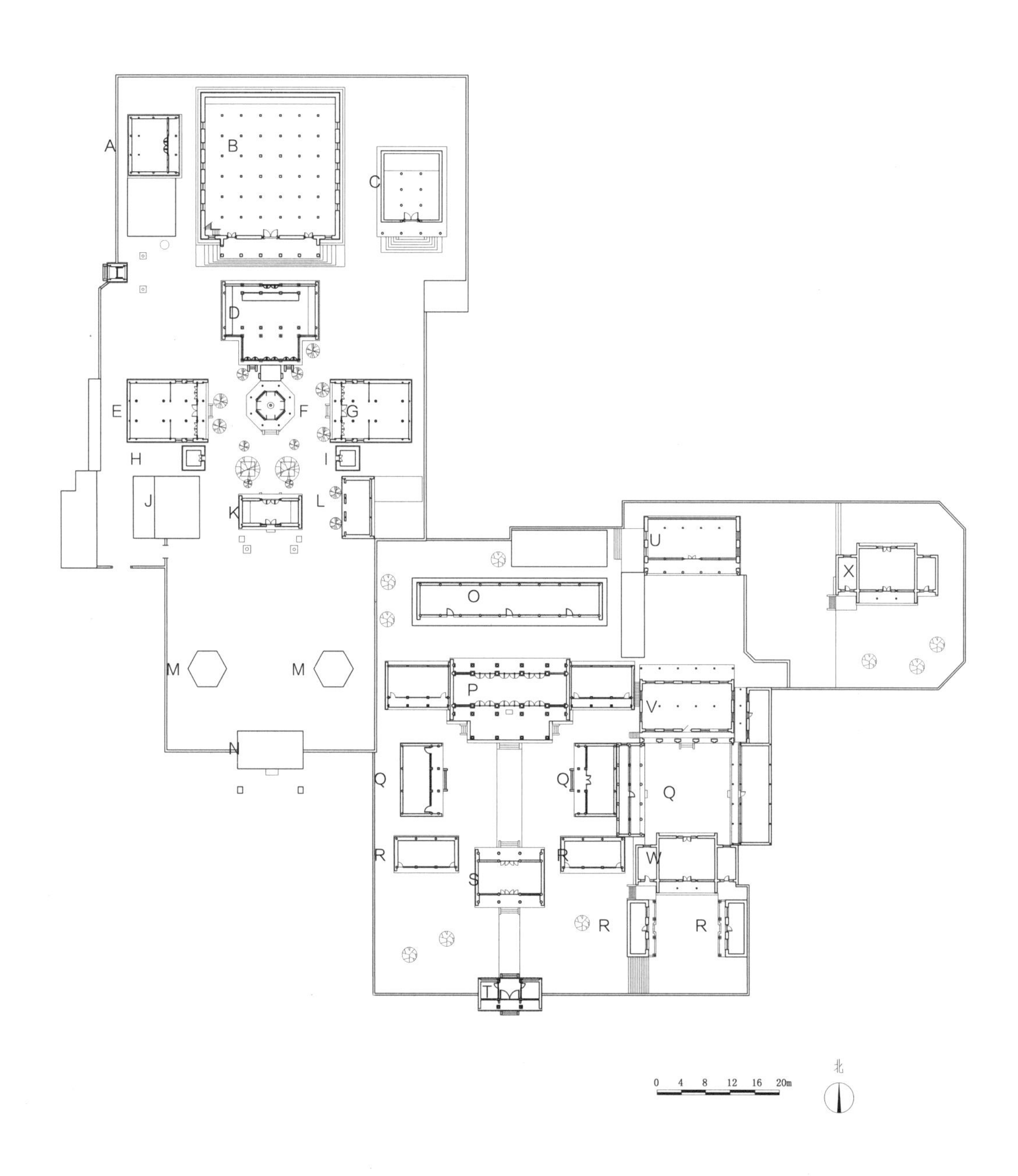

A.观世音殿
B.大雄宝殿
C.阿拉善大殿
D.三世佛殿
E.白哈五王殿
F.转经阁
G.吉祥天女殿
H.鼓 楼
I.钟 楼
J.办公室
K.四大天王殿
L.厢 房
M.亭
N.山 门
O.后罩房
P.迎恩堂
Q.厢 房
R.配 殿
S.大 堂
T.大 门
U.东路后殿
V.东路大殿
W.东路前殿
X.寝 殿

衙门庙总平面图

2.1 衙门庙·三世佛殿

单位:毫米

建筑名称	汉语正式名称	三世佛殿	俗称	三世佛殿													
概述	初建年	康熙五十八年（1719年）	建筑朝向	南	建筑层数	一											
	建筑简要描述	汉藏结合砖木混合结构体系，汉藏结合装饰风格															
	重建重修记载	分别于1987年、1992年、2006年修缮															
		信息来源	寺庙喇嘛口述														
结构规模	结构形式	砖木混合	相连的建筑	南向与转经阁以台基相连	室内天井	无											
	建筑平面形式	凸字形	外廊形式	无廊													
	通面阔	16395	开间数	5间	明间	3250	次间	3100	梢间	3100	次梢间	——	尽间	——			
	通进深	15920	进深数	4间	进深尺寸（前→后）	3800→3000→5480→1650											
	柱子数量	——	柱子间距	横向尺寸	——	（藏式建筑结构体系填写此栏，不含廊柱）											
				纵向尺寸	——												
	其他	——															
建筑主体（大木作）（石作）（瓦作）	屋顶	屋顶形式	卷棚结合硬山屋顶	瓦作	布瓦												
	外墙	主体材料	青砖	材料规格	260×70	饰面颜色	灰色										
		墙体收分	无	边玛檐墙	无	边玛材料	——										
	斗栱、梁架	斗栱	有	平身科斗口尺寸	60	梁架关系	五檩无廊、六檩无廊										
	柱、柱式（前廊柱）	形式	汉式	柱身断面形状	圆形	断面尺寸	直径D=265	（在没有前廊柱的情况下，填写室内柱及其特征）									
		柱身材料	木材	柱身收分	有	栌斗、托木	无	雀替	无								
		柱础	有	柱础形状	圆形	柱础尺寸	直径D=400										
	台基	台基类型	普通台基	台基高度	450	台基地面铺设材料	石材										
	其他	——															
装修（小木作）（彩画）	门(正面)	隔扇	门楣	无	堆经	无	门帘	无									
	窗（正面）	槛窗	窗楣	无	窗套	无	窗帘	无									
	室内隔扇	隔扇	无	隔扇位置	——												
	室内地面、楼面	地面材料及规格	木板2050×120	楼面材料及规格	——												
	室内楼梯	楼梯	无	楼梯位置	——	楼梯材料	——	楼段宽度	——								
	天花、藻井	天花	无	天花类型	——	藻井	无	藻井类型	——								
	彩画	柱头	有	柱身	无	梁架	有	走马板	无								
		门、窗	有	天花	——	藻井	——	其他彩画	——								
	其他	悬塑	无	佛龛	有	匾额	无										
装饰	室内	帷幔	无	幕帘彩绘	无	壁画	不祥	唐卡	有								
		经幡	有	经幢	有	柱毯	有	其他	——								
	室外	玛尼轮	有	苏勒德	无	宝顶	有	祥麟法轮	无								
		四角经幢	无	经幡	无	铜饰	无	石刻、砖雕	有								
		仙人走兽	无	壁画	有	其他	——										
陈设	室内	主佛像	三世佛	佛像基座	莲花瓣基座												
		法座	有	藏经橱	无	经床	有	诵经桌	有	法鼓	有	玛尼轮	无	坛城	无	其他	——
	室外	旗杆	无	苏勒德	无	狮子	无	经幡	无	玛尼轮	无	香炉	无	五供	无	其他	——
	其他	——															
备注	——																
调查日期	2010/08/16	调查人员	付瑞峰	整理日期	2010/08/16	整理人员	付瑞峰、李国保										

三世佛殿基本概况表1

衙门庙 · 三世佛殿 · 档案照片

照片名称	斜前方	照片名称	侧立面	照片名称	斜后方
照片名称	翼角1	照片名称	翼角2	照片名称	正门
照片名称	隔扇	照片名称	槛窗	照片名称	室内正面
照片名称	室内柱廊	照片名称	室内侧面	照片名称	室内柱子
照片名称	室内柱头	照片名称	室内梁架	照片名称	牌匾
备注	——				
摄影日期	2010/08/22	摄影人员	高旭、王志强		

三世佛殿基本概况表2

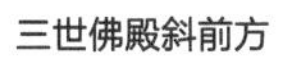

三世佛殿斜前方

三世佛殿斜后方（左图）
三世佛殿室内局部
（右图）

三世佛殿侧面（左图）
三世佛殿局部（右图）

三世佛殿平面图

三世佛殿正立面图

三世佛殿背立面图

三世佛殿侧立面图（左图）

三世佛殿剖面图（右图）

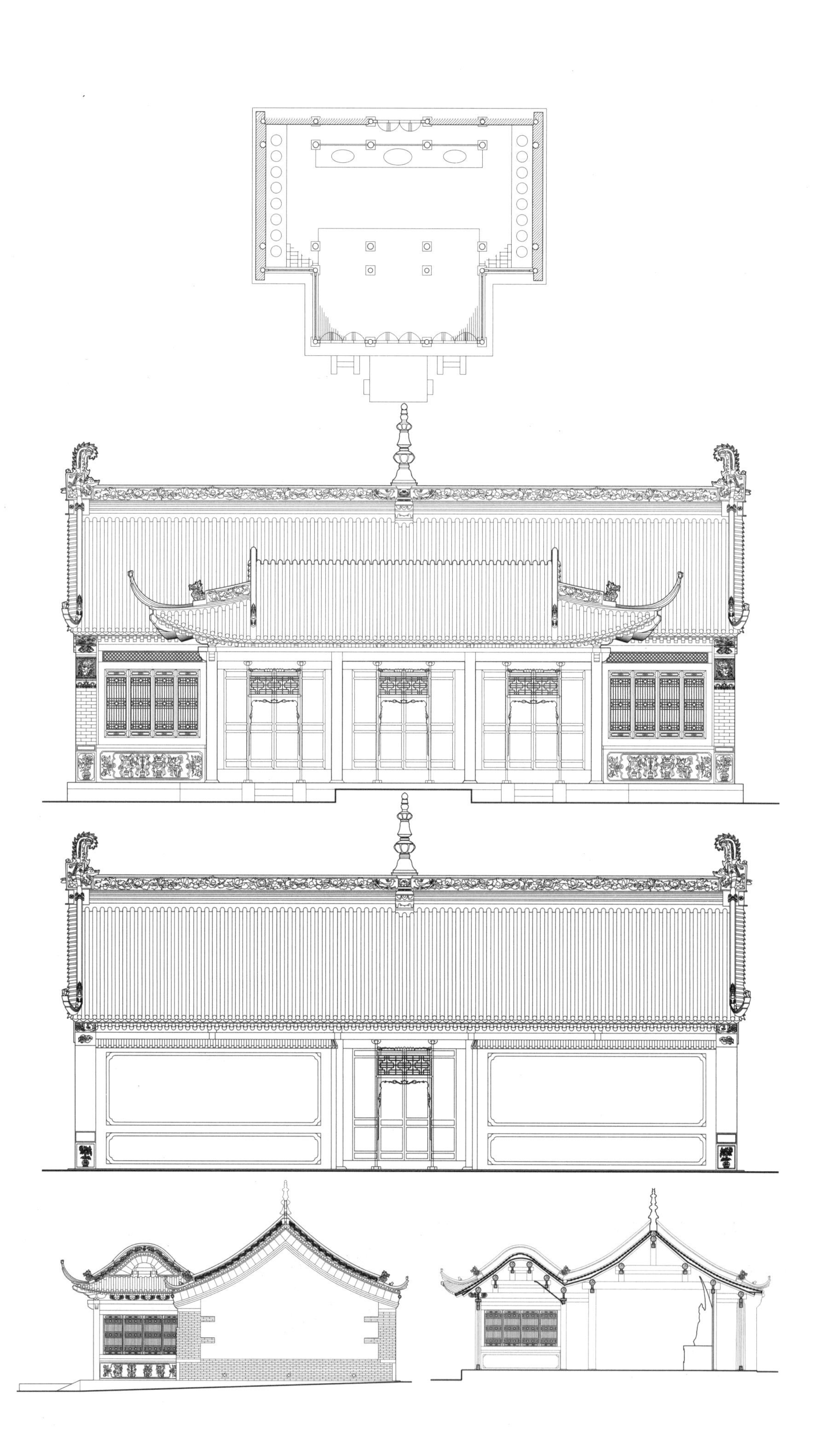

2.2 衙门庙·大雄宝殿

单位:毫米

建筑名称	汉语正式名称	大雄宝殿	俗称	大殿													
概述	初建年	1741年	建筑朝向	南	建筑层数	三											
	建筑简要描述	纯藏式结构体系，汉藏结合的装饰风格															
	重建重修记载	“文化大革命”期间被破坏，1990年重新修缮															
		信息来源	德勒格.内蒙古喇嘛教史.内蒙古人民出版社，1998;寺庙喇嘛口述														
结构规模	结构形式	砖木混合	相连的建筑	无	室内天井	都纲法式											
	建筑平面形式	凸字形	外廊形式	前廊													
	通面阔	24910	开间数	7间	明间	3190	次间	3190	梢间	3190	次梢间	3190	尽间	——			
	通进深	28520	进深数	8间	进深尺寸（前→后）	2800→3240→3240→3240→324→3240→3240→3240											
	柱子数量	36	柱子间距	横向尺寸	——	（藏式建筑结构体系填写此栏，不含廊柱）											
				纵向尺寸	——												
	其他	——															
建筑主体（大木作）（石作）（瓦作）	屋顶	屋顶形式	歇山结合藏式密肋平屋顶	瓦作	黄琉璃瓦												
	外墙	主体材料	青砖	材料规格	330×170×60	饰面颜色	青色，红色										
		墙体收分	有	边玛檐墙	有	边玛材料	边玛草										
	斗栱、梁架	斗栱	无	平身科斗口尺寸	——	梁架关系	吊顶，不详										
	柱、柱式（前廊柱）	形式	藏式	柱身断面形状	八棱柱	断面尺寸	240×240			（在没有前廊柱的情况下，填写室内柱及其特征）							
		柱身材料	木材	柱身收分	有	栌斗、托木	有	雀替	无								
		柱础	有	柱础形状	方形	柱础尺寸	500×500										
	台基	台基类型	须弥座	台基高度	1360	台基地面铺设材料	石材										
	其他	——															
装修（小木作）（彩画）	门(正面)	板门	门楣	有	堆经	有	门帘	无									
	窗（正面）	藏式明窗	窗楣	有	窗套	有	窗帘	无									
	室内隔扇	隔扇	有	隔扇位置	二楼室内空间分割												
	室内地面、楼面	地面材料及规格	石材800×140	楼面材料及规格	木材，规格不均												
	室内楼梯	楼梯	有	楼梯位置	进正门左侧	楼梯材料	木材	楼段宽度	1050								
	天花、藻井	天花	有	天花类型	井口天花	藻井	无	藻井类型	——								
	彩画	柱头	不详（柱毯）	柱身	不详（柱毯）	梁架	有	走马板	无								
		门、窗	有	天花	有	藻井	——	其他彩画	——								
	其他	悬塑	无	佛龛	有	匾额	无										
装饰	室内	帷幔	有	幕帘彩绘	有	壁画	有	唐卡	有								
		经幡	有	经幢	有	柱毯	有	其他	——								
	室外	玛尼轮	无	苏勒德	无	宝顶	有	祥麟法轮	有								
		四角经幢	有	经幡	无	铜饰	有	石刻、砖雕	有								
		仙人走兽	2	壁画	有	其他	——										
陈设	室内	主佛像	释迦牟尼	佛像基座	须弥座												
		法座	有	藏经橱	有	经床	有	诵经桌	有	法鼓	有	玛尼轮	有	坛城	有	其他	——
	室外	旗杆	无	苏勒德	无	狮子	无	经幡	无	玛尼轮	无	香炉	无	五供	无	其他	——
	其他	——															
备注	——																
调查日期	2010/08/16	调查人员	付瑞峰	整理日期	2010/08/16	整理人员	付瑞峰、李国保										

大雄宝殿基本概况表1

衙门庙·大雄宝殿·档案照片

照片名称	斜前方1	照片名称	斜前方2	照片名称	斜后方

照片名称	室外局部	照片名称	侧立面一层窗户	照片名称	边玛墙

照片名称	正门	照片名称	前廊柱	照片名称	室内正面

照片名称	室内空间1	照片名称	室内柱子	照片名称	室内楼梯

照片名称	室内天井	照片名称	室内局部	照片名称	室内空间2

备注	——				
摄影日期	2010/08/22	摄影人员	高旭、王志强		

大雄宝殿基本概况表2

大雄宝殿斜后方

大雄宝殿斜前方1（左图）
大雄宝殿侧面（右图）

大雄宝殿斜前方2（左图）
大雄宝殿殿门（右图）

大雄宝殿一层平面图
（左图）
大雄宝殿正立面
（右图）

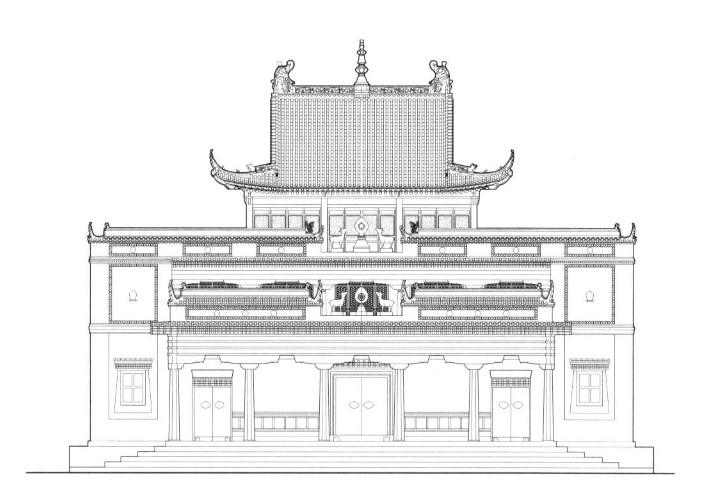

大雄宝殿二层平面图
（左图）
大雄宝殿侧立面
（右图）

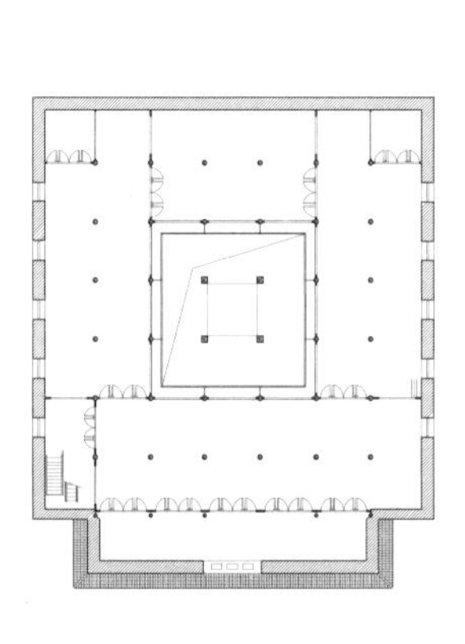

大雄宝殿三层平面图
（左图）
大雄宝殿剖面图
（右图）

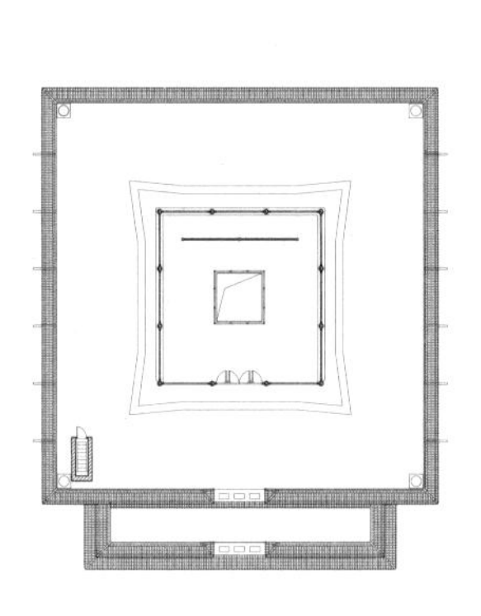

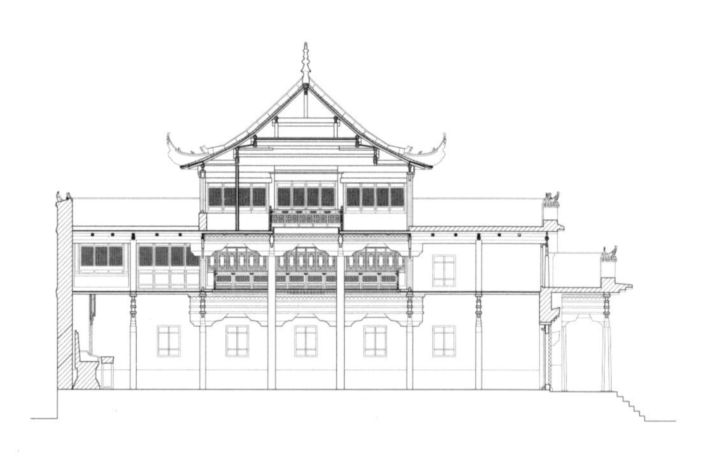

2.3 衙门庙 · 吉祥天女殿

单位:毫米

建筑名称	汉语正式名称	吉祥天女殿						俗称	——						
概述	初建年	不详						建筑朝向	西		建筑层数	一			
	建筑简要描述	汉藏结合结构体系，汉藏结合装饰风格													
	重建重修记载	不详													
		信息来源	——												
结构规模	结构形式	砖木混合		相连的建筑	无				室内天井		都纲法式				
	建筑平面形式	长字形		外廊形式	前廊										
	通面阔	10160	开间数	3间	明间	3220	次间	2910	梢间	——	次梢间	——	尽间	——	
	通进深	13555	进深数	4间	进深尺寸（前→后）			1040→1900→3680→4470→1120							
	柱子数量	——	柱子间距	横向尺寸	——			（藏式建筑结构体系填写此栏，不含廊柱）							
				纵向尺寸	——										
	其他	——													
建筑主体（大木作）（石作）（瓦作）	屋顶	屋顶形式	藏式密肋平屋顶结合经堂歇山屋顶、佛殿硬山屋顶					瓦作	布瓦						
	外墙	主体材料	青砖	材料规格	280×140×60			饰面颜色	灰色						
		墙体收分	有	边玛檐墙	有			边玛材料	边玛草						
	斗栱、梁架	斗栱	有	平身科斗口尺寸		60		梁架关系	见备注						
	柱、柱式（前廊柱）	形式	汉式	柱身断面形状	圆形	断面尺寸	直径 D=240			（在没有前廊柱的情况下，填写室内柱及其特征）					
		柱身材料	木材	柱身收分	有	栌斗、托木	无	雀替	无						
		柱础	有	柱础形状	圆形	柱础尺寸	直径 D=290								
	台基	台基类型	普通台基	台基高度	130	台基地面铺设材料		石材，规格不均							
	其他	——													
装修（小木作）（彩画）	门(正面)	板门		门楣	有	堆经	有	门帘	无						
	窗（正面）	槛窗		窗楣	无	窗套	无	窗帘	无						
	室内隔扇	隔扇	无	隔扇位置	——										
	室内地面、楼面	地面材料及规格		水泥砂浆抹平		楼面材料及规格		——							
	室内楼梯	楼梯	无	楼梯位置	——	楼梯材料	——	楼段宽度	——						
	天花、藻井	天花	有	天花类型	井口天花	藻井	无	藻井类型	——						
	彩画	柱头	有	柱身	无	梁架	有	走马板	有						
		门、窗	有	天花	无	藻井	——	其他彩画	——						
	其他	悬塑	无	佛龛	有	匾额	无								
装饰	室内	帷幔	有	幕帘彩绘	无	壁画	无	唐卡	有						
		经幡	有	经幢	有	柱毯	有	其他	——						
	室外	玛尼轮	无	苏勒德	无	宝顶	有	祥麟法轮	有						
		四角经幢	无	经幡	无	铜饰	有	石刻、砖雕	有						
		仙人走兽	无	壁画	无	其他	——								
陈设	室内	主佛像	吉祥天女			佛像基座	无								
		法座	有	藏经橱	无	经床	有	诵经桌	有	法鼓	有	玛尼轮	无	坛城	无
		其他	——												
	室外	旗杆	无	苏勒德	无	狮子	无	经幡	无	玛尼轮	无	香炉	有	五供	无
		其他	——												
	其他	——													
备注	梁架关系：前面为藏式密肋平屋顶结合歇山顶，后面为硬山屋顶，内部吊顶，不详														
调查日期	2010/08/17	调查人员	付瑞峰	整理日期	2010/08/17	整理人员	付瑞峰、李国保								

吉祥天女殿基本概况表1

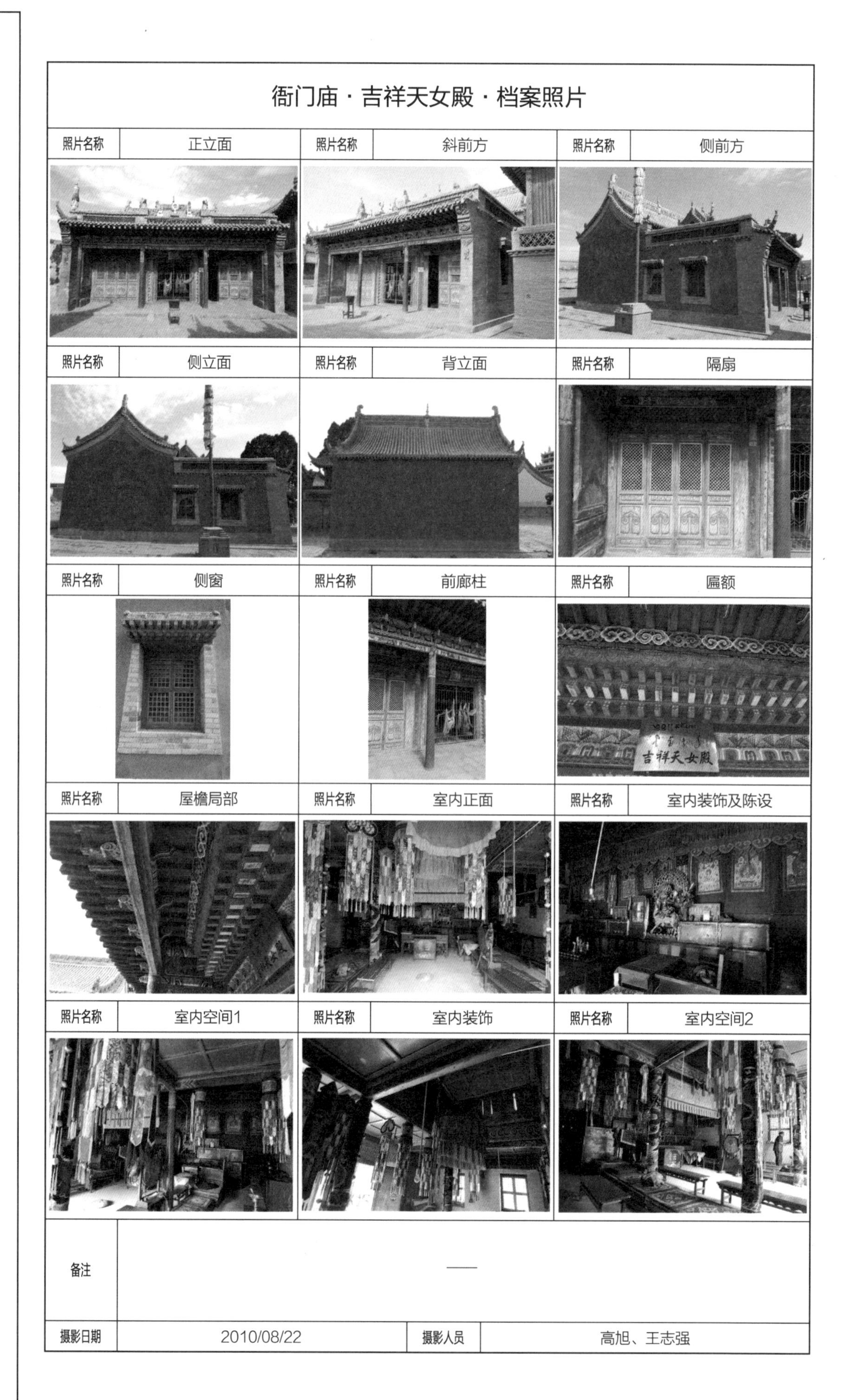

衙门庙 · 吉祥天女殿 · 档案照片

照片名称	正立面	照片名称	斜前方	照片名称	侧前方
照片名称	侧立面	照片名称	背立面	照片名称	隔扇
照片名称	侧窗	照片名称	前廊柱	照片名称	匾额
照片名称	屋檐局部	照片名称	室内正面	照片名称	室内装饰及陈设
照片名称	室内空间1	照片名称	室内装饰	照片名称	室内空间2
备注	——				
摄影日期	2010/08/22	摄影人员	高旭、王志强		

吉祥天女殿基本概况表2

吉祥天女殿正前方

吉祥天女殿室内正面（左图）
吉祥天女殿斜前方（右图）

吉祥天女殿窗（左图）
吉祥天女殿背面（右图）

2.4 衔门庙·白哈五王殿

单位:毫米

建筑名称	汉语正式名称	白哈五王殿	俗称	——															
概述	初建年	不详	建筑朝向	东	建筑层数	一													
	建筑简要描述	汉藏砖木混合结构体系，汉藏结合装饰风格																	
	重建重修记载	不详																	
		信息来源	——																
结构规模	结构形式	砖木混合	相连的建筑	无	室内天井	都纲法式													
	建筑平面形式	长方形	外廊形式	前廊															
	通面阔	10160	开间数	3间	明间	3220	次间	2910	梢间	——	次梢间	——	尽间	——					
	通进深	13555	进深数	5间	进深尺寸（前→后）	1040→1900→3680→4470→1120													
	柱子数量	6	柱子间距	横向尺寸	——	（藏式建筑结构体系填写此栏，不含廊柱）													
				纵向尺寸	——														
	其他	——																	
建筑主体（大木作）（石作）（瓦作）	屋顶	屋顶形式	藏式密肋平屋顶结合经堂歇山屋顶、佛殿硬山屋顶	瓦作	布瓦														
	外墙	主体材料	青砖	材料规格	270×130×60	饰面颜色	灰色												
		墙体收分	有	边玛檐墙	有	边玛材料	青砖、红色涂料												
	斗栱、梁架	斗栱	有	平身科斗口尺寸	60	梁架关系	见备注												
	柱、柱式（前廊柱）	形式	汉式	柱身断面形状	圆形	断面尺寸	直径D=240	（在没有前廊柱的情况下，填写室内柱及其特征）											
		柱身材料	木材	柱身收分	有	栌斗、托木	无	雀替	无										
		柱础	有	柱础形状	腹盆	柱础尺寸	直径D=360												
	台基	台基类型	普通台基	台基高度	300	台基地面铺设材料	石材、规格不均												
	其他	——																	
装修（小木作）（彩画）	门(正面)	板门	门楣	有	堆经	有	门帘	无											
	窗（正面）	槛窗	窗楣	无	窗套	无	窗帘	无											
	室内隔扇	隔扇	无	隔扇位置	——														
	室内地面、楼面	地面材料及规格	大理石方400×400	楼面材料及规格	——														
	室内楼梯	楼梯	无	楼梯位置	——	楼梯材料	——	楼段宽度	——										
	天花、藻井	天花	有	天花类型	井口天花	藻井	无	藻井类型	——										
	彩画	柱头	有	柱身	无	梁架	有	走马板	无										
		门、窗	有	天花	有	藻井	——	其他彩画	——										
	其他	悬塑	无	佛龛	有	匾额	无												
装饰	室内	帷幔	有	幕帘彩绘	无	壁画	无	唐卡	有										
		经幡	有	经幢	有	柱毯	有	其他	——										
	室外	玛尼轮	无	苏勒德	无	宝顶	有	祥麟法轮	有										
		四角经幢	无	经幡	无	铜饰	无	石刻、砖雕	有										
		仙人走兽	无	壁画	有	其他													
陈设	室内	主佛像	五道佛	佛像基座	无														
		法座	无	藏经橱	无	经床	无	诵经桌	有	法鼓	有	玛尼轮	无	坛城	无	其他	——		
	室外	旗杆	无	苏勒德	无	狮子	无	经幡	无	玛尼轮	无	香炉	有	五供	无	其他	——		
	其他	——																	
备注	梁架关系：前面为藏式密肋平屋顶结合歇山顶，后面为硬山屋顶，内部吊顶，不详																		
调查日期	2010/08/17	调查人员	付瑞峰	整理日期	2010/08/17	整理人员	付瑞峰、李国保												

白哈五王殿基本概况表1

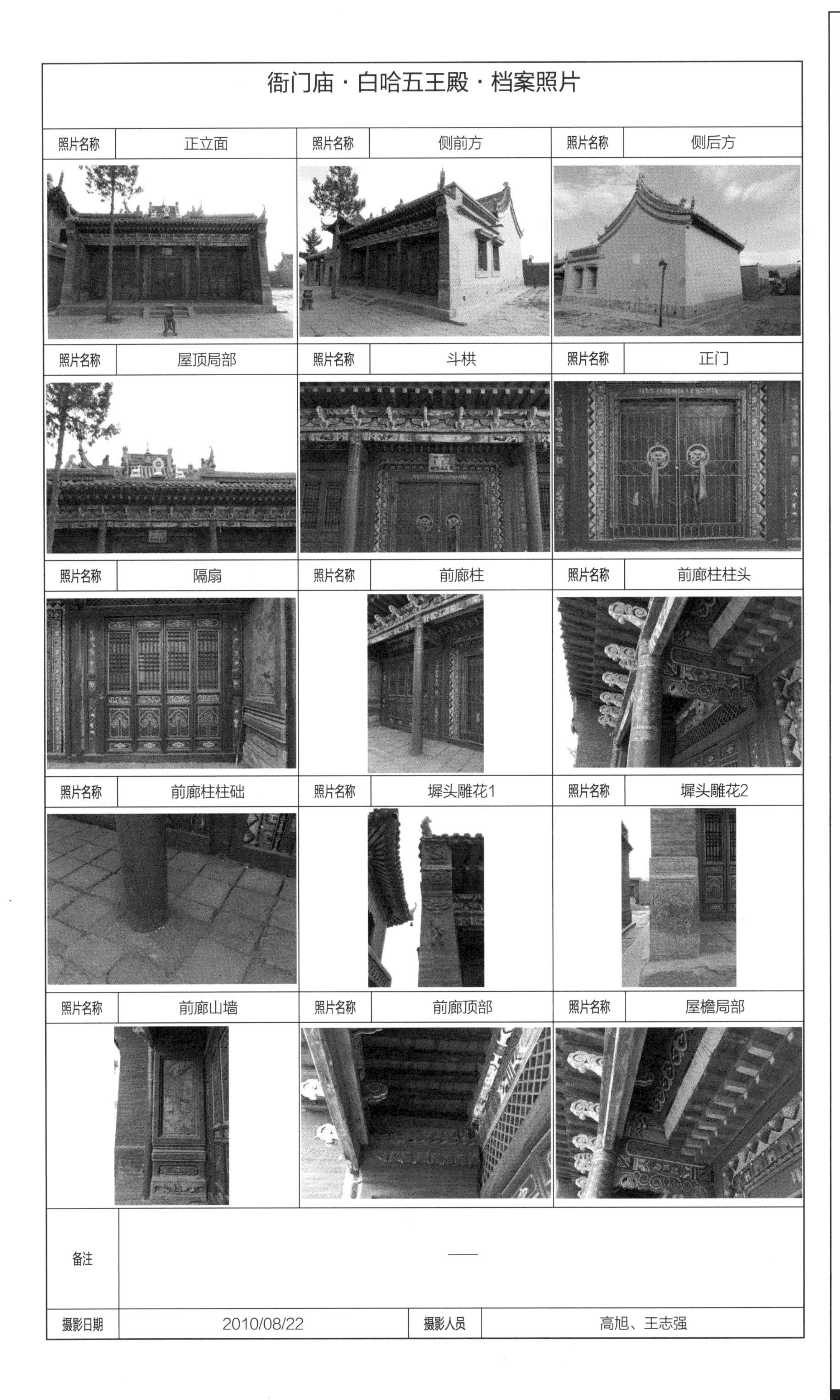

衙门庙·白哈五王殿·档案照片

照片名称	正立面	照片名称	侧前方	照片名称	侧后方
照片名称	屋顶局部	照片名称	斗栱	照片名称	正门
照片名称	隔扇	照片名称	前廊柱	照片名称	前廊柱柱头
照片名称	前廊柱柱础	照片名称	墀头雕花1	照片名称	墀头雕花2
照片名称	前廊山墙	照片名称	前廊顶部	照片名称	屋檐局部

备注	——		
摄影日期	2010/08/22	摄影人员	高旭、王志强

白哈五王殿基本概况表2

白哈五王殿正前方

白哈五王殿斜前方
（左图）
白哈五王殿室外柱子
（右图）

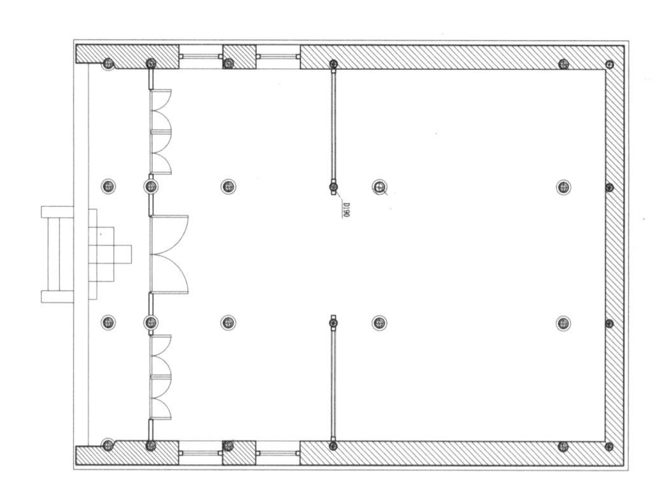

白哈五王殿平面图

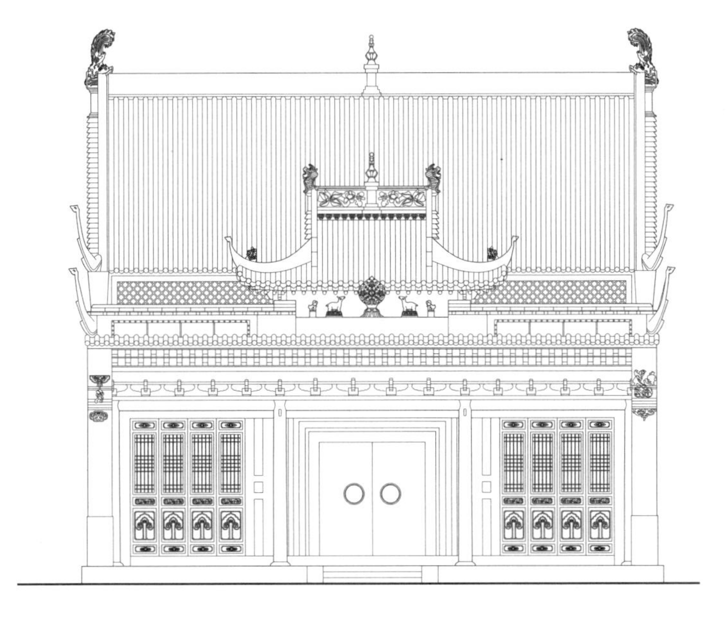

白哈五王殿立面图

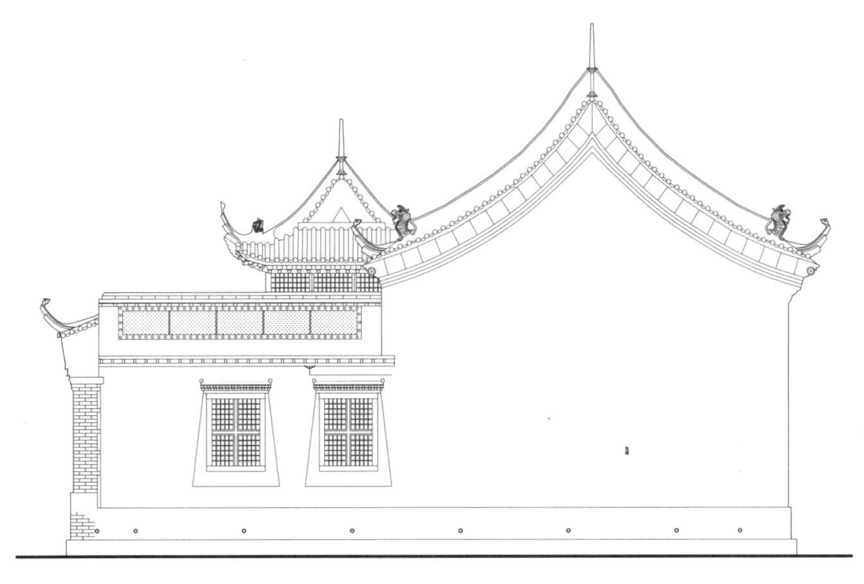

白哈五王殿侧立面

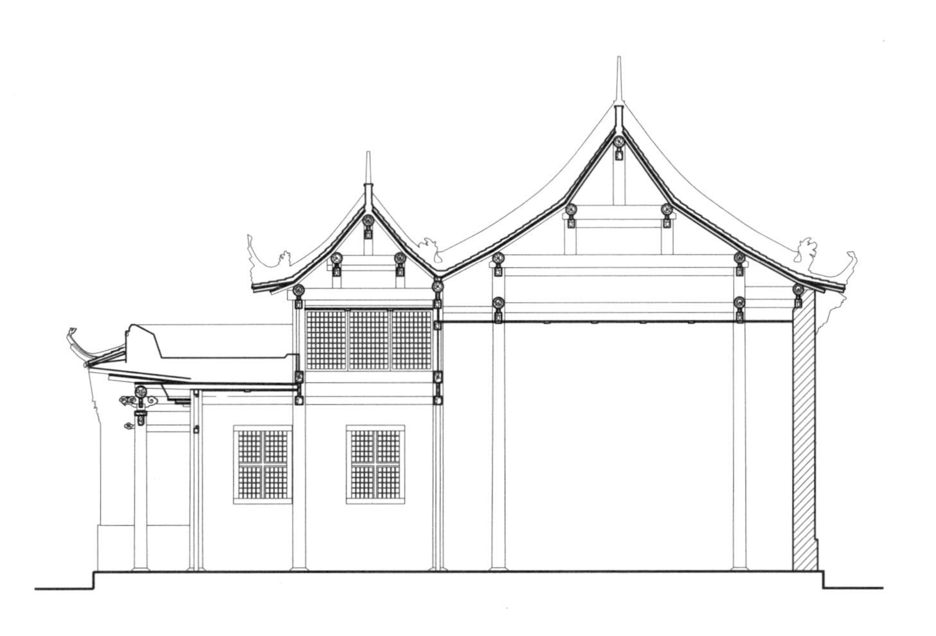

白哈五王殿剖面图

2.5 衙门庙·阿拉善大殿

单位:毫米

建筑名称	汉语正式名称	阿拉善大殿	俗称	阿拉善													
概述	初建年	1741年	建筑朝向	南	建筑层数	一											
	建筑简要描述	汉藏结合建筑结构体系，汉藏结合装饰风格															
	重建重修记载	“文化大革命”期间被毁，2003年重新修建															
		信息来源	寺庙喇嘛口述														
结构规模	结构形式	砖木混合	相连的建筑	无	室内天井	都纲法式											
	建筑平面形式	长字形	外廊形式	前廊													
	通面阔	11450	开间数	3间	明间	3190	次间	3155	梢间	——	次梢间	——	尽间	——			
	通进深	14350	进深数	4间	进深尺寸（前→后）	1950→2500→2500→4000											
	柱子数量	6	柱子间距	横向尺寸	——	（藏式建筑结构体系填写此栏，不含廊柱）											
				纵向尺寸	——												
	其他	——															
建筑主体（大木作）（石作）（瓦作）	屋顶	屋顶形式	密肋平顶结合歇山屋顶	瓦作	黄琉璃瓦												
	外墙	主体材料	红砖	材料规格	240×120×53	饰面颜色	灰色										
		墙体收分	有	边玛檐墙	有	边玛材料	泥土、红色涂料										
	斗栱、梁架	斗栱	无	平身科斗口尺寸	50	梁架关系	6柱17梁										
	柱、柱式（前廊柱）	形式	藏式	柱身断面形状	小八角	断面尺寸	200×200	（在没有前廊柱的情况下，填写室内柱及其特征）									
		柱身材料	木材	柱身收分	有	栌斗、托木	无	雀替	无								
		柱础	有	柱础形状	莲花瓣	柱础尺寸	直径 D=300										
	台基	台基类型	普通台基	台基高度	770	台基地面铺设材料	大理石、规格不均										
	其他	——															
装修（小木作）（彩画）	门(正面)	板门	门楣	有	堆经	有	门帘	无									
	窗（正面）	藏式盲窗、槛窗	窗楣	有	窗套	有	窗帘	无									
	室内隔扇	隔扇	无	隔扇位置	——												
	室内地面、楼面	地面材料及规格	水泥抹平	楼面材料及规格	——												
	室内楼梯	楼梯	无	楼梯位置	——	楼梯材料	——	楼段宽度	——								
	天花、藻井	天花	无	天花类型	——	藻井	无	藻井类型	——								
	彩画	柱头	无	柱身	无	梁架	无	走马板	有								
		门、窗	有	天花	无	藻井	——	其他彩画	——								
	其他	悬塑	无	佛龛	有	匾额	无										
装饰	室内	帷幔	有	幕帘彩绘	无	壁画	无	唐卡	有								
		经幡	无	经幢	有	柱毯	有	其他	——								
	室外	玛尼轮	无	苏勒德	无	宝顶	有	祥麟法轮	有								
		四角经幢	有	经幡	无	铜饰	有	石刻、砖雕	有								
		仙人走兽	3	壁画	有	其他	——										
陈设	室内	主佛像	阿拉善地方神	佛像基座	须弥座												
		法座	有	藏经橱	无	经床	有	诵经桌	有	法鼓	有	玛尼轮	无	坛城	无	其他	——
	室外	旗杆	有	苏勒德	无	狮子	有	经幡	无	玛尼轮	无	香炉	有	五供	无	其他	——
	其他																
备注	——																
调查日期	2010/08/17	调查人员	付瑞峰	整理日期	2010/08/17	整理人员	付瑞峰、李国保										

阿拉善大殿基本概况表1

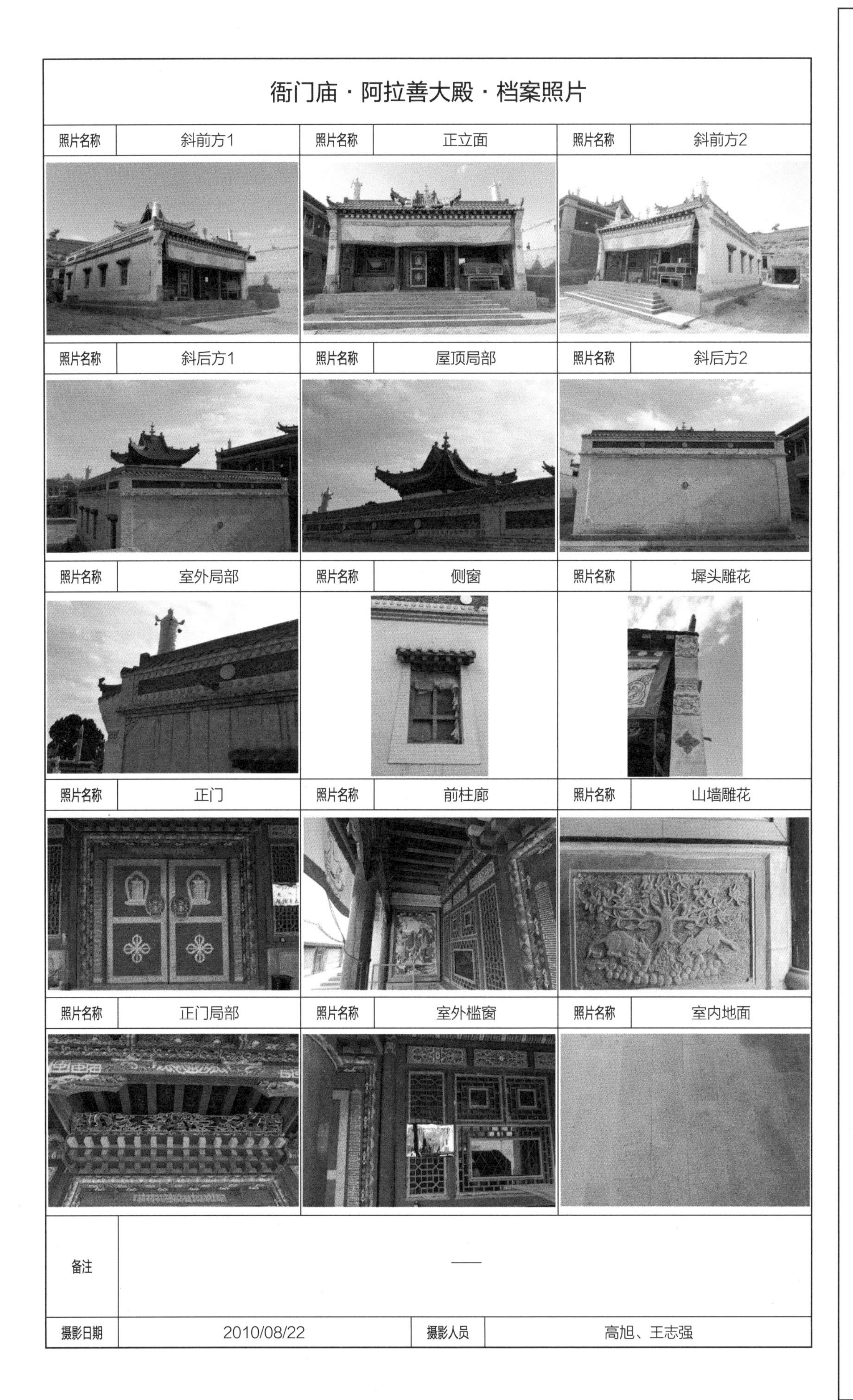

衙门庙·阿拉善大殿·档案照片

照片名称	斜前方1	照片名称	正立面	照片名称	斜前方2
照片名称	斜后方1	照片名称	屋顶局部	照片名称	斜后方2
照片名称	室外局部	照片名称	侧窗	照片名称	墀头雕花
照片名称	正门	照片名称	前柱廊	照片名称	山墙雕花
照片名称	正门局部	照片名称	室外槛窗	照片名称	室内地面
备注	——				
摄影日期	2010/08/22	摄影人员	高旭、王志强		

阿拉善大殿正前方

阿拉善大殿斜前方（左图）
阿拉善大殿正门（右图）

2.6 衙门庙 · 山门

山门正前方(左图)
山门室外柱子（右图）

2.7 衙门庙 · 四大天王殿

四大天王殿正前方

四大天王殿正门（左上图）
四大天王殿侧面（左下图）
四大天王殿斜后方（右图）

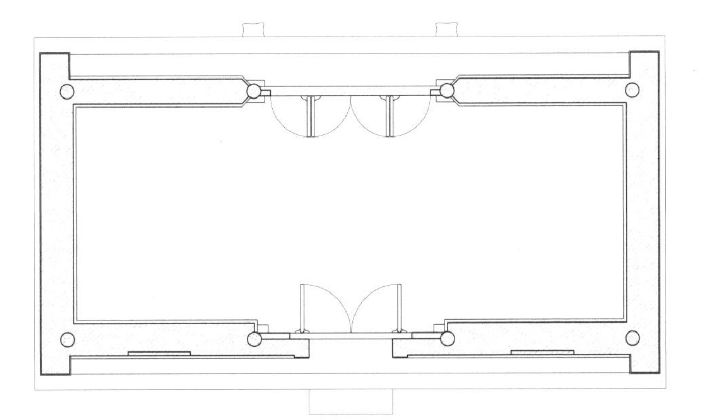

四大天王殿一层平面图

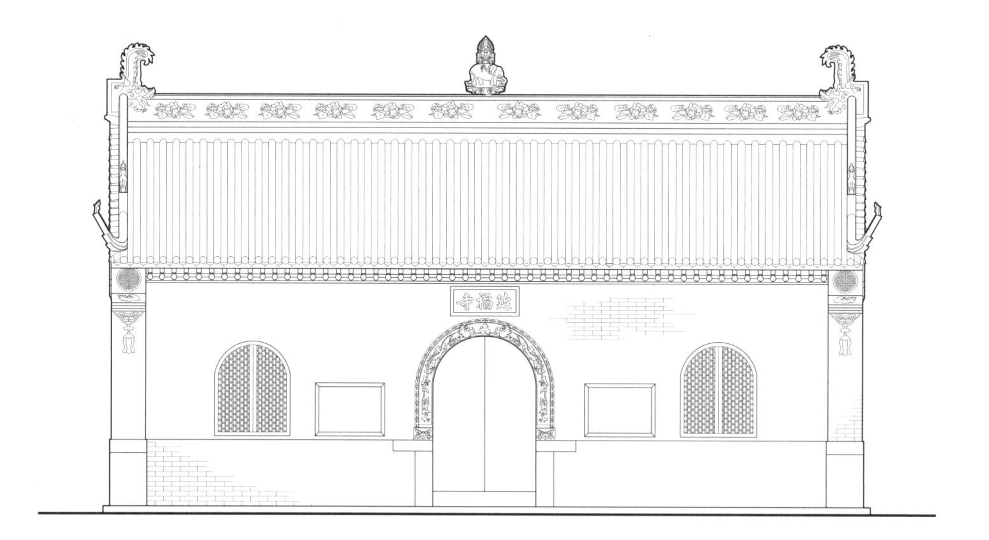

四大天王殿正立面图

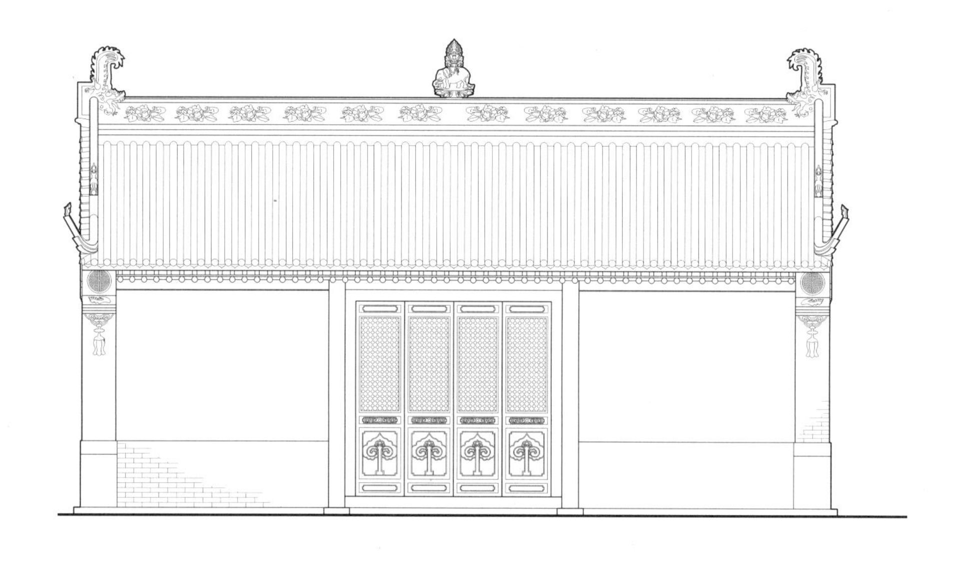

四大天王殿背立面图

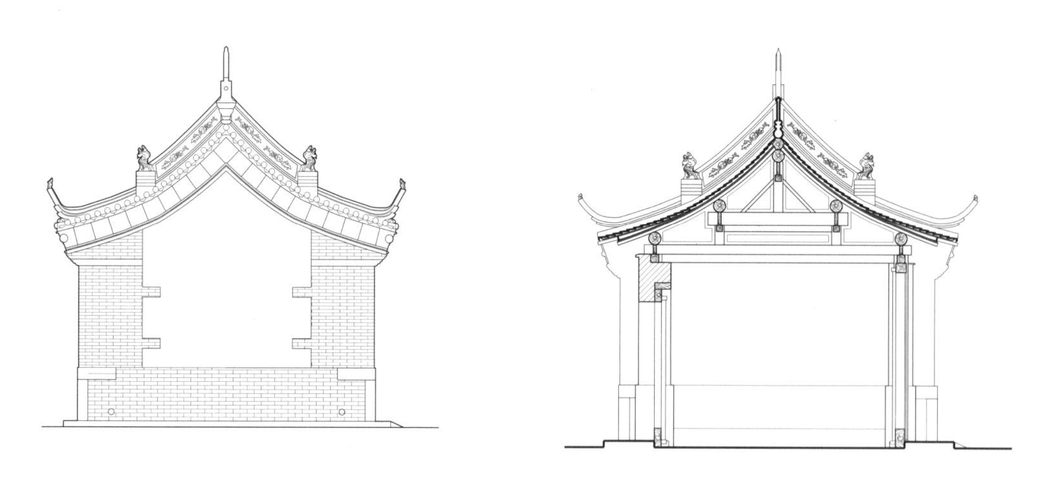

四大天王殿侧立面图
（左图）
四大天王殿剖面图
（右图）

2.8 衙门庙 · 钟鼓楼及转经阁

转经阁正前方

钟楼室外柱子（左图）

钟楼斜前方（左图）
钟楼侧面（中图）
钟楼正前方（右图）

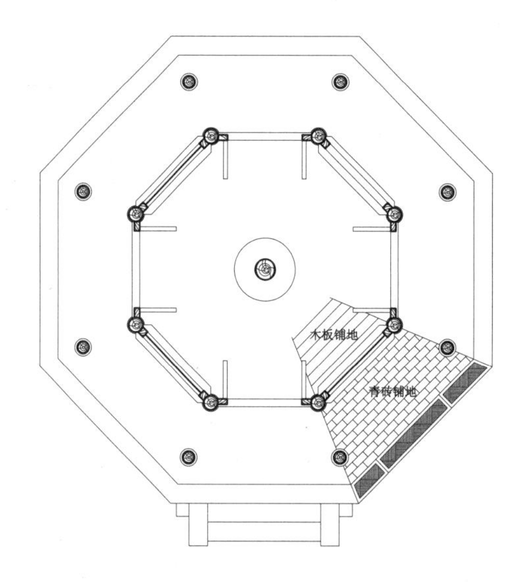

转经阁平面图

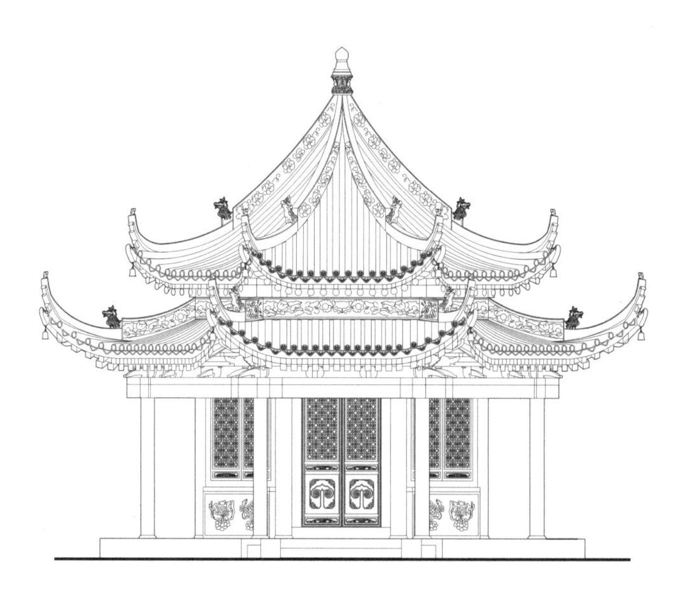

转经阁正立面图

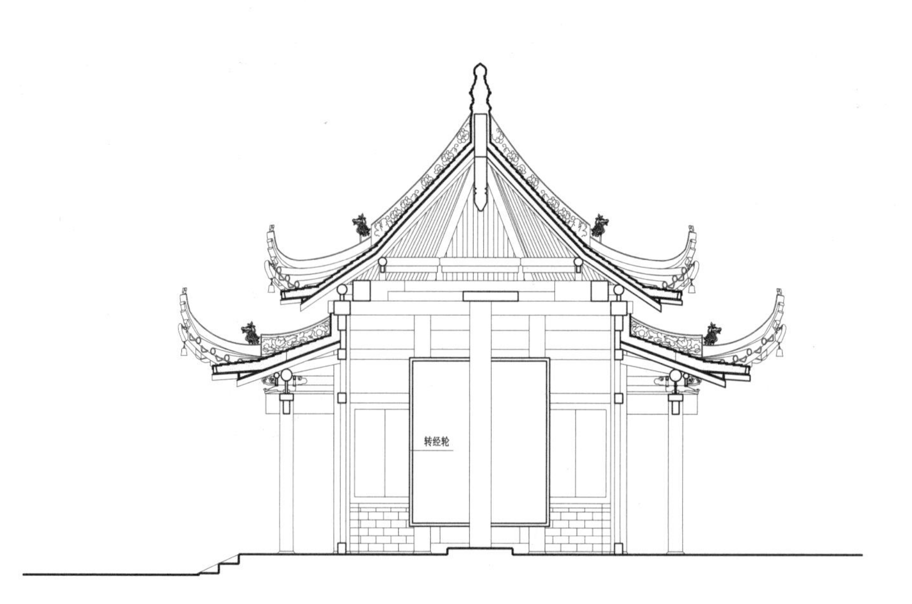

转经阁剖面图

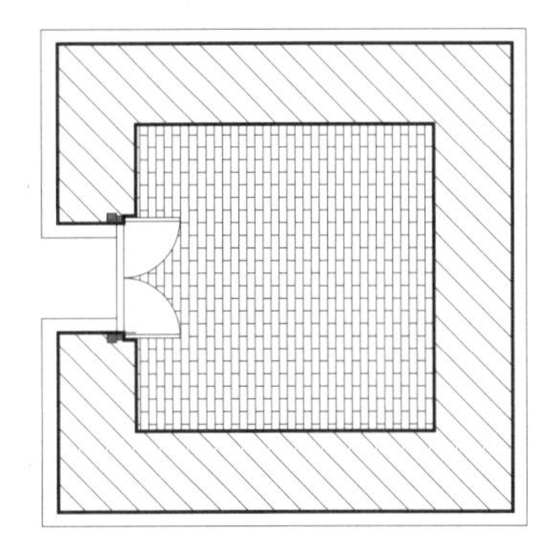

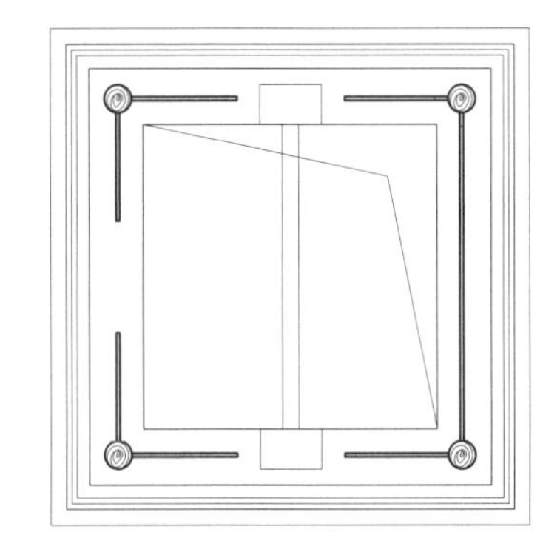

钟楼底层平面图（左图）
钟楼上层平面图（右图）

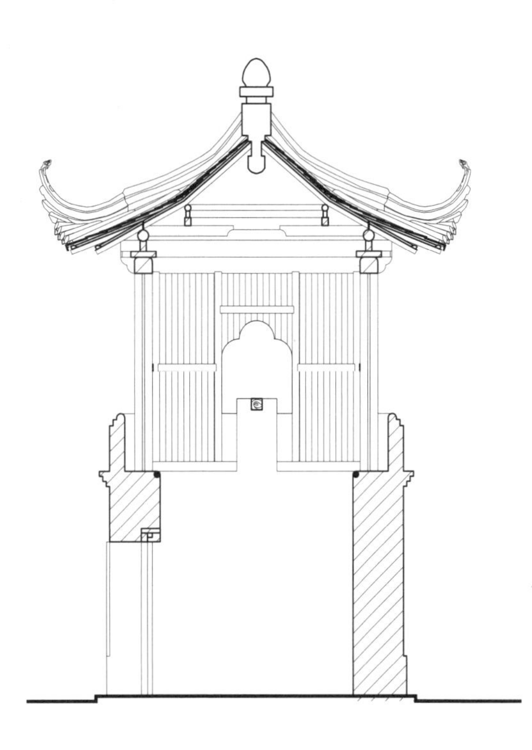

钟楼正立面图（左图）
钟楼剖面图（右图）

2.9 衙门庙·观世音殿

观世音殿斜前方（左图）
观世音殿墀头（右图）

观世音殿匾额（左图）
观世音殿室外柱子（右图）

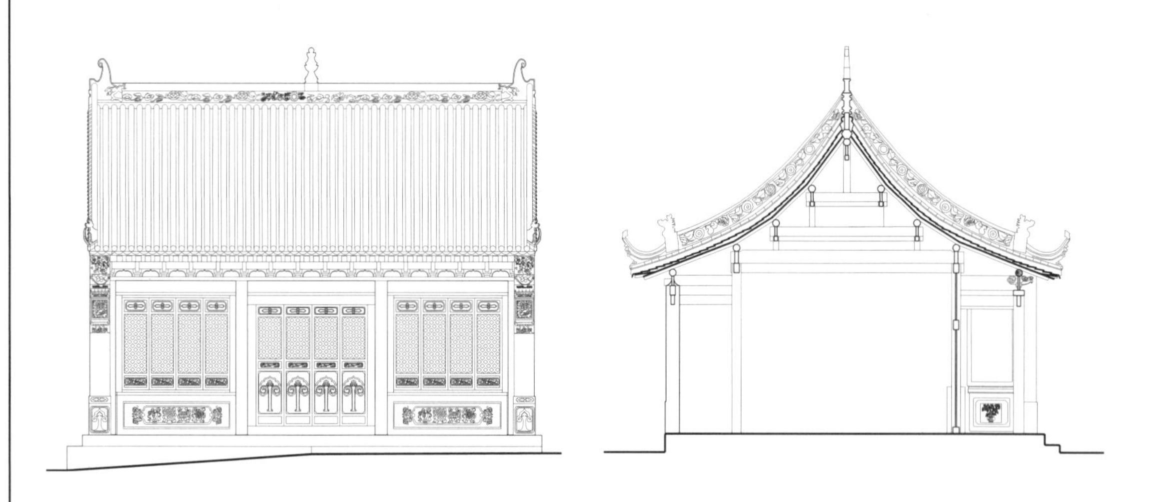

观世音殿立面图（左图）
观世音殿剖面图（右图）

2.10 衙门庙 · 财神殿

财神殿正前方

财神殿室内侧面（左图）
财神殿斜前方（右图）

财神殿室内柱子（左图）
财神殿侧面（右图）

巴丹吉林庙

Badajiiren Temple

3 巴丹吉林庙 Badaijiren Temple

巴丹吉林庙建筑群

巴丹吉林庙为原阿拉善和硕特旗寺庙，系衙门庙11座属庙之一。

关于该庙的创建有两种不同的观点。其一为，乾隆五十六年（1791年）堪布玛尼其喇嘛在巴丹吉林沙漠中新建一座寺庙，供奉其从家乡卫拉特带来的玛沁尼布塔格佛像，后得巴丹吉林庙之名。其二为，1830—1855年，由第二世玛尼其喇嘛罗布桑利格登与一名地方官员及额日博黑庙几名喇嘛在巴丹吉林沙漠忠伊和热之地新建寺庙。最初只搭建几顶蒙古包，后新建一座藏式小殿与两间膳房，因寺庙地处巴丹吉林沙漠深处，故称巴丹吉林庙，寺庙所处地方被更名为苏莫吉林。

寺庙建筑风格为汉藏结合式建筑。至民国3年（1914年），寺庙有青瓦顶汉式大雄宝殿、藏式密宗殿、藏式护法殿、青瓦顶汉式观音殿、法轮殿、3间汉式玛尼其喇嘛拉布隆等5座殿宇1座活佛府，寺庙东南及东侧有2座庙仓，东北侧有1座膳房。大雄宝殿外围有近400平方米的大院。“文化大革命”中寺庙严重受损，大雄宝殿因被用作嘎查仓库，故幸免于难。1992年起，修缮寺庙，正式恢复法会。此庙现为阿拉善地区保存完好的一座寺庙。

参考文献:

[1] 斯·苏雅拉图.神奇的巴丹吉林（蒙古文、汉文）.呼和浩特:内蒙古人民出版社,2011，8.

[2] 勃儿吉斤·道尔格.阿拉善和硕特（上册）（蒙古文）.海拉尔:内蒙古文化出版社,2002，5.

巴丹吉林庙大雄宝殿

巴丹吉林庙·基本概况 1

寺院蒙古语藏语名称	蒙古语	[illegible]	寺院汉语名称	汉语正式名称	巴丹吉林庙
	藏语	དགའ་ལྡན་ཕུན་ཚོགས་རབ་རྒྱས་གླིང་།		俗称	巴丹吉林庙
	汉语语义	巴丹吉林沙漠中的庙	寺院汉语名称的由来		依据所处地方命名

所在地	阿拉善盟阿拉善右旗巴丹吉林沙漠腹地	东经	102° 25′	北纬	39° 48′

初建年	乾隆二十年(1755年)	保护等级	自治区级文物保护单位
盛期时间	不详	盛期喇嘛僧/现有喇嘛僧	近160/2人

历史沿革

1830—1855年间，在巴丹吉林沙漠始建寺庙。
1896年，同治回民起义中寺庙被烧毁,之后搭建蒙古包延续法会。
1903—1906年，该寺部分喇嘛经3年准备工作，备齐建庙材料。
1906年，新建以大雄宝殿为主的部分建筑。
1906—1914年，建全了寺庙所有建筑。
1951年，因地震大雄宝殿受损，后修缮殿堂。
“文化大革命”中寺庙严重受损，仅存大雄宝殿。
1999年，整体修缮大雄宝殿室内装饰。
2007年，整体加固大雄宝殿外墙，更换室内木地板

资料来源

[1] 斯·苏雅拉图.神奇的巴丹吉林（蒙古文、汉文）.呼和浩特:内蒙古人民出版社,2011.8.
[2] 勃儿吉斤·道尔格.阿拉善和硕特（上册）（蒙古文）.海拉尔:内蒙古文化出版社,2002.5.

现状描述	巴丹吉林庙现有建筑皆为历史遗存，大雄宝殿和活佛府均以院落围合，其建筑的结构体系基本完好，汉藏式的建筑风格	描述时间	2010/08/22
		描述人	付瑞峰

调查日期	2010/8/22	调查人员	高旭、李国保、付瑞峰、宝山、王志强

巴林吉林基本概况表1

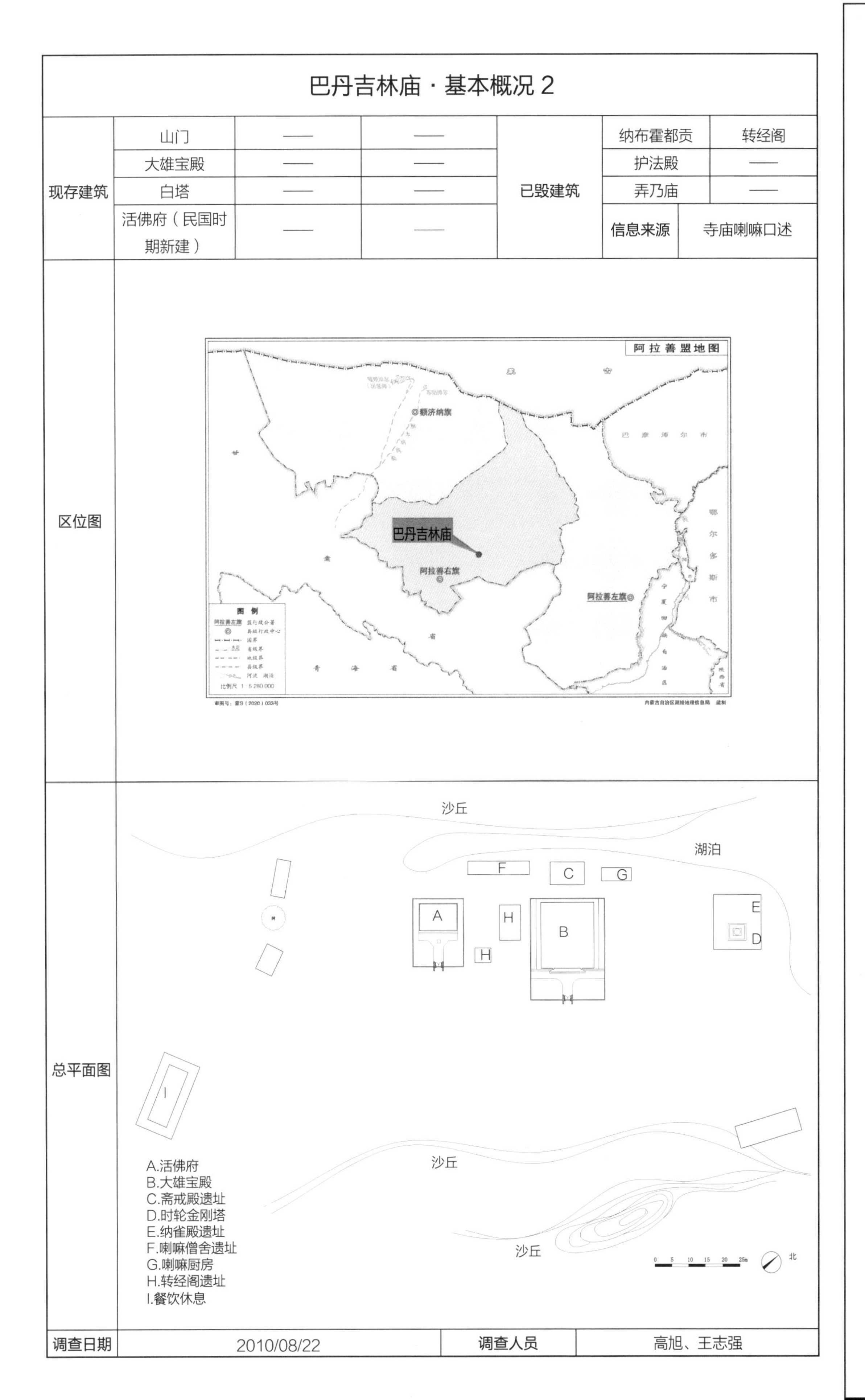

巴丹吉林庙 · 基本概况 2

现存建筑				已毁建筑		
现存建筑	山门	——	——	已毁建筑	纳布霍都贡	转经阁
	大雄宝殿	——	——		护法殿	——
	白塔	——	——		弄乃庙	——
	活佛府（民国时期新建）	——	——		信息来源	寺庙喇嘛口述

区位图

总平面图

调查日期	2010/08/22	调查人员	高旭、王志强

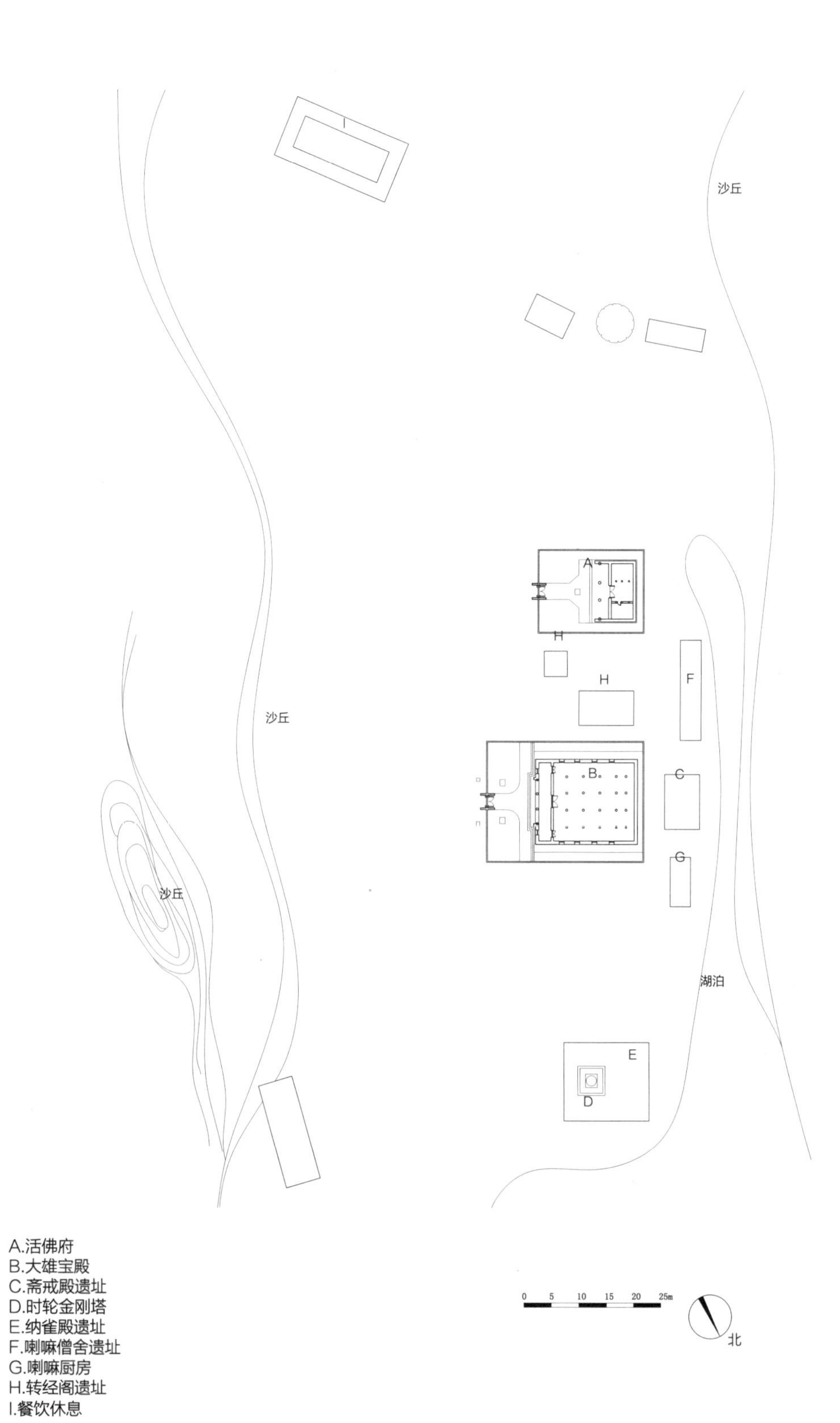
沙丘
沙丘
沙丘
湖泊
A
B
C
D
E
F
G
H
H
I
A.活佛府
B.大雄宝殿
C.斋戒殿遗址
D.时轮金刚塔
E.纳雀殿遗址
F.喇嘛僧舍遗址
G.喇嘛厨房
H.转经阁遗址
I.餐饮休息
0 5 10 15 20 25m
北

巴林吉林总平面图

3.1 巴丹吉林庙·大雄宝殿

单位:毫米

建筑名称	汉语正式名称	——	俗称	大殿													
概述	初建年	乾隆二十年（1755年）	建筑朝向	东偏南约25°	建筑层数	二											
	建筑简要描述	汉藏混合结构体系，藏式装饰风格															
	重建重修记载	1951年因地震重新修缮，1991年室内装修。2007年整体修缮，外墙加固															
		信息来源	寺庙喇嘛口述														
结构规模	结构形式	黏土砖木混合	相连的建筑	无	室内天井	都纲法式											
	建筑平面形式	长方形	外廊形式	前廊													
	通面阔	9110	开间数	5间	明间	3070	次间	2970	梢间	3010	次梢间	——	尽间	——			
	通进深	16920	进深数	7间	进深尺寸（前→后）	2200→2420→3011→2970→3000→1800→1520											
	柱子数量	——	柱子间距	横向尺寸	——	（藏式建筑结构体系填写此栏，不含廊柱）											
				纵向尺寸	——												
	其他	——															
建筑主体（大木作）（石作）（瓦作）	屋顶	屋顶形式	藏式密肋平顶、汉式歇山顶	瓦作	布瓦												
	外墙	主体材料	青砖	材料规格	260×120×60	饰面颜色	白色										
		墙体收分	有	边玛檐墙	有	边玛材料	边玛草										
	斗栱、梁架	斗栱	无	平身科斗口尺寸	——	梁架关系	吊顶不详										
	柱、柱式（前廊柱）	形式	藏式	柱身断面形状	方形	断面尺寸	210×210	（在没有前廊柱的情况下，填写室内柱及其特征）									
		柱身材料	木材	柱身收分	有	栌斗、托木	有	雀替	无								
		柱础	无	柱础形状	——	柱础尺寸	——										
	台基	台基类型	普通式台基	台基高度	470	台基地面铺设材料	石材（500×500）										
	其他	——															
装修（小木作）（彩画）	门(正面)	板门	门楣	有	堆经	有	门帘	有									
	窗（正面）	藏式明窗	窗楣	有	窗套	有	窗帘	无									
	室内隔扇	隔扇	无	隔扇位置	——												
	室内地面、楼面	地面材料及规格	木材（1475×宽不均）	楼面材料及规格	木板，规格不均												
	室外楼梯	楼梯	有	楼梯位置	正门左侧	楼梯材料	木板	梯段宽度	700								
	天花、藻井	天花	有	天花类型	井口天花	藻井	无	藻井类型	——								
	彩画	柱头	有	柱身	无	梁架	有	走马板	无								
		门、窗	有	天花	有	藻井	——	其他彩画	无								
	其他	悬塑	无	佛龛	有	匾额	无										
装饰	室内	帷幔	无	幕帘彩绘	有	壁画	无	唐卡	有								
		经幡	有	经幢	有	柱毯	有	其他	——								
	室外	玛尼轮	无	苏勒德	无	宝顶	有	祥麟法轮	有								
		四角经幢	有	经幡	无	铜饰	无	石刻、砖雕	有								
		仙人走兽	无	壁画	有	其他	——										
陈设	室内	主佛像	释迦牟尼	佛像基座	须弥座												
		法座	有	藏经橱	有	经床	有	诵经桌	有	法鼓	有	玛尼轮	有	坛城	无	其他	—
	室外	旗杆	有	苏勒德	无	狮子	有	经幡	无	玛尼轮	无	香炉	有	五供	无	其他	—
	其他	——															
备注	——																
调查日期	2010/08/22	调查人员	付瑞峰	整理日期	2010/08/24	整理人员	付瑞峰、李国保										

大雄宝殿基本概况表1

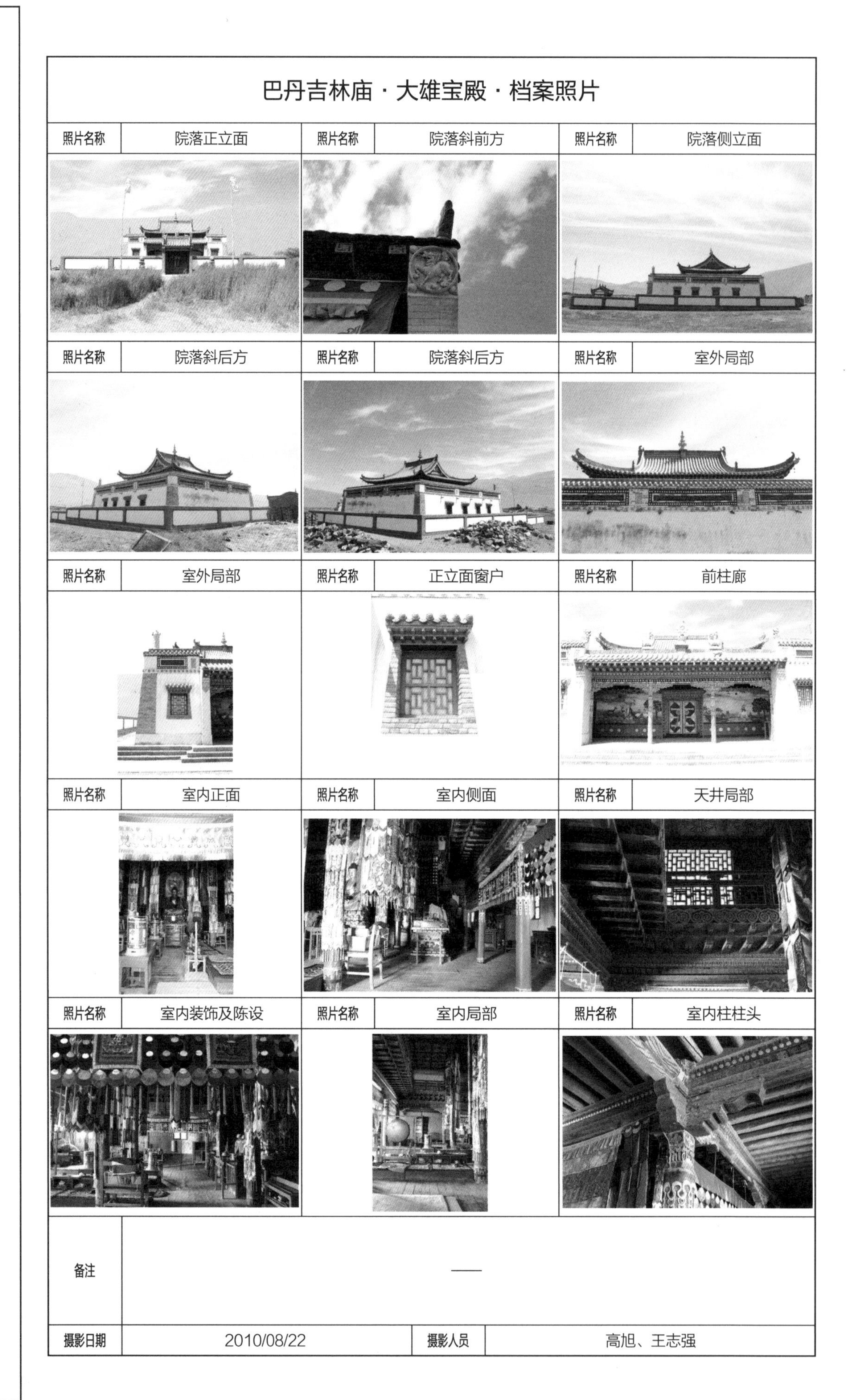

巴丹吉林庙·大雄宝殿·档案照片

照片名称	院落正立面	照片名称	院落斜前方	照片名称	院落侧立面
照片名称	院落斜后方	照片名称	院落斜后方	照片名称	室外局部
照片名称	室外局部	照片名称	正立面窗户	照片名称	前柱廊
照片名称	室内正面	照片名称	室内侧面	照片名称	天井局部
照片名称	室内装饰及陈设	照片名称	室内局部	照片名称	室内柱柱头
备注	——				
摄影日期	2010/08/22	摄影人员	高旭、王志强		

大雄宝殿基本概况表2

大雄宝殿院落正视图

大雄宝殿前廊（左图）
大雄宝殿正面门（右图）

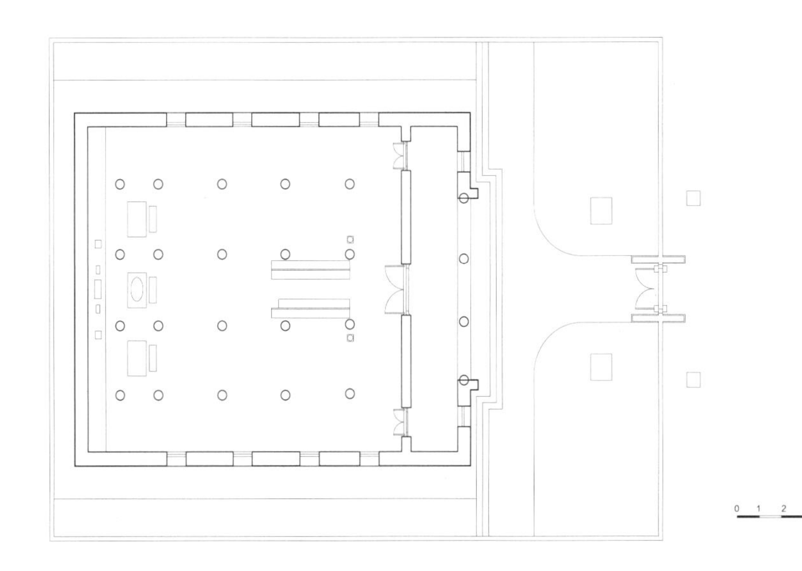

大雄宝殿一层平面图

3.2 巴丹吉林庙·活佛住所

活佛住所院落斜前方

活佛住所前廊（左图）
活佛住所斜前方（右图）

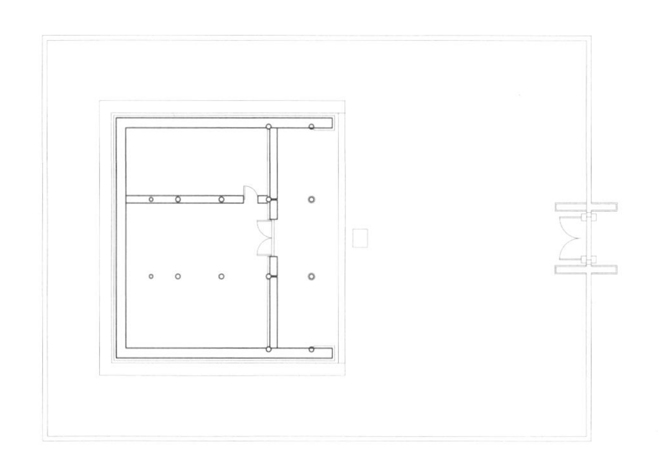

活佛住所平面图

巴彦淖尔市地区

Bayannaoer City

底图来源：内蒙古自治区自然资源厅官网 内蒙古地图
审图号：蒙S（2017）028号

巴彦淖尔市辖乌拉特前旗、乌拉特中旗、乌拉特后旗、杭锦后旗4旗，五原县、磴口县2县及临河市1区。该市前身为巴彦淖尔盟，2004年撤盟设为巴彦淖尔市。巴彦淖尔盟建立于1956年，现辖区由清时乌兰察布盟乌拉特三公旗、阿拉善厄鲁特旗、伊克昭盟部分地区组成。市辖区内曾有60余座藏传佛教寺庙，现存10余座已恢复重建或尚有建筑遗存的寺庙，课题组实地调研9座寺庙。

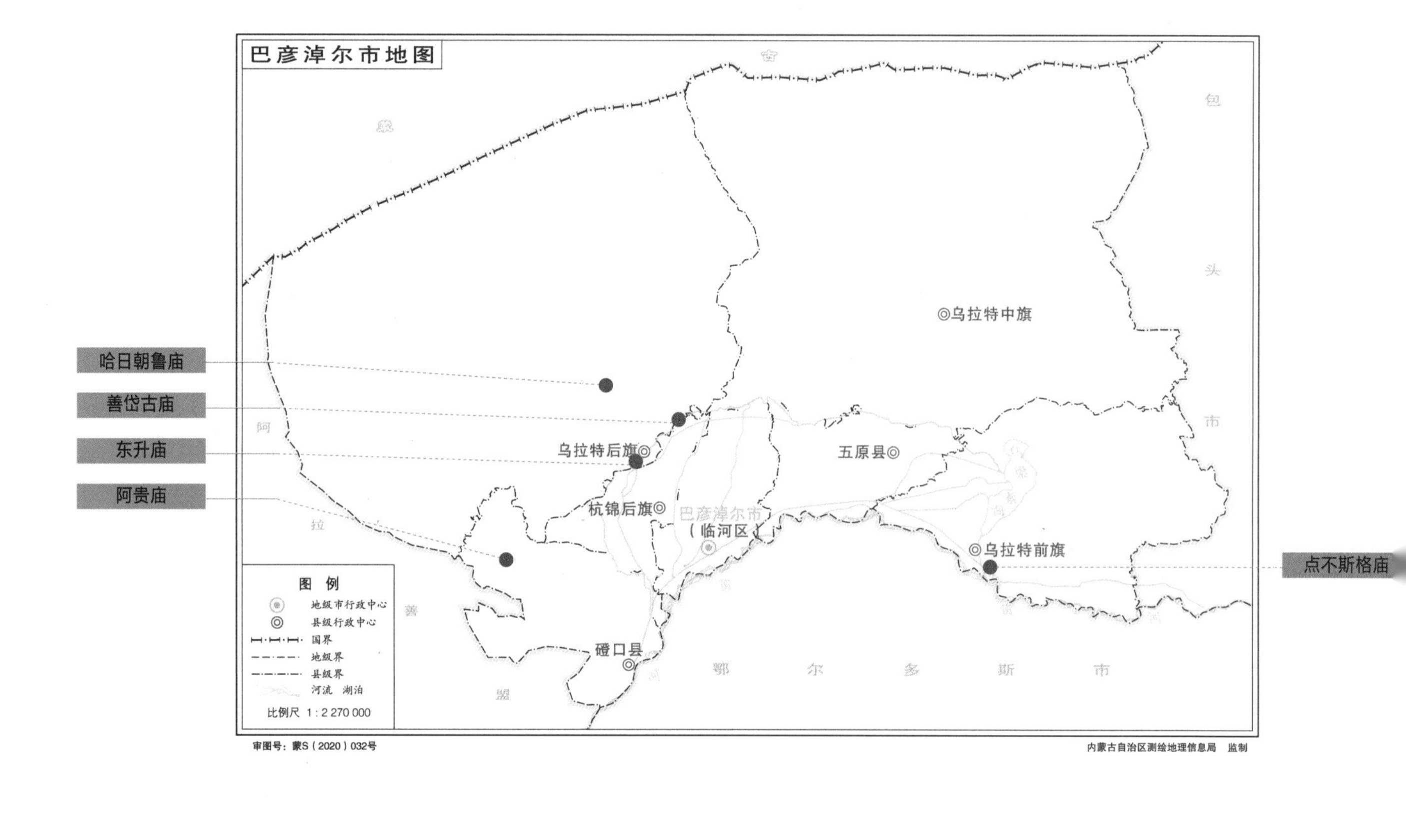
巴彦淖尔市地图
乌拉特中旗
哈日朝鲁庙
善岱古庙
东升庙
阿贵庙
乌拉特后旗
五原县
杭锦后旗
巴彦淖尔市
（临河区）
乌拉特前旗
点不斯格庙
磴口县
图例
地级市行政中心
县级行政中心
国界
地级界
县级界
河流 湖泊
比例尺 1：2 270 000
鄂 尔 多 斯 市
审图号：蒙S（2020）032号
内蒙古自治区测绘地理信息局 监制

点不斯格庙

Debseg Temple

1 点不斯格庙 Debseg Temple

点不斯格庙大雄宝殿

点不斯格庙为原乌兰察布盟乌拉特西公旗寺庙，系该旗哈日诺特氏苏木的寺庙。咸丰五年（1855年），清廷赐满、蒙古、汉、藏四体“寿华寺”匾额。该庙以蒙古语诵经而著称。

据嘉庆十六年（1811年）著《乌拉特寺庙名录》记载，乌拉特西公旗公爷庙（迁址前的梅日更庙）喇嘛罗布桑却日格以自己的财产始建于西拉点布斯格之地。寺庙始建年为康熙四十年（1701年）。

寺庙建筑风格为汉藏结合式建筑。寺庙占地面积2000多平方米，有大雄宝殿等建筑。

在“文化大革命”中，寺庙严重受损，仅存大雄宝殿与一处庙仓。

参考文献:

[1] 乌拉特寺庙名录.内蒙古师范大学图书馆藏古籍（编号 02077）.

[2]（清）葛尔丹旺楚克多尔济.梅日更召创建史（蒙古文）.巴 · 孟和校注.海拉尔:内蒙古文化出版社,1994,4.

[3] 倪玉明.图说巴彦诺尔（汉文）.呼和浩特:远方出版社,2007,12.

点不斯格庙建筑群

<table>
<tr><th colspan="8">点不斯格庙 · 基本概况 1</th></tr>
<tr><td rowspan="3">寺院蒙古语藏语名称</td><td>蒙古语</td><td colspan="2">[illegible]</td><td rowspan="2">寺院汉语名称</td><td>汉语正式名称</td><td colspan="2">寿华寺</td></tr>
<tr><td>藏语</td><td colspan="2">——</td><td>俗称</td><td colspan="2">点不斯格庙</td></tr>
<tr><td>汉语语义</td><td colspan="2">高岗地的寺庙</td><td colspan="2">寺院汉语名称的由来</td><td colspan="2">咸丰五年(1855年)
清廷赐名“寿华寺”</td></tr>
<tr><td>所在地</td><td colspan="3">内蒙古自治区巴彦淖尔盟乌拉特前旗东北山脉中20公里处</td><td>东经</td><td>108° 51′</td><td>北纬</td><td>40° 41′</td></tr>
<tr><td>初建年</td><td colspan="2">康熙四十年（1701年）</td><td>保护等级</td><td colspan="4">——</td></tr>
<tr><td>盛期时间</td><td colspan="2">不详</td><td colspan="2">盛期喇嘛僧/现有喇嘛僧数</td><td colspan="3">300余人/34人</td></tr>
<tr><td rowspan="2">历史沿革</td><td colspan="7">1701年，始建寺庙。
“文化大革命”中严重受损，仅存一座大雄宝殿与一处庙仓</td></tr>
<tr><td>资料来源</td><td colspan="6">[1] 乌拉特寺庙名录.内蒙古师范大学图书馆藏古籍（编号 02077）.
[2]（清）噶尔丹旺楚克多尔济，梅日更召创建史（蒙古文）.巴 · 孟和校注.海拉尔:内蒙古文化出版社,1994，4.
[3] 倪玉明.图说巴彦诺尔（汉文）.呼和浩特:远方出版社,2007，12.</td></tr>
<tr><td rowspan="2">现状描述</td><td colspan="4" rowspan="2">寺庙依山而建，现仅存殿宇一座和庙仓一处。殿宇建筑风格为汉藏结合，屋顶形式复杂，前柱廊装饰精美</td><td>描述时间</td><td colspan="2">2010/10/31</td></tr>
<tr><td>描述人</td><td colspan="2">付瑞峰</td></tr>
<tr><td>调查日期</td><td colspan="2">2010/10/31</td><td>调查人员</td><td colspan="4">李国保、宝山、乔恩懋、付瑞峰</td></tr>
</table>

点不斯格庙基本概况表1

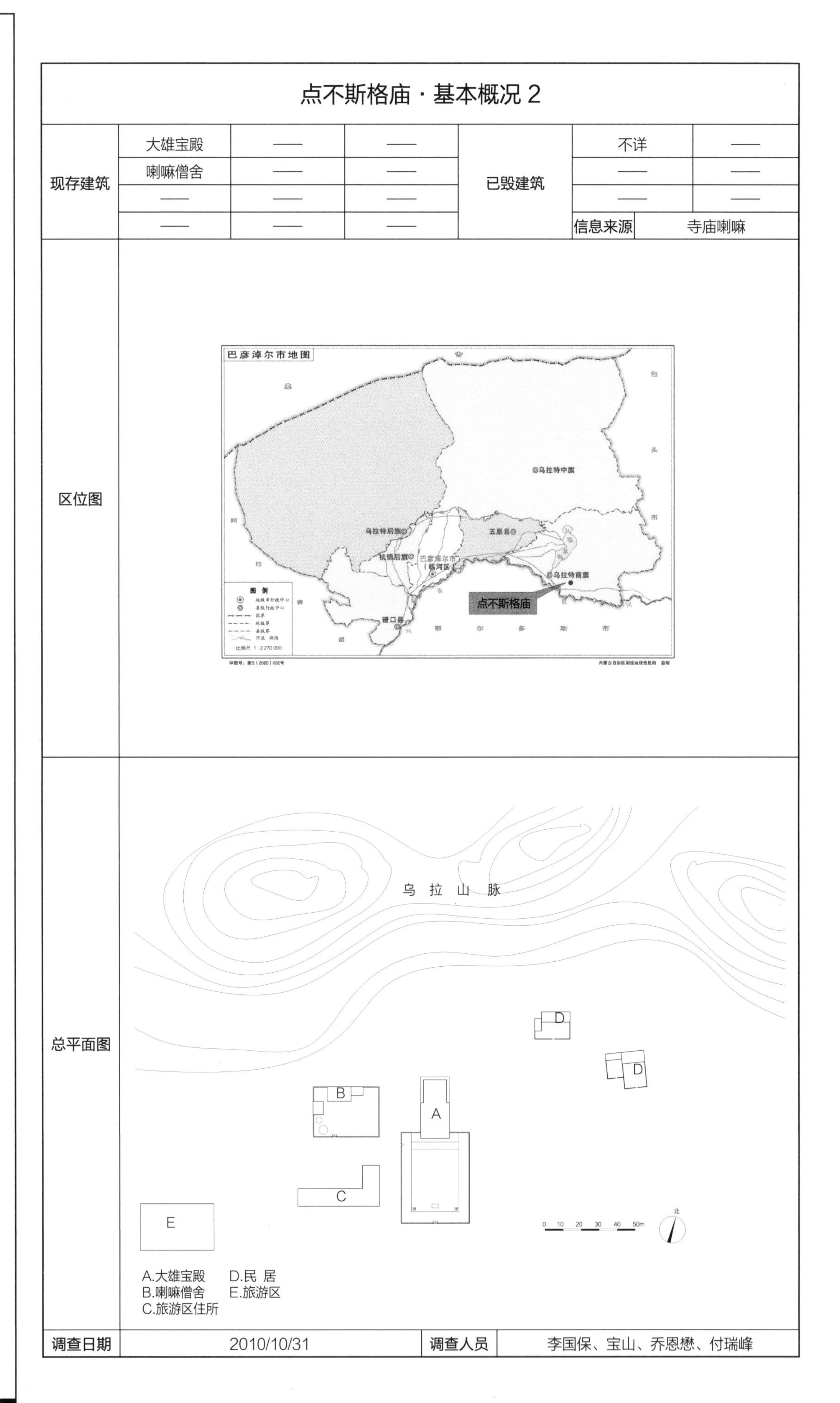

点不斯格庙·基本概况 2

现存建筑	大雄宝殿	——	——	已毁建筑	不详	——
	喇嘛僧舍	——	——		——	——
	——	——	——		——	——
	——	——	——		信息来源	寺庙喇嘛
区位图						
总平面图						
调查日期	2010/10/31		调查人员	李国保、宝山、乔恩懋、付瑞峰		

点不斯格庙基本概况表2

1.1 点不斯格庙 · 大雄宝殿

大雄宝殿斜前方

大雄宝殿前廊（左图）

大雄宝殿正前方（右上图）

大雄宝殿二层檐部（右下图）

1.2 点不斯格庙 · 喇嘛住所

喇嘛住所院落斜前方（左图）

喇嘛住所斜后方（右图）

2 阿贵庙 Agui Temple

2 阿贵庙 Agui Temple

阿贵庙建筑群

阿贵庙为原阿拉善和硕特旗寺庙，系阿拉善八大寺庙之一，衙门庙11座属庙之一。民国元年（1912年），民国政府蒙藏事务局赐予蒙古、汉、满、藏四体“宗乘寺”匾额。该庙曾有红塔寺等几座属庙。阿贵庙为内蒙古地区现存唯一的一座按照宁玛派仪轨活动的喇嘛教红教派寺庙。

据传说，印度佛教密宗大师巴达玛桑布巴，蒙古语称为罗本钦布，汉语称为莲花生，于公元774年来到现阿贵庙所在地，并在此修法9个多月。当时此地有五个山洞，并有一母所生的5个仙女（杭瑞玛）住在其中。故此，莲花生为山洞起名“亚吉拉杭瑞玛阿贵”。嘉庆三年（1798年），喀尔喀梅日更王旗查干乌拉庙五世道格新诺颜呼图克图阿格旺罗布桑丹赞若布萨勒到阿拉善旗巡锡弘法，向阿拉善扎萨克旺钦班巴日申请一处修行之地。经准许起初在贡呼洞之地搭建蒙古包修行，后在南寺两位呼图克图的建议下迁至哈如纳阿贵（一说杭瑞玛阿贵之变音）之地，新建寺庙，阿拉善旗衙赐名“达西日布甘丹令”，习称罗本钦布阿贵或阿贵庙。

寺庙建筑风格为藏式建筑。至清末时该寺仍有大雄宝殿、扎仓殿等2座经殿共64间、庙仓12座72间、僧舍1000余间。该庙有五大山洞，或曰石窟，分别为莲花生洞、上乐金刚洞、亥母洞、吉祥天女洞、珈蓝神洞，其中莲花生洞最大，长约25米，宽约15米，高4米，内供莲花生佛像。

在“文化大革命”中，寺庙严重受损。20世纪80年代开始修缮、复建寺庙建筑，经20余年的建设，已建全大雄宝殿、金刚亥母殿等5座主要殿堂与1座活佛拉布隆的中等规模的寺庙。

参考文献:

[1] 苏瓦迪.莲花生洞（蒙古文）.北京:民族出版社,2008，4.
[2] 政协内蒙古自治区委员会文史资料委员会.内蒙古喇嘛教纪例（第四十五辑）,1997，1.

阿贵庙 · 基本概况 1

<table>
<tr><td rowspan="3">寺院蒙古语藏语名称</td><td>蒙古语</td><td>[illegible]</td><td rowspan="2">寺院汉语名称</td><td>汉语正式名称</td><td colspan="3">宗乘寺</td></tr>
<tr><td>藏语</td><td>དམ་པ་དར་རྒྱས་གླིང་།</td><td>俗称</td><td colspan="3">阿贵庙</td></tr>
<tr><td>汉语语义</td><td>洞庙</td><td colspan="2">寺院汉语名称的由来</td><td colspan="3">清廷赐名；另一说为1912年民国政府赐匾（参考《莲花生洞》）</td></tr>
<tr><td>所在地</td><td colspan="4">巴彦淖尔盟磴口县沙金套海苏木境内</td><td>东经 106° 21′</td><td colspan="2">北纬 40° 43′</td></tr>
<tr><td>初建年</td><td colspan="2">嘉庆八年（1803年）</td><td>保护等级</td><td colspan="4">自治区级文物保护单位</td></tr>
<tr><td>盛期时间</td><td colspan="2">1986年</td><td colspan="2">盛期喇嘛僧/现有喇嘛僧数</td><td colspan="3">近400人/27人</td></tr>
<tr><td rowspan="2">历史沿革</td><td colspan="7">1798年，阿格旺罗布桑丹赞若布萨勒新建寺庙，同年回喀尔喀部。
1801年，阿格旺罗布桑丹赞若布萨勒从喀尔喀蒙古回本寺视察寺庙修建工程时带来藏语甘珠尔经1部、5尺高莲花生像及一些查玛用具。
“文化大革命”时寺庙严重受损，仅存5个天然山洞，大雄宝殿仅存框架。
20世纪80年代起，修缮寺庙建筑。
1982年，修缮大雄宝殿，并逐步修缮其他殿宇。
1990年，迪鲁瓦呼图克图驾临该寺，并出资修建殿堂、佛塔及庙仓。
1998年，修缮时轮殿、金刚亥母殿、护法殿。
2006年，新建多闻天王殿，修缮大雄宝殿。
2008—2009年，维修莲花洞台阶，并加建两个凉亭。
2010年，修缮观音殿、亥母殿、白马金刚殿</td></tr>
<tr><td>资料来源</td><td colspan="6">［1］苏瓦迪.莲花生洞（蒙古文）.北京:民族出版社,2008.4.
［2］政协内蒙古自治区委员会文史资料委员会.内蒙古喇嘛教纪例（第四十五辑）.1997，1.
［3］调研访谈记录.</td></tr>
<tr><td rowspan="2">现状描述</td><td colspan="4" rowspan="2">“文化大革命”期间寺庙受损严重，仅存5个天然洞和大雄宝殿框架。20世纪80年代开始重建，现存殿宇5座，天然岩洞5个，殿宇分别是汉藏结合的结构体系和纯藏式结构体系</td><td>描述时间</td><td colspan="2">2010/08/27</td></tr>
<tr><td>描述人</td><td colspan="2">付瑞峰</td></tr>
<tr><td>调查日期</td><td colspan="2">2010/08/27</td><td>调查人员</td><td colspan="4">李国保、高旭、宝山、王志强、付瑞峰</td></tr>
</table>

阿贵庙基本概况表1

阿贵庙·基本概况 2

现存建筑	时轮金刚殿	金刚亥母殿	扎嘎生布窑	已毁建筑	看门洞	——
	大雄宝殿	寺庙管理用房	达日额柯窑		修行洞	——
	护法殿	喇嘛僧舍	额尔登珠窑		观音洞	——
	财神殿	阿贵洞	桑布嘎日布窑		信息来源	寺庙大喇嘛口述

区位图

总平面图

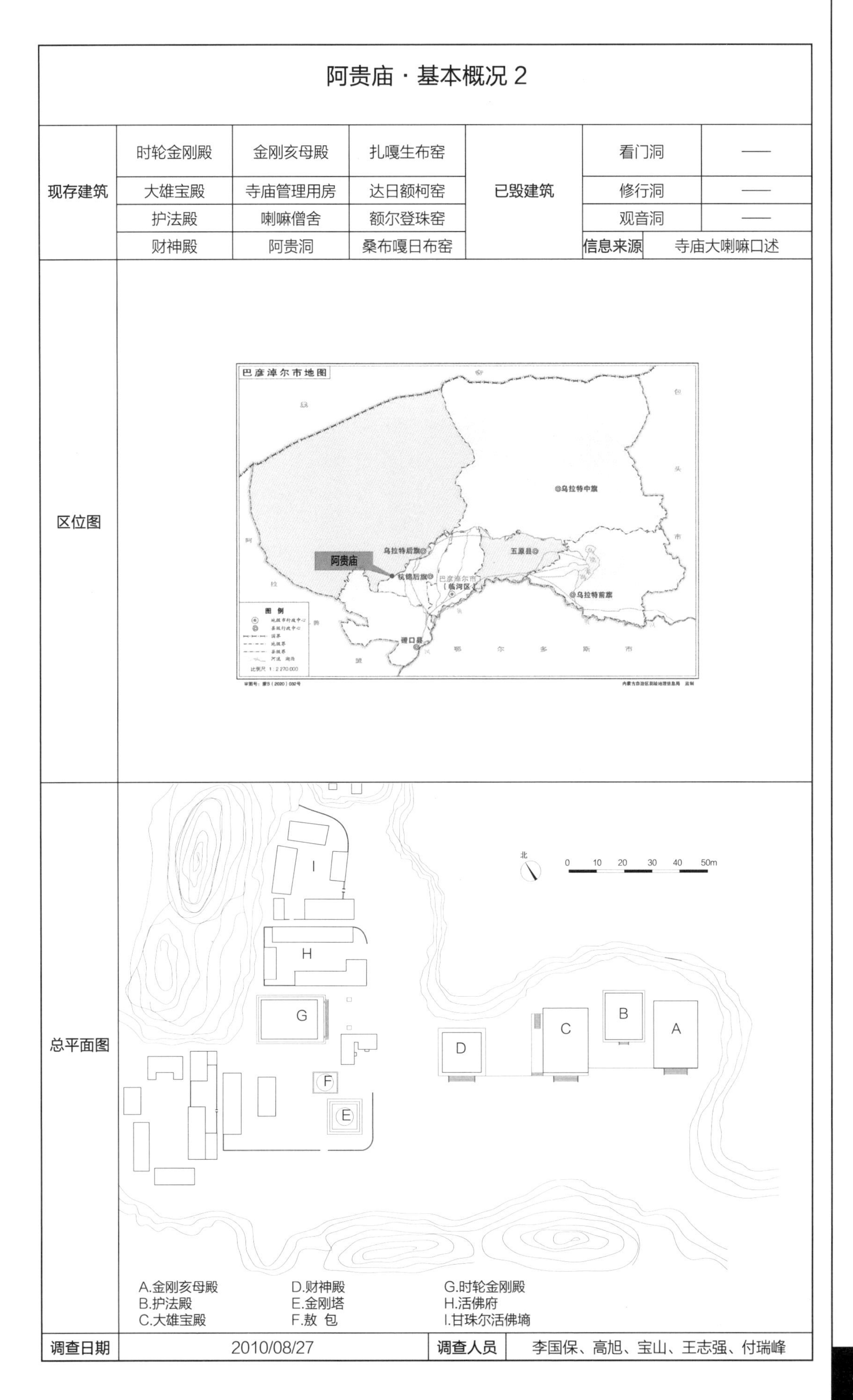

A.金刚亥母殿
B.护法殿
C.大雄宝殿
D.财神殿
E.金刚塔
F.敖 包
G.时轮金刚殿
H.活佛府
I.甘珠尔活佛墒

调查日期	2010/08/27	调查人员	李国保、高旭、宝山、王志强、付瑞峰

阿贵庙基本概况表2

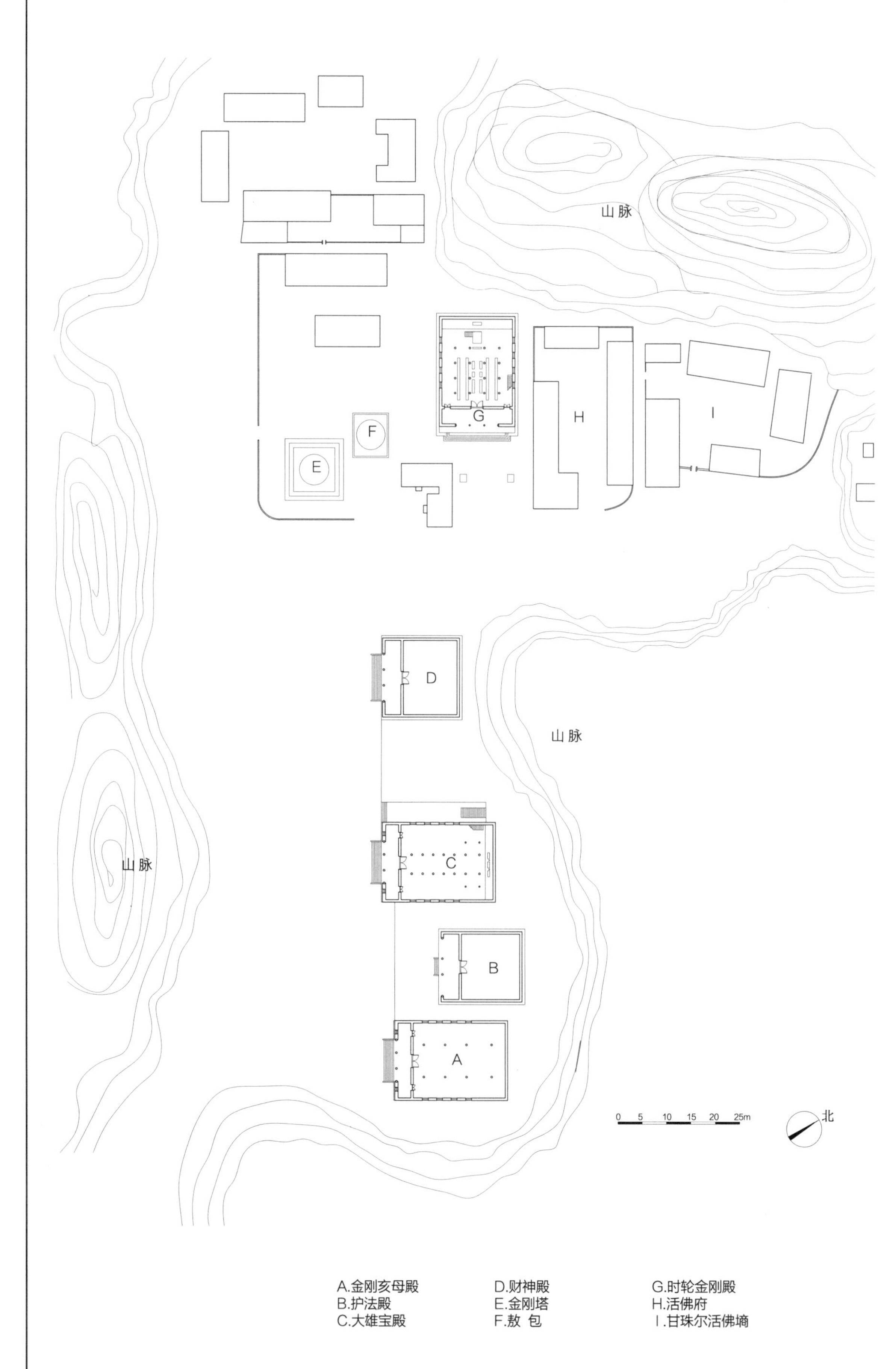

阿贵庙总平面图

A.金刚亥母殿
B.护法殿
C.大雄宝殿
D.财神殿
E.金刚塔
F.敖 包
G.时轮金刚殿
H.活佛府
I.甘珠尔活佛塥

2.1 阿贵庙 · 大雄宝殿

单位:毫米

建筑名称	汉语正式名称	会功殿	俗称	大雄宝殿、朝格钦独贡													
概述	初建年	嘉庆三年（1798年）	建筑朝向	南偏西约27°	建筑层数	二											
	建筑简要描述	藏式结构体系，藏式装饰风格															
	重建重修记载	“文化大革命”期间被毁，仅存框架，1982年修缮，2006年再次修缮，2008年室内修缮															
		信息来源	寺庙喇嘛口述														
结构规模	结构形式	混凝土结构	相连的建筑	无	室内天井	佛殿都纲法式											
	建筑平面形式	长方形	外廊形式	前廊													
	通面阔	15400	开间数	3间	明间	——	次间	——	梢间	——	次梢间	——	尽间	——			
	通进深	22530	进深数	9间	进深尺寸（前→后）	——											
	柱子数量	18	柱子间距	横向尺寸	2540、3150	（藏式建筑结构体系填写此栏，不含廊柱）											
				纵向尺寸	2850、2200、2250、2300												
	其他	——															
建筑主体（大木作）（石作）（瓦作）	屋顶	屋顶形式	藏式密肋平屋顶	瓦作	——												
	外墙	主体材料	石材	材料规格	规格不均	饰面颜色	白色										
		墙体收分	有	边玛檐墙	有	边玛材料	红色涂料粉刷										
	斗栱、梁架	斗栱	无	平身科斗口尺寸	——	梁架关系	不详										
	柱、柱式（前廊柱）	形式	汉式	柱身断面形状	圆形	断面尺寸	周长C=1260	（在没有前廊柱的情况下，填写室内柱及其特征）									
		柱身材料	木材	柱身收分	有	栌斗、托木	无	雀替	有								
		柱础	无	柱础形状	——	柱础尺寸	——										
	台基	台基类型	普通式台基	台基高度	1500	台基地面铺设材料	青砖240×120										
	其他	——															
装修（小木作）（彩画）	门(正面)	板门	门楣	无	堆经	有	门帘	无									
	窗（正面）	藏式明窗	窗楣	有	窗套	有	窗帘	有									
	室内隔扇	隔扇	无	隔扇位置	——												
	室内地面、楼面	地面材料及规格	青砖240×120	楼面材料及规格	木板，规格不均												
	室内楼梯	楼梯	有	楼梯位置	佛殿左右两侧	楼梯材料	木材	楼段宽度	930								
	天花、藻井	天花	有	天花类型	海漫天花	藻井	无	藻井类型	——								
	彩画	柱头	无	柱身	有	梁架	有	走马板	无								
		门、窗	有	天花	有	藻井	——	其他彩画	——								
	其他	悬塑	无	佛龛	有	匾额	无										
装饰	室内	帷幔	无	幕帘彩绘	有	壁画	无	唐卡	有								
		经幡	有	经幢	有	柱毯	有	其他	——								
	室外	玛尼轮	无	苏勒德	无	宝顶	有	祥麟法轮	有								
		四角经幢	有	经幡	无	铜饰	有	石刻、砖雕	有								
		仙人走兽	无	壁画	有	其他	——										
陈设	室内	主佛像	莲花生大师	佛像基座	莲花座												
		法座	有	藏经橱	有	经床	有	诵经桌	有	法鼓	有	玛尼轮	无	坛城	无	其他	——
	室外	旗杆	无	苏勒德	无	狮子	无	经幡	无	玛尼轮	无	香炉	有	五供	无	其他	——
	其他	——															
备注	——																
调查日期	2010/08/27	调查人员	付瑞峰	整理日期	2010/08/27	整理人员	李国保、付瑞峰										

大雄宝殿基本概况表1

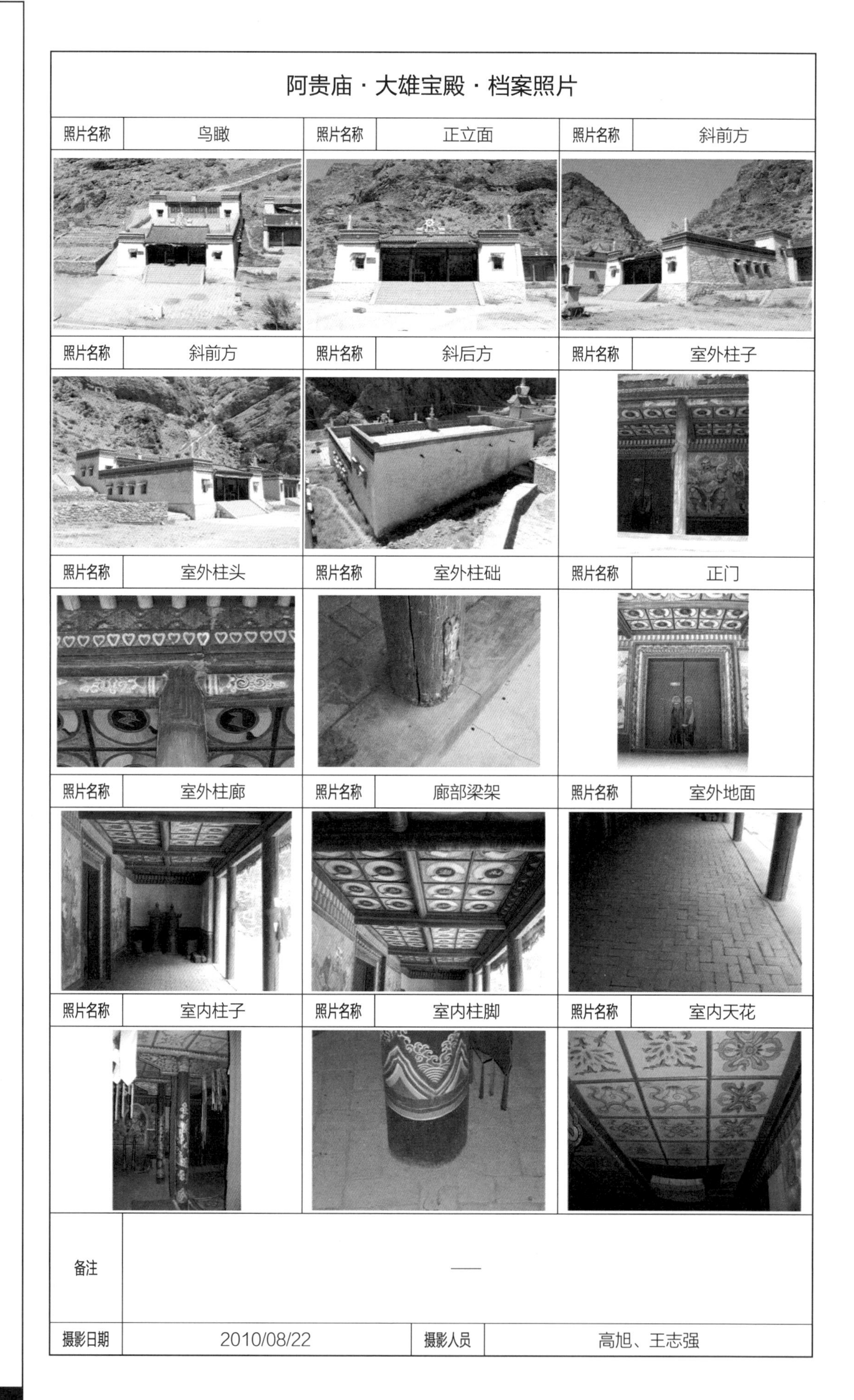

阿贵庙·大雄宝殿·档案照片

照片名称	鸟瞰	照片名称	正立面	照片名称	斜前方
照片名称	斜前方	照片名称	斜后方	照片名称	室外柱子
照片名称	室外柱头	照片名称	室外柱础	照片名称	正门
照片名称	室外柱廊	照片名称	廊部梁架	照片名称	室外地面
照片名称	室内柱子	照片名称	室内柱脚	照片名称	室内天花
备注	—				
摄影日期	2010/08/22	摄影人员	高旭、王志强		

大雄宝殿基本概况表2

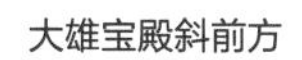

大雄宝殿斜前方

大雄宝殿正前方

大雄宝殿室外柱头（左图）

大雄宝殿斜后方（右图）

2.2 阿贵庙·护法殿

护法殿正前方1（左图）

护法殿正前方2（左图）
护法殿室外柱子（右图）

护法殿背面（左图）
护法殿门（右图）

2.3 阿贵庙 · 金刚亥母殿

金刚亥母殿正前方（左上图）
金刚亥母殿室外柱头（左下图）
金刚亥母殿室外柱头（右图）

2.4 阿贵庙 · 财神殿

财神殿鸟瞰图（左图）
财神殿（右上图）
财神殿斜后方（右下图）

2.5 阿贵庙 · 时轮金刚殿

时轮金刚殿正前方（左上图）
时轮金刚殿室内正面（左下图）
时轮金刚殿鸟瞰图（右图）

2.6 阿贵庙 · 其他建筑

阿贵洞（左上图）
阿贵洞门（左下图）
白塔（右图）

善岱古庙

Shaletew Temple

3 善岱古庙 Shaletew Temple

善岱古庙建筑群

善岱古庙为原伊克昭盟鄂尔多斯左翼后旗（俗称达拉特旗）寺庙，系达拉特旗境内布勒吉庙建在乌兰察布盟乌拉特西公旗的一座寺庙，为西公旗境内修建最早的一座寺庙。康熙五十九年（1720年），清廷御赐满、蒙古、汉、藏四体“咸化寺”匾额。寺庙管辖戈壁庙、舍贵腾庙、阿贵庙等属庙。

寺庙始建于康熙十五年（1676年），原址位于巴音敖包之地（原临河县巴音敖包乡），后因洪水的影响，不久迁至夏日占布拉圪台（今乌拉特后旗乌盖苏木所在地）重建殿宇。关于寺庙始建年代另有三种说法，分别为1723年、1738年及1902年。依据清廷赐匾年代推断，寺庙至少始建于1720年前，而寺庙迁址与扩建时期应在雍正与乾隆年间。寺庙位于阴山南麓，依据所处地名，俗称萨拉达巴庙。寺庙建成后屡遭兵灾匪患及火灾，几次被毁，几次重建。

寺庙建筑风格以藏式建筑为主，兼有汉式建筑。该庙占地面积为2平方公里，建筑面积达1000平方米。至1958年仍有80间双层大雄宝殿、60间单层大雄宝殿、双层千神殿、经学殿、劳林殿、乌兰拉布隆等14座殿堂及西仓、庙仓、教书院、医药房、查玛用具房、凉亭等多处建筑。寺庙有僧舍120余间，共有108塔，其中有特大佛塔3座。

“文化大革命”中寺庙严重受损，仅存大雄宝殿、东配殿、西配殿、东厢院、西厢院等大小五座庙宇。1980年起开始修缮、重建寺庙，恢复法事。现已成为乌拉特后旗全旗喇嘛教活动中心。该寺为乌拉特后旗乃至巴彦淖尔市保存最完整的藏传佛教寺庙。

参考文献:

[1] 巴图苏和.关于善岱古庙文史情况的调研报告.乌拉特后旗政协文件,2012，2.

[2] 乔吉.内蒙古寺庙（汉文）.呼和浩特:内蒙古人民出版社,1994，8.

[3] 倪玉明.图说巴彦诺尔（汉文）.呼和浩特:远方出版社,2007，12.

[4] 调研记录——寺庙碑文记载,2010.

善岱古庙大雄宝殿（左图）
善岱古庙密宗殿（右图）

善岱古庙·基本概况 1

<table>
<tr><td rowspan="3">寺院蒙古语藏语名称</td><td>蒙古语</td><td>[illegible]</td><td rowspan="2">寺院汉语名称</td><td>汉语正式名称</td><td>咸化寺</td></tr>
<tr><td>藏语</td><td>ཀུན་བདེ་གླིང་།</td><td>俗称</td><td>善岱古庙、乌盖庙</td></tr>
<tr><td>汉语语义</td><td>——</td><td colspan="2">寺院汉语名称的由来</td><td>清廷赐名</td></tr>
<tr><td>所在地</td><td colspan="4">巴彦淖尔盟乌拉特后旗</td><td>东经 107° 04′ 北纬 41° 06′</td></tr>
<tr><td>初建年</td><td colspan="2">雍正元年（1723年）</td><td>保护等级</td><td colspan="2">自治区级文物保护单位</td></tr>
<tr><td>盛期时间</td><td colspan="2">清中晚期</td><td colspan="2">盛期喇嘛僧/现有喇嘛僧数</td><td>不详/14人</td></tr>
<tr><td rowspan="2">历史沿革</td><td colspan="5">1677年，新建寺庙。
1895年，重修大雄宝殿。
“文化大革命”中寺庙严重受损，仅存3座殿宇和一座活佛府。
1980年，修葺部分殿宇。
2003年，修缮乌兰拉布隆殿</td></tr>
<tr><td>资料来源</td><td colspan="4">[1] 倪玉明.图说巴彦诺尔（汉文）.呼和浩特：远方出版社, 2007，12.
[2] 乔吉.内蒙古寺庙.呼和浩特：内蒙古人民出版社,1994，8.
[3] 寺庙碑文记载.
[4] 调研访谈记录.</td></tr>
<tr><td rowspan="2">现状描述</td><td colspan="3" rowspan="2">寺庙依山而建，整体布局延轴线对称，并随山势级级递升。现存五座殿皆为历史遗存，均属纯藏式平屋顶建筑</td><td>描述时间</td><td>2010/10/29</td></tr>
<tr><td>描述人</td><td>付瑞峰</td></tr>
<tr><td>调查日期</td><td colspan="2">2010/10/29</td><td>调查人员</td><td colspan="2">李国保、宝山、乔恩懋、付瑞峰</td></tr>
</table>

善岱古庙基本概况表1

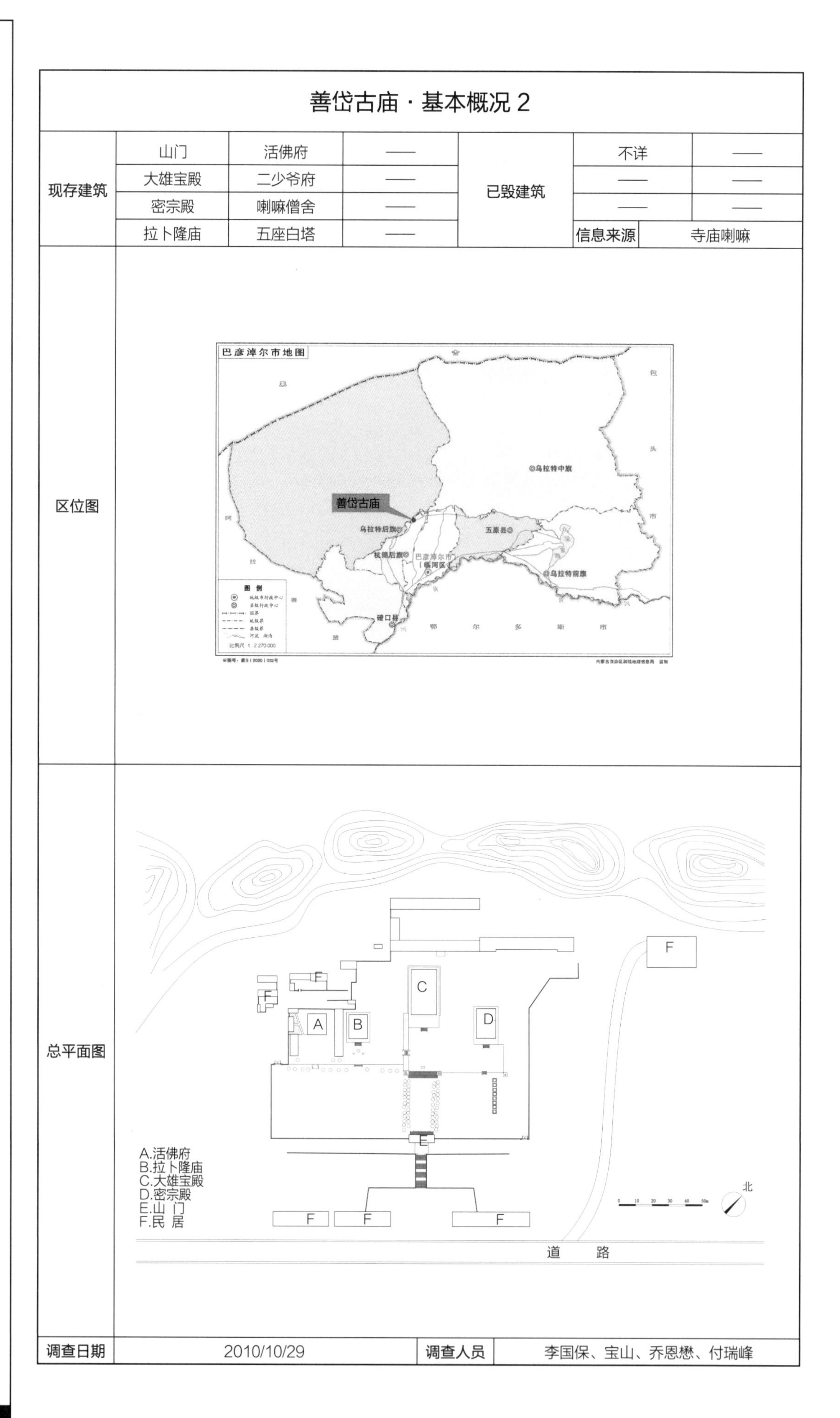

善岱古庙 · 基本概况 2

现存建筑				已毁建筑		
现存建筑	山门	活佛府	——	已毁建筑	不详	——
	大雄宝殿	二少爷府	——		——	——
	密宗殿	喇嘛僧舍	——		——	——
	拉卜隆庙	五座白塔	——		信息来源	寺庙喇嘛
区位图						
总平面图						
调查日期	2010/10/29			调查人员	李国保、宝山、乔恩懋、付瑞峰	

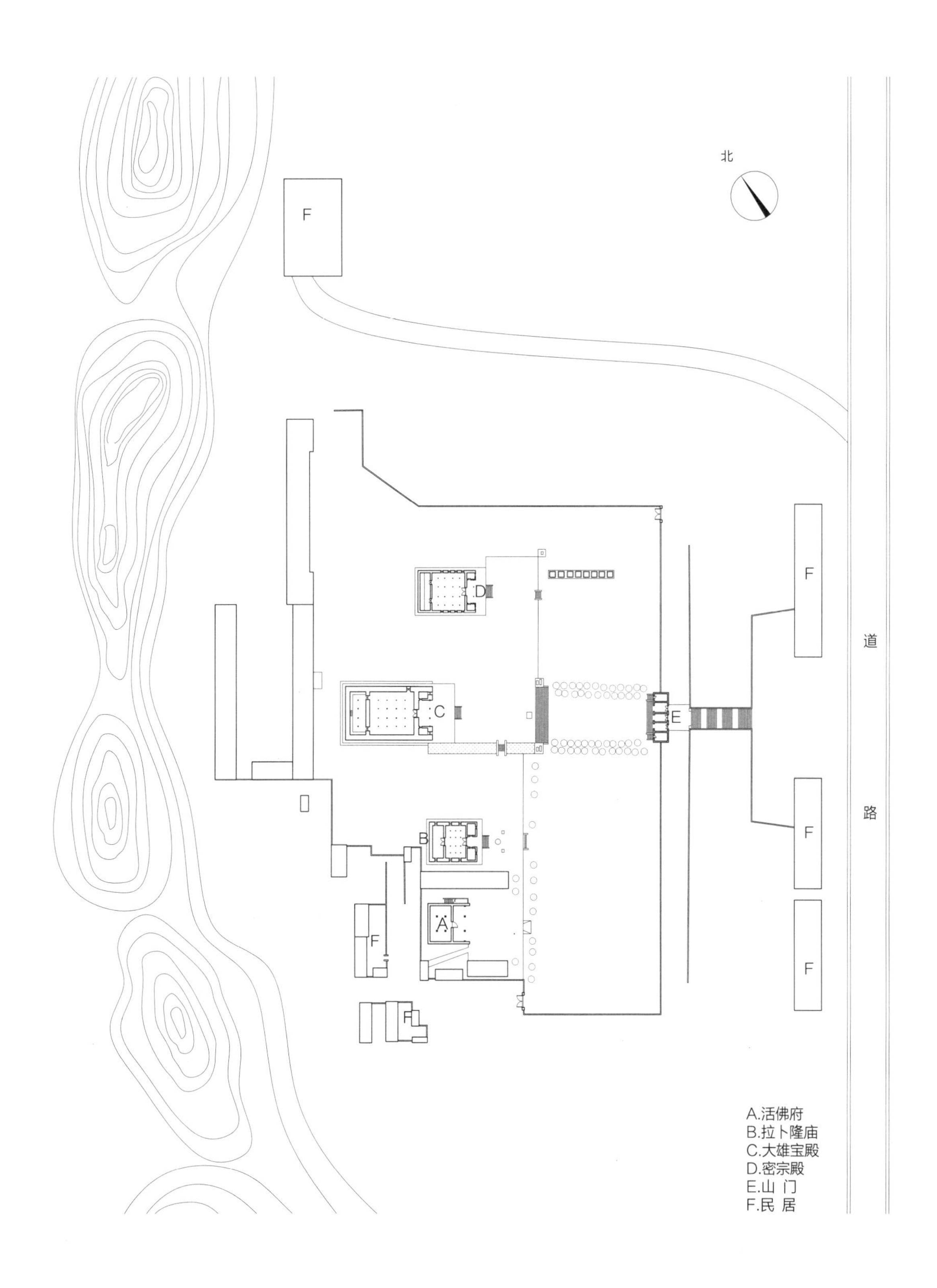

善岱古庙总平面图

3.1 善岱古庙・大雄宝殿

单位:毫米

建筑名称	汉语正式名称	大雄宝殿	俗称	朝格钦													
概述	初建年	不详	建筑朝向	南偏东约45°	建筑层数	三											
	建筑简要描述	纯藏式砖木混合结构体系，汉藏结合装饰风格															
	重建重修记载	1980年政府出资修缮，1997年外表面粉刷装修															
		信息来源	寺庙喇嘛（噶拉桑旺其格，75岁）口述														
结构规模	结构形式	砖木混合	相连的建筑	无	室内天井	都纲法式											
	建筑平面形式	长方形	外廊形式	前廊													
	通面阔	14710	开间数	5间	明间	——	次间	——	梢间	——	次梢间	——	尽间	——			
	通进深	25690	进深数	8间	进深尺寸（前→后）												
	柱子数量	经堂16，佛殿2	柱子间距	横向尺寸	2850、2570	（藏式建筑结构体系填写此栏，不含廊柱）											
				纵向尺寸	2900、2710												
	其他	——															
建筑主体（大木作）（石作）（瓦作）	屋顶	屋顶形式	藏式密肋平屋顶	瓦作	——												
	外墙	主体材料	青砖	材料规格	290×130×60	饰面颜色	白色										
		墙体收分	有	边玛檐墙	有	边玛材料	青砖，红色涂料										
	斗栱、梁架	斗栱	无	平身科斗口尺寸	——	梁架关系	16柱28梁（经堂）										
	柱、柱式（前廊柱）	形式	藏式	柱身断面形状	小八棱柱	断面尺寸	195×40×130	（在没有前廊柱的情况下，填写室内柱及其特征）									
		柱身材料	木材	柱身收分	有	栌斗、托木	有	雀替	无								
		柱础	无	柱础形状	——	柱础尺寸	——										
	台基	台基类型	普通台基	台基高度	900	台基地面铺设材料	方砖（300×300）										
	其他	——															
装修（小木作）（彩画）	门(正面)	板门	门楣	无	堆经	有	门帘	无									
	窗（正面）	藏式明窗	窗楣	有	窗套	有	窗帘	有									
	室内隔扇	隔扇	有	隔扇位置	经堂二楼室内外空间分割												
	室内地面、楼面	地面材料及规格	水泥砂浆抹平	楼面材料及规格	土坯抹平												
	室内楼梯	楼梯	有	楼梯位置	柱廊右侧暗厢中	楼梯材料	木材	楼段宽度	800								
	天花、藻井	天花	无	天花类型	——	藻井	无	藻井类型	——								
	彩画	柱头	有	柱身	无	梁架	有	走马板	无								
		门、窗	有	天花	——	藻井	——	其他彩画	无								
	其他	悬塑	有	佛龛	有	匾额	无										
装饰	室内	帷幔	有	幕帘彩绘	有	壁画	有	唐卡	有								
		经幡	有	经幢	有	柱毯	无	其他	——								
	室外	玛尼轮	有	苏勒德	无	宝顶	有	祥麟法轮	有								
		四角经幢	有	经幡	无	铜饰	有	石刻、砖雕	有								
		仙人走兽	无	壁画	有	其他	无										
陈设	室内	主佛像	释迦牟尼	佛像基座	普通台基												
		法座	有	藏经橱	有	经床	有	诵经桌	有	法鼓	有	玛尼轮	无	坛城	有	其他	——
	室外	旗杆	无	苏勒德	有	狮子	有	经幡	有	玛尼轮	有	香炉	有	五供	无	其他	——
	其他	——															
备注	——																
调查日期	2010/10/29	调查人员	付瑞峰	整理日期	2010/10/29	整理人员	李国保、付瑞峰										

大雄宝殿基本概况表1

善岱古庙·大雄宝殿·档案照片					
照片名称	正立面	照片名称	斜前方	照片名称	侧立面
照片名称	斜后方	照片名称	背立面	照片名称	室外柱子
照片名称	室外柱头	照片名称	室外柱础	照片名称	正门
照片名称	室内正面1	照片名称	室内正面2	照片名称	室内侧面
照片名称	室内天花	照片名称	佛像	照片名称	室内悬塑
备注	—				
摄影日期	2010/10/29	摄影人员	乔恩懋		

大雄宝殿正前方

大雄宝殿侧面(左图)
大雄宝殿室外柱头(右图)

大雄宝殿佛殿前廊
（左图）
大雄宝殿二层屋面
（右上图）
大雄宝殿前廊正前方
（右中图）

大雄宝殿室内正面
（左图）
大雄宝殿室内悬塑
（右下图）

3.2 善岱古庙 · 密宗殿

单位:毫米

建筑名称	汉语正式名称	密宗殿	俗称	居德巴独贡													
概述	初建年	不详	建筑朝向	南偏东约45°	建筑层数	二											
	建筑简要描述	纯藏式砖木混合结构体系，汉藏结合装饰风格															
	重建重修记载	1980年政府出资修缮，1997年外表面粉刷装修															
		信息来源	寺庙喇嘛（噶拉桑旺其格，75岁）口述														
结构规模	结构形式	砖木混合	相连的建筑	无	室内天井	都纲法式											
	建筑平面形式	长方形	外廊形式	前廊													
	通面阔	12080	开间数	5间	明间	——	次间	——	梢间	——	次梢间	——	尽间	——			
	通进深	17190	进深数	7间	进深尺寸（前→后）	2400→1940→2130→2550→2080→2210→3880											
	柱子数量	16	柱子间距	横向尺寸	2980、2560	（藏式建筑结构体系填写此栏，											
				纵向尺寸	2130、2550、2080	不含廊柱）											
	其他	——															
建筑主体（大木作）（石作）（瓦作）	屋顶	屋顶形式	藏式密肋平屋顶	瓦作	——												
	外墙	主体材料	青砖	材料规格	260×135×60	饰面颜色	白色										
		墙体收分	有	边玛檐墙	有	边玛材料	青砖（红色涂料粉刷）										
	斗栱、梁架	斗栱	无	平身科斗口尺寸	——	梁架关系	16柱24梁										
	柱、柱式（前廊柱）	形式	藏式	柱身断面形状	圆形	断面尺寸	周长C=560	（在没有前廊柱的情况下，填写室内柱及其特征）									
		柱身材料	木材	柱身收分	有	栌斗、托木	有	雀替	无								
		柱础	无	柱础形状	——	柱础尺寸	——										
	台基	台基类型	普通台基	台基高度	820	台基地面铺设材料	方砖（300×300）										
	其他	——															
装修（小木作）（彩画）	门(正面)	板门	门楣	无	堆经	有	门帘	无									
	窗（正面）	藏式明窗	窗楣	有	窗套	有	窗帘	无									
	室内隔扇	隔扇	无	隔扇位置	——												
	室内地面、楼面	地面材料及规格	方砖(265×265)	楼面材料及规格	——												
	室内楼梯	楼梯	有	楼梯位置	柱廊右侧暗厢中	楼梯材料	木材	楼段宽度	770								
	天花、藻井	天花	有	天花类型	海漫（布）	藻井	无	藻井类型	——								
	彩画	柱头	有	柱身	无	梁架	有	走马板	无								
		门、窗	无	天花	无	藻井	无	其他彩画	无								
	其他	悬塑	无	佛龛	有	匾额	无										
装饰	室内	帷幔	无	幕帘彩绘	有	壁画	有	唐卡	有								
		经幡	有	经幢	有	柱毯	无	其他	——								
	室外	玛尼轮	无	苏勒德	无	宝顶	有	祥麟法轮	无								
		四角经幢	无	经幡	有	铜饰	有	石刻、砖雕	有								
		仙人走兽	无	壁画	有	其他	——										
陈设	室内	主佛像	宗喀巴	佛像基座	须弥座												
		法座	无	藏经橱	无	经床	无	诵经桌	无	法鼓	无	玛尼轮	无	坛城	无	其他	——
	室外	旗杆	无	苏勒德	无	狮子	无	经幡	有	玛尼轮	无	香炉	无	五供	无	其他	——
	其他	——															
备注	——																
调查日期	2010/10/29	调查人员	付瑞峰	整理日期	2010/10/29	整理人员	付瑞峰、李国保										

密宗殿基本概况表1

善岱古庙·密宗殿·档案照片

照片名称	正立面	照片名称	斜前方	照片名称	侧立面
照片名称	斜后方	照片名称	背立面	照片名称	室外柱子
照片名称	室外柱头	照片名称	室外柱础	照片名称	正门
照片名称	室内正面	照片名称	经堂室内侧面	照片名称	佛殿室内侧面
照片名称	室内柱子	照片名称	室内柱头	照片名称	佛像
备注	——				
摄影日期	2010/10/29	摄影人员	乔恩懋		

密宗殿基本概况表2

密宗殿正前方

密宗殿室内正前方（左图）
密宗殿正门（右图）

密宗殿室内侧面（左图）
密宗殿室内天井（右上图）
密宗殿室内佛像（右下图）

3.3 善岱古庙·拉卜隆庙

单位:毫米

建筑名称	汉语正式名称	——	俗称	拉卜隆庙		
概述	初建年	不详	建筑朝向	南偏东约45°	建筑层数	二
	建筑简要描述	纯藏式砖木混合结构体系，汉藏结合装饰风格				
	重建重修记载	1980年政府出资修缮，1997年外表面粉刷装修				
		信息来源	寺庙喇嘛（噶拉桑旺其格，75岁）口述			

结构规模														
结构规模	结构形式	砖木混合	相连的建筑	无	室内天井	都纲法式								
	建筑平面形式	长方形	外廊形式	前廊										
	通面阔	11070	开间数	5间	明间	——	次间	——	梢间	——	次梢间	——	尽间	——
	通进深	14220	进深数	6间	进深尺寸（前→后）	3040→1800→1600→1610→1870→4300								
	柱子数量	12	柱子间距	横向尺寸	2100、2590	（藏式建筑结构体系填写此栏，不含廊柱）								
				纵向尺寸	1610、1600									
	其他	——												

建筑主体（大木作）（石作）（瓦作）										
建筑主体（大木作）（石作）（瓦作）	屋顶	屋顶形式	藏式密肋平屋顶	瓦作	——					
	外墙	主体材料	青砖	材料规格	265×140×70	饰面颜色	白色			
		墙体收分	有	边玛檐墙	有	边玛材料	青砖（红色涂料粉刷）			
	斗栱、梁架	斗栱	无	平身科斗口尺寸	——	梁架关系	12柱15梁			
	柱、柱式（前廊柱）	形式	藏式	柱身断面形状	小八棱柱	断面尺寸	65×40×70			（在没有前廊柱的情况下，填写室内柱及其特征）
		柱身材料	石材	柱身收分	有	栌斗、托木	有	雀替	无	
		柱础	无	柱础形状	——	柱础尺寸	——			
	台基	台基类型	普通台基	台基高度	770	台基地面铺设材料	青砖(280×130)			
	其他	——								

装修（小木作）（彩画）									
装修（小木作）（彩画）	门(正面)	板门		门楣	无	堆经	无	门帘	有
	窗（正面）	藏式明窗		窗楣	有	窗套	有	窗帘	有
	室内隔扇	隔扇	无	隔扇位置	——				
	室内地面、楼面	地面材料及规格		方砖（270×130）		楼面材料及规格		不详	
	室内楼梯	楼梯	有	楼梯位置	柱廊右侧暗厢中	楼梯材料	木材	楼段宽度	760
	天花、藻井	天花	无	天花类型	——	藻井	无	藻井类型	——
	彩画	柱头	有	柱身	无	梁架	有	走马板	无
		门、窗	无	天花	——	藻井	——	其他彩画	无
	其他	悬塑	无	佛龛	有	匾额	无		

装饰									
装饰	室内	帷幔	有	幕帘彩绘	有	壁画	有	唐卡	有
		经幡	有	经幢	有	柱毯	有	其他	——
	室外	玛尼轮	无	苏勒德	无	宝顶	有	祥麟法轮	无
		四角经幢	无	经幡	有	铜饰	有	石刻、砖雕	有
		仙人走兽	无	壁画	有	其他	无		

陈设																	
陈设	室内	主佛像	无量寿佛					佛像基座	须弥座								
		法座	无	藏经橱	无	经床	无	诵经桌	无	法鼓	无	玛尼轮	无	坛城	无	其他	——
	室外	旗杆	无	苏勒德	有	狮子	无	经幡	无	玛尼轮	无	香炉	有	五供	无	其他	——
	其他	——															

备注	——

调查日期	2010/10/29	调查人员	付瑞峰	整理日期	2010/10/29	整理人员	李国保、付瑞峰

拉卜隆庙基本概况表1

善岱古庙 · 拉卜隆庙 · 档案照片

照片名称 	正立面	照片名称 	斜前方	照片名称 	侧立面
照片名称 	斜后方	照片名称 	正门	照片名称 	室外柱子
照片名称 	室外柱头	照片名称 	室外柱础	照片名称 	藏式窗户
照片名称 	室内正面	照片名称 	经堂室内侧面	照片名称 	佛殿室内侧面
照片名称 	室内柱子	照片名称 	室内柱头	照片名称 	佛像
备注	——				
摄影日期	2010/10/29	摄影人员	乔恩懋		

拉卜隆庙斜前方(左图)
拉卜隆庙正面门（右上图）
拉卜隆庙室内侧面（右下图）

3.4 善岱古庙 · 活佛府

活佛府正前方（左图）
活佛府室外柱头（右图）

活佛府室内正面（左图）
活佛府室内侧面（右图）

3.5 善岱古庙 · 二少爷府

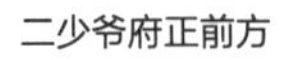

二少爷府正前方

山门正前方（左图）
二少爷府门窗（右图）

喇嘛僧舍正前方(左图）
二少爷府前廊（右图）

鄂尔多斯市地区

Erdos City

底图来源：内蒙古自治区自然资源厅官网 内蒙古地图
审图号：蒙S（2017）028号

鄂尔多斯市辖达拉特旗、准格尔旗、伊金霍洛旗、乌审旗、杭锦旗、鄂托克旗、鄂托克前旗7旗，东胜区1区。该市前身为伊克昭盟，2001年撤盟设地级鄂尔多斯市。现辖地由清顺治年间设立的鄂尔多斯6旗，即左翼中旗（郡王旗）、左翼前旗（准格尔旗）、左翼后旗（达拉特旗）、右翼中旗（鄂托克旗）、右翼前旗（乌审旗）、右翼后旗（杭锦旗）及乾隆年间增设的右翼前末旗（扎萨克旗）7旗组成，七旗会盟于王爱召，故称伊克昭盟。市辖区内曾有320余座藏传佛教寺庙，现存20余座已恢复重建或尚有建筑遗存的寺庙，课题组实地调研18座寺庙。

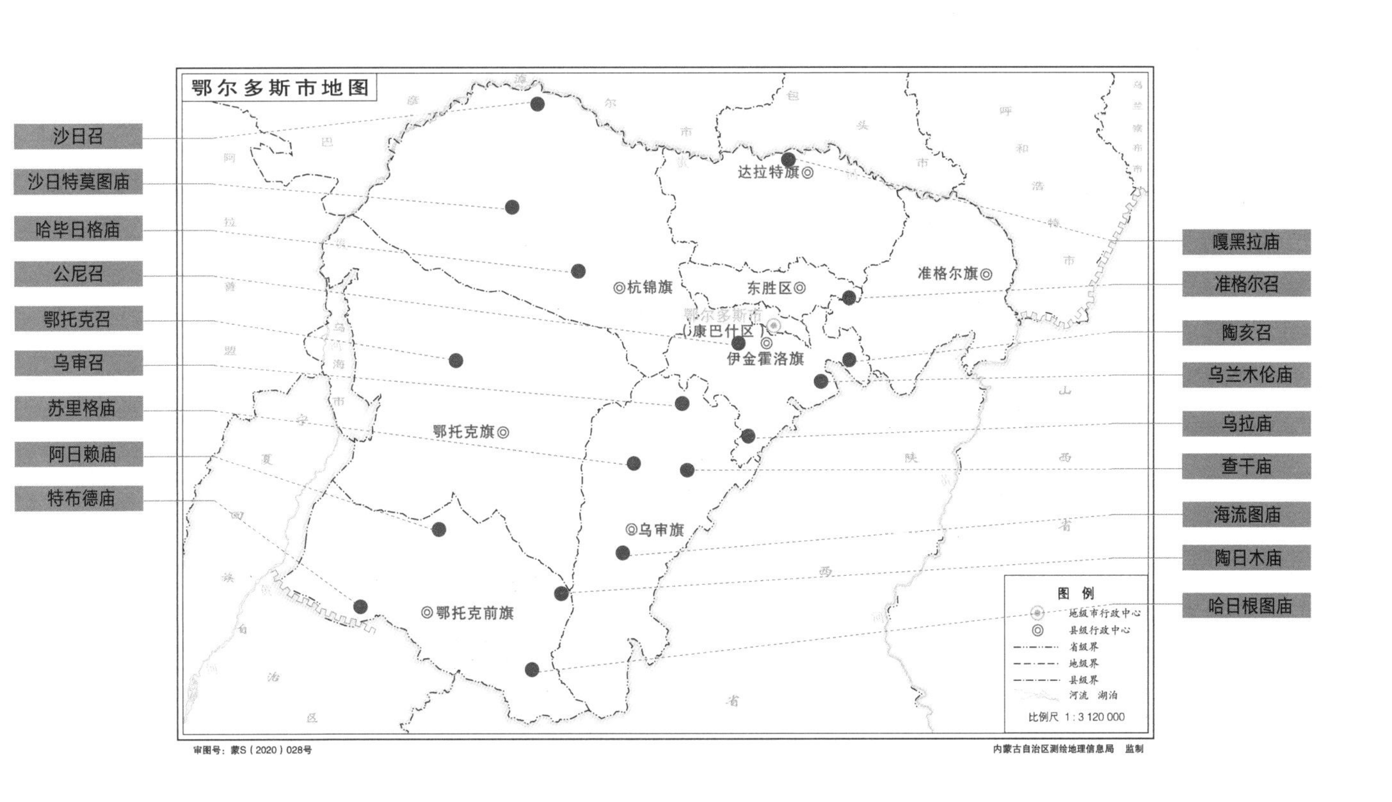
鄂尔多斯市地图
沙日召
沙日特莫图庙
哈毕日格庙
公尼召
鄂托克召
乌审召
苏里格庙
阿日赖庙
特布德庙
嘎黑拉庙
准格尔召
陶亥召
乌兰木伦庙
乌拉庙
查干庙
海流图庙
陶日木庙
哈日根图庙
达拉特旗
杭锦旗
东胜区
准格尔旗
鄂尔多斯市
（康巴什区）
伊金霍洛旗
鄂托克旗
乌审旗
鄂托克前旗
图例
比例尺 1:3 120 000
审图号：蒙S（2020）028号
内蒙古自治区测绘地理信息局 监制

1 准格尔召 Zhungarzhao Temple

1 准格尔召 Zhungarzhao Temple

准格尔召建筑群

准格尔召为原伊克昭盟鄂尔多斯左翼前旗（俗称准格尔旗）寺庙，系鄂尔多斯建造最早的藏传佛教格鲁派寺庙及准格尔旗规模最为宏大的寺庙。明廷赐名“宝藏寺”，民国12年（1923年）民国政府赐名“宝堂寺”。寺庙管辖拉白乌拉庙、西巴日台庙、寨子阿贵、希日嘎庙、宝日汗图阿贵、巴音图库木庙、西敖萨拉庙、东敖萨拉庙、钦达木尼庙、药师庙、渥巴锡诺颜庙、浩亚日乌孙庙、呼仁高勒庙、巴日庙、达诺颜赞康庙等15座属庙。

明天启三年（1623年），准格尔旗第一任扎萨克斯仁之祖父明盖岱青、图日布洪台吉等人从陕西神木请来工匠，在乌力吉图山之地新建黄绿色琉璃瓦重檐殿宇，初建该寺。最初称该庙为西拉召、衮额日格召、贝子召。据《白银鉴·准格尔召史》记载，此庙与土默特阿拉坦汗所建美岱召、鄂尔多斯博硕克图济农所建伊克西拉召三座寺庙式样规模十分相似。

寺庙建筑风格为汉、藏建筑并存。至民国末年寺庙有65间大雄宝殿、42间释迦牟尼殿、8间弥勒殿、8间莲花生殿等铺有黄绿色琉璃瓦的三重檐双层汉式殿宇，此4座殿宇外围有方形大院。此外，有8间双层藏式法轮殿、35间藏式舍利殿、24间双层释迦牟尼殿、3间护法殿、35间藏式五大神殿、35间藏式显宗院殿、25间密宗殿、3间天王殿等殿宇，显宗殿加院落共计96间、3间北京喇嘛殿、3间文殊菩萨殿、7间观音殿、9间藏式大黑天殿、2间大庙仓骑狮护法神殿、3间博格达喇嘛殿、3间赞丹舍利殿、5间药师殿、3间堪布喇嘛殿、8间双层阿日扎嘎尔殿、4间塔殿、1间灵藏室共4座、9间根皮殿、6间官仓殿等大小30余座殿宇及官仓11间、学部仓11间、大庙仓26间、22座佛塔，僧舍218座。

“文化大革命”中寺庙遭严重损毁，仅存大雄宝殿、释迦牟尼殿等10余座空殿堂。1980年起，重点修缮大雄宝殿等4座殿宇，设为准格尔旗佛教活动点之一。

参考文献：

［1］萨·那日松，特木尔巴特尔.鄂尔多斯寺院（蒙古文）.海拉尔:内蒙古文化出版社,2000，5.

［2］萨·那日松.鄂尔多斯人历史文献集（第四辑）.内蒙古伊克昭盟档案馆,1984，9.

准格尔召释迦牟尼殿

<table>
<tr><th colspan="7">准格尔召·基本概况 1</th></tr>
<tr><td rowspan="3">寺院蒙古语藏语名称</td><td>蒙古语</td><td>[illegible]</td><td rowspan="2">寺院汉语名称</td><td>汉语正式名称</td><td colspan="2">宝堂寺</td></tr>
<tr><td>藏语</td><td>དགའ་ལྡན་བཤད་སྒྲུབ་དར་རྒྱས་གླིང་།</td><td>俗称</td><td colspan="2">准格尔召、西召</td></tr>
<tr><td>汉语语义</td><td>准格尔召</td><td colspan="2">寺院汉语名称的由来</td><td colspan="2">民国政府赐名</td></tr>
<tr><td>所在地</td><td colspan="4">鄂尔多斯市准格尔旗准格尔召苏木</td><td>东经 110° 08′</td><td>北纬 39° 36′</td></tr>
<tr><td>初建年</td><td colspan="2">天启三年（1623年）</td><td>保护等级</td><td colspan="3">自治区级文物保护单位</td></tr>
<tr><td>盛期时间</td><td colspan="2">不详</td><td colspan="2">盛期喇嘛僧/现有喇嘛僧数</td><td colspan="2">1000余人／30余人</td></tr>
<tr><td rowspan="2">历史沿革</td><td colspan="6">1622年，始建寺庙。1623年，台吉土格日布始建本召大雄宝殿，经过3年时间完成。
1633年，察哈尔林丹汗兵败西逃途中路经此庙，破坏了寺庙。
1741年，增建殿宇与僧舍，并上奏理藩院，规定僧人常住寺庙。
1742年，六世达赖驾临寺庙。
1751年，创建显宗学部。
1785年，修缮寺庙。
1819年，创建密宗学部，传授拉占巴学位。
1871年，修缮寺庙，并将大雄宝殿扩至65间，释迦牟尼殿扩至42间，弥勒殿扩至8间，莲花生殿扩至8间，并加层于此四座殿宇。
1920—1922年，修缮大雄宝殿、释迦牟尼殿等殿宇，装饰成三重檐双层殿宇。
1980年，恢复宗教活动，并开始修缮召庙。
1985年，被列为内蒙古自治区重点文物保护单位。
1999年，大规模修缮准格尔召。2002年后，陆续将各个殿堂的佛像塑全，恢复原貌。
2006年，建成白塔</td></tr>
<tr><td>资料来源</td><td colspan="5">［1］韩福海,韩钧宇.美丽的准格尔召.呼和浩特:内蒙古人民出版社,2008，4.
［2］萨·那日松，特木尔巴特尔.鄂尔多斯寺院（蒙古文）.海拉尔:内蒙古文化出版社,2005，5.</td></tr>
<tr><td rowspan="2">现状描述</td><td colspan="4" rowspan="2">准格尔召建筑群，现存建筑大雄宝殿、观音庙、舍利殿、五道庙、六臂护法殿、千佛殿均为遗存建筑，其中大雄宝殿建造年代最久，在整个召庙建筑群中具有核心地位。整个建筑群体保存完好，每年仍有各种宗教活动在此开展，是鄂尔多斯地区藏传佛教东路发展的典型代表寺庙</td><td>描述时间</td><td>2010/11/08</td></tr>
<tr><td>描述人</td><td>杜娟</td></tr>
<tr><td>调查日期</td><td colspan="2">2010/11/08</td><td>调查人员</td><td colspan="3">宝山</td></tr>
</table>

准格尔召基本概况表1

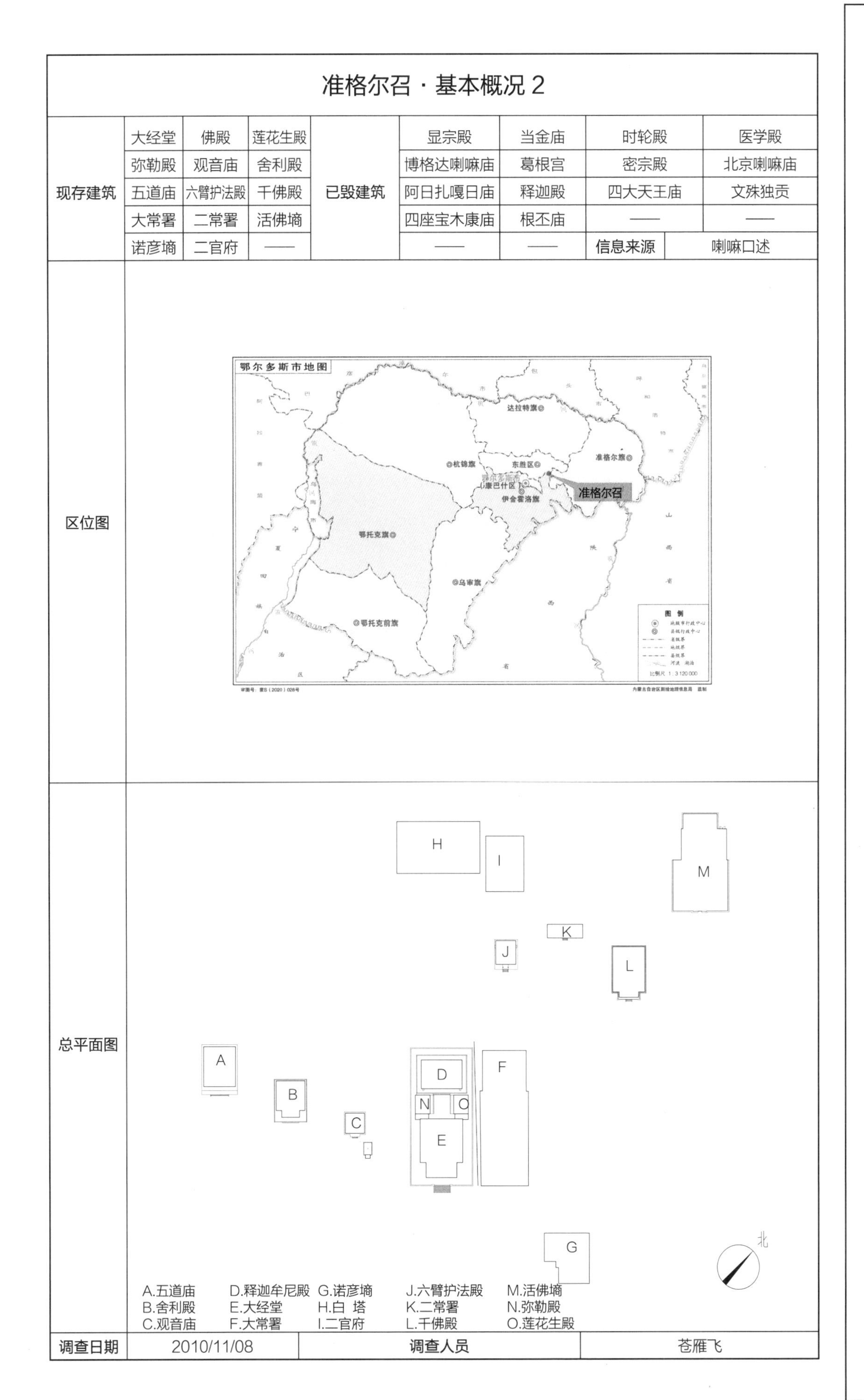

准格尔召·基本概况 2

现存建筑				已毁建筑				
现存建筑	大经堂	佛殿	莲花生殿	已毁建筑	显宗殿	当金庙	时轮殿	医学殿
	弥勒殿	观音庙	舍利殿		博格达喇嘛庙	葛根宫	密宗殿	北京喇嘛庙
	五道庙	六臂护法殿	千佛殿		阿日扎嘎日庙	释迦殿	四大天王庙	文殊独贡
	大常署	二常署	活佛墒		四座宝木康庙	根丕庙	——	——
	诺彦墒	二官府	——		——	——	信息来源	喇嘛口述

区位图

总平面图

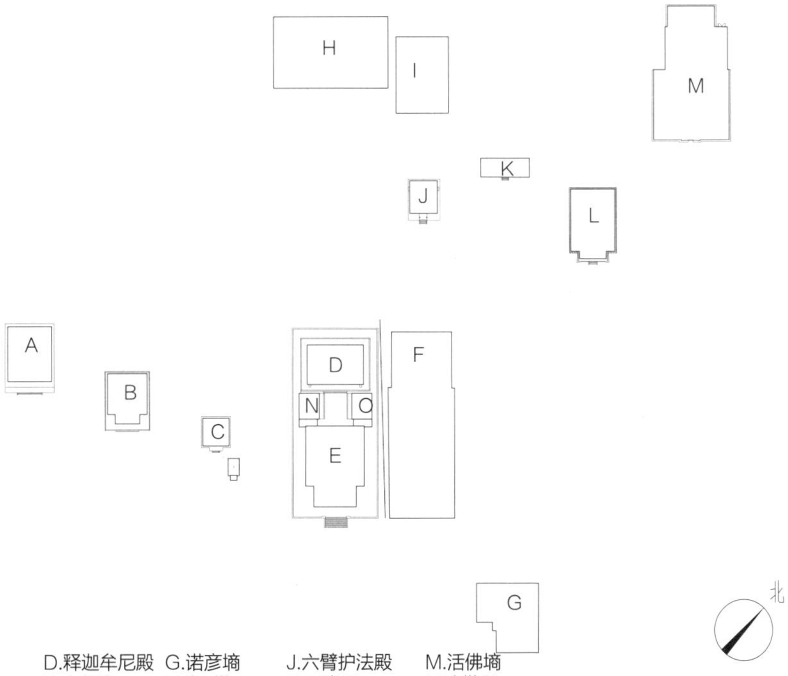

A.五道庙　D.释迦牟尼殿　G.诺彦墒　J.六臂护法殿　M.活佛墒
B.舍利殿　E.大经堂　H.白 塔　K.二常署　N.弥勒殿
C.观音庙　F.大常署　I.二官府　L.千佛殿　O.莲花生殿

调查日期	2010/11/08	调查人员	苍雁飞

准格尔召基本概况表2

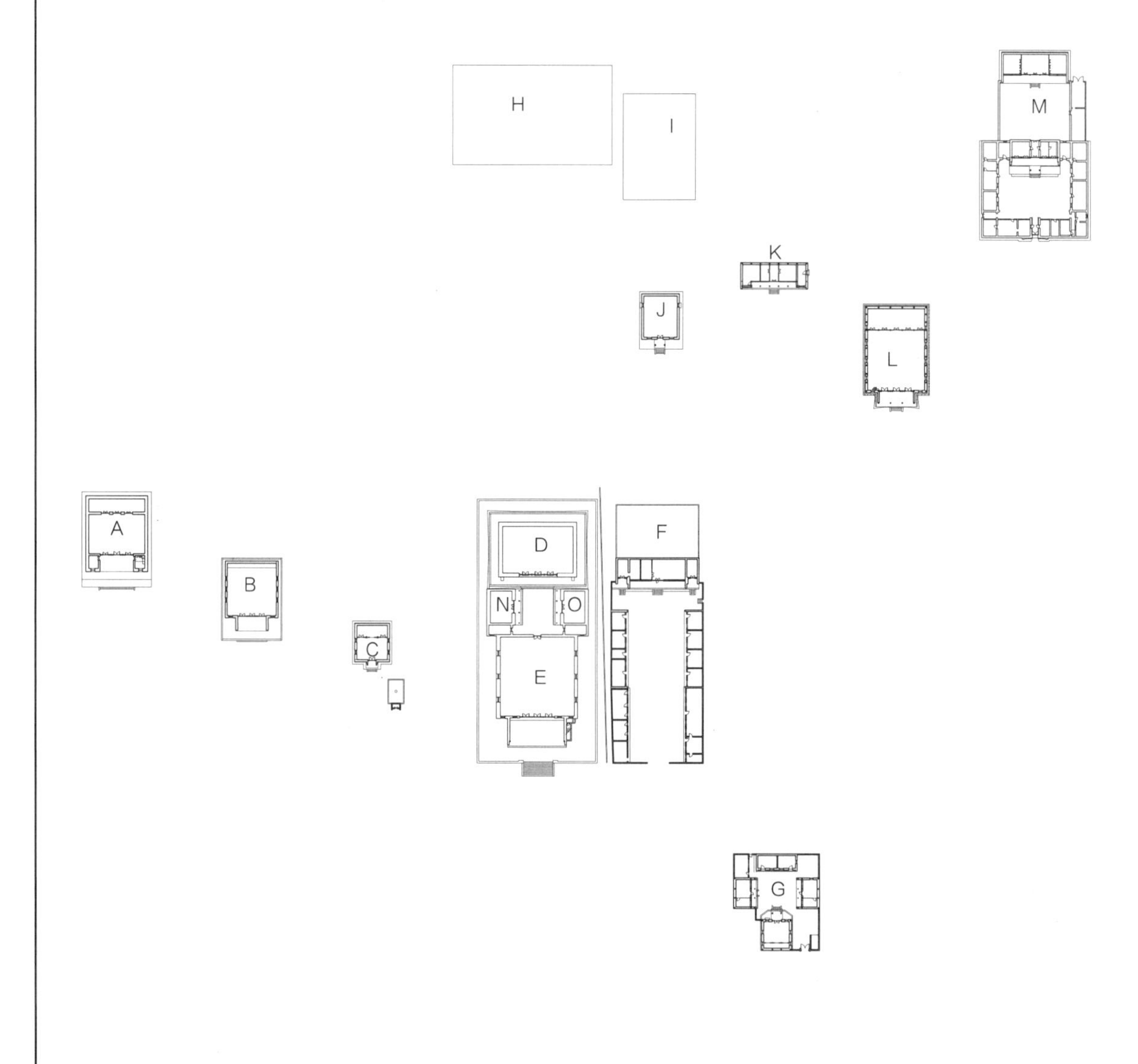

A.五道庙　D.释迦牟尼殿　G.诺彦璫　J.六臂护法殿　M.活佛璫
B.舍利殿　E.大经堂　H.白 塔　K.二常署　N.弥勒殿
C.观音庙　F.大常署　I.二官府　L.千佛殿　O.莲花生殿

准格尔召总平面图

1.1 准格尔召·大经堂

单位:毫米

建筑名称	汉语正式名称	大经堂			俗称	大殿			
概述	初建年	天启三年（1623年）			建筑朝向	南		建筑层数	二
	建筑简要描述	汉藏结合式建筑							
	重建重修记载	1750年（乾隆十五年），由本旗第五代札萨克贝子贝勒那木吉勒道尔吉进行了一次维修和扩建。1871年九代札萨克扎那嘎如迪修过一次寺庙，并扩建。1920—1922年，十一代札萨克阿拉坦敖其尔及其子宝音巴达日呼、泰吉那顺达来等献银三万两，加上泰吉官布扎布等群众的布施，将大雄宝殿修成三层檐的两层寺庙，并加上金银油漆							
		信息来源	准格尔召庙志						
结构规模	结构形式	砖木		相连的建筑	与莲花生殿、弥勒殿相连接		室内天井	无	
	建筑平面形式	凸字形		外廊形式	前廊				
	通面阔	21340	开间数	7	明间 3090	次间 3100	梢间 ——	次梢间 ——	尽间 ——
	通进深	21630	进深数	7	进深尺寸（前→后）		3025→3100→3090→3090→3100→3120→3105		
	柱子数量	——	柱子间距	横向尺寸	——		（藏式建筑结构体系填写此栏，不含廊柱）		
				纵向尺寸	——				
	其他	——							
建筑主体（大木作）（石作）（瓦作）	屋顶	屋顶形式	重檐歇山、歇山与藏式平屋顶相结合				瓦作	绿琉璃黄剪边	
	外墙	主体材料	砖	材料规格	150×310×70		饰面颜色	白	
		墙体收分	无	边玛檐墙	无		边玛材料	——	
	斗栱、梁架	斗栱	无	平身科斗口尺寸		——	梁架关系	十五檩有前廊	
	柱、柱式（前廊柱）	形式	汉式	柱身断面形状	圆	断面尺寸	直径D=260		（在没有前廊柱的情况下，填写室内柱及其特征）
		柱身材料	木材	柱身收分	无	栌斗、托木	无	雀替 无	
		柱础	有	柱础形状	方础上置圆础	柱础尺寸	直径D=400		
	台基	台基类型	带砖墙普通式台基	台基高度	1760	台基地面铺设材料		石材	
	其他	——							
装修（小木作）（彩画）	门(正面)	板门		门楣	无	堆经	无	门帘	有
	窗（正面）	藏式明窗		窗楣	无	窗套	无	窗帘	无
	室内隔扇	隔扇	无	隔扇位置	——				
	室内地面、楼面	地面材料及规格		砖250×250		楼面材料及规格		——	
	室内楼梯	楼梯	无	楼梯位置	——	楼梯材料	——	梯段宽度	——
	天花、藻井	天花	无	天花类型	——	藻井	无	藻井类型	——
	彩画	柱头	无	柱身	无	梁架	有	走马板	无
		门、窗	无	天花	——	藻井	——	其他彩画	——
	其他	悬塑	无	佛龛	有	匾额	有		
装饰	室内	帷幔	无	幕帘彩绘	无	壁画	无	唐卡	有
		经幡	有	经幢	有	柱毯	有	其他	——
	室外	玛尼轮	无	苏勒德	无	宝顶	有	祥麟法轮	有
		四角经幢	有	经幡	无	铜饰	无	石刻、砖雕	有
		仙人走兽	1+3	壁画	有	其他	——		
陈设	室内	主佛像	——			佛像基座	莲花座		
		法座 有	藏经橱 无	经床 有	诵经桌 有	法鼓 有	玛尼轮 无	坛城 无	其他 ——
	室外	旗杆 无	苏勒德 有	狮子 有	经幡 有	玛尼轮 无	香炉 有	五供 无	其他 ——
	其他	——							
备注	匾额上书噶丹夏珠达尔杰朗。经堂中无主佛像								
调查日期	2010/08/11	调查人员	杜娟	整理日期	2010/08/12	整理人员	杜娟		

大经堂基本概况表1

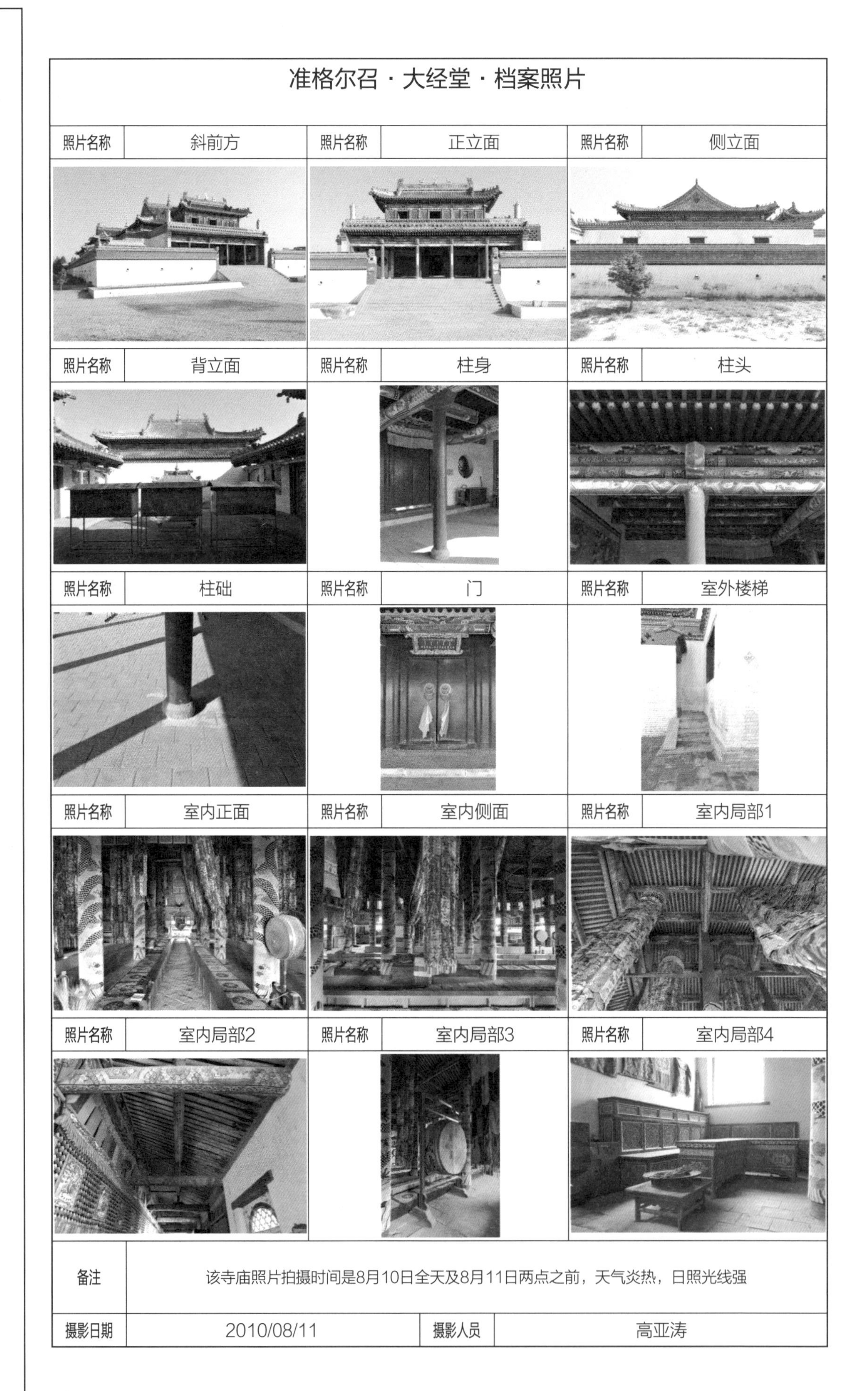

准格尔召・大经堂・档案照片

照片名称	斜前方	照片名称	正立面	照片名称	侧立面
照片名称	背立面	照片名称	柱身	照片名称	柱头
照片名称	柱础	照片名称	门	照片名称	室外楼梯
照片名称	室内正面	照片名称	室内侧面	照片名称	室内局部1
照片名称	室内局部2	照片名称	室内局部3	照片名称	室内局部4

备注	该寺庙照片拍摄时间是8月10日全天及8月11日两点之前，天气炎热，日照光线强		
摄影日期	2010/08/11	摄影人员	高亚涛

大经堂基本概况表2

大经堂正前方

大经堂侧面

大经堂室内（左图）
大经堂柱头（右图）

1.2 准格尔召·释迦牟尼殿

单位:毫米

建筑名称	汉语正式名称	释迦牟尼殿	俗称	大雄宝殿													
概述	初建年	天启三年（1623年）	建筑朝向	南	建筑层数	一											
	建筑简要描述	汉式建筑															
	重建重修记载	1750年（乾隆十五年），由本旗第五代札萨克贝子贝勒那木吉勒道尔吉进行了一次维修和扩建。1871年，九代札萨克扎那嘎如迪修过一次寺庙，并扩建。1920—1922年，十一代札萨克阿拉坦敖其尔及其子宝音巴达日呼、泰吉那顺达来等献银三万两，加上泰吉官布扎布等群众的布施，将大雄宝殿修成三层檐的两层寺庙，并加上金银油漆															
		信息来源	准格尔召庙志														
结构规模	结构形式	砖木	相连的建筑	无	室内天井	无											
	建筑平面形式	长方形	外廊形式	四周回廊													
	通面阔	20430	开间数	7	明间	3360	次间	3340	梢间	——	次梢间	——	尽间	——			
	通进深	13430	进深数	4	进深尺寸（前→后）	3380→3370→3320→3360											
	柱子数量	——	柱子间距	横向尺寸	——	（藏式建筑结构体系填写此栏，不含廊柱）											
				纵向尺寸	——												
	其他	——															
建筑主体（大木作）（石作）（瓦作）	屋顶	屋顶形式	重檐歇山	瓦作	绿琉璃黄剪边												
	外墙	主体材料	砖	材料规格	160×310×70	饰面颜色	土黄										
		墙体收分	有	边玛檐墙	无	边玛材料	——										
	斗栱、梁架	斗栱	有	平身科斗口尺寸	——	梁架关系	十一檩有回廊										
	柱、柱式（前廊柱）	形式	藏式	柱身断面形状	八楞	断面尺寸	300×300	（在没有前廊柱的情况下，填写室内柱及其特征）									
		柱身材料	木材	柱身收分	有	栌斗、托木	有	雀替	无								
		柱础	有	柱础形状	方础上置圆础	柱础尺寸	直径 D=580										
	台基	台基类型	普通式台基	台基高度	480	台基地面铺设材料	方砖										
	其他	——															
装修（小木作）（彩画）	门(正面)	格栅	门楣	无	堆经	无	门帘	无									
	窗（正面）	无	窗楣	——	窗套	——	窗帘	——									
	室内隔扇	隔扇	无	隔扇位置	——												
	室内地面、楼面	地面材料及规格	条形木板、规格不一	楼面材料及规格	——												
	室内楼梯	楼梯	无	楼梯位置	——	楼梯材料	——	梯段宽度	——								
	天花、藻井	天花	有	天花类型	井口天花	藻井	有	藻井类型	四方变八方								
	彩画	柱头	无	柱身	无	梁架	有	走马板	有								
		门、窗	无	天花	有	藻井	有	其他彩画	——								
	其他	悬塑	无	佛龛	有	匾额	有										
装饰	室内	帷幔	无	幕帘彩绘	无	壁画	无	唐卡	无								
		经幡	有	经幢	有	柱毯	有	其他	——								
	室外	玛尼轮	无	苏勒德	无	宝顶	有	祥麟法轮	无								
		四角经幢	无	经幡	无	铜饰	无	石刻、砖雕	有								
		仙人走兽	1仙人+3走兽	壁画	无	其他	——										
陈设	室内	主佛像	释迦牟尼佛	佛像基座	莲花座												
		法座	无	藏经橱	无	经床	无	诵经桌	无	法鼓	无	玛尼轮	无	坛城	无	其他	——
	室外	旗杆	无	苏勒德	无	狮子	无	经幡	无	玛尼轮	有	香炉	有	五供	无	其他	——
	其他	——															
备注	室内正在做装修，匾额上书宝堂寺。佛殿为全木结构，据说没有使用过一颗铁钉，均为纯木块相契的榫卯结构，历经数百年仍然完好无损地屹立着																
调查日期	2010/08/11	调查人员	杜娟	整理日期	2010/08/12	整理人员	杜娟										

释迦牟尼殿基本概况表1

准格尔召·释迦牟尼殿·档案照片

照片名称 	正立面	照片名称 	侧立面	照片名称 	斜后方
照片名称 	背立面	照片名称 	柱身	照片名称 	柱头及斗栱
照片名称 	室外局部1	照片名称 	室外局部2	照片名称 	室外局部3
照片名称 	室外局部4	照片名称 	室内天花	照片名称 	室内地面
照片名称 	室内柱础	照片名称 	室内藻井	照片名称 	室内局部
备注	佛殿室内正在进行佛龛、佛像的制作，室内照片拍摄困难				
摄影日期	2010/08/11	摄影人员	高亚涛		

释迦牟尼殿基本概况表2

释迦牟尼殿正前方(左图)

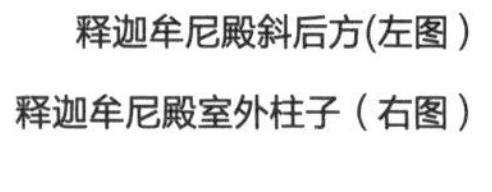

释迦牟尼殿斜后方(左图)

释迦牟尼殿室外柱子(右图)

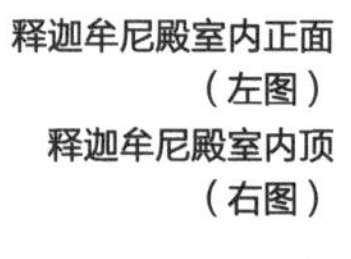

释迦牟尼殿室内正面
(左图)
释迦牟尼殿室内顶
(右图)

释迦牟尼殿室内侧面
(左图)
释迦牟尼殿室内柱
(右图)

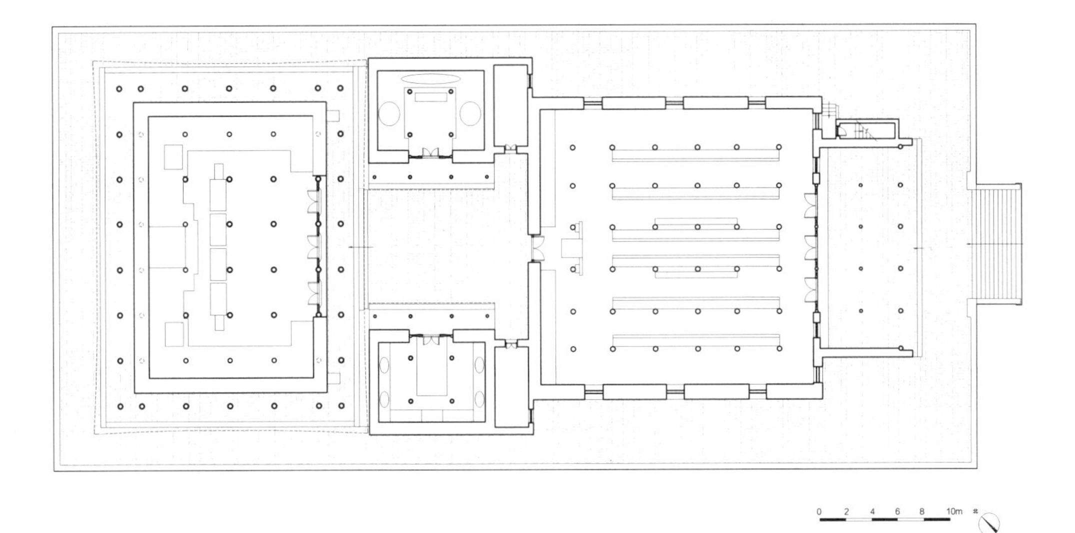

释迦牟尼殿一层平面图

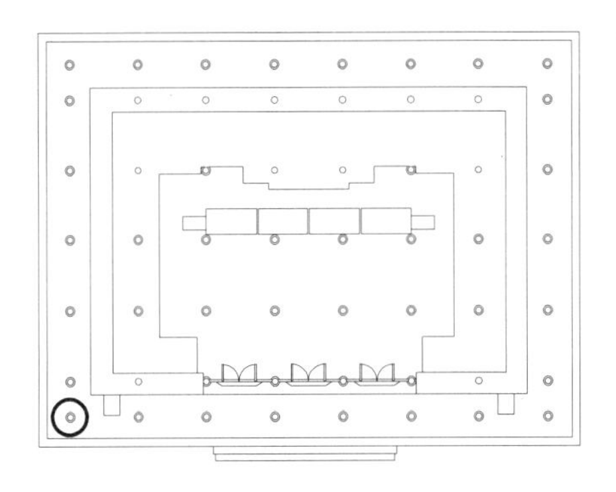

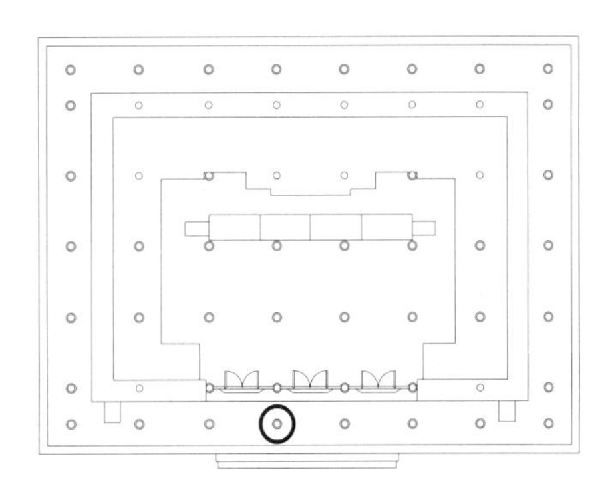

释迦牟尼殿角柱平面图（左图）
释迦牟尼殿檐柱平面图（右图）

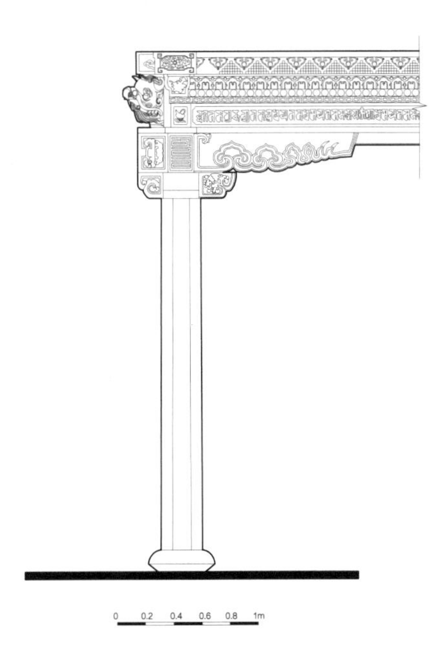

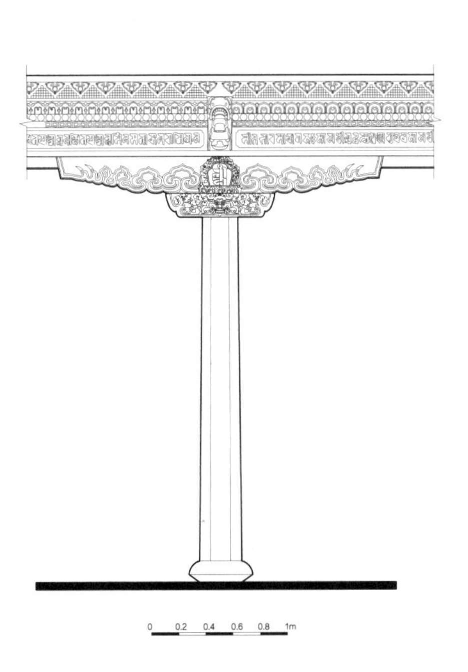

释迦牟尼殿角柱立面图（左图）
释迦牟尼殿檐柱立面图（右图）

1.3 准格尔召·莲花生殿

单位:毫米

建筑名称	汉语正式名称	莲花生殿	俗称	——															
概述	初建年	不详	建筑朝向	西	建筑层数	一													
	建筑简要描述	汉藏结合式建筑																	
	重建重修记载	1922年（民国11年）原址上重建																	
		信息来源	准格尔召庙志																
结构规模	结构形式	砖木	相连的建筑	南侧与大经堂相连接	室内天井	无													
	建筑平面形式	长方形	外廊形式	前廊															
	通面阔	8530	开间数	3	明间	3120	次间	2705	梢间	——	次梢间	——	尽间	——					
	通进深	6800	进深数	3	进深尺寸（前→后）	1620→3180→2000													
	柱子数量	——	柱子间距	横向尺寸	——	（藏式建筑结构体系填写此栏，不含廊柱）													
				纵向尺寸	——														
	其他	——																	
建筑主体（大木作）（石作）（瓦作）	屋顶	屋顶形式	硬山顶与歇山顶相结合	瓦作	绿琉璃黄剪边														
	外墙	主体材料	砖	材料规格	160×320×60	饰面颜色	白												
		墙体收分	无	边玛檐墙	无	边玛材料	——												
	斗栱、梁架	斗栱	无	平身科斗口尺寸	——	梁架关系	——												
	柱、柱式（前廊柱）	形式	汉式	柱身断面形状	圆柱	断面尺寸	直径D=190	（在没有前廊柱的情况下，填写室内柱及其特征）											
		柱身材料	木材	柱身收分	无	栌斗、托木	无	雀替	有										
		柱础	有	柱础形状	方础上置圆础	柱础尺寸	直径D=380												
	台基	台基类型	普通式台基	台基高度	150	台基地面铺设材料	条石、方砖												
	其他	——																	
装修（小木作）（彩画）	门(正面)	格栅	门楣	无	堆经	无	门帘	无											
	窗（正面）	无	窗楣	——	窗套	——	窗帘	——											
	室内隔扇	隔扇	无	隔扇位置	——														
	室内地面、楼面	地面材料及规格	砖150×310、210×210	楼面材料及规格	——														
	室内楼梯	楼梯	无	楼梯位置	——	楼梯材料	——	梯段宽度	——										
	天花、藻井	天花	有	天花类型	井口天花	藻井	有	藻井类型	四方										
	彩画	柱头	无	柱身	无	梁架	有	走马板	有										
		门、窗	有	天花	有	藻井	有	其他彩画	——										
	其他	悬塑	有	佛龛	无	匾额	无												
装饰	室内	帷幔	无	幕帘彩绘	无	壁画	无	唐卡	无										
		经幡	有	经幢	有	柱毯	无	其他											
	室外	玛尼轮	无	苏勒德	无	宝顶	有	祥麟法轮	无										
		四角经幢	无	经幡	无	铜饰	无	石刻、砖雕	有										
		仙人走兽	1仙人+3走兽	壁画	有	其他	——												
陈设	室内	主佛像	莲花生大师	佛像基座	莲花座														
		法座	无	藏经橱	无	经床	无	诵经桌	无	法鼓	无	玛尼轮	无	坛城	无	其他	——		
	室外	旗杆	无	苏勒德	无	狮子	无	经幡	无	玛尼轮	无	香炉	有	五供	无	其他	——		
	其他	——																	
备注	莲花生殿内主要供奉的是西藏密宗红教开山祖师莲花生大师的塑像，塑像正下方是马头明王，莲花生大师塑像后方供奉宁玛八尊金刚护法。殿内的墙壁是用泥沙雕塑的佛陀世界，是准格尔召内的一绝																		
调查日期	2010/08/11	调查人员	杜娟	整理日期	2010/08/12	整理人员	杜娟												

莲花生殿基本概况表1

准格尔召·莲花生殿·档案照片

照片名称	斜前方	照片名称	正立面	照片名称	柱身
照片名称	柱头	照片名称	柱础	照片名称	门
照片名称	室外局部1	照片名称	室外局部2	照片名称	室外局部3
照片名称	室内正面	照片名称	室内侧面	照片名称	室内天花
照片名称	室内地面	照片名称	室内柱础	照片名称	壁画

备注	——		
摄影日期	2010/08/11	摄影人员	高亚涛

莲花生殿斜前方

莲花生殿室内顶棚（左图）
莲花生殿室内柱子（右图）

莲花生殿室内正面

1.4 准格尔召 · 弥勒殿

单位:毫米

建筑名称	汉语正式名称	弥勒殿	俗称	——													
概述	初建年	不详	建筑朝向	东	建筑层数	一											
	建筑简要描述	汉式建筑															
	重建重修记载	1922年（民国11年）原址上重建															
		信息来源	准格尔召庙志														
结构规模	结构形式	砖木	相连的建筑	与南侧大经堂相连接	室内天井	无											
	建筑平面形式	长方形	外廊形式	前廊													
	通面阔	8530	开间数	3	明间	3120	次间	2705	梢间	——	次梢间	——	尽间	——			
	通进深	6800	进深数	3	进深尺寸（前→后）	1620→3180→2000											
	柱子数量	——	柱子间距	横向尺寸	——	（藏式建筑结构体系填写此栏，不含廊柱）											
				纵向尺寸	——												
	其他	——															
建筑主体（大木作）（石作）（瓦作）	屋顶	屋顶形式	硬山顶与歇山顶相结合	瓦作	——												
	外墙	主体材料	砖	材料规格	160×320×60	饰面颜色	白										
		墙体收分	无	边玛檐墙	无	边玛材料	——										
	斗栱、梁架	斗栱	无	平身科斗口尺寸	——	梁架关系	——										
	柱、柱式（前廊柱）	形式	汉式	柱身断面形状	圆	断面尺寸	直径D=190	（在没有前廊柱的情况下，填写室内柱及其特征）									
		柱身材料	木材	柱身收分	无	栌斗、托木	无	雀替	有								
		柱础	有	柱础形状	圆形	柱础尺寸	直径D=380										
	台基	台基类型	普通式台基	台基高度	150	台基地面铺设材料	方砖+条石										
	其他	——															
装修（小木作）（彩画）	门(正面)	格栅	门楣	无	堆经	无	门帘	无									
	窗（正面）	无	窗楣	——	窗套	——	窗帘	——									
	室内隔扇	隔扇	无	隔扇位置	——												
	室内地面、楼面	地面材料及规格	条形木板、规格不一，砖330×160	楼面材料及规格	——												
	室内楼梯	楼梯	无	楼梯位置	——	楼梯材料	——	梯段宽度	——								
	天花、藻井	天花	有	天花类型	平棋	藻井	有	藻井类型	方形								
	彩画	柱头	无	柱身	无	梁架	有	走马板	有								
		门、窗	有	天花	有	藻井	有	其他彩画	——								
	其他	悬塑	无	佛龛	有	匾额	无										
装饰	室内	帷幔	无	幕帘彩绘	有	壁画	有	唐卡	无								
		经幡	无	经幢	无	柱毯	有	其他									
	室外	玛尼轮	无	苏勒德	无	宝顶	有	祥麟法轮	无								
		四角经幢	无	经幡	无	铜饰	无	石刻、砖雕	有								
		仙人走兽	1仙人+3走兽	壁画	有	其他	——										
陈设	室内	主佛像	弥勒佛	佛像基座	莲花座												
		法座	无	藏经橱	有	经床	无	诵经桌	无	法鼓	无	玛尼轮	无	坛城	无	其他	——
	室外	旗杆	无	苏勒德	无	狮子	无	经幡	无	玛尼轮	无	香炉	有	五供	无	其他	——
	其他	——															
备注	因该殿内存放大量经书，有时也称其为藏经阁																
调查日期	2010/08/11	调查人员	杜娟	整理日期	2010/08/12	整理人员	杜娟										

弥勒殿基本概况表1

准格尔召·弥勒殿·档案照片

照片名称 	斜前方	照片名称 	正立面	照片名称 	斜后方
照片名称 	背立面	照片名称 	柱身	照片名称 	柱头
照片名称 	柱础	照片名称 	门	照片名称 	室外布局
照片名称 	室内正面	照片名称 	室内侧面	照片名称 	室内天花
照片名称 	室内柱身	照片名称 	室内局部	照片名称 	室外局部
备注	——				
摄影日期	2010/08/12	摄影人员	高亚涛		

弥勒殿基本概况表2

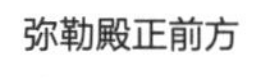

弥勒殿正前方

弥勒殿背面（左图）
弥勒殿室内侧面（右图）

弥勒殿佛像（左图）
弥勒殿室内顶棚
（右上图）
弥勒殿室内柱子
（右下图）

1.5 准格尔召·观音庙

单位:毫米

建筑名称	汉语正式名称	观音庙				俗称	——					
概述	初建年	不详				建筑朝向	南		建筑层数	一		
	建筑简要描述	汉藏结合式建筑										
	重建重修记载	不详										
		信息来源	——									
结构规模	结构形式	砖木		相连的建筑	无				室内天井	无		
	建筑平面形式	凸字形		外廊形式	前廊							
	通面阔	9320	开间数	3	明间	3250	次间	3035	梢间	——	次梢间 ——	尽间 ——
	通进深	9680	进深数	4	进深尺寸（前→后）		1550→1450→3237→3438					
	柱子数量	——	柱子间距	横向尺寸	——		（藏式建筑结构体系填写此栏，不含廊柱）					
				纵向尺寸	——							
	其他	——										
建筑主体（大木作）（石作）（瓦作）	屋顶	屋顶形式	歇山顶与藏式平屋顶相结合				瓦作	灰色布瓦				
	外墙	主体材料	砖	材料规格	150×320×60		饰面颜色	白				
		墙体收分	有	边玛檐墙	有		边玛材料	砖（刷红）				
	斗栱、梁架	斗栱	无	平身科斗口尺寸		——	梁架关系	五檩无廊				
	柱、柱式（前廊柱）	形式	汉式	柱身断面形状	方	断面尺寸	230×230		（在没有前廊柱的情况下，填写室内柱及其特征）			
		柱身材料	木材	柱身收分	无	栌斗、托木	无	雀替	有			
		柱础	有	柱础形状	方	柱础尺寸	280×280					
	台基	台基类型	普通式台基	台基高度	330	台基地面铺设材料		砖、条石				
	其他	——										
装修（小木作）（彩画）	门(正面)	板门		门楣	无	堆经	无	门帘	无			
	窗（正面）	无		窗楣	——	窗套	——	窗帘	——			
	室内隔扇	隔扇	有	隔扇位置	——							
	室内地面、楼面	地面材料及规格		方砖300×300		楼面材料及规格		——				
	室内楼梯	楼梯	无	楼梯位置	——	楼梯材料	——	梯段宽度	——			
	天花、藻井	天花	无	天花类型	——	藻井	无	藻井类型	——			
	彩画	柱头	无	柱身	无	梁架	有	走马板	有			
		门、窗	无	天花	——	藻井	——	其他彩画	——			
	其他	悬塑	无	佛龛	无	匾额	无					
装饰	室内	帷幔	无	幕帘彩绘	有	壁画	有	唐卡	无			
		经幡	有	经幢	无	柱毯	无	其他	——			
	室外	玛尼轮	无	苏勒德	无	宝顶	有	祥麟法轮	有			
		四角经幢	有	经幡	无	铜饰	无	石刻、砖雕	无			
		仙人走兽	1仙人+3走兽	壁画	无	其他	——					
陈设	室内	主佛像	千手千眼观音			佛像基座	莲花座					
		法座 无	藏经橱 无	经床 无	诵经桌 无	法鼓 无	玛尼轮 有	坛城 无	其他 ——			
	室外	旗杆 无	苏勒德 无	狮子 无	经幡 无	玛尼轮 无	香炉 无	五供 无	其他 ——			
	其他	——										
备注	——											
调查日期	2010/08/11	调查人员	杜娟	整理日期	2010/08/12	整理人员	杜娟					

观音庙基本概况表1

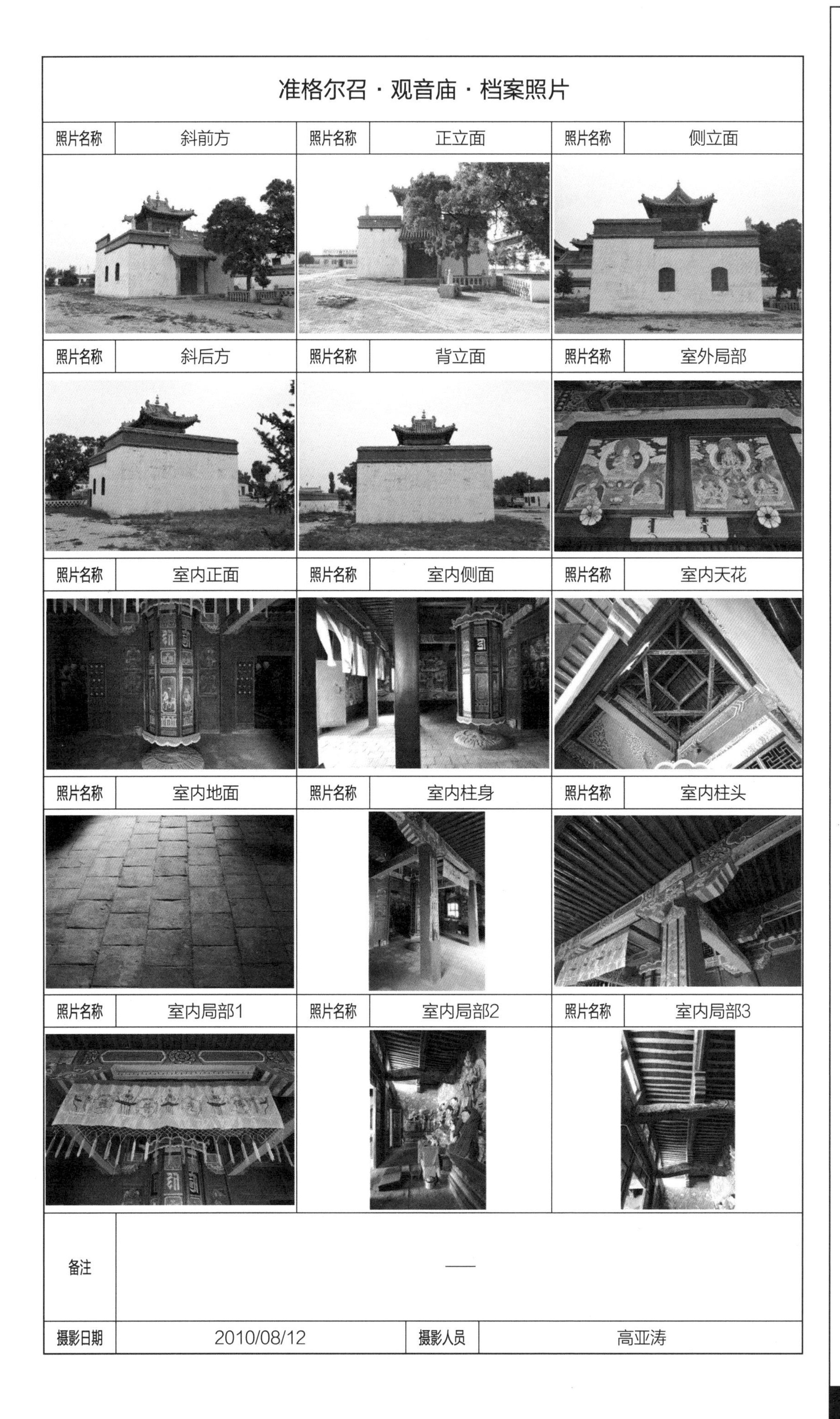

准格尔召·观音庙·档案照片

照片名称	斜前方	照片名称	正立面	照片名称	侧立面
照片名称	斜后方	照片名称	背立面	照片名称	室外局部
照片名称	室内正面	照片名称	室内侧面	照片名称	室内天花
照片名称	室内地面	照片名称	室内柱身	照片名称	室内柱头
照片名称	室内局部1	照片名称	室内局部2	照片名称	室内局部3

备注	——		
摄影日期	2010/08/12	摄影人员	高亚涛

观音庙斜前方

观音庙斜后方（左图）
观音庙室内佛像（右图）

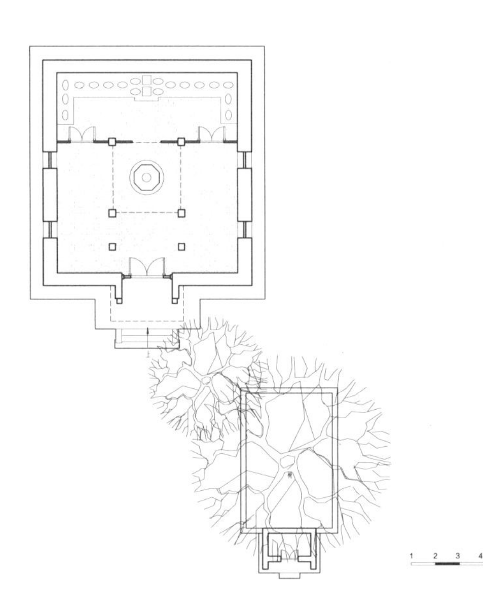

观音庙一层平面图(左图)
观音庙室内柱子（右图）

1.6 准格尔召 · 舍利殿

单位:毫米

建筑名称	汉语正式名称	舍利殿	俗称	——															
概述	初建年	不详	建筑朝向	南	建筑层数	一													
	建筑简要描述	汉藏结合式建筑																	
	重建重修记载	不详																	
		信息来源	——																
结构规模	结构形式	砖木	相连的建筑	无	室内天井	无													
	建筑平面形式	凸字形	外廊形式	前廊															
	通面阔	14450	开间数	5	明间	2860	次间	2855	梢间	2940	次梢间	——	尽间	——					
	通进深	14340	进深数	5	进深尺寸（前→后）	2810→2870→2860→2860→2940													
	柱子数量	——	柱子间距	横向尺寸	——	（藏式建筑结构体系填写此栏，不含廊柱）													
				纵向尺寸	——														
	其他	——																	
建筑主体（大木作）（石作）（瓦作）	屋顶	屋顶形式	歇山	瓦作	灰色布瓦														
	外墙	主体材料	砖	材料规格	160×320×70	饰面颜色	白												
		墙体收分	无	边玛檐墙	无	边玛材料	——												
	斗栱、梁架	斗栱	无	平身科斗口尺寸	——	梁架关系	十一檩有廊												
	柱、柱式（前廊柱）	形式	汉式	柱身断面形状	圆	断面尺寸	直径 D=200		（在没有前廊柱的情况下，填写室内柱及其特征）										
		柱身材料	木材	柱身收分	无	栌斗、托木	无	雀替	无										
		柱础	有	柱础形状	圆形	柱础尺寸	直径 D=280												
	台基	台基类型	普通式台基	台基高度	300	台基地面铺设材料	方砖												
	其他	——																	
装修（小木作）（彩画）	门(正面)	板门	门楣	无	堆经	无	门帘	无											
	窗（正面）	藏式明窗	窗楣	无	窗套	无	窗帘	无											
	室内隔扇	隔扇	无	隔扇位置	——														
	室内地面、楼面	地面材料及规格	条木、规格不一	楼面材料及规格	——														
	室内楼梯	楼梯	无	楼梯位置	——	楼梯材料	——	梯段宽度	——										
	天花、藻井	天花	无	天花类型	——	藻井	无	藻井类型	——										
	彩画	柱头	无	柱身	无	梁架	有	走马板	有										
		门、窗	无	天花	——	藻井	——	其他彩画	——										
	其他	悬塑	无	佛龛	无	匾额	无												
装饰	室内	帷幔	无	幕帘彩绘	无	壁画	有	唐卡	有										
		经幡	有	经幢	有	柱毯	有	其他	——										
	室外	玛尼轮	无	苏勒德	无	宝顶	有	祥麟法轮	无										
		四角经幢	无	经幡	无	铜饰	无	石刻、砖雕	无										
		仙人走兽	1+5	壁画	有	其他	——												
陈设	室内	主佛像	无	佛像基座	——														
		法座	无	藏经橱	无	经床	无	诵经桌	无	法鼓	无	玛尼轮	无	坛城	无	其他	——		
	室外	旗杆	无	苏勒德	无	狮子	无	经幡	无	玛尼轮	无	香炉	无	五供	无	其他	——		
	其他	——																	
备注	准格尔召的舍利殿中供奉着两座佛塔，一座银塔是准格尔活佛的舍利骨灰塔，另一座铜塔是甘肃省天祝县祝贡寺活佛的舍利骨灰塔。两座舍利塔的后方从左至右供奉着金刚瑜伽母、时轮金刚、胜乐金刚、密集金刚、大威德金刚、大轮金刚、欢喜金刚																		
调查日期	2010/08/11	调查人员	杜娟	整理日期	2010/08/12	整理人员	杜娟												

舍利殿基本概况表1

准格尔召·舍利殿·档案照片

照片名称 	斜前方	照片名称 	正立面	照片名称 	正侧面
照片名称 	斜后方	照片名称 	背立面	照片名称 	柱头
照片名称 	柱础	照片名称 	室外柱子	照片名称 	室内正面
照片名称 	室内侧面	照片名称 	室内地面	照片名称 	室内局部1
照片名称 	室内局部2	照片名称 	室内局部3	照片名称 	室内局部4
备注	——				
摄影日期	2010/08/22	摄影人员	高雅涛		

舍利殿基本概况表2

舍利殿正前方

舍利殿侧面

舍利殿室内柱子（左图）
舍利殿室内梁架（右图）

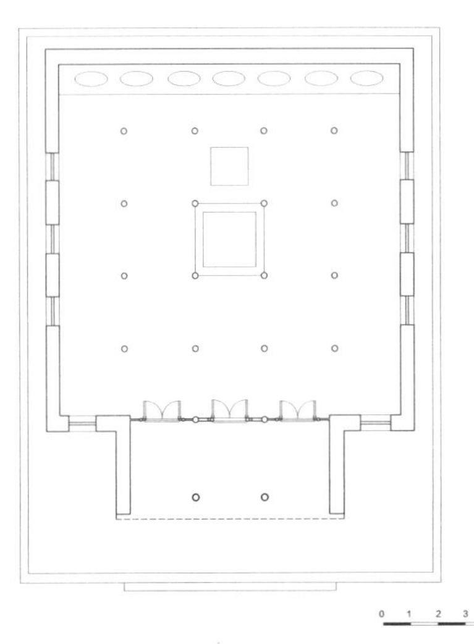

舍利殿一层平面图（左图）
舍利殿室内陈设（右图）

1.7 准格尔召 · 五道庙

单位:毫米

建筑名称	汉语正式名称	护法五王殿	俗称	五道庙													
概述	初建年	约清同治年间	建筑朝向	南	建筑层数	二											
	建筑简要描述	汉藏结合式建筑															
	重建重修记载	不详															
		信息来源	——														
结构规模	结构形式	砖木	相连的建筑	无	室内天井	都纲法式											
	建筑平面形式	凹字形	外廊形式	前廊													
	通面阔	16250	开间数	5	明间	3210	次间	3000	梢间	3250	次梢间	——	尽间	——			
	通进深	19295	进深数	7	进深尺寸（前→后）	3942→1968→1930→1630→1650→1970→1980											
	柱子数量	——	柱子间距	横向尺寸	——	（藏式建筑结构体系填写此栏，不含廊柱）											
				纵向尺寸	——												
	其他	——															
建筑主体（大木作）（石作）（瓦作）	屋顶	屋顶形式	歇山顶与藏式平屋顶相结合	瓦作	——												
	外墙	主体材料	砖	材料规格	240×300×60	饰面颜色	白										
		墙体收分	有	边玛檐墙	有	边玛材料	砖（刷红）										
	斗栱、梁架	斗栱	无	平身科斗口尺寸	——	梁架关系	五檩前廊										
	柱、柱式（前廊柱）	形式	藏式	柱身断面形状	方	断面尺寸	200×200	（在没有前廊柱的情况下，填写室内柱及其特征）									
		柱身材料	木材	柱身收分	无	栌斗、托木	有	雀替	无								
		柱础	有	柱础形状	方形	柱础尺寸	360×360										
	台基	台基类型	普通式台基	台基高度	400	台基地面铺设材料	方砖										
	其他	——															
装修（小木作）（彩画）	门(正面)	板门	门楣	无	堆经	有	门帘	无									
	窗（正面）	藏式明窗	窗楣	无	窗套	无	窗帘	无									
	室内隔扇	隔扇	有	隔扇位置	——												
	室内地面、楼面	地面材料及规格	条木、规格不一	楼面材料及规格	——												
	室内楼梯	楼梯	无	楼梯位置	——	楼梯材料	——	梯段宽度	——								
	天花、藻井	天花	无	天花类型	——	藻井	无	藻井类型	——								
	彩画	柱头	无	柱身	无	梁架	有	走马板	无								
		门、窗	无	天花	——	藻井	——	其他彩画	——								
	其他	悬塑	无	佛龛	有	匾额	无										
装饰	室内	帷幔	有	幕帘彩绘	无	壁画	无	唐卡	无								
		经幡	有	经幢	有	柱毯	无	其他									
	室外	玛尼轮	无	苏勒德	无	宝顶	有	祥麟法轮	有								
		四角经幢	有	经幡	无	铜饰	有	石刻、砖雕	无								
		仙人走兽	无	壁画	无	其他	——										
陈设	室内	主佛像	护法五王	佛像基座	莲花座												
		法座	无	藏经橱	无	经床	有	诵经桌	有	法鼓	有	玛尼轮	无	坛城	无	其他	——
	室外	旗杆	无	苏勒德	无	狮子	有	经幡	无	玛尼轮	无	香炉	有	五供	无	其他	——
	其他	——															
备注	——																
调查日期	2010/08/11	调查人员	杜娟	整理日期	2010/08/12	整理人员	杜娟										

五道庙基本概况表1

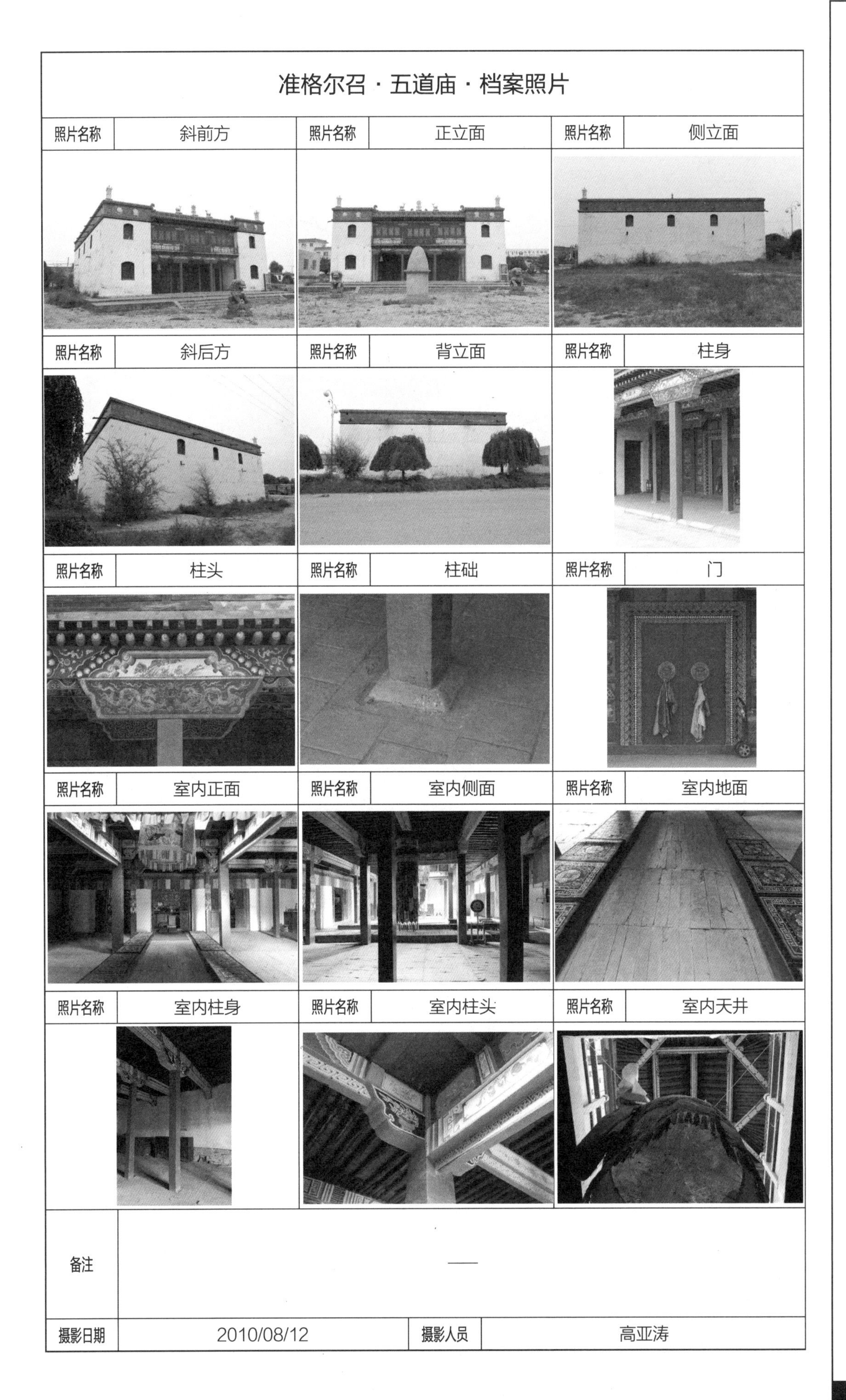

准格尔召 · 五道庙 · 档案照片

照片名称	斜前方	照片名称	正立面	照片名称	侧立面
照片名称	斜后方	照片名称	背立面	照片名称	柱身
照片名称	柱头	照片名称	柱础	照片名称	门
照片名称	室内正面	照片名称	室内侧面	照片名称	室内地面
照片名称	室内柱身	照片名称	室内柱头	照片名称	室内天井
备注	—				
摄影日期	2010/08/12	摄影人员	高亚涛		

五道庙基本概况表2

五道庙正面

五道庙斜后方（左图）
五道庙正面门（右图）

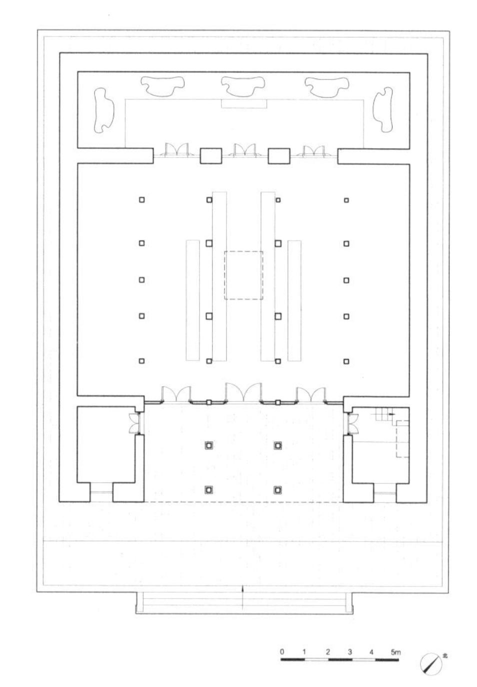

五道庙室外柱头（左上图）
五道庙室内正面（左下图）
五道庙一层平面图（右图）

1.8 准格尔召·六臂护法殿

单位:毫米

建筑名称	汉语正式名称	大黑天护法殿	俗称	六臂护法殿													
概述	初建年	不详	建筑朝向	南	建筑层数	一											
	建筑简要描述	藏式建筑															
	重建重修记载	不详															
		信息来源	——														
结构规模	结构形式	砖木	相连的建筑	无	室内天井	无											
	建筑平面形式	长方形	外廊形式	前廊													
	通面阔	9855	开间数	3	明间	——	次间	——	梢间	——	次梢间	——	尽间	——			
	通进深	11500	进深数	5	进深尺寸（前→后）	——											
	柱子数量	8	柱子间距	横向尺寸	3075、3390	（藏式建筑结构体系填写此栏，不含廊柱）											
				纵向尺寸	2105、1780、1825												
	其他	——															
建筑主体（大木作）（石作）（瓦作）	屋顶	屋顶形式	藏式平屋顶	瓦作	——												
	外墙	主体材料	砖	材料规格	150×315×60	饰面颜色	白										
		墙体收分	有	边玛檐墙	有	边玛材料	砖（刷红）										
	斗栱、梁架	斗栱	无	平身科斗口尺寸	——	梁架关系	梁纵排架										
	柱、柱式（前廊柱）	形式	藏式	柱身断面形状	圆	断面尺寸	直径 D=185	（在没有前廊柱的情况下，填写室内柱及其特征）									
		柱身材料	木材	柱身收分	无	栌斗、托木	有	雀替	无								
		柱础	有	柱础形状	方形	柱础尺寸	500×500										
	台基	台基类型	普通式台基	台基高度	680	台基地面铺设材料	方砖										
	其他	——															
装修（小木作）（彩画）	门(正面)	板门	门楣	无	堆经	无	门帘	无									
	窗（正面）	藏式明窗	窗楣	无	窗套	无	窗帘	无									
	室内隔扇	隔扇	无	隔扇位置	——												
	室内地面、楼面	地面材料及规格	方砖300×300	楼面材料及规格	——												
	室内楼梯	楼梯	无	楼梯位置	——	楼梯材料	——	梯段宽度	——								
	天花、藻井	天花	无	天花类型	——	藻井	无	藻井类型	——								
	彩画	柱头	无	柱身	无	梁架	有	走马板	有								
		门、窗	无	天花	——	藻井	——	其他彩画	——								
	其他	悬塑	无	佛龛	无	匾额	无										
装饰	室内	帷幔	无	幕帘彩绘	无	壁画	有	唐卡	无								
		经幡	无	经幢	无	柱毯	无	其他									
	室外	玛尼轮	无	苏勒德	无	宝顶	有	祥麟法轮	有								
		四角经幢	有	经幡	有	铜饰	有	石刻、砖雕	无								
		仙人走兽	无	壁画	无	其他	——										
陈设	室内	主佛像	六臂护法大黑天	佛像基座	莲花座												
		法座	无	藏经橱	无	经床	无	诵经桌	无	法鼓	无	玛尼轮	无	坛城	无	其他	——
	室外	旗杆	无	苏勒德	无	狮子	无	经幡	有	玛尼轮	无	香炉	无	五供	无	其他	——
	其他	——															
备注	——																
调查日期	2010/08/11	调查人员	杜娟	整理日期	2010/08/12	整理人员	杜娟										

六臂护法殿基本概况表1

准格尔召·六臂护法殿·档案照片

照片名称 	斜前方	照片名称 	正立面	照片名称 	侧立面
照片名称 	斜后方	照片名称 	柱础	照片名称 	台基
照片名称 	门	照片名称 	室外局部	照片名称 	窗户
照片名称 	室内正面	照片名称 	室内地面	照片名称 	室内柱身
照片名称 	室内柱头	照片名称 	室内柱础	照片名称 	室内局部
备注	——				
摄影日期	2010/08/12	摄影人员	高亚涛		

六臂护法殿基本概况表2

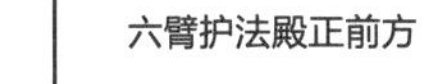

六臂护法殿正前方

六臂护法殿侧面（左图）
六臂护法殿室内正面（右图）

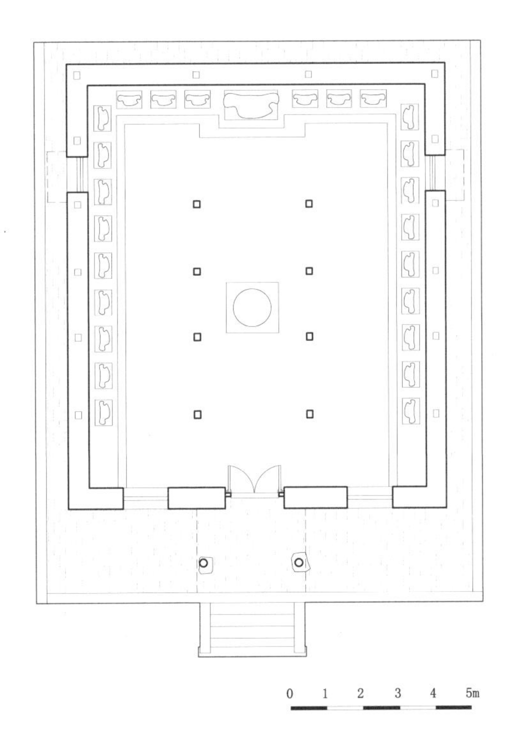

六臂护法殿室内柱子（左图）
六臂护法一层平面图（右图）

1.9 准格尔召·千佛殿

单位:毫米

建筑名称	汉语正式名称	千佛殿				俗称	——							
概述	初建年	不详				建筑朝向	南		建筑层数	一				
	建筑简要描述	汉藏结合式建筑												
	重建重修记载	不详												
		信息来源	——											
结构规模	结构形式	砖木	相连的建筑	无					室内天井	天井				
	建筑平面形式	方形	外廊形式	前廊										
	通面阔	16210	开间数	5	明间	3200	次间	3220	梢间	3285	次梢间	——	尽间	——
	通进深	24220	进深数	7	进深尺寸（前→后）				2290→2290→2280→2290→2280→2280→2540					
	柱子数量	——	柱子间距	横向尺寸	——				（藏式建筑结构体系填写此栏，不含廊柱）					
				纵向尺寸	——									
	其他	——												
建筑主体（大木作）（石作）（瓦作）	屋顶	屋顶形式	卷棚、硬山与藏式平屋顶相结合				瓦作	灰布瓦						
	外墙	主体材料	砖	材料规格	160×300×60		饰面颜色	白						
		墙体收分	无	边玛檐墙	有		边玛材料	砖（刷红）						
	斗栱、梁架	斗栱	无	平身科斗口尺寸		——	梁架关系	梁纵排架						
	柱、柱式（前廊柱）	形式	藏式	柱身断面形状	方	断面尺寸	210×210			（在没有前廊柱的情况下，填写室内柱及其特征）				
		柱身材料	混凝土	柱身收分	无	栌斗、托木	有	雀替	无					
		柱础	有	柱础形状	方	柱础尺寸	300×300							
	台基	台基类型	普通式台基	台基高度	580	台基地面铺设材料		方砖						
	其他	——												
装修（小木作）（彩画）	门(正面)	板门		门楣	无	堆经	有	门帘	无					
	窗（正面）	藏式明窗		窗楣	无	窗套	无	窗帘	无					
	室内隔扇	隔扇	有	隔扇位置	——									
	室内地面、楼面	地面材料及规格		方砖285×285		楼面材料及规格		——						
	室内楼梯	楼梯	无	楼梯位置	——	楼梯材料	——	梯段宽度	——					
	天花、藻井	天花	有	天花类型	井口天花	藻井	无	藻井类型	——					
	彩画	柱头	无	柱身	无	梁架	有	走马板	有					
		门、窗	无	天花	——	藻井	——	其他彩画	——					
	其他	悬塑	无	佛龛	有	匾额	无							
装饰	室内	帷幔	无	幕帘彩绘	无	壁画	无	唐卡	无					
		经幡	无	经幢	无	柱毯	无	其他						
	室外	玛尼轮	无	苏勒德	无	宝顶	有	祥麟法轮	有					
		四角经幢	有	经幡	无	铜饰	有	石刻、砖雕	无					
		仙人走兽	1仙人+3走兽	壁画	有	其他	——							
陈设	室内	主佛像	宗喀巴大师			佛像基座	莲花座							
		法座	有	藏经橱	无	经床	有	诵经桌	无	法鼓	无	玛尼轮	无	坛城	无	其他	——
	室外	旗杆	无	苏勒德	无	狮子	无	经幡	无	玛尼轮	无	香炉	有	五供	无	其他	——
	其他	——												
备注	千佛殿，原为扎仓庙，是喇嘛讲经的地方。2004年请进了1000尊宗喀巴大师的塑像，因此被称为千佛殿													
调查日期	2010/08/11	调查人员	杜娟	整理日期	2010/08/12	整理人员	杜娟							

千佛殿基本概况表1

准格尔召 · 千佛殿 · 档案照片

照片名称	斜前方	照片名称	正立面	照片名称	斜后方
照片名称	柱身	照片名称	柱础	照片名称	台基
照片名称	门	照片名称	室内正面	照片名称	室内侧面
照片名称	室内天花	照片名称	室内柱身	照片名称	室内柱头
照片名称	室内柱础	照片名称	室内局部1	照片名称	室内局部2

备注	——		
摄影日期	2010/08/12	摄影人员	高亚涛

千佛殿斜前方

千佛殿斜后方

千佛殿室内正面
（左图）
千佛殿佛殿内佛像
（右图）

千佛殿佛殿室内侧面
（左图）
千佛殿经堂室内侧面
（右图）

乌审召

Wushenzhao Temple

2 乌审召 Wushenzhao Temple

乌审召建筑群

乌审召为原伊克昭盟鄂尔多斯右翼前旗（乌审旗）寺庙，系该旗规模最为宏大的寺庙。寺庙管辖巴音陶勒盖庙、哈西雅图诵经会、查干庙、布日都庙、努恒庙、新庙、陶格其纳日诵经会、拉布隆庙、曼哈图庙、巴日松古庙、呼吉日图庙、乌兰陶勒盖庙、梅林庙、达日罕喇嘛庙、乌日图诵经会、嘎鲁图庙、海流图庙、陶日木庙等18座属庙与诵经会。

乾隆年间，安多藏区僧人囊苏喇嘛云游至乌审旗东部，并结识乌审旗扎萨克固山贝子若西斯仁。二人商定在乌审旗建庙弘法，并于乾隆元年（1736年）至乾隆五年（1740年）间，在善达河源塔本哈日陶勒盖之地新建德都庙，囊素喇嘛驻锡于西殿，将器具存放于东殿。该庙初建时称囊苏庙，后改称甘珠尔庙。光绪四年（1878年），乌审旗阿格公察克图尔斯仁等为该寺献一尊鎏金释迦牟尼像，此后寺庙被称为乌审召。

寺庙建筑风格为汉藏结合式建筑。“文化大革命”前寺庙有49间双层大雄宝殿、49间双层参尼殿、44间显宗殿、25间双层德都庙、25间护法殿、25间时轮殿、9间无量寿佛殿、9间普明佛殿、9间大黑天殿、9间胜乐金刚殿、9间密集金刚殿、9间弥勒殿、9间白伞盖佛母殿、4间龙王殿、4间法轮殿、3间度母殿、3间吉祥天女殿、3间大密九尊马头明王殿、3间罗汉殿、3间药师殿、3间女神殿、3间土地神殿、3间罗本喇嘛大黑天殿等大小25座殿堂，18座庙仓、大小203座佛塔。

“文化大革命”中寺庙严重损毁。1984年该寺僧人清理修缮殿宇，正式恢复法会。至2006年，经二十余年的建设，修复并建全了该庙主要建筑。

参考文献：

[1] 莫·哈斯苏度.乌审召简史（蒙古文）.“阿拉腾甘德尔”期刊专辑,1989，3.

[2] 阿·哈斯朝格图,额尔克固特·巴布.乌审召—历史悠久的乌审召暨乌审召修缮记（蒙古文）.呼和浩特，阿儿含只文化有限责任公司,2007，6.

[3] 色·那日松,和日布忠乃.博格多活佛与乌审召（蒙古文）.呼和浩特:内蒙古人民出版社,2008，12.

乌审召·基本概况 1

<table>
<tr><td rowspan="3">寺院蒙古语藏语名称</td><td>蒙古语</td><td>ᠦᠦᠰᠢᠨ ᠵᠤᠤ</td><td rowspan="2">寺院汉语名称</td><td>汉语正式名称</td><td colspan="3">甘珠尔庙</td></tr>
<tr><td>藏语</td><td>བདེ་ཆེན་དམ་ཆོས་བཀའ་གསལ་གླིང་།</td><td>俗称</td><td colspan="3">乌审召</td></tr>
<tr><td>汉语语义</td><td>乌审召</td><td colspan="2">寺院汉语名称的由来</td><td colspan="3">清廷赐名</td></tr>
<tr><td>所在地</td><td colspan="4">鄂尔多斯市乌审旗乌审召镇</td><td>东经 109° 01′</td><td colspan="2">北纬 39° 09′</td></tr>
<tr><td>初建年</td><td colspan="2">1736—1740年</td><td>保护等级</td><td colspan="4">自治区级文物保护单位</td></tr>
<tr><td>盛期时间</td><td colspan="2">不详</td><td colspan="2">盛期喇嘛僧/现有喇嘛僧数</td><td colspan="3">3000人／100人</td></tr>
<tr><td rowspan="2">历史沿革</td><td colspan="7">1734年，新建30间双层大雄宝殿。
1777年，新建显宗殿。
1829年，新建时轮金刚殿。
1868—1871年，同治年间回民起义中寺庙多次受损。
1874年，重建49间双层大雄宝殿。
1888年，重建49间双层显宗殿。
1940—1942年，在九世班禅的建议下新建扎荣卡修尔佛塔一座。相传该塔在世界上仅有三座。
1943年，寺庙被国民党二十六师部分士兵洗劫。
至1958年，乌审召仍有宗教活动。
1966—1968年，寺庙建筑被拆毁，仅存德都庙、时轮殿、吉祥天女殿、活佛仓佛殿等四座殿宇与一座佛塔。
1984年，乌审召恢复宗教活动。
1989—1991年，修缮乌审召。
2005—2006年，修缮并复建乌审召殿堂与院落</td></tr>
<tr><td>资料来源</td><td colspan="6">[1]莫·哈斯苏度.乌审召简史（蒙古文）.“阿拉腾甘德尔”期刊专辑,1989，3.
[2]阿·哈斯朝格图,额尔克固特·巴布.乌审召——历史悠久的乌审召暨乌审召修缮记（蒙古文）.呼和浩特:阿儿含只文化有限责任公司,2007，6 .
[3]色·那日松，和日布忠乃编著.博格多活佛与乌审召（蒙古文）.呼和浩特:内蒙古人民出版社,2008，12.
[4]调研访谈记录.</td></tr>
<tr><td rowspan="2">现状描述</td><td colspan="4" rowspan="2">召庙规模较大，主要单体建筑十余座，以一严整院落围合，其中有德都庙、时轮金刚殿、吉祥天女殿及活佛仓和一座佛塔为历史遗存。2008年新建了药王殿，其复建的扎荣卡修尔佛塔是国内仅有。召庙周边喇嘛住所形成的古聚落较有特色</td><td>描述时间</td><td colspan="2">2010/07/31</td></tr>
<tr><td>描述人</td><td colspan="2">杜娟</td></tr>
<tr><td>调查日期</td><td colspan="2">2010/07/31</td><td>调查人员</td><td colspan="4">宝山</td></tr>
</table>

乌审召基本概况表1

<table>
<tr><td colspan="7">乌审召・基本概况 2</td></tr>
<tr><td rowspan="4">现存建筑</td><td>大经堂</td><td>弥勒殿</td><td>时轮金刚殿</td><td rowspan="4">已毁建筑</td><td>却伊拉扎仓殿</td><td>甘珠尔庙（古庙）</td></tr>
<tr><td>钟楼庙</td><td>度母殿</td><td>德都殿</td><td>满巴扎仓</td><td>卓德巴扎仓</td></tr>
<tr><td>长寿佛殿</td><td>药师佛殿</td><td>活佛墒</td><td>丁科尔扎仓</td><td>——</td></tr>
<tr><td>法王殿</td><td>扎荣噶沙尔</td><td>六臂护法殿</td><td>信息来源</td><td>《鄂尔多斯寺院》</td></tr>
<tr><td>区位图</td><td colspan="6">鄂尔多斯市地图
达拉特旗 杭锦旗 东胜区 准格尔旗 康巴什区 伊金霍洛旗 乌审召 鄂托克旗 乌审旗 鄂托克前旗</td></tr>
<tr><td>总平面图</td><td colspan="6">A.法王殿
B.扎荣噶沙尔
C.六臂护法殿
D.白塔
E.度母殿
F.德都殿
G.长寿佛殿
H.活佛墒
I.喇嘛僧舍
J.时轮学院
K.钟楼庙
L.大经堂
M.闻思学院
N.弥勒殿
O.山门
P.苏勒德殿
Q.药王殿</td></tr>
<tr><td>调查日期</td><td colspan="3">2010/07/30</td><td>调查人员</td><td colspan="2">苍雁飞</td></tr>
</table>

乌审召基本概况表2

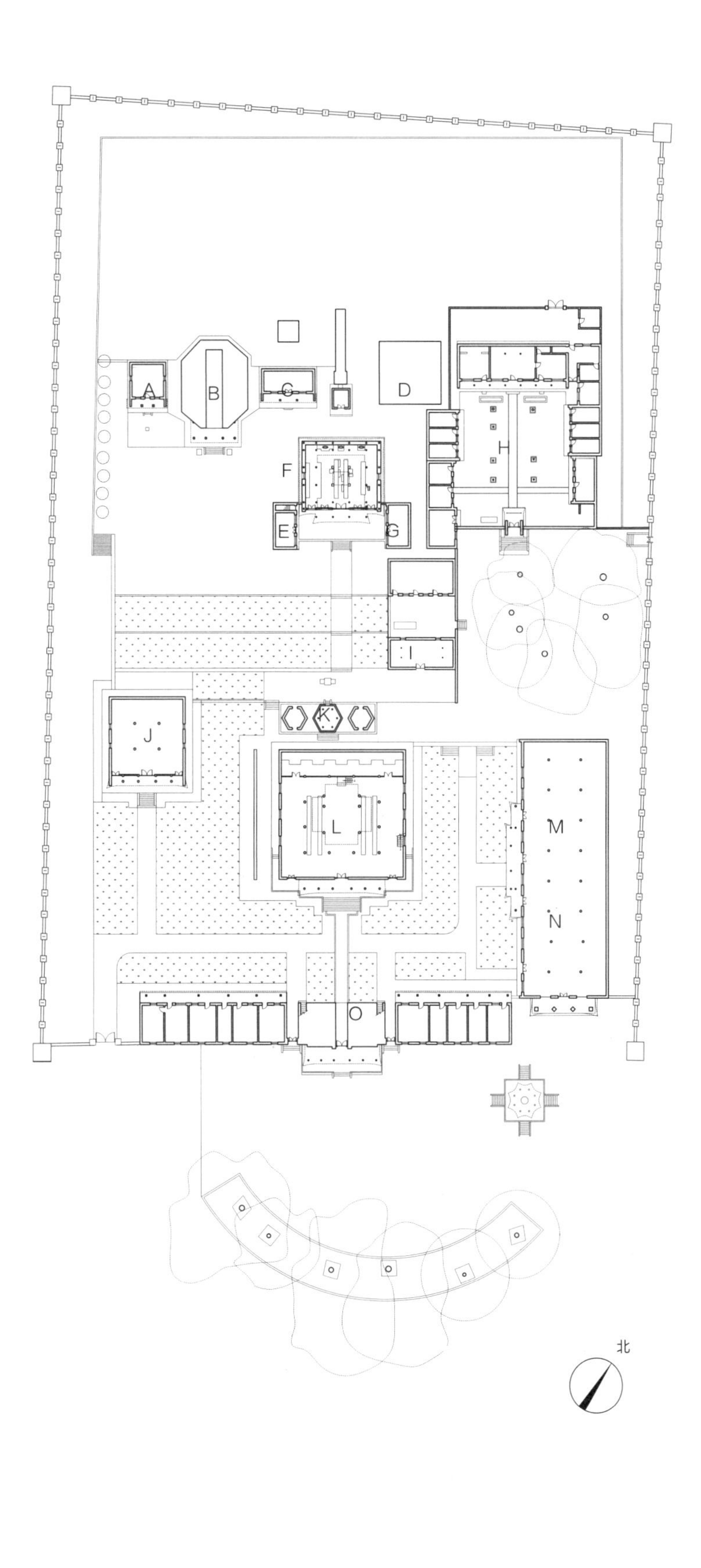

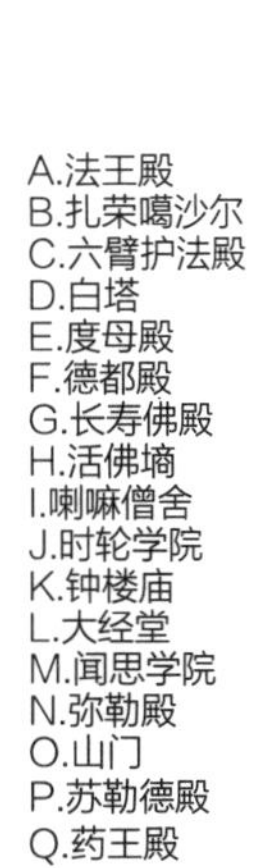

A.法王殿
B.扎荣噶沙尔
C.六臂护法殿
D.白塔
E.度母殿
F.德都殿
G.长寿佛殿
H.活佛墒
I.喇嘛僧舍
J.时轮学院
K.钟楼庙
L.大经堂
M.闻思学院
N.弥勒殿
O.山门
P.苏勒德殿
Q.药王殿

乌审召总平面图

2.1 乌审召·大经堂

单位:毫米

建筑名称	汉语正式名称	大经堂				俗称	朝格钦独贡										
概述	初建年	清朝乾隆年间				建筑朝向	南		建筑层数	三							
	建筑简要描述	汉藏结合式															
	重建重修记载	藏译为朝格钦独贡，建于清朝乾隆年间，由乌审旗第五任王爷热西斯仁出资修建。新建的大经堂是乌审召最大的建筑，占地面积453平方米，三层，是乌审召举行综合性法事活动的场所															
		信息来源	殿外牌匾														
结构规模	结构形式	钢筋混凝土	相连的建筑	无	室内天井	都纲法式											
	建筑平面形式	长方形	外廊形式	前廊													
	通面阔	21050	开间数	5	明间	4170	次间	4240	梢间	4200	次梢间	——	尽间	——			
	通进深	20910	进深数	5	进深尺寸（前→后）	4180→4230→4090→4230→4180											
	柱子数量	——	柱子间距	横向尺寸	——	（藏式建筑结构体系填写此栏，不含廊柱）											
				纵向尺寸	——												
	其他	——															
建筑主体（大木作）（石作）（瓦作）	屋顶	屋顶形式	重檐歇山+藏式平屋顶	瓦作	赭石色琉璃瓦屋面												
	外墙	主体材料	混凝土	材料规格	——	饰面颜色	白										
		墙体收分	无	边玛檐墙	无	边玛材料	——										
	斗栱、梁架	斗栱	（一层）无	平身科斗口尺寸	——	梁架关系	——										
	柱、柱式（前廊柱）	形式	藏式	柱身断面形状	圆	断面尺寸	直径D=230		（在没有前廊柱的情况下，填写室内柱及其特征）								
		柱身材料	木材	柱身收分	有	栌斗、托木	有	雀替	无								
		柱础	有	柱础形状	圆	柱础尺寸	直径D=300										
	台基	台基类型	带勾栏式台基	台基高度	1650	台基地面铺设材料	水磨石										
	其他	——															
装修（小木作）（彩画）	门(正面)	板门	门楣	有	堆经	有	门帘	有									
	窗（正面）	槛窗	窗楣	无	窗套	无	窗帘	无									
	室内隔扇	隔扇	有	隔扇位置	经堂后方												
	室内地面、楼面	地面材料及规格	砖600×400	楼面材料及规格	混凝土												
	室内楼梯	楼梯	右	楼梯位置	进正门右侧	楼梯材料	木材	梯段宽度	1000								
	天花、藻井	天花	有	天花类型	井口天花	藻井	有	藻井类型	四方变八方变圆								
	彩画	柱头	有	柱身	有	梁架	有	走马板	——								
		门、窗	有	天花	有	藻井	有	其他彩画	——								
	其他	悬塑	无	佛龛	有	匾额	无										
装饰	室内	帷幔	无	幕帘彩绘	无	壁画	有	唐卡	有								
		经幡	有	经幢	有	柱毯	有	其他									
	室外	玛尼轮	无	苏勒德	无	宝顶	有	祥麟法轮	有								
		四角经幢	无	经幡	有	铜饰	无	石刻、砖雕	无								
		仙人走兽	无	壁画	有	其他	——										
陈设	室内	主佛像	释迦牟尼佛	佛像基座	莲花座												
		法座	有	藏经橱	有	经床	有	诵经桌	有	法鼓	有	玛尼轮	无	坛城	无	其他	——
	室外	旗杆	无	苏勒德	无	狮子	无	经幡	无	玛尼轮	无	香炉	有	五供	无	其他	——
	其他	——															
备注	——																
调查日期	2010/07/31	调查人员	杜娟	整理日期	2010/08/02	整理人员	杜娟										

大经堂基本概况表1

乌审召·大经堂·档案照片

照片名称	斜前方	照片名称	正立面	照片名称	斜后方
照片名称	背立面	照片名称	柱头	照片名称	柱础
照片名称	台基	照片名称	门	照片名称	室外局部
照片名称	室内正面	照片名称	室内侧面	照片名称	室内天花
照片名称	室内柱身	照片名称	室内楼梯	照片名称	室内局部
备注	该寺庙照片拍摄时间在7月30日傍晚及7月31日下午两点之前，7月31日阴天，光线较暗，无云				
摄影日期	2010/07/31	摄影人员	高亚涛		

大经堂斜前方

大经堂背面（左图）
大经堂正门（右图）

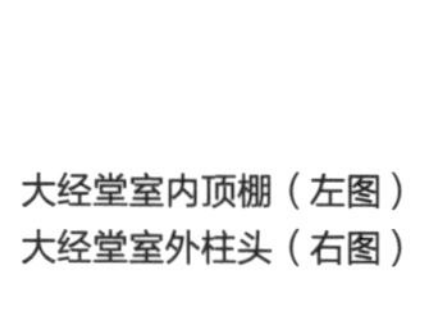

大经堂室内顶棚（左图）
大经堂室外柱头（右图）

大经堂室内正面（左图）
大经堂室内侧面（右图）

2.2 乌审召·德格都苏莫殿

单位:毫米

建筑名称	汉语正式名称	不详		俗称	德格都苏莫殿		
概述	初建年	不详		建筑朝向	南	建筑层数	二
	建筑简要描述	汉式建筑					
	重建重修记载	德格都苏莫殿，蒙古语意为上庙热西热布占楞，是乌审召最早建造的庙之一，也是保存情况较好、历史价值较高的一个寺庙					
		信息来源	殿外牌匾				

结构规模											
结构形式	砖木结构		相连的建筑	东西侧与度母殿、长寿佛殿以墙相连接				室内天井	都纲法式		
建筑平面形式	长方形		外廊形式	前廊							
通面阔	12778	开间数	5	明间	2800	次间	2855	梢间	2136	次梢间	——	尽间	——
通进深	10775	进深数	5	进深尺寸（前→后）		980→2900→2920→2815→1160					
柱子数量	——	柱子间距	横向尺寸	——		（藏式建筑结构体系填写此栏，不含廊柱）					
			纵向尺寸	——							
其他	——										

建筑主体（大木作）（石作）（瓦作）								
屋顶	屋顶形式	歇山			瓦作	布瓦		
外墙	主体材料	砖	材料规格	400×300×190	饰面颜色	青灰		
	墙体收分	无	边玛檐墙	无	边玛材料	——		
斗栱、梁架	斗栱	无	平身科斗口尺寸	——	梁架关系	六檩前廊		
柱、柱式（前廊柱）	形式	汉式	柱身断面形状	圆	断面尺寸	直径D=170		（在没有前廊柱的情况下，填写室内柱及其特征）
	柱身材料	木材	柱身收分	无	栌斗、托木	无	雀替	有
	柱础	有	柱础形状	圆	柱础尺寸	直径D=260		
台基	台基类型	普通式台基	台基高度	140	台基地面铺设材料		砖	
其他	——							

装修（小木作）（彩画）								
门(正面)	板门		门楣	无	堆经	无	门帘	无
窗（正面）	槛窗		窗楣	无	窗套	无	窗帘	无
室内隔扇	隔扇	无	隔扇位置	——				
室内地面、楼面	地面材料及规格		砖240×120×60，条形木板（规格不一）		楼面材料及规格		条形木板、规格不一	
室内楼梯	楼梯	有	楼梯位置	进正门右侧	楼梯材料	木材	梯段宽度	650
天花、藻井	天花	无	天花类型	——	藻井	无	藻井类型	——
彩画	柱头	有	柱身	无	梁架	无	走马板	——
	门、窗	有	天花	无	藻井	无	其他彩画	——
其他	悬塑	无	佛龛	无	匾额	无		

装饰								
室内	帷幔	无	幕帘彩绘	有	壁画	有	唐卡	有
	经幡	有	经幢	有	柱毯	有	其他	——
室外	玛尼轮	无	苏勒德	无	宝顶	有	祥麟法轮	无
	四角经幢	无	经幡	无	铜饰	无	石刻、砖雕	无
	仙人走兽	无	壁画	无	其他	——		

陈设																
室内	主佛像	宗喀巴、释迦牟尼佛、未来佛					佛像基座	莲花座								
	法座	无	藏经橱	有	经床	有	诵经桌	有	法鼓	有	玛尼轮	无	坛城	无	其他	——
室外	旗杆	无	苏勒德	无	狮子	无	经幡	无	玛尼轮	无	香炉	有	五供	无	其他	——
其他	——															

备注	初建年大约在300多年前，后因“文化大革命”期间作为库房而留存至今						
调查日期	2010/07/31	调查人员	杜娟	整理日期	2010/08/02	整理人员	杜娟

德格都苏莫殿基本概况表1

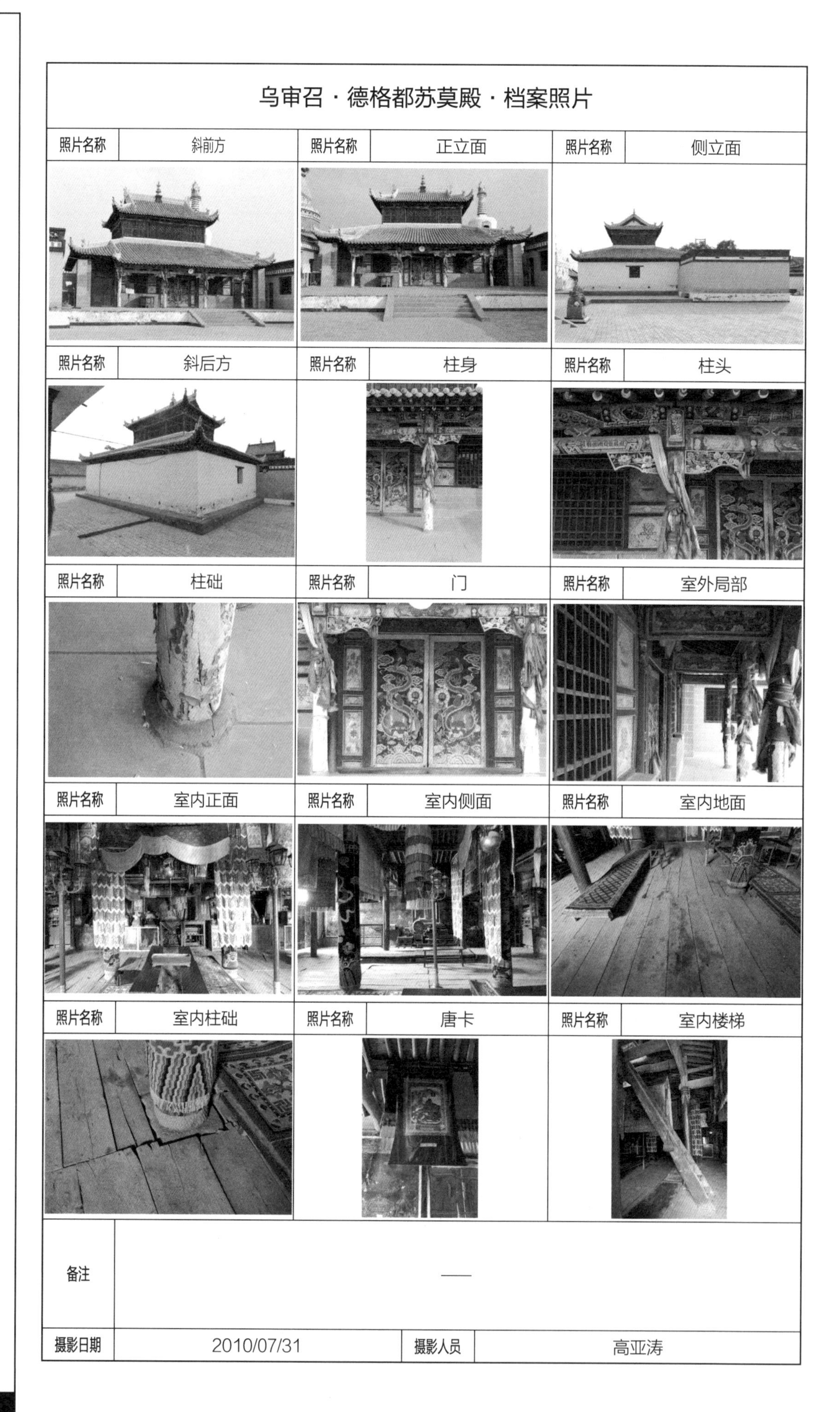

乌审召·德格都苏莫殿·档案照片

照片名称	斜前方	照片名称	正立面	照片名称	侧立面
照片名称	斜后方	照片名称	柱身	照片名称	柱头
照片名称	柱础	照片名称	门	照片名称	室外局部
照片名称	室内正面	照片名称	室内侧面	照片名称	室内地面
照片名称	室内柱础	照片名称	唐卡	照片名称	室内楼梯
备注	——				
摄影日期	2010/07/31	摄影人员	高亚涛		

德格都苏莫殿基本概况
表2

德格都苏莫殿正前方

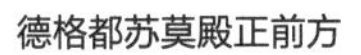

德格都苏莫殿室外柱头（左图）
德格都苏莫殿前廊（中图）
德格都苏莫殿正门（右图）

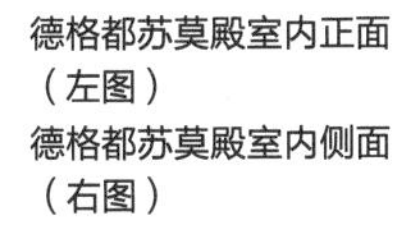

德格都苏莫殿室内正面（左图）
德格都苏莫殿室内侧面（右图）

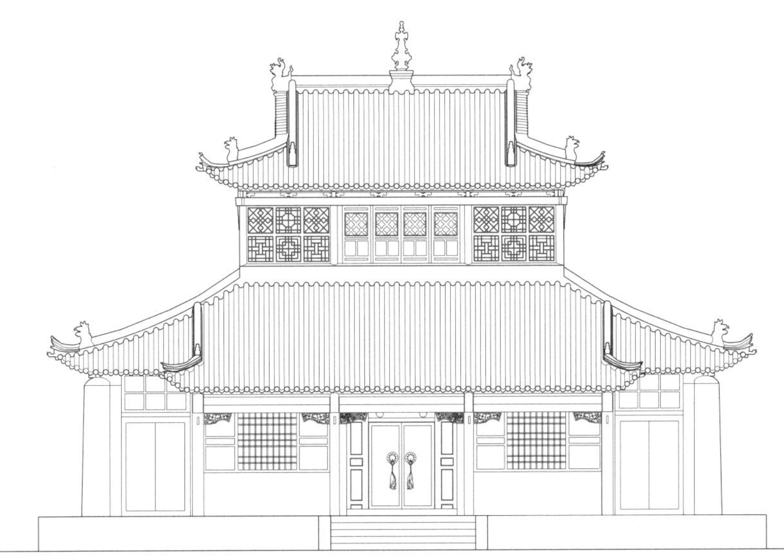

德格都苏莫殿正立面图

德格都苏莫殿一层平面图
（左图）
德格都苏莫殿二层平面图
（右图）

德格都苏莫殿剖面图1

注:本图由日本东京大学包慕萍研究员主持的“内蒙古自治区、青海藏传佛教文化遗产调查”项目测绘并提供。

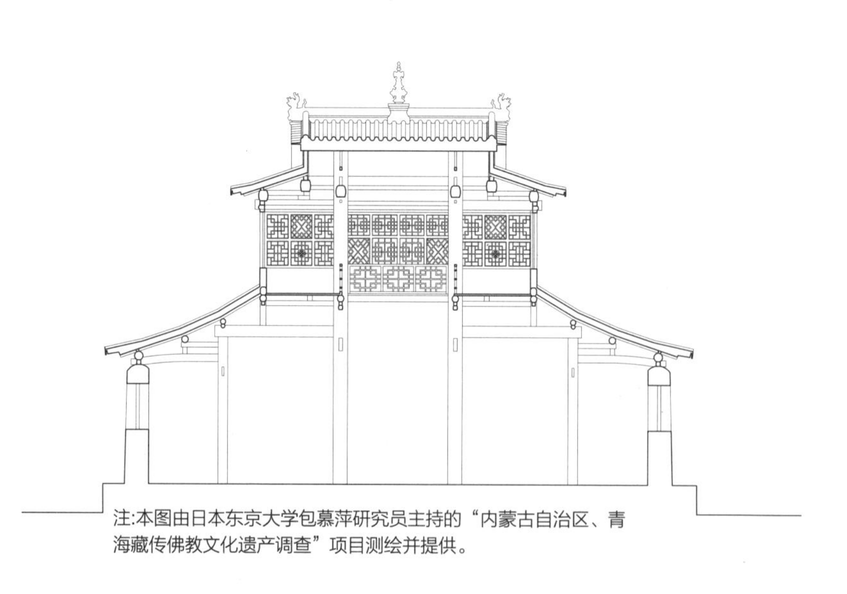

德格都苏莫殿剖面图2

注:本图由日本东京大学包慕萍研究员主持的“内蒙古自治区、青海藏传佛教文化遗产调查”项目测绘并提供。

2.3 乌审召·时轮金刚殿

单位:毫米

建筑名称	汉语正式名称	时轮金刚殿	俗称	东科尔殿、东科尔扎仓													
概述	初建年	道光八年（1828年）	建筑朝向	南	建筑层数	二
	建筑简要描述	汉藏结合式
	重建重修记载	藏译为东科尔扎仓，主要学习讲授数学、天文、地理、艺术、占卜、历法等，于1828年由协理陶岱修建，现有建筑面积是156平方米
		信息来源	殿外牌匾
结构规模	结构形式	钢筋混凝土	相连的建筑	无	室内天井	都纲法式
	建筑平面形式	方形	外廊形式	前廊		
	通面阔	12560	开间数	3	明间	4100	次间	4230	梢间	——	次梢间	——	尽间	——
	通进深	12600	进深数	3	进深尺寸（前→后）	4200→4200→4200
	柱子数量	——	柱子间距	横向尺寸	——	（藏式建筑结构体系填写此栏，不含廊柱）
				纵向尺寸	——	
	其他	——
建筑主体（大木作）（石作）（瓦作）	屋顶	屋顶形式	歇山与藏式平屋顶相结合	瓦作	布瓦
	外墙	主体材料	混凝土	材料规格	——	饰面颜色	朱红
		墙体收分	无	边玛檐墙	有	边玛材料	砖（刷红）
	斗栱、梁架	斗栱	无	平身科斗口尺寸	——	梁架关系	五檩有廊
	柱、柱式（前廊柱）	形式	汉式	柱身断面形状	圆	断面尺寸	直径D=200	（在没有前廊柱的情况下，填写室内柱及其特征）
		柱身材料	木材	柱身收分	无	栌斗、托木	无	雀替	有
		柱础	有	柱础形状	圆础上置莲花础	柱础尺寸	直径D=350
	台基	台基类型	普通式台基	台基高度	1105	台基地面铺设材料	砖（水泥抹平）
	其他	——
装修（小木作）（彩画）	门(正面)	板门	门楣	无	堆经	无	门帘	有
	窗（正面）	槛窗	窗楣	无	窗套	无	窗帘	无
	室内隔扇	隔扇	无	隔扇位置	——
	室内地面、楼面	地面材料及规格	瓷砖400×400	楼面材料及规格	水泥抹平
	室内楼梯	楼梯	有	楼梯位置	进正门右侧	楼梯材料	木材	梯段宽度	870
	天花、藻井	天花	无	天花类型	——	藻井	无	藻井类型	——
	彩画	柱头	有	柱身	有	梁架	有	走马板	有
		门、窗	有	天花	——	藻井	——	其他彩画	——
	其他	悬塑	无	佛龛	有	匾额	无
装饰	室内	帷幔	无	幕帘彩绘	无	壁画	有	唐卡	有
		经幡	有	经幢	有	柱毯	无	其他	
	室外	玛尼轮	无	苏勒德	有	宝顶	有	祥麟法轮	有
		四角经幢	有	经幡	有	铜饰	无	石刻、砖雕	无
		仙人走兽	无	壁画	有	其他	——
陈设	室内	主佛像	时轮金刚	佛像基座	无
		法座	有	藏经橱	无	经床	有	诵经桌	有	法鼓	有	玛尼轮	无	坛城	无	其他	——
	室外	旗杆	无	苏勒德	无	狮子	无	经幡	无	玛尼轮	无	香炉	有	五供	无	其他	——
	其他	——
备注	——
调查日期	2010/07/31	调查人员	杜娟	整理日期	2010/08/03	整理人员	杜娟

时轮金刚殿基本概况表1

乌审召·时轮金刚殿·档案照片

照片名称	斜前方	照片名称	正立面	照片名称	侧立面
照片名称	斜后方	照片名称	背立面	照片名称	柱身
照片名称	柱头	照片名称	柱础	照片名称	门
照片名称	室外台基	照片名称	室内柱子	照片名称	正面窗
照片名称	檐部 局部	照片名称	二层斜前方	照片名称	室外局部
备注	该殿全天举行法会，室内无法拍照				
摄影日期	2010/07/31	摄影人员	高亚涛		

时轮金刚殿基本概况表2

时轮金刚殿正前方

时轮金刚殿斜前方
（右图）

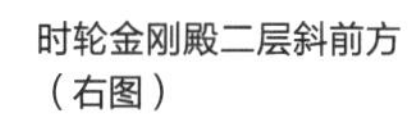

时轮金刚殿二层斜前方
（右图）

时轮金刚殿室外柱子
（左图）
时轮金刚殿背面
（右图）

2.4 乌审召·药师佛殿

单位:毫米

建筑名称	汉语正式名称	药师佛殿	俗称	敖特其独贡													
概述	初建年	2008年	建筑朝向	南	建筑层数	二											
	建筑简要描述	汉藏结合式															
	重建重修记载	2008年开始修建，并于当年完工															
		信息来源	喇嘛口述														
结构规模	结构形式	砖木	相连的建筑	无	室内天井	都纲法式											
	建筑平面形式	凹字形	外廊形式	前廊													
	通面阔	9300	开间数	5	明间	3300	次间	3000	梢间	3000	次梢间	——	尽间	——			
	通进深	12000	进深数	4	进深尺寸（前→后）	3000→3000→3000→3000											
	柱子数量	——	柱子间距	横向尺寸	——	（藏式建筑结构体系填写此栏，不含廊柱）											
				纵向尺寸	——												
	其他	——															
建筑主体（大木作）（石作）（瓦作）	屋顶	屋顶形式	歇山与藏式平屋顶相结合	瓦作	布瓦												
	外墙	主体材料	砖	材料规格	240×120×60	饰面颜色	灰										
		墙体收分	无	边玛檐墙	有	边玛材料	泥（刷红）										
	斗栱、梁架	斗栱	无	平身科斗口尺寸	——	梁架关系	——										
	柱、柱式（前廊柱）	形式	藏式	柱身断面形状	十二楞	断面尺寸	450×450	（在没有前廊柱的情况下，填写室内柱及其特征）									
		柱身材料	木材	柱身收分	无	栌斗、托木	有	雀替	无								
		柱础	有	柱础形状	圆形	柱础尺寸	直径 D=550										
	台基	台基类型	普通式台基	台基高度	550	台基地面铺设材料	砖										
	其他	——															
装修（小木作）（彩画）	门(正面)	板门	门楣	有	堆经	有	门帘	无									
	窗（正面）	藏式明窗	窗楣	有	窗套	有	窗帘	有									
	室内隔扇	隔扇	有	隔扇位置	经堂后方												
	室内地面、楼面	地面材料及规格	条形木板,规格不一	楼面材料及规格	条形木板，规格不一												
	室内楼梯	楼梯	无（活动爬梯）	楼梯位置	——	楼梯材料	——	梯段宽度	——								
	天花、藻井	天花	有	天花类型	井口天花	藻井	无	藻井类型	——								
	彩画	柱头	有	柱身	无	梁架	有	走马板	——								
		门、窗	有	天花	有	藻井	——	其他彩画	——								
	其他	悬塑	无	佛龛	有	匾额	有										
装饰	室内	帷幔	无	幕帘彩绘	有	壁画	有	唐卡	有								
		经幡	有	经幢	有	柱毯	无	其他	——								
	室外	玛尼轮	无	苏勒德	有	宝顶	有	祥麟法轮	有								
		四角经幢	有	经幡	有	铜饰	有	石刻、砖雕	无								
		仙人走兽	0+3	壁画	有	其他	——										
陈设	室内	主佛像	八药师佛	佛像基座	莲花座												
		法座	有	藏经橱	有	经床	有	诵经桌	有	法鼓	有	玛尼轮	无	坛城	无	其他	——
	室外	旗杆	无	苏勒德	有	狮子	有	经幡	有	玛尼轮	无	香炉	有	五供	无	其他	——
	其他	——															
备注	匾额上题吉祥昌盛医明学院（2008年11月题）																
调查日期	2010/07/31	调查人员	杜娟	整理日期	2010/08/03	整理人员	杜娟										

药师佛殿基本概况1

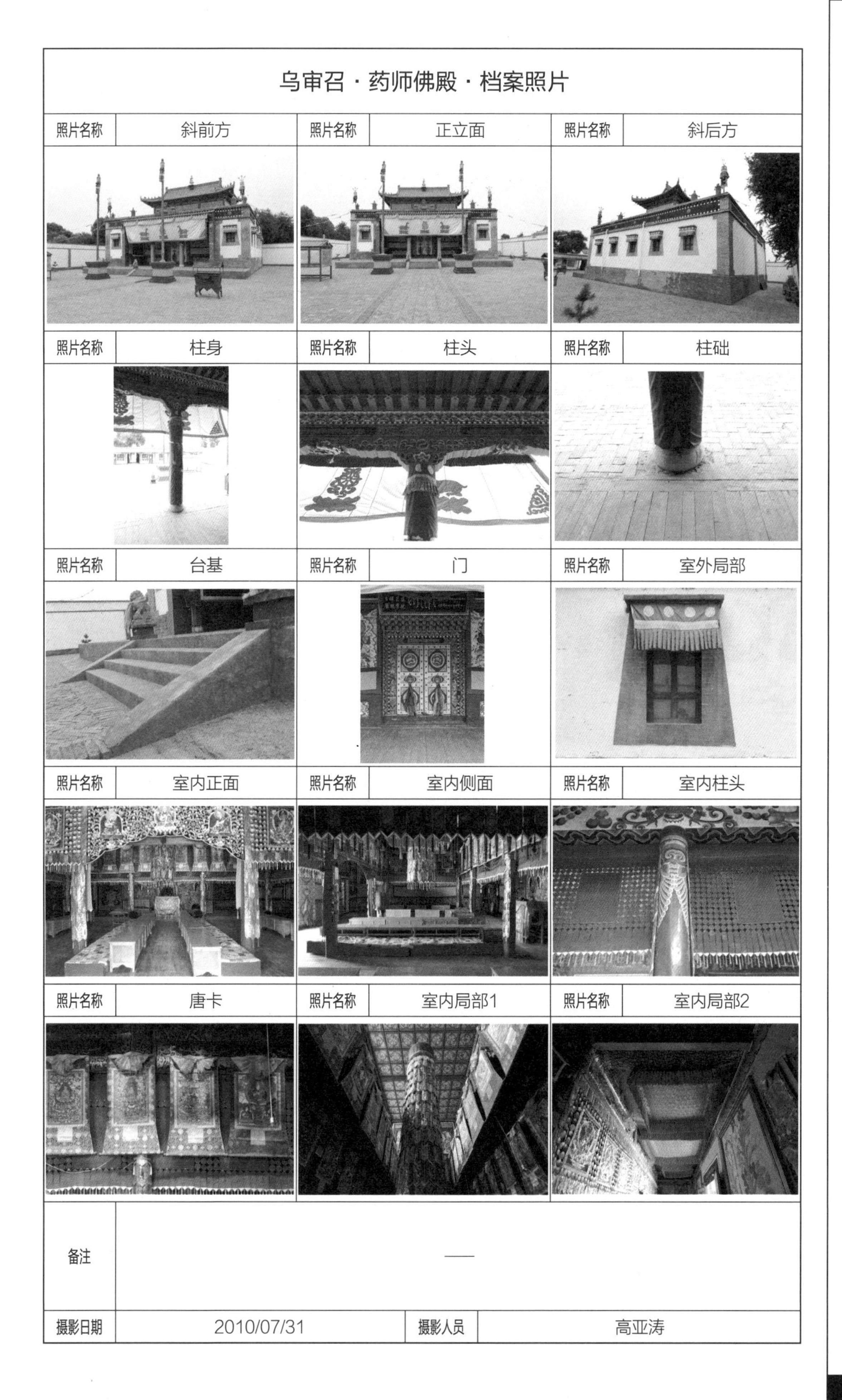

乌审召·药师佛殿·档案照片

照片名称	斜前方	照片名称	正立面	照片名称	斜后方
照片名称	柱身	照片名称	柱头	照片名称	柱础
照片名称	台基	照片名称	门	照片名称	室外局部
照片名称	室内正面	照片名称	室内侧面	照片名称	室内柱头
照片名称	唐卡	照片名称	室内局部1	照片名称	室内局部2
备注	—				
摄影日期	2010/07/31	摄影人员	高亚涛		

药师佛殿正前方

药师佛殿室内正面（左图）
药师佛殿正门（右图）

药师佛殿室内侧面（左图）
药师佛殿经幢（右图）

药师佛殿匾额

海流图庙

Hailiut Temple

3 海流图庙 Hailiut Temple

海流图庙建筑群

海流图庙为原伊克昭盟鄂尔多斯右翼前旗（乌审旗）寺庙，系乌审召18座属庙之一。

康熙十年（1671年），乌审旗笔帖式嘎日玛将海流图河附近的两个诵经会合并为一个，建两间小庙，取名拉西登诵经会。后嘎日玛剃发为僧，取僧名为萨木丹嘉木苏，专心于弘法利生。相传他到拉萨朝圣，归来时带回一件49间大雄宝殿的红檀木模型及佛像、经文等珍贵物品。康熙五十四年（1715年），乌审旗台吉巴岱及萨木丹嘉木苏在旗衙的支持下扩建寺庙。在萨木丹嘉木苏圆寂后，台吉巴岱将寺庙继续扩建成具有18座殿堂的寺庙，故该庙又称台吉巴岱庙。因萨木丹嘉木苏曾被清廷赐予国师称号，该庙也称国师庙，当地汉民音译为古神庙。

寺庙建筑风格以汉式建筑为主，兼有窑洞式建筑。光绪元年（1875年）经重建后的寺庙建筑规模为:25间双层大雄宝殿、12间砖窑式显宗殿、6间双层天王殿及龙王殿、贡嘎日佛殿、土地神殿、释迦牟尼殿、密集金刚殿、胜乐金刚殿、普明佛殿、文殊菩萨殿、五大神殿、莲花生与骑狮护法神殿、会庙、白伞盖佛母殿、法轮殿等14座砖窑式佛殿，每殿为3间，佛殿间各有一座佛塔。寺庙周边有9座佛塔，东仓、中仓与西仓等3座活佛仓及僧舍200余间。

“文化大革命”中寺庙严重受损，仅存11座窑洞式佛殿。1983年寺庙在乌审召第五世博格达活佛的主持下正式恢复法会。

参考文献:

[1] 阿拉腾巴嘎那.乌审海流图庙（蒙古文）.呼和浩特:内蒙古人民出版社,2007，12.

[2] 萨·那日松，特木尔巴特尔.鄂尔多斯寺院（蒙古文）.海拉尔:内蒙古文化出版社,2000，5.

海流图庙释迦牟尼殿

海流图庙 · 基本概况 1

寺院蒙古语藏语名称	蒙古语	ᠬᠠᠶᠢᠯᠤᠲᠤ ᠶᠢᠨ ᠰᠦᠮᠡ	寺院汉语名称	汉语正式名称	——
	藏语	དམ་པ་དར་རྒྱས་གླིང་།		俗称	海流图庙、古神庙
	汉语语义	水獭地的庙	寺院汉语名称的由来		——
所在地	鄂尔多斯市乌审旗巴音柴达木苏木		东经 108° 39′		北纬 38° 20′
初建年	1671年		保护等级	自治区级文物保护单位	
盛期时间	不详		盛期喇嘛僧/现有喇嘛僧数		200余人 / 18人

历史沿革

1715年，扩建寺庙，新建49间两层大雄宝殿、两处窑洞经堂，班禅大师赐名“召丹巴丹日吉令”，成为乌审召属庙。

1868年，49间大雄宝殿毁于回民起义。

1875年，重建寺庙，新建25间大雄宝殿。

1943—1946年，寺庙遭受国民党26师的破坏，除大雄宝殿与14座砖窑之外其余建筑均被烧毁。

1950—1953年，重建寺庙。

1967年，寺庙严重受损，仅存12座砖窑式佛殿。

1983年，寺庙恢复法会。

1989年，被上级指定为宗教事务场所

资料来源

[1] 阿拉腾巴嘎那.乌审海流图庙（蒙古文）.呼和浩特:内蒙古人民出版社,2007,12.

[2] 萨·那日松，特木尔巴特尔.鄂尔多斯寺院（蒙古文）.海拉尔:内蒙古文化出版社,2000,5.

[3] 调研访谈记录.

现状描述	寺庙内留存的11座窑洞式佛殿均为遗存建筑，并且建筑形制统一，保存完好。原天王殿已被完全拆掉，原址上已重建一天王殿，正在做粉刷装饰；山门和塔皆为新建	描述时间	2010/08/01
		描述人	杜娟
调查日期	2010/08/01	调查人员	宝山

海流图庙基本概况表1

海流图庙 · 基本概况 2

现存建筑	白伞盖天母殿	护法殿	观世音菩萨殿	胜乐殿	已毁建筑	拉西登诵经会	大雄宝殿
	大威德金刚殿	大日如来殿	文殊菩萨殿	五神殿		——	——
	密集金刚殿	释迦牟尼殿	药师佛殿	四大天王殿		——	——
	喇嘛处	山门	白塔	——		信息来源	《鄂托克寺庙》

区位图

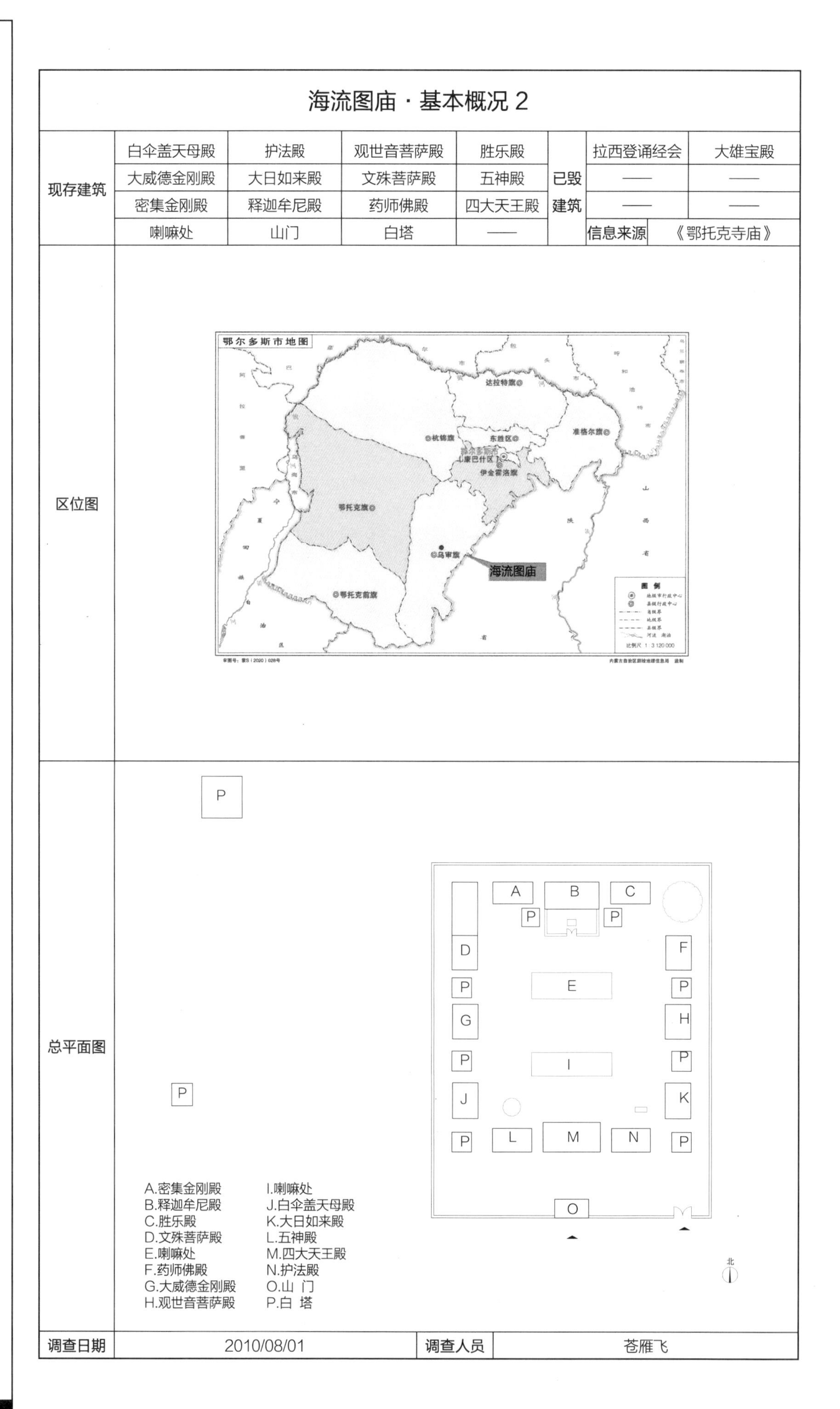

总平面图

A.密集金刚殿　B.释迦牟尼殿　C.胜乐殿　D.文殊菩萨殿　E.喇嘛处　F.药师佛殿　G.大威德金刚殿　H.观世音菩萨殿　I.喇嘛处　J.白伞盖天母殿　K.大日如来殿　L.五神殿　M.四大天王殿　N.护法殿　O.山 门　P.白 塔

调查日期	2010/08/01	调查人员	苍雁飞

海流图庙基本概况表2

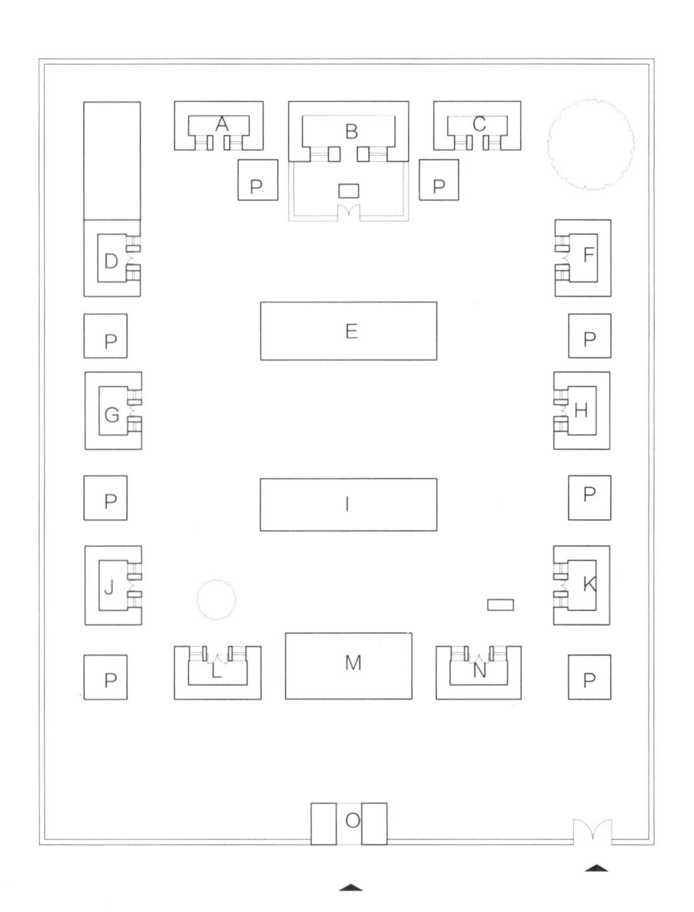

A.密集金刚殿
B.释迦牟尼殿
C.胜乐殿
D.文殊菩萨殿
E.喇嘛处
F.药师佛殿
G.大威德金刚殿
H.观世音菩萨殿
I.喇嘛处
J.白伞盖天母殿
K.大日如来殿
L.五神殿
M.四大天王殿
N.护法殿
O.山 门
P.白 塔

北

海流图庙总平面图

3.1 海流图庙·释迦牟尼殿

释迦牟尼殿院落斜前方

释迦牟尼殿正前方（左图）
释迦牟尼殿院门（右图）

释迦牟尼殿室内空间（左图）
释迦牟尼殿正面门（右图）

3.2 海流图庙·白伞盖天母殿

白伞盖天母殿侧面

白伞盖天母殿门（左图）
白伞盖天母殿窗（右图）

白伞盖天母殿正面门

3.3 海流图庙 · 喇嘛处

喇嘛处斜前方（左图）
喇嘛处斜后方（右图）

3.4 海流图庙 · 山门

山门正立面

3.5 海流图庙 · 白塔

佛塔（左图）
庙塔斜前方（中图）
寺庙院墙（右图）

4 陶亥召

Taohaizhao Temple

4 陶亥召 Taohaizhao Temple

陶亥召建筑群

陶亥召为原伊克昭盟鄂尔多斯左翼中旗（郡王旗）寺庙，寺庙管辖修古日庙、脑干宝拉格庙、吉如和庙等3座属庙。

约顺治九年（1652年），君王旗台吉若西遵从五世达赖喇嘛之示意，在苏布日干陶亥之地新建一座小庙，1740年绘《鄂尔多斯七旗地图》上标注该寺为“台吉若西庙”。该庙仅有1座佛殿、2座佛塔及几间僧房，后因失火而烧毁。乾隆十七年（1752年），郡王旗协理哈汗宝等人听取青海塔尔寺呼比勒汗喇嘛的指教，在敖努音陶亥之地重建寺庙，俗称陶亥召。当地汉民称原庙，即苏布日干陶亥召为旧庙;称新建寺庙，即敖努音陶亥召为新庙。

寺庙建筑风格为藏汉式建筑并存。至20世纪50年代该庙有49间大雄宝殿、49间三层释迦牟尼殿（也称伊克召）、49间显宗殿、25间时轮殿、密宗殿等5大殿宇。有占地面积分别为3000平方米的显宗院、二十一度母殿院、松巴堪布仓及占地4000平方米的活佛府三进院落等4大院落。有五大神殿、大小两座天王殿、大小两座法轮殿、布如木钦布殿、大黑天殿、巴格殿、龙王殿、斋戒殿、大威德殿等大小殿堂10余座，松巴堪布府、公尼召活佛府、活佛府等3座拉布隆，9座白塔、6座庙仓共计100间，僧舍200余间。陶亥召时轮殿北有一座寺庙，名为伊和乌孙庙，该寺原在浩赖河盘，后由于寺庙所处草地被卖给神木地主，故将寺庙移至陶亥召。

“文化大革命”中寺庙严重受损，而大雄宝殿等殿宇由于被用作存放粮食与物资的仓库，留存至今。1986年起原陶亥召数十名喇嘛修缮几处殿堂，正式恢复了法会。

参考文献：
[1] 哈斯朝鲁.陶亥召（蒙古文）.呼和浩特:内蒙古人民出版社, 2008，9.
[2] 彻·哈斯毕力格图.陶亥召——新庙（蒙古文）（内部资料）,2002，10.

陶亥召大雄宝殿

陶亥召 · 基本概况 1

寺院蒙古语藏语名称	蒙古语	[illegible]	寺院汉语名称	汉语正式名称	吉祥如意寺
	藏语	བཀྲ་ཤིས་ལྷུན་ཚོགས་གླིང་།		俗称	陶亥召、新庙、拉西庙
	汉语语义	湾庙	寺院汉语名称的由来		不详
所在地	鄂尔多斯市伊金霍洛旗纳林陶亥镇新庙村		东经 110° 23′	北纬 39° 23′	
初建年	1654年		保护等级	不详	
盛期时间	不详		盛期喇嘛僧/现有喇嘛僧	500余人／—	
历史沿革	1652年，在苏布日干陶亥之地始建寺庙。 1752年，在敖努音陶亥之地重建寺庙。 1870年，新建伊克召殿。该殿顶为绿色，故称绿殿，其二层为25间，三层为9间。 同治年间，回民起义中寺庙严重受损。 民国时期，被国民党军队占领使用。 1976年，伊克召殿被拆毁。 1986年，修缮部分残存殿堂，正式恢复法会。 2001年，新建一座佛塔				
	资料来源	[1] 哈斯朝鲁.陶亥召（蒙古文）.呼和浩特:内蒙古人民出版社,2008，9. [2] 彻 · 哈斯毕力格图.陶亥召——新庙（蒙古文）（内部资料）.2002，10. [3] 调研访谈记录.			
现状描述	大雄宝殿、时轮僧院、法相僧院、龙王庙、观音殿、佛爷府属于遗存建筑，主体未被破坏，近年来对其进行了一定的修缮及翻新。其他建筑均属新建建筑。调研时，该召庙正在新建3层汉式密乘僧院。召庙整体规模较为完整，遗存建筑保存较为完好		描述时间	2010/07/29	
			描述人	杜娟	
调查日期	2010/07/29		调查人员	宝山	

陶亥召基本概况表1

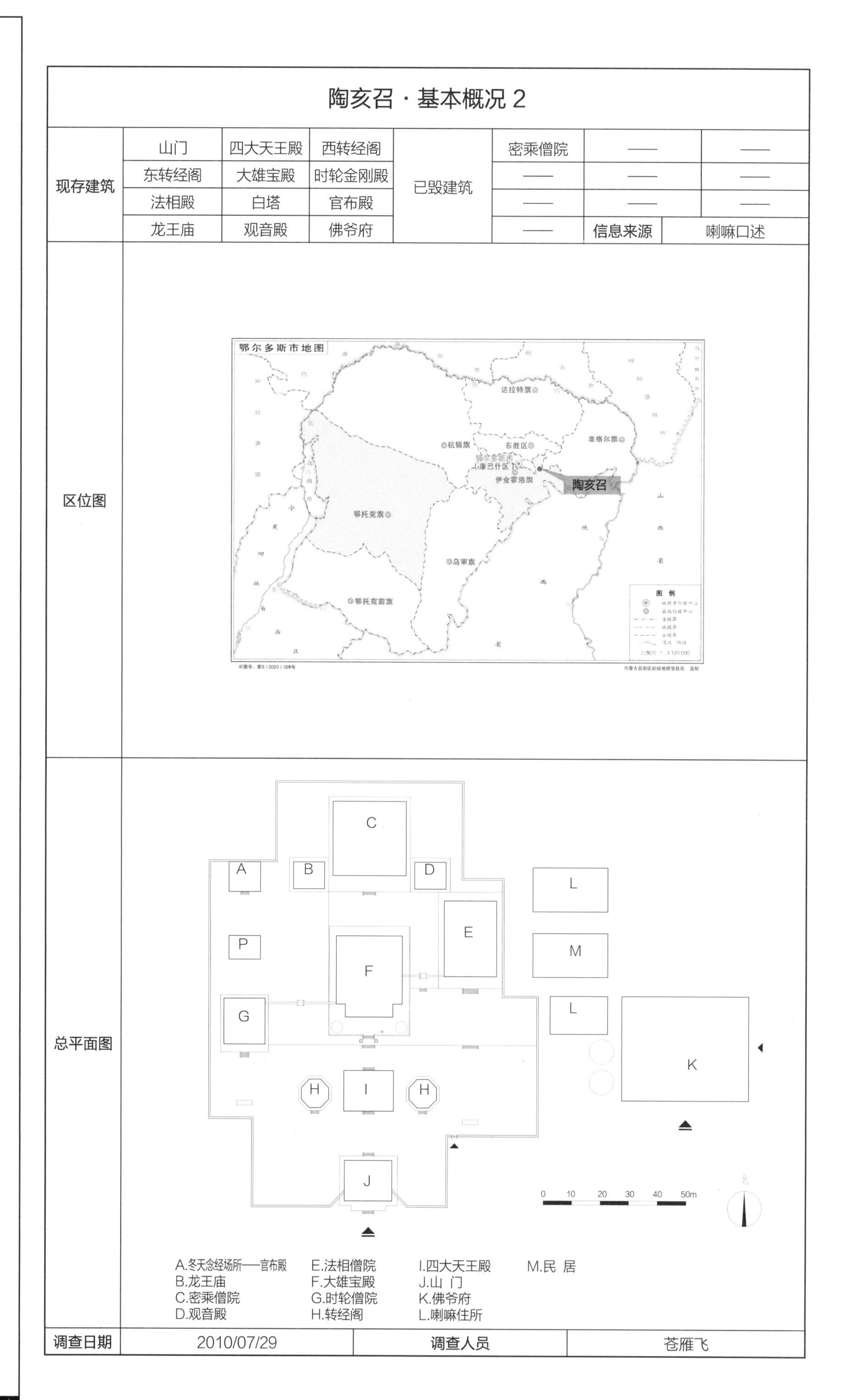

陶亥召·基本概况 2

现存建筑	山门	四大天王殿	西转经阁	已毁建筑	密乘僧院	——	——
	东转经阁	大雄宝殿	时轮金刚殿		——	——	——
	法相殿	白塔	官布殿		——	——	——
	龙王庙	观音殿	佛爷府		——	信息来源	喇嘛口述

区位图

总平面图

A.冬天念经场所——官布殿
B.龙王庙
C.密乘僧院
D.观音殿
E.法相僧院
F.大雄宝殿
G.时轮僧院
H.转经阁
I.四大天王殿
J.山 门
K.佛爷府
L.喇嘛住所
M.民 居

调查日期	2010/07/29	调查人员	苍雁飞

陶亥召基本概况表2

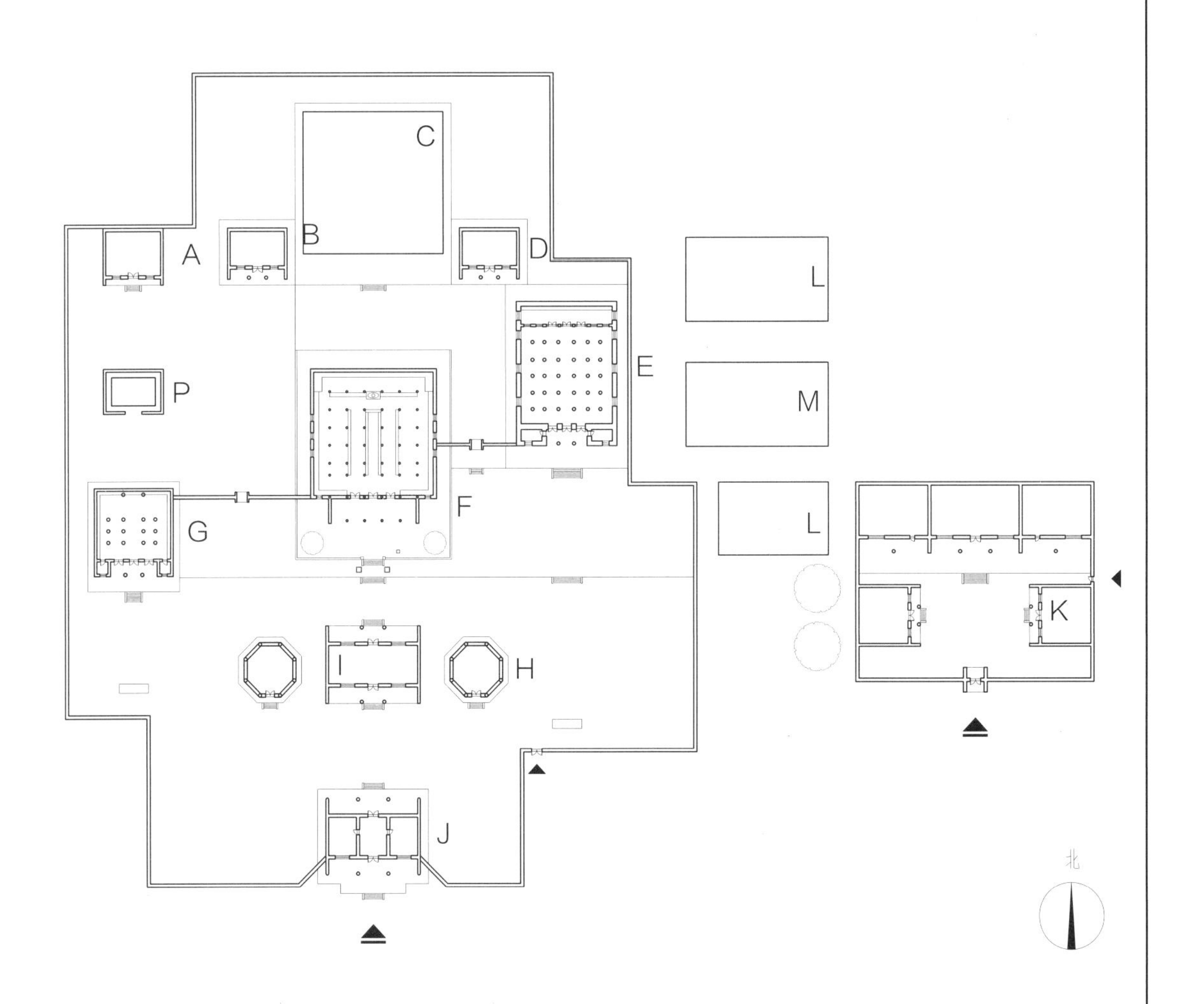

A.冬天念经场所——官布殿
B.龙王庙
C.密乘僧院
D.观音殿
E.法相僧院
F.大雄宝殿
G.时轮僧院
H.转经阁
I.四大天王殿
J.山 门
K.佛爷府
L.喇嘛住所
M.民 居

陶亥召总平面图

4.1 陶亥召 · 大雄宝殿

单位:毫米

建筑名称	汉语正式名称	大雄宝殿	俗称	朝格钦独贡													
概述	初建年	乾隆十七年（1752年）	建筑朝向	南	建筑层数	一											
	建筑简要描述	汉藏结合式															
	重建重修记载	20世纪80年代重修															
		信息来源	住持口述														
结构规模	结构形式	砖木结构	相连的建筑	无	室内天井	无											
	建筑平面形式	凹字形	外廊形式	前廊													
	通面阔	21900	开间数	7间	明间	3200	次间	3200	梢间	3200	次梢间	——	尽间	2950			
	通进深	23100	进深数	7间	进深尺寸（前→后）	2600→3160→3190→3150→3200→3200→4600											
	柱子数量	——	柱子间距	横向尺寸	——	（藏式建筑结构体系填写此栏，不含廊柱）											
				纵向尺寸	——												
	其他	主体结构均未破坏，只在一定部位做过相应的粉刷装修															
建筑主体（大木作）（石作）（瓦作）	屋顶	屋顶形式	卷棚顶、歇山顶与藏式平屋顶相结合	瓦作	灰色布瓦												
	外墙	主体材料	青砖	材料规格	160×310×70	饰面颜色	白										
		墙体收分	无	边玛檐墙	有	边玛材料	砖（刷红）										
	斗栱、梁架	斗栱	有	平身科斗口尺寸	无法测量	梁架关系	九檩前廊										
	柱、柱式（前廊柱）	形式	汉式	柱身断面形状	圆	断面尺寸	直径 D=200	（在没有前廊柱的情况下，填写室内柱及其特征）									
		柱身材料	木材	柱身收分	无	栌斗、托木	无	雀替	有								
		柱础	有	柱础形状	方形础上置莲花础	柱础尺寸	400×400										
	台基	台基类型	带砖墙式普通台基	台基高度	960	台基地面铺设材料	砖										
	其他	——															
装修（小木作）（彩画）	门(正面)	板门	门楣	无	堆经	有	门帘	有									
	窗（正面）	藏式盲窗	窗楣	有	窗套	无	窗帘	无									
	室内隔扇	隔扇	无	隔扇位置	——												
	室内地面、楼面	地面材料及规格	瓷砖（600×600），砖（240×120），条形木板	楼面材料及规格	——												
	室内楼梯	楼梯	无	楼梯位置	——	楼梯材料	——	楼段宽度	——								
	天花、藻井	天花	无	天花类型	——	藻井	无	藻井类型	——								
	彩画	柱头	有	柱身	无	梁架	有	走马板	有								
		门、窗	无	天花	——	藻井	——	其他彩画	——								
	其他	悬塑	无	佛龛	有	匾额	有										
装饰	室内	帷幔	有	幕帘彩绘	无	壁画	有	唐卡	有								
		经幡	有	经幢	有	柱毯	无	其他	——								
	室外	玛尼轮	无	苏勒德	无	宝顶	有	祥麟法轮	有								
		四角经幢	有	经幡	有	铜饰	有	石刻、砖雕	无								
		仙人走兽	无	壁画	有	其他	——										
陈设	室内	主佛像	释迦牟尼佛	佛像基座	莲花座												
		法座	有	藏经橱	有	经床	有	诵经桌	有	法鼓	有	玛尼轮	无	坛城	无	其他	——
	室外	旗杆	无	苏勒德	无	狮子	有	经幡	无	玛尼轮	无	香炉	无	五供	无	其他	——
	其他	——															
备注	壁画以及墙上的彩绘均为较古老的作品																
调查日期	2010/07/29	调查人员	杜娟	整理日期	2010/07/29	整理人员	杜娟										

大雄宝殿基本概况表1

陶亥召 · 大雄宝殿 · 档案照片

照片名称	斜前方	照片名称	正立面	照片名称	侧立面
照片名称	斜后方	照片名称	柱础	照片名称	台基
照片名称	门	照片名称	柱头	照片名称	室外局部
照片名称	室内正面	照片名称	室内侧面	照片名称	室内地面
照片名称	佛像	照片名称	壁画	照片名称	室内局部
备注	—				
摄影日期	2010/07/29	摄影人员	高亚涛		

大雄宝殿正前方

大雄宝殿侧面（左图）

大雄宝殿斜前方（左图）
大雄宝殿经幢（右图）

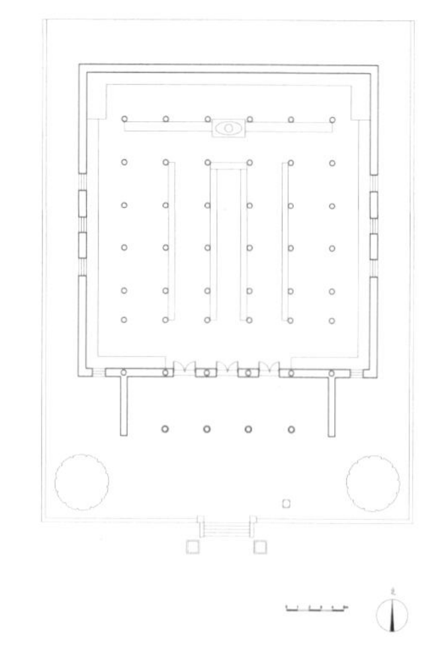

大雄宝殿室内正前方（左图）
大雄宝殿一层平面图（右图）

4.2 陶亥召·时轮金刚殿

单位:毫米

建筑名称	汉语正式名称	时轮金刚殿	俗称	东科尔独贡				
概述	初建年	同治九年（1870年）	建筑朝向	南	建筑层数	一		
	建筑简要描述	藏式建筑						
	重建重修记载	20世纪80年代做过一定维修						
		信息来源	住持口述					
结构规模	结构形式	砖木结构	相连的建筑	无	室内天井	无		
	建筑平面形式	凹字形	外廊形式	前廊				
	通面阔	13500	开间数	5间	明间 3650	次间 2775	梢间 2200	次梢间 —— / 尽间 ——
	通进深	15050	进深数	5间	进深尺寸（前→后）	3000→2900→2050→2100→5000		
	柱子数量	12	柱子间距	横向尺寸	2775→3650→2775	（藏式建筑结构体系填写此栏，不含廊柱）		
				纵向尺寸	2900→2050→2100			
	其他	——						
建筑主体（大木作）（石作）（瓦作）	屋顶	屋顶形式	藏式平屋顶		瓦作	——		
	外墙	主体材料	青砖	材料规格	310×160×60	饰面颜色	白	
		墙体收分	无	边玛檐墙	有	边玛材料	砖（刷红）	
	斗栱、梁架	斗栱	无	平身科斗口尺寸	——	梁架关系	梁横纵排架	
	柱、柱式（前廊柱）	形式	藏式	柱身断面形状	小八角	断面尺寸	200×200	（在没有前廊柱的情况下，填写室内柱及其特征）
		柱身材料	木材	柱身收分	无	栌斗、托木	有	雀替 无
		柱础	有	柱础形状	圆形	柱础尺寸	直径 D=350	
	台基	台基类型	普通式台基	台基高度	930	台基地面铺设材料	砖	
	其他	——						
装修（小木作）（彩画）	门(正面)	板门		门楣	无	堆经	有	门帘 无
	窗（正面）	藏式明窗		窗楣	无	窗套	无	窗帘 无
	室内隔扇	隔扇	无	隔扇位置	——			
	室内地面、楼面	地面材料及规格		砖（240×120×60）		楼面材料及规格	——	
	室内楼梯	楼梯	无	楼梯位置	——	楼梯材料	——	楼段宽度 ——
	天花、藻井	天花	无	天花类型	——	藻井	无	藻井类型 ——
	彩画	柱头	无	柱身	无	梁架	有	走马板 有
		门、窗	无	天花	无	藻井	无	其他彩画 ——
	其他	悬塑	无	佛龛	有	匾额	无	
装饰	室内	帷幔	无	幕帘彩绘	有	壁画	无	唐卡 有
		经幡	有	经幢	有	柱毯	有	其他
	室外	玛尼轮	无	苏勒德	有	宝顶	无	祥麟法轮 无
		四角经幢	无	经幡	无	铜饰	无	石刻、砖雕 无
		仙人走兽	无	壁画	无	其他	——	
陈设	室内	主佛像	千手观音			佛像基座	无	
		法座 无	藏经橱 无	经床 有	诵经桌 有	法鼓 无	玛尼轮 无	坛城 无 / 其他 ——
	室外	旗杆 无	苏勒德 有	狮子 无	经幡 无	玛尼轮 无	香炉 无	五供 无 / 其他 ——
	其他	——						
备注	——							
调查日期	2010/07/29	调查人员	杜娟	整理日期	2010/07/29	整理人员	杜娟	

时轮金刚殿基本概况表1

陶亥召·时轮金刚殿·档案照片					
照片名称 	斜前方	照片名称 	正立面	照片名称 	侧立面
照片名称 	柱身	照片名称 	柱头	照片名称 	柱础
照片名称 	台基	照片名称 	门	照片名称 	窗户
照片名称 	室内柱身	照片名称 	唐卡	照片名称 	室内局部
照片名称 	室内经幢	照片名称 	室内唐卡	照片名称 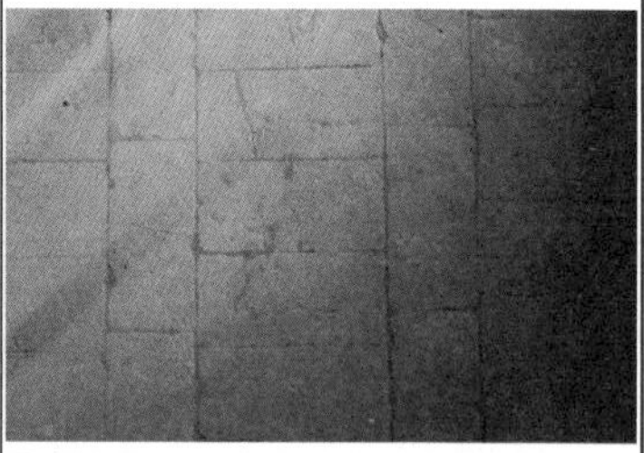	室内地面
备注	—				
摄影日期	2010/07/29	摄影人员	高亚涛		

时轮金刚殿基本概况表2

时轮金刚殿正前方

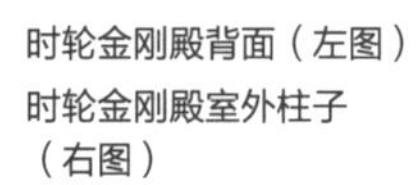

时轮金刚殿背面（左图）

时轮金刚殿室外柱子（右图）

时轮金刚殿室内梁柱（左图）

时轮金刚殿正面门（右图）

时轮金刚殿室内顶棚

时轮金刚殿室内唐卡
（左图）
时轮金刚殿室内地面
（右图）

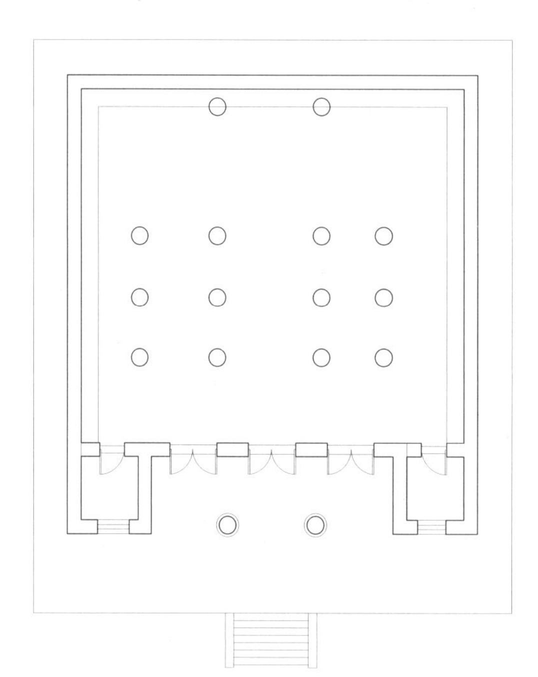

时轮金刚殿一层平面图

4.3 陶亥召·法相殿

单位:毫米

建筑名称	汉语正式名称	法相殿	俗称	却仁独贡													
概述	初建年	不详	建筑朝向	南	建筑层数	一											
	建筑简要描述	藏式建筑															
	重建重修记载	20世纪80年代做过一定维修															
		信息来源	住持口述														
结构规模	结构形式	砖木结构	相连的建筑	无	室内天井	天井											
	建筑平面形式	凹字形	外廊形式	前廊													
	通面阔	16700	开间数	7间	明间	——	次间	——	梢间	——	次梢间	——	尽间	——			
	通进深	19800	进深数	7间	进深尺寸（前→后）	——											
	柱子数量	36	柱子间距	横向尺寸	2135、2550、2830	（藏式建筑结构体系填写此栏，不含廊柱）											
				纵向尺寸	2800、2900、3000、3150												
	其他	——															
建筑主体（大木作）（石作）（瓦作）	屋顶	屋顶形式	藏式平屋顶	瓦作	——												
	外墙	主体材料	青砖	材料规格	310×160×60	饰面颜色	白										
		墙体收分	无	边玛檐墙	有	边玛材料	砖（刷红）										
	斗栱、梁架	斗栱	无	平身科斗口尺寸	——	梁架关系	面向天井纵排架										
	柱、柱式（前廊柱）	形式	藏式	柱身断面形状	方	断面尺寸	160×160	（在没有前廊柱的情况下，填写室内柱及其特征）									
		柱身材料	木材	柱身收分	有	栌斗、托木	有	雀替	无								
		柱础	有	柱础形状	圆形	柱础尺寸	直径 D=160										
	台基	台基类型	普通式台基	台基高度	600	台基地面铺设材料	砖										
	其他	——															
装修（小木作）（彩画）	门(正面)	板门	门楣	无	堆经	有	门帘	无									
	窗（正面）	藏式明窗	窗楣	有	窗套	无	窗帘	无									
	室内隔扇	隔扇	有	隔扇位置	经堂后侧												
	室内地面、楼面	地面材料及规格	瓷砖（600×600）	楼面材料及规格	——												
	室内楼梯	楼梯	无	楼梯位置	——	楼梯材料	——	梯段宽度	——								
	天花、藻井	天花	无	天花类型	——	藻井	无	藻井类型	——								
	彩画	柱头	无	柱身	无	梁架	有	走马板	有								
		门、窗	无	天花	——	藻井	——	其他彩画	——								
	其他	悬塑	无	佛龛	有	匾额	无										
装饰	室内	帷幔	无	幕帘彩绘	有	壁画	有	唐卡	有								
		经幡	有	经幢	有	柱毯	无	其他	——								
	室外	玛尼轮	无	苏勒德	有	宝顶	有	祥麟法轮	无								
		四角经幢	无	经幡	无	铜饰	无	石刻、砖雕	无								
		仙人走兽	无	壁画	有	其他	——										
陈设	室内	主佛像	千手观音	佛像基座	莲花座												
		法座	有	藏经橱	无	经床	有	诵经桌	有	法鼓	无	玛尼轮	无	坛城	无	其他	——
	室外	旗杆	无	苏勒德	有	狮子	无	经幡	无	玛尼轮	无	香炉	无	五供	无	其他	——
	其他	——															
备注	——																
调查日期	2010/07/29	调查人员	杜娟	整理日期	2010/07/29	整理人员	杜娟										

法相殿基本概况表1

陶亥召·法相殿·档案照片					
照片名称 	斜前方	照片名称 	正立面	照片名称 	柱身
照片名称 	柱头	照片名称 	柱础	照片名称 	台基
照片名称 	门	照片名称 	窗户	照片名称 	室外地面
照片名称 	室内正面	照片名称 	室内侧面	照片名称 	室内柱身
照片名称 	室内局部1	照片名称 	唐卡	照片名称 	室内局部2
备注	——				
摄影日期	2010/07/29	摄影人员	高亚涛		

法相殿基本概况表2

法相殿正前方

法相殿斜前方

法相殿室外柱子（左图）
法相殿室内唐卡（右图）

法相殿室内正面

法相殿室内柱（左图）
法相殿室内柱头（右图）

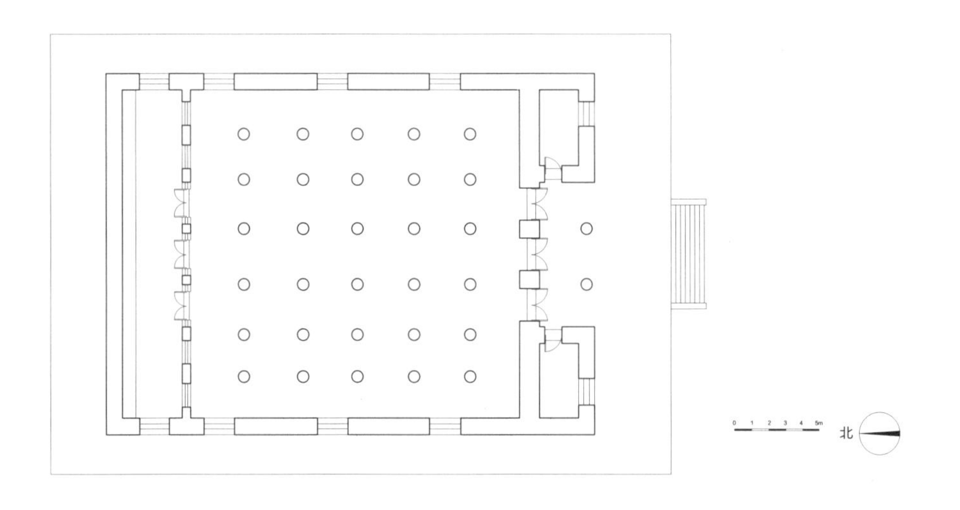

法相殿一层平面图

5 特布德庙

Tubed Temple

5 特布德庙 Tubed Temple

特布德庙建筑群

特布德庙为原伊克昭盟鄂尔多斯右翼中旗（鄂托克旗）寺庙。最初为上海庙属庙，后成为鄂托克召属庙。

相传，在寺庙所在地曾有一名藏族僧人苦练经法，该僧坐化于一处岩洞中，并留下“用流沙将岩洞填满并在其上修建大黑天庙”的遗言。地方施主遵照其言，新建了大黑天殿及僧舍。民国9年（1920年），扩建了寺庙。蒙古语称“藏族”为“特布德”，故该庙被称为特布德庙。

寺庙建筑为汉藏结合式建筑。“文化大革命”前寺庙有49间汉藏结合式双层大雄宝殿、12间大黑天殿、9间呼比勒汗喇嘛拉布隆、4座佛塔、7间大庙仓房舍、3间膳房及僧舍15座共70余间。

“文化大革命”中寺庙严重受损。1982年起修缮寺庙，正式恢复法会。

参考文献：

［1］阿日宾巴雅尔，曹纳木.鄂托克寺庙（蒙古文）.海拉尔:内蒙古文化出版社,1998,8.

特布德庙大雄宝殿（左图）
特布德庙建筑群（右图）

特布德庙 · 基本概况 1

<table>
<tr><td rowspan="3">寺院蒙古语藏语名称</td><td>蒙古语</td><td>ᠲᠦᠪᠡᠳ ᠰᠦᠮ᠎ᠡ</td><td rowspan="2">寺院汉语名称</td><td>汉语正式名称</td><td colspan="4">吉祥大乘寺</td></tr>
<tr><td>藏语</td><td>བཀྲ་ཤིས་ཐེག་ཆེན་གླིང་།</td><td>俗称</td><td colspan="4">特布德庙</td></tr>
<tr><td>汉语语义</td><td>藏庙</td><td colspan="2">寺院汉语名称的由来</td><td colspan="4">藏名汉译</td></tr>
<tr><td>所在地</td><td colspan="4">鄂尔多斯市鄂前旗上海庙镇特布德嘎查</td><td>东经</td><td>106° 54′</td><td>北纬</td><td>38° 14′</td></tr>
<tr><td>初建年</td><td colspan="2">1920年</td><td>保护等级</td><td colspan="5">——</td></tr>
<tr><td>盛期时间</td><td colspan="2">不详</td><td colspan="2">盛期喇嘛僧/现有喇嘛僧数</td><td colspan="4">不详 / 15人</td></tr>
<tr><td rowspan="2">历史沿革</td><td colspan="8">1920年，扩建寺庙。
1926年，扩建寺庙。
“文化大革命”中寺庙严重受损，仅存大雄宝殿。
1988年，修复大雄宝殿</td></tr>
<tr><td>资料来源</td><td colspan="7">[1] 阿日宾巴雅尔,曹纳木.鄂托克寺庙（蒙古文）.海拉尔:内蒙古文化出版社,1998.8.
[2] 调研访谈记录.</td></tr>
<tr><td rowspan="2">现状描述</td><td colspan="4" rowspan="2">大雄宝殿为遗存建筑，保存情况较为良好，现召庙内只有大雄宝殿、佛爷墒、喇嘛住所及旧塔遗址留存</td><td colspan="2">描述时间</td><td colspan="2">2010/08/03</td></tr>
<tr><td colspan="2">描述人</td><td colspan="2">杜娟</td></tr>
<tr><td>调查日期</td><td colspan="2">2010/08/03</td><td>调查人员</td><td colspan="5">宝山</td></tr>
</table>

特布德庙基本概况表1

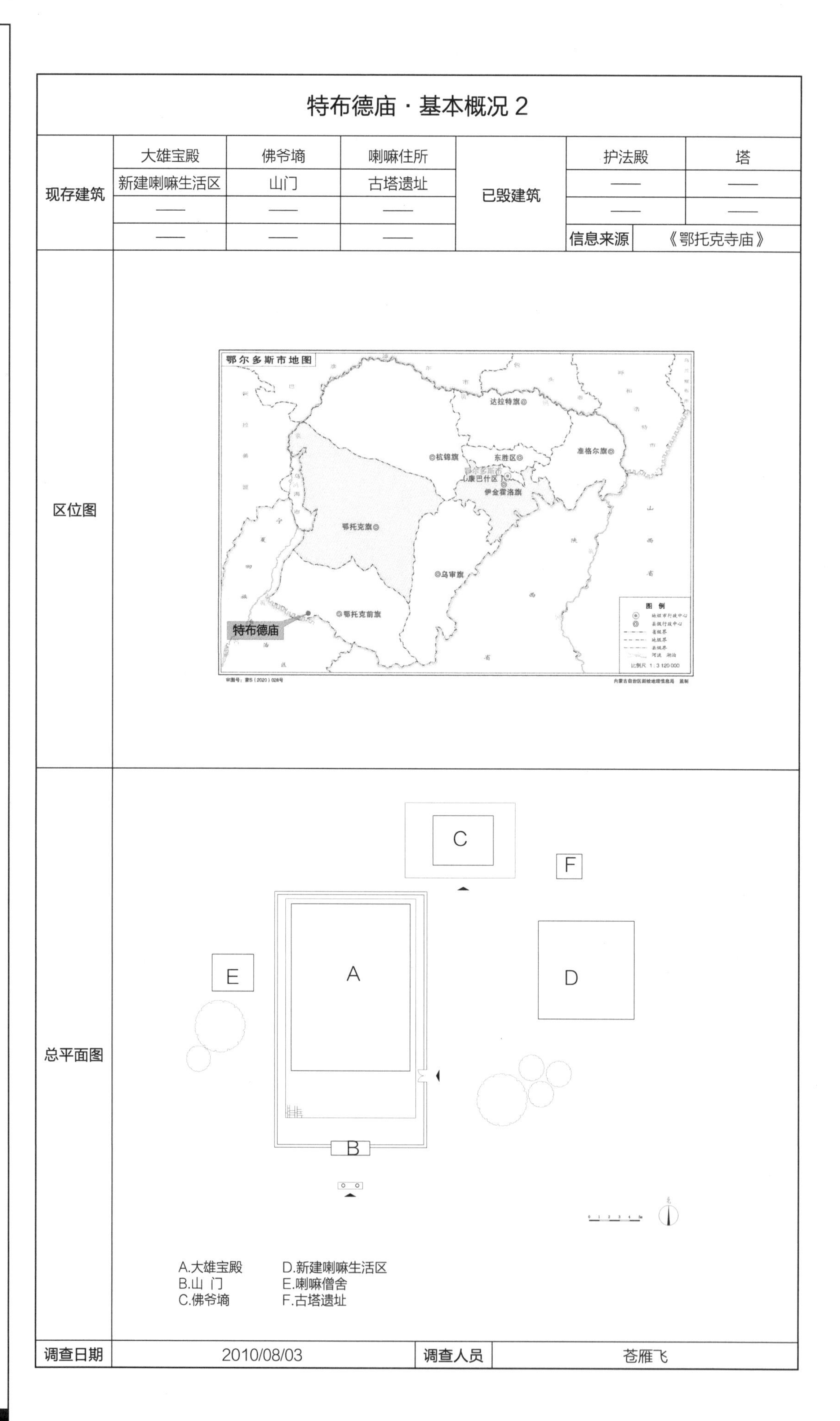

特布德庙·基本概况 2

现存建筑				已毁建筑		
现存建筑	大雄宝殿	佛爷滳	喇嘛住所	已毁建筑	护法殿	塔
	新建喇嘛生活区	山门	古塔遗址		——	——
	——	——	——		——	——
	——	——	——		信息来源	《鄂托克寺庙》
区位图						
总平面图						
调查日期	2010/08/03		调查人员	苍雁飞		

特布德庙基本概况表2

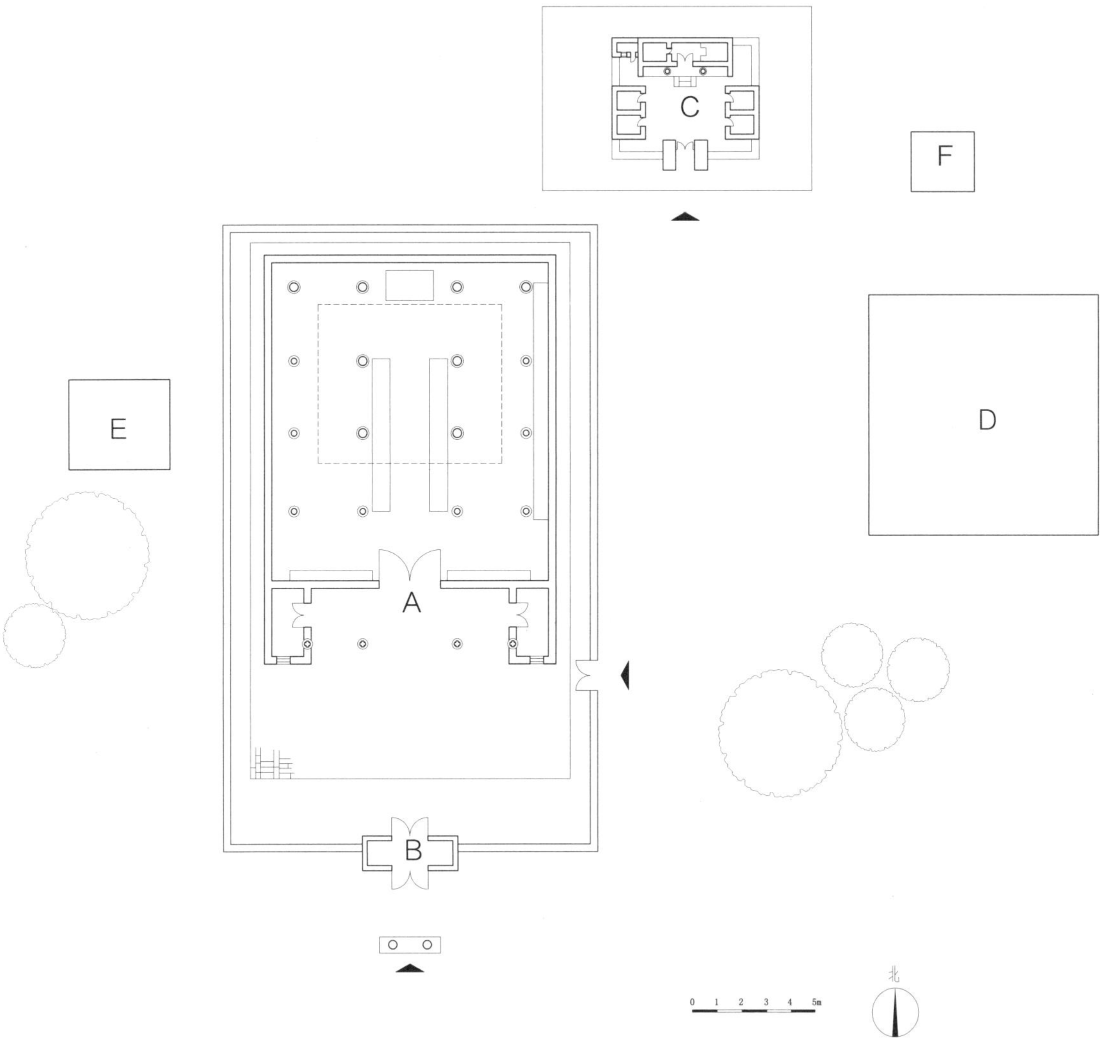

A.大雄宝殿　　D.新建喇嘛生活区
B.山 门　　E.喇嘛僧舍
C.佛爷墒　　F.古塔遗址

特布德庙总平面图

5.1 特布德庙·大雄宝殿

单位:毫米

建筑名称	汉语正式名称	大雄宝殿		俗称	朝格钦独贡												
概述	初建年	不详		建筑朝向	南	建筑层数	——										
	建筑简要描述	汉藏结合式															
	重建重修记载	20世纪80年代曾经修葺过															
		信息来源	喇嘛口述														
结构规模	结构形式	砖木	相连的建筑	无	室内天井	无											
	建筑平面形式	凹字形	外廊形式	前廊													
	通面阔	11250	开间数	5	明间	3250	次间	2950	梢间	1050	次梢间	——	尽间	——			
	通进深	12650	进深数	5	进深尺寸（前→后）	2550→2950→2900→3000→1250											
	柱子数量	16	柱子间距	横向尺寸	2950→3250→2950	（藏式建筑结构体系填写此栏，不含廊柱）											
				纵向尺寸	2950→2900→3000												
	其他	该寺庙名称是藏庙，填表人观测其建筑形制是汉藏结合式															
建筑主体（大木作）（石作）（瓦作）	屋顶	屋顶形式	庑殿、硬山与藏式平屋顶相结合	瓦作	赭石色琉璃瓦屋面												
	外墙	主体材料	砖	材料规格	260×120×60	饰面颜色	白										
		墙体收分	无	边玛檐墙	无	边玛材料	——										
	斗栱、梁架	斗栱	无	平身科斗口尺寸	——	梁架关系	十一檩前廊										
	柱、柱式（前廊柱）	形式	藏式	柱身断面形状	八瓣形	断面尺寸	240×240	（在没有前廊柱的情况下，填写室内柱及其特征）									
		柱身材料	木材	柱身收分	有	栌斗、托木	有	雀替	无								
		柱础	有	柱础形状	方础上置圆础	柱础尺寸	450×450										
	台基	台基类型	无	台基高度	——	台基地面铺设材料	——										
	其他	——															
装修（小木作）（彩画）	门(正面)	板门	门楣	无	堆经	有	门帘	无									
	窗（正面）	槛窗	窗楣	无	窗套	无	窗帘	无									
	室内隔扇	隔扇	无	隔扇位置	——												
	室内地面、楼面	地面材料及规格	水泥抹平	楼面材料及规格	——												
	室内楼梯	楼梯	无	楼梯位置	——	楼梯材料	——	梯段宽度	——								
	天花、藻井	天花	无	天花类型	——	藻井	无	藻井类型	——								
	彩画	柱头	有	柱身	有	梁架	有	走马板	——								
		门、窗	无	天花	——	藻井	——	其他彩画	——								
	其他	悬塑	无	佛龛	有	匾额	有										
装饰	室内	帷幔	无	幕帘彩绘	无	壁画	无	唐卡	有								
		经幡	无	经幢	有	柱毯	无	其他	——								
	室外	玛尼轮	无	苏勒德	有	宝顶	有	祥麟法轮	有								
		四角经幢	无	经幡	无	铜饰	无	石刻、砖雕	无								
		仙人走兽	2走兽	壁画	无	其他	——										
陈设	室内	主佛像	宗喀巴大师	佛像基座	无												
		法座	有	藏经橱	有	经床	有	诵经桌	有	法鼓	有	玛尼轮	无	坛城	无	其他	——
	室外	旗杆	无	苏勒德	有	狮子	无	经幡	无	玛尼轮	无	香炉	无	五供	无	其他	——
	其他	——															
备注	——																
调查日期	2010/08/03	调查人员	杜娟	整理日期	2010/08/04	整理人员	杜娟										

大雄宝殿基本概况表1

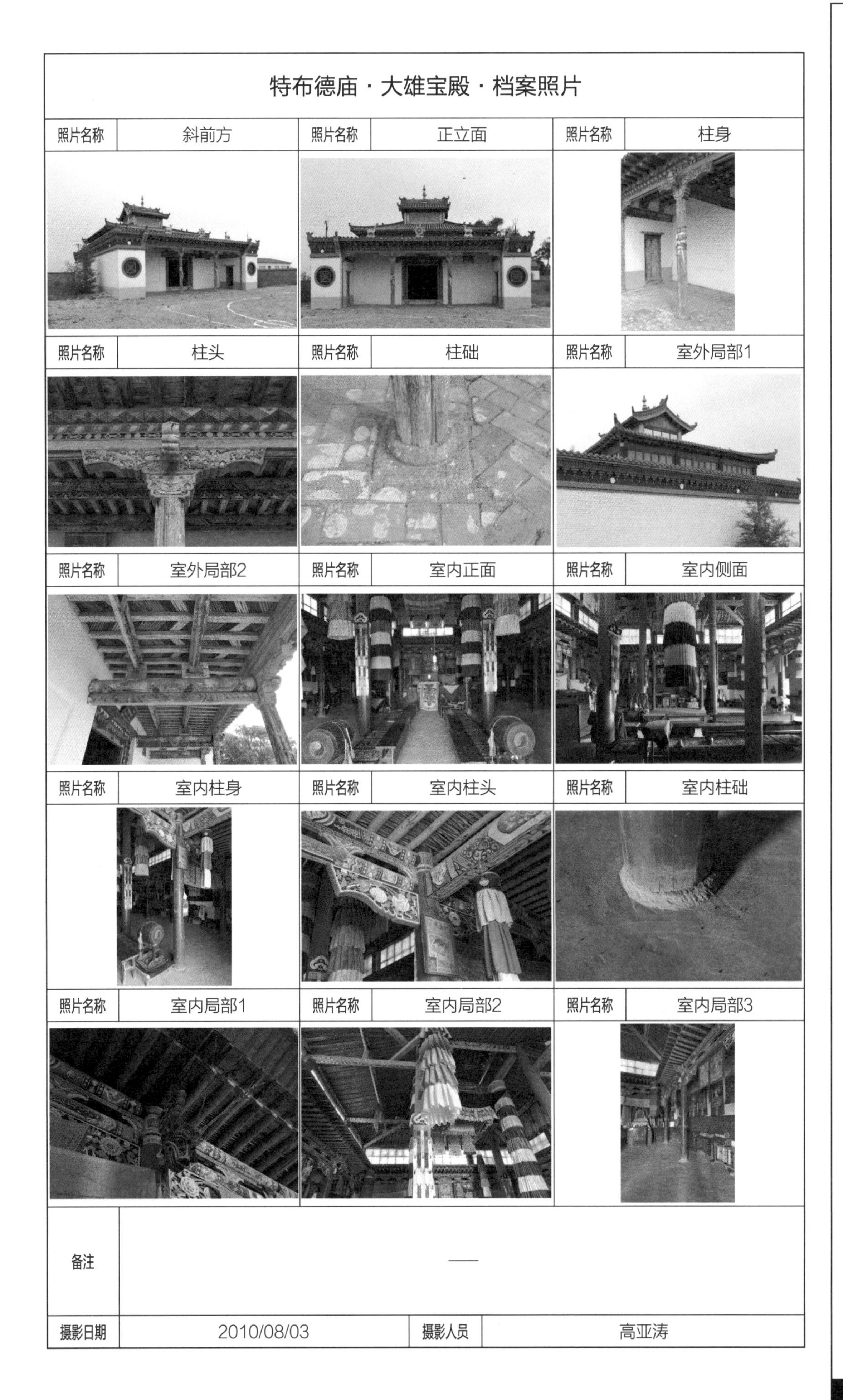

特布德庙·大雄宝殿·档案照片

照片名称	斜前方	照片名称	正立面	照片名称	柱身
照片名称	柱头	照片名称	柱础	照片名称	室外局部1
照片名称	室外局部2	照片名称	室内正面	照片名称	室内侧面
照片名称	室内柱身	照片名称	室内柱头	照片名称	室内柱础
照片名称	室内局部1	照片名称	室内局部2	照片名称	室内局部3
备注	—				
摄影日期	2010/08/03	摄影人员	高亚涛		

大雄宝殿正前方

大雄宝殿侧面

大雄宝殿室外柱子（左图）
大雄宝殿室外陈设（右图）

大雄宝殿室内正面

大雄宝殿室内柱子（左图）
大雄宝殿室内侧面（右上图）
大雄宝殿室内陈设（右中图）
大雄宝殿室内顶棚（右下图）

包头市地区

Baotou City

底图来源：内蒙古自治区自然资源厅官网 内蒙古地图
审图号：蒙S（2017）028号

包头市辖昆都仑区、青山区、东河区、九原区、石拐区、白云鄂博矿区6个市辖区、固阳县1县、土默特右旗、达尔罕茂明安联合旗2旗。现辖区由清时乌兰察布盟乌拉特三公旗、喀尔喀右翼旗、茂明安旗及归化城土默特旗部分地区组成。市辖区内现有15座已恢复重建或尚有建筑遗存的寺庙，课题组实地调研7座寺庙。

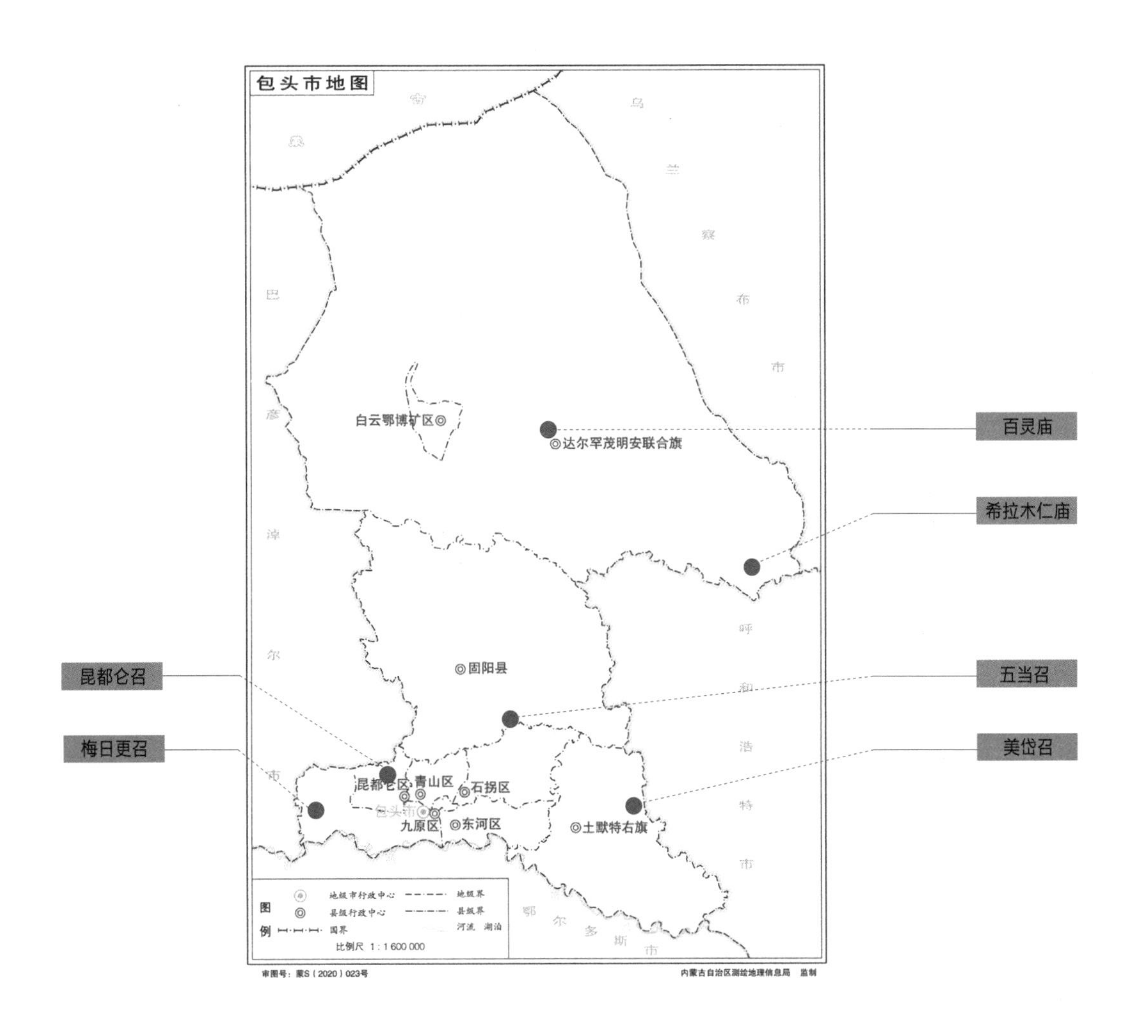
包头市地图
百灵庙
希拉木仁庙
昆都仑召
五当召
梅日更召
美岱召
白云鄂博矿区
达尔罕茂明安联合旗
固阳县
昆都仑区
青山区
石拐区
包头市
九原区
东河区
土默特右旗
蒙
古
乌
兰
察
布
市
巴
彦
淖
尔
市
呼
和
浩
特
市
鄂
尔
多
斯
市
图例
地级市行政中心
县级行政中心
国界
地级界
县级界
河流 湖泊
比例尺 1：1 600 000
审图号：蒙S（2020）023号
内蒙古自治区测绘地理信息局 监制

1 美岱召 Maidarzhao Temple

1 美岱召 Maidarzhao Temple

美岱召建筑群

美岱召为呼和浩特八小召之一,系内蒙古地区建造最早的藏传佛教格鲁派寺庙之一。该寺在明代曾为阿拉坦汗的家庙，或称汗庙，并位于金国之都大板升城。因此，寺庙具有城寺合一、政教一体的独特形制。后随政治中心的东移以及政权的更替，城堡逐步改扩成为寺庙，清时降为由呼和浩特喇嘛印务处管辖的召庙。该寺原名灵觉寺，乾隆五十二年（1787年），清廷赐名“寿灵寺”。寺中珍藏巨型纪实壁画“蒙古贵族礼佛图”及美岱召现存唯一的文字资料——泰和门石匾等珍贵文物。

明隆庆六年（1572年），土默特部阿拉坦汗始建于大板升城，即福化城内。万历三十四年（1606年），阿拉坦汗嫡孙大成台吉妻乌兰妣吉用金银宝石铸造一尊弥勒（蒙古语称迈达日宝日汗）佛像，请达赖喇嘛派往蒙古掌教的迈达日呼图克图主持开光仪式，并留此庙掌教。由此蒙古人将该庙称为“迈达日召”，又译作“迈达里召”，汉语简略为“美岱召”。从此，召名替代城名，该城以美岱召闻名。

寺庙建筑以汉式建筑为主。寺庙有大雄宝殿、西万佛殿、东万佛殿、八角庙、护法神殿、两廊庙、天王殿、吉瓦殿、白马天神殿、佛爷府、公爷府、太后庙、达赖庙等殿堂、院落。建筑外围有城墙，城略呈方形，方城墙体四角有重檐歇山顶角楼。因由城改为寺庙的独特建寺历程，宫殿、府邸与寺庙建筑相互融合，构成一体。有学者称西万佛殿为最初所建土默特汗庙，大雄宝殿与太后庙分别为阿拉坦汗与三娘子的陵宫。寺庙现存明代建筑10座，256间。

“文化大革命”中寺庙严重受损，部分寺庙被拆除。但因寺庙先后被用作果园、军用粮库，故其建筑保存至今。1985年起，大规模维修美岱召。

参考文献：

[1] 包头市人民政府专家顾问组.塞外城寺——美岱召.呼和浩特:内蒙古人民出版社,2009,9.
[2] 土默特左旗土默特志编纂委员会办公室.土默特史料（第十六集），1985,2.
[3] 金峰.大青山下的美岱召.呼和浩特:内蒙古人民出版社,2011,7.

美岱召泰和门

美岱召 · 基本概况 1

寺院蒙古语藏语名称	蒙古语	[illegible]	寺院汉语名称	汉语正式名称	灵觉寺
	藏语	——		俗称	美岱召、麦达力召
	汉语语义	弥勒佛庙	寺院汉语名称的由来		明廷赐名
所在地	包头市土默特右旗美岱召村		东经	110° 42′	北纬 40° 36′
初建年	隆庆六年（1572年）		保护等级	全国重点文物保护单位	
盛期时间	——		盛期喇嘛僧/现有喇嘛僧数	——/ 0	
历史沿革	1565年，阿拉坦汗仿元大都，改扩建大板升城。 1571年，明廷赐名大板升城为福化城。 1572年，史料所记最早建寺的年代。寺名灵觉寺，为现存的西万佛殿。 1581年，归化城建成，大明金国的政治中心逐渐东移。 1585年，三娘子筑城别居，离开大板升城。 1586年，三世达赖至美岱召为阿拉坦汗举行第二次安葬仪式。 1606年，乌兰妣吉修泰和门，迎迈达里活佛为大雄宝殿内银制弥勒佛像举行开光仪式。 1627年，迈达里活佛东行，活佛府后改称护法神庙。 1849年，对着城门口新建大照壁。 1958年，美岱召生产大队将美岱召改为果园。 “文化大革命”中，寺庙严重受损，天王殿、白马天神殿等建筑被拆除。 1969年，寺庙建筑被用作战备粮库，建筑受到空前破坏。 1979年，美岱召文物保管所成立，美岱召由文物部门管理。 1985年，大规模维修寺庙					
	资料来源	[1] 包头市人民政府专家顾问组.塞外城寺——美岱召.呼和浩特:内蒙古人民出版社,2009,9. [2] 土默特左旗土默特志编纂委员会办公室.土默特史料（第十六集）.1985,2. [3] 金峰.大青山下的美岱召.呼和浩特:内蒙古人民出版社,2011,7.			
现状描述	美岱召皆为遗存建筑，城寺合一的寺庙布局		描述时间	2010/08/09	
			描述人	萨日朗	
调查日期	2010/08/06		调查人员	白丽燕、萨日朗	

美岱召基本概况表1

美岱召 · 基本概况 2

现存建筑	大雄宝殿	琉璃殿	太后庙	已毁建筑	泰合门城门楼（原）	罗汉殿（原）
	乃琼庙	八角庙	西万佛殿		观音殿（原）	东万佛殿
	达赖庙	泰合门城楼	四个角楼		——	——
	观音殿	罗汉堂	——		信息来源	[1] 调研访谈记录. [2] 乔吉,马永真.内蒙古古城.呼和浩特：内蒙古人民出版社,2003.3.
区位图						
总平面图						
调查日期	2010/08/06		调查人员	白丽燕、萨日朗		

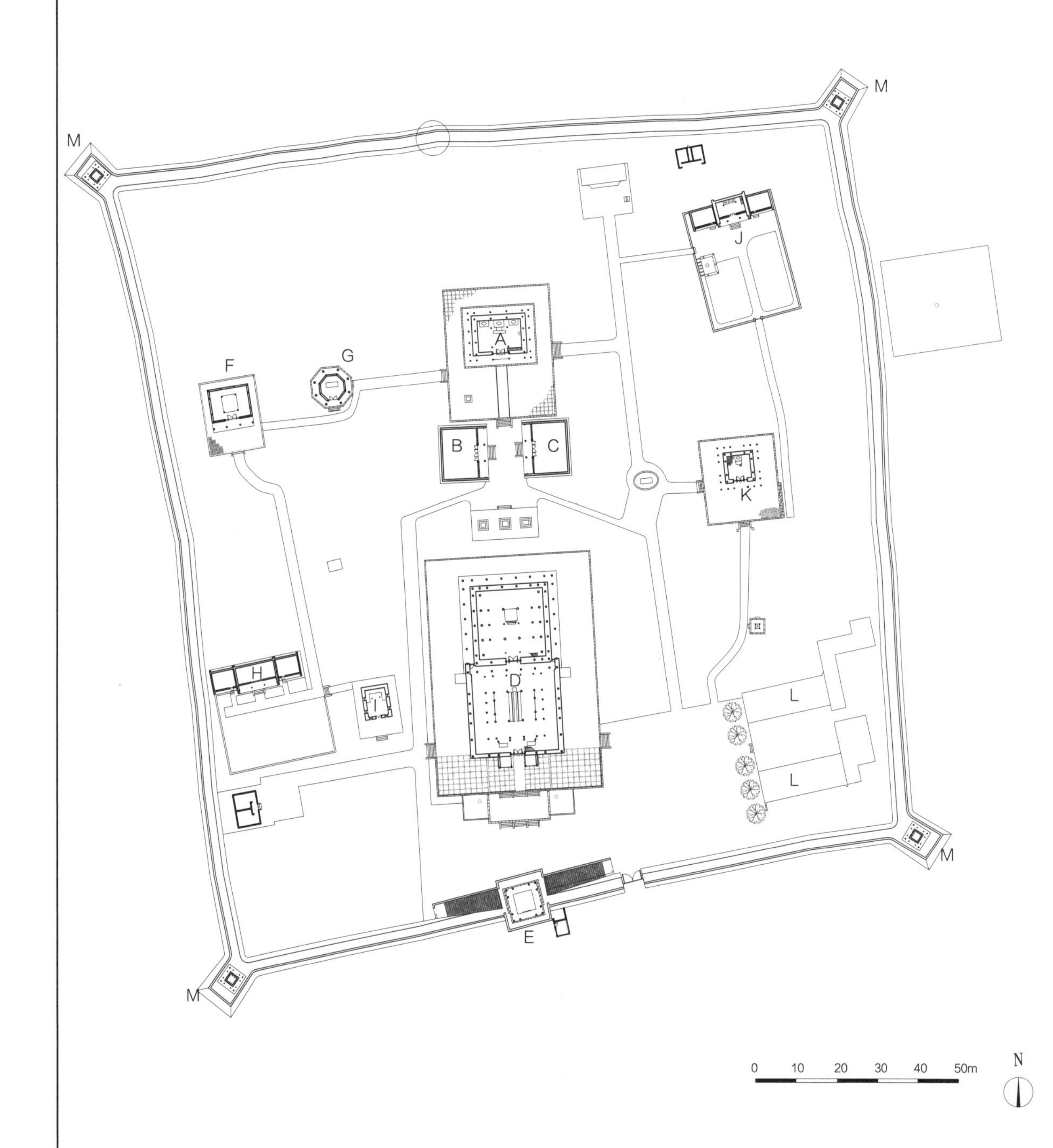

A.琉璃殿
B.法物流通处
C.罗汉堂
D.大雄宝殿
E.泰和门城楼
F.西万佛殿
G.八角庙
H.佛爷府
I.乃琼庙
J.达赖庙
K.太后庙
L.游客中心
M.角 楼

美岱召总平面图

1.1 美岱召·大雄宝殿

单位:毫米

建筑名称	汉语正式名称	大雄宝殿				俗称	朝克沁独贡				
概述	初建年	明代				建筑朝向	南		建筑层数	三	
	建筑简要描述	汉藏结合式，重檐歇山屋顶，汉式结构体系，外部看为三层阁楼式建筑									
	重建重修记载	——									
		信息来源	——								
结构规模	结构形式	砖木混合		相连的建筑	无			室内天井	都纲法式		
	建筑平面形式	凸字形		外廊形式	前廊、半回廊				无		
	通面阔	22000	开间数	7	明间	3550	次间	3550	梢间	2000	次梢间 —— 尽间 3400
	通进深	41970	进深数	7	进深尺寸（前→后）		3700→3850→2300→2300→2300→3850→2700				
	柱子数量	45	柱子间距	横向尺寸	——						
				纵向尺寸	——						
	其他	——									
建筑主体（大木作）（石作）（瓦作）	屋顶	屋顶形式	重檐歇山			瓦作	布瓦				
	外墙	主体材料	砖	材料规格	290×140×50		饰面颜色	白			
		墙体收分	有	边玛檐墙	有		边玛材料	砖			
	斗栱、梁架	斗栱	有	平身科斗口尺寸		不详	梁架关系	不详（有吊顶）			
	柱、柱式（前廊柱）	形式	汉式	柱身断面形状	圆	断面尺寸	直径 $D=300$			（在没有前廊柱的情况下，填写室内柱及其特征）	
		柱身材料	木	柱身收分	无	栌斗、托木	有	雀替	有		
		柱础	有	柱础形状	方上圆	柱础尺寸	450×450 直径 $D=390$				
	台基	台基类型	普通台基	台基高度	730、1030	台基地面铺设材料		条石、鹅卵石			
	其他	——									
装修（小木作）（彩画）	门(正面)	汉式板门		门楣	——	堆经	——	门帘	——		
	窗（正面）	无		窗楣	——	窗套	——	窗帘	——		
	室内隔扇	隔扇	无	隔扇位置	——						
	室内地面、楼面	地面材料及规格		木地板260×2100		楼面材料及规格		木地板（规格不均）			
	室内楼梯	楼梯	有	楼梯位置	进入殿门左侧	楼梯材料	木	梯段宽度	700		
	天花、藻井	天花	有	天花类型	井口	藻井	有	藻井类型	八边形		
	彩画	柱头	有	柱身	有	梁架	有	走马板	无		
		门、窗	无	天花	有	藻井	有	其他彩画	——		
	其他	悬塑	无	佛龛	无	匾额	寿灵寺				
装饰	室内	帷幔	有	幕帘彩绘	无	壁画	有	唐卡	无		
		经幡	有	经幢	无	柱毯	无	其他	——		
	室外	玛尼轮	无	苏勒德	无	宝顶	有	祥麟法轮	无		
		四角经幢	有	经幡	无	铜饰	有	石刻、砖雕	有		
		仙人走兽	无	壁画	有	其他	——				
陈设	室内	主佛像	弥勒佛像			佛像基座	莲花座				
		法座 无	藏经橱 无	经床 无	诵经桌 无	法鼓 无	玛尼轮 无	坛城 无	其他 ——		
	室外	旗杆 无	苏勒德 无	狮子 无	经幡 有	玛尼轮 无	香炉 有	五供 无	其他 ——		
	其他	——									
备注	——										
调查日期	2010/08/06	调查人员	白丽燕、萨日朗	整理日期	2010/08/09	整理人员	萨日朗				

大雄宝殿基本概况表1

美岱召 ·大雄宝殿·档案照片					
照片名称	斜前方1	照片名称	正立面	照片名称	斜前方2

照片名称	斜前方3	照片名称	斜后方	照片名称	北立面

照片名称	柱身	照片名称	柱头	照片名称	柱础

照片名称	经堂室内正面	照片名称	经堂室内壁画	照片名称	经堂室内唐卡

照片名称	佛殿室内正面	照片名称	佛殿室内侧面	照片名称	佛殿室内天花

备注	—				
摄影日期	2010/08/06	摄影人员	孟祎军		

大雄宝殿基本概况表2

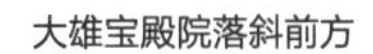

大雄宝殿院落斜前方

大雄宝殿正面（左图）
大雄宝殿背面（右图）

大雄宝殿室内正面（左图）

大雄宝殿室内壁画（右图）

大雄宝殿正立面图

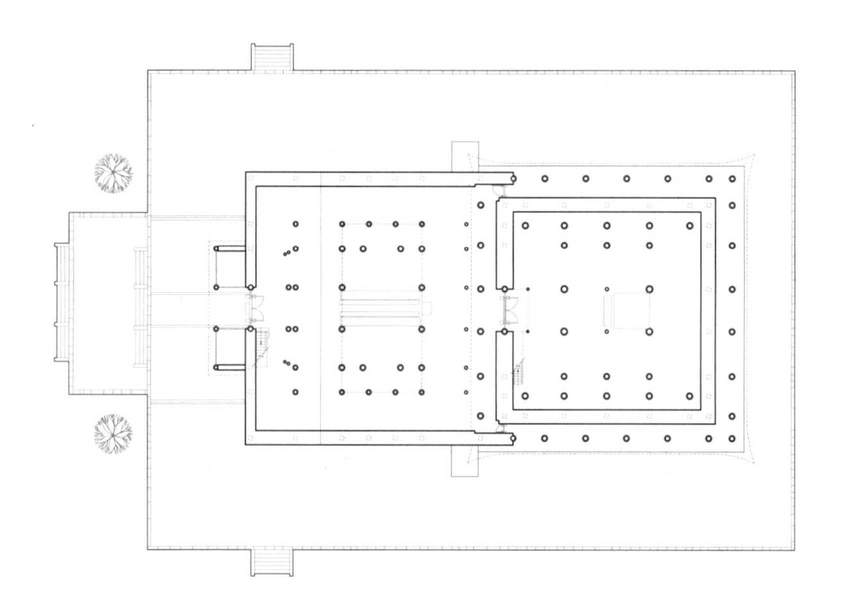

大雄宝殿一层平面图

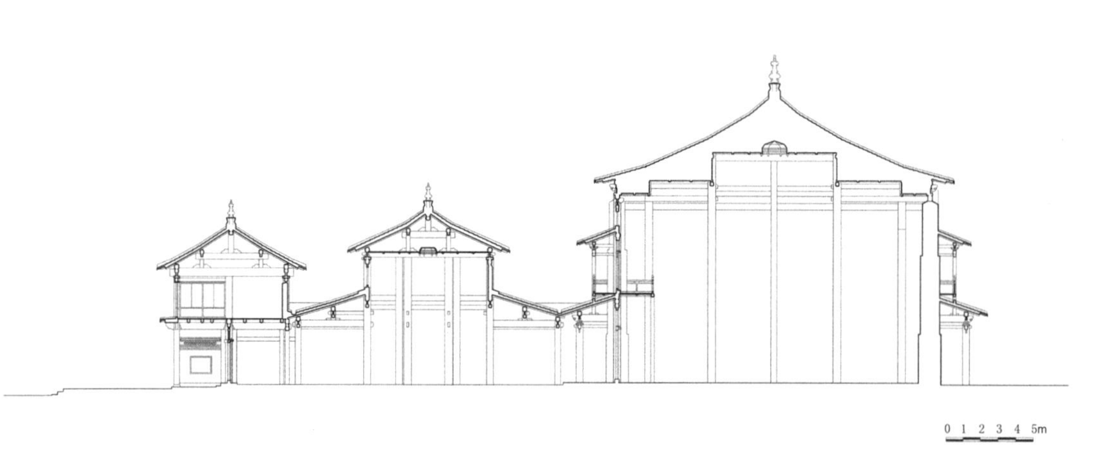

大雄宝殿剖面图

1.2 美岱召 · 琉璃殿

单位:毫米

建筑名称	汉语正式名称		俗称	琉璃殿										
概述	初建年	嘉靖四十五年（1566年）	建筑朝向	南	建筑层数	三								
	建筑简要描述	汉式建筑，重檐歇山屋顶，汉式结构体系												
	重建重修记载	——												
		信息来源	——											
结构规模	结构形式	砖木混合	相连的建筑	无	室内天井	无								
	建筑平面形式	长方	外廊形式	回廊										
	通面阔	13300	开间数	3	明间	3900	次间	3350	梢间	——	次梢间	——	尽间	——
	通进深	9600	进深数	3	进深尺寸（前→后）	1050→4570→1870								
	柱子数量	——	柱子间距	横向尺寸	——	（藏式建筑结构体系填写此栏，不含廊柱）								
				纵向尺寸	——									
	其他	——												
建筑主体（大木作）（石作）（瓦作）	屋顶	屋顶形式	重檐歇山	瓦作	绿色琉璃瓦屋面									
	外墙	主体材料	砖抹灰	材料规格	270×130×50	饰面颜色	白							
		墙体收分	有	边玛檐墙	无	边玛材料	——							
	斗栱、梁架	斗栱	有	平身科斗口尺寸	不详	梁架关系	不详（有吊顶）							
	柱、柱式（前廊柱）	形式	汉式	柱身断面形状	圆	断面尺寸	直径$D=280$	（在没有前廊柱的情况下，填写室内柱及其特征）						
		柱身材料	木	柱身收分	无	栌斗、托木	无	雀替	有					
		柱础	有	柱础形状	圆	柱础尺寸	直径$D=360$							
	台基	台基类型	普通台基	台基高度	1070	台基地面铺设材料	条石、卵石							
	其他	——												
装修（小木作）（彩画）	门(正面)	板门	门楣	无	堆经	无	门帘	无						
	窗（正面）	槛窗	窗楣	无	窗套	有	窗帘	无						
	室内隔扇	隔扇	无	隔扇位置	——									
	室内地面、楼面	地面材料及规格	石块930×1800	楼面材料及规格	木板（规格不均）									
	室内楼梯	楼梯	有	楼梯位置	进入殿门左侧	楼梯材料	木	梯段宽度	500					
	天花、藻井	天花	有	天花类型	井口	藻井	无	藻井类型	——					
	彩画	柱头	无	柱身	无	梁架	有	走马板	无					
		门、窗	无	天花	有	藻井	——	其他彩画	——					
	其他	悬塑	无	佛龛	无	匾额	无							
装饰	室内	帷幔	无	幕帘彩绘	无	壁画	有	唐卡	无					
		经幡	无	经幢	有	柱毯	无	其他	——					
	室外	玛尼轮	无	苏勒德	无	宝顶	有	祥麟法轮	无					
		四角经幢	无	经幡	有	铜饰	无	石刻、砖雕	无					
		仙人走兽	无	壁画	无	其他	——							

陈设	室内	主佛像	弥勒佛像	佛像基座	莲花座												
		法座	有	藏经橱	无	经床	无	诵经桌	无	法鼓	无	玛尼轮	无	坛城	无	其他	——
	室外	旗杆	无	苏勒德	无	狮子	无	经幡	有	玛尼轮	无	香炉	有	五供	无	其他	——
	其他	——															

备注	——						
调查日期	2010/08/06	调查人员	白丽燕、萨日朗	整理日期	2010/08/09	整理人员	萨日朗

美岱召 ·琉璃殿·档案照片					
照片名称 	斜前方	照片名称 	正立面	照片名称 	侧立面
照片名称 	斜后方	照片名称 	北立面	照片名称 	柱身
照片名称 	柱头	照片名称 	柱础	照片名称 	柱廊
照片名称 	门	照片名称 	二层隔扇窗	照片名称 	台基
照片名称 	室内佛像	照片名称 	室内壁画	照片名称 	室内天花
备注	——				
摄影日期	2010/08/06	摄影人员	孟祎军		

琉璃殿基本概况表2

琉璃殿正前方

琉璃殿斜前方（左图）
琉璃殿檐部（右图）

琉璃殿斜后方（左图）
琉璃殿室外柱子（右图）

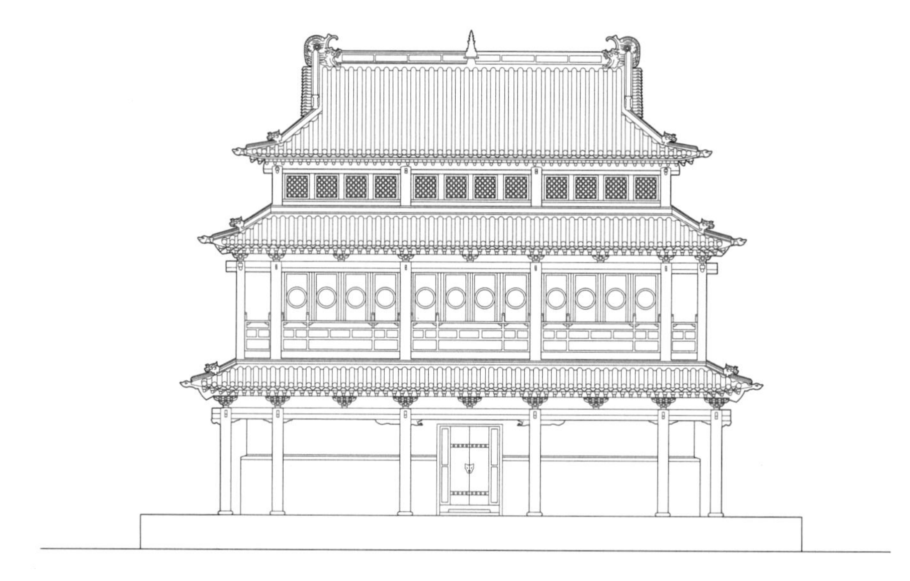

琉璃殿正立面图

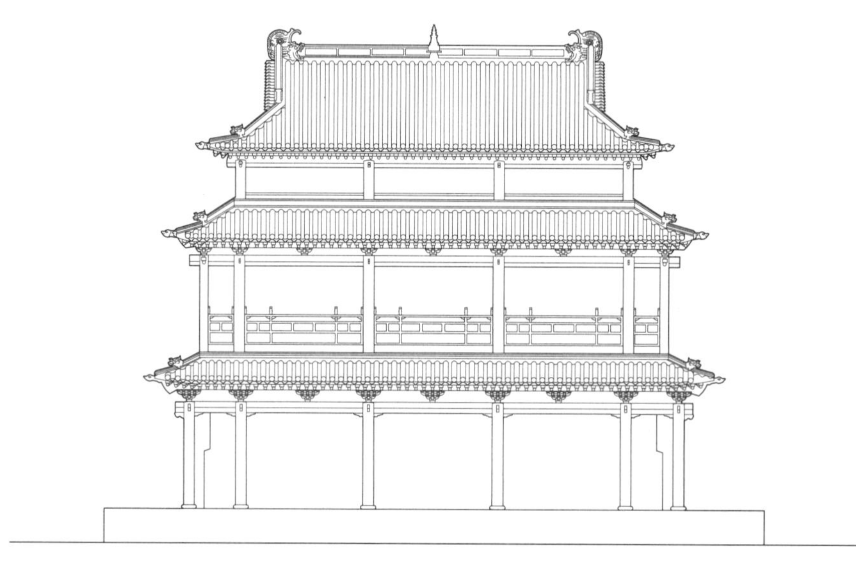

琉璃殿背立面图

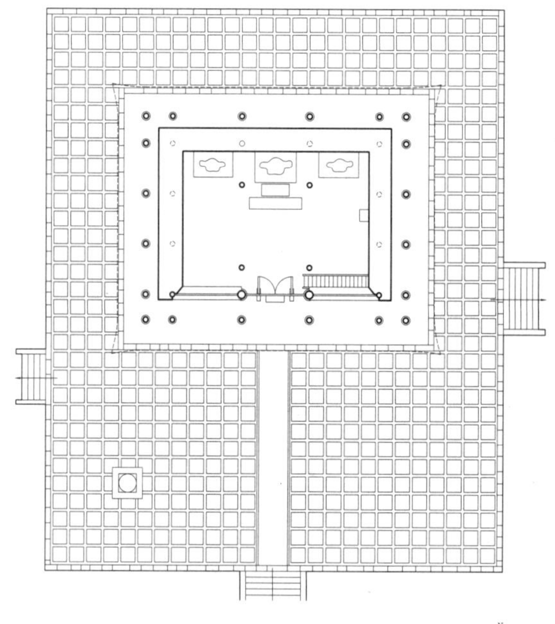

琉璃殿一层平面图

1.3 美岱召·乃琼庙

单位:毫米

建筑名称	汉语正式名称	乃琼殿								俗称	护法神殿						
概述	初建年		建筑朝向	南	建筑层数	二											
	建筑简要描述	藏式建筑、藏式平屋顶															
	重建重修记载	不详															
		信息来源	不详														
结构规模	结构形式	砖木混合	相连的建筑	无	室内天井	无											
	建筑平面形式	方形	外廊形式	无													
	通面阔	7000	开间数	3	明间	2000	次间	1620	梢间	——	次梢间	——	尽间	——			
	通进深	7850	进深数	2	进深尺寸(前→后)	3400→2960											
	柱子数量	2	柱子间距	横向尺寸	2000	(藏式建筑结构体系填写此栏,不含廊柱)											
				纵向尺寸	——												
	其他	——															
建筑主体(大木作)(石作)(瓦作)	屋顶	屋顶形式	藏式平屋顶	瓦作	布瓦												
	外墙	主体材料	砖	材料规格	290×140×50	饰面颜色	白										
		墙体收分	有	边玛檐墙	有	边玛材料	红砖										
	斗栱、梁架	斗栱	无	平身科斗口尺寸	——	梁架关系	不详(有吊顶)										
	柱、柱式(前廊柱)	形式	汉式	柱身断面形状	圆	断面尺寸	直径 $D=200$	(在没有前廊柱的情况下,填写室内柱及其特征)									
		柱身材料	木	柱身收分	无	栌斗、托木	无	雀替	无								
		柱础	有	柱础形状	圆	柱础尺寸	直径 $D=350$										
	台基	台基类型	普通台基	台基高度	260	台基地面铺设材料	条石、方砖										
	其他	——															
装修(小木作)(彩画)	门(正面)	藏式门	门楣	有	堆经	无	门帘	无									
	窗(正面)	藏式明窗	窗楣	无	窗套	有	窗帘	无									
	室内隔扇	隔扇	无	隔扇位置	——												
	室内地面、楼面	地面材料及规格	水泥地面	楼面材料及规格	木板(规格不均)												
	室内楼梯	楼梯	有	楼梯位置	进入殿门左侧	楼梯材料	木	梯段宽度	500								
	天花、藻井	天花	有	天花类型	不详	藻井	无	藻井类型	——								
	彩画	柱头	无	柱身	无	梁架	无	走马板	无								
		门、窗	无	天花	无	藻井	——	其他彩画	——								
	其他	悬塑	无	佛龛	无	匾额	无										
装饰	室内	帷幔	有	幕帘彩绘	无	壁画	有	唐卡	无								
		经幡	无	经幢	有	柱毯	无	其他	——								
	室外	玛尼轮	无	苏勒德	无	宝顶	无	祥麟法轮	有								
		四角经幢	无	经幡	无	铜饰	无	石刻、砖雕	无								
		仙人走兽	无	壁画	无	其他	——										
陈设	室内	主佛像	一层:五尊护法像 二层:麦达力坐像	佛像基座	须弥座												
		法座	无	藏经橱	无	经床	无	诵经桌	无	法鼓	无	玛尼轮	无	坛城	无	其他	——
	室外	旗杆	无	苏勒德	无	狮子	无	经幡	无	玛尼轮	无	香炉	有	五供	无	其他	——
	其他	——															
备注	——																
调查日期	2010/08/06	调查人员	白丽燕、萨日朗	整理日期	2010/08/09	整理人员	萨日朗										

乃琼庙基本概况表1

美岱召 ·乃琼庙·档案照片					
照片名称 	斜前方	照片名称 	正立面	照片名称 	侧立面
照片名称 	祥麟法轮	照片名称 	边玛檐墙	照片名称 	局部
照片名称 	入口	照片名称 	台基	照片名称 	窗
照片名称 	室内正面	照片名称 	室内柱	照片名称 	盲窗
照片名称 	室内佛像1	照片名称 	室内佛像2	照片名称 	室内佛像3
备注	——				
摄影日期	2010/08/06	摄影人员	孟祎军		

乃琼庙基本概况表2

乃琼庙斜前方

乃琼庙室内正面（左图）
乃琼庙室外局部（右图）

乃琼庙侧面（左图）
乃琼庙窗（右图）

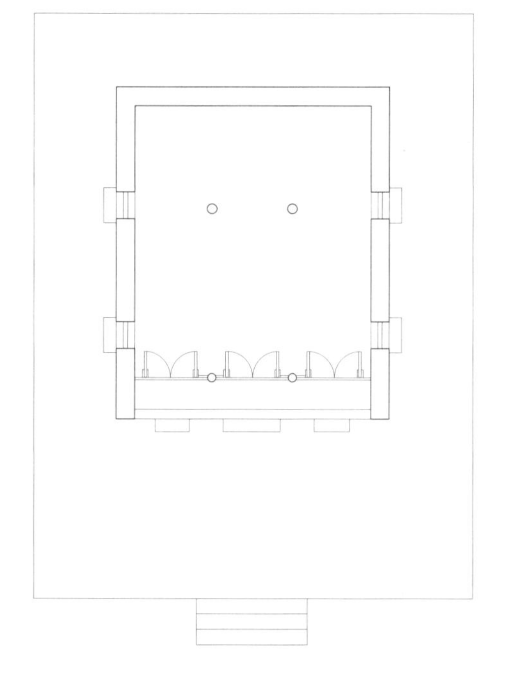

乃琼庙一层平面图

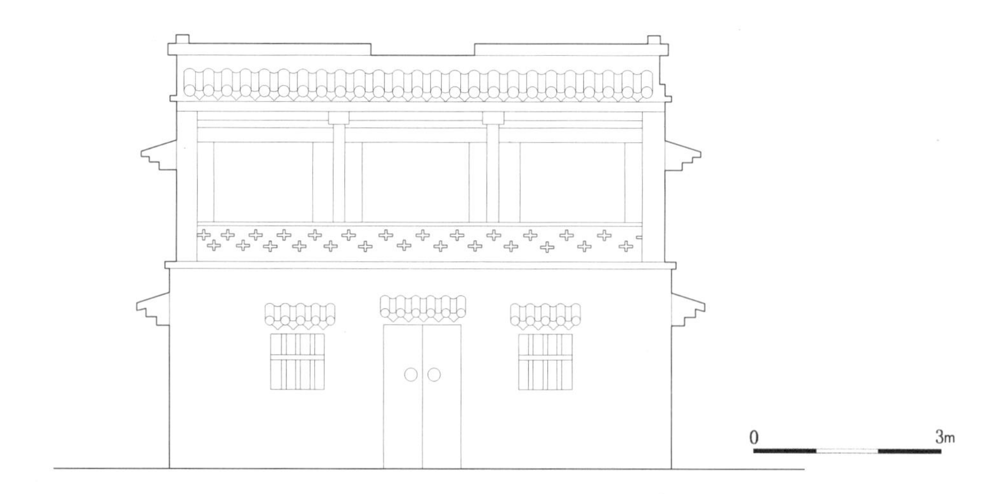

乃琼庙立面图

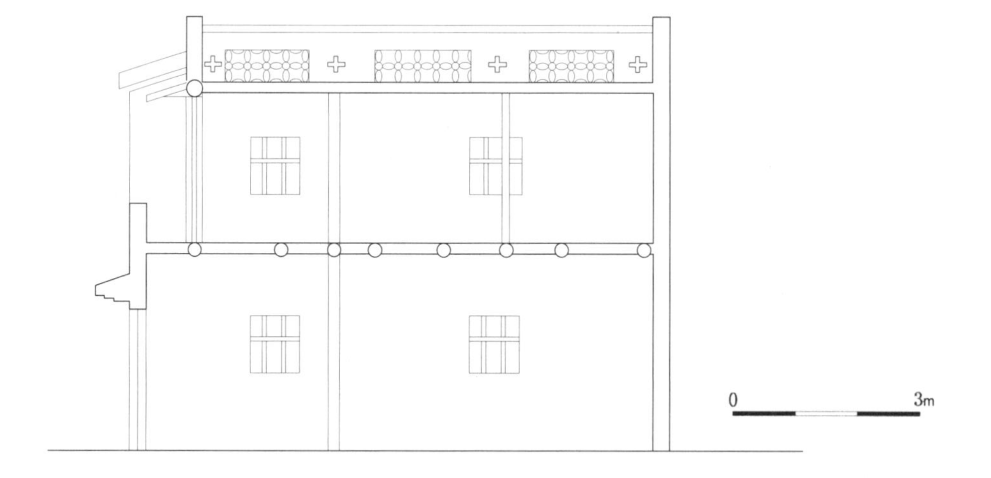

乃琼庙剖面图

1.4 美岱召·西万佛殿

单位:毫米

建筑名称	汉语正式名称	西万佛殿		俗称	——												
概述	初建年	不详		建筑朝向	南	建筑层数	一										
	建筑简要描述	汉式建筑，室内为木悬山式风格，塑众多佛像															
	重建重修记载	——															
		信息来源	——														
结构规模	结构形式	砖木混合		相连的建筑	无		室内天井	无									
	建筑平面形式	长方		外廊形式	前廊												
	通面阔	9800	开间数	3	明间	3420	次间	3160	梢间	——	次梢间	——	尽间	——			
	通进深	7500	进深数	3	进深尺寸（前→后）		1925→3745→1840										
	柱子数量	——	柱子间距	横向尺寸	——		（藏式建筑结构体系填写此栏，不含廊柱）										
				纵向尺寸	——												
	其他	——															
建筑主体（大木作）（石作）（瓦作）	屋顶	屋顶形式	硬山			瓦作	布瓦										
	外墙	主体材料	砖抹灰	材料规格	270×130×50		饰面颜色	白									
		墙体收分	无	边玛檐墙	无		边玛材料	——									
	斗栱、梁架	斗栱	有	平身科斗口尺寸		不详	梁架关系	不详（有吊顶）									
	柱、柱式（前廊柱）	形式	汉式	柱身断面形状	圆	断面尺寸	直径 $D=220$			（在没有前廊柱的情况下，填写室内柱及其特征）							
		柱身材料	木	柱身收分	无	栌斗、托木	无	雀替	有								
		柱础	有	柱础形状	圆	柱础尺寸	直径 $D=350$										
	台基	台基类型	普通台基	台基高度	3200	台基地面铺设材料		条石、卵石									
	其他	——															
装修（小木作）（彩画）	门(正面)	板门		门楣	无	堆经	无	门帘	无								
	窗（正面）	槛窗		窗楣	无	窗套	有	窗帘	无								
	室内隔扇	隔扇	无	隔扇位置	——												
	室内地面、楼面	地面材料及规格		方砖240×240		楼面材料及规格		无									
	室内楼梯	楼梯	——	楼梯位置	——	楼梯材料	——	梯段宽度	——								
	天花、藻井	天花	无	天花类型	——	藻井	无	藻井类型	——								
	彩画	柱头	无	柱身	无	梁架	有	走马板	无								
		门、窗	无	天花	——	藻井	——	其他彩画	——								
	其他	悬塑	有	佛龛	无	匾额	无										
装饰	室内	帷幔	无	幕帘彩绘	无	壁画	无	唐卡	无								
		经幡	无	经幢	无	柱毯	无	其他	——								
	室外	玛尼轮	无	苏勒德	无	宝顶	无	祥麟法轮	无								
		四角经幢	无	经幡	有	铜饰	无	石刻、砖雕	无								
		仙人走兽	无	壁画	无	其他	——										
陈设	室内	主佛像	释迦牟尼佛			佛像基座	莲花座										
		法座	无	藏经橱	无	经床	无	诵经桌	无	法鼓	无	玛尼轮	无	坛城	无	其他	——
	室外	旗杆	无	苏勒德	无	狮子	无	经幡	有	玛尼轮	无	香炉	无	五供	无	其他	——
	其他	——															
备注	——																
调查日期	2010/08/06	调查人员	白丽燕、萨日朗	整理日期	2010/08/09	整理人员	萨日朗										

西万佛殿基本概况表1

美岱召 ·西万佛殿 · 档案照片

照片名称	斜前方	照片名称	正立面	照片名称	侧立面
照片名称	斜后方	照片名称	柱身	照片名称	柱础
照片名称	门	照片名称	窗	照片名称	台基
照片名称	斗栱	照片名称	柱廊	照片名称	地面
照片名称	室内正面	照片名称	室内局部	照片名称	室内佛像
备注	—				
摄影日期	2010/08/06	摄影人员	孟祎军		

西万佛殿基本概况表2

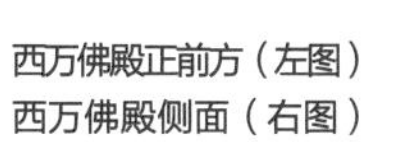

西万佛殿正前方（左图）
西万佛殿侧面（右图）

西万佛殿背面（左图）
西万佛殿前廊（右图）

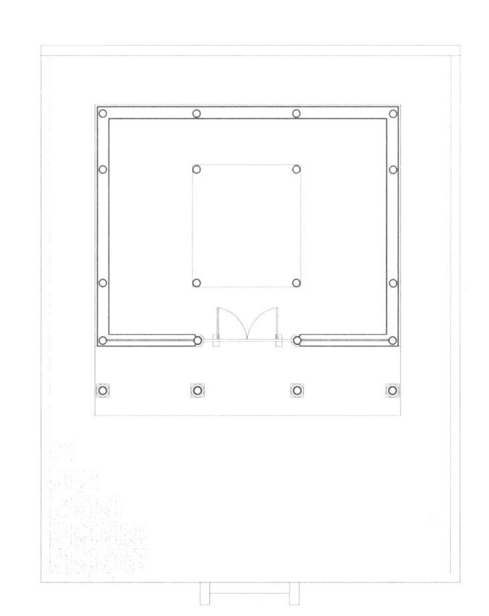

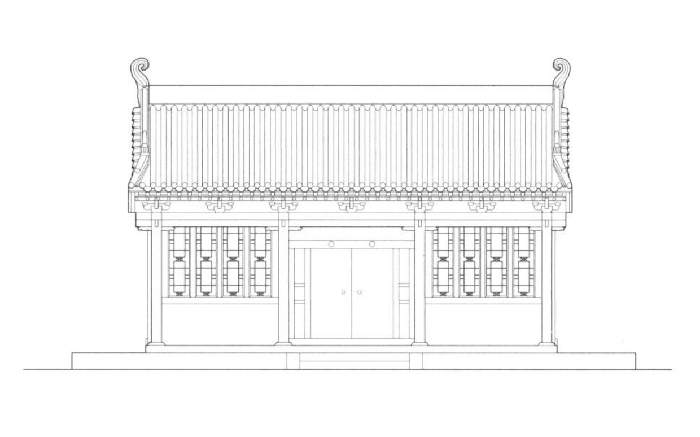

西万佛殿一层平面图（左图）

西万佛殿立面图（右图）

1.5 美岱召·八角庙

单位:毫米

建筑名称	汉语正式名称	八角庙				俗称		老君庙									
概述	初建年	1585年始建				建筑朝向		南		建筑层数		一					
	建筑简要描述	——															
	重建重修记载	——															
		信息来源	——														
结构规模	结构形式	砖木混合		相连的建筑		无				室内天井		无					
	建筑平面形式	八角形		外廊形式		回廊											
	通面阔	6000	开间数	3	明间	3300	次间	2300	梢间	——	次梢间	——	尽间	——			
	通进深	6000	进深数	3	进深尺寸（前→后）		2300→3300→2300										
	柱子数量	——	柱子间距	横向尺寸	——		（藏式建筑结构体系填写此栏，不含廊柱）										
				纵向尺寸	——												
	其他	——															
建筑主体（大木作）（石作）（瓦作）	屋顶	屋顶形式	八角攒尖顶			瓦作		布瓦									
	外墙	主体材料	砖	材料规格	290×140×50	饰面颜色		白、灰									
		墙体收分	无	边玛檐墙	——	边玛材料		——									
	斗栱、梁架	斗栱	无	平身科斗口尺寸	——	梁架关系		抬梁式									
	柱、柱式（前廊柱）	形式	汉式	柱身断面形状	圆	断面尺寸	直径$D=200$			（在没有前廊柱的情况下，填写室内柱及其特征）							
		柱身材料	木	柱身收分	无	栌斗、托木	无	雀替	无								
		柱础	有	柱础形状	圆	柱础尺寸	直径$D=270$										
	台基	台基类型	普通台基	台基高度	280	台基地面铺设材料		条石、方砖									
	其他	——															
装修（小木作）（彩画）	门(正面)	板门		门楣	无	堆经	无	门帘	无								
	窗（正面）	无		窗楣	——	窗套	——	窗帘	——								
	室内隔扇	隔扇	无	隔扇位置	——												
	室内地面、楼面	地面材料及规格		方砖280×280		楼面材料及规格		无									
	室内楼梯	楼梯	——	楼梯位置	——	楼梯材料	——	梯段宽度	——								
	天花、藻井	天花	无	天花类型	——	藻井	无	藻井类型	——								
	彩画	柱头	有	柱身	无	梁架	有	走马板	无								
		门、窗	无	天花	——	藻井	——	其他彩画	——								
	其他	悬塑	无	佛龛	无	匾额	“清凉境”										
装饰	室内	帷幔	无	幕帘彩绘	无	壁画	有	唐卡	无								
		经幡	无	经幢	无	柱毯	无	其他	——								
	室外	玛尼轮	无	苏勒德	无	宝顶	无	祥麟法轮	无								
		四角经幢	无	经幡	无	铜饰	无	石刻、砖雕	无								
		仙人走兽	无	壁画	无	其他	——										
陈设	室内	主佛像	大威德金刚尊神			佛像基座	莲花座										
		法座	无	藏经橱	无	经床	无	诵经桌	无	法鼓	无	玛尼轮	无	坛城	无	其他	——
	室外	旗杆	无	苏勒德	无	狮子	无	经幡	无	玛尼轮	无	香炉	无	五供	无	其他	——
	其他	——															
备注	——																
调查日期	2010/08/06	调查人员	白丽燕、萨日朗	整理日期	2010/08/09	整理人员	萨日朗										

八角庙基本概况表1

美岱召 ·八角庙·档案照片					
照片名称	斜前方	照片名称	正前方	照片名称	侧立面
照片名称	北立面	照片名称	柱身	照片名称	柱头
照片名称	柱础	照片名称	入口	照片名称	台基
照片名称	室内正面	照片名称	左墙壁画	照片名称	右墙壁画
照片名称	室内屋顶梁架	照片名称	佛像	照片名称	地面
备注	——				
摄影日期	2010/08/06	摄影人员	孟祎军		

八角庙基本概况表2

八角庙正前方

八角庙背面（左图）
八角庙室外柱子（右图）

八角庙室内佛像（左图）
八角庙正面门（右图）

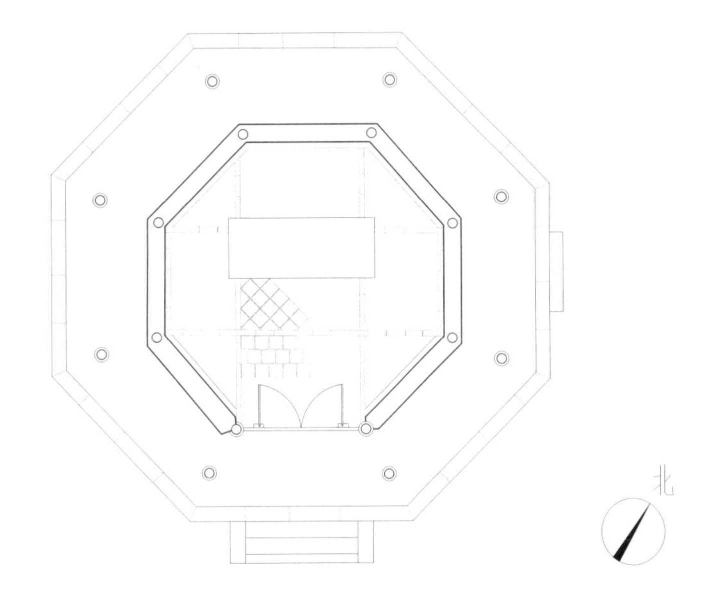

八角庙平面图

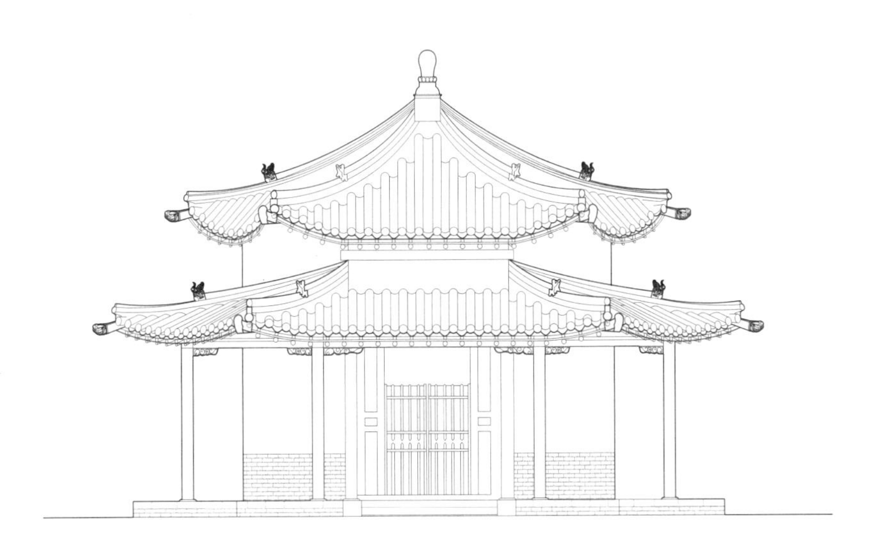

八角庙正立面图

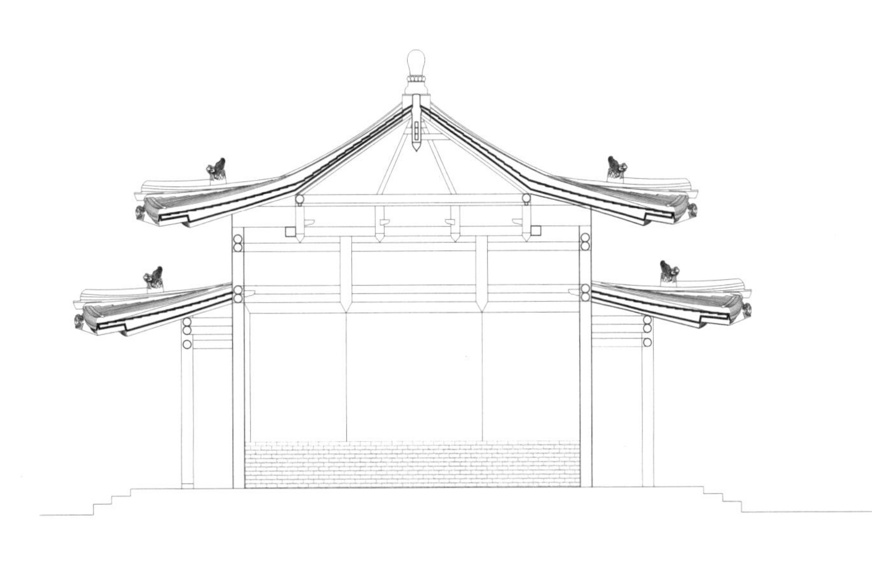

八角庙剖面图

1.6 美岱召·太后庙

单位:毫米

建筑名称	汉语正式名称	太后庙		俗称	三娘子庙		
概述	初建年	——		建筑朝向	南	建筑层数	—
	建筑简要描述	汉式建筑					
	重建重修记载	——					
		信息来源	——				
结构规模	结构形式	砖木混合	相连的建筑	无		室内天井	无
	建筑平面形式	方	外廊形式	回廊			
	通面阔	7600	开间数	5	明间 3300 次间 1650 梢间 1650 次梢间 —— 尽间 ——		
	通进深	7400	进深数	5	进深尺寸（前→后）	1650→1650→3160→1650→1650	
	柱子数量	——	柱子间距	横向尺寸	——	（藏式建筑结构体系填写此栏，不含廊柱）	
				纵向尺寸	——		
	其他	——					
建筑主体（大木作）（石作）（瓦作）	屋顶	屋顶形式	重檐歇山		瓦作	布瓦	
	外墙	主体材料	砖	材料规格	290×140×50	饰面颜色	白
		墙体收分	有	边玛檐墙	无	边玛材料	——
	斗栱、梁架	斗栱	无	平身科斗口尺寸	——	梁架关系	不详（有吊顶）
	柱、柱式（前廊柱）	形式	汉式	柱身断面形状	圆	断面尺寸	直径 $D=260$
		柱身材料	木	柱身收分	无	栌斗、托木	无　雀替　有
		柱础	有	柱础形状	圆	柱础尺寸	直径 $D=300$
		（在没有前廊柱的情况下，填写室内柱及其特征）					
	台基	台基类型	普通台基	台基高度	600	台基地面铺设材料	条石、方砖
	其他	——					
装修（小木作）（彩画）	门(正面)	板门	门楣	无	堆经	无	门帘　无
	窗（正面）	无	窗楣	——	窗套	——	窗帘　——
	室内隔扇	隔扇	无	隔扇位置	——		
	室内地面、楼面	地面材料及规格		方砖280×280	楼面材料及规格	无	
	室内楼梯	楼梯	——	楼梯位置	——	楼梯材料	—— 梯段宽度 ——
	天花、藻井	天花	有	天花类型	井口	藻井	无 藻井类型 ——
	彩画	柱头	有	柱身	无	梁架	有 走马板 有
		门、窗	无	天花	有	藻井	无 其他彩画 ——
	其他	悬塑	无	佛龛	无	匾额	钟宝夫人享堂
装饰	室内	帷幔	有	幕帘彩绘	无	壁画	有 唐卡 无
		经幡	无	经幢	无	柱毯	无 其他 ——
	室外	玛尼轮	无	苏勒德	无	宝顶	有 祥麟法轮 无
		四角经幢	无	经幡	无	铜饰	无 石刻、砖雕 无
		仙人走兽	无	壁画	无	其他	——
陈设	室内	主佛像	三娘子骨灰塔		佛像基座	莲花座	
		法座 无　藏经橱 无　经床 无	诵经桌 无	法鼓 无	玛尼轮 无	坛城 无	其他 ——
	室外	旗杆 无　苏勒德 无　狮子 有	经幡 无	玛尼轮 无	香炉 无	五供 无	其他 ——
	其他	——					
备注	——						
调查日期	2010/08/06	调查人员	白丽燕、萨日朗	整理日期	2010/08/09	整理人员	萨日朗

太后庙基本概况表1

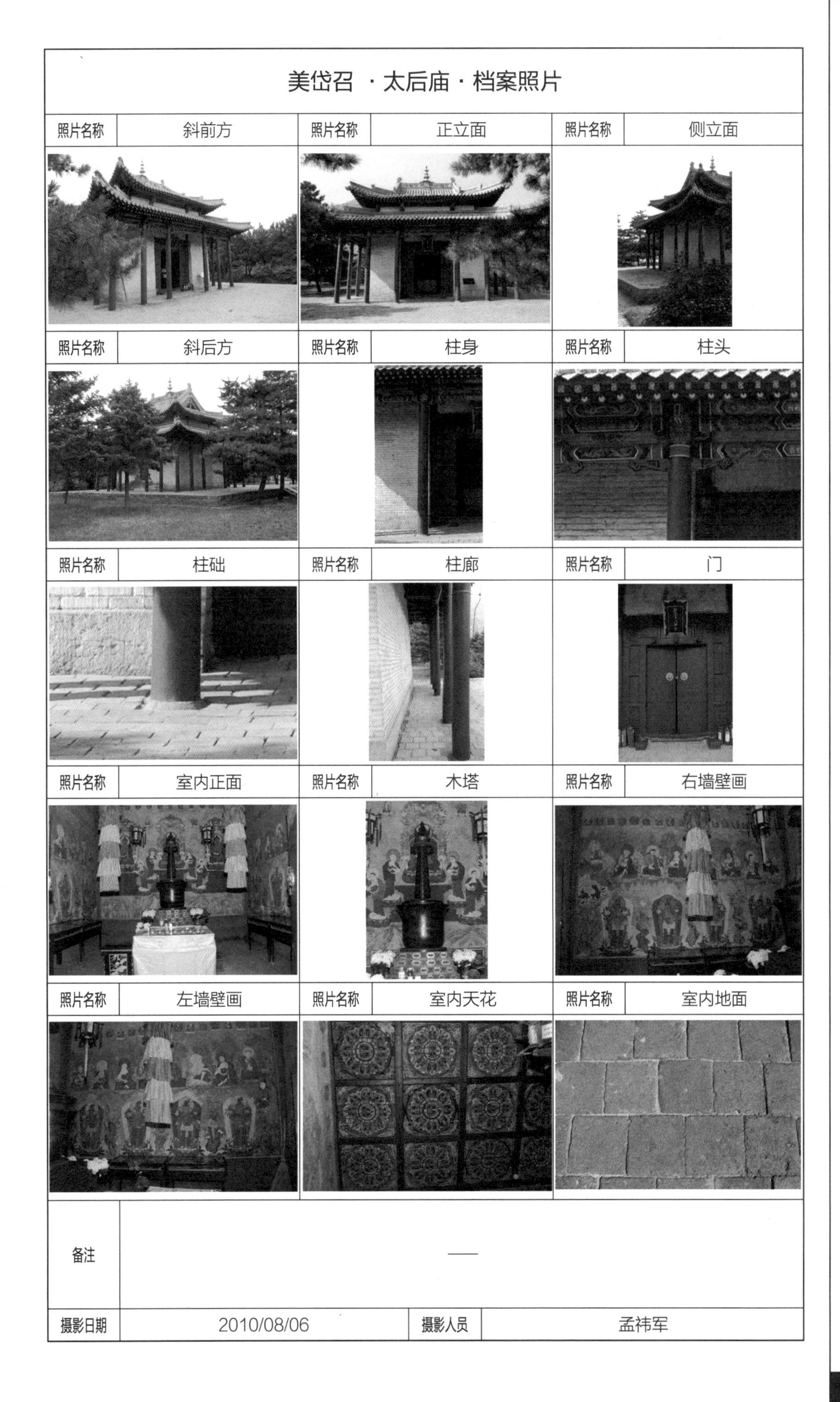

美岱召 ·太后庙·档案照片
照片名称 斜前方 照片名称 正立面 照片名称 侧立面
照片名称 斜后方 照片名称 柱身 照片名称 柱头
照片名称 柱础 照片名称 柱廊 照片名称 门
照片名称 室内正面 照片名称 木塔 照片名称 右墙壁画
照片名称 左墙壁画 照片名称 室内天花 照片名称 室内地面
备注 —
摄影日期 2010/08/06 摄影人员 孟祎军

太后庙基本概况表2

太后庙正前方

太后庙斜后方（左图）
太后庙侧面（右图）

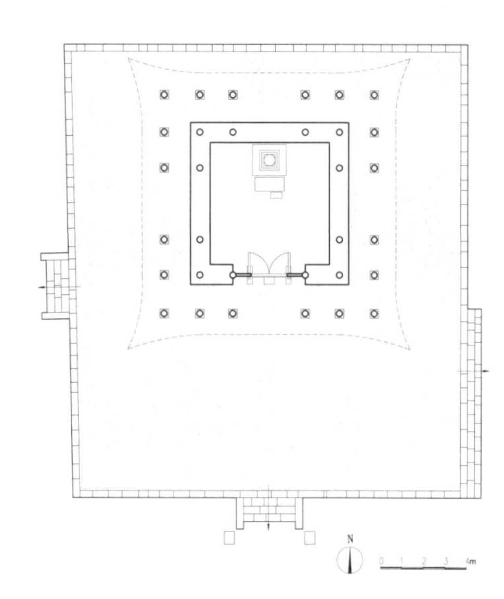

太后庙室内正面（左图）
太后庙一层平面图（右图）

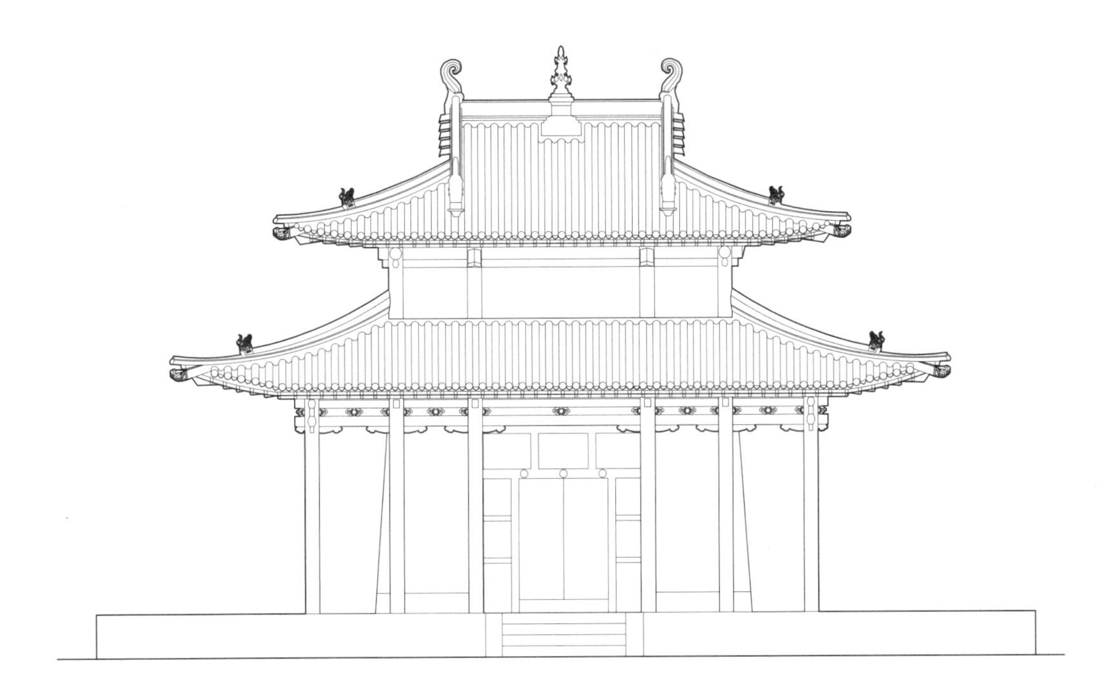

太后庙正立面图

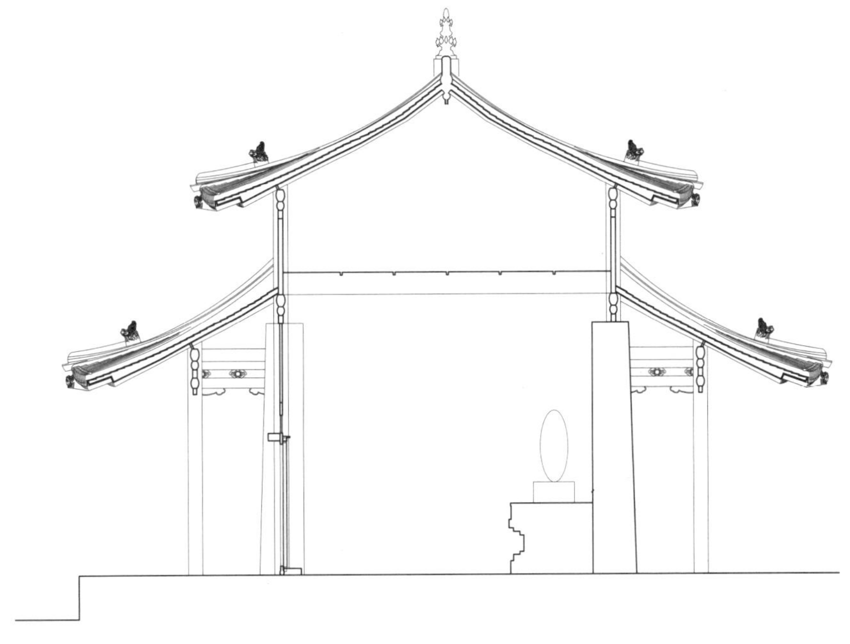

太后庙剖面图

1.7 美岱召·达赖庙

单位:毫米

建筑名称	汉语正式名称	达赖庙	俗称	——													
概述	初建年	——	建筑朝向	南	建筑层数	二
	建筑简要描述	汉式建筑，二层硬山式小楼，东西各连有耳房
	重建重修记载	——
		信息来源	——
结构规模	结构形式	砖木混合	相连的建筑	东西各有耳房相连	室内天井	无
	建筑平面形式	长方	外廊形式	前廊
	通面阔	8350	开间数	3	明间	3000	次间	2200	梢间	——	次梢间	——	尽间	——
	通进深	5000	进深数	3	进深尺寸（前→后）	——
	柱子数量	——	柱子间距	横向尺寸	——	（藏式建筑结构体系填写此栏，不含廊柱）
				纵向尺寸	——
	其他	——
建筑主体（大木作）（石作）（瓦作）	屋顶	屋顶形式	硬山	瓦作	布瓦
	外墙	主体材料	砖	材料规格	260×120×55	饰面颜色	灰
		墙体收分	无	边玛檐墙	无	边玛材料	——
	斗栱、梁架	斗栱	无	平身科斗口尺寸	——	梁架关系	不详（有吊顶）
	柱、柱式（前廊柱）	形式	汉式	柱身断面形状	圆	断面尺寸	直径$D=200$	（在没有前廊柱的情况下，填写室内柱及其特征）
		柱身材料	木	柱身收分	无	栌斗、托木	无	雀替	有
		柱础	有	柱础形状	圆	柱础尺寸	直径$D=300$
	台基	台基类型	普通	台基高度	730	台基地面铺设材料	条石、方砖、长砖
	其他	——
装修（小木作）（彩画）	门(正面)	汉式板门	门楣	无	堆经	无	门帘	无
	窗（正面）	槛窗	窗楣	无	窗套	有	窗帘	无
	室内隔扇	隔扇	无	隔扇位置	——
	室内地面、楼面	地面材料及规格	方砖	楼面材料及规格	木板、规格不均
	室内楼梯	楼梯	有	楼梯位置	入口左侧	楼梯材料	木	梯段宽度	600
	天花、藻井	天花	有	天花类型	不详	藻井	无	藻井类型	——
	彩画	柱头	有	柱身	无	梁架	有	走马板	无
		门、窗	无	天花	无	藻井	——	其他彩画	——
	其他	悬塑	无	佛龛	无	匾额	无
装饰	室内	帷幔	有	幕帘彩绘	无	壁画	无	唐卡	无
		经幡	无	经幢	无	柱毯	无	其他	——
	室外	玛尼轮	无	苏勒德	无	宝顶	无	祥麟法轮	无
		四角经幢	无	经幡	无	铜饰	无	石刻、砖雕	无
		仙人走兽	无	壁画	有	其他	——
陈设	室内	主佛像	宗喀巴	佛像基座	莲花座
		法座	无	藏经橱	无	经床	无	诵经桌	无	法鼓	无	玛尼轮	无	坛城	无	其他	——
	室外	旗杆	无	苏勒德	无	狮子	无	经幡	无	玛尼轮	无	香炉	无	五供	无	其他	——
	其他	——
备注	——
调查日期	2010/08/06	调查人员	白丽燕、萨日朗	整理日期	2010/08/09	整理人员	萨日朗

达赖庙基本概况表1

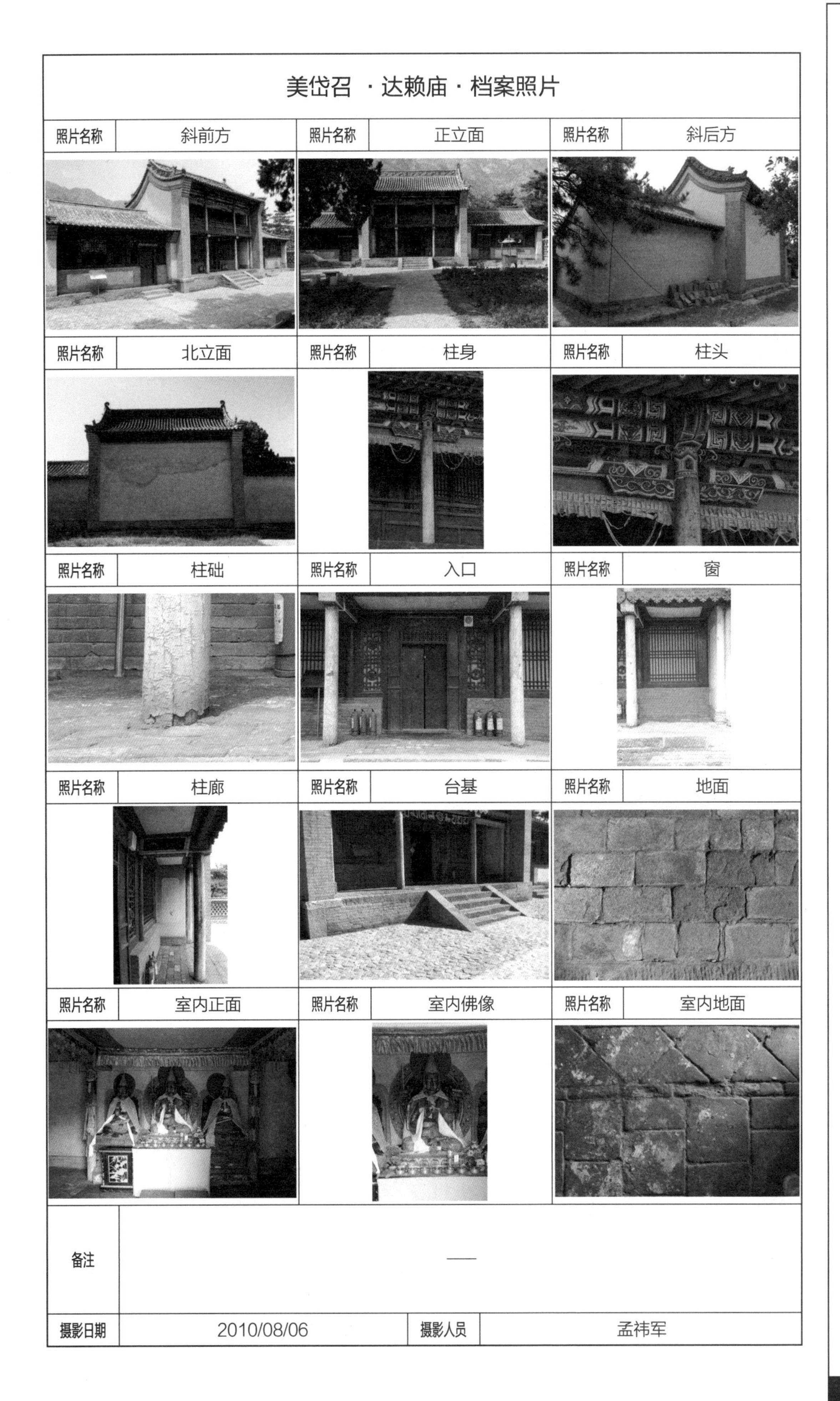

美岱召 ·达赖庙·档案照片

照片名称	斜前方	照片名称	正立面	照片名称	斜后方
照片名称	北立面	照片名称	柱身	照片名称	柱头
照片名称	柱础	照片名称	入口	照片名称	窗
照片名称	柱廊	照片名称	台基	照片名称	地面
照片名称	室内正面	照片名称	室内佛像	照片名称	室内地面

备注	——		
摄影日期	2010/08/06	摄影人员	孟祎军

达赖庙正前方

达赖庙斜前方（左图）
达赖庙院门（右图）

达赖庙佛像（左图）
达赖庙斜后方（右图）

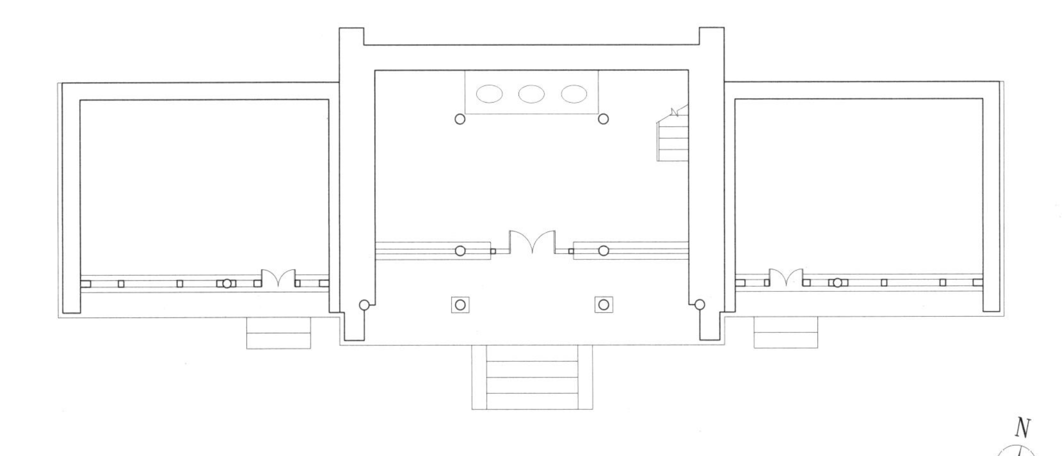

达赖庙一层平面图

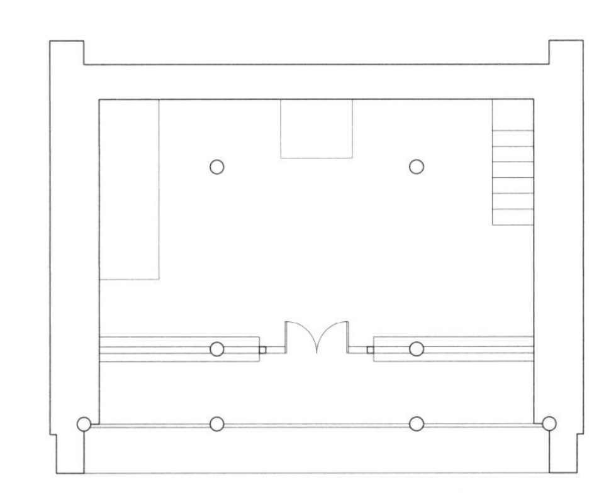

达赖庙二层平面图

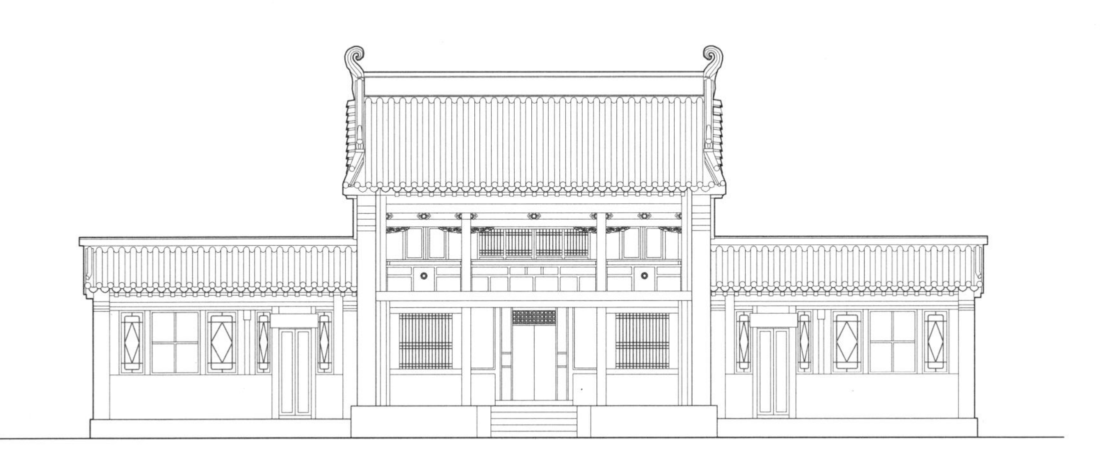

达赖庙立面图

1.8 美岱召·罗汉堂

罗汉堂正前方

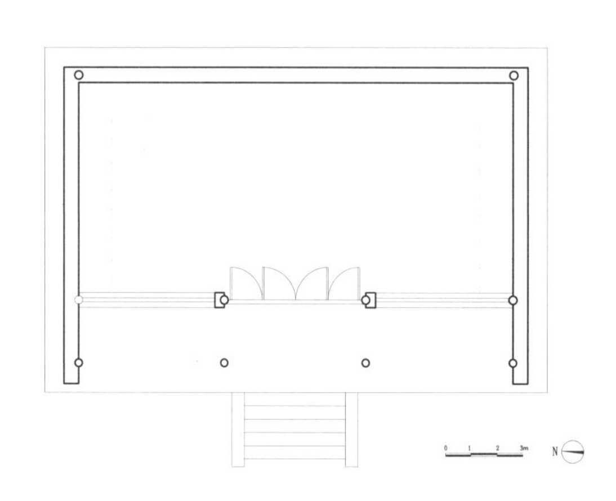

罗汉堂斜前方
（左上图）
罗汉堂一层平面
（左下图）
罗汉堂室外柱子
（右图）

1.9 美岱召・泰和门

泰和门斜前方

泰和门侧面（右图）
泰和门斜后方（左图）

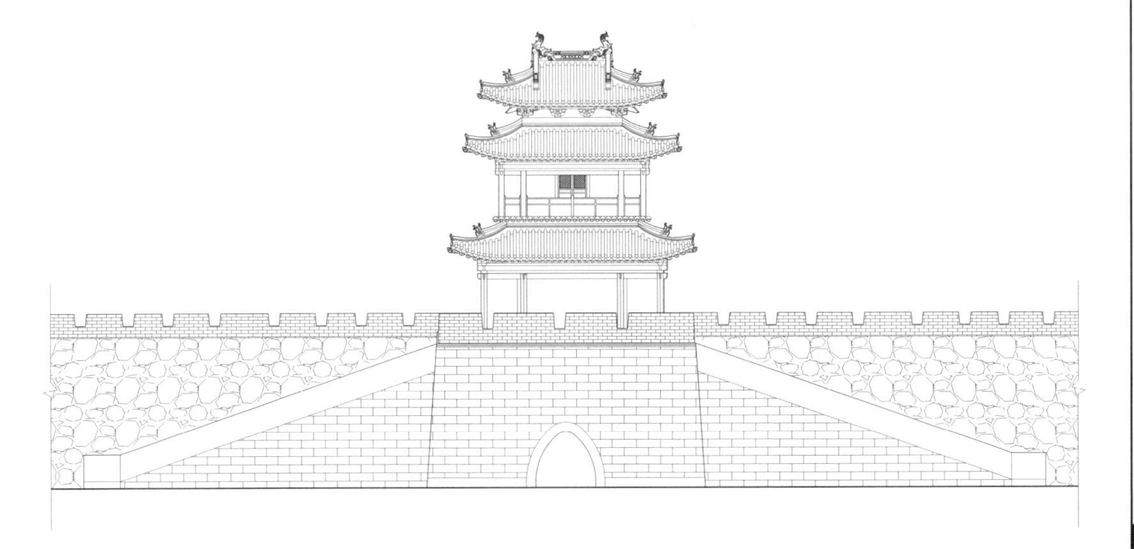

泰和门正立面图

五当召

Badgar Temple

2 五当召 Badgar Temple

五当召建筑群

五当召地处原乌拉特东公旗所辖之地，系章嘉活佛属庙，蒙古地区最知名的几座学问寺之一。乾隆二十一年（1756年），清廷御赐满、蒙古、汉、藏四体“广觉寺”匾额。该庙无属庙，亦不受任何盟旗与喇嘛印务处管理，寺庙喇嘛管事会在寺庙辖区内实施政教合一的管理制度。寺庙初由理藩院管理，嘉庆四年（1799年），交由绥远将军监管。

乾隆十四年（1749年），高僧阿格旺曲日莫在吉忽伦图山始建寺庙。该僧曾受教于多伦甘珠尔活佛，并在西藏哲蚌寺、扎什伦布寺精研显宗学。七世达赖赐予其“洞阔尔班迪达”称号，故在蒙古之地弘扬时轮经法成为其意愿，寺庙最初创建的经殿即为时轮殿。据传，鄂尔多斯左翼前旗扎萨克王纳木吉勒道尔吉最初出资建造了该庙。七世达赖赐名“巴达格尔禅林”，俗称五当召、宝唐图庙、吉忽伦图庙。

寺庙建筑风格为藏式建筑。该寺有双层时轮殿、双层护法殿（又称黄庙）、81间三层大雄宝殿、阿会殿、显宗殿、菩提道学殿等6座大殿，洞阔尔活佛府、章嘉活佛府、甘珠尔活佛府等3座拉布隆及1座供奉历世活佛舍利塔的苏卜盖陵，僧舍94排。寺庙建筑共计2500余间，占地面积近20公顷，其中不包括已拆毁的远方殿、斋戒殿、根坯庙等殿宇。寺庙建筑因逐年扩建，殿堂式样各异，但以西藏扎什伦布寺等名寺为效仿蓝图，寺庙建筑又构成一个有机体系。

“文化大革命”中寺庙资产被大量损毁，但建筑保存完好。1973年起，国家陆续拨款修缮、扩建寺庙，使该庙保持了原有建筑规模，成为自治区内少有的保存完好、体系完整、运转较早的藏传佛教寺庙。

参考文献：

[1] 色力和扎布，勒 · 乌云毕力格.五当召（蒙古文）.赤峰: 内蒙古科学技术出版社, 1991，3.

[2] 青格勒扎布.五当召三百年（蒙古文）.呼和浩特: 内蒙古人民出版社,2003，11.

[3] 杨 · 道尔基.阴山五当召（蒙古文、汉文）.呼和浩特:内蒙古人民出版社,2008，9.

五当召转经筒

<table>
<tr><td colspan="8">五当召·基本概况 1</td></tr>
<tr><td rowspan="3">寺院蒙古语藏语名称</td><td>蒙古语</td><td>[illegible]</td><td rowspan="2">寺院汉语名称</td><td>汉语正式名称</td><td colspan="3">广觉寺</td></tr>
<tr><td>藏语</td><td>——</td><td>俗称</td><td colspan="3">五当召</td></tr>
<tr><td>汉语语义</td><td>白色莲花召</td><td colspan="2">寺院汉语名称的由来</td><td colspan="3">清廷赐名</td></tr>
<tr><td>所在地</td><td colspan="4">包头市石拐区</td><td>东经 110° 18′</td><td colspan="2">北纬 40° 47′</td></tr>
<tr><td>初建年</td><td colspan="2">乾隆十四年（1749年）</td><td>保护等级</td><td colspan="4">全国重点文物保护单位</td></tr>
<tr><td>盛期时间</td><td colspan="2">乾隆至同治时期</td><td colspan="2">盛期喇嘛僧/现有喇嘛僧数</td><td colspan="3">1700人/30余人</td></tr>
<tr><td>历史沿革</td><td colspan="7">1749年，始建寺庙，最先新建活佛居住地——色木沁宫，即现今的苏布盖林。
1749年，始建第一座经堂——时轮殿。
1750年，始建护法殿。
约1750—1757年，在西拉哈达山僻静之处，新建根丕庙。
1757年，始建大雄宝殿。
1784年，始建洞阔尔活佛拉布隆。
1800年，始建阿会殿，为寺庙唯一坐西朝东的佛殿。
1835年，始建显宗殿。
1842年，新建章嘉活佛府。
1932年，九世班禅驾临寺庙。
1949年，修缮显宗殿。
1952年，活佛拉布隆二层失火，整个建筑被烧毁。
1958年，寺庙停止法事活动，大批喇嘛还俗。
1966年，封闭寺庙所有殿堂、仓库。根丕庙等殿宇被拆毁。
1968年，寺庙大量财产遗失，但建筑由吉忽伦图公社民兵看管，得以保存。
1972年，被列为内蒙古自治区级重点文物保护单位。
1973年，开始修缮寺庙，在此后十余年国家陆续拨款维修、扩建寺庙。
1986年，成立“内蒙古佛学院”，两年后转至乌素图召。
1996年，被列为国家级重点文物保护单位。同年，包头地区发生强烈地震，对千佛殿、甘珠尔活佛府、苏布盖林进行了维修。
2001年，新建根丕活佛府、仁布切殿、展旦召（即安居殿）及格日乌拉黄庙。
2002年，维修僧舍。
2003年，在原址上新建一座斋戒殿</td></tr>
<tr><td></td><td>资料来源</td><td colspan="6">[1] 色力和扎布,勒·乌云毕力格.五当召（蒙古文）.赤峰:内蒙古科学技术出版社,1991，3.
[2] 青格勒扎布.五当召三百年（蒙古文）.呼和浩特:内蒙古人民出版社,2003，11.
[3] 杨·道尔基.阴山五当召（蒙古文,汉文）.呼和浩特:内蒙古人民出版社,2008，9.</td></tr>
<tr><td rowspan="2">现状描述</td><td colspan="4" rowspan="2">五当召(广觉寺)的主体建筑，现由六大经堂、三座活佛府、一幢安放本召历世活佛舍利塔的灵堂,以及94栋(共两千余间，现存四十余栋)喇嘛住宿的白色藏式小土楼组成，占地三百多亩。其中大雄宝殿、显宗殿、时轮殿、当圪希德殿、菩提道殿、阿会殿、金刚护法殿、苏卜盖陵、洞阔尔活佛府、章嘉活佛府、甘珠尔活佛府为历史遗迹，农乃殿为新建建筑，庚毗庙已毁，仅存遗址,其他喇嘛僧舍多数为历史建筑,其建筑风格主要是纯藏式建筑，结构形式以石木混合为主</td><td>描述时间</td><td colspan="2">2010/10/10</td></tr>
<tr><td>描述人</td><td colspan="2">卢文娟</td></tr>
<tr><td>调查日期</td><td colspan="2">2010/10/09</td><td>调查人员</td><td colspan="4">韩瑛、卢文娟</td></tr>
</table>

五当召基本概况表1

五当召·基本概况 2

现存建筑	大雄宝殿	喇弥仁殿	甘珠尔活佛府	已毁建筑	郝拉银殿	斋戒殿
	显宗殿	斋戒殿（新建）	洞阔尔活佛府		——	——
	时轮殿	苏卜盖陵	章嘉活佛府		信息来源	王磊义，姚桂轩，郭建中.藏传佛教寺院美岱召五当召调查与研究.北京：中国藏学出版社,2009,4.
	当圪希德殿	阿会殿	——			

区位图

总平面图

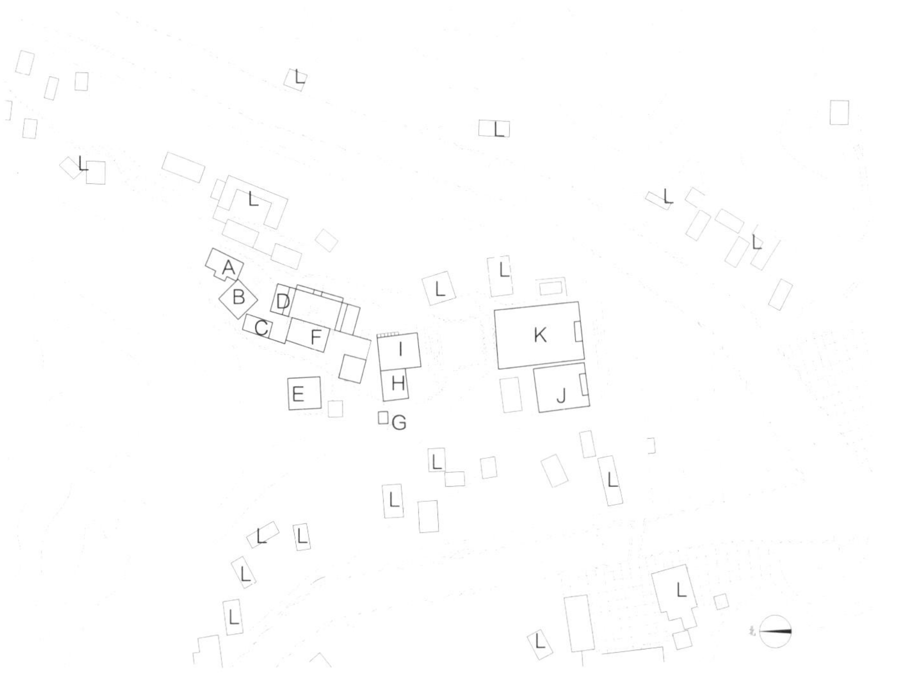

A.苏卜盖陵
B.阿会殿
C.章嘉活佛府
D.洞阔尔活佛府
E.喇弥仁独贡
F.甘珠尔活佛府
G.小黄庙
H.当圪希德殿
I.时轮殿
J.显宗殿
K.大雄宝殿
L.僧 舍

调查日期	2010/10/09	调查人员	韩瑛、卢文娟

A.苏卜盖陵
B.阿会殿
C.章嘉活佛府
D.洞阔尔活佛府
E.喇弥仁独贡
F.甘珠尔活佛府
G.小黄庙
H.当圪希德殿
I.时轮殿
J.显宗殿
K.大雄宝殿
L.僧 舍

五当召总平面图

2.1 五当召·大雄宝殿

单位:毫米

建筑名称	汉语正式名称	大雄宝殿	俗称	苏古沁大殿		
概述	初建年	乾隆二十二年（1757年）	建筑朝向	南	建筑层数	二
	建筑简要描述	藏式结构体系，藏式风格装饰，前廊为藏式柱廊，经堂内部为都纲法式空间格局				
	重建重修记载	不详				
		信息来源	——			

结构规模														
	结构形式	石木混合	相连的建筑	无	室内天井	都纲法式								
	建筑平面形式	凹	外廊形式	前廊										
	通面阔	36330	开间数	——	明间	——	次间	——	梢间	——	次梢间	——	尽间	——
	通进深	28500	进深数	——	进深尺寸（前→后）	——								
	柱子数量	80	柱子间距	横向尺寸	2800→3200→3000→2720	（藏式建筑结构体系填写此栏，不含廊柱）								
				纵向尺寸	2930→2950→2970→2980									
	其他													

建筑主体（大木作）（石作）（瓦作）										
	屋顶	屋顶形式	密肋平顶	瓦作	抹灰屋面					
	外墙	主体材料	土坯	材料规格	不规则	饰面颜色	白			
		墙体收分	有	边玛檐墙	有	边玛材料	边玛草			
	斗栱、梁架	斗栱	无	平身科斗口尺寸	——	梁架关系	——			
	柱、柱式（前廊柱）	形式	藏式	柱身断面形状	十二楞柱	断面尺寸	390×390			（在没有前廊柱的情况下，填写室内柱及其特征）
		柱身材料	木材	柱身收分	有	栌斗、托木	有	雀替	无	
		柱础	有	柱础形状	圆	柱础尺寸	直径 $D=360$			
	台基	台基类型	普通台基	台基高度	2860	台基地面铺设材料	石材			
	其他	无								

装修（小木作）（彩画）									
	门(正面)	板门		门楣	无	堆经	有	门帘	无
	窗（正面）	藏式明窗		窗楣	有	窗套	有	窗帘	有
	室内隔扇	隔扇	有	隔扇位置	经堂和佛殿之间				
	室内地面、楼面	地面材料及规格		不规则天然石材		楼面材料及规格		木材	
	室内楼梯	楼梯	有	楼梯位置	佛殿左侧	楼梯材料	木材	梯段宽度	不详
	天花、藻井	天花	有	天花类型	井口天花	藻井	无	藻井类型	无
	彩画	柱头	有	柱身	有	梁架	有	走马板	无
		门、窗	无	天花	有	藻井	——	其他彩画	无
	其他	悬塑	无	佛龛	无	匾额	无		

装饰									
	室内	帷幔	有	幕帘彩绘	无	壁画	有	唐卡	有
		经幡	无	经幢	无	柱毯	有	其他	无
	室外	玛尼轮	无	苏勒德	有	宝顶	有	祥麟法轮	无
		四角经幢	有	经幡	有	铜饰	有	石刻、砖雕	有
		仙人走兽	无	壁画	有	其他	无		

陈设																	
	室内	主佛像		三世佛				佛像基座		不详							
		法座	有	藏经橱	有	经床	有	诵经桌	有	法鼓	无	玛尼轮	无	坛城	无	其他	无
	室外	旗杆	有	苏勒德	无	狮子	无	经幡	有	玛尼轮	无	香炉	有	五供	无	其他	无
	其他	——															

备注	——

调查日期	2010/10/09	调查人员	韩瑛、卢文娟	整理日期	2010/10/09	整理人员	卢文娟

大雄宝殿基本概况表1

五当召·大雄宝殿·档案照片					
照片名称	鸟瞰图	照片名称	正立面	照片名称	斜后方
照片名称	柱廊	照片名称	正门	照片名称	室外柱头
照片名称	室外柱础	照片名称	台基石	照片名称	经幢
照片名称	窗户	照片名称	室内正面	照片名称	都纲法式
照片名称	室内柱身	照片名称	室内天花	照片名称	壁画
备注	——				
摄影日期	2010/10/09	摄影人员	孟祎军		

大雄宝殿基本概况表2

大雄宝殿正前方

大雄宝殿室内正面（左图）

大雄宝殿侧面（右图）

大雄宝殿室内空间（左图）

大雄宝殿正面门（右图）

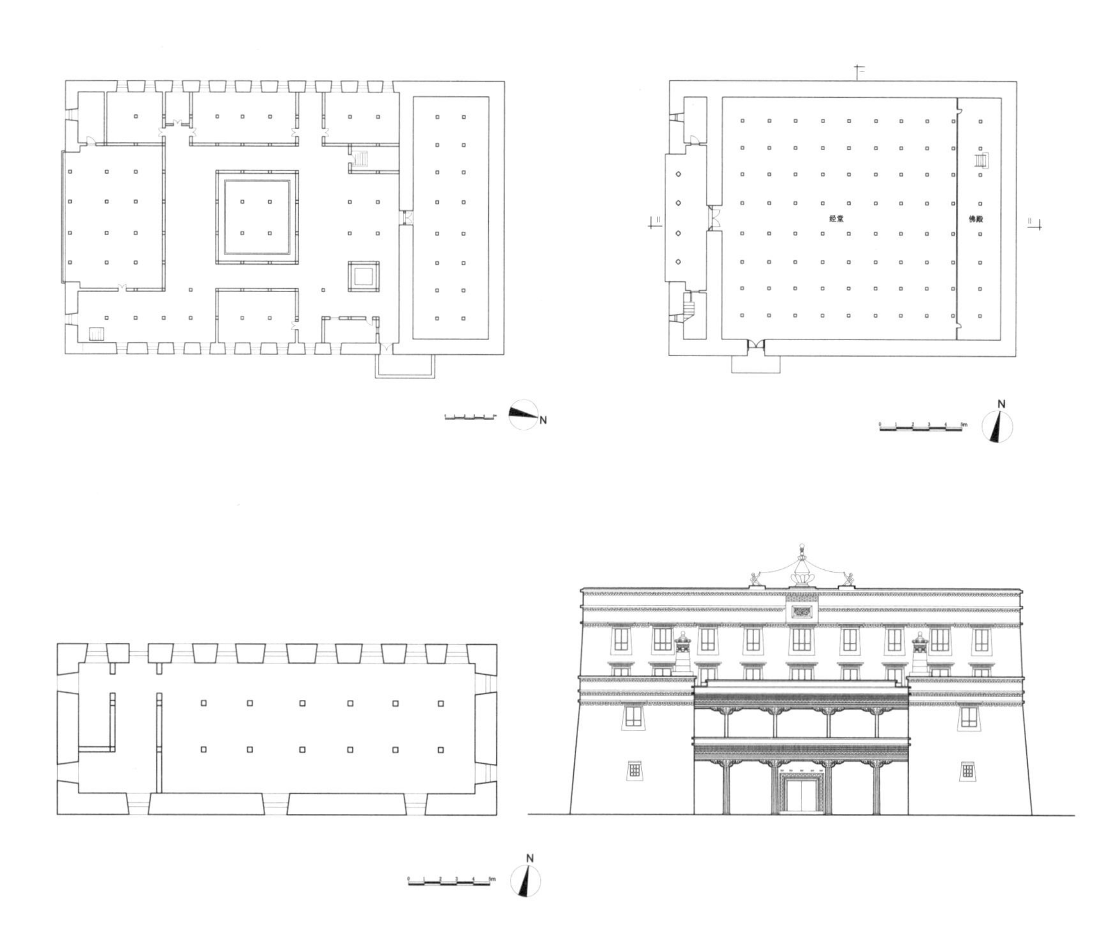

大雄宝殿一层平面图
（左图）
大雄宝殿二层平面图
（右图）

大雄宝殿三层平面图
（左图）
大雄宝殿正立面图
（右图）

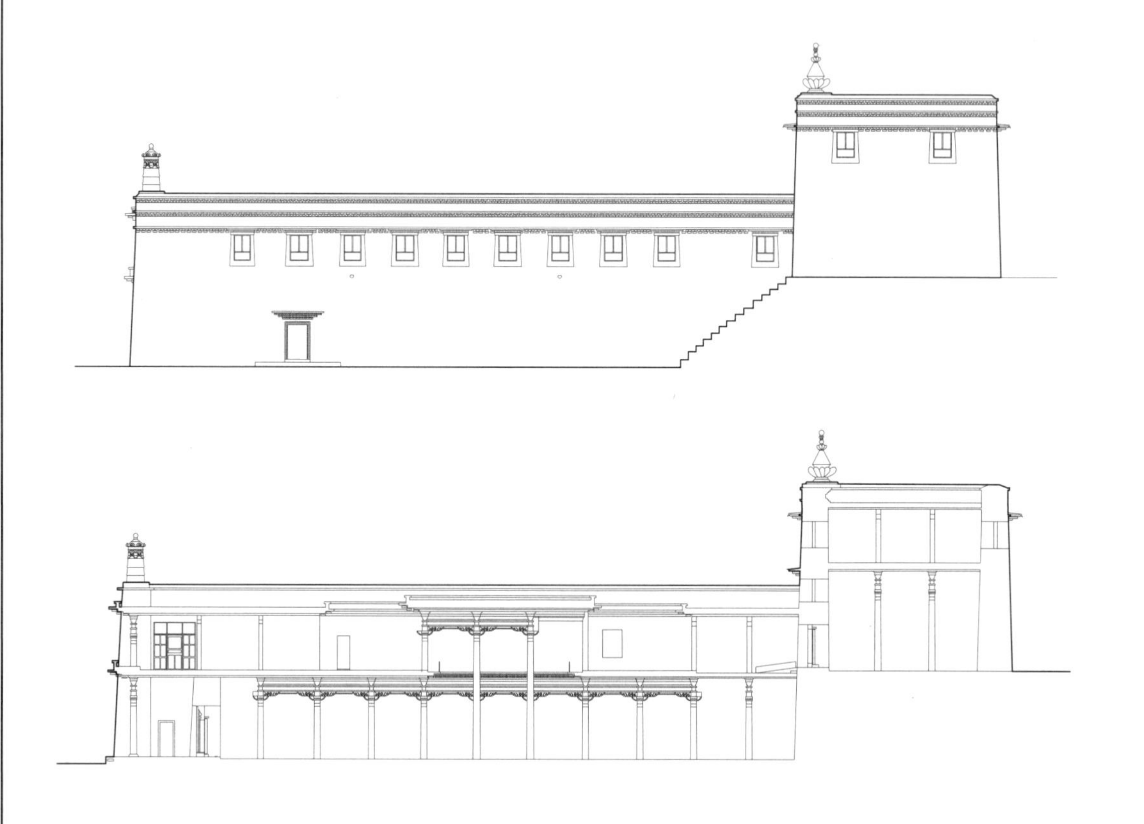

大雄宝殿侧立面图

大雄宝殿剖面图

2.2 五当召·显宗殿

单位:毫米

建筑名称	汉语正式名称	法相殿、显宗殿	俗称	却依拉殿													
概述	初建年	道光十五年（1835年）	建筑朝向	南	建筑层数	二											
	建筑简要描述	藏式结构体系，藏式风格装饰，前廊为藏式柱廊															
	重建重修记载	不详															
		信息来源	——														
结构规模	结构形式	石木混合	相连的建筑	无	室内天井	有											
	建筑平面形式	凹字形	外廊形式	前廊													
	通面阔	23060	开间数	——	明间	——	次间	——	梢间	——	次梢间	——	尽间	——			
	通进深	28580	进深数	——	进深尺寸（前→后）	——											
	柱子数量	40	柱子间距	横向尺寸	2880→3090→2980→2880	（藏式建筑结构体系填写此栏，不含廊柱）											
				纵向尺寸	2930→2950→2970→2980												
	其他	无															
建筑主体（大木作）（石作）（瓦作）	屋顶	屋顶形式	密肋平顶	瓦作	抹灰屋面												
	外墙	主体材料	土坯	材料规格	不详	饰面颜色	黑、白、红										
		墙体收分	有	边玛檐墙	有	边玛材料	红泥抹面										
	斗栱、梁架	斗栱	无	平身科斗口尺寸	——	梁架关系	面向天井纵排架										
	柱、柱式（前廊柱）	形式	藏式	柱身断面形状	十二楞柱	断面尺寸	360×200	（在没有前廊柱的情况下，填写室内柱及其特征）									
		柱身材料	木材	柱身收分	有	栌斗、托木	有	雀替	无								
		柱础	无	柱础形状	——	柱础尺寸	——										
	台基	台基类型	普通台基	台基高度	1350	台基地面铺设材料	不规则天然石材										
	其他	无															
装修（小木作）（彩画）	门(正面)	板门	门楣	有	堆经	无	门帘	无									
	窗（正面）	藏式明窗	窗楣	有	窗套	有	窗帘	有									
	室内隔扇	隔扇	无	隔扇位置	——												
	室内地面、楼面	地面材料及规格	不规则天然石材	楼面材料及规格	阿嘎土												
	室内楼梯	楼梯	有	楼梯位置	入口东侧	楼梯材料	木材	梯段宽度	865								
	天花、藻井	天花	有	天花类型	井口天花	藻井	无	藻井类型	——								
	彩画	柱头	有	柱身	有	梁架	有	走马板	无								
		门、窗	有	天花	有	藻井	——	其他彩画	无								
	其他	悬塑	无	佛龛	有	匾额	无										
装饰	室内	帷幔	有	幕帘彩绘	有	壁画	有	唐卡	有								
		经幡	无	经幢	有	柱毯	有	其他	无								
	室外	玛尼轮	无	苏勒德	有	宝顶	无	祥麟法轮	有								
		四角经幢	有	经幡	有	铜饰	有	石刻、砖雕	无								
		仙人走兽	无	壁画	有	其他	无										
陈设	室内	主佛像	弥勒佛	佛像基座	莲花座												
		法座	无	藏经橱	无	经床	有	诵经桌	有	法鼓	无	玛尼轮	无	坛城	无	其他	无
	室外	旗杆	无	苏勒德	有	狮子	无	经幡	有	玛尼轮	无	香炉	有	五供	无	其他	无
	其他	无															
备注	——																
调查日期	2010/10/09	调查人员	韩瑛、卢文娟	整理日期	2010/10/09	整理人员	卢文娟										

显宗殿基本概况表1

五当召·显宗殿·档案照片

照片名称 	正立面	照片名称 	侧立面	照片名称 	侧面楼梯
照片名称 	正门	照片名称 	柱廊侧面	照片名称 	室外壁画
照片名称 	祥麟法轮	照片名称 	经幢	照片名称 	苏勒德
照片名称 	室内正面	照片名称 	室内楼梯	照片名称 	室内柱头1
照片名称 	室内柱头2	照片名称 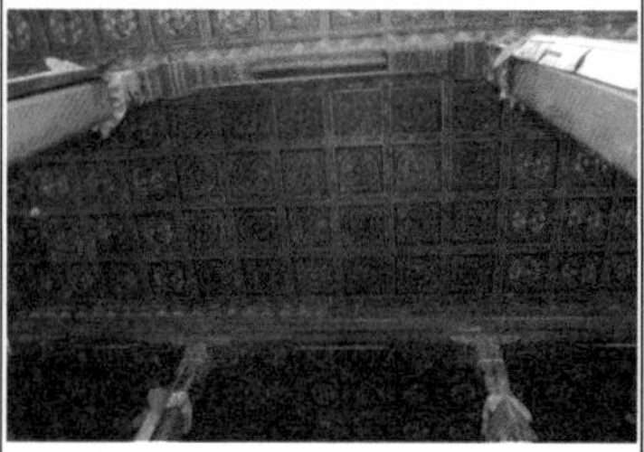	室内天花	照片名称 	室内地面
备注	——				
摄影日期	2010/10/09	摄影人员	孟祎军		

显宗殿基本概况表1

显宗殿正前方

显宗殿局部
（左图）
显宗殿正门
（右上图）
显宗殿室内柱
（右中图）
显宗殿室内正面
（右下图）

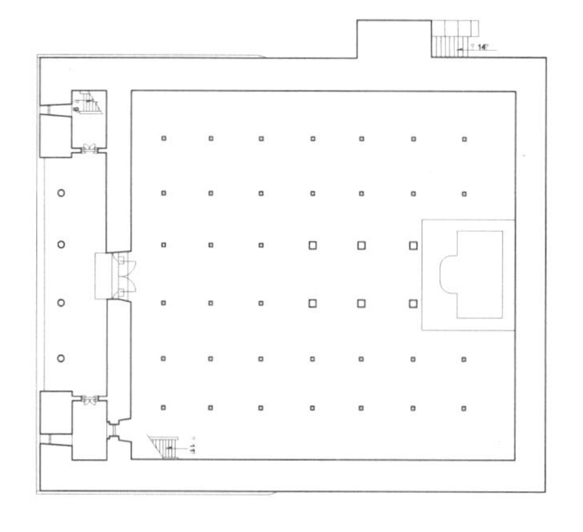

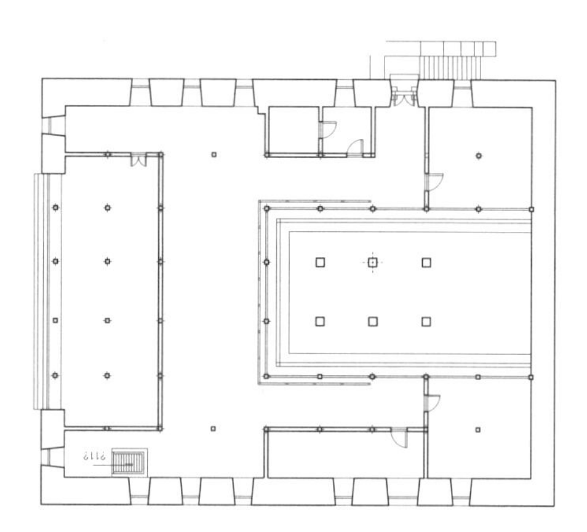

显宗殿一层平面图
（左图）
显宗殿二层平面图
（右图）

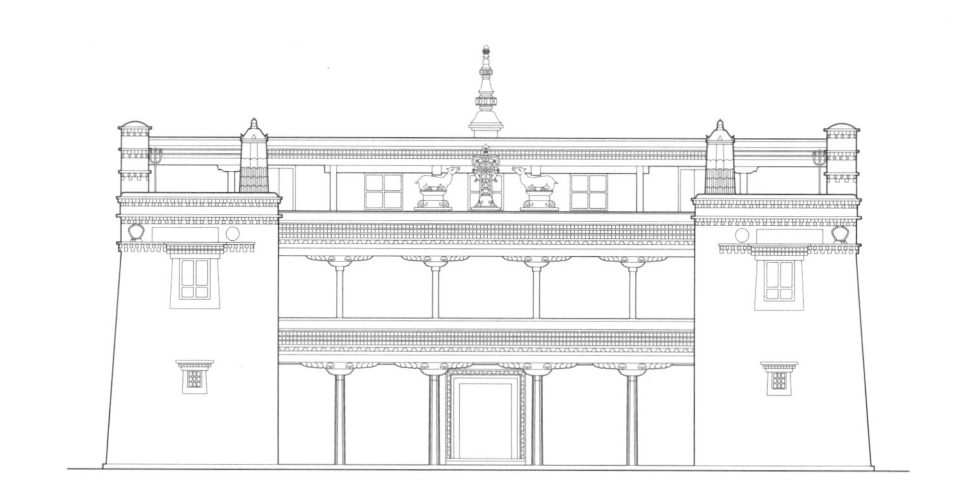

显宗殿正立面图

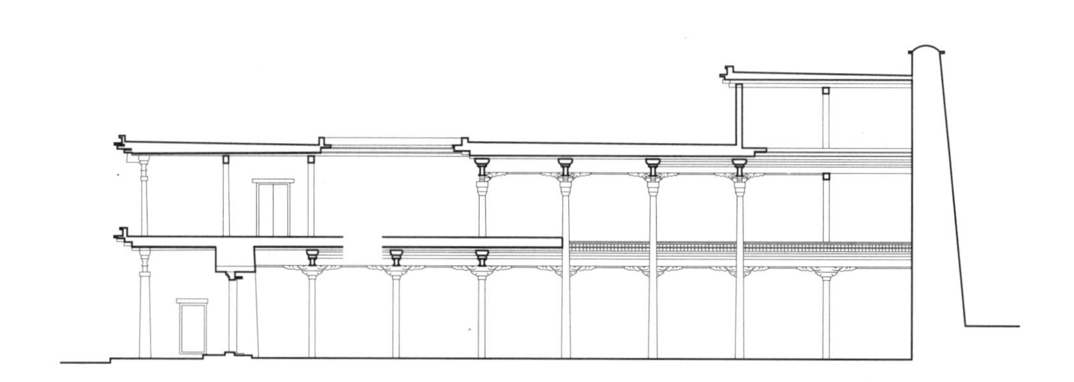

显宗殿剖面图

2.3 五当召·时轮殿

单位:毫米

建筑名称	汉语正式名称	时轮金刚殿	俗称	洞阔尔殿													
概述	初建年	乾隆十四年（1749年）	建筑朝向	南	建筑层数	三											
	建筑简要描述	藏式结构体系，藏式风格装饰，前廊为藏式柱廊															
	重建重修记载	不详															
		信息来源	——														
结构规模	结构形式	土木混合	相连的建筑	无	室内天井	有											
	建筑平面形式	凸字形	外廊形式	前廊													
	通面阔	19350	开间数	——	明间	——	次间	——	梢间	——	次梢间	——	尽间	——			
	通进深	23500	进深数	——	进深尺寸（前→后）	——											
	柱子数量	24	柱子间距	横向尺寸	2340→2640→2360→2320	（藏式建筑结构体系填写此栏，不含廊柱）											
				纵向尺寸	2310→2340→2350→2380												
	其他	无															
建筑主体（大木作）（石作）（瓦作）	屋顶	屋顶形式	密肋平顶	瓦作	——												
	外墙	主体材料	土坯	材料规格	不详	饰面颜色	白										
		墙体收分	有	边玛檐墙	有	边玛材料	砖、边玛草										
	斗栱、梁架	斗栱	无	平身科斗口尺寸	——	梁架关系	不详										
	柱、柱式（前廊柱）	形式	藏式	柱身断面形状	十二楞柱	断面尺寸	455×455	（在没有前廊柱的情况下，填写室内柱及其特征）									
		柱身材料	木材	柱身收分	有	栌斗、托木	有	雀替	无								
		柱础	有	柱础形状	不规则	柱础尺寸	不详										
	台基	台基类型	普通台基	台基高度	4000	台基地面铺设材料	不规则天然石材										
	其他	无															
装修（小木作）（彩画）	门(正面)	板门	门楣	无	堆经	有	门帘	无									
	窗（正面）	藏式明窗	窗楣	有	窗套	有	窗帘	有									
	室内隔扇	隔扇	无	隔扇位置	——												
	室内地面、楼面	地面材料及规格	不规则天然石材	楼面材料及规格	不详												
	室内楼梯	楼梯	有	楼梯位置	佛殿左侧	楼梯材料	木材	梯段宽度	不详								
	天花、藻井	天花	有	天花类型	井口天花	藻井	无	藻井类型	——								
	彩画	柱头	有	柱身	有	梁架	有	走马板	无								
		门、窗	无	天花	有	藻井	——	其他彩画	无								
	其他	悬塑	无	佛龛	无	匾额	有										
装饰	室内	帷幔	无	幕帘彩绘	无	壁画	有	唐卡	有								
		经幡	有	经幢	有	柱毯	有	其他	无								
	室外	玛尼轮	无	苏勒德	有	宝顶	有	祥麟法轮	有								
		四角经幢	有	经幡	无	铜饰	有	石刻、砖雕	有								
		仙人走兽	无	壁画	有	其他	无										
陈设	室内	主佛像	无	佛像基座	——												
		法座	有	藏经橱	无	经床	有	诵经桌	有	法鼓	无	玛尼轮	无	坛城	无	其他	无
	室外	旗杆	有	苏勒德	无	狮子	无	经幡	无	玛尼轮	无	香炉	无	五供	无	其他	无
	其他	无															
备注	——																
调查日期	2010/10/09	调查人员	韩瑛、卢文娟	整理日期	2010/10/09	整理人员	卢文娟										

时轮殿基本概况表1

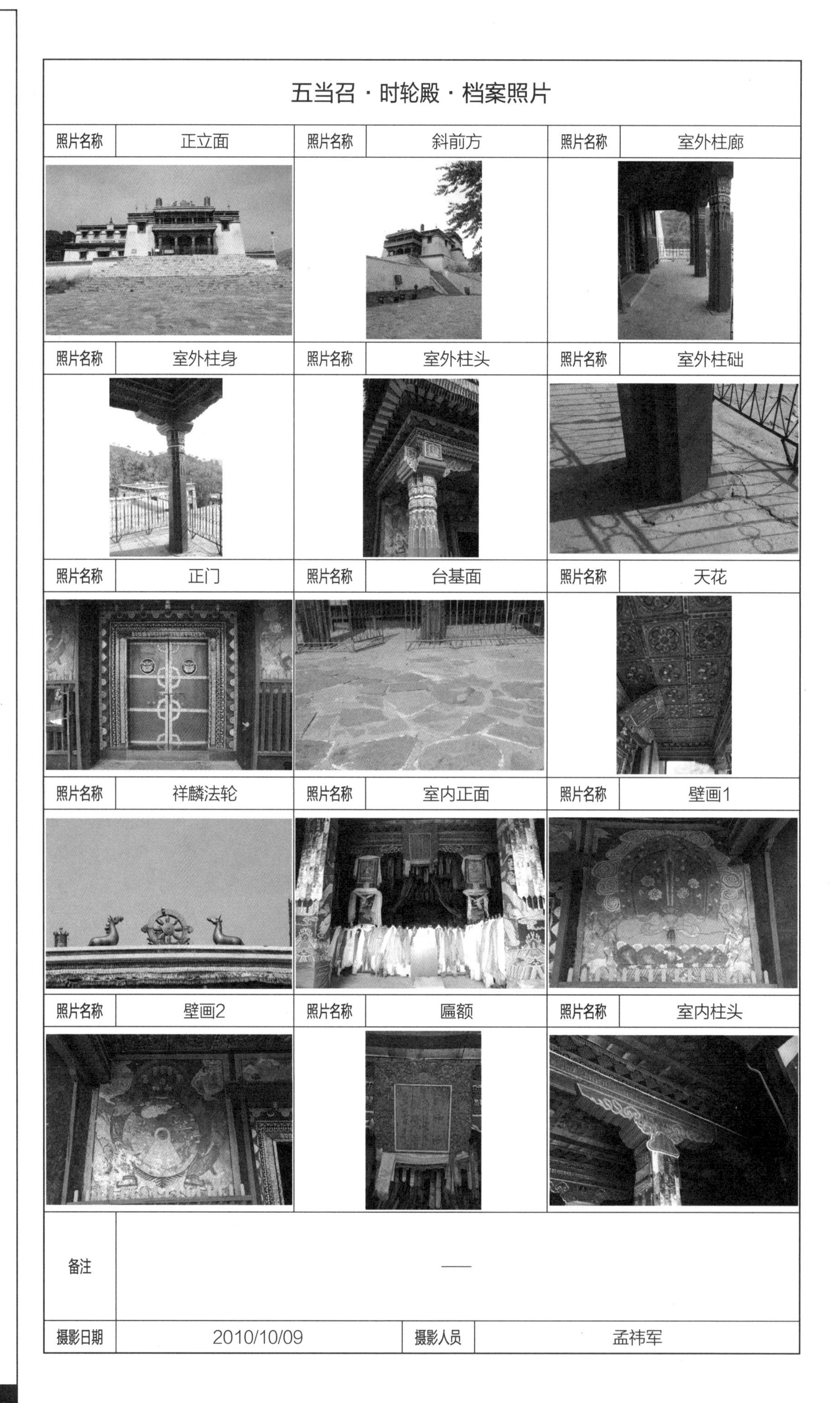

五当召·时轮殿·档案照片					
照片名称	正立面	照片名称	斜前方	照片名称	室外柱廊
照片名称	室外柱身	照片名称	室外柱头	照片名称	室外柱础
照片名称	正门	照片名称	台基面	照片名称	天花
照片名称	祥麟法轮	照片名称	室内正面	照片名称	壁画1
照片名称	壁画2	照片名称	匾额	照片名称	室内柱头
备注	—				
摄影日期	2010/10/09	摄影人员	孟祎军		

时轮殿基本概况表2

时轮殿斜前方

显宗殿室内正面（左上图）
显宗殿室外柱头（左下图）
显宗殿前廊（右图）

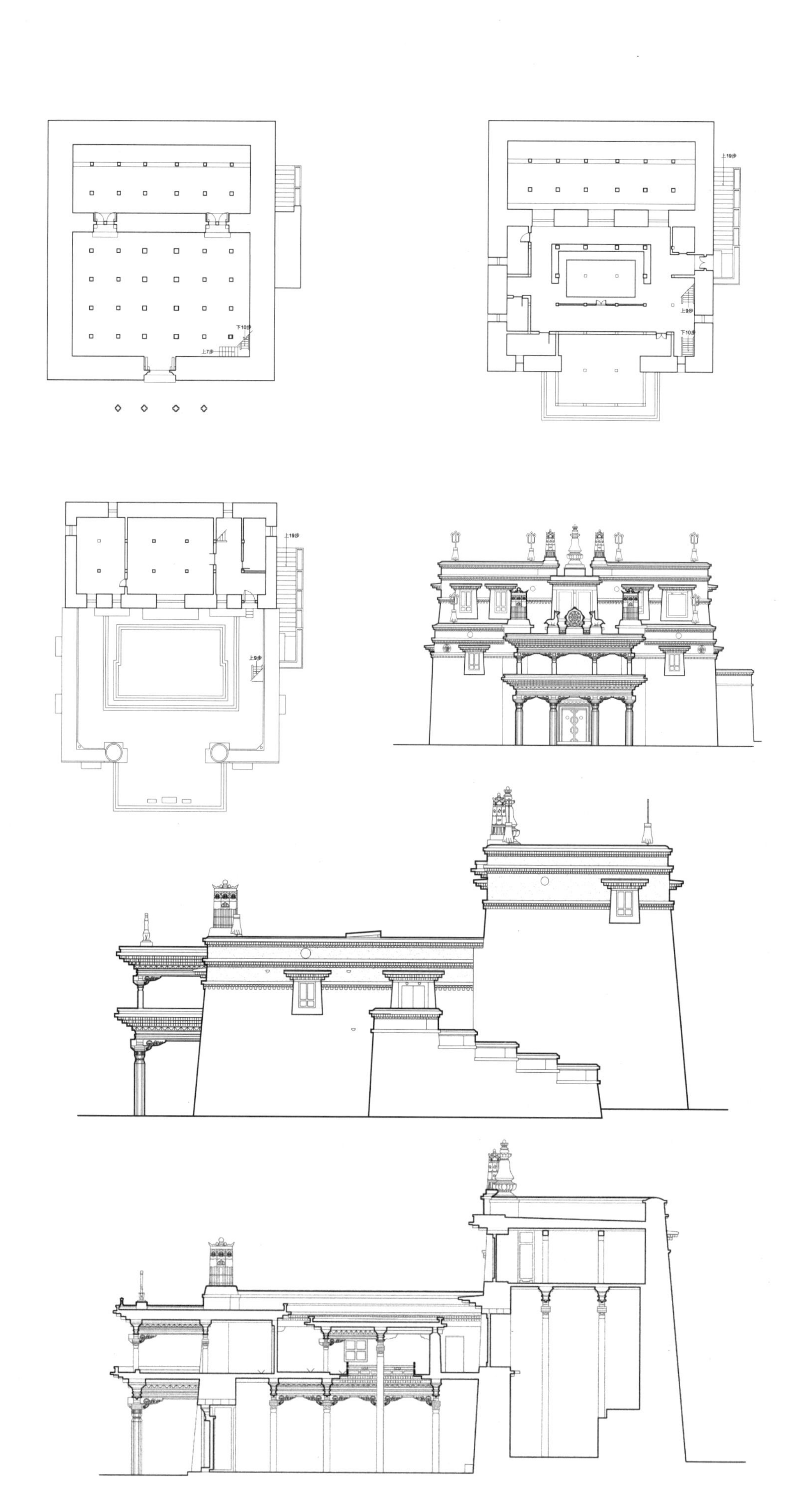

时轮殿一层平面图
（左图）

时轮殿二层平面图
（右图）

时轮殿三层平面图
（左图）

时轮殿正立面图
（右图）

时轮殿侧立面图

时轮殿剖面图

2.4 五当召·当圪希德殿

单位:毫米

建筑名称	汉语正式名称	护法殿	俗称	当圪希德殿													
概述	初建年	乾隆十五年（1750年）	建筑朝向	南	建筑层数	一											
	建筑简要描述	藏式结构体系，藏式风格装饰，形同藏式民居，级别较低															
	重建重修记载	不详															
		信息来源	——														
结构规模	结构形式	土木混合	相连的建筑	时轮殿	室内天井	有											
	建筑平面形式	方形	外廊形式	无外廊（内测廊）													
	通面阔	14170	开间数	——	明间	——	次间	——	梢间	——	次梢间	——	尽间	——			
	通进深	10495	进深数	——	进深尺寸（前→后）	——											
	柱子数量	4	柱子间距	横向尺寸	2920→3100→2960	（藏式建筑结构体系填写此栏，不含廊柱）											
				纵向尺寸	——												
	其他	无															
建筑主体（大木作）（石作）（瓦作）	屋顶	屋顶形式	密肋平顶	瓦作	——												
	外墙	主体材料	土坯	材料规格	不详	饰面颜色	白										
		墙体收分	有	边玛檐墙	有	边玛材料	边玛草										
	斗栱、梁架	斗栱	无	平身科斗口尺寸	——	梁架关系	面向天井纵排架										
	柱、柱式（前廊柱）	形式	藏式	柱身断面形状	方形	断面尺寸	250×250	（在没有前廊柱的情况下，填写室内柱及其特征）									
		柱身材料	木材	柱身收分	有	栌斗、托木	有	雀替	无								
		柱础	无	柱础形状	——	柱础尺寸	——										
	台基	台基类型	普通台基	台基高度	900	台基地面铺设材料	不规则天然石材										
	其他	无															
装修（小木作）（彩画）	门(正面)	板门	门楣	有	堆经	无	门帘	有									
	窗（正面）	藏式明窗	窗楣	有	窗套	有	窗帘	有									
	室内隔扇	隔扇	无	隔扇位置	——												
	室内地面、楼面	地面材料及规格	不规则天然石材	楼面材料及规格	无												
	室内楼梯	楼梯	无	楼梯位置	——	楼梯材料	——	梯段宽度	——								
	天花、藻井	天花	有	天花类型	海墁天花	藻井	无	藻井类型	——								
	彩画	柱头	有	柱身	有	梁架	有	走马板	无								
		门、窗	无	天花	有	藻井	——	其他彩画	窗眉								
	其他	悬塑	无	佛龛	无	匾额	无										
装饰	室内	帷幔	有	幕帘彩绘	有	壁画	有	唐卡	无								
		经幡	无	经幢	无	柱毯	有	其他	有大量哈达								
	室外	玛尼轮	无	苏勒德	有	宝顶	有	祥麟法轮	无								
		四角经幢	有	经幡	无	铜饰	无	石刻、砖雕	无								
		仙人走兽	无	壁画	无	其他	无										
陈设	室内	主佛像	大维德金刚	佛像基座	莲花座												
		法座	无	藏经橱	无	经床	无	诵经桌	无	法鼓	无	玛尼轮	无	坛城	无	其他	无
	室外	旗杆	无	苏勒德	无	狮子	无	经幡	无	玛尼轮	无	香炉	无	五供	无	其他	无
	其他	无															
备注	——																
调查日期	2010/10/09	调查人员	韩瑛、卢文娟	整理日期	2010/10/09	整理人员	卢文娟										

当圪希德殿基本概况表1

五当召・当圪希德殿・档案照片

照片名称	正立面	照片名称	与时轮殿相接处	照片名称	正门

照片名称	窗户1	照片名称	窗户2	照片名称	宝顶

照片名称	经幢	照片名称	苏勒德	照片名称	室内正面

照片名称	室内天花	照片名称	室内柱身	照片名称	
				—	
照片名称		照片名称		照片名称	
—		—		—	
备注	—				
摄影日期	2010/10/09	摄影人员	孟祎军		

当圪希德殿基本概况表2

当圪希德殿正前方

当圪希德殿殿与时轮殿衔接处（左图）

当圪希德殿室内柱子（右上图）

当圪希德殿殿室内佛殿正门（右下图）

当圪希德殿一层平面图
（左图）
当圪希德殿二层平面图
（右图）

当圪希德殿正立面图

当圪希德殿侧立面图

当圪希德殿剖面图

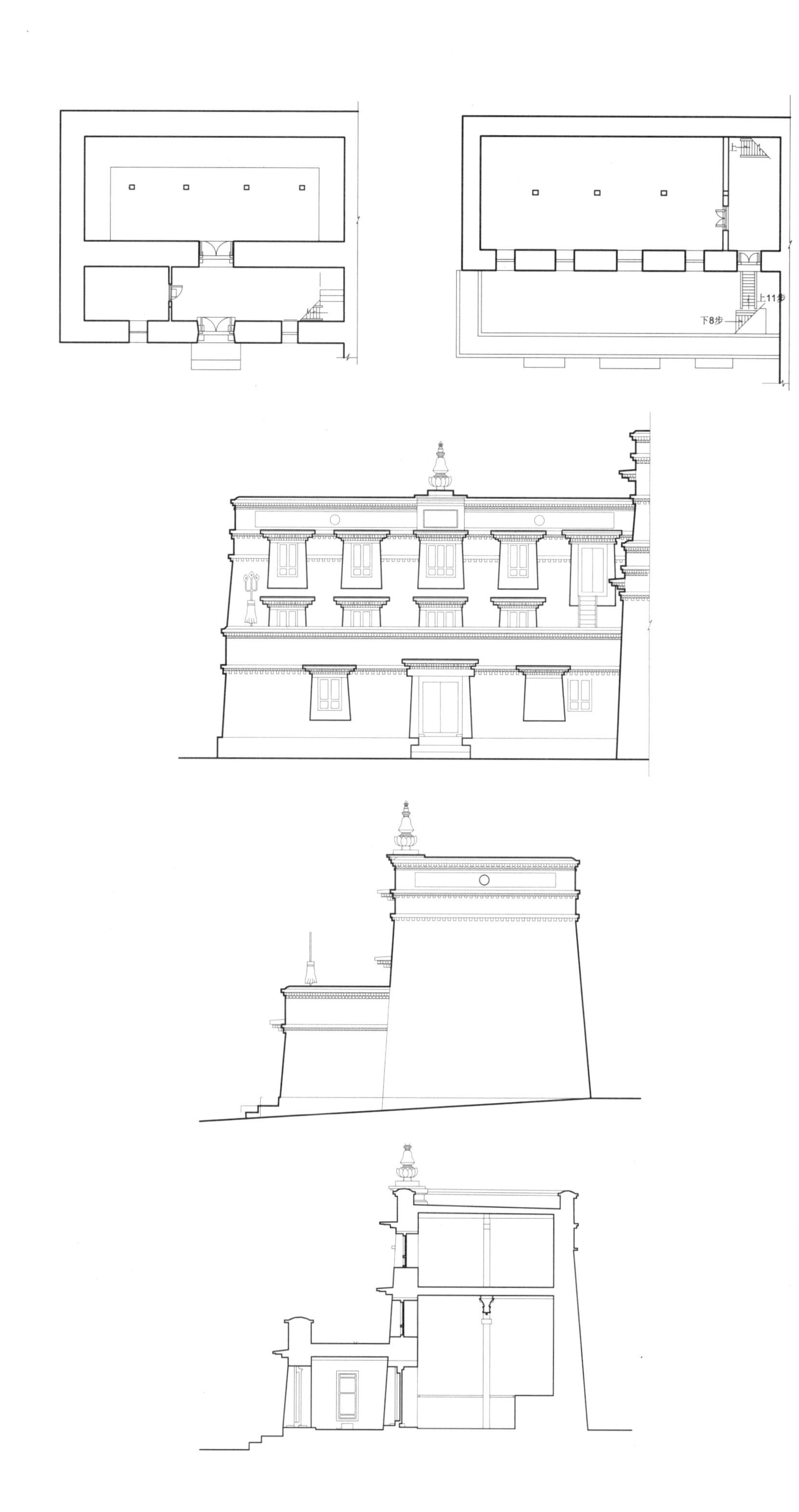

2.5 五当召·喇弥仁殿

单位:毫米

建筑名称	汉语正式名称	法相殿	俗称	喇弥仁殿													
概述	初建年	光绪十八年（1892年）	建筑朝向	南	建筑层数	二											
	建筑简要描述	藏式结构体系，藏式风格装饰，前廊为藏式柱廊，经堂内部为都纲法式空间格局															
	重建重修记载	不详															
		信息来源	——														
结构规模	结构形式	土木混合	相连的建筑	无	室内天井	都纲法式											
	建筑平面形式	凸字形	外廊形式	前廊													
	通面阔	16100	开间数	——	明间	——	次间	——	梢间	——	次梢间	——	尽间	——			
	通进深	11500	进深数	——	进深尺寸（前→后）	——											
	柱子数量	16	柱子间距	横向尺寸	2600→2620→2600	（藏式建筑结构体系填写此栏，不含廊柱）											
				纵向尺寸	3250→2560→2600												
	其他	无															
建筑主体（大木作）（石作）（瓦作）	屋顶	屋顶形式	密肋平顶	瓦作	——												
	外墙	主体材料	土坯	材料规格	不详	饰面颜色	白										
		墙体收分	有	边玛檐墙	有	边玛材料	砖、边玛草										
	斗栱、梁架	斗栱	无	平身科斗口尺寸	——	梁架关系	面向天井纵排架										
	柱、柱式（前廊柱）	形式	藏式	柱身断面形状	十二楞柱、八楞柱	断面尺寸	440×440、205×205	（在没有前廊柱的情况下，填写室内柱及其特征）									
		柱身材料	木材	柱身收分	有	栌斗、托木	有	雀替	无								
		柱础	有	柱础形状	圆形	柱础尺寸	直径$D=400$										
	台基	台基类型	普通台基	台基高度	1200	台基地面铺设材料	不规则天然石材										
	其他	无															
装修（小木作）（彩画）	门(正面)	板门	门楣	无	堆经	有	门帘	无									
	窗（正面）	藏式明窗	窗楣	有	窗套	有	窗帘	有									
	室内隔扇	隔扇	无	隔扇位置	——												
	室内地面、楼面	地面材料及规格	不规则天然石材	楼面材料及规格	不详												
	室内楼梯	楼梯	有	楼梯位置	后墙右侧	楼梯材料	木	梯段宽度	860								
	天花、藻井	天花	有	天花类型	井口	藻井	无	藻井类型	——								
	彩画	柱头	有	柱身	有	梁架	有	走马板	无								
		门、窗	无	天花	有	藻井	无	其他彩画	无								
	其他	悬塑	无	佛龛	有	匾额	无										
装饰	室内	帷幔	无	幕帘彩绘	无	壁画	有	唐卡	有								
		经幡	无	经幢	无	柱毯	有	其他	无								
	室外	玛尼轮	无	苏勒德	有	宝顶	无	祥麟法轮	有								
		四角经幢	有	经幡	无	铜饰	有	石刻、砖雕	有								
		仙人走兽	无	壁画	有	其他	无										
陈设	室内	主佛像	宗喀巴	佛像基座	莲花座												
		法座	无	藏经橱	无	经床	有	诵经桌	无	法鼓	无	玛尼轮	无	坛城	无	其他	无
	室外	旗杆	无	苏勒德	无	狮子	无	经幡	无	玛尼轮	无	香炉	有	五供	无	其他	无
	其他	无															
备注	——																
调查日期	2010/10/09	调查人员	韩瑛、卢文娟	整理日期	2010/10/09	整理人员	卢文娟										

喇弥仁殿基本概况表1

五当召·喇弥仁殿·档案照片

照片名称	正立面	照片名称	斜前方	照片名称	正门
照片名称	室外柱头	照片名称	室外柱础	照片名称	台基
照片名称	室外装饰	照片名称	柱廊天花	照片名称	室内正面
照片名称	室内侧面	照片名称	室内柱身	照片名称	室内天花
照片名称	室内柱头	照片名称	室内窗	照片名称	壁画
备注	——				
摄影日期	2010/10/09	摄影人员	孟祎军		

喇弥仁殿基本概况表2

喇弥仁殿斜前方

喇弥仁殿正前方

喇弥仁殿正门（左图）
喇弥仁殿室外柱头
（右图）

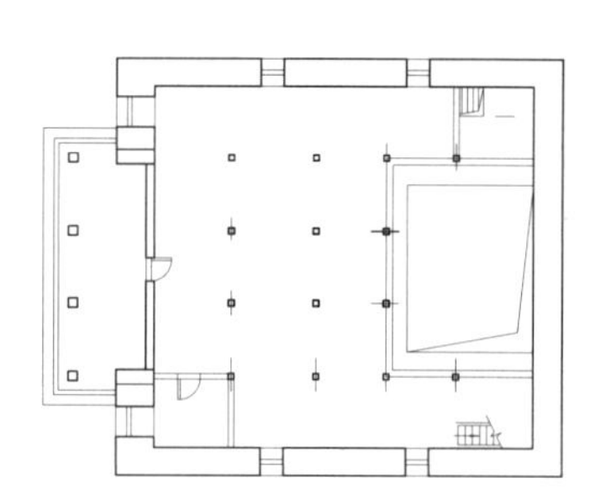

喇弥仁殿一层平面图
（左图）
喇弥仁殿二层平面图
（右图）

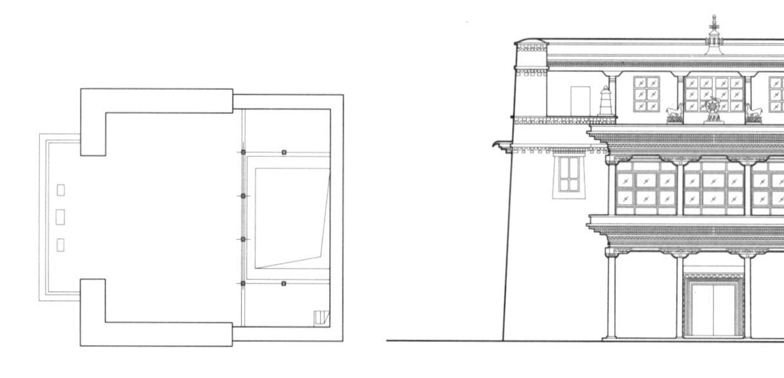

喇弥仁殿三层平面图
（左图）
喇弥仁殿正立面图
（右图）

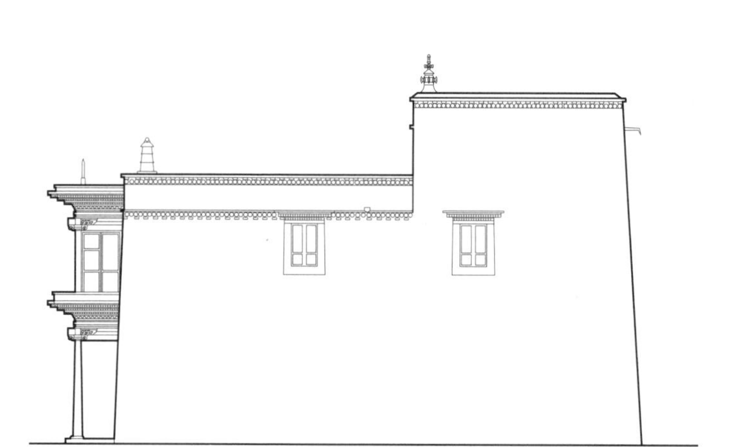

喇弥仁殿侧立面图

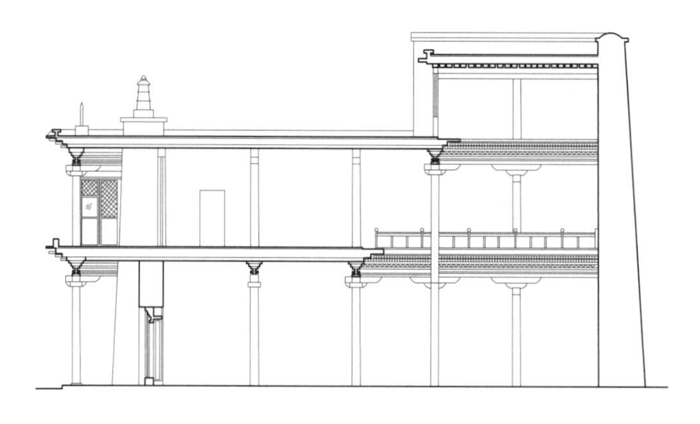

喇弥仁殿剖面图

2.6 五当召·斋戒殿

单位:毫米

建筑名称	汉语正式名称	金科殿	俗称	农乃殿													
概述	初建年	不详	建筑朝向	南	建筑层数	二											
	建筑简要描述	藏式结构体系，藏式风格装饰，前廊为藏式柱廊，在原努乃殿的遗址上建造															
	重建重修记载	原建筑已毁，现在原址上新建一座面阔七开间的二层建筑，更名金科殿															
		信息来源	王磊义，姚桂轩，郭建中.藏传佛教寺院美岱召五当召调查与研究.北京：中国藏学出版社，2009，4.														
结构规模	结构形式	框架结构	相连的建筑	无	室内天井	无											
	建筑平面形式	方形	外廊形式	前廊													
	通面阔	——	开间数	——	明间	——	次间	——	梢间	——	次梢间	——	尽间	——			
	通进深	——	进深数	——	进深尺寸（前→后）	——											
	柱子数量	——	柱子间距	横向尺寸	——	（藏式建筑结构体系填写此栏，不含廊柱）											
				纵向尺寸	——												
	其他	无															
建筑主体（大木作）（石作）（瓦作）	屋顶	屋顶形式	密肋平顶	瓦作	——												
	外墙	主体材料	土坯	材料规格	不详	饰面颜色	白										
		墙体收分	有	边玛檐墙	有	边玛材料	砖、边玛草										
	斗栱、梁架	斗栱	无	平身科斗口尺寸	——	梁架关系	不详										
	柱、柱式（前廊柱）	形式	藏式	柱身断面形状	十二楞柱	断面尺寸	370×370	（在没有前廊柱的情况下，填写室内柱及其特征）									
		柱身材料	木材	柱身收分	有	栌斗、托木	有	雀替	无								
		柱础	无	柱础形状	——	柱础尺寸	——										
	台基	台基类型	普通台基	台基高度	1440	台基地面铺设材料	木材、石材										
	其他	无															
装修（小木作）（彩画）	门(正面)	板门	门楣	无	堆经	有	门帘	无									
	窗（正面）	藏式明窗	窗楣	有	窗套	有	窗帘	有									
	室内隔扇	隔扇	无	隔扇位置	——												
	室内地面、楼面	地面材料及规格	石材，500×500	楼面材料及规格	不详												
	室内楼梯	楼梯	有	楼梯位置	入口左侧	楼梯材料	钢筋混凝土	梯段宽度	1260								
	天花、藻井	天花	有	天花类型	井口天花	藻井	无	藻井类型	——								
	彩画	柱头	有	柱身	有	梁架	有	走马板	有								
		门、窗	无	天花	有	藻井	无	其他彩画	无								
	其他	悬塑	无	佛龛	有	匾额	无										
装饰	室内	帷幔	有	幕帘彩绘	无	壁画	有	唐卡	无								
		经幡	无	经幢	有	柱毯	有	其他	无								
	室外	玛尼轮	无	苏勒德	有	宝顶	无	祥麟法轮	有								
		四角经幢	有	经幡	无	铜饰	有	石刻、砖雕	有								
		仙人走兽	无	壁画	有	其他	门上有镏金										
陈设	室内	主佛像	三世佛	佛像基座	莲花座												
		法座	有	藏经橱	无	经床	有	诵经桌	有	法鼓	无	玛尼轮	无	坛城	无	其他	无
	室外	旗杆	无	苏勒德	无	狮子	无	经幡	无	玛尼轮	无	香炉	有	五供	无	其他	无
	其他	无															
备注	——																
调查日期	2010/10/09	调查人员	韩瑛、卢文娟	整理日期	2010/10/09	整理人员	卢文娟										

斋戒殿基本概况表1

五当召·斋戒殿·档案照片					
照片名称 	正立面	照片名称 	斜前方	照片名称 	斜前方
照片名称 	侧立面	照片名称 	室外柱廊	照片名称 	柱廊天花
照片名称 	柱廊	照片名称 	室外柱身	照片名称 	室外柱头
照片名称 	四角经幢	照片名称 	正门	照片名称 	室内门
照片名称 	室内正面	照片名称 	室内天花	照片名称 	壁画
备注	—				
摄影日期	2010/10/09	摄影人员	孟祎军		

斋戒殿基本概况表2

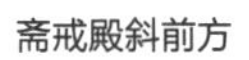

斋戒殿斜前方

斋戒殿廊柱（左图）
斋戒殿室外柱头（右图）

斋戒殿室内正面（左图）
斋戒殿正面门（右图）

2.7 五当召·苏卜盖陵

单位:毫米

建筑名称	汉语正式名称	——	俗称	苏卜盖陵													
概述	初建年	1763年	建筑朝向	东	建筑层数	二											
	建筑简要描述	藏式结构体系，藏式风格装饰，前廊为藏式柱廊															
	重建重修记载	不详															
		信息来源	——														
结构规模	结构形式	土木混合	相连的建筑	无	室内天井	无											
	建筑平面形式	不规则	外廊形式	侧廊													
	通面阔	5030	开间数	——	明间	——	次间	——	梢间	——	次梢间	——	尽间	——			
	通进深	8590	进深数	——	进深尺寸（前→后）	——											
	柱子数量	2	柱子间距	横向尺寸	2120	（藏式建筑结构体系填写此栏，不含廊柱）											
				纵向尺寸	——												
	其他	无															
建筑主体（大木作）（石作）（瓦作）	屋顶	屋顶形式	密肋平顶	瓦作	抹灰屋面												
	外墙	主体材料	土坯	材料规格	不详	饰面颜色	白										
		墙体收分	有	边玛檐墙	无	边玛材料	——										
	斗栱、梁架	斗栱	无	平身科斗口尺寸	——	梁架关系	不详										
	柱、柱式（前廊柱）	形式	藏式	柱身断面形状	方形	断面尺寸	180×180	（在没有前廊柱的情况下，填写室内柱及其特征）									
		柱身材料	木材	柱身收分	无	栌斗、托木	有	雀替	无								
		柱础	无	柱础形状	——	柱础尺寸	——										
	台基	台基类型	普通台基	台基高度	600	台基地面铺设材料	木材										
	其他	无															
装修（小木作）（彩画）	门(正面)	板门	门楣	无	堆经	无	门帘	有									
	窗（正面）	藏式暗窗	窗楣	有	窗套	有	窗帘	有									
	室内隔扇	隔扇	有	隔扇位置	——												
	室内地面、楼面	地面材料及规格	不规则天然石材	楼面材料及规格	不详												
	室内楼梯	楼梯	有	楼梯位置	入口左侧	楼梯材料	木材	梯段宽度	600								
	天花、藻井	天花	有	天花类型	海漫天花	藻井	无	藻井类型	——								
	彩画	柱头	有	柱身	有	梁架	有	走马板	无								
		门、窗	无	天花	有	藻井	无	其他彩画	无								
	其他	悬塑	无	佛龛	有	匾额	有										
装饰	室内	帷幔	无	幕帘彩绘	无	壁画	无	唐卡	无								
		经幡	有	经幢	无	柱毯	无	其他	无								
	室外	玛尼轮	有	苏勒德	有	宝顶	有	祥麟法轮	无								
		四角经幢	有	经幡	无	铜饰	无	石刻、砖雕	无								
		仙人走兽	无	壁画	有	其他	无										
陈设	室内	主佛像	长寿佛	佛像基座	不详												
		法座	无	藏经橱	无	经床	无	诵经桌	无	法鼓	无	玛尼轮	无	坛城	无	其他	无
	室外	旗杆	无	苏勒德	无	狮子	无	经幡	无	玛尼轮	有	香炉	无	五供	无	其他	无
	其他	无															
备注	——																
调查日期	2010/10/09	调查人员	韩瑛、卢文娟	整理日期	2010/10/09	整理人员	卢文娟										

苏卜盖陵基本概况表1

五当召 · 苏卜盖陵 · 档案照片

照片名称	正立面	照片名称	斜前方	照片名称	侧立面
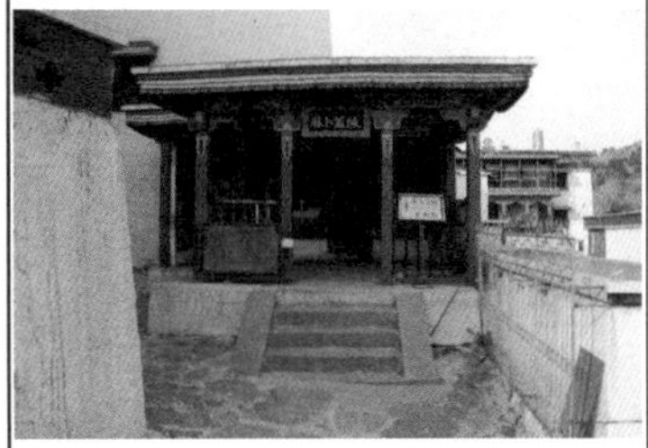		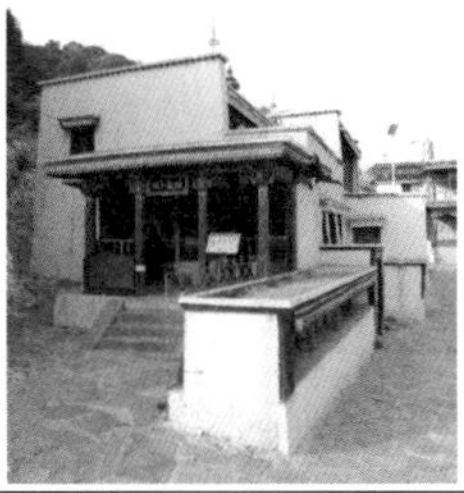		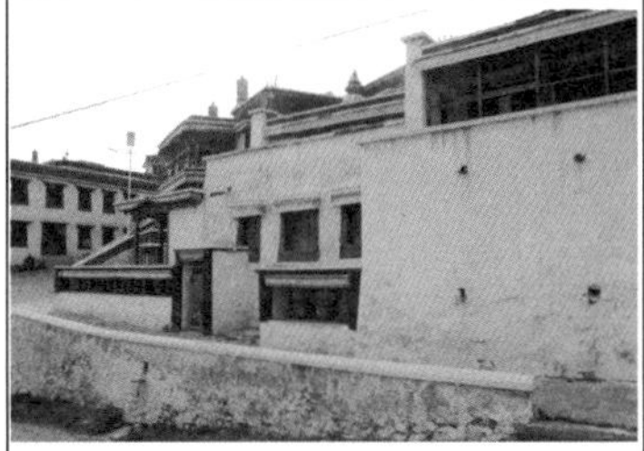	
照片名称	玛尼轮	照片名称	室外柱身	照片名称	室外柱头
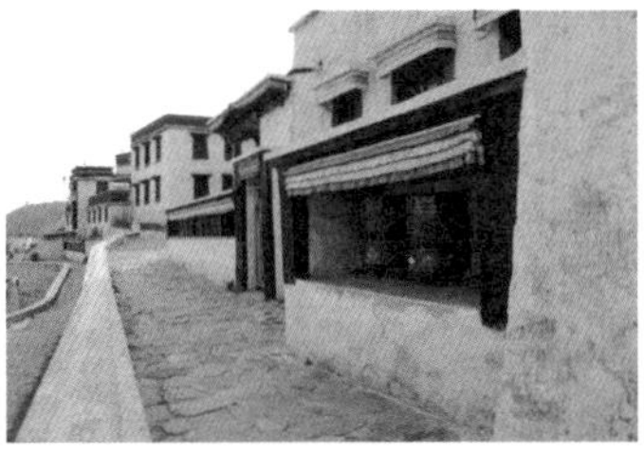		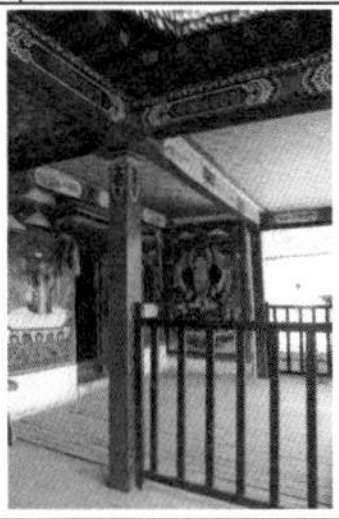		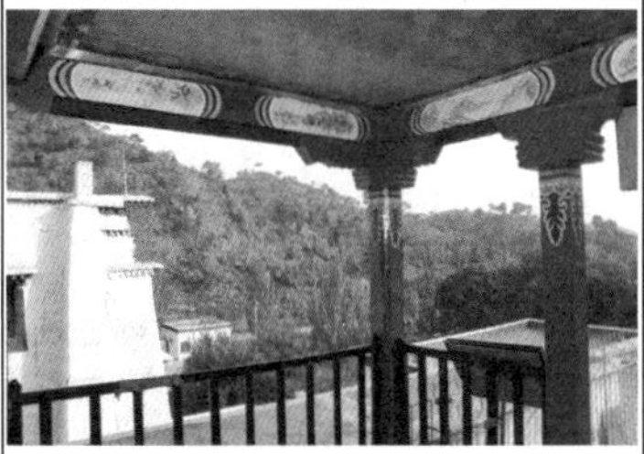	
照片名称	苏勒德	照片名称	台基	照片名称	正门
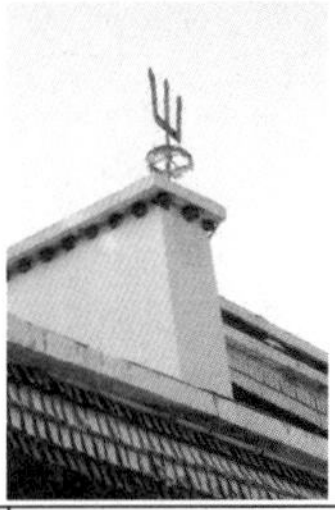		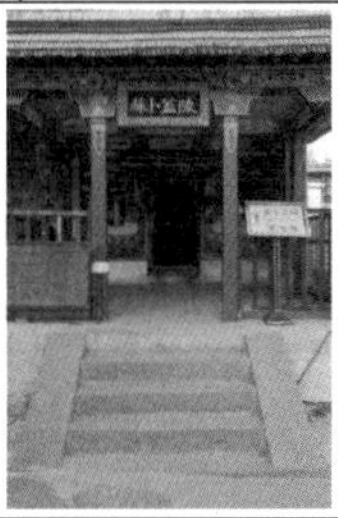		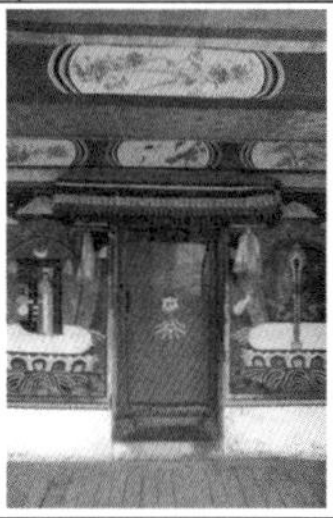	
照片名称	室内正面	照片名称	楼梯	照片名称	唐卡
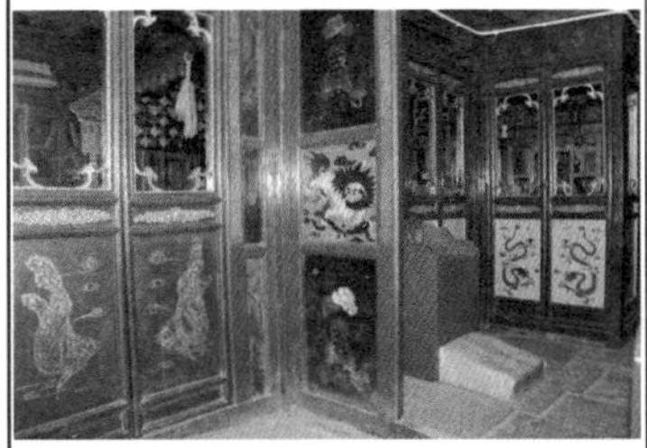		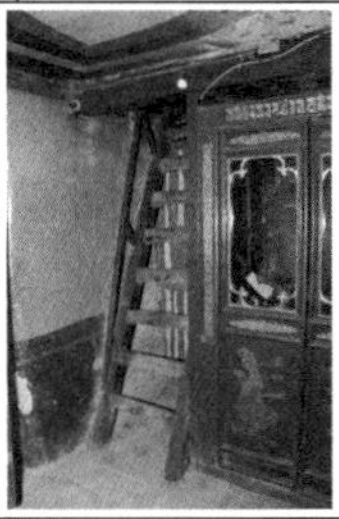		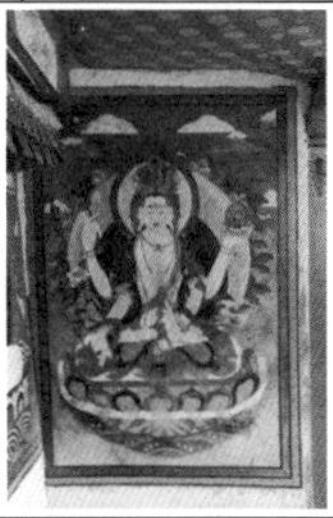	
照片名称	室内佛像	照片名称	室内局部	照片名称	室内陈设
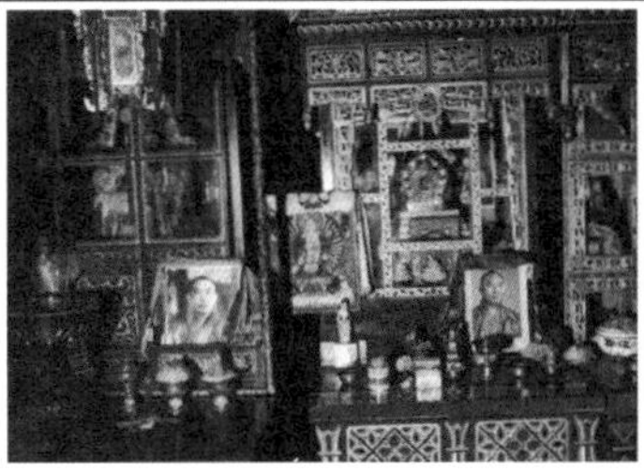		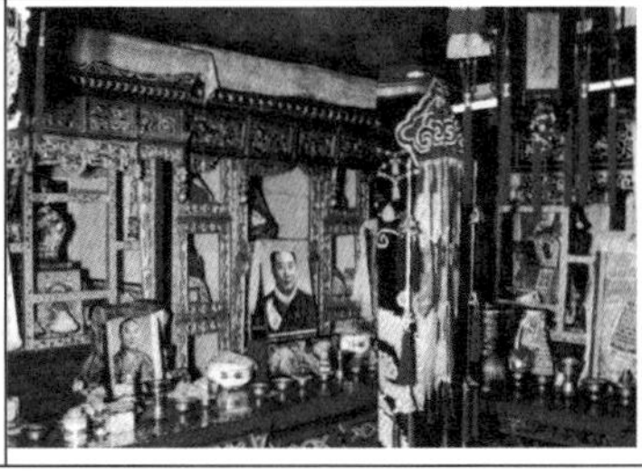		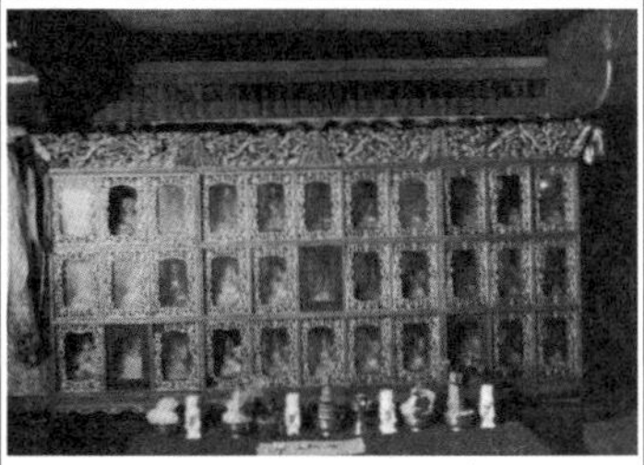	
备注	——				
摄影日期	2010/10/09	摄影人员	孟祎军		

苏卜盖陵基本概况表2

苏卜盖陵斜前方（左图）
苏卜盖陵室内局部
（右上图）
苏卜盖陵室内梯子
（右下图）

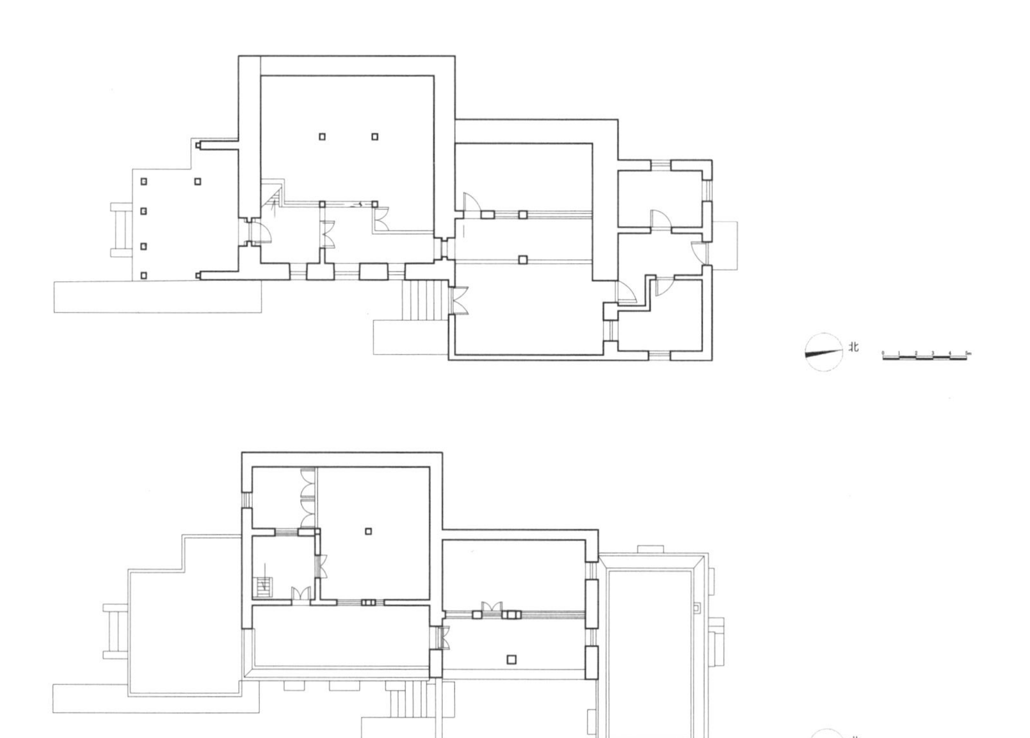

苏卜盖陵一层平面图

苏卜盖陵二层平面图

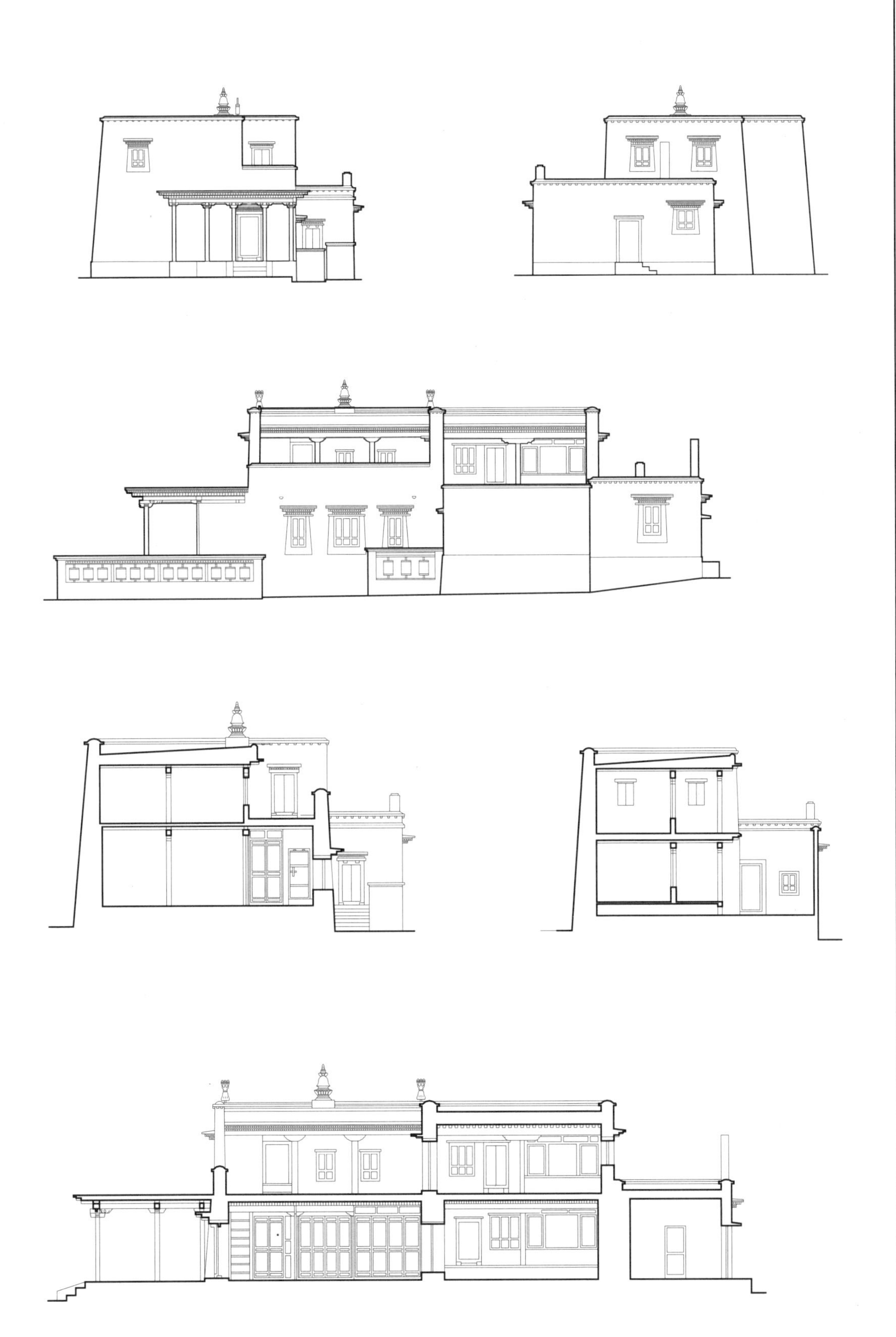

苏卜盖陵立面图

苏卜盖陵剖面图

2.8 五当召 · 阿会殿

单位:毫米

建筑名称	汉语正式名称	——	俗称	阿会殿													
概述	初建年	1800年	建筑朝向	东	建筑层数	一											
	建筑简要描述	藏式结构体系，藏式风格装饰，前廊为藏式柱廊															
	重建重修记载	不详															
		信息来源	——														
结构规模	结构形式	土木混合	相连的建筑	无	室内天井	有											
	建筑平面形式	方形	外廊形式	前廊													
	通面阔	16180	开间数	——	明间	——	次间	——	梢间	——	次梢间	——	尽间	——			
	通进深	10500	进深数	——	进深尺寸（前→后）	——											
	柱子数量	8	柱子间距	横向尺寸	2630→2630→2780	（藏式建筑结构体系填写此栏，不含廊柱）											
				纵向尺寸	2440												
	其他	无															
建筑主体（大木作）（石作）（瓦作）	屋顶	屋顶形式	密肋平顶	瓦作	抹灰屋面												
	外墙	主体材料	土坯	材料规格	不详	饰面颜色	白										
		墙体收分	有	边玛檐墙	有	边玛材料	水泥、边玛草										
	斗栱、梁架	斗栱	无	平身科斗口尺寸	——	梁架关系	八柱十四梁										
	柱、柱式（前廊柱）	形式	藏式	柱身断面形状	八楞柱	断面尺寸	290×290	（在没有前廊柱的情况下，填写室内柱及其特征）									
		柱身材料	木材	柱身收分	有	栌斗、托木	有	雀替	无								
		柱础	有	柱础形状	方形	柱础尺寸	周长 $C=1600$										
	台基	台基类型	普通台基	台基高度	2210	台基地面铺设材料	石材										
	其他	无															
装修（小木作）（彩画）	门(正面)	板门	门楣	无	堆经	有	门帘	无									
	窗（正面）	藏式明窗	窗楣	有	窗套	有	窗帘	有									
	室内隔扇	隔扇	有	隔扇位置	——												
	室内地面、楼面	地面材料及规格	石材 1180×500	楼面材料及规格	不详												
	室内楼梯	楼梯	无	楼梯位置	——	楼梯材料	——	梯段宽度	——								
	天花、藻井	天花	有	天花类型	井口天花	藻井	无	藻井类型	——								
	彩画	柱头	有	柱身	有	梁架	有	走马板	无								
		门、窗	无	天花	有	藻井	无	其他彩画	无								
	其他	悬塑	有	佛龛	无	匾额	无										
装饰	室内	帷幔	无	幕帘彩绘	无	壁画	有	唐卡	有								
		经幡	无	经幢	有	柱毯	有	其他	无								
	室外	玛尼轮	无	苏勒德	有	宝顶	有	祥麟法轮	无								
		四角经幢	有	经幡	无	铜饰	有	石刻、砖雕	无								
		仙人走兽	无	壁画	有	其他	无										
陈设	室内	主佛像	八世活佛	佛像基座	不详												
		法座	有	藏经橱	有	经床	有	诵经桌	有	法鼓	无	玛尼轮	无	坛城	无	其他	无
	室外	旗杆	无	苏勒德	无	狮子	无	经幡	无	玛尼轮	无	香炉	有	五供	无	其他	无
	其他	无															
备注	——																
调查日期	2010/10/09	调查人员	韩瑛、卢文娟	整理日期	2010/10/09	整理人员	卢文娟										

阿会殿基本概况表1

五当召·阿会殿·档案照片

照片名称	正立面	照片名称	台基	照片名称	外廊
照片名称	正门	照片名称	外窗	照片名称	柱身
照片名称	室内正面	照片名称	室内柱身	照片名称	壁画
照片名称	室内局部	照片名称	室内装饰	照片名称	室内陈设
照片名称	室内佛像1	照片名称	室内佛像2	照片名称	
				——	
备注	——				
摄影日期	2010/10/09	摄影人员	孟祎军		

阿会殿正前方

阿会殿室内正面（左上图）

阿会殿室内柱子（左中图）

阿会殿室内陈设（左下图）
阿会殿室外柱子（右图）

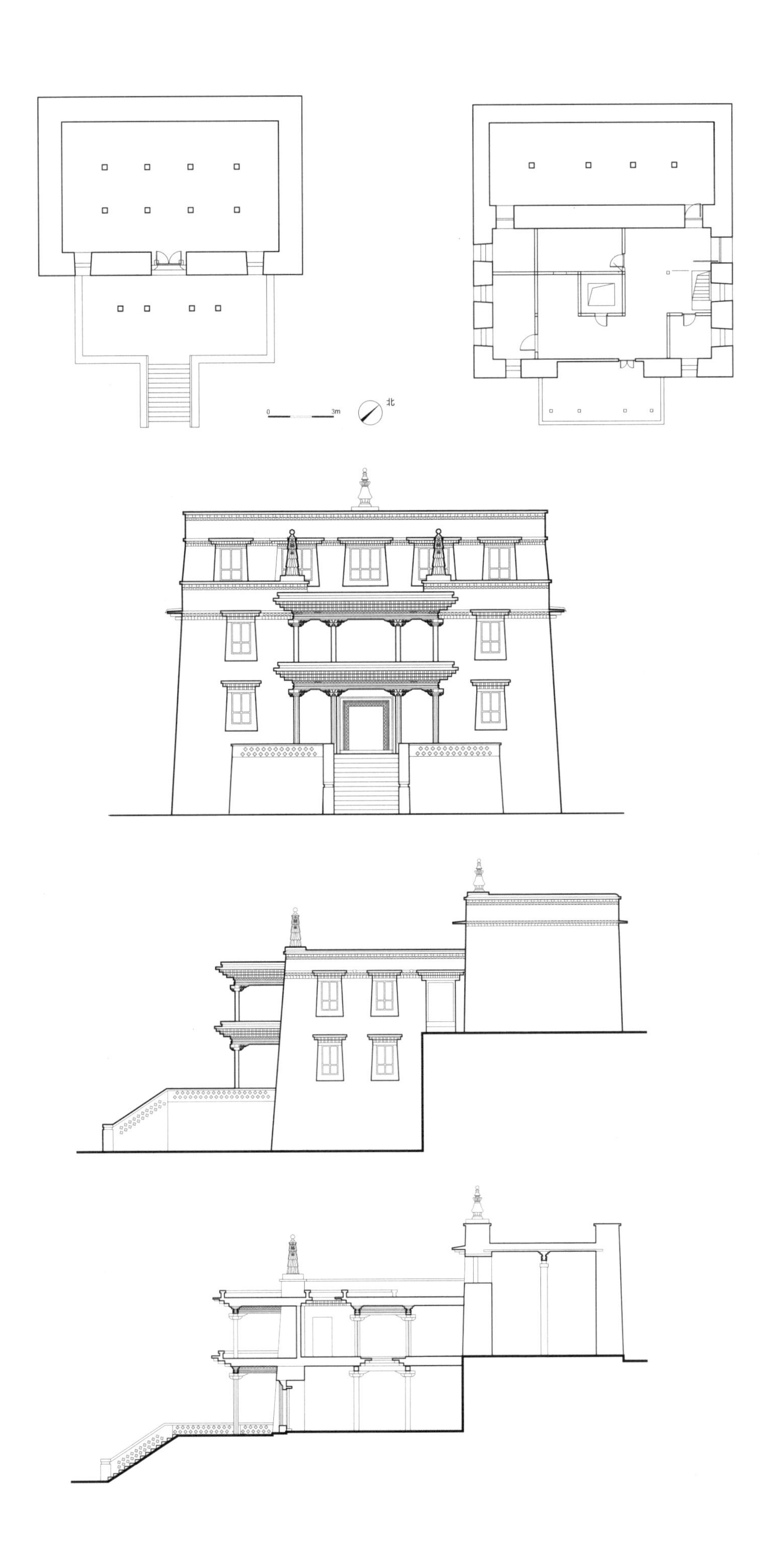

阿会殿一层平面图（左图）

阿会殿二层平面图（右图）

阿会殿正立面图

阿会殿侧立面图

阿会殿剖面图

2.9 五当召·甘珠尔活佛府

单位:毫米

建筑名称	汉语正式名称	甘珠尔活佛府								俗称	拉卜隆						
概述	初建年	1842年								建筑朝向	东		建筑层数	二			
	建筑简要描述	藏式结构体系，藏式风格装饰，前廊为藏式柱廊															
	重建重修记载	不详															
		信息来源	——														
结构规模	结构形式	土木混合		相连的建筑	无					室内天井	有						
	建筑平面形式	凹形		外廊形式	前廊												
	通面阔	14960	开间数	——	明间	——	次间	——	梢间	——	次梢间	——	尽间	——			
	通进深	9150	进深数	——	进深尺寸（前→后）			——									
	柱子数量	6	柱子间距	横向尺寸	2520			（藏式建筑结构体系填写此栏，不含廊柱）									
				纵向尺寸	1940、2910												
	其他	无															
建筑主体（大木作）（石作）（瓦作）	屋顶	屋顶形式	密肋平顶					瓦作	——								
	外墙	主体材料	土坯	材料规格	不详			饰面颜色	白								
		墙体收分	有	边玛檐墙	有			边玛材料	砖								
	斗栱、梁架	斗栱	无	平身科斗口尺寸		——		梁架关系	二柱三梁								
	柱、柱式（前廊柱）	形式	藏式	柱身断面形状	十二楞柱	断面尺寸	255×255			（在没有前廊柱的情况下，填写室内柱及其特征）							
		柱身材料	木材	柱身收分	有	栌斗、托木	有	雀替	无								
		柱础	有	柱础形状	圆形	柱础尺寸	直径 $D=315$										
	台基	台基类型	普通台基	台基高度	400	台基地面铺设材料		石材									
	其他	无															
装修（小木作）（彩画）	门(正面)	风门		门楣	无	堆经	无	门帘	无								
	窗（正面）	支摘窗		窗楣	无	窗套	无	窗帘	无								
	室内隔扇	隔扇	无	隔扇位置	——												
	室内地面、楼面	地面材料及规格		木板、规格不详		楼面材料及规格		不详									
	室内楼梯	楼梯	有	楼梯位置	不详	楼梯材料	不详	梯段宽度	不详								
	天花、藻井	天花	有	天花类型	海漫天花	藻井	无	藻井类型									
	彩画	柱头	有	柱身	有	梁架	有	走马板	有								
		门、窗	有	天花	无	藻井	无	其他彩画	无								
	其他	悬塑	无	佛龛	无	匾额	无										
装饰	室内	帷幔	有	幕帘彩绘	有	壁画	有	唐卡	有								
		经幡	无	经幢	无	柱毯	无	其他	无								
	室外	玛尼轮	无	苏勒德	有	宝顶	有	祥麟法轮	无								
		四角经幢	无	经幡	无	铜饰	无	石刻、砖雕	无								
		仙人走兽	无	壁画	有	其他	无										
陈设	室内	主佛像	甘珠尔活佛					佛像基座	——								
		法座	有	藏经橱	无	经床	无	诵经桌	无	法鼓	无	玛尼轮	无	坛城	无	其他	无
	室外	旗杆	无	苏勒德	无	狮子	无	经幡	无	玛尼轮	无	香炉	无	五供	无	其他	无
	其他	无															
备注	——																
调查日期	2010/10/09	调查人员	韩瑛、卢文娟	整理日期	2010/10/09	整理人员	卢文娟										

甘珠尔活佛府基本概况
表1

五当召 · 甘珠尔活佛府 · 档案照片					
照片名称	正立面	照片名称	斜前方	照片名称	外廊
照片名称	正门	照片名称	台基	照片名称	室外柱身
照片名称	室外柱头	照片名称	室外柱础	照片名称	四角经幢
照片名称	苏勒德	照片名称	室内正面	照片名称	室内侧面
照片名称	室内柱头	照片名称	彩绘	照片名称	室内唐卡
备注	—				
摄影日期	2010/10/09	摄影人员	孟祎军		

甘珠尔活佛府基本概况表2

甘珠尔活佛府正前方

甘珠尔活佛府室外柱子（左图）

甘珠尔活佛府窗（右图）

甘珠尔活佛府室内正面（左图）

甘珠尔活佛府室内唐卡（右图）

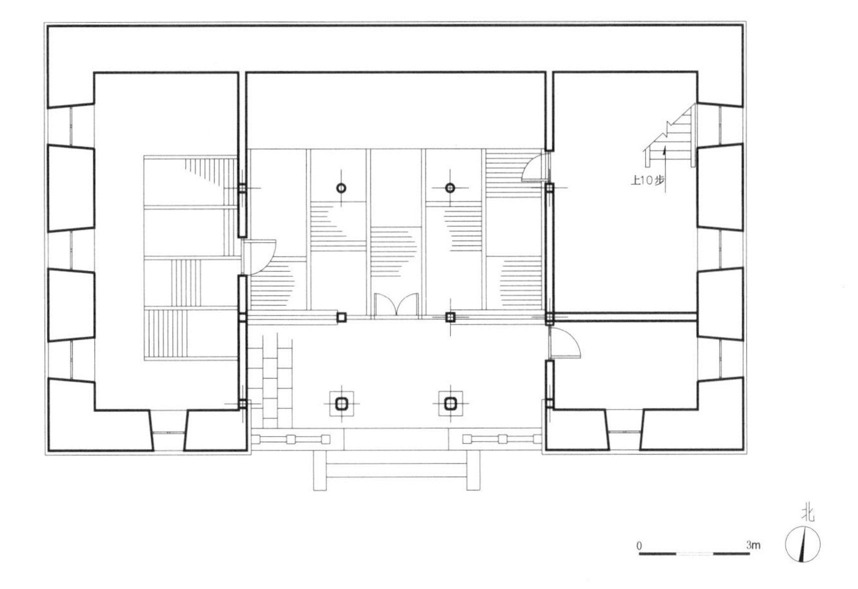

甘珠尔活佛府一层平面图

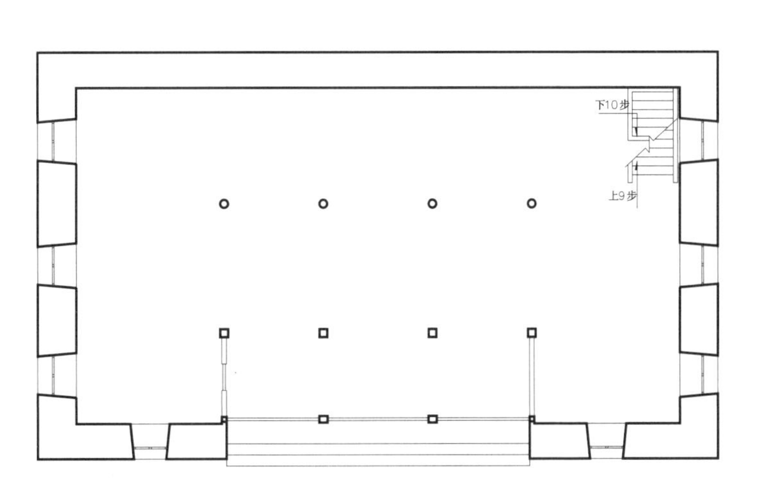

甘珠尔活佛府二层平面图

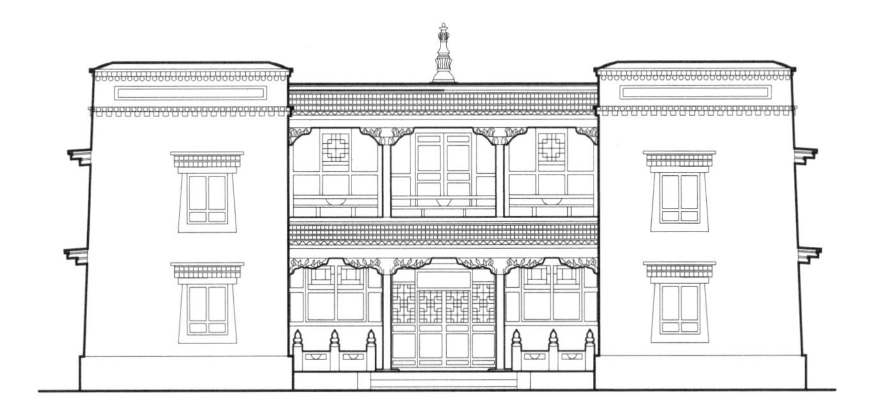

甘珠尔活佛府正立面图

3 梅日更召 Merigen Temple

3 梅日更召 Merigen Temple

梅日更召建筑群1

梅日更召为原乌兰察布盟乌拉特西公旗寺庙，系该旗旗庙及乌拉特三大名寺之一。乾隆三十八年（1773年），清廷御赐满、蒙古、汉、藏四体“昌梵寺”匾额。寺庙管辖吉日嘎朗图庙、席勒庙、额尔顿宝拉格庙、根还庙、查干庙共5座属庙，同时作为旗庙，管理旗境内24座寺庙的法事活动。梅日更召及其管辖的西公旗寺庙以翻译藏文经典，并以蒙古语诵经而著称。寺庙曾出现梅日更活佛罗布桑丹毕贾拉森及乌格里衮达赉、诺门达赉、丹毕贾拉森等著名学者。

康熙十一年（1672年）起乌拉特西公旗扎萨克镇国公诺门在本旗建寺弘法，在海日图之地新建一座寺庙，请内齐托音呼图克图之弟子梅日更迪彦其迪努瓦主持法事。迪努瓦即第一世梅日更活佛，梅日更之名源自迪努瓦之名号。康熙四十年（1701年），扎萨克达日玛希日出资在浩亚日胡都格之地另建一座寺庙，于康熙四十四年（1705年）请梅日更活佛丹津嘉木苏在新寺坐床，由此将新寺称为梅日更庙，将原海日图庙作为镇国公家庙，称为公音庙。

寺庙建筑风格为汉藏结合式建筑。寺庙在其最盛时占地面积约4520平方米，有弥勒殿、大雄宝殿等4座大殿，大雄宝殿南有查玛舞场，东侧为膳食房，西侧为会管房，南有天王殿。梅日更活佛拉布隆及佛仓占据寺庙中心区位，其西侧为大雄宝殿，北侧有大庙仓院、西达喇嘛仓两进院，其东侧有乔尔吉上师仓三进院、固始上师仓两进院，寺庙最北端为席勒喇嘛仓两进院。寺庙有8座佛塔。

“文化大革命”中寺庙严重受损，后有部队驻扎于该庙。1984年起寺庙正式恢复法会。

参考文献：

[1] 巴·孟和等.梅日更召（蒙古文）.海拉尔:内蒙古文化出版社,1996，10.
[2]（清）葛尔丹旺楚克多尔济.梅日更召创建史（蒙古文）.海拉尔:内蒙古文化出版社,1994，4.
[3] 巴·孟和.梅日更葛根罗桑丹毕坚赞研究（蒙古文）.海拉尔:内蒙古文化出版社,1995，5.

梅日更召建筑群2

<table>
<tr><th colspan="8">梅日更召 · 基本概况 1</th></tr>
<tr><td rowspan="3">寺院蒙古语藏语名称</td><td>蒙古语</td><td>ᠮᠡᠷᠭᠡᠨ ᠵᠣᠣ</td><td rowspan="2">寺院汉语名称</td><td>汉语正式名称</td><td colspan="3">广法寺</td></tr>
<tr><td>藏语</td><td>——</td><td>俗称</td><td colspan="3">梅日更召</td></tr>
<tr><td>汉语语义</td><td>聪明、智慧</td><td colspan="2">寺院汉语名称的由来</td><td colspan="3">清廷赐名</td></tr>
<tr><td>所在地</td><td colspan="4">包头市昆都仑区西月三十五公里处</td><td>东经 110° 02′</td><td colspan="2">北纬 40° 42′</td></tr>
<tr><td>初建年</td><td colspan="2">初建于康熙十六年（1677年）</td><td colspan="2">保护等级</td><td colspan="3">自治区级文物保护单位</td></tr>
<tr><td>盛期时间</td><td colspan="2">乾隆时期</td><td colspan="2">盛期喇嘛僧/现有喇嘛僧数</td><td colspan="3">500人/5余人</td></tr>
<tr><td rowspan="2">历史沿革</td><td colspan="7">1677年，在海日图之地新建一座寺庙，后称公音庙。
1702年，在浩亚日胡都格之地新建一座寺庙，称梅日更庙</td></tr>
<tr><td>资料来源</td><td colspan="6">[1] 巴 · 孟和等.梅日更召（蒙古文）.海拉尔:内蒙古文化出版社,1996，10.
[2]（清）葛尔丹旺楚克多尔济.巴 · 孟和校注.梅日更召创建史（蒙古文）.海拉尔:内蒙古文化出版社,1994，4.
[3] 巴 · 孟和.梅日更葛根罗桑丹毕坚赞研究（蒙古文）.海拉尔:内蒙古文化出版社,1995，5.</td></tr>
<tr><td rowspan="2">现状描述</td><td colspan="4" rowspan="2">梅日更召寺庙主要建筑由四大天王殿、护法殿、经堂佛殿组成，另外还有葛根仓、葛根住所、活佛舍利殿、乔尔吉仓、固始仓、四大喇嘛仓、大甲坝以及一系列厢房和喇嘛住所，其中除西大天王殿和护法殿新建外，其余均属历史建筑，主要建筑风格为汉、藏混合，结构形式为砖、木混合结构</td><td>描述时间</td><td colspan="2">2010/10/08</td></tr>
<tr><td>描述人</td><td colspan="2">卢文娟</td></tr>
<tr><td>调查日期</td><td colspan="2">2010/10/08</td><td colspan="2">调查人员</td><td colspan="3">韩瑛、卢文娟</td></tr>
</table>

梅日更召基本概况表1

梅日更召 · 基本概况 2

现存建筑	经堂	四大天王殿（新建）	护法殿	已毁建筑	四大天王殿（原）	护法殿
	佛殿	葛根仓	葛根舍利殿		西勒喇嘛仓	呼毕勒汗仓
	葛根住所	乔尔吉仓	固始仓		药师仓	——
	西大喇嘛仓	——	——		信息来源	寺庙喇嘛口述

区位图

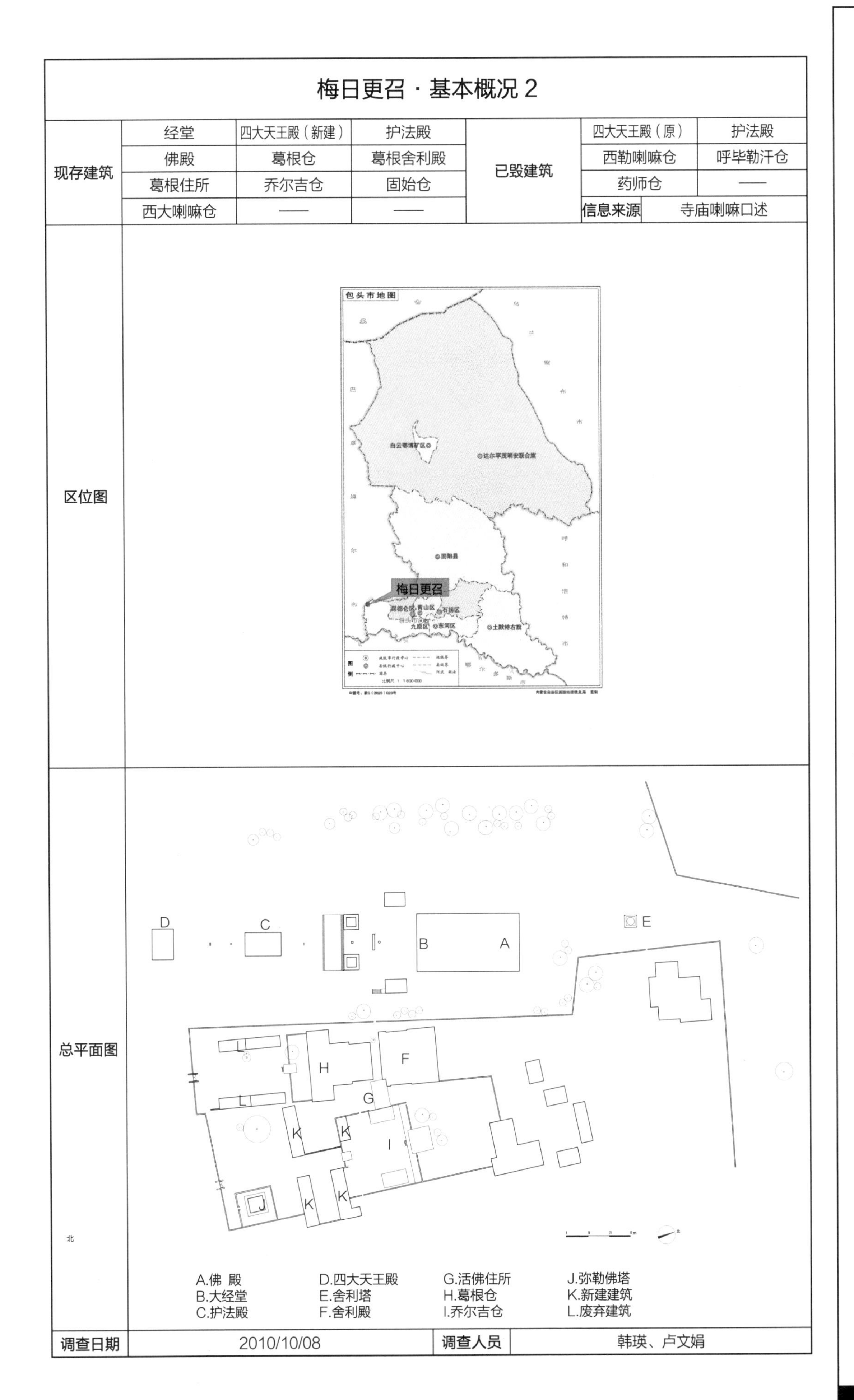

总平面图

A.佛 殿
B.大经堂
C.护法殿
D.四大天王殿
E.舍利塔
F.舍利殿
G.活佛住所
H.葛根仓
I.乔尔吉仓
J.弥勒佛塔
K.新建建筑
L.废弃建筑

调查日期	2010/10/08	调查人员	韩瑛、卢文娟

梅日更召基本概况表2

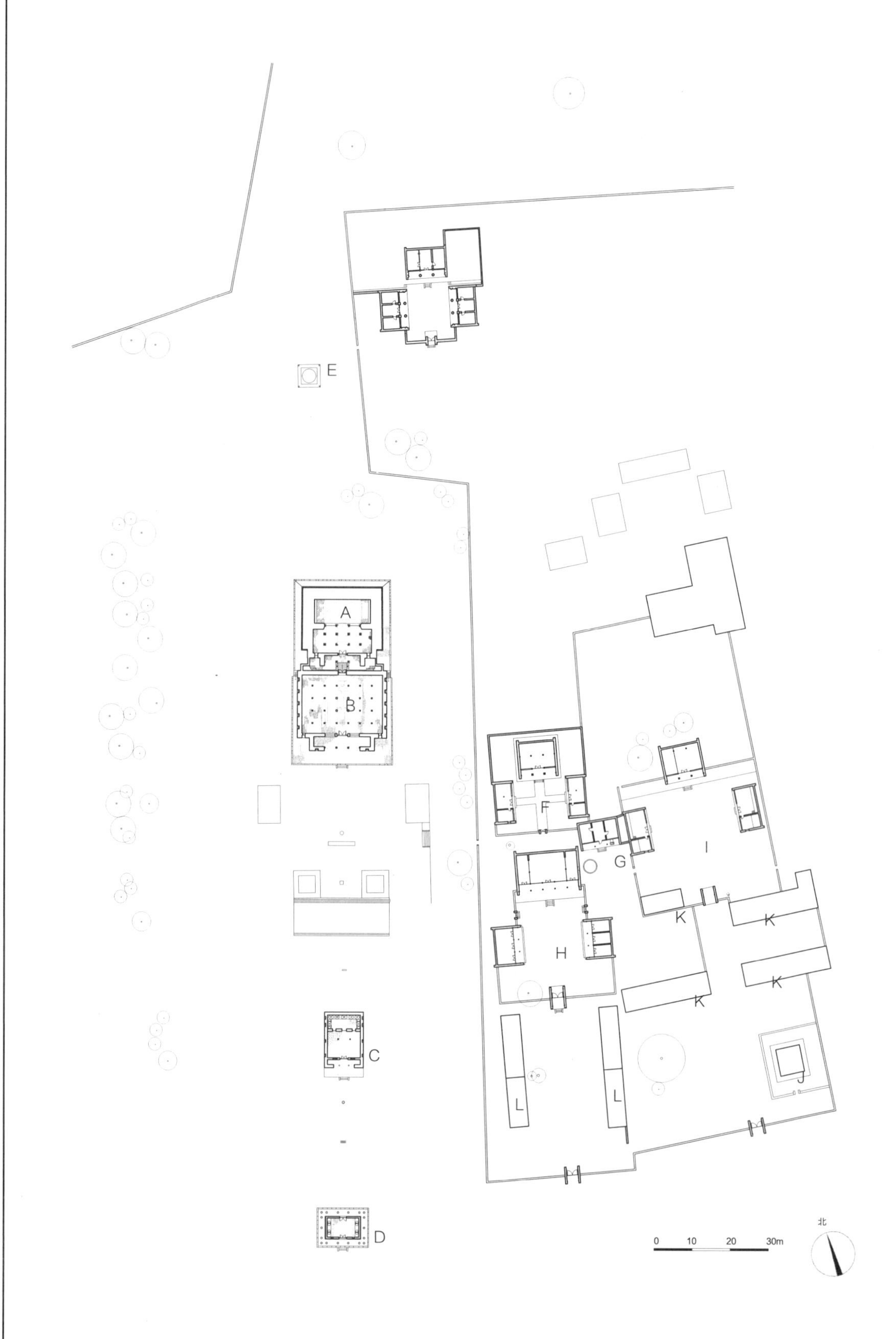

梅日更召总平面图

A.佛 殿
B.大经堂
C.护法殿
D.四大天王殿
E.舍利塔
F.舍利殿
G.活佛住所
H.葛根仓
I.乔尔吉仓
J.弥勒佛塔
K.新建建筑
L.废弃建筑

3.1 梅日更召 · 大雄宝殿（经堂）

单位:毫米

建筑名称	汉语正式名称	大雄宝殿（经堂）	俗称	——													
概述	初建年	康熙十六年（1677年）	建筑朝向	南	建筑层数	二											
	建筑简要描述	汉藏混合，藏式结构体系，局部歇山屋顶，藏式风格装饰，藏式柱廊，经堂内部为都纲法式															
	重建重修记载	不详															
		信息来源	——														
结构规模	结构形式	砖木混合	相连的建筑	北面与佛殿以墙体连接，中间以台阶过渡	室内天井	都纲法式											
	建筑平面形式	凹字形	外廊形式	前廊													
	通面阔	12678	开间数	——	明间	——	次间	——	梢间	——	次梢间	——	尽间	——			
	通进深	9830	进深数	——	进深尺寸（前→后）	——											
	柱子数量	24	柱子间距	横向尺寸	3168→3100→3200→3210→3152	（藏式建筑结构体系填写此栏，不含廊柱）											
				纵向尺寸	3190→3155→3185												
	其他	——															
建筑主体（大木作）（石作）（瓦作）	屋顶	屋顶形式	密肋平顶，中间歇山顶	瓦作	抹灰屋面												
	外墙	主体材料	青砖	材料规格	300×175×90	饰面颜色	土黄										
		墙体收分	有	边玛檐墙	有	边玛材料	砖										
	斗栱、梁架	斗栱	无	平身科斗口尺寸	——	梁架关系	面向天井纵排架										
	柱、柱式（前廊柱）	形式	藏式	柱身断面形状	十二楞柱	断面尺寸	260×260	（在没有前廊柱的情况下，填写室内柱及其特征）									
		柱身材料	木材	柱身收分	有	栌斗、托木	有	雀替	无								
		柱础	有	柱础形状	方形	柱础尺寸	650×650										
	台基	台基类型	普通台基	台基高度	5800	台基地面铺设材料	石材										
	其他	——															
装修（小木作）（彩画）	门(正面)	板门	门楣	无	堆经	有	门帘	无									
	窗（正面）	藏式明窗	窗楣	有	窗套	无	窗帘	无									
	室内隔扇	隔扇	无	隔扇位置	——												
	室内地面、楼面	地面材料及规格	方砖 300×300	楼面材料及规格	木材、规格不详												
	室内楼梯	楼梯	无	楼梯位置	——	楼梯材料	——	梯段宽度	——								
	天花、藻井	天花	有	天花类型	井口天花	藻井	无	藻井类型	——								
	彩画	柱头	有	柱身	有	梁架	有	走马板	无								
		门、窗	无	天花	有	藻井	——	其他彩画	窗眉								
	其他	悬塑	无	佛龛	无	匾额	有										
装饰	室内	帷幔	有	幕帘彩绘	无	壁画	有	唐卡	无								
		经幡	有	经幢	有	柱毯	有	其他									
	室外	玛尼轮	有	苏勒德	有	宝顶	有	祥麟法轮	有								
		四角经幢	无	经幡	无	铜饰	有	石刻、砖雕	有								
		仙人走兽	无	壁画	有	其他	——										
陈设	室内	主佛像	——	佛像基座	——												
		法座	有	藏经橱	无	经床	有	诵经桌	有	法鼓	无	玛尼轮	无	坛城	无	其他	——
	室外	旗杆	无	苏勒德	无	狮子	无	经幡	无	玛尼轮	有	香炉	有	五供	无	其他	——
	其他	——															
备注	——																
调查日期	2010/10/08	调查人员	韩瑛、卢文娟	整理日期	2010/10/08	整理人员	卢文娟										

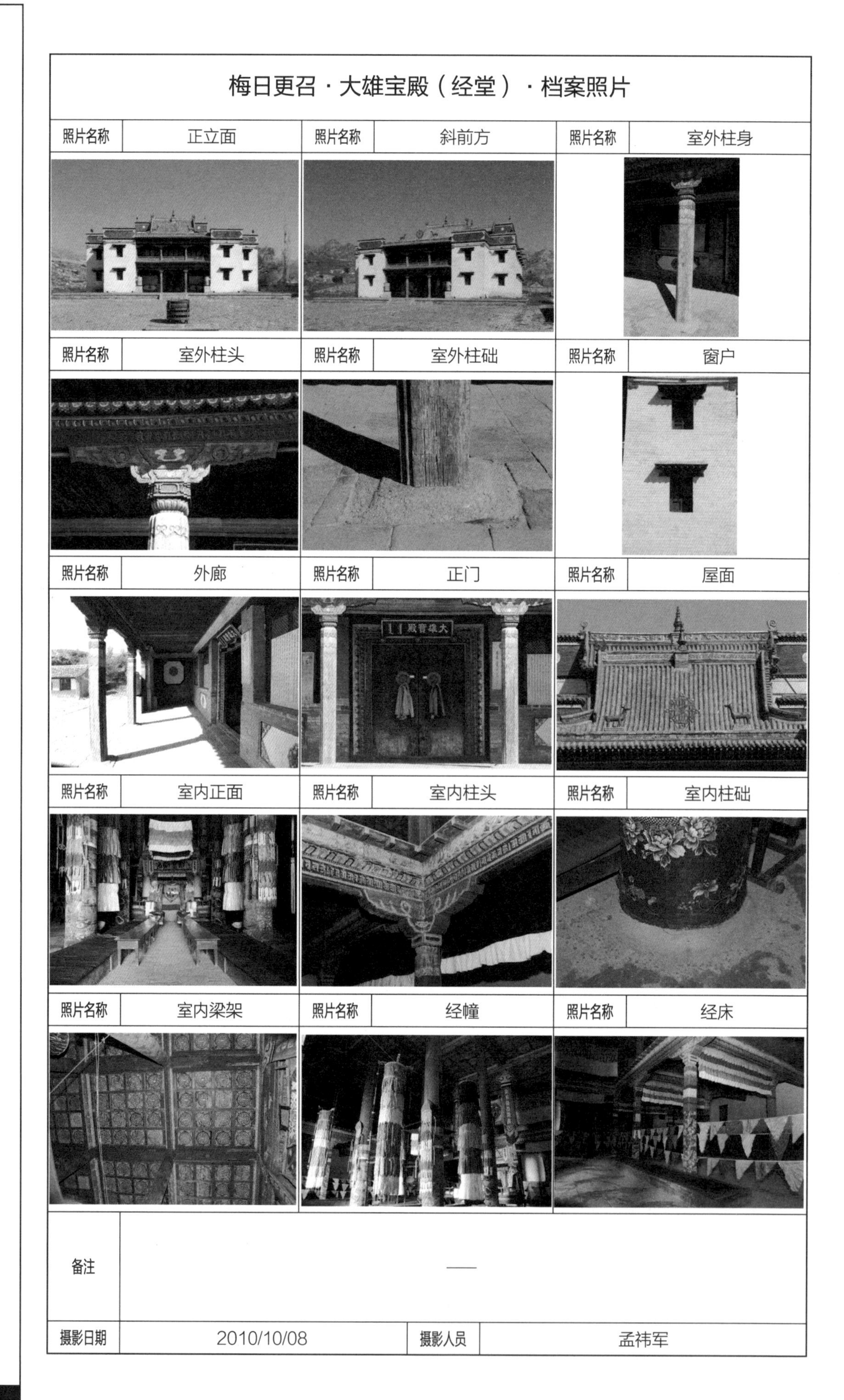

梅日更召·大雄宝殿（经堂）·档案照片

照片名称	正立面	照片名称	斜前方	照片名称	室外柱身
照片名称	室外柱头	照片名称	室外柱础	照片名称	窗户
照片名称	外廊	照片名称	正门	照片名称	屋面
照片名称	室内正面	照片名称	室内柱头	照片名称	室内柱础
照片名称	室内梁架	照片名称	经幢	照片名称	经床

备注	——		
摄影日期	2010/10/08	摄影人员	孟祎军

大雄宝殿（经堂）
基本概况表2

梅日更召·大雄宝殿（佛殿）

单位:毫米

建筑名称	汉语正式名称	朝格钦独贡、弥勒佛殿	俗称	大雄宝殿（佛殿）													
概述	初建年	不详	建筑朝向	南	建筑层数	二											
	建筑简要描述	汉藏混合，藏式结构体系，藏式风格装饰，藏式柱廊															
	重建重修记载	不详															
		信息来源	——														
结构规模	结构形式	砖木混合	相连的建筑	南面与经堂以墙体连接，中间以台阶过渡	室内天井	有											
	建筑平面形式	凹字形	外廊形式	前廊													
	通面阔	9570	开间数	——	明间	——	次间	——	梢间	——	次梢间	——	尽间	——			
	通进深	4855	进深数	——	进深尺寸（前→后）	——											
	柱子数量	10	柱子间距	横向尺寸	3160→3220→3190	（藏式建筑结构体系填写此栏，不含廊柱）											
				纵向尺寸	2630→2225												
	其他	——															
建筑主体（大木作）（石作）（瓦作）	屋顶	屋顶形式	密肋平顶	瓦作	抹灰屋面												
	外墙	主体材料	青砖	材料规格	300×135×60	饰面颜色	土黄										
		墙体收分	有	边玛檐墙	有	边玛材料	砖										
	斗栱、梁架	斗栱	无	平身科斗口尺寸	——	梁架关系	二柱四梁										
	柱、柱式（前廊柱）	形式	藏式	柱身断面形状	十二楞柱	断面尺寸	275×275	（在没有前廊柱的情况下，填写室内柱及其特征）									
		柱身材料	木材	柱身收分	有	栌斗、托木	有	雀替	无								
		柱础	有	柱础形状	圆形	柱础尺寸	直径 $D=360$										
	台基	台基类型	普通台基	台基高度	440	台基地面铺设材料	方砖										
	其他	——															
装修（小木作）（彩画）	门(正面)	板门	门楣	无	堆经	有	门帘	无									
	窗（正面）	槛窗	窗楣	无	窗套	无	窗帘	无									
	室内隔扇	隔扇	无	隔扇位置	——												
	室内地面、楼面	地面材料及规格	方砖300×700	楼面材料及规格	不详												
	室内楼梯	楼梯	无	楼梯位置	——	楼梯材料	——	梯段宽度	——								
	天花、藻井	天花	有	天花类型	海漫天花	藻井	无	藻井类型	——								
	彩画	柱头	有	柱身	有	梁架	有	走马板	无								
		门、窗	无	天花	无	藻井	——	其他彩画	无								
	其他	悬塑	无	佛龛	无	匾额	有										
装饰	室内	帷幔	有	幕帘彩绘	无	壁画	无	唐卡	有								
		经幡	无	经幢	有	柱毯	有	其他	——								
	室外	玛尼轮	无	苏勒德	有	宝顶	有	祥麟法轮	有								
		四角经幢	无	经幡	无	铜饰	有	石刻、砖雕	有								
		仙人走兽	无	壁画	无	其他	——										
陈设	室内	主佛像	弥勒佛	佛像基座	莲花座												
		法座	无	藏经橱	无	经床	无	诵经桌	无	法鼓	无	玛尼轮	无	坛城	无	其他	——
	室外	旗杆	无	苏勒德	无	狮子	无	经幡	无	玛尼轮	无	香炉	无	五供	无	其他	——
	其他	——															
备注	——																
调查日期	2010/10/08	调查人员	韩瑛、卢文娟	整理日期	2010/10/08	整理人员	卢文娟										

大雄宝殿（佛殿）
基本概况表1

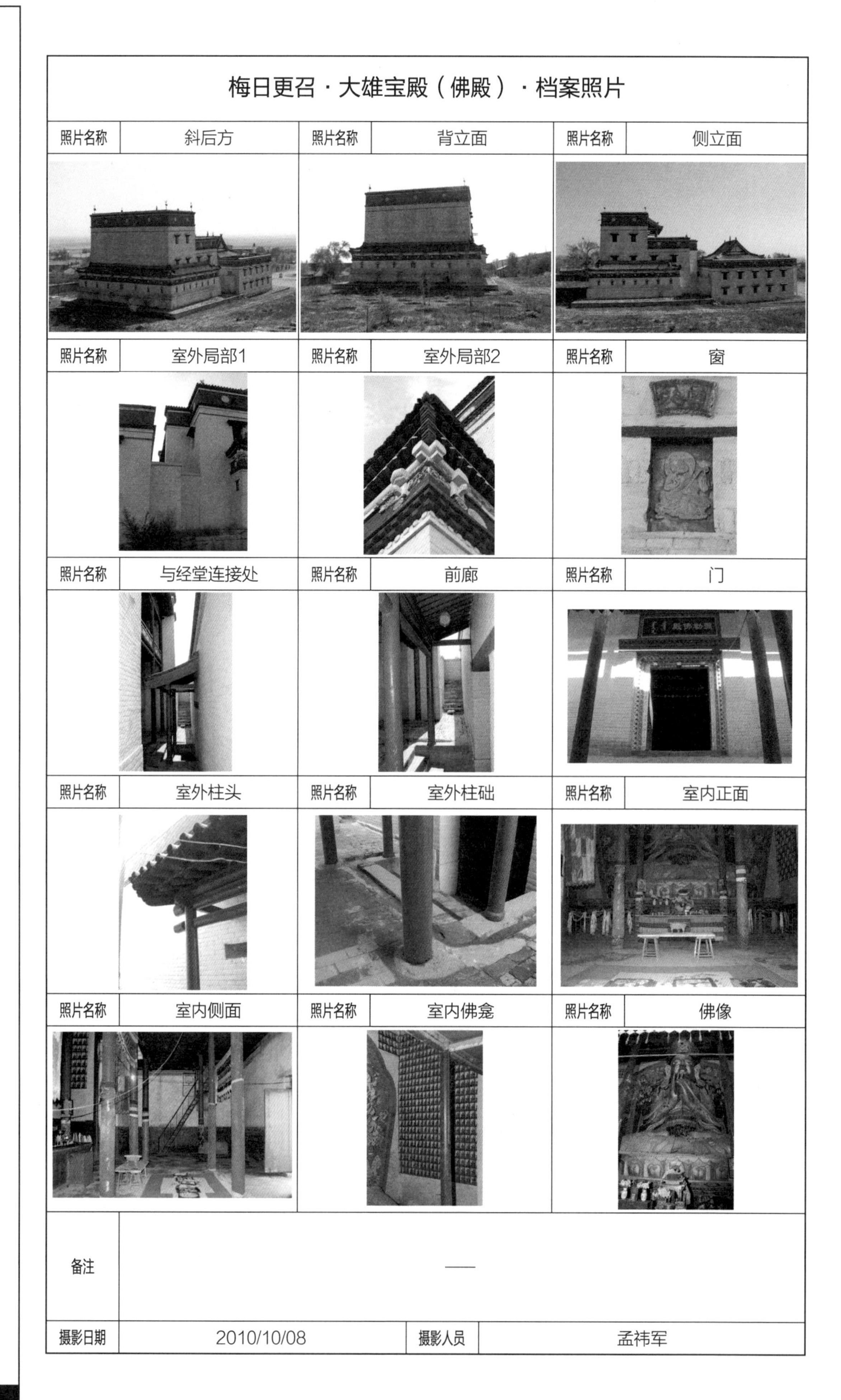

梅日更召·大雄宝殿（佛殿）·档案照片

照片名称	斜后方	照片名称	背立面	照片名称	侧立面
照片名称	室外局部1	照片名称	室外局部2	照片名称	窗
照片名称	与经堂连接处	照片名称	前廊	照片名称	门
照片名称	室外柱头	照片名称	室外柱础	照片名称	室内正面
照片名称	室内侧面	照片名称	室内佛龛	照片名称	佛像
备注	——				
摄影日期	2010/10/08	摄影人员	孟祎军		

大雄宝殿（佛殿）斜后方

大雄宝殿正前方

大雄宝殿（经堂）室内空间（左图）
大雄宝殿（经堂）室内陈设（右图）

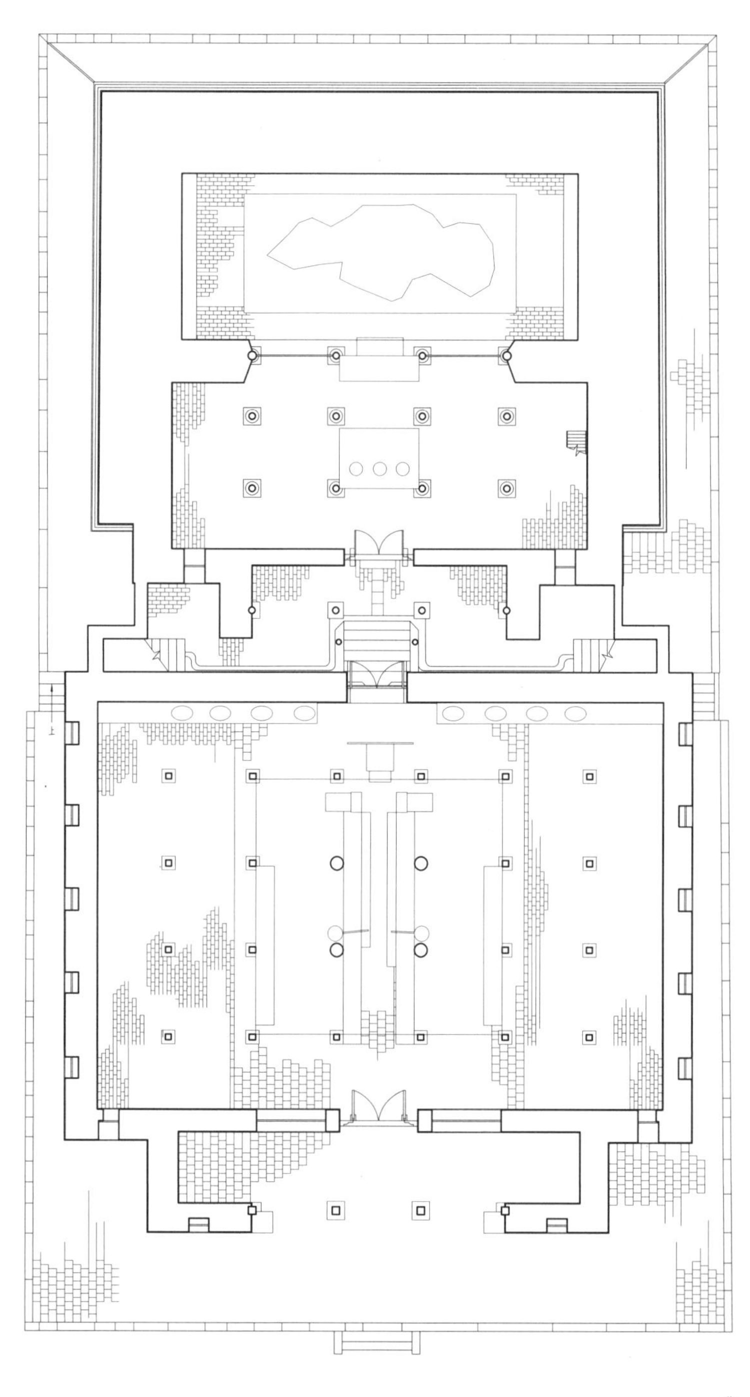

大雄宝殿一层平面图

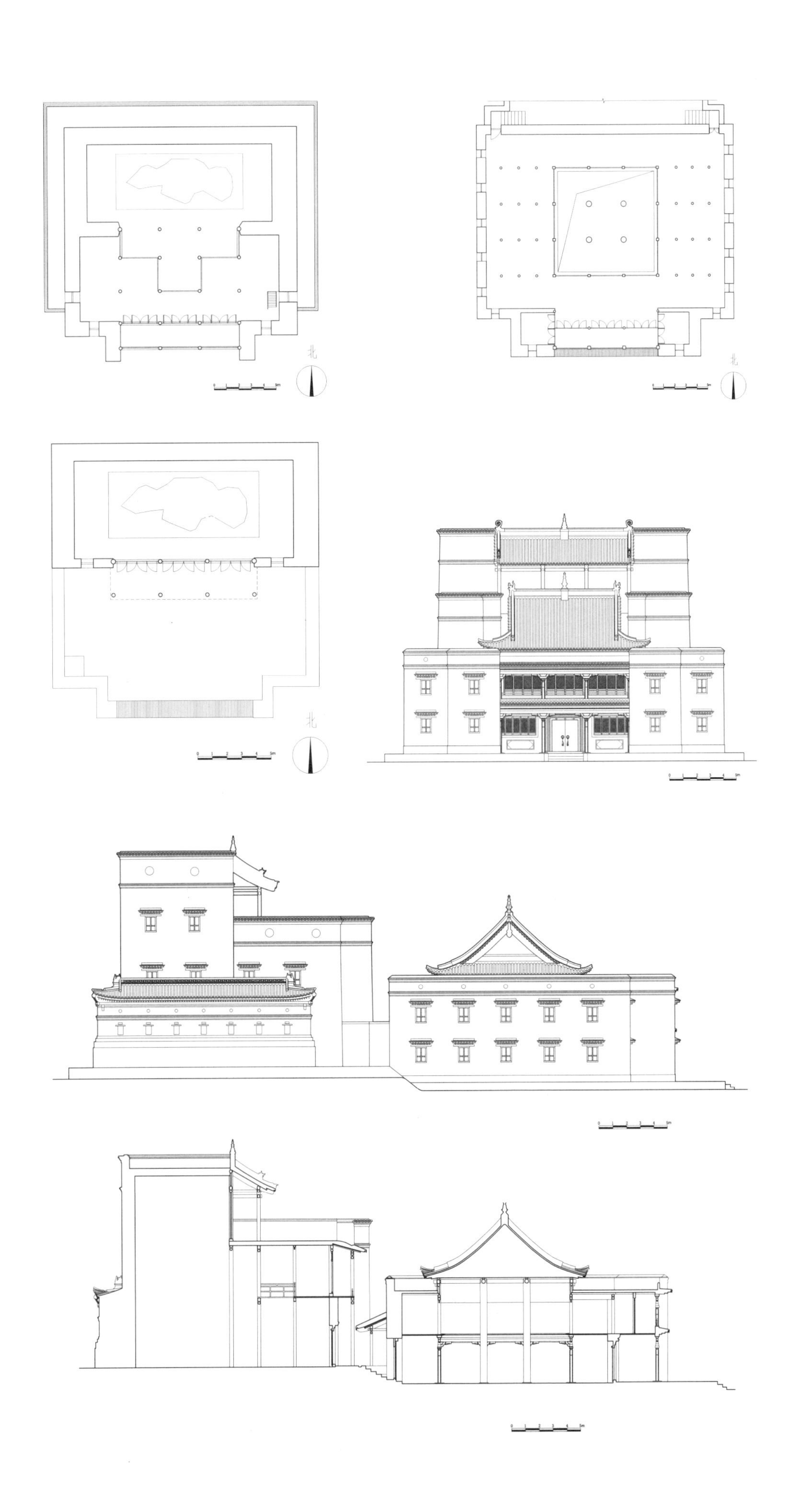

大雄宝殿（佛殿）
二层平面图（左图）

大雄宝殿（经堂）
二层平面图（右图）

大雄宝殿（佛殿）
三层平面图（左图）

大雄宝殿南立面图
（右图）

大雄宝殿侧立面图

大雄宝殿剖面图

3.2 梅日更召 · 护法殿

单位:毫米

建筑名称	汉语正式名称	护法殿		俗称	——								
概述	初建年	不详		建筑朝向	南	建筑层数	一						
	建筑简要描述	藏式结构体系，藏式风格装饰，前廊为藏式柱廊											
	重建重修记载	不详											
		信息来源	——										
结构规模	结构形式	砖木混合		相连的建筑	南面与经堂以墙体连接，中间以台阶过渡		室内天井	无					
	建筑平面形式	凹字形		外廊形式	前廊								
	通面阔	2550	开间数	——	明间	——	次间	——	梢间	——	次梢间	——	尽间 ——
	通进深	——	进深数	——	进深尺寸（前→后）	——							
	柱子数量	2	柱子间距	横向尺寸	2550	（藏式建筑结构体系填写此栏，不含廊柱）							
				纵向尺寸	——								
	其他	——											
建筑主体（大木作）（石作）（瓦作）	屋顶	屋顶形式	密肋平顶			瓦作	抹灰屋面						
	外墙	主体材料	青砖	材料规格	300×175×90	饰面颜色	土黄						
		墙体收分	有	边玛檐墙	有	边玛材料	砖						
	斗栱、梁架	斗栱	无	平身科斗口尺寸	——	梁架关系	面向天井纵排架						
	柱、柱式（前廊柱）	形式	汉式	柱身断面形状	圆	断面尺寸	直径 D=260		（在没有前廊柱的情况下，填写室内柱及其特征）				
		柱身材料	木材	柱身收分	有	栌斗、托木	无	雀替	有				
		柱础	有	柱础形状	方形	柱础尺寸	600×600						
	台基	台基类型	普通台基	台基高度	1100	台基地面铺设材料	方砖 150×300						
	其他	——											
装修（小木作）（彩画）	门(正面)	板门		门楣	无	堆经	有	门帘	无				
	窗（正面）	藏式明窗		窗楣	有	窗套	无	窗帘	无				
	室内隔扇	隔扇	无	隔扇位置	——								
	室内地面、楼面	地面材料及规格		方砖150×300		楼面材料及规格	木材						
	室内楼梯	楼梯	有	楼梯位置	入口右侧	楼梯材料	钢	梯段宽度	800				
	天花、藻井	天花	无	天花类型	——	藻井	无	藻井类型	——				
	彩画	柱头	无	柱身	无	梁架	无	走马板	无				
		门、窗	无	天花	无	藻井	——	其他彩画	无				
	其他	悬塑	无	佛龛	无	匾额	有						
装饰	室内	帷幔	无	幕帘彩绘	无	壁画	有	唐卡	有				
		经幡	有	经幢	有	柱毯	无	其他	——				
	室外	玛尼轮	无	苏勒德	有	宝顶	有	祥麟法轮	有				
		四角经幢	无	经幡	无	铜饰	无	石刻、砖雕	有				
		仙人走兽	无	壁画	无	其他	——						
陈设	室内	主佛像	大维德金刚			佛像基座	莲花座						
		法座	无	藏经橱	无	经床	有	诵经桌	有	法鼓	无	玛尼轮	无
		坛城	无	其他	——								
	室外	旗杆	无	苏勒德	无	狮子	无	经幡	无	玛尼轮	无	香炉	有
		五供	无	其他	——								
	其他	——											
备注	——												
调查日期	2010/10/08	调查人员	韩瑛、卢文娟	整理日期	2010/10/08	整理人员	卢文娟						

护法殿基本概况表1

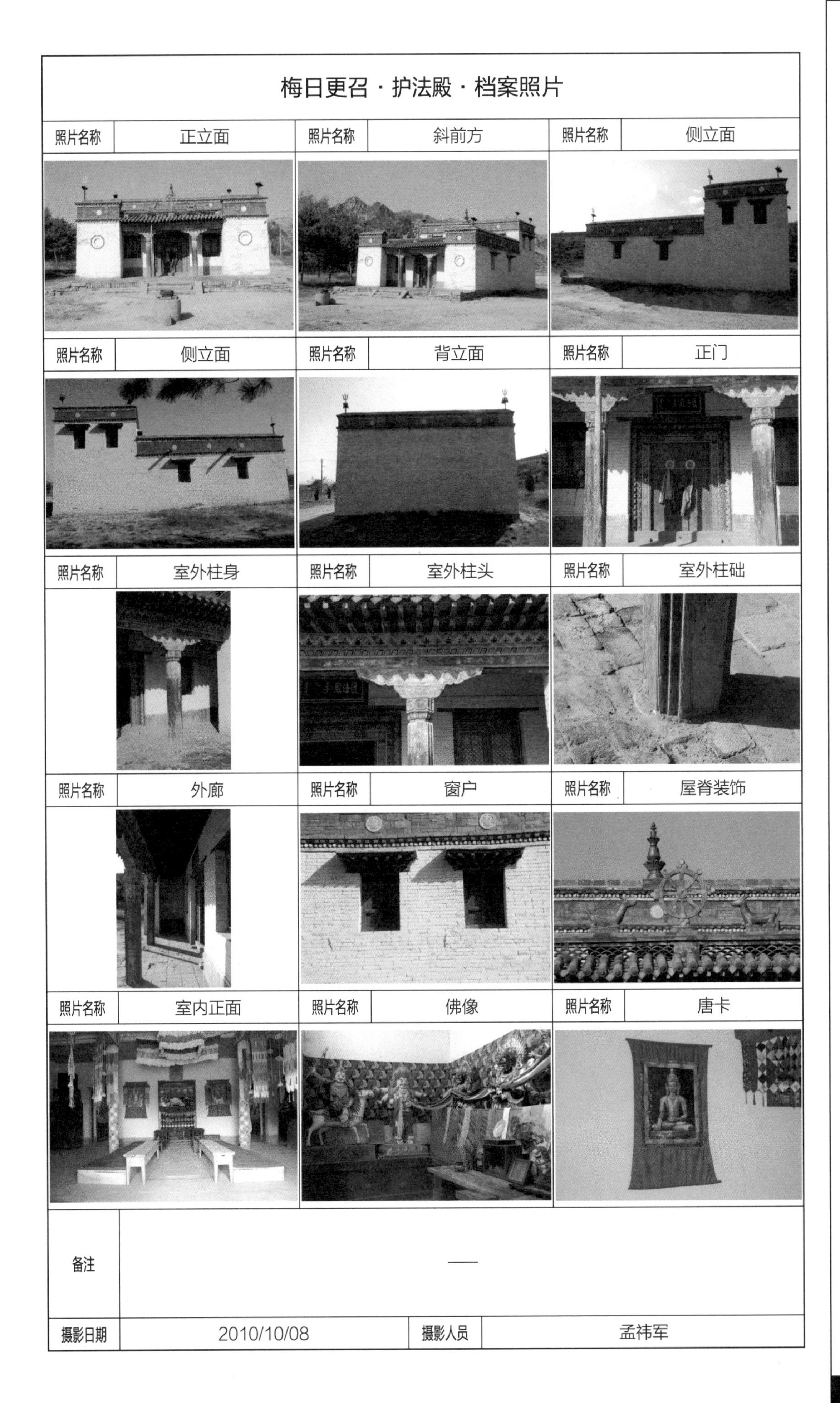

梅日更召・护法殿・档案照片

照片名称	正立面	照片名称	斜前方	照片名称	侧立面
照片名称	侧立面	照片名称	背立面	照片名称	正门
照片名称	室外柱身	照片名称	室外柱头	照片名称	室外柱础
照片名称	外廊	照片名称	窗户	照片名称	屋脊装饰
照片名称	室内正面	照片名称	佛像	照片名称	唐卡

备注	——		
摄影日期	2010/10/08	摄影人员	孟祎军

护法殿正前方

护法殿斜前方（左图）
护法殿室外前廊（右图）

护法殿鸟瞰图（左图）
护法殿室内柱子（右图）

3.3 梅日更召·四大天王殿

单位:毫米

建筑名称	汉语正式名称	四大天王殿	俗称	天王殿													
概述	初建年	不详	建筑朝向	南	建筑层数	一											
	建筑简要描述	汉式结构体系，前廊为汉式柱廊，建筑内外为汉式装饰															
	重建重修记载	不详															
		信息来源	——														
结构规模	结构形式	砖木混合	相连的建筑	无	室内天井	无											
	建筑平面形式	长方形	外廊形式	回廊													
	通面阔	13850	开间数	5	明间	2795	次间	3020	梢间	1290	次梢间	——	尽间	——			
	通进深	7790	进深数	4	进深尺寸（前→后）	1290→2600→2630→1270											
	柱子数量	——	柱子间距	横向尺寸	——	（藏式建筑结构体系填写此栏，不含廊柱）											
				纵向尺寸	——												
	其他	——															
建筑主体（大木作）（石作）（瓦作）	屋顶	屋顶形式	歇山顶	瓦作	布瓦												
	外墙	主体材料	青砖	材料规格	300×135×60	饰面颜色	青砖色										
		墙体收分	无	边玛檐墙	无	边玛材料	——										
	斗栱、梁架	斗栱	无	平身科斗口尺寸	无	梁架关系	五檩前后廊										
	柱、柱式（前廊柱）	形式	汉式	柱身断面形状	圆形	断面尺寸	直径D=260	（在没有前廊柱的情况下，填写室内柱及其特征）									
		柱身材料	木材	柱身收分	有	栌斗、托木	有	雀替	无								
		柱础	有	柱础形状	圆形	柱础尺寸	直径D=340										
	台基	台基类型	普通台基	台基高度	650	台基地面铺设材料	石材										
	其他	——															
装修（小木作）（彩画）	门(正面)	板门	门楣	无	堆经	有	门帘	无									
	窗（正面）	牖窗	窗楣	无	窗套	有	窗帘	无									
	室内隔扇	隔扇	无	隔扇位置	——												
	室内地面、楼面	地面材料及规格	方砖 300×300	楼面材料及规格	不详												
	室内楼梯	楼梯	无	楼梯位置	——	楼梯材料	——	梯段宽度	——								
	天花、藻井	天花	无	天花类型	——	藻井	无	藻井类型	——								
	彩画	柱头	有	柱身	有	梁架	有	走马板	无								
		门、窗	有	天花	无	藻井	——	其他彩画	——								
	其他	悬塑	无	佛龛	无	匾额	有										
装饰	室内	帷幔	无	幕帘彩绘	无	壁画	有	唐卡	无								
		经幡	无	经幢	无	柱毯	无	其他	——								
	室外	玛尼轮	无	苏勒德	无	宝顶	无	祥麟法轮	无								
		四角经幢	无	经幡	无	铜饰	有	石刻、砖雕	无								
		仙人走兽	无	壁画	无	其他	——										
陈设	室内	主佛像	四大天王	佛像基座	——												
		法座	无	藏经橱	无	经床	无	诵经桌	无	法鼓	无	玛尼轮	无	坛城	无	其他	——
	室外	旗杆	有	苏勒德	无	狮子	有	经幡	无	玛尼轮	无	香炉	无	五供	无	其他	——
	其他	——															
备注	——																
调查日期	2010/10/08	调查人员	韩瑛、卢文娟	整理日期	2010/10/08	整理人员	卢文娟										

四大天王殿基本概况表1

梅日更召·四大天王殿·档案照片

照片名称	正立面	照片名称	斜前方	照片名称	背立面
照片名称	侧立面1	照片名称	侧立面2	照片名称	正门
照片名称	室外柱身	照片名称	室外柱头	照片名称	室外柱础
照片名称	外廊	照片名称	台基	照片名称	佛像
照片名称	壁画	照片名称	室内梁架	照片名称	室内地面
备注	——				
摄影日期	2010/10/08	摄影人员	孟祎军		

四大天王殿基本概况表2

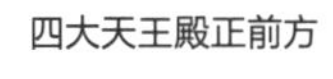

四大天王殿正前方

四大天王殿侧立面（左图）

四大天王殿室外柱子（右图）

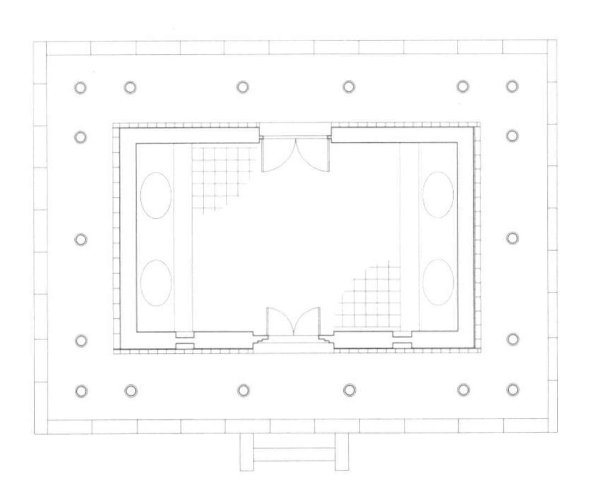

四大天王殿一层平面图（左图）

四大天王殿佛像（右图）

3.4 梅日更召·舍利殿

舍利殿正前方

舍利殿东厢房正前方（左图）
舍利殿西厢房斜前方（右图）

3.5 梅日更召·活佛住所

活佛住所正前方（左图）
活佛住所侧面（右图）

活佛住所门（左图）
活佛住所前廊（右图）

3.6 梅日更召 · 乔尔吉仓

乔尔吉仓正前方（左上图）
乔尔吉仓东厢房（左下图）
乔尔吉仓室外柱（右图）

3.7 梅日更召 · 西达喇嘛仓

西达喇嘛仓院落斜前方

西达喇嘛仓正前方（左图）
西达喇嘛仓院门（右图）

西达喇嘛仓斜前方

西达喇嘛仓西配殿正前方
（左图）
西达喇嘛仓配殿室外柱子
（右图）

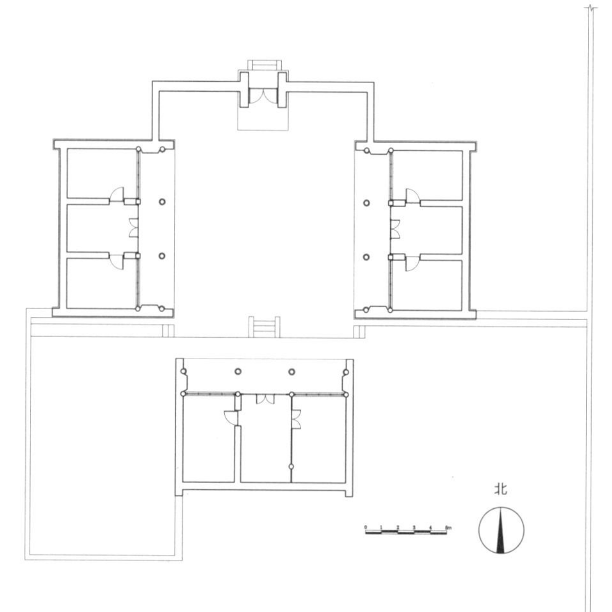

西达喇嘛仓平面图

3.8 梅日更召·固始仓

固始仓斜前方

固始仓配殿斜前方（左图）
固始仓斜后方（右图）

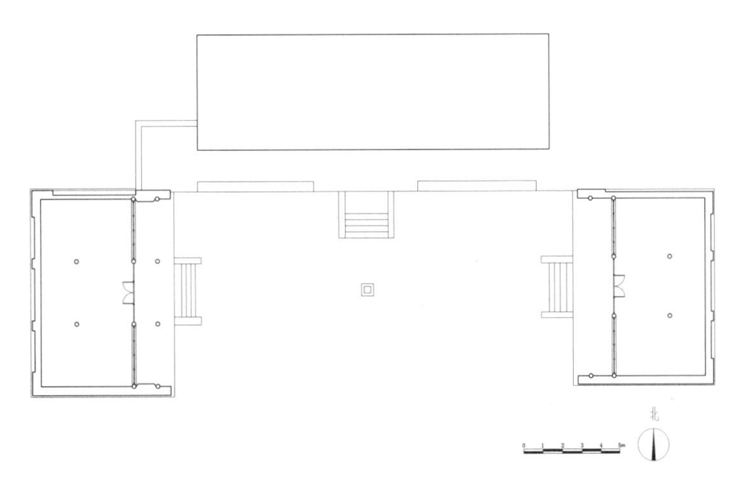

固始仓平面图

3.9 梅日更召·葛根仓

葛根仓正前方

葛根仓院落（左图）
葛根仓东配殿室外柱子（右图）

葛根仓东配殿正前方（左图）
葛根仓院门正面（右图）

3.10 梅日更召・白塔

白塔

白弥勒佛塔（左图）
白舍利塔（右上图）
白塔立面图（右下图）

4 昆都仑召 Hundele Temple

4 昆都仑召 Hundele Temple

昆都仑召建筑群

昆都仑召为原乌兰察布盟乌拉特中公旗寺庙，系该旗旗庙及乌拉特三大名寺之一。据说乾隆年间清廷御赐“法禧寺”匾额。昆都仑召为原乌拉特中公旗寺庙中唯一用藏语诵经的寺庙。历史名著《水晶鉴》作者金巴道尔吉出自该寺。

雍正七年（1729年），两名青海游僧在地方信众的支持下初建该庙，称吉日嘎朗图庙。游僧为叔侄二人，侄嘉木桑桑宝在寺庙建成后到多伦诺尔研习佛法十余年，回到本寺后经乌拉特中公旗王公及寺庙僧人的选定，成为寺庙活佛。在其主持下，经二十余年的建设，该寺成为规模宏大的寺庙。因寺庙位于昆都仑河河畔，故俗称昆都勒庙，也称昆都仑召。

寺庙建筑风格为藏式建筑与汉式建筑并存。至中华人民共和国成立前该庙建筑群占地面积近11公顷，81间大雄宝殿、天王殿、小黄庙、斋戒殿等殿宇23余座，僧舍近60间，活佛府2座，佛塔3座。除此之外，有供旗扎萨克王公祭拜休憩的公爷府、小殿、八角亭等建筑及1座专供祭拜乌拉特部先祖哈布图哈撒尔的汗撒尔殿。该寺活佛高僧闭关修行的根坯庙位于寺北四里处阴山下，为一处四合院，有1座佛殿与2座配殿。

“文化大革命”中寺庙严重受损，但存大雄宝殿、寺庙最早的建筑一小黄庙等10余座殿堂与50余间僧舍。从“丑牛之乱”至中华人民共和国成立初期，在多次兵灾匪患中该寺部分殿宇虽遭受损毁，但整体建筑却幸存下来，成为原乌拉特东公旗唯一一座保存较为完好的寺庙。2010年寺庙进行大规模扩建。

参考文献：

[1] 傲特恒贺希格，勒 · 乌云毕力格.昆都仑召（内部资料）（蒙古文）.包头市少数民族古籍整理办公室，1991，12.

[2] 莫德力图.乌兰察布史略（第十一辑）.政协乌兰察布盟委员会文史资料研究委员会，1997，10.

昆都仑召四大天王殿

<table>
<tr><th colspan="9">昆都仑召·基本概况 1</th></tr>
<tr><td rowspan="3">寺院蒙古语藏语名称</td><td>蒙古语</td><td>[illegible]</td><td rowspan="2">寺院汉语名称</td><td>汉语正式名称</td><td colspan="4">法禧寺</td></tr>
<tr><td>藏语</td><td>——</td><td>俗称</td><td colspan="4">昆都仑召</td></tr>
<tr><td>汉语语义</td><td>昆都仑(横）河庙</td><td colspan="2">寺院汉语名称的由来</td><td colspan="4">清廷赐名</td></tr>
<tr><td>所在地</td><td colspan="4">包头市昆都仑区西北乌拉山南麓山下的昆都仑河右岸</td><td>东经</td><td>109° 47′</td><td>北纬</td><td>40° 42′</td></tr>
<tr><td>初建年</td><td colspan="2">雍正七年或康熙四十年（1729年）</td><td>保护等级</td><td colspan="5">自治区级文物保护单位</td></tr>
<tr><td>盛期时间</td><td colspan="2">清康熙二十六年（1687年）或清光绪时期</td><td colspan="2">盛期喇嘛僧/现有喇嘛僧数</td><td colspan="4">1000余人/近20人</td></tr>
<tr><td rowspan="2">历史沿革</td><td colspan="8">1729年，新建寺庙，初建时只有小黄庙一座。
1913年，丑牛之乱中乌拉特中公旗南部寺庙遭受战乱洗劫，部分僧人迁至该庙。该庙轻微受损。
“文化大革命”期间寺庙严重受损，仅剩10余间殿堂与50余间僧舍。
2010年，大规模修缮、扩建寺庙</td></tr>
<tr><td>资料来源</td><td colspan="7">[1] 傲特恒贺希格,勒·乌云毕力格.昆都仑召（内部资料）（蒙古文）.包头市少数民族古籍整理办公室，1991，12.
[2] 莫德力图.乌兰察布史略（第十一辑）.政协乌兰察布盟委员会文史资料研究委员会，1997，10.</td></tr>
<tr><td rowspan="2">现状描述</td><td colspan="4" rowspan="2">昆都仑召寺庙现存建筑主要有大雄宝殿、度母殿、天王殿、时轮殿、小黄庙、白塔、王爷府、东、西活佛府以及部分喇嘛住所等。以上建筑均属历史遗存建筑，其主要建筑风格以藏式和汉、藏混合两种形式为主。目前正在大规模新建，新建建筑形式和特点暂不详</td><td colspan="2">描述时间</td><td colspan="2">2010/10/10</td></tr>
<tr><td colspan="2">描述人</td><td colspan="2">卢文娟</td></tr>
<tr><td>调查日期</td><td colspan="2">2010/10/10</td><td>调查人员</td><td colspan="5">韩瑛、卢文娟</td></tr>
</table>

昆都仑召基本概况表1

昆都仑召・基本概况 2

现存建筑				已毁建筑		
现存建筑	大雄宝殿	时轮殿	王爷府	已毁建筑	楞木横	小独贡
	小黄庙	东活佛府	白塔		小塔楼	八角过亭
	四大天王殿	西活佛府	——		大甲巴	山门
	度母殿	哈萨尔殿	——		信息来源	——

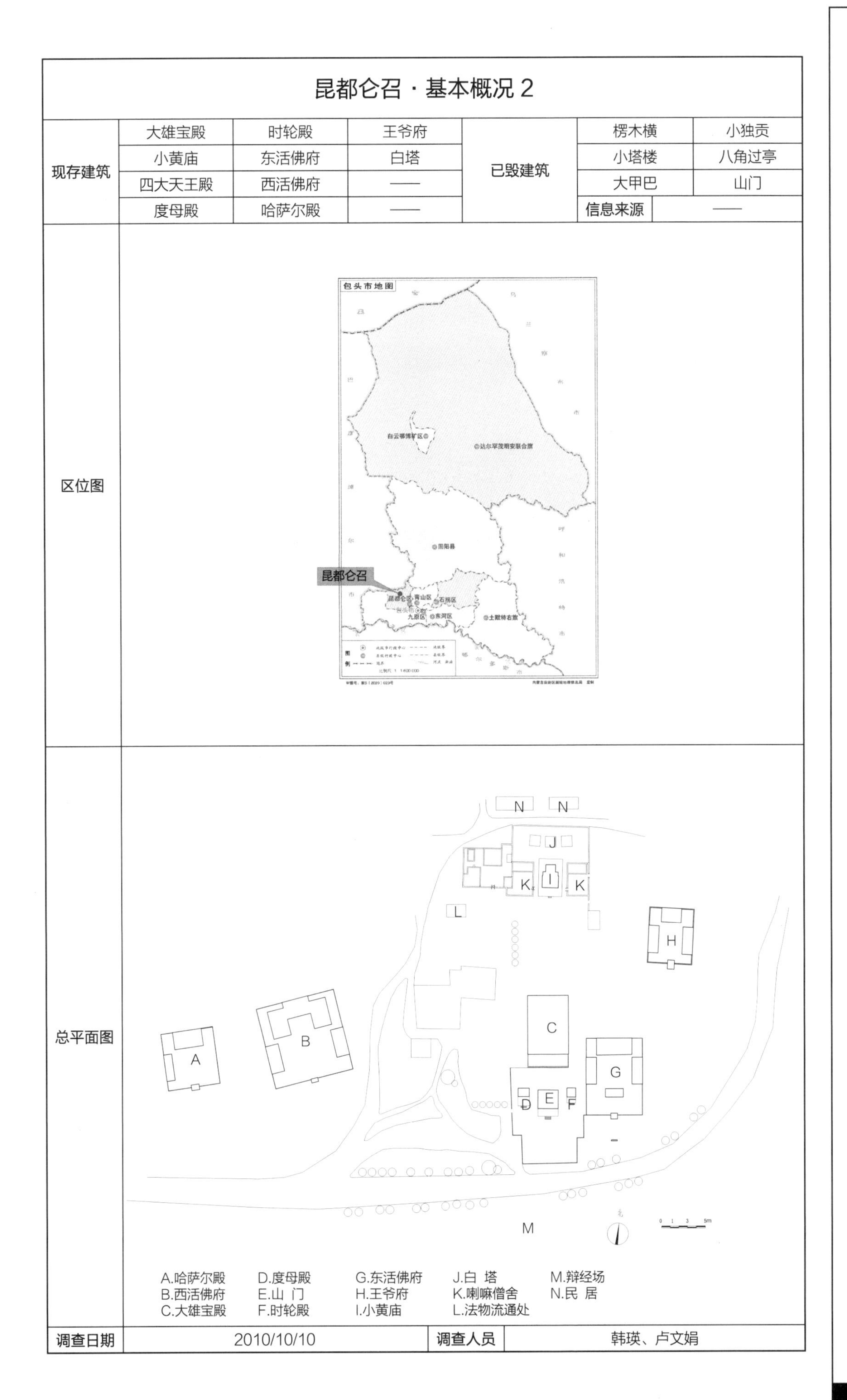

调查日期	2010/10/10	调查人员	韩瑛、卢文娟

昆都仑召基本概况表2

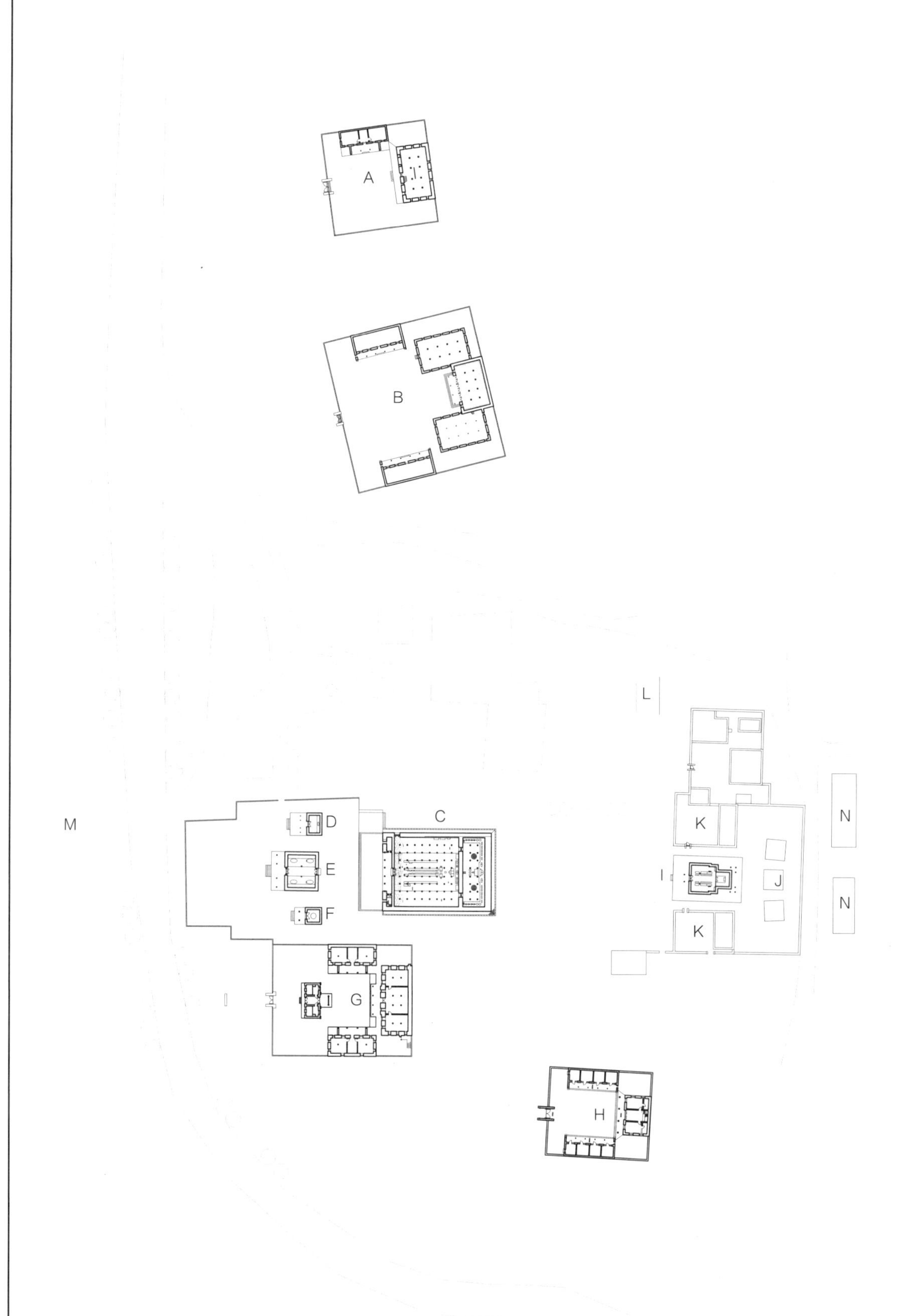
A
B
C
D
E
F
G
H
I
J
K
K
L
M
N
N
A.哈萨尔殿
B.西活佛府
C.大雄宝殿
D.度母殿
E.山 门
F.时轮殿
G.东活佛府
H.王爷府
I.小黄庙
J.白 塔
K.喇嘛僧舍
L.法物流通处
M.辩经场
N.民 居
0 1 3 5m
北

昆都仑召总平面图

4.1 昆都仑召·大雄宝殿

单位:毫米

建筑名称	汉语正式名称	大雄宝殿	俗称	朝克沁独贡													
概述	初建年	不详	建筑朝向	南	建筑层数	二											
	建筑简要描述	纯藏式建筑形制，布局为前经堂后佛殿式，经堂两层，佛殿通高三层。大殿各方向均开九间，含明柱61根，总面积达1161平方米，是昆都仑举行大型法事活动和集会的场所															
	重建重修记载	不详															
		信息来源	实地调研														
结构规模	结构形式	砖木混合	相连的建筑	无	室内天井	都刚法式											
	建筑平面形式	凹字形	外廊形式	前廊													
	通面阔		开间数	——	明间	——	次间	——	梢间	——	次梢间	——	尽间	——			
	通进深		进深数	——	进深尺寸（前→后）	——											
	柱子数量	80	柱子间距	横向尺寸	2755→2750→3170→2755	（藏式建筑结构体系填写此栏，不含廊柱）											
				纵向尺寸	2755→2750→3170→2755												
	其他	无															
建筑主体（大木作）（石作）（瓦作）	屋顶	屋顶形式	密肋平顶	瓦作	抹灰屋面												
	外墙	主体材料	青砖	材料规格	不详	饰面颜色	白色										
		墙体收分	有	边玛檐墙	有	边玛材料	砖										
	斗栱、梁架	斗栱	无	平身科斗口尺寸	——	梁架关系	面向天井纵排架										
	柱、柱式（前廊柱）	形式	藏式	柱身断面形状	八楞柱	断面尺寸	350×350	（在没有前廊柱的情况下，填写室内柱及其特征）									
		柱身材料	木材	柱身收分	有	栌斗、托木	有	雀替	无								
		柱础	有	柱础形状	方形	柱础尺寸	周长 $C=2520$										
	台基	台基类型	普通台基	台基高度	830	台基地面铺设材料	石材										
	其他	无															
装修（小木作）（彩画）	门(正面)	板门	门楣	无	堆经	有	门帘	无									
	窗（正面）	藏式明窗	窗楣	无	窗套	有	窗帘	有									
	室内隔扇	隔扇	无	隔扇位置	——												
	室内地面、楼面	地面材料及规格	石材 450×450	楼面材料及规格	不详												
	室内楼梯	楼梯	有	楼梯位置	右侧小房子内	楼梯材料	不详	梯段宽度	不详								
	天花、藻井	天花	有	天花类型	彻上明造	藻井	无	藻井类型	——								
	彩画	柱头	有	柱身	有	梁架	有	走马板	无								
		门、窗	无	天花	无	藻井	——	其他彩画	无								
	其他	悬塑	无	佛龛	无	匾额	无										
装饰	室内	帷幔	无	幕帘彩绘	无	壁画	有	唐卡	有								
		经幡	无	经幢	有	柱毯	有	其他	有华盖								
	室外	玛尼轮	无	苏勒德	有	宝顶	无	祥麟法轮	有								
		四角经幢	有	经幡	有	铜饰	有	石刻、砖雕	有								
		仙人走兽	无	壁画	有	其他	无										
陈设	室内	主佛像	三世佛	佛像基座	莲花座												
		法座	有	藏经橱	有	经床	有	诵经桌	有	法鼓	有	玛尼轮	无	坛城	无	其他	无
	室外	旗杆	无	苏勒德	无	狮子	无	经幡	有	玛尼轮	无	香炉	有	五供	无	其他	无
	其他	无															
备注	——																
调查日期	2010/10/10	调查人员	韩瑛、卢文娟	整理日期	2010/10/10	整理人员	卢文娟										

大雄宝殿基本概况表1

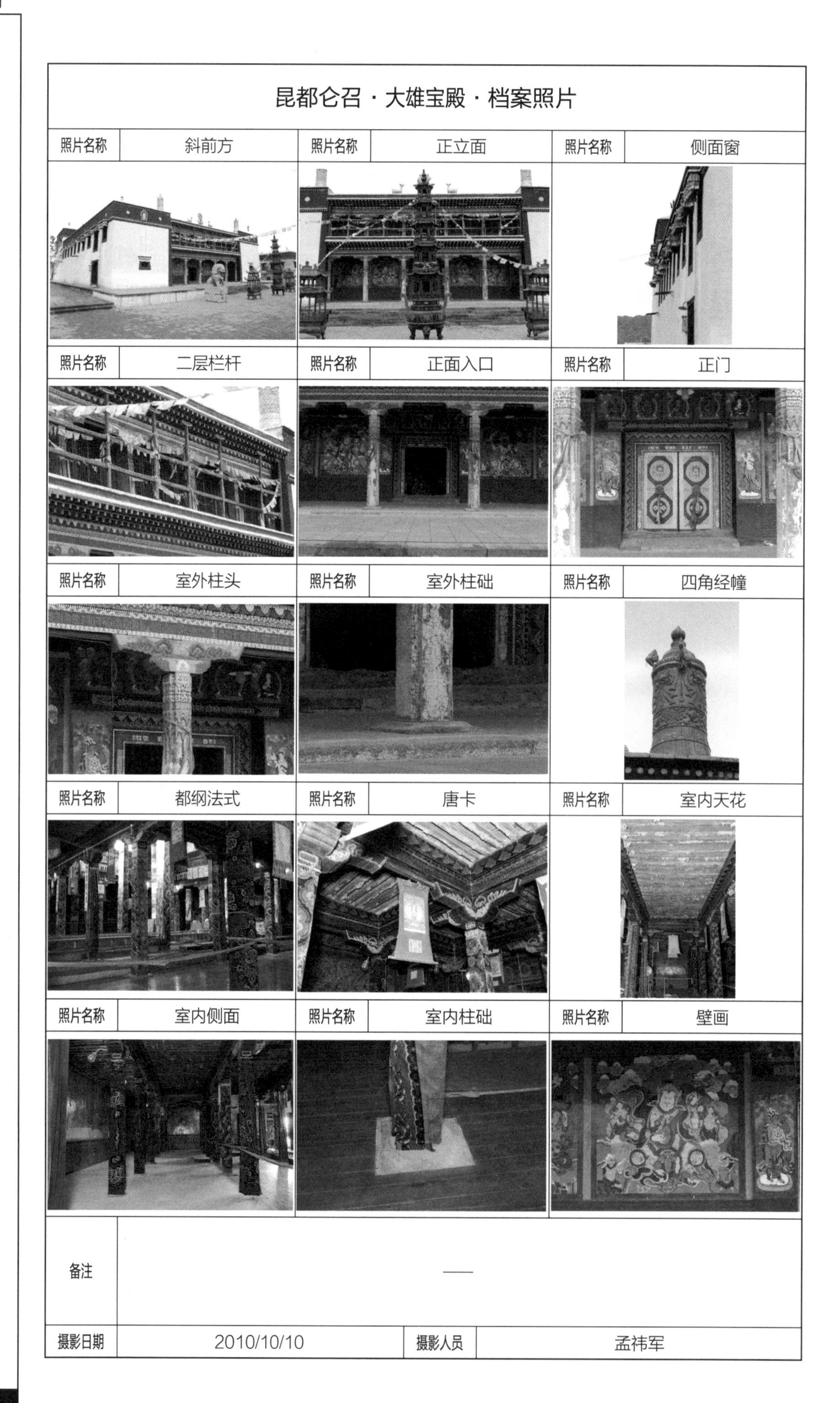

昆都仑召·大雄宝殿·档案照片					
照片名称	斜前方	照片名称	正立面	照片名称	侧面窗
照片名称	二层栏杆	照片名称	正面入口	照片名称	正门
照片名称	室外柱头	照片名称	室外柱础	照片名称	四角经幢
照片名称	都纲法式	照片名称	唐卡	照片名称	室内天花
照片名称	室内侧面	照片名称	室内柱础	照片名称	壁画
备注	——				
摄影日期	2010/10/10	摄影人员	孟祎军		

大雄宝殿基本概况表2

大雄宝殿建筑群

大雄宝殿斜前方

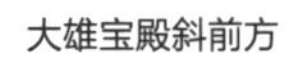

大雄宝殿正面正门（左图）

大雄宝殿室外柱头（中图）

大雄宝殿胜利经幢（右图）

大雄宝殿室内空间

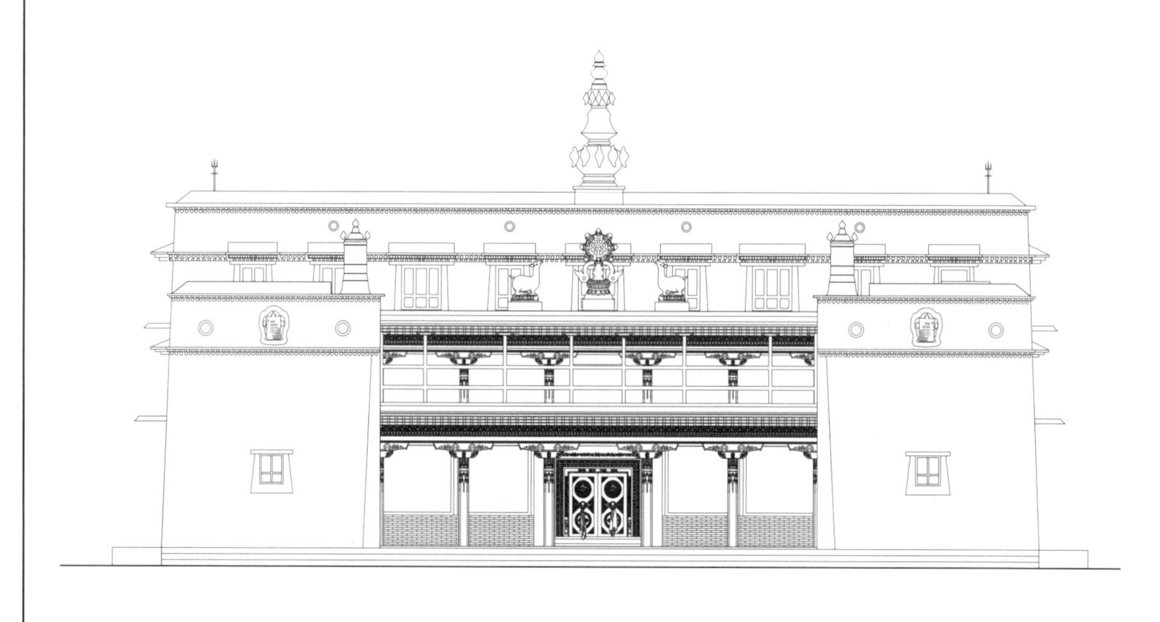
大雄宝殿正立面图

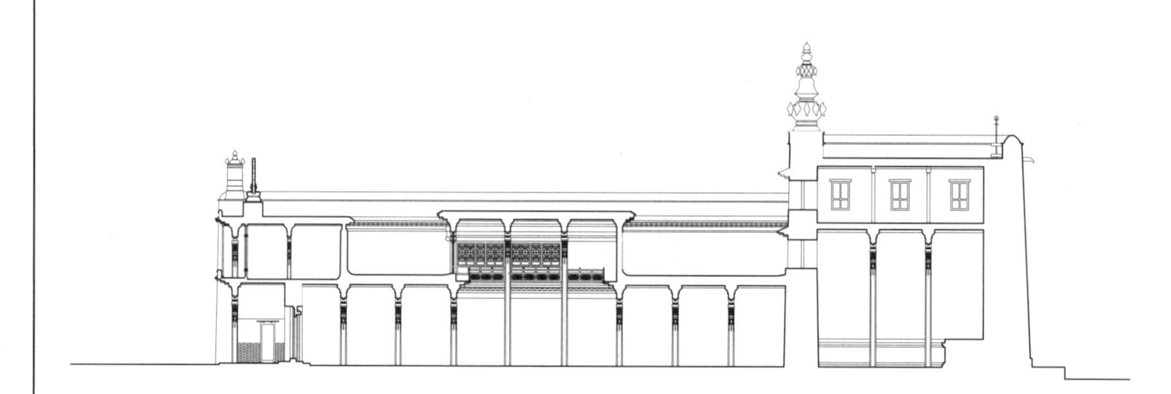
大雄宝殿剖面图

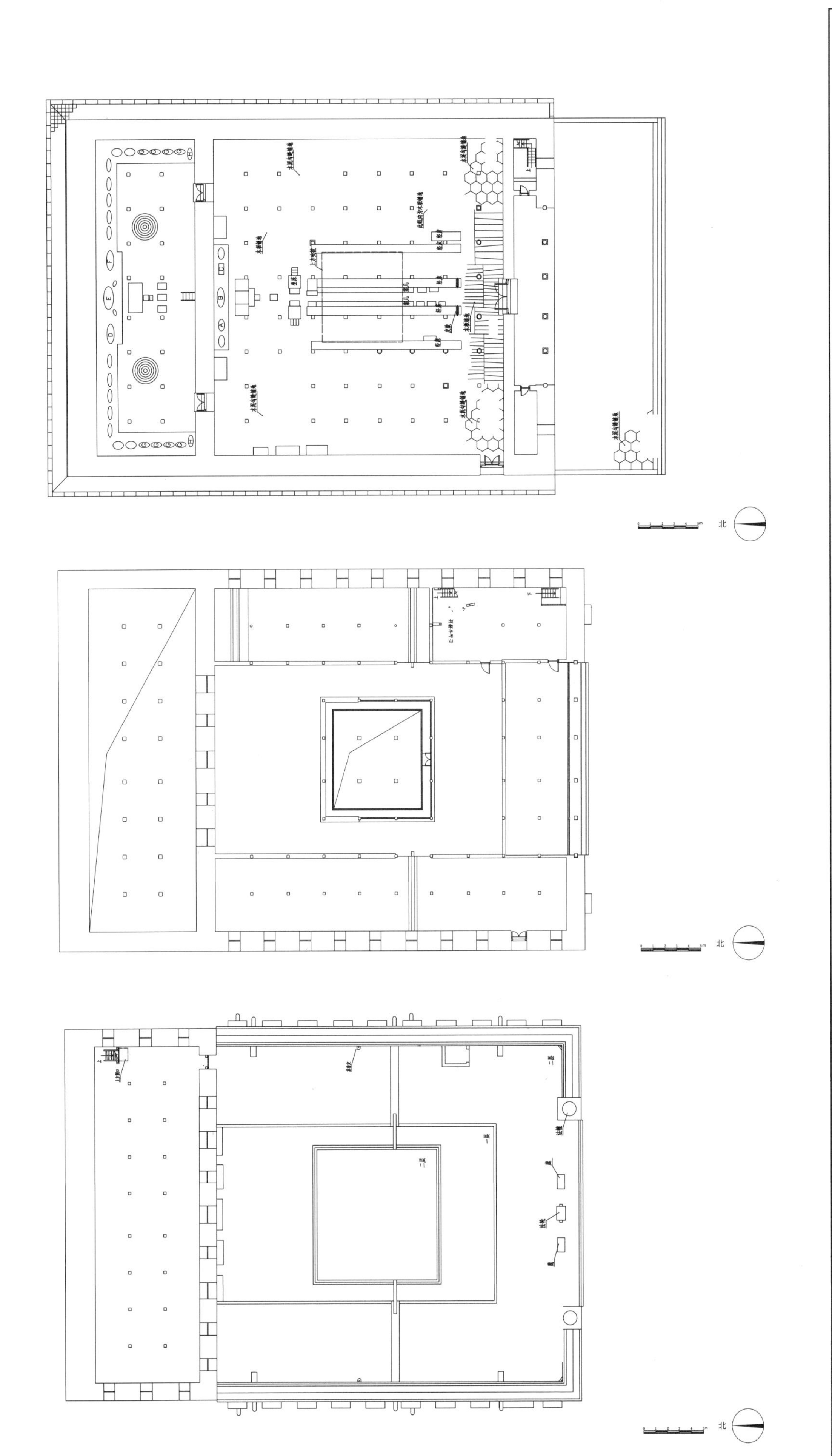

大雄宝殿一层平面图

大雄宝殿二层平面图

大雄宝殿三层平面图

4.2 昆都仑召·四大天王殿

单位:毫米

建筑名称	汉语正式名称	四大天王殿	俗称	——													
概述	初建年	不详	建筑朝向	南	建筑层数	一											
	建筑简要描述	汉藏混合，藏式结构体系，前廊为汉式柱廊，前廊部分汉式歇山顶与建筑主体相接															
	重建重修记载	无															
		信息来源	实地调研														
结构规模	结构形式	砖木混合	相连的建筑	无	室内天井	无											
	建筑平面形式	长方形	外廊形式	前廊													
	通面阔	——	开间数	——	明间	——	次间	——	梢间	——	次梢间	——	尽间	——			
	通进深	——	进深数	——	进深尺寸（前→后）	——											
	柱子数量	2	柱子间距	横向尺寸	2770	（藏式建筑结构体系填写此栏，不含廊柱）											
				纵向尺寸	——												
	其他	无															
建筑主体（大木作）（石作）（瓦作）	屋顶	屋顶形式	密肋平顶	瓦作	抹灰屋面												
	外墙	主体材料	青砖	材料规格	不详	饰面颜色	白色										
		墙体收分	有	边玛檐墙	有	边玛材料	红砖										
	斗栱、梁架	斗栱	无	平身科斗口尺寸	无	梁架关系	二柱三梁										
	柱、柱式（前廊柱）	形式	汉式	柱身断面形状	圆形	断面尺寸	直径 $D=220$	（在没有前廊柱的情况下，填写室内柱及其特征）									
		柱身材料	木材	柱身收分	无	栌斗、托木	无	雀替	有								
		柱础	无	柱础形状	——	柱础尺寸	——										
	台基	台基类型	普通台基	台基高度	870	台基地面铺设材料	石材										
	其他	无															
装修（小木作）（彩画）	门(正面)	板门	门楣	无	堆经	有	门帘	无									
	窗（正面）	无	窗楣	——	窗套	——	窗帘	——									
	室内隔扇	隔扇	无	隔扇位置	——												
	室内地面、楼面	地面材料及规格	石材450×450	楼面材料及规格	——												
	室内楼梯	楼梯	无	楼梯位置	——	楼梯材料	——	梯段宽度	——								
	天花、藻井	天花	有	天花类型	海漫天花	藻井	无	藻井类型	——								
	彩画	柱头	有	柱身	有	梁架	有	走马板	无								
		门、窗	无	天花	无	藻井	——	其他彩画	无								
	其他	悬塑	无	佛龛	无	匾额	无										
装饰	室内	帷幔	无	幕帘彩绘	有	壁画	有	唐卡	无								
		经幡	无	经幢	无	柱毯	无	其他									
	室外	玛尼轮	无	苏勒德	无	宝顶	有	祥麟法轮	无								
		四角经幢	无	经幡	有	铜饰	无	石刻、砖雕	有								
		仙人走兽	一兽	壁画	有	其他	无										
陈设	室内	主佛像	四大天王	佛像基座	——												
		法座	无	藏经橱	无	经床	无	诵经桌	无	法鼓	无	玛尼轮	无	坛城	无	其他	无
	室外	旗杆	有	苏勒德	无	狮子	无	经幡	有	玛尼轮	无	香炉	有	五供	无	其他	无
	其他	无															
备注	——																
调查日期	2010/10/10	调查人员	韩瑛、卢文娟	整理日期	2010/10/10	整理人员	卢文娟										

四大天王殿基本概况表1

昆都仑召·四大天王殿·档案照片					
照片名称	正立面	照片名称	斜前方	照片名称	背立面
照片名称	斜后方	照片名称	正门	照片名称	室外柱础
照片名称	柱廊	照片名称	台基	照片名称	香炉
照片名称	屋顶装饰	照片名称	宝顶	照片名称	室内柱头
照片名称	室内佛像1	照片名称	室内佛像2	照片名称	室内壁画
备注	—				
摄影日期	2010/10/10	摄影人员	孟祎军		

四大天王殿斜前方

四大天王殿正前方

四大天王殿斜后方（左图）
四大天王殿室内柱头（右图）

4.3 昆都仑召·小黄庙

单位:毫米

建筑名称	汉语正式名称	——	俗称	小黄庙													
概述	初建年	清雍正七年（1729年）	建筑朝向	南	建筑层数	一
	建筑简要描述	小黄庙建筑为前经堂后佛殿式。经堂是典型的藏式风格，门楼及佛殿部分是典型的汉式佛寺建筑风格
	重建重修记载	不详
		信息来源	实地调研
结构规模	结构形式	砖木混合	相连的建筑	无	室内天井	无
	建筑平面形式	方形	外廊形式	前后廊		
	通面阔		开间数	——	明间	——	次间	——	梢间	——	次梢间	——	尽间	——
	通进深		进深数	——	进深尺寸（前→后）	——
	柱子数量	10	柱子间距	横向尺寸	3000、700、1410	（藏式建筑结构体系填写此栏，不含廊柱）
				纵向尺寸	2740、440、1610	
	其他	无
建筑主体（大木作）（石作）（瓦作）	屋顶	屋顶形式	密肋平顶，局部歇山顶	瓦作	抹灰屋面、布瓦
	外墙	主体材料	青砖	材料规格	300×140×70	饰面颜色	土黄
		墙体收分	有	边玛檐墙	有	边玛材料	砖
	斗栱、梁架	斗栱	有	平身科斗口尺寸	不详	梁架关系	梁架关系很复杂，无法确定
	柱、柱式（前廊柱）	形式	汉式	柱身断面形状	圆形	断面尺寸	直径 $D=250$	（在没有前廊柱的情况下，填写室内柱及其特征）
		柱身材料	木材	柱身收分	有	栌斗、托木	无	雀替	无	
		柱础	有	柱础形状	方形	柱础尺寸	周长 $C=1520$	
	台基	台基类型	普通台基	台基高度	570	台基地面铺设材料	石材
	其他	无
装修（小木作）（彩画）	门(正面)	板门	门楣	无	堆经	无	门帘	无
	窗（正面）	牖窗	窗楣	无	窗套	有	窗帘	无
	室内隔扇	隔扇	有	隔扇位置	中部
	室内地面、楼面	地面材料及规格	木板、规格不详	楼面材料及规格	——
	室内楼梯	楼梯	无	楼梯位置	——	楼梯材料	——	楼段宽度	——
	天花、藻井	天花	有	天花类型	彻上明造	藻井	有	藻井类型	八方型
	彩画	柱头	有	柱身	有	梁架	有	走马板	无
		门、窗	无	天花	无	藻井	无	其他彩画	无
	其他	悬塑	无	佛龛	有	匾额	——
装饰	室内	帷幔	无	幕帘彩绘	无	壁画	有	唐卡	有
		经幡	无	经幢	有	柱毯	无	其他	无
	室外	玛尼轮	无	苏勒德	无	宝顶	有	祥麟法轮	无
		四角经幢	无	经幡	无	铜饰	无	石刻、砖雕	有
		仙人走兽	1+2	壁画	无	其他	——
陈设	室内	主佛像	四臂观音	佛像基座	莲花座
		法座	无	藏经橱	无	经床	有	诵经桌	有	法鼓	无	玛尼轮	无	坛城	无	其他	无
	室外	旗杆	无	苏勒德	无	狮子	无	经幡	无	玛尼轮	无	香炉	无	五供	无	其他	无
	其他	——
备注	——
调查日期	2010/10/10	调查人员	韩瑛、卢文娟	整理日期	2010/10/10	整理人员	卢文娟

小黄庙基本概况表1

昆都仑召 · 小黄庙 · 档案照片

照片名称	正立面	照片名称	侧立面	照片名称	斜后方

照片名称	背立面	照片名称	正门	照片名称	室外角柱

照片名称	室外柱础	照片名称	宝顶	照片名称	室内正面

照片名称	室内天花	照片名称	室内柱头	照片名称	佛龛

照片名称	藻井	照片名称	佛像	照片名称	壁画

备注	——				
摄影日期	2010/10/10	摄影人员	孟祎军		

小黄庙基本概况表2

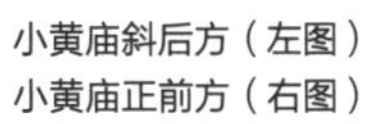

小黄庙斜后方（左图）
小黄庙正前方（右图）

小黄庙室内正面（左上图）
小黄庙吻兽（左下图）
小黄庙宝顶（左中图）
小黄庙佛像（右图）

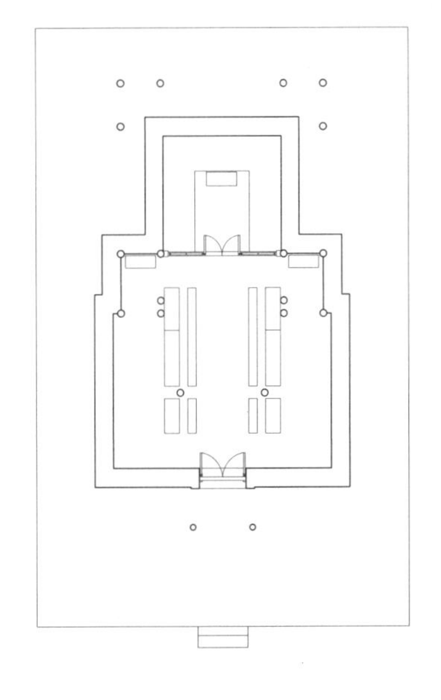

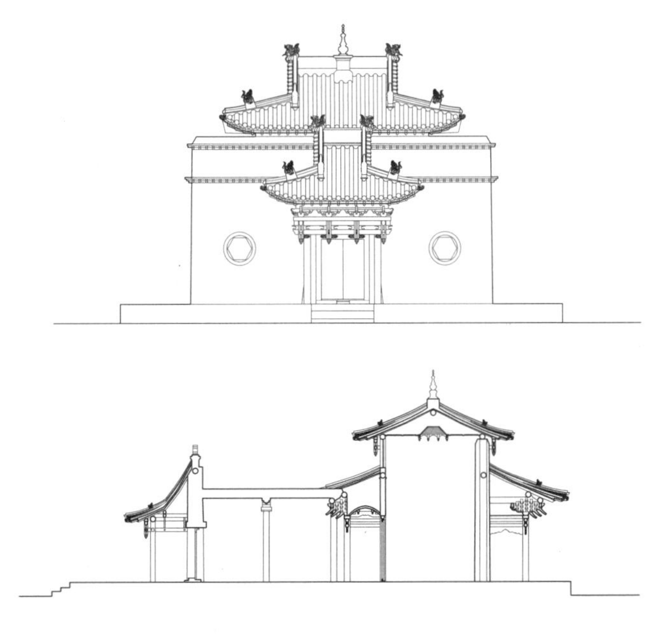

小黄庙一层平面图
（左图）
小黄庙正立面图
（右上图）
小黄庙剖面图
（右下图）

4.4 昆都仑召·时轮殿

时轮殿斜前方

4.5 昆都仑召·度母殿

度母殿斜前方

4.6 昆都仑召 · 东活佛府

东活佛府斜前方

东活佛府前廊

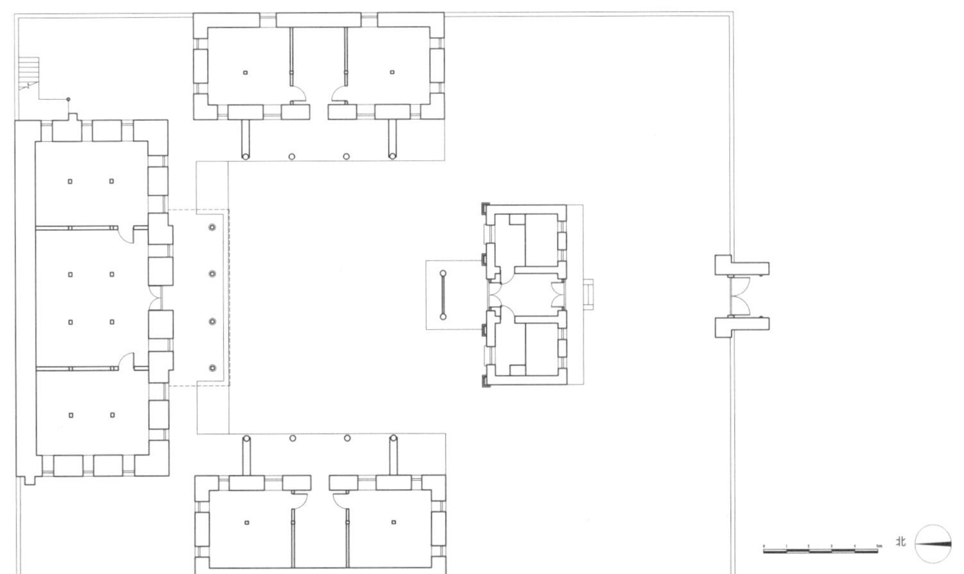

东活佛府平面图

东活佛府正立面图

4.7 昆都仑召·王爷府

王爷府院落

4.8 昆都仑召·白塔

白塔

5 百灵庙

Batahalaga Temple

5 百灵庙 Batahalaga Temple

百灵庙建筑群

百灵庙为原乌兰察布盟喀尔喀右翼旗（俗称达尔罕贝勒旗）寺庙，系该旗旗庙。康熙年间清廷御赐满、蒙古、汉、藏四体“广福寺”匾额。该寺地处漠南通往漠北大库伦及新疆的交通要道，商旅云集。近代内蒙古史上的重要历史事件多发生于此庙。

康熙三十九年（1700年），喀尔喀右翼部第一任扎萨克和硕达尔罕亲王本塔尔之子诺内到五台山朝圣，请回文殊菩萨像一尊及佛经一部。并邀请多伦诺尔甘珠尔活佛甄选庙址，请归化城席力图召第四世活佛阿格旺若布坦设计殿宇式样与规模，于康熙四十二年（1703年），大兴土木，在本塔尔亲王冬营盘乌力吉陶亥之地，按照归化城小召式样新建寺庙。最初，作为诺内亲王家庙，只有一座大雄宝殿，后陆续扩建为规模宏大的旗庙。据传，康熙帝称乌力吉陶亥为地势险峻的要塞通道，故改称此地为巴图哈拉嘎，由此习称此庙为巴图哈拉嘎庙，又因诺内之子执政时清廷免去其亲王爵位，改封为贝勒，故此庙又称贝勒庙，汉译为白林庙或百灵庙。

寺庙建筑风格以汉式建筑为主，兼有藏式建筑。至中华人民共和国成立前该庙有三层大雄宝殿、显宗殿、密宗殿、医药殿、时轮殿5大殿宇，班禅行宫等8座活佛拉布隆，此外有白伞盖佛母殿、天王殿、甘珠尔殿、甘珠尔殿、药师殿及清廷特派达喇嘛住所——孟克斋戒殿等多座殿宇。 寺庙香灯师房舍8座，膳房5座，庙仓5座，由全旗四个佐领各建10间僧舍共计40间以及后来僧人自建的房舍360余座，蒙古包300余座。

“文化大革命”中寺庙严重受损，仅存大雄宝殿、显宗殿、丹珠尔殿、甘珠尔殿、天王殿、密宗殿、医学府7座殿宇。1980年起重新修缮寺庙建筑，经9年建设，于1989年10月21日，正式恢复法会。

参考文献：

[1] 白春花等.百灵庙（蒙古文）.呼和浩特:内蒙古人民出版社,1997，12.

[2] 满都麦,莫德尔图.乌兰察布寺院（蒙古文）.海拉尔:内蒙古文化出版社,1996，5.

百灵庙坛城

<table>
<tr><th colspan="8">百灵庙 · 基本概况 1</th></tr>
<tr><td rowspan="3">寺院蒙古语藏语名称</td><td>蒙古语</td><td>[illegible]</td><td rowspan="2">寺院汉语名称</td><td>汉语正式名称</td><td colspan="3">广福寺</td></tr>
<tr><td>藏语</td><td>——</td><td>俗称</td><td colspan="3">百灵庙</td></tr>
<tr><td>汉语语义</td><td>贝勒庙</td><td colspan="2">寺院汉语名称的由来</td><td colspan="3">清廷赐名</td></tr>
<tr><td>所在地</td><td colspan="4">包头市达尔罕茂明安联合旗百灵庙镇</td><td>东经 109° 44′</td><td colspan="2">北纬 41° 30′</td></tr>
<tr><td>初建年</td><td colspan="2">清康熙四十二年（1702年）</td><td>保护等级</td><td colspan="4">自治区级文物保护单位</td></tr>
<tr><td>盛期时间</td><td colspan="2">清道光三十年（1850年）</td><td colspan="2">盛期喇嘛僧/现有喇嘛僧数</td><td colspan="3">约1500/25</td></tr>
<tr><td rowspan="2">历史沿革</td><td colspan="7">1701年，从阴山、大库伦等地驮运建材，从归化、山西等地请来工匠。
1703年，正式开工建设寺庙。
1705年，经三年零两月的时间建成大雄宝殿。
1738年，首次扩建寺庙，新建显宗殿。
1740年，建立密宗学部与医药学部。
1742年，新建密宗殿。
1864年，新建医药殿。
1872年，新建时轮殿。
1911年后，丑牛之乱中6座殿堂被革命军烧毁。
1914—1926年，修复5座殿堂。
1932—1933年，九世班禅驻跸于此庙。
1936—1937年，国民党军队掠夺寺庙财产，炸毁部分建筑。寺庙受损后旗衙将五大学部分别部署于旗内五座寺庙，维持了法事的延续。
1942年，新建2座佛塔。
1966年，寺庙严重受损，仅存大雄宝殿等大小7座殿宇。
1980年，开始修缮寺庙。
1987年左右，拆除密宗殿与医学府。
1989年，正式恢复法会。
2003年，新建东西配殿，东西侧门及两座白塔</td></tr>
<tr><td>资料来源</td><td colspan="6">[1] 白春花等.百灵庙（蒙古文）.呼和浩特:内蒙古人民出版社,1997，12.
[2] 满都麦,莫德尔图.乌兰察布寺院（蒙古文）.海拉尔:内蒙古文化出版社,1996,5.
[3] 调研访谈记录 .</td></tr>
<tr><td rowspan="2">现状描述</td><td colspan="4" rowspan="2">现百灵庙大雄宝殿（大雄宝殿）与显宗殿、丹珠尔殿、甘珠尔殿、四大天王殿基本完好局部有破损，其中两座佛塔，东西配殿、东西侧门等为现代复建</td><td>描述时间</td><td colspan="2">2010/07/28</td></tr>
<tr><td>描述人</td><td colspan="2">张宇</td></tr>
<tr><td>调查日期</td><td colspan="2">2010/07/28</td><td>调查人员</td><td colspan="4">张宇、何鑫、卢文娟</td></tr>
</table>

百灵庙基本概况表1

百灵庙·基本概况 2

<table>
<tr><td rowspan="4">现存建筑</td><td>大雄宝殿</td><td>显宗殿</td><td>丹珠尔殿</td><td rowspan="4">已毁建筑</td><td>朱德布殿（密宗殿）</td><td colspan="2">吉如海殿</td></tr>
<tr><td>甘珠尔殿</td><td>九世班禅府邸</td><td>四大天王殿</td><td>时轮殿</td><td colspan="2">门巴殿</td></tr>
<tr><td>新东配殿</td><td>新西配殿</td><td>东西侧门</td><td>医学府</td><td colspan="2">——</td></tr>
<tr><td>两座白塔</td><td>——</td><td>——</td><td>信息来源</td><td colspan="2">《百灵庙修缮工程修复方案》、大喇嘛口述</td></tr>
<tr><td>区位图</td><td colspan="7"></td></tr>
<tr><td>总平面图</td><td colspan="7">
A.显宗殿　D.甘珠尔殿　G.服务用房　J.喇嘛僧舍
B.大雄宝殿　E.丹珠尔殿　H.四大天王殿　K.已毁院落
C.白 塔　F.管理用房　I.九世班禅府邸</td></tr>
<tr><td>调查日期</td><td colspan="3">2010/07/29</td><td>调查人员</td><td colspan="3">张宇、卢文娟</td></tr>
</table>

百灵庙基本概况表2

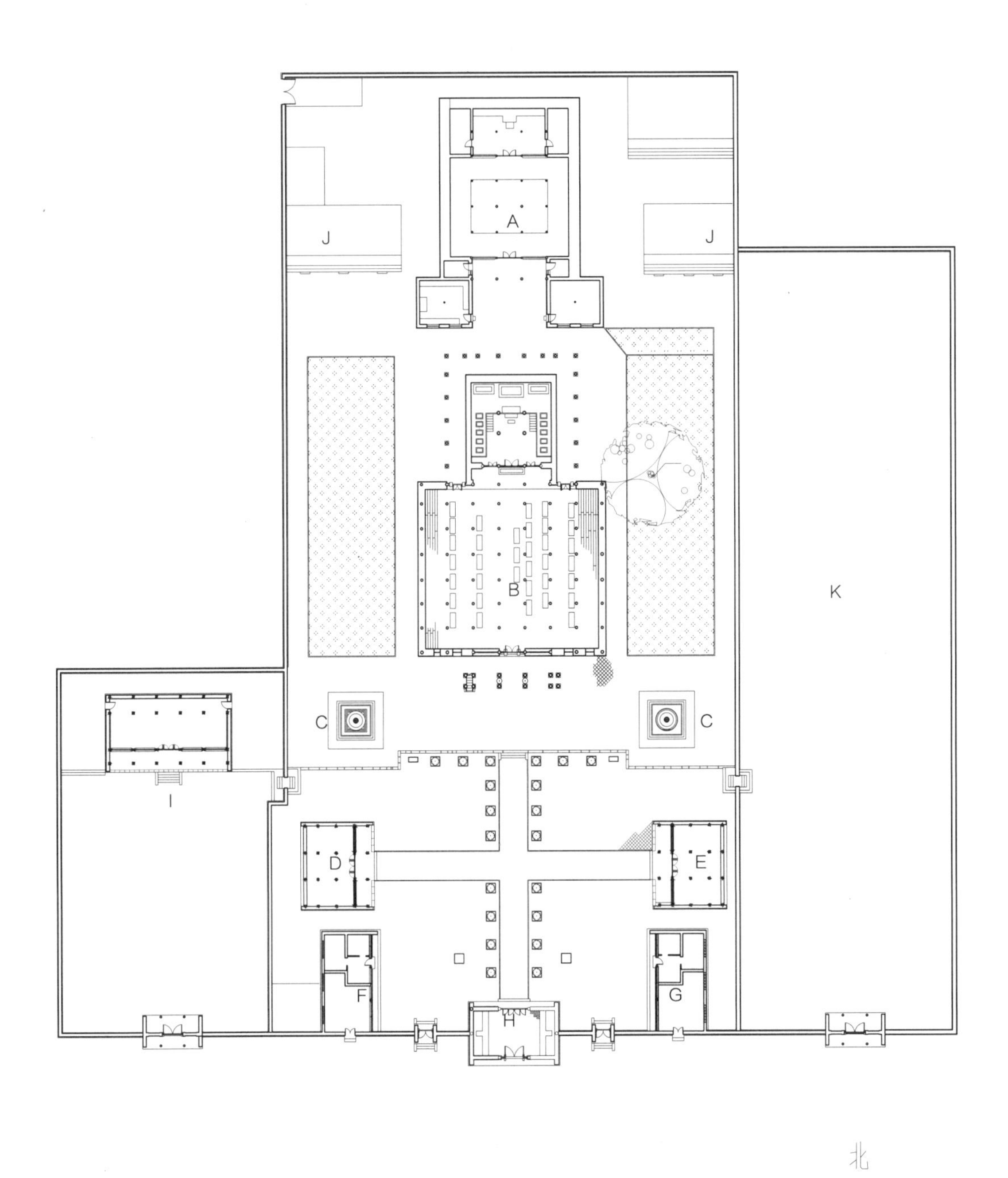

A.显宗殿　D.甘珠尔殿　G.服务用房　J.喇嘛僧舍
B.大雄宝殿　E.丹珠尔殿　H.四大天王殿　K.已毁院落
C.白 塔　F.管理用房　I.九世班禅府邸

百灵庙总平面图

5.1 百灵庙·大雄宝殿

单位:毫米

建筑名称	汉语正式名称	大雄宝殿	俗称	大雄宝殿													
概述	初建年	不详	建筑朝向	南	建筑层数	二											
	建筑简要描述	汉藏结合，由三大空间组成，二层歇山顶由南至北逐次升高															
	重建重修记载	2003—2006年，室内壁画、顶棚进行了维护与修缮															
		信息来源	大喇嘛口述														
结构规模	结构形式	砖木混合	相连的建筑	无	室内天井	经堂都纲法式											
	建筑平面形式	凸字形	外廊形式	前后廊													
	通面阔	23040	开间数	7	明间	3150	次间	3150	梢间	3150	次梢间	——	尽间	3150			
	通进深	33630	进深数	14	进深尺寸（前→后）	2950→2875→2800→3040→3060→3000→2300→2165→2750→2700→2750→2200											
	柱子数量	经堂36根 佛殿4根	柱子间距	横向尺寸	——	（藏式建筑结构体系填写此栏，不含廊柱）											
				纵向尺寸	——												
	其他	——															
建筑主体（大木作）（石作）（瓦作）	屋顶	屋顶形式	佛殿为重檐歇山、南面两殿为一层藏式平顶二层歇山	瓦作	布瓦												
	外墙	主体材料	青砖	材料规格	300×140×65	饰面颜色	红、白										
		墙体收分	有	边玛檐墙	有	边玛材料	砖										
	斗栱、梁架	斗栱	无	平身科斗口尺寸	无	梁架关系	不详										
	柱、柱式（前廊柱）	形式	汉式	柱身断面形状	圆	断面尺寸	直径$D=350$	（在没有前廊柱的情况下，填写室内柱及其特征）									
		柱身材料	木材	柱身收分	有	栌斗、托木	无	雀替	有								
		柱础	有	柱础形状	圆	柱础尺寸	直径$D=500$										
	台基	台基类型	普通台基	台基高度	600	台基地面铺设材料	方砖										
	其他	——															
装修（小木作）（彩画）	门(正面)	板门	门楣	无	堆经	无	门帘	无									
	窗（正面）	槛窗/藏式盲窗	窗楣	无	窗套	无	窗帘	有									
	室内隔扇	隔扇	有	隔扇位置	在经堂与佛殿之间												
	室内地面、楼面	地面材料及规格	木板3950×（220~340）	楼面材料及规格	——												
	室外楼梯	楼梯	有	楼梯位置	南面偏东	楼梯材料	木材	梯段宽度	1280								
	天花、藻井	天花	有	天花类型	井口	藻井	有	藻井类型	四方变八方								
	彩画	柱头	有	柱身	有	梁架	有	走马板	无								
		门、窗	有	天花	有	藻井	有	其他彩画	——								
	其他	悬塑	无	佛龛	无	匾额	广福寺										
装饰	室内	帷幔	有	幕帘彩绘	无	壁画	有	唐卡	有								
		经幡	有	经幢	有	柱毯	无	其他									
	室外	玛尼轮	无	苏勒德	有	宝顶	有	祥麟法轮	有								
		四角经幢	有	经幡	有	铜饰	有	石刻、砖雕	无								
		仙人走兽	无仙人+3	壁画	无	其他	——										
陈设	室内	主佛像	三世佛	佛像基座	弥须座												
		法座	有	藏经橱	无	经床	有	诵经桌	有	法鼓	有	玛尼轮	无	坛城	有	其他	——
	室外	旗杆	有	苏勒德	无	狮子	有	经幡	有	玛尼轮	有	香炉	有	五供	有	其他	——
	其他	——															
备注	——																
调查日期	2010/07/30	调查人员	张宇、何鑫	整理日期	2010/07/31	整理人员	张宇、卢文娟										

大雄宝殿基本概况表1

百灵庙·大雄宝殿·档案照片

照片名称	鸟瞰图	照片名称	斜前方	照片名称	正立面
照片名称	前廊侧面	照片名称	佛殿侧立面	照片名称	室外柱头
照片名称	后廊柱	照片名称	台基	照片名称	正门
照片名称	室内天花	照片名称	室内局部	照片名称	室内柱身
照片名称	藻井	照片名称	佛像	照片名称	坛城
备注	—				
摄影日期	2010/07/31	摄影人员	高旭		

大雄宝殿斜前方
（左图）
大雄宝殿室外梯子
（右图）

大雄宝殿室内空间
（左图）
大雄宝殿室外柱子
（右图）

大雄宝殿室内陈设
（左图）
大雄宝殿匾额（右上图）
大雄宝殿藻井（右下图）

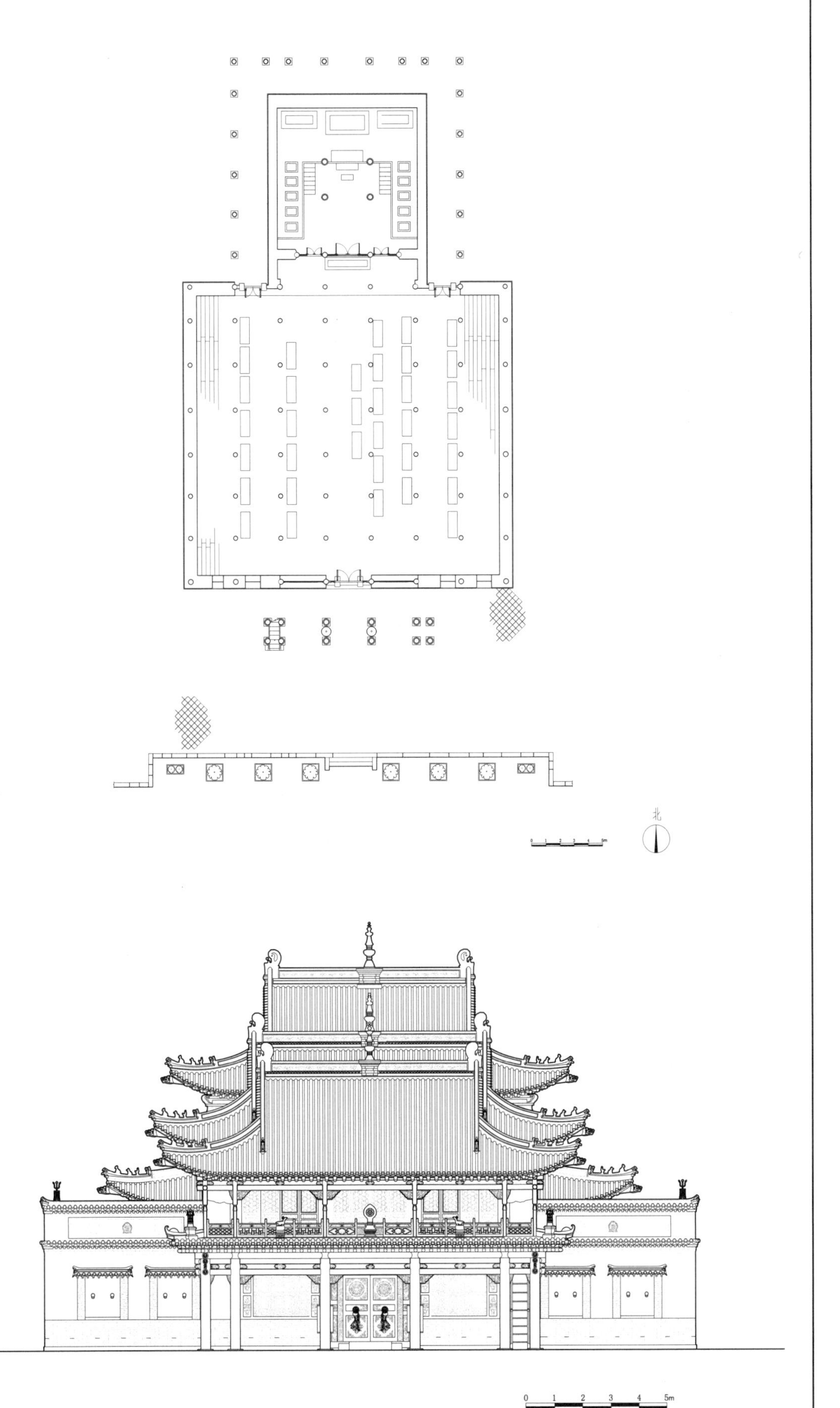

大雄宝殿一层平面图

大雄宝殿南立面图

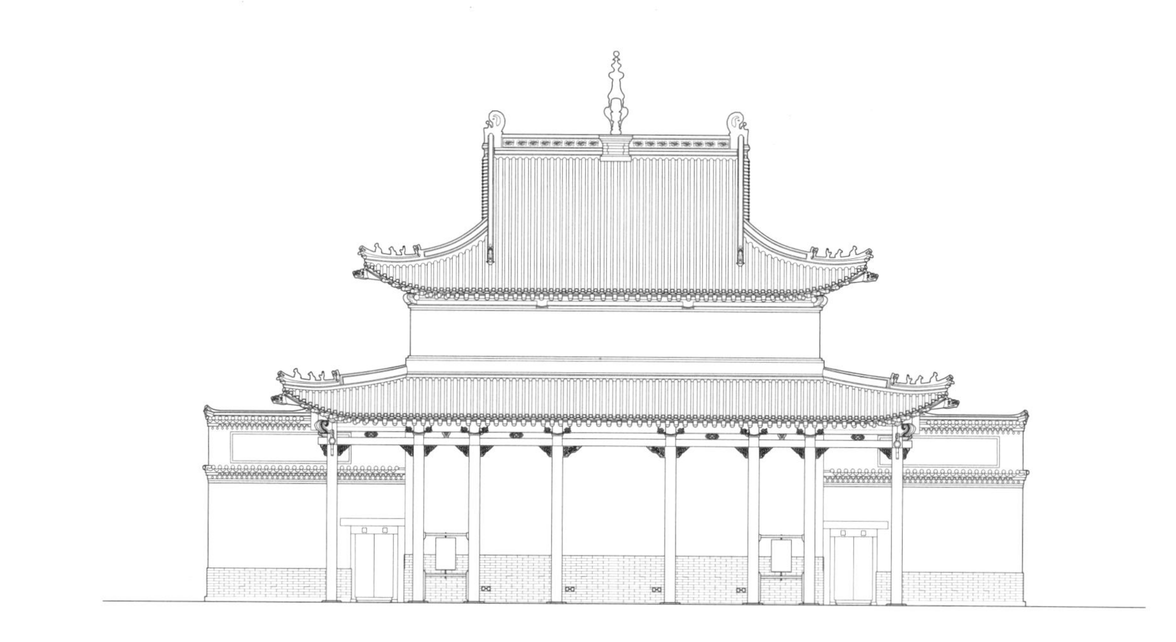

大雄宝殿北立面图

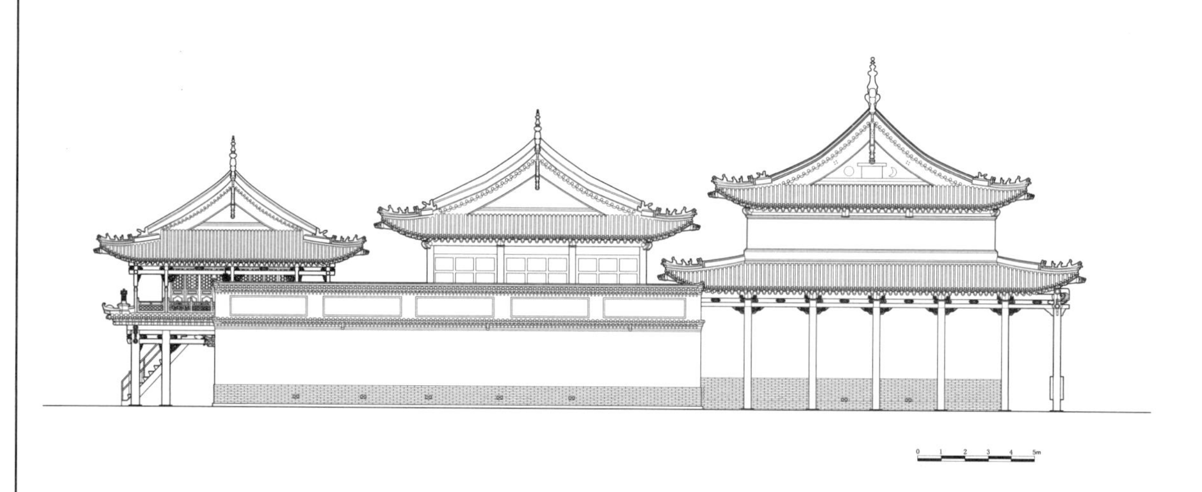

大雄宝殿侧立面图

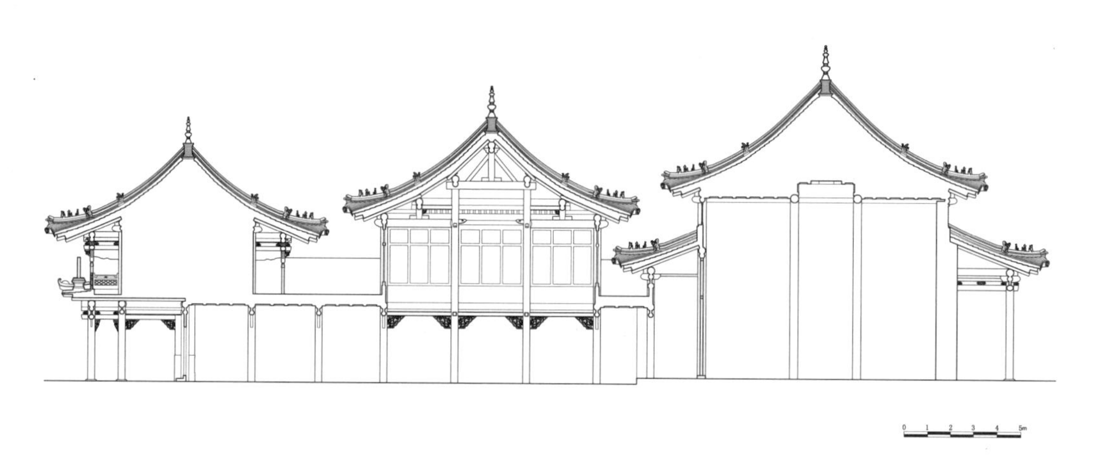

大雄宝殿剖面图

5.2 百灵庙 · 显宗殿

单位:毫米

建筑名称	汉语正式名称	法相殿			俗称	却伊拉殿					
概述	初建年	不详			建筑朝向	南		建筑层数	一		
	建筑简要描述	纯藏式风格、前经堂后佛殿									
	重建重修记载	2003—2006年间，室内壁画、顶棚进行了维护与修缮									
		信息来源	大喇嘛口述								
结构规模	结构形式	砖木混合		相连的建筑	无			室内天井	都刚法式		
	建筑平面形式	凹字形		外廊形式	前廊						
	通面阔	22800	开间数	7	明间	——	次间	——	梢间	——	次梢间 —— 尽间 ——
	通进深	27340	进深数	9	进深尺寸（前→后）			——			
	柱子数量	经堂12根 佛殿2根	柱子间距	横向尺寸	3000			（藏式建筑结构体系填写此栏，不含廊柱）			
				纵向尺寸	2770、2530、3078、3040						
	其他	——									
建筑主体（大木作）（石作）（瓦作）	屋顶	屋顶形式	密肋平顶			瓦作	布瓦				
	外墙	主体材料	青砖	材料规格	280×135×60		饰面颜色	白			
		墙体收分	有	边玛檐墙	有		边玛材料	砖			
	斗栱、梁架	斗栱	无	平身科斗口尺寸		无	梁架关系	不详			
	柱、柱式（前廊柱）	形式	藏式	柱身断面形状	十二楞柱	断面尺寸	225×225			（在没有前廊柱的情况下，填写室内柱及其特征）	
		柱身材料	木材	柱身收分	无	栌斗、托木	有	雀替	无		
		柱础	不详	柱础形状	不详	柱础尺寸	——				
	台基	台基类型	无	台基高度	——	台基地面铺设材料		——			
	其他	——									
装修（小木作）（彩画）	门(正面)	格栅门		门楣	无	堆经	无	门帘	无		
	窗（正面）	槛窗/藏式窗		窗楣	有	窗套	无	窗帘	有		
	室内隔扇	隔扇	有	隔扇位置	经堂与佛殿之间						
	室内地面、楼面	地面材料及规格		木板3800×（250~360）		楼面材料及规格		无			
	室内楼梯	楼梯	无	楼梯位置	——	楼梯材料	——	梯段宽度	——		
	天花、藻井	天花	有	天花类型	井口	藻井	无	藻井类型	——		
	彩画	柱头	有	柱身	有	梁架	有	走马板	无		
		门、窗	有	天花	有	藻井	无	其他彩画	——		
	其他	悬塑	无	佛龛	无	匾额	无				
装饰	室内	帷幔	有	幕帘彩绘	有	壁画	有	唐卡	有		
		经幡	无	经幢	有	柱毯	无	其他	哈达		
	室外	玛尼轮	无	苏勒德	有	宝顶	有	祥麟法轮	有		
		四角经幢	有	经幡	无	铜饰	无	石刻、砖雕	有		
		仙人走兽	无	壁画	无	其他	——				
陈设	室内	主佛像	弥勒佛			佛像基座	莲花座				
		法座 有	藏经橱 有	经床 有	诵经桌 有	法鼓 无	玛尼轮 有	坛城 无	其他 ——		
	室外	旗杆 无	苏勒德 无	狮子 无	经幡 无	玛尼轮 无	香炉 有	五供 有	其他 水缸		
	其他	——									
备注	——										
调查日期	2010/07/30	调查人员	张宇、何鑫	整理日期	2010/07/30	整理人员	张宇、卢文娟				

显宗殿基本概况表1

百灵庙·显宗殿·档案照片					
照片名称 	鸟瞰图	照片名称 	斜前方	照片名称 	正立面局部
照片名称 	柱廊	照片名称 	室外柱身	照片名称 	室外柱头
照片名称 	南立面藏式窗	照片名称 	正门	照片名称 	窗户
照片名称 	室内正面	照片名称 	室内侧面	照片名称 	经堂与佛殿柱
照片名称 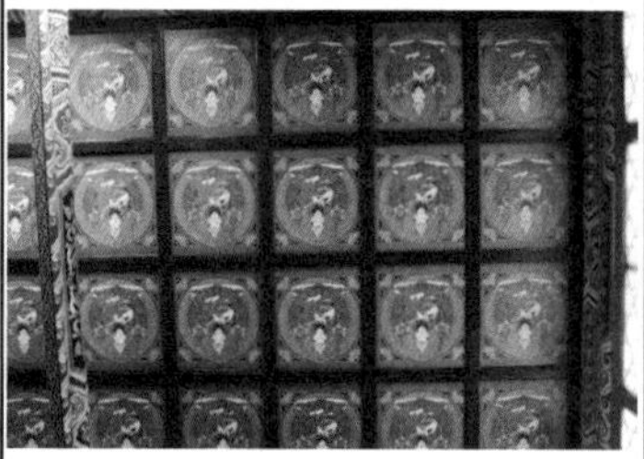	室内天花	照片名称 	室内高窗	照片名称 	主佛像
备注	——				
摄影日期	2010/07/30	摄影人员	高旭		

显宗殿基本概况表2

显宗殿前廊正面

显宗殿室内空间（左上图）
显宗殿侧面（右上图）

显宗殿室外柱子（左下图）
显宗殿都纲法式（右下图）

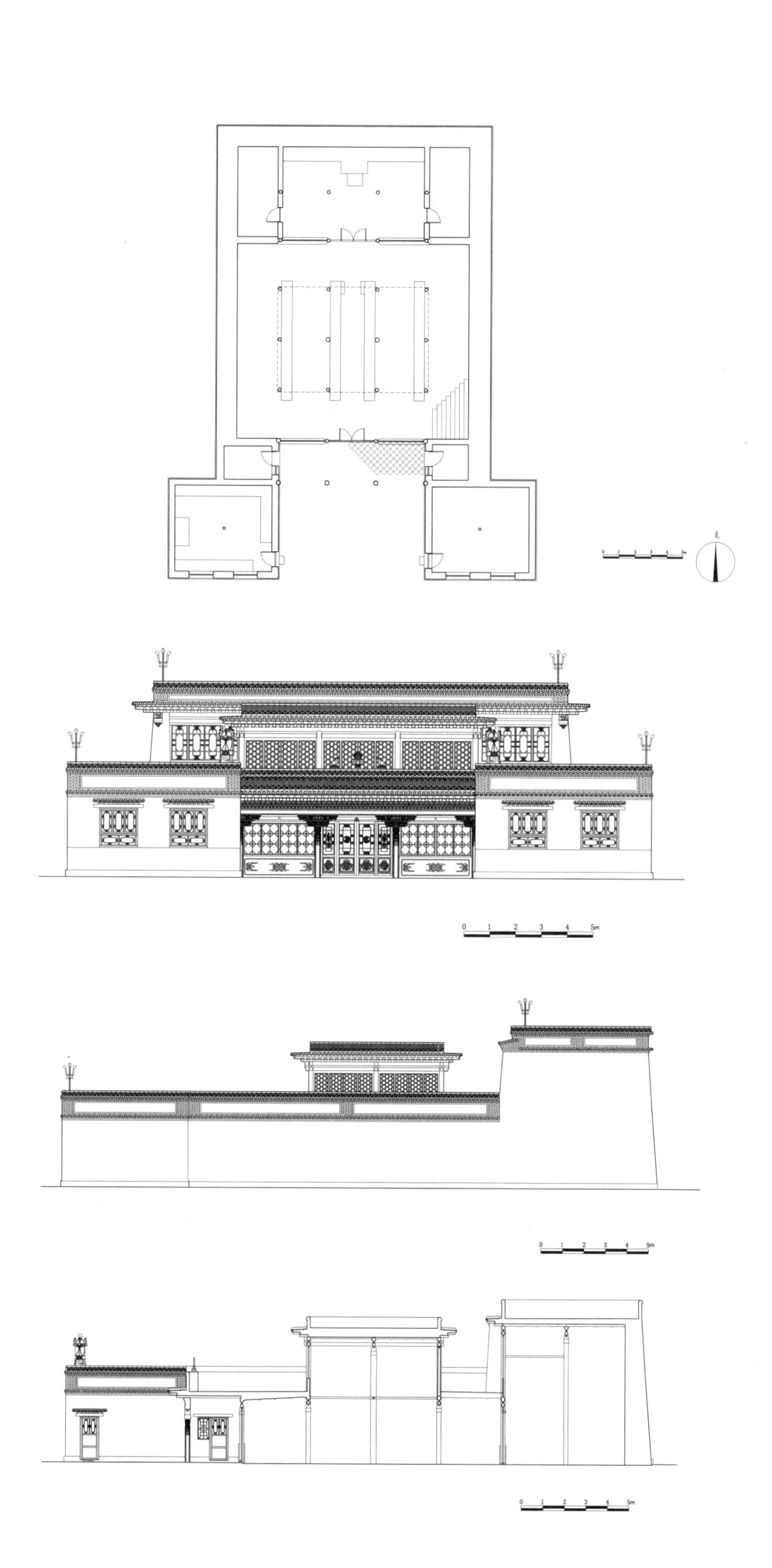

显宗殿一层平面图

显宗殿正立面图

显宗殿侧立面图

显宗殿剖面图

5.3 百灵庙·四大天王殿

单位:毫米

建筑名称	汉语正式名称	四大天王殿	俗称	——													
概述	初建年	约清康熙四十二年（1703年）	建筑朝向	南	建筑层数	一											
	建筑简要描述	汉式建筑，是该寺中保存原貌最完好者，今作为四大天王殿兼山门使用															
	重建重修记载	2003年彩画翻新															
		信息来源	大喇嘛														
结构规模	结构形式	木结构	相连的建筑	无	室内天井	无											
	建筑平面形式	长方形	外廊形式	无													
	通面阔	10860	开间数	3	明间	3600	次间	3400	梢间	——	次梢间	——	尽间	——			
	通进深	7390	进深数	2	进深尺寸（前→后）	2875→2925											
	柱子数量	——	柱子间距	横向尺寸	——	（藏式建筑结构体系填写此栏，不含廊柱）											
				纵向尺寸	——												
	其他	——															
建筑主体（大木作）（石作）（瓦作）	屋顶	屋顶形式	硬山	瓦作	布瓦												
	外墙	主体材料	青砖	材料规格	270×135×55	饰面颜色	灰										
		墙体收分	无	边玛檐墙	无	边玛材料	——										
	斗栱、梁架	斗栱	无	平身科斗口尺寸	——	梁架关系	五檩无廊										
	柱、柱式（前廊柱）	形式	汉式	柱身断面形状	圆	断面尺寸	直径$D=360$	（在没有前廊柱的情况下，填写室内柱及其特征）									
		柱身材料	木材	柱身收分	有	栌斗、托木	无	雀替	无								
		柱础	不详	柱础形状	不详	柱础尺寸	不详										
	台基	台基类型	普通台基	台基高度		台基地面铺设材料	条石600×290、方砖60×60										
	其他	——															
装修（小木作）（彩画）	门(正面)	板门	门楣	无	堆经	无	门帘	无									
	窗（正面）	无	窗楣	无	窗套	无	窗帘	无									
	室内隔扇	隔扇	无	隔扇位置	——												
	室内地面、楼面	地面材料及规格	红砖 240×150	楼面材料及规格	——												
	室内楼梯	楼梯	无	楼梯位置	——	楼梯材料	——	梯段宽度	——								
	天花、藻井	天花	无	天花类型	——	藻井	无	藻井类型	——								
	彩画	柱头	有	柱身	无	梁架	有	走马板	无								
		门、窗	有	天花	无	藻井	——	其他彩画	——								
	其他	悬塑	无	佛龛	有	匾额	无										
装饰	室内	帷幔	有	幕帘彩绘	有	壁画	无	唐卡	无								
		经幡	有	经幢	无	柱毯	无	其他	——								
	室外	玛尼轮	无	苏勒德	无	宝顶	有	祥麟法轮	无								
		四角经幢	无	经幡	有	铜饰	无	石刻、砖雕	无								
		仙人走兽	无	壁画	有	其他	——										
陈设	室内	主佛像	四大天王	佛像基座	——												
		法座	无	藏经橱	无	经床	无	诵经桌	无	法鼓	无	玛尼轮	无	坛城	无	其他	——
	室外	旗杆	无	苏勒德	无	狮子	无	经幡	有	玛尼轮	无	香炉	无	五供	无	其他	——
	其他	——															
备注	——																
调查日期	2010/07/31	调查人员	何鑫	整理日期	2010/07/31	整理人员	张宇、何鑫、卢文娟										

四大天王殿基本概况表1

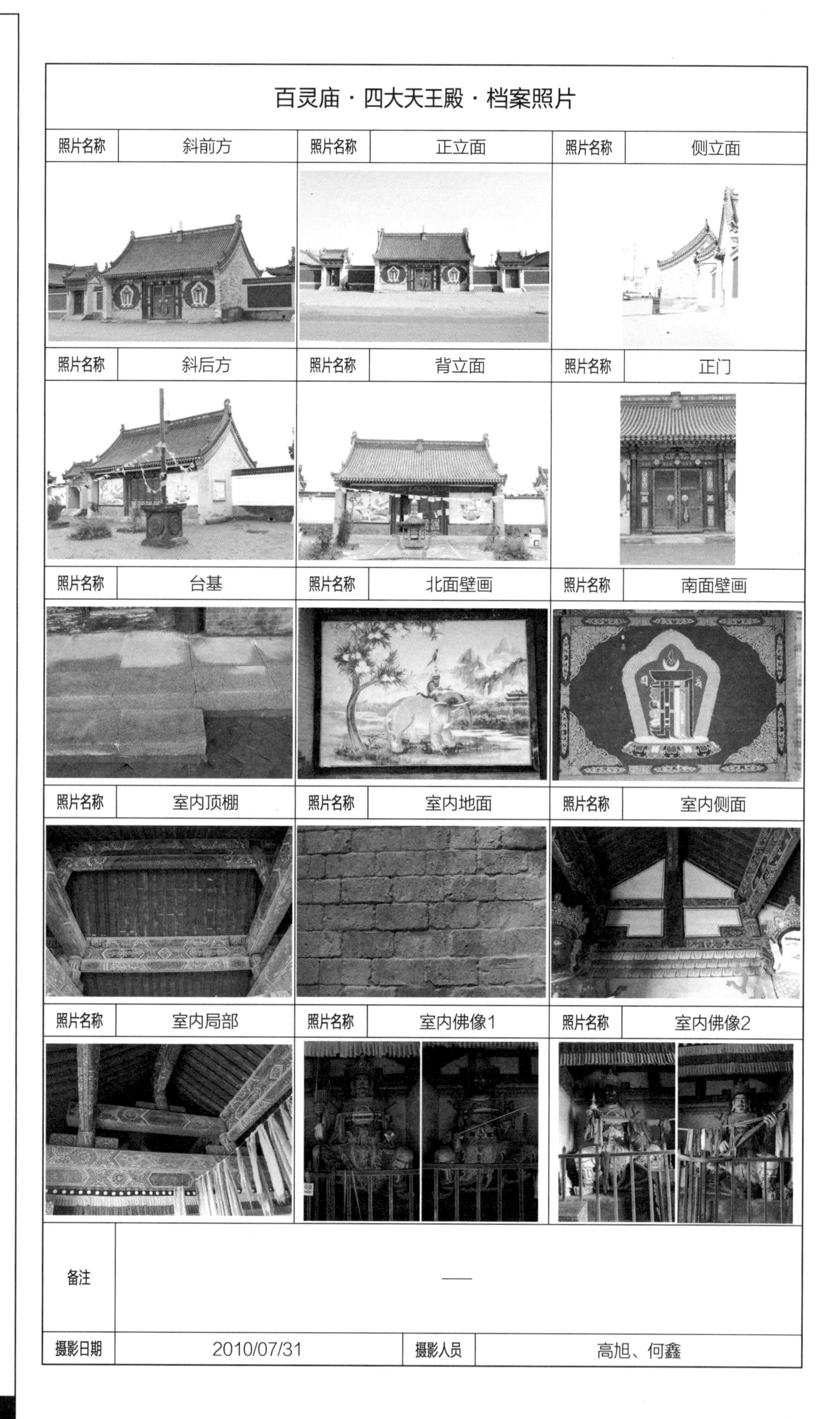

百灵庙·四大天王殿·档案照片					
照片名称	斜前方	照片名称	正立面	照片名称	侧立面
照片名称	斜后方	照片名称	背立面	照片名称	正门
照片名称	台基	照片名称	北面壁画	照片名称	南面壁画
照片名称	室内顶棚	照片名称	室内地面	照片名称	室内侧面
照片名称	室内局部	照片名称	室内佛像1	照片名称	室内佛像2
备注	——				
摄影日期	2010/07/31	摄影人员	高旭、何鑫		

四大天王殿基本概况表2

四大天王殿正前方

四大天王殿室外壁画（左上图）
四大天王殿室内梁架（左下图）
四大天王殿室外柱子（右图）

四大天王殿正立面图

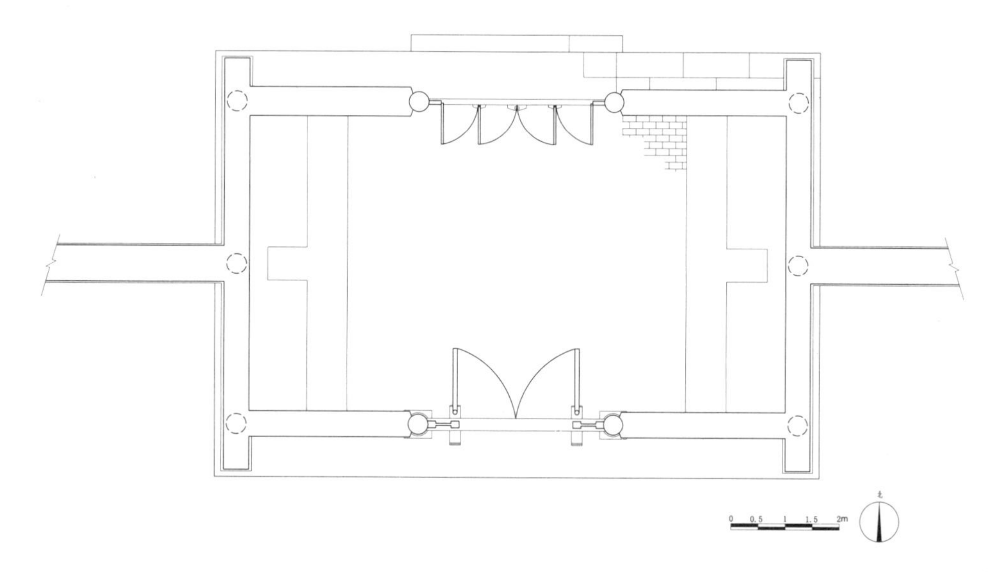

四大天王殿平面图

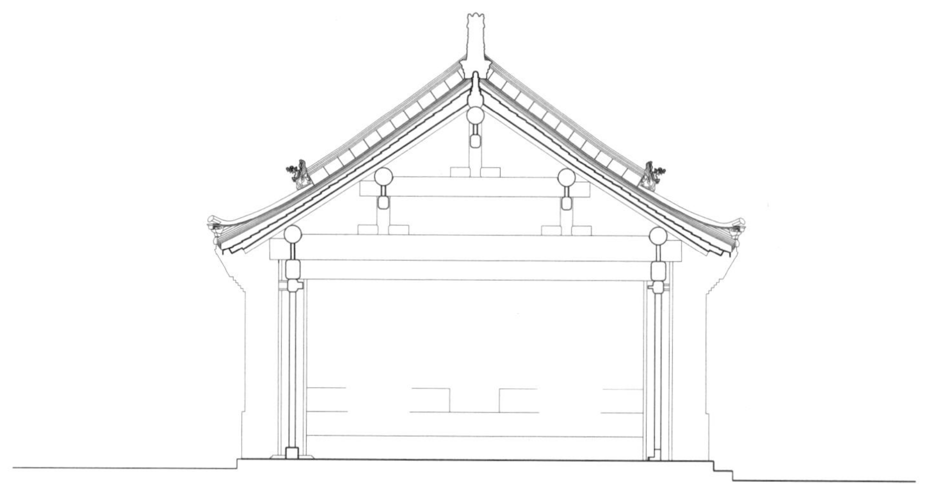

四大天王殿剖面图

5.4 百灵庙·甘珠尔殿

单位:毫米

建筑名称	汉语正式名称	甘珠尔殿	俗称	西藏经阁													
概述	初建年	约清康熙四十二年（1703年）	建筑朝向	东	建筑层数	—											
	建筑简要描述	汉式结构体系，汉式装饰，级别较低															
	重建重修记载	2010年室内装修															
		信息来源	现场观察及大喇嘛口述														
结构规模	结构形式	木结构	相连的建筑	无	室内天井	无											
	建筑平面形式	长方形	外廊形式	前廊													
	通面阔	10390	开间数	3	明间	3150	次间	3220	梢间	——	次梢间	——	尽间	——			
	通进深	8780	进深数	4	进深尺寸（前→后）	1580→2230→2230→1580											
	柱子数量	——	柱子间距	横向尺寸	——	（藏式建筑结构体系填写此栏，不含廊柱）											
				纵向尺寸	——												
	其他	——															
建筑主体（大木作）（石作）（瓦作）	屋顶	屋顶形式	硬山	瓦作	布瓦												
	外墙	主体材料	青砖	材料规格	270×135×55	饰面颜色	灰										
		墙体收分	无	边玛檐墙	无	边玛材料	——										
	斗栱、梁架	斗栱	无	平身科斗口尺寸	——	梁架关系	不详										
	柱、柱式（前廊柱）	形式	汉式	柱身断面形状	圆	断面尺寸	直径D=90	（在没有前廊柱的情况下，填写室内柱及其特征）									
		柱身材料	木材	柱身收分	有	栌斗、托木	无	雀替	无								
		柱础	有	柱础形状	不详	柱础尺寸	不详										
	台基	台基类型	普通台基	台基高度	400	台基地面铺设材料	条石（800×350）、方砖（310×310）										
	其他	——															
装修（小木作）（彩画）	门(正面)	隔扇	门楣	无	堆经	无	门帘	无									
	窗（正面）	槛门	窗楣	无	窗套	无	窗帘	无									
	室内隔扇	隔扇	无	隔扇位置	——												
	室内地面、楼面	地面材料及规格	不详	楼面材料及规格	——												
	室内楼梯	楼梯	无	楼梯位置	——	楼梯材料	——	楼段宽度	——								
	天花、藻井	天花	有	天花类型	井口	藻井	无	藻井类型	——								
	彩画	柱头	有	柱身	无	梁架	有	走马板	有								
		门、窗	有	天花	无	藻井	——	其他彩画									
	其他	悬塑	无	佛龛	有	匾额	无										
装饰	室内	帷幔	无	幕帘彩绘	无	壁画	无	唐卡	无								
		经幡	无	经幢	无	柱毯	无	其他	——								
	室外	玛尼轮	无	苏勒德	无	宝顶	有	祥麟法轮	无								
		四角经幢	无	经幡	无	铜饰	无	石刻、砖雕	无								
		仙人走兽	无	壁画	有	其他	——										
陈设	室内	主佛像	无	佛像基座	无												
		法座	无	藏经橱	无	经床	无	诵经桌	无	法鼓	无	玛尼轮	无	坛城	无	其他	——
	室外	旗杆	无	苏勒德	无	狮子	无	经幡	无	玛尼轮	无	香炉	无	五供	无	其他	——
	其他	——															
备注	——																
调查日期	2010/07/31	调查人员	张宇、何鑫	整理日期	2010/07/31	整理人员	张宇、卢文娟										

百灵庙·甘珠尔殿·档案照片

照片名称	斜前方1	照片名称	正立面	照片名称	侧立面
照片名称	斜前方2	照片名称	室外柱身	照片名称	室外柱头
照片名称	室外柱础	照片名称	正门	照片名称	窗户
照片名称	台基	照片名称	砖雕	照片名称	墙体材料
照片名称	南侧壁画	照片名称	北侧壁画	照片名称	——
					—
备注	——				
摄影日期	2010/07/31	摄影人员	高旭、何鑫		

甘珠尔殿基本概况表2

甘珠尔殿正前方

甘珠尔殿斜前方（左上图）
甘珠尔殿侧面（左下图）
甘珠尔殿室外柱子（右图）

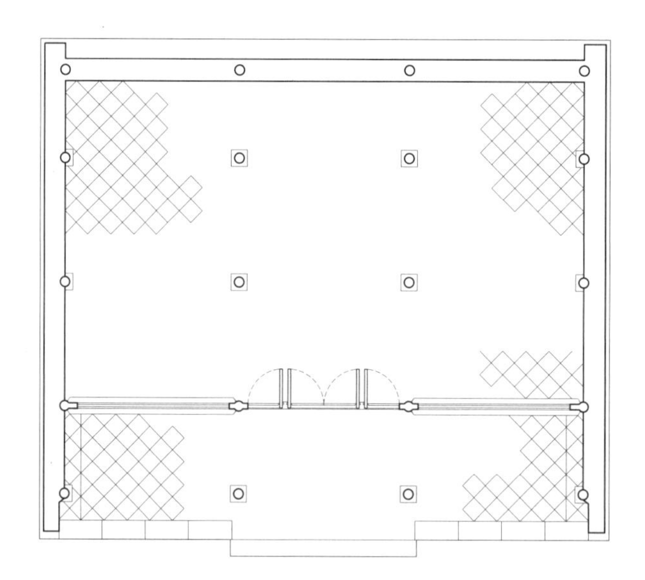

甘珠尔殿一层平面图

甘珠尔殿正立面图

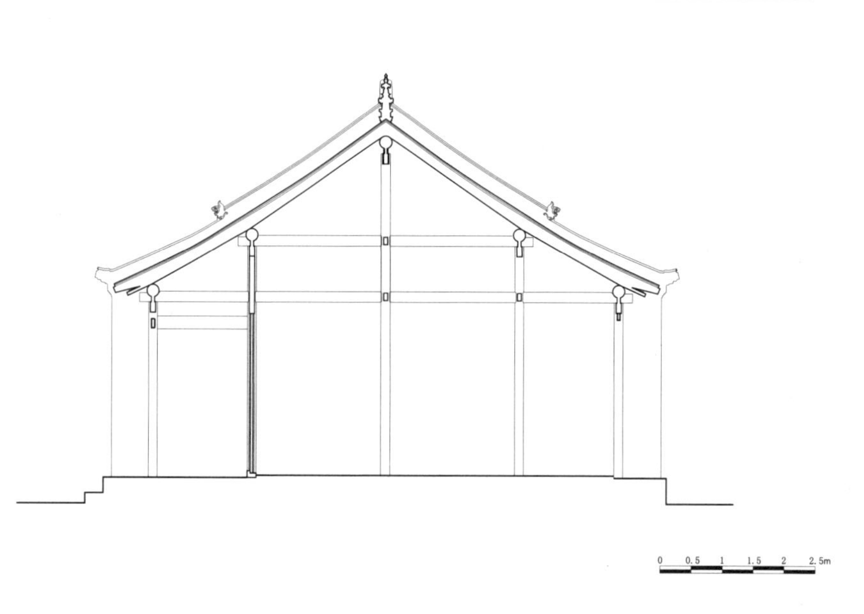

甘珠尔殿剖面图

5.5 百灵庙·九世班禅府邸

单位:毫米

建筑名称	汉语正式名称	九世班禅府邸		俗称	乌兰拉卜隆、红宫		
概述	初建年	不详		建筑朝向	南	建筑层数	一
	建筑简要描述	汉式结构					
	重建重修记载	2003年重建					
		信息来源	大喇嘛口述				

结构规模													
结构形式	木结构		相连的建筑	无				室内天井	无				
建筑平面形式	青砖		外廊形式	前廊									
通面阔	15240	开间数	5	明间	2930	次间	2850	梢间	2875	次梢间	——	尽间	——
通进深	9100	进深数	3	进深尺寸（前→后）		1655→4350→1910							
柱子数量	——	柱子间距	横向尺寸	——		（藏式建筑结构体系填写此栏，不含廊柱）							
			纵向尺寸	——									
其他	——												

建筑主体（大木作）（石作）（瓦作）								
屋顶	屋顶形式	硬山				瓦作	布瓦	
外墙	主体材料	青砖	材料规格	280×135×65		饰面颜色	灰	
	墙体收分	无	边玛檐墙	无		边玛材料	——	
斗栱、梁架	斗栱	无	平身科斗口尺寸		——	梁架关系	六檩前廊	
柱、柱式（前廊柱）	形式	汉式	柱身断面形状	圆	断面尺寸	直径 $D=105$		（在没有前廊柱的情况下，填写室内柱及其特征）
	柱身材料	木材	柱身收分	有	栌斗、托木	无	雀替 有	
	柱础	不详	柱础形状	不详	柱础尺寸	——		
台基	台基类型	普通台基	台基高度	960	台基地面铺设材料		条形砖（600×300） 方砖（600×600）	
其他	——							

装修（小木作）（彩画）								
门(正面)	格栅门		门楣	无	堆经	无	门帘	无
窗（正面）	格栅窗		窗楣	无	窗套	无	窗帘	无
室内隔扇	隔扇	有	隔扇位置	位于两侧尽间与梢间之间				
室内地面、楼面	地面材料及规格		不详		楼面材料及规格		——	
室内楼梯	楼梯	无	楼梯位置	——	楼梯材料	——	梯段宽度	——
天花、藻井	天花	有	天花类型	井口	藻井	无	藻井类型	无
彩画	柱头	有	柱身	无	梁架	有	走马板	有
	门、窗	有	天花	有	藻井	——	其他彩画	——
其他	悬塑	无	佛龛	无	匾额	无		

装饰								
室内	帷幔	无	幕帘彩绘	无	壁画	无	唐卡	无
	经幡	无	经幢	无	柱毯	无	其他	——
室外	玛尼轮	无	苏勒德	无	宝顶	有	祥麟法轮	无
	四角经幢	无	经幡	无	铜饰	无	石刻、砖雕	有
	仙人走兽	3走兽	壁画	有	其他	——		

陈设																
室内	主佛像	无							佛像基座	——						
	法座	无	藏经橱	无	经床	无	诵经桌	无	法鼓	无	玛尼轮	无	坛城	无	其他	——
室外	旗杆	无	苏勒德	无	狮子	无	经幡	无	玛尼轮	无	香炉	无	五供	无	其他	——
其他	——															

备注	——						
调查日期	2010/07/29	调查人员	张宇、何鑫	整理日期	2010/07/29	整理人员	张宇、卢文娟

九世班禅府邸基本概况表1

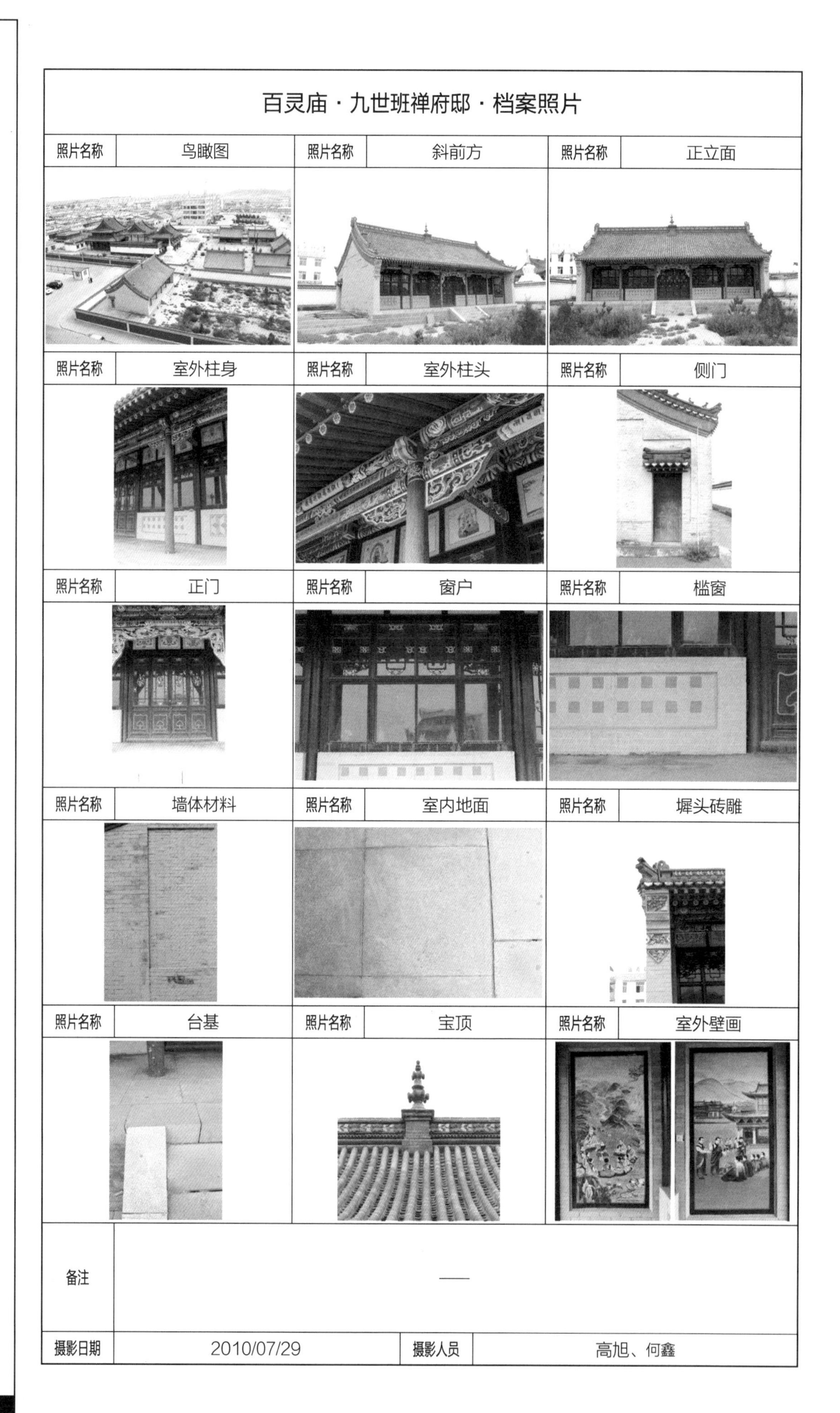

百灵庙·九世班禅府邸·档案照片

照片名称	鸟瞰图	照片名称	斜前方	照片名称	正立面
照片名称	室外柱身	照片名称	室外柱头	照片名称	侧门
照片名称	正门	照片名称	窗户	照片名称	槛窗
照片名称	墙体材料	照片名称	室内地面	照片名称	墀头砖雕
照片名称	台基	照片名称	宝顶	照片名称	室外壁画
备注	——				
摄影日期	2010/07/29	摄影人员	高旭、何鑫		

九世班禅府邸基本概况
表2

九世班禅府邸斜前方

九世班禅府邸正前方
（左上图）
九世班禅府邸墀头
（左下图）
九世班禅府邸室外柱子
（右图）

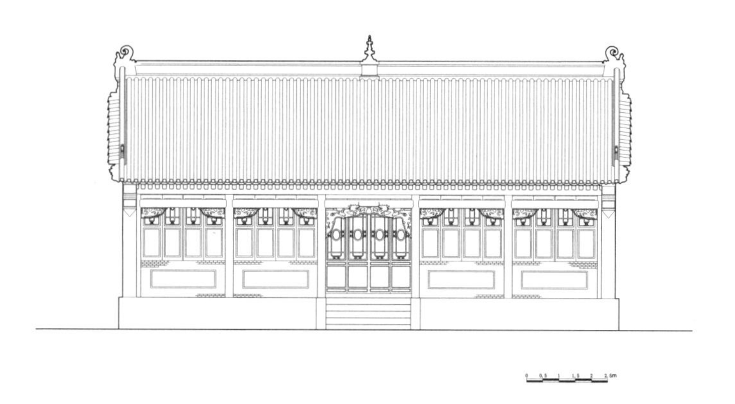

九世班禅府邸正立面图

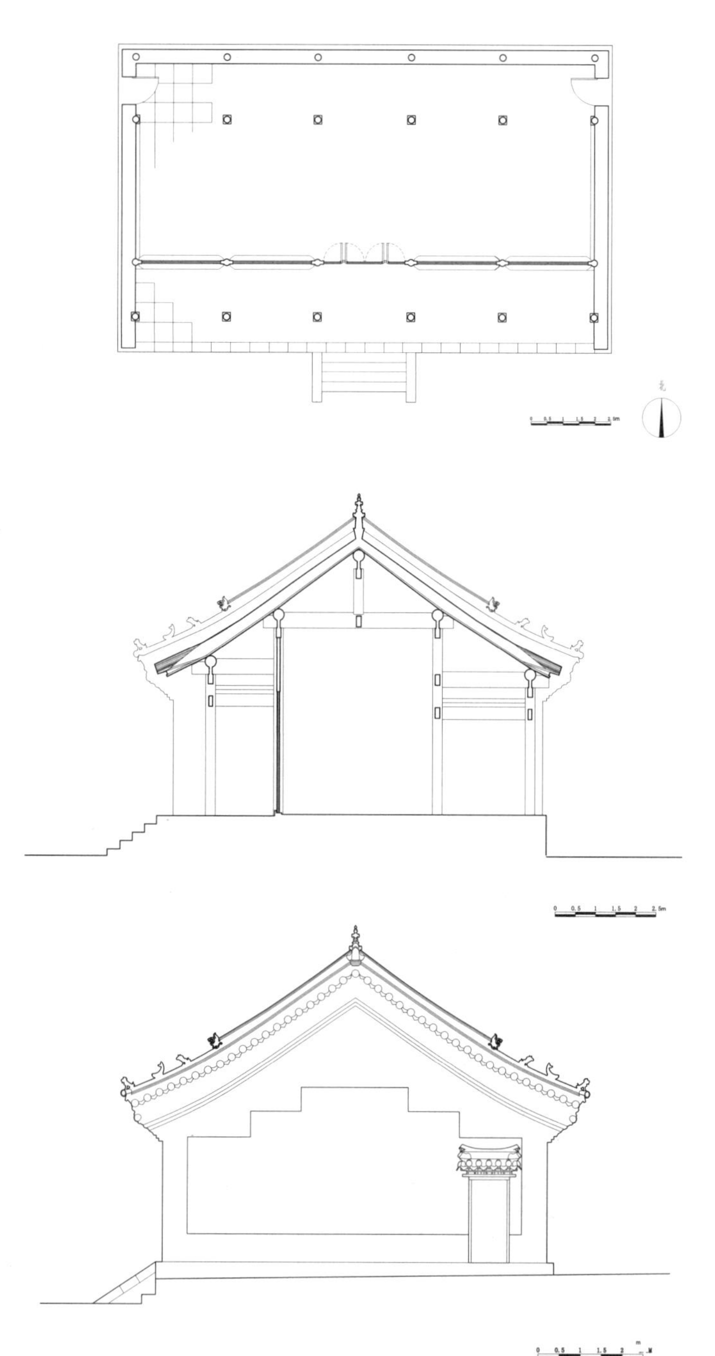

九世班禅府邸一层平面图

九世班禅府邸剖面图

九世班禅府邸侧立面图

希拉木仁庙

Xiaramuren Temple

6 希拉木仁庙 Xiaramuren Temple

希拉木仁庙建筑群

希拉木仁庙为原归化城土默特旗寺庙，系席力图召4座属庙之一及避暑庙。乾隆三十四年（1769年），清廷御赐满、蒙古、汉、藏四体“普会寺”匾额。

乾隆三十四年（1769年），时任呼和浩特掌印扎萨克达喇嘛的第六世席力图呼图克图阿格旺罗布桑达瓦以私财于归化城北一百五十里处希拉木仁之地（属土默特旗大青山北四苏木牧区，习称土默特旗第七区）新建寺庙，习称席力图希拉木仁庙、北席力图召。大雄宝殿仿效西藏扎什伦布寺而建。寺庙由席力图召管理并由主庙委派达喇嘛管理该庙。

寺庙建筑风格为汉藏结合式建筑，至中华人民共和国成立前该庙有三层大雄宝殿、天王殿、5间玛尼仓、5间经会仓、膳房、审讯厅、呼图克图拉布隆、舍利殿（内供第六世席力图呼图克图舍利）等建筑及11座庙仓。

“文化大革命”中寺庙严重受损，仅存大雄宝殿等部分殿宇。1984年起修缮寺庙，第十一世席力图呼图克图主持开光，正式恢复法会。

参考文献：

［1］拉·乌云毕力格.希拉穆仁庙（蒙古文）.呼和浩特：内蒙古人民出版社,1999,12．
［2］满都麦,莫德尔图.乌兰察布寺院（蒙古文）.海拉尔：内蒙古文化出版社,1996,5.
［3］荣祥，荣赓麟.土默特沿革（征求意见稿）.1981,6.

希拉木仁庙建匾额（左图）
希拉木仁庙转经桶（中图）
希拉木仁庙僧房（右图）

希拉木仁庙 · 基本概况 1

<table>
<tr><td rowspan="3">寺院蒙古语藏语名称</td><td>蒙古语</td><td>[illegible]</td><td rowspan="2">寺院汉语名称</td><td>汉语正式名称</td><td colspan="4">兴源寺</td></tr>
<tr><td>藏语</td><td>ཀུན་མཐུན་གླིང་།</td><td>俗称</td><td colspan="4">席力图希拉木仁召</td></tr>
<tr><td>汉语语义</td><td>黄河召</td><td colspan="2">寺院汉语名称的由来</td><td colspan="4">清廷赐名</td></tr>
<tr><td>所在地</td><td colspan="4">包头市达茂旗希拉慕仁镇，阴山北麓，希拉木仁河畔</td><td>东经</td><td>111° 14′</td><td>北纬</td><td>41° 19′</td></tr>
<tr><td>初建年</td><td colspan="2">乾隆三十四年（1769年）</td><td>保护等级</td><td colspan="5">——</td></tr>
<tr><td>盛期时间</td><td colspan="2">不详</td><td colspan="2">盛期喇嘛僧/现有喇嘛僧数</td><td colspan="4">300余人/</td></tr>
<tr><td rowspan="2">历史沿革</td><td colspan="8">1769年，始建寺庙。
1919年，该寺达喇嘛在庙中创办学堂，招收僧俗13名学生。
“文化大革命”中寺庙严重受损。
1940年，希拉木仁庙管辖地被编为席力图旗，该寺萨木丹达喇嘛任旗长。
1984年，修缮大雄宝殿、天王殿</td></tr>
<tr><td>资料来源</td><td colspan="7">[1] 拉 · 乌云毕力格.希拉穆仁庙（蒙古文）.呼和浩特:内蒙古人民出版社,1999,12.
[2] 满都麦,莫德尔图.乌兰察布寺院（蒙古文）.海拉尔:内蒙古文化出版社,1996,5.
[3] 调研访谈记录.</td></tr>
<tr><td rowspan="2">现状描述</td><td colspan="4" rowspan="2">希拉木仁召目前现存建筑主要有天王殿、大雄宝殿、两侧配殿、护法殿、中院佛爷府和西院六世活佛供殿等建筑群。四大天王殿建筑年代不详，其余建筑均属历史遗存建筑,其主要建筑风格以汉藏结合和汉式建筑为主</td><td colspan="2">描述时间</td><td colspan="2">2010/10/30</td></tr>
<tr><td colspan="2">描述人</td><td colspan="2">卢文娟</td></tr>
<tr><td>调查日期</td><td colspan="2">2010/10/30</td><td>调查人员</td><td colspan="5">韩瑛、卢文娟</td></tr>
</table>

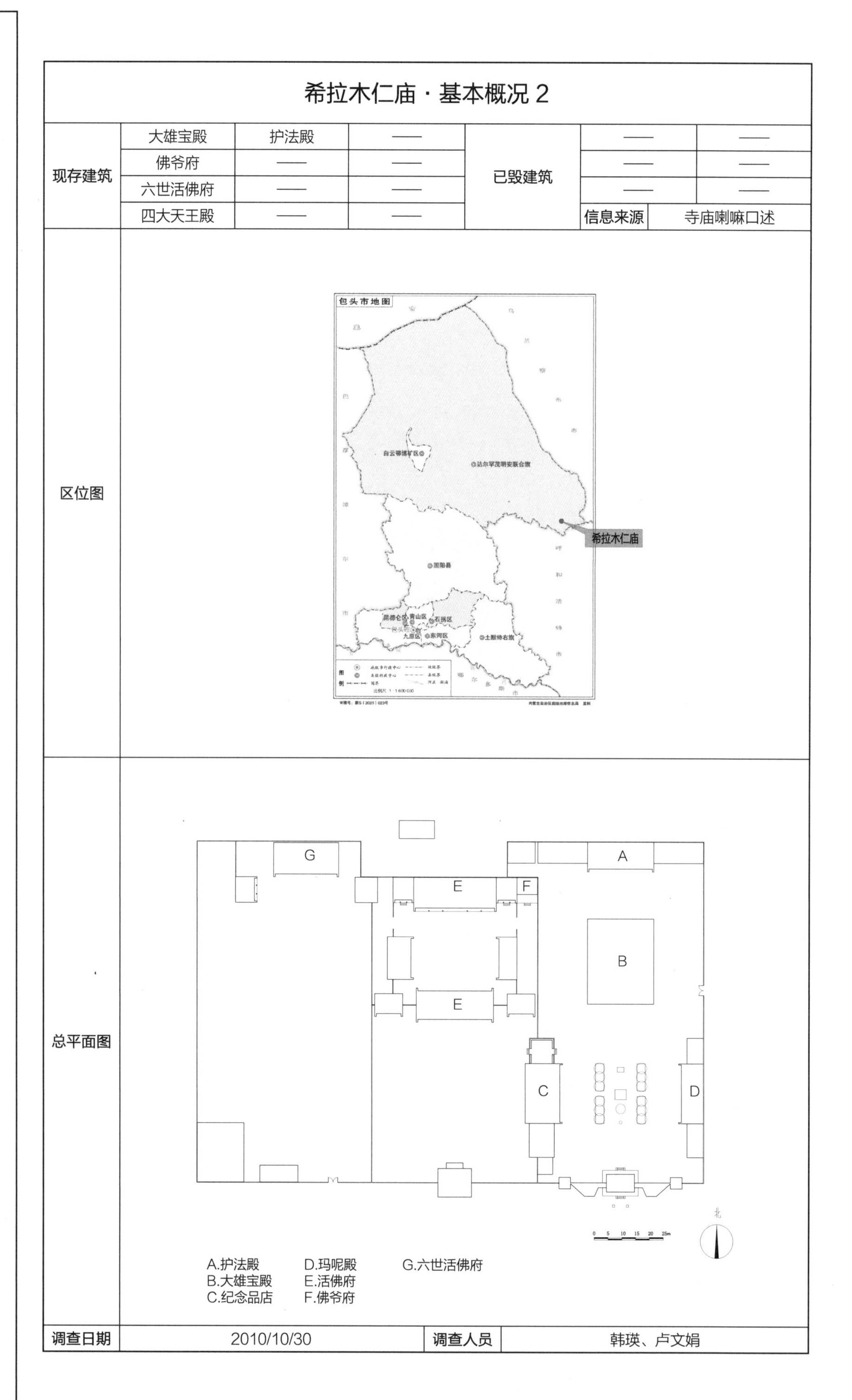

希拉木仁庙·基本概况 2

现存建筑	大雄宝殿	护法殿	——	已毁建筑	——	——
	佛爷府	——	——		——	——
	六世活佛府	——	——		——	——
	四大天王殿	——	——		信息来源	寺庙喇嘛口述
区位图						
总平面图						
调查日期	2010/10/30		调查人员	韩瑛、卢文娟		

希拉木仁庙基本概况表2

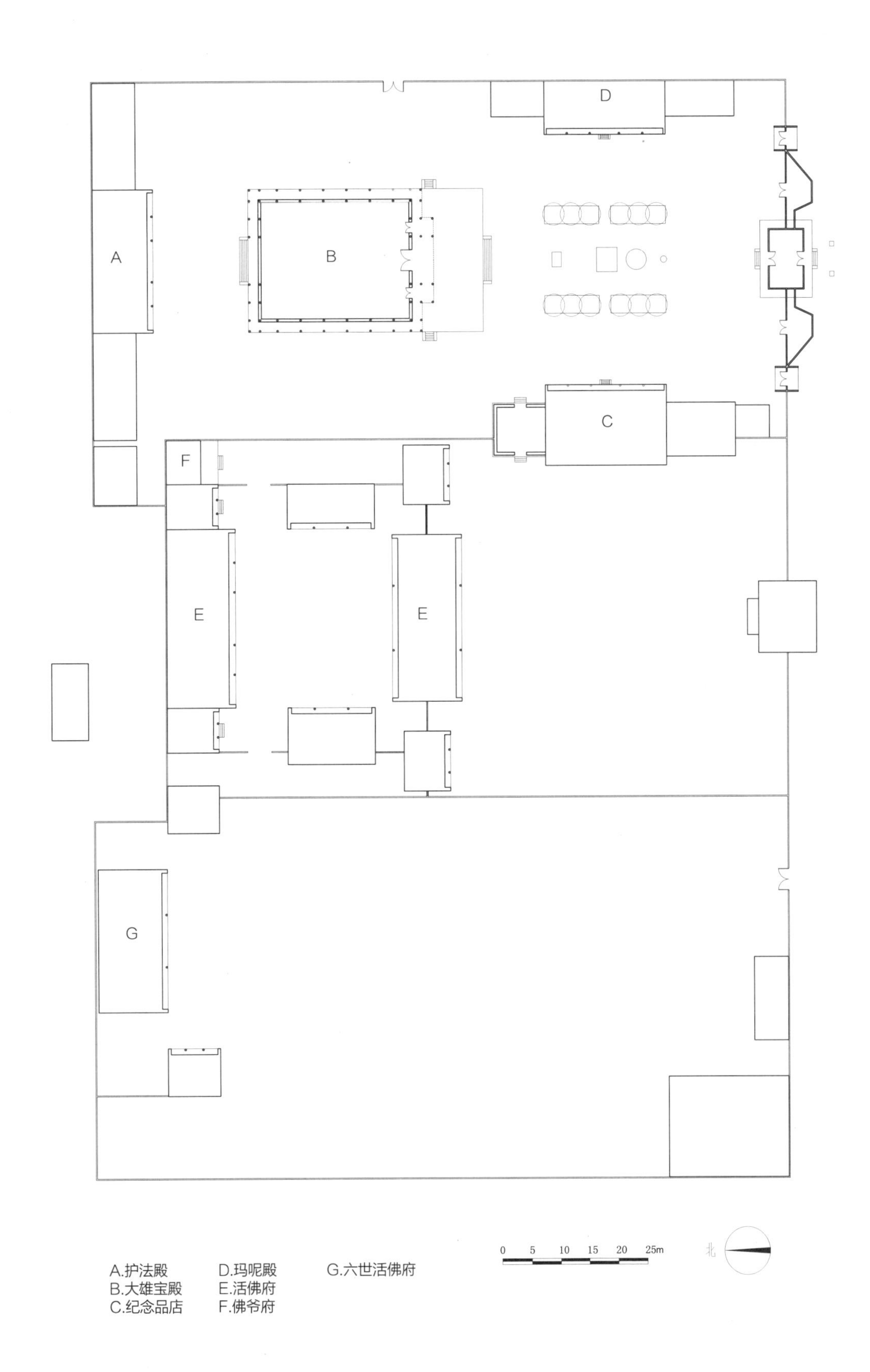

A
B
D
C
F
E
E
G
0 5 10 15 20 25m
北
A.护法殿
B.大雄宝殿
C.纪念品店
D.玛呢殿
E.活佛府
F.佛爷府
G.六世活佛府

希拉木仁庙总平面图

6.1 希拉木仁庙·大雄宝殿

单位:毫米

建筑名称	汉语正式名称	大雄宝殿	俗称	朝格钦独贡													
概述	初建年	乾隆三十四年（1769年）	建筑朝向	南	建筑层数	二
	建筑简要描述	汉藏风格巧妙结合、藏式柱廊、藏式装饰风格
	重建重修记载	不详
		信息来源	——
结构规模	结构形式	砖木混合	相连的建筑	无	室内天井	都刚法式
	建筑平面形式	长方形	外廊形式	回廊		
	通面阔	——	开间数	——	明间	——	次间	——	梢间	——	次梢间	——	尽间	——
	通进深	——	进深数	——	进深尺寸（前→后）	——
	柱子数量	20	柱子间距	横向尺寸	——	（藏式建筑结构体系填写此栏，不含廊柱）
				纵向尺寸	——	
	其他	——
建筑主体（大木作）（石作）（瓦作）	屋顶	屋顶形式	密肋平顶，局部歇山顶	瓦作	抹灰
	外墙	主体材料	青砖	材料规格	300×140×55	饰面颜色	白色、灰色
		墙体收分	有	边玛檐墙	有	边玛材料	边玛草
	斗栱、梁架	斗栱	无	平身科斗口尺寸	——	梁架关系	面向天井纵排架
	柱、柱式（前廊柱）	形式	藏式	柱身断面形状	圆形	断面尺寸	直径 $D=380$			（在没有前廊柱的情况下，填写室内柱及其特征）
		柱身材料	木材	柱身收分	无	栌斗、托木	有	雀替	无	
		柱础	有	柱础形状	莲花	柱础尺寸	600×600			
	台基	台基类型	普通台基	台基高度	920	台基地面铺设材料	石材
	其他	——
装修（小木作）（彩画）	门(正面)	板门	门楣	有	堆经	有	门帘	无
	窗（正面）	无(隔栅)	窗楣	有	窗套	有	窗帘	有
	室内隔扇	隔扇	有	隔扇位置	佛殿和经堂之间
	室内地面、楼面	地面材料及规格	木板、规格不详	楼面材料及规格	不详
	室内楼梯	楼梯	有	楼梯位置	入口右侧	楼梯材料	木材	梯段宽度	1100
	天花、藻井	天花	有	天花类型	彻上明造	藻井	无	藻井类型	——
	彩画	柱头	有	柱身	有	梁架	有	走马板	无
		门、窗	无	天花	无	藻井	——	其他彩画	——
	其他	悬塑	无	佛龛	无	匾额	普会寺
装饰	室内	帷幔	有	幕帘彩绘	有	壁画	有	唐卡	有
		经幡	有	经幢	有	柱毯	有	其他	——
	室外	玛尼轮	有	苏勒德	无	宝顶	有	祥麟法轮	有
		四角经幢	有	经幡	有	铜饰	有	石刻、砖雕	有
		仙人走兽	一仙人两兽	壁画	有	其他	——
陈设	室内	主佛像	三世佛	佛像基座	莲花座
		法座	有	藏经橱	无	经床	有	诵经桌	有	法鼓	有	玛尼轮	无	坛城	无	其他	——
	室外	旗杆	有	苏勒德	无	狮子	无	经幡	有	玛尼轮	有	香炉	有	五供	无	其他	——
	其他	——
备注	——
调查日期	2010/10/30	调查人员	韩瑛、卢文娟	整理日期	2010/10/30	整理人员	卢文娟

大雄宝殿基本概况表1

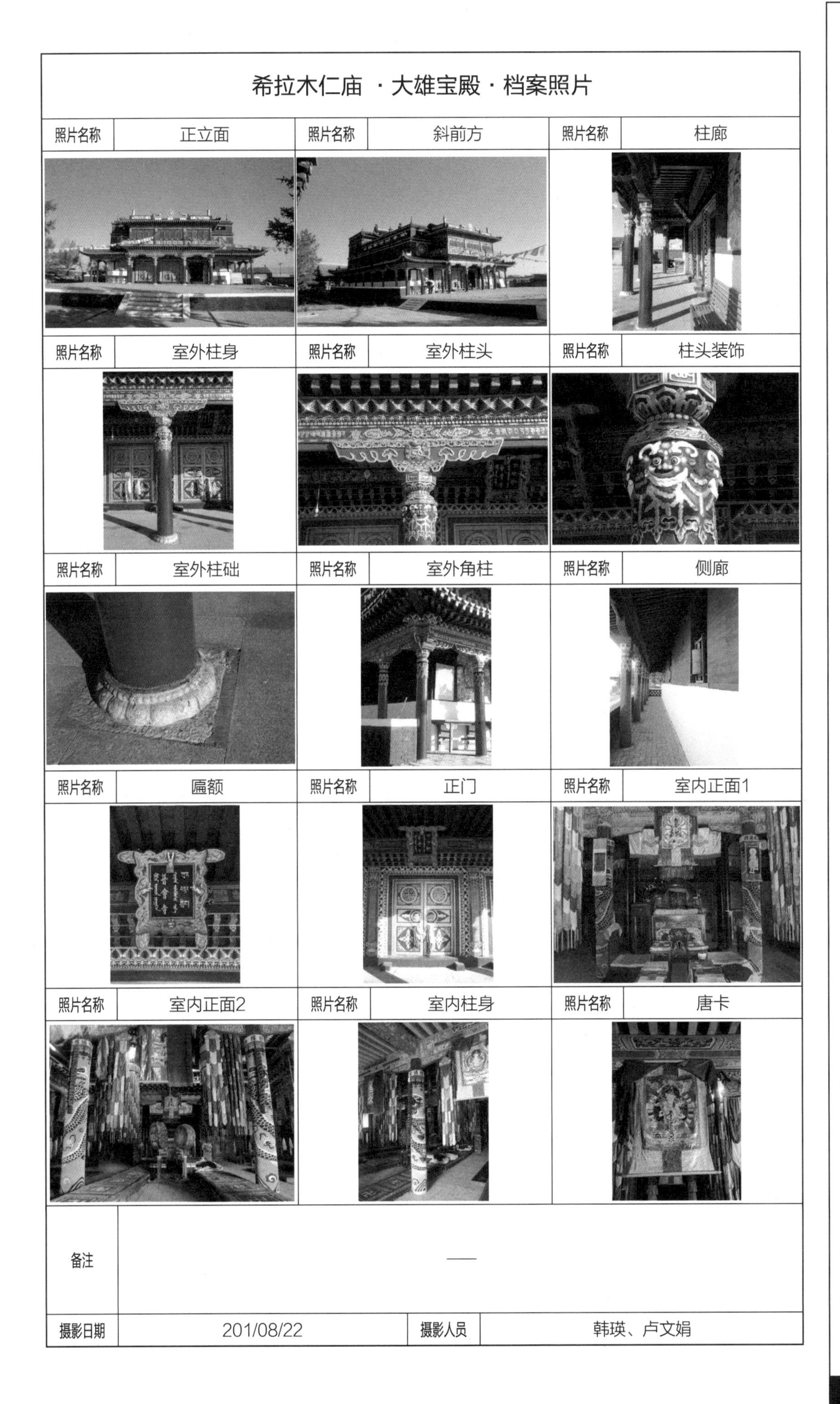

希拉木仁庙 ·大雄宝殿·档案照片

照片名称	正立面	照片名称	斜前方	照片名称	柱廊
照片名称	室外柱身	照片名称	室外柱头	照片名称	柱头装饰
照片名称	室外柱础	照片名称	室外角柱	照片名称	侧廊
照片名称	匾额	照片名称	正门	照片名称	室内正面1
照片名称	室内正面2	照片名称	室内柱身	照片名称	唐卡
备注	——				
摄影日期	201/08/22	摄影人员	韩瑛、卢文娟		

大雄宝殿斜前方

大雄宝殿正前方（左图）

大雄宝殿前廊（右图）

大雄宝殿室内正面（左上图）

大雄宝殿室内侧面（左下图）

大雄宝殿室外柱子（右图）

6.2 希拉木仁庙·四大天王殿

单位:毫米

建筑名称	汉语正式名称	四大天王殿	俗称	天王殿											
概述	初建年	不详	建筑朝向	南	建筑层数	一									
	建筑简要描述	汉式结构体系，汉式装饰风格，歇山屋顶形式													
	重建重修记载	不详													
		信息来源	——												
结构规模	结构形式	砖木混合	相连的建筑	无	室内天井	无									
	建筑平面形式	长方形	外廊形式	无廊											
	通面阔	——	开间数	3	明间	——	次间	——	梢间	——	次梢间	——	尽间	——	
	通进深	——	进深数	2	进深尺寸（前→后）	——									
	柱子数量	——	柱子间距	横向尺寸	——	（藏式建筑结构体系填写此栏，不含廊柱）									
				纵向尺寸	——										
	其他	——													
建筑主体（大木作）（石作）（瓦作）	屋顶	屋顶形式	歇山顶	瓦作	布瓦屋面										
	外墙	主体材料	青砖	材料规格	280×140×55	饰面颜色	灰色								
		墙体收分	无	边玛檐墙	无	边玛材料	——								
	斗栱、梁架	斗栱	无	平身科斗口尺寸	无	梁架关系	三檩无廊								
	柱、柱式（前廊柱）	形式	汉式	柱身断面形状	圆形	断面尺寸	直径 $D=260$	（在没有前廊柱的情况下，填写室内柱及其特征）							
		柱身材料	木材	柱身收分	无	栌斗、托木	无	雀替	无						
		柱础	无	柱础形状	——	柱础尺寸	——								
	台基	台基类型	普通台基	台基高度	540	台基地面铺设材料	石材								
	其他	——													
装修（小木作）（彩画）	门(正面)	板门	门楣	有	堆经	有	门帘	无							
	窗（正面）	隔栅	窗楣	无	窗套	无	窗帘	无							
	室内隔扇	隔扇	无	隔扇位置	——										
	室内地面、楼面	地面材料及规格	水泥地面	楼面材料及规格	——										
	室内楼梯	楼梯	无	楼梯位置	——	楼梯材料	——	梯段宽度	——						
	天花、藻井	天花	有	天花类型	井口天花	藻井	无	藻井类型	——						
	彩画	柱头	有	柱身	无	梁架	有	走马板	无						
		门、窗	无	天花	有	藻井	——	其他彩画	——						
	其他	悬塑	无	佛龛	无	匾额	无								
装饰	室内	帷幔	无	幕帘彩绘	有	壁画	无	唐卡	无						
		经幡	无	经幢	无	柱毯	无	其他	——						
	室外	玛尼轮	无	苏勒德	无	宝顶	有	祥麟法轮	无						
		四角经幢	无	经幡	无	铜饰	无	石刻、砖雕	有						
		仙人走兽	1仙+1兽	壁画	无	其他	——								
陈设	室内	主佛像	四大天王	佛像基座	——										
		法座	无	藏经橱	无	经床	无	诵经桌	无	法鼓	无	玛尼轮	无	坛城	无
		其他	——												
	室外	旗杆	无	苏勒德	无	狮子	有	经幡	无	玛尼轮	无	香炉	无	五供	无
		其他	——												
	其他	——													
备注	——														
调查日期	2010/10/30	调查人员	韩瑛、卢文娟	整理日期	2010/10/30	整理人员	卢文娟								

四大天王殿基本概况表1

希拉木仁庙·四大天王殿·档案照片

照片名称 	正立面	照片名称 	斜前方1	照片名称 	斜前方2
照片名称 	侧立面	照片名称 	背立面	照片名称 	屋顶装饰
照片名称 	室外柱头1	照片名称 	室外柱头2	照片名称 	正门
照片名称 	台基	照片名称 	天花	照片名称 	佛像1
照片名称 	佛像2	照片名称 	佛像3	照片名称 	佛像4
备注	—				
摄影日期	2010/08/22		摄影人员	韩瑛、卢文娟	

四大天王殿基本概况表2

四大天王殿斜前方

四大天王殿正前方（左图）
四大天王殿室内佛像（右图）

四大天王殿背面（左图）
四大天王殿室外柱头（右图）

6.3 希拉木仁庙·佛爷府

佛爷府院落正前方

佛爷府室外柱子（左图）
佛爷府斜前方（右上图）
佛爷府正前方（右中图）
佛爷府侧面（右下图）

呼和浩特市地区

Hohhot City

底图来源：内蒙古自治区自然资源厅官网 内蒙古地图
审图号：蒙S（2017）028号

呼和浩特市辖新城区、赛罕区、回民区、玉泉区4区，托克托县、清水河县、和林格尔县、武川县4县，土默特左旗1旗。现辖区由清时归化、绥远二城,天聪年间设立的土默特二都统旗及雍正至乾隆年间设置的归化城厅、绥远城厅、萨拉齐厅、清水河厅、和林格尔厅、托克托城厅等诸厅组成。呼和浩特地区曾被誉为拥有“七大召、八小召、七十二个绵绵召”的召城。市辖区内曾有39座召庙，现存10余座已恢复重建或尚有建筑遗存的召庙，课题组实地调研7座寺庙。

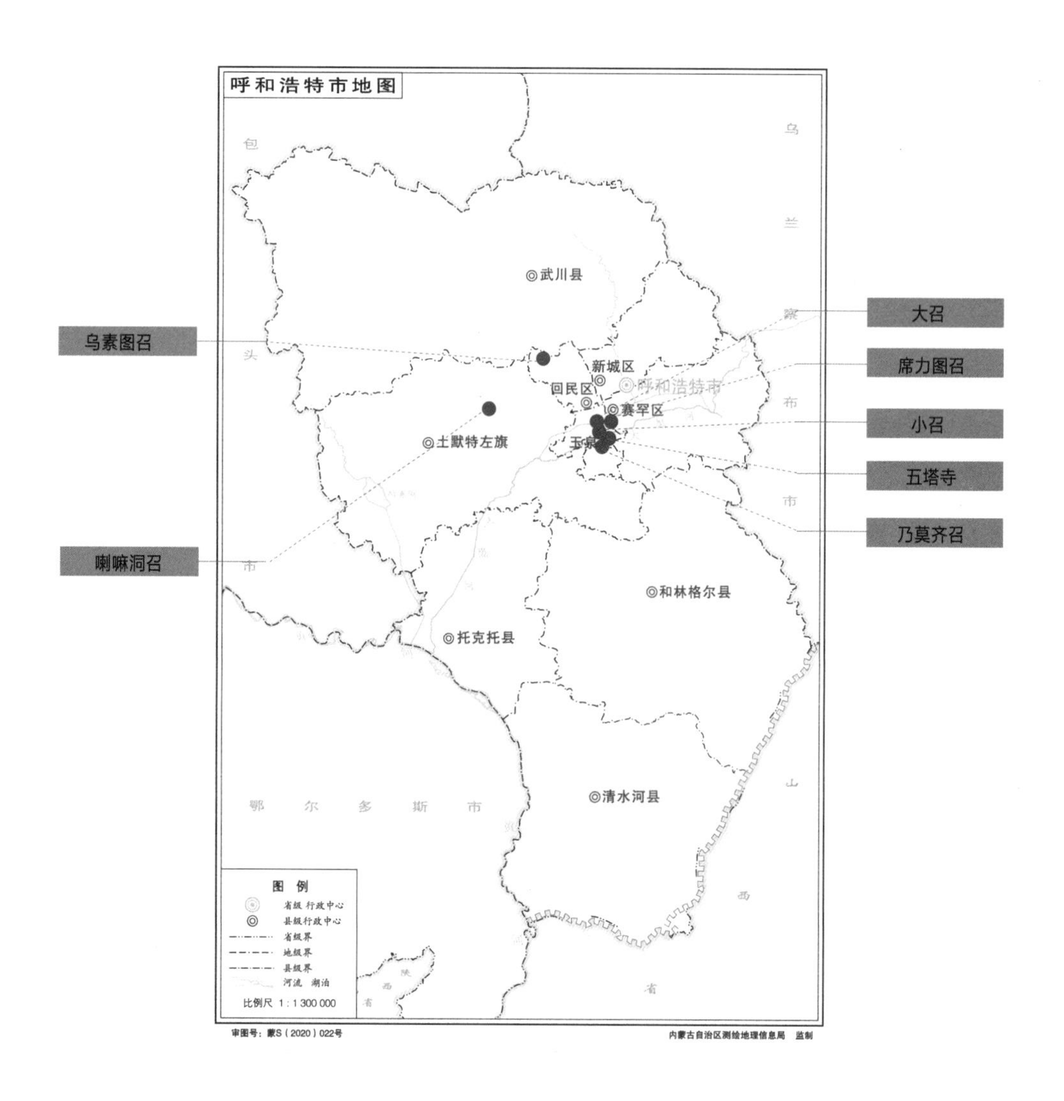

呼和浩特市地图
乌素图召
喇嘛洞召
大召
席力图召
小召
五塔寺
乃莫齐召
武川县
新城区
回民区
呼和浩特市
赛罕区
玉泉区
土默特左旗
和林格尔县
托克托县
清水河县
包头市
乌兰察布市
鄂尔多斯市
山西省
图例
省级行政中心
县级行政中心
省级界
地级界
县级界
河流 湖泊
比例尺 1:1 300 000
审图号：蒙S（2020）022号
内蒙古自治区测绘地理信息局 监制

大召

Dazhao Temple

1 大召 Dazhao Temple

大召全景图

大召为呼和浩特七大召之一，系漠南蒙古地区第一座藏传佛教格鲁派寺庙。明万历年间，明廷赐名“弘慈寺”；清崇德五年（1640年），清太宗命土默特都统古禄格楚琥尔扩修寺庙，御赐蒙古、汉、满三体“无量寺”匾额。清代呼和浩特喇嘛印务处设于大召内。寺庙珍藏龙雕、壁画、明天启年间铸造的背部有蒙古文铭文的铁狮子等明代寺庙古建艺术品及现已展存内蒙古博物院的月明楼绘画等珍贵文物。寺前有玉泉井。

明万历六年（1578年），以土默特阿拉坦汗为首的蒙古右翼诸部首领于青海湖畔新建的恰布恰庙（仰华寺）会见三世达赖喇嘛索南嘉措，阿拉坦汗从青海返回后，于次年（1579年）在归化城南门外0.5公里处新建寺庙，习称阿拉坦汗庙、葛根汗庙。因寺中供奉银制释迦牟尼像，故称银佛寺。17世纪初，蒙古译经师们在大召将甘珠尔经译成蒙古文，故也称甘珠尔庙。寺庙建成后对左右翼蒙古、漠北喀尔喀以及卫拉特的弘法事业影响至深，漠北喀尔喀部第一座格鲁派寺院额尔德尼召即采用了大召的式样。

寺庙建筑风格为汉藏结合式建筑。寺院由正院、东仓、西仓三大院落单元组成，东西两仓各有仓门，北端互通构成环绕寺庙的甬道。正院内有天王殿、菩提过殿、大雄宝殿、九间楼4座主殿及钟楼、鼓楼、无量佛殿、长寿佛殿、集密佛殿、胜乐佛殿、老道房、东西耳房及其前东西配殿等殿宇楼阁。东仓内有菩萨庙、公中仓、喇嘛印务处、五间亭等建筑，西仓内有乃春庙、五间楼、东西配房等建筑。

寺庙在民国年间已破败不堪，“文化大革命”中又丢失大量文物。至20世纪80年代五间楼、菩萨庙等部分建筑被拆除。1983年起开始恢复法事，历经20余年的修缮与复建，基本恢复了寺庙往日的宏伟规模。

参考文献：

[1] 金峰整理注释.呼和浩特召庙（蒙古文）.呼和浩特：内蒙古人民出版社，1982,11.

[2] 政协内蒙古自治区委员会文史资料委员会.内蒙古喇嘛教纪例（第四十五辑）,1997,1.

[3] 金峰.大青山下的美岱召.呼和浩特：内蒙古人民出版社，2011,7.

大召序列侧面图（左图）
大雄宝殿正面图（右图）

<table>
<tr><th colspan="6">大召 · 基本概况 1</th></tr>
<tr><td rowspan="3">寺院蒙古语藏语名称</td><td>蒙古语</td><td>ᠶᠡᢈᠡ ᠵᠤᠤ</td><td rowspan="2">寺院汉语名称</td><td>汉语正式名称</td><td>无量寺</td></tr>
<tr><td>藏语</td><td>དཔག་མེད་གླིང་།</td><td>俗称</td><td>大召</td></tr>
<tr><td>汉语语义</td><td>大寺院</td><td colspan="2">寺院汉语名称的由来</td><td>清廷赐名</td></tr>
<tr><td>所在地</td><td colspan="3">呼和浩特市玉泉区</td><td colspan="2">东经 111° 38′ 北纬 40° 47′</td></tr>
<tr><td>初建年</td><td colspan="2">万历七年（1579年）</td><td>保护等级</td><td colspan="2">国家级重点文物保护单位</td></tr>
<tr><td>盛期时间</td><td colspan="2">1586—1700年</td><td colspan="2">盛期喇嘛僧/现有喇嘛僧数</td><td>约300人/60人</td></tr>
<tr><td rowspan="2">历史沿革</td><td colspan="5">1579年，始建寺庙，1580年竣工。
1586年，三世达赖亲临大召，主持了银佛开光法会。
1602—1607年，蒙古译经师们将甘珠尔经最先在该寺译成蒙古文。
1632年，清太宗皇太极讨伐林丹汗途径呼和浩特下榻该寺。
1640年，清太宗命土默特都统古禄格楚琥尔扩修寺庙。
1652年，五世达赖进京，途经归化城驻锡于大召。
1698年，内齐托音二世动用小召庙仓，将该寺殿顶换铺黄琉璃瓦，改为“帝庙”。
1878年，大召喇嘛曾在召中粉壁上添绘佛像，绘功粗劣，破坏了原有的朴素风格。
1893年，大召已非常破旧，佛像壁画都剥落不全。
1904年，大召扎萨克达喇嘛凯穆楚克募缘重修大召。此后至中华人民共和国成立前夕，寺庙遭连续破坏，召庙的前院和山门前先后被辟为商业广场。
1959年，为迎接十世班禅大师重绘佛殿建筑，并在经堂佛殿内安装了电灯。
1966年起，停止对外开放和各种宗教活动。
1967年，被呼和浩特市友谊服装厂占用。
1980年，对经堂门面进行了彩画。
1983年，友谊服装厂迁出，修葺山门和九间楼，开始恢复宗教活动。
1985—1986年，修缮大雄宝殿。
1986年，成为自治区重点文物保护单位。
1990—1997年，修缮菩提过殿和前、后东西配殿，恢复重建了钟鼓楼和大召广场。
2003年，全面维修大召经堂、佛殿、配殿。
2004—2005年，对大召后院进行地面硬化，并修建了8座汉白玉塔，修缮乃春庙、藏经阁。
2006年，将大殿前东西殿配置铜鎏金“千佛殿”。
2007年，恢复重建玉佛殿、菩萨殿。
2008年，恢复重建弥勒佛殿，添置玉佛殿、菩萨殿。
2009年，恢复重建大白伞盖殿、公中仓</td></tr>
<tr><td>资料来源</td><td colspan="4">[1] 金峰整理注释.呼和浩特召庙（蒙古文）.呼和浩特:内蒙古人民出版社,1982，11.
[2] 政协内蒙古自治区委员会文史资料委员会.内蒙古喇嘛教纪例（第四十五辑），1997，1.
[3] 金峰.大青山下的美岱召.呼和浩特:内蒙古人民出版社,2011，7.
[4] 调研访谈记录， 2010，8.</td></tr>
<tr><td rowspan="2">现状描述</td><td colspan="3" rowspan="2">大召是呼和浩特现存最大、最完整的木结构建筑，坐北朝南，三院串联，东西设两个侧院，有20余间殿宇，玉佛殿、菩萨殿、钟鼓楼、弥勒佛殿、大白傅韵霏经多次修葺，现状保存良好。现有喇嘛60名,庙内保持传统宗教活动，已成为呼和浩特市旅游中心</td><td>描述时间</td><td>2010/05/12</td></tr>
<tr><td>描述人</td><td>萨日朗</td></tr>
<tr><td>调查日期</td><td colspan="2">2010/05/12</td><td>调查人员</td><td colspan="2">白丽燕、萨日朗</td></tr>
</table>

大召基本概况表1

大召·基本概况 2

现存建筑					已毁建筑		
牌楼	山门	天王殿	钟鼓楼	菩提过殿		钟鼓楼（原）	菩萨殿（原）
大雄宝殿	九间楼	天王殿(西院)	乃春庙	藏经阁		庇佑殿（原）	玉佛殿（原）
庇佑殿	公中仓佛殿	天王殿（东院）	菩萨殿	玉佛殿		公中仓佛殿（原）	弥勒佛殿（原）等
弥勒佛殿	——	——	——	——		信息来源	大召大事记

区位图

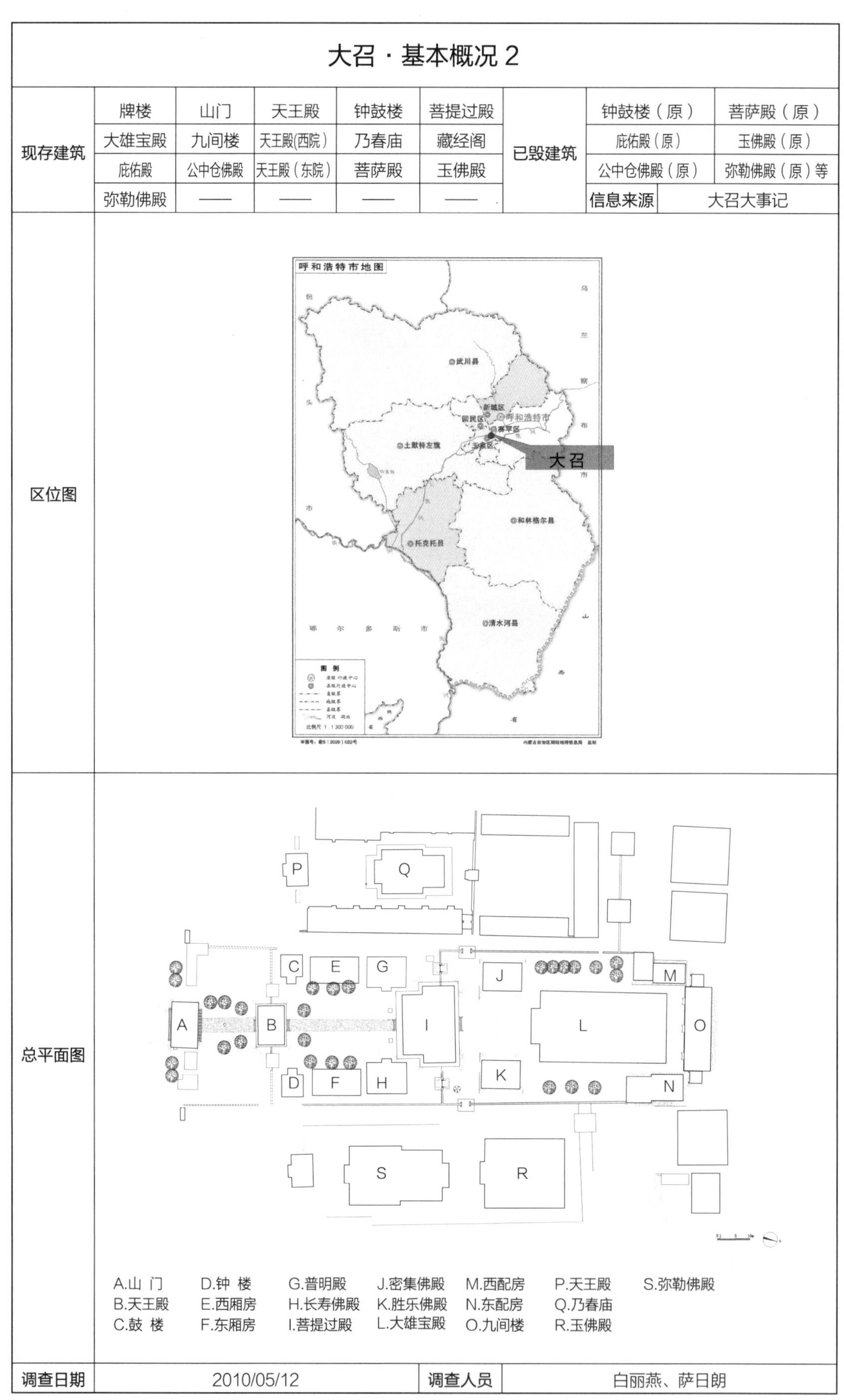

总平面图

调查日期	2010/05/12	调查人员	白丽燕、萨日朗

大召基本概况表2

注：本召庙测绘图由天津大学王其亨教授为首的国家精品课程《古建筑测绘》教研组合作测绘并提供。

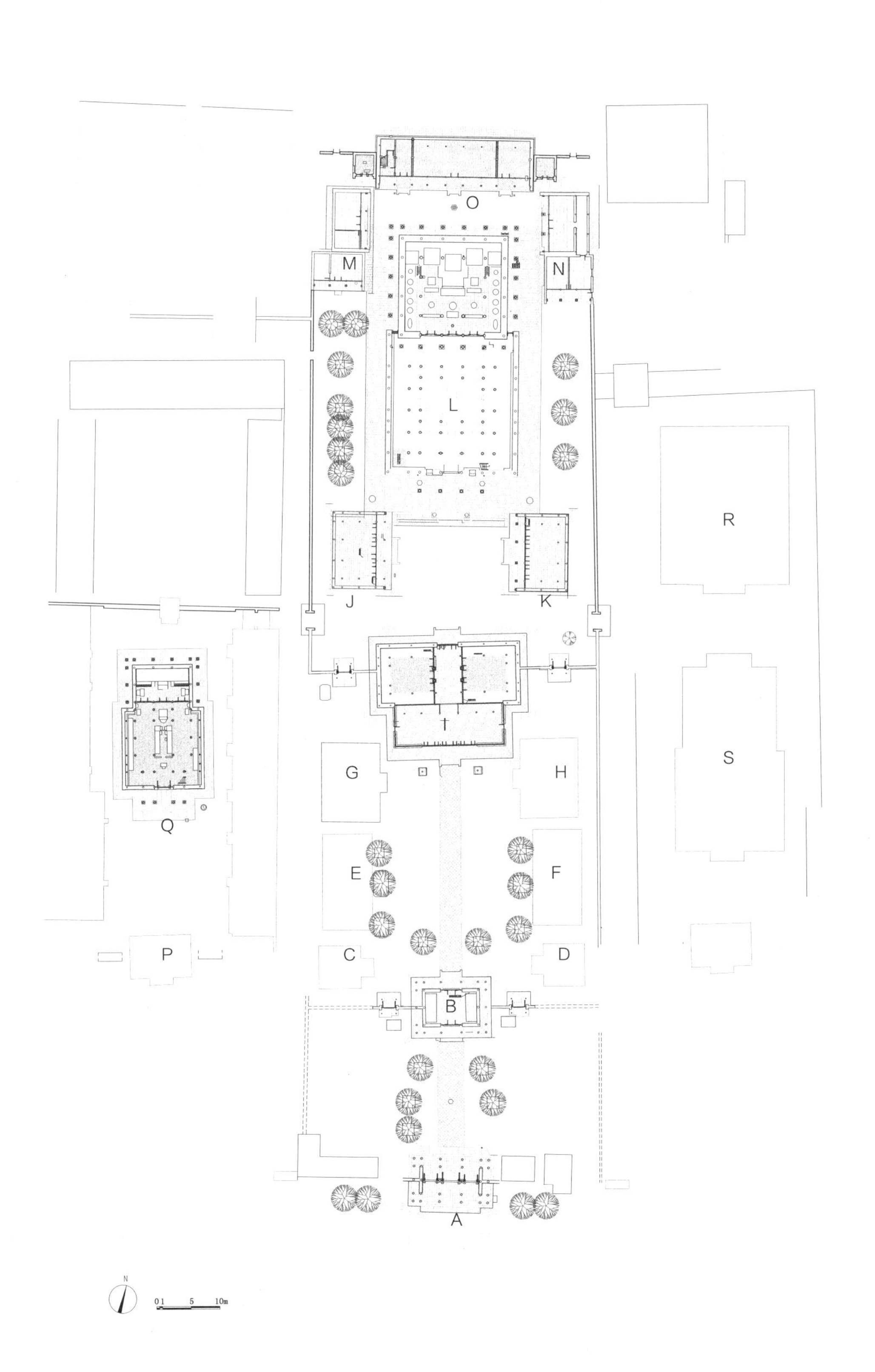

A.山 门　D.钟 楼　G.普明殿　J.密集佛殿　M.西配房　P.天王殿　S.弥勒佛殿
B.天王殿　E.西厢房　H.长寿佛殿　K.胜乐佛殿　N.东配房　Q.乃春庙
C.鼓 楼　F.东厢房　I.菩提过殿　L.大雄宝殿　O.九间楼　R.玉佛殿

大召总平面图

1.1 大召 · 山门

单位:毫米

建筑名称	汉语正式名称	山门	俗称	——													
概述	初建年	万历十五年（1587年）始建	建筑朝向	南	建筑层数	一											
	建筑简要描述	汉式建筑，歇山屋顶，汉式结构体系															
	重建重修记载	——															
		信息来源	——														
结构规模	结构形式	砖木混合	相连的建筑	无	室内天井	无											
	建筑平面形式	方形	外廊形式	回廊													
	通面阔	12970	开间数	5	明间	3850	次间	3250	梢间	1310	次梢间	——	尽间	——			
	通进深	7420	进深数	4	进深尺寸（前→后）	1400→2360→2360→1300											
	柱子数量	——	柱子间距	横向尺寸	——	（藏式建筑结构体系填写此栏，不含廊柱）											
				纵向尺寸													
	其他	——															
建筑主体	屋顶	屋顶形式	歇山	瓦作	琉璃瓦												
	外墙	主体材料	青砖	材料规格	300×150×60	饰面颜色	红										
		墙体收分	无	边玛檐墙	无	边玛材料	——										
	斗栱、梁架	斗栱	有	平身科斗口尺寸	不详	梁架关系	不详(有吊顶）										
	柱、柱式（前廊柱）	形式	汉	柱身断面形状	圆	断面尺寸	直径 $D=330$	（在没有前廊柱的情况下，填写室内柱及其特征）									
		柱身材料	木	柱身收分	有	栌斗、托木	无	雀替	有								
		柱础	有	柱础形状	圆	柱础尺寸	直径 $D=420$										
	台基	台基类型	普通台基	台基高度	590	台基地面铺设材料	石材										
	其他	——															
装修（小木作）（彩画）	门(正面)	板门	门楣	无	堆经	无	门帘	无									
	窗（正面）	无	窗楣	无	窗套	无	窗帘	无									
	室内隔扇	隔扇	无	隔扇位置	——												
	室内地面、楼面	地面材料及规格	砖，规格不详	楼面材料及规格	——												
	室内楼梯	楼梯	无	楼梯位置	——	楼梯材料	——	梯段宽度	——								
	天花、藻井	天花	有	天花类型	井口	藻井	无	藻井类型	——								
	彩画	柱头	有	柱身	无	梁架	有	走马板	无								
		门、窗	无	天花	有	藻井	——	其他彩画	——								
	其他	悬塑	无	佛龛	无	匾额	无量寺，丁亥年佛诞										
装饰	室内	帷幔	无	幕帘彩绘	无	壁画	无	唐卡	无								
		经幡	无	经幢	无	柱毯	无	其他	——								
	室外	玛尼轮	无	苏勒德	无	宝顶	无	祥麟法轮	无								
		四角经幢	无	经幡	无	铜饰	无	石刻、砖雕	无								
		仙人走兽	1+4	壁画	无	其他	——										
陈设	室内	主佛像	无	佛像基座	无												
		法座	无	藏经橱	无	经床	无	诵经桌	无	法鼓	无	玛尼轮	无	坛城	无	其他	——
	室外	旗杆	无	苏勒德	无	狮子	有	经幡	无	玛尼轮	无	香炉	无	五供	无	其他	——
	其他	——															
备注	——																
调查日期	2010/05/01	调查人员	白丽燕、萨日朗	整理日期	2010/05/12	整理人员	萨日朗										

山门基本概况表1

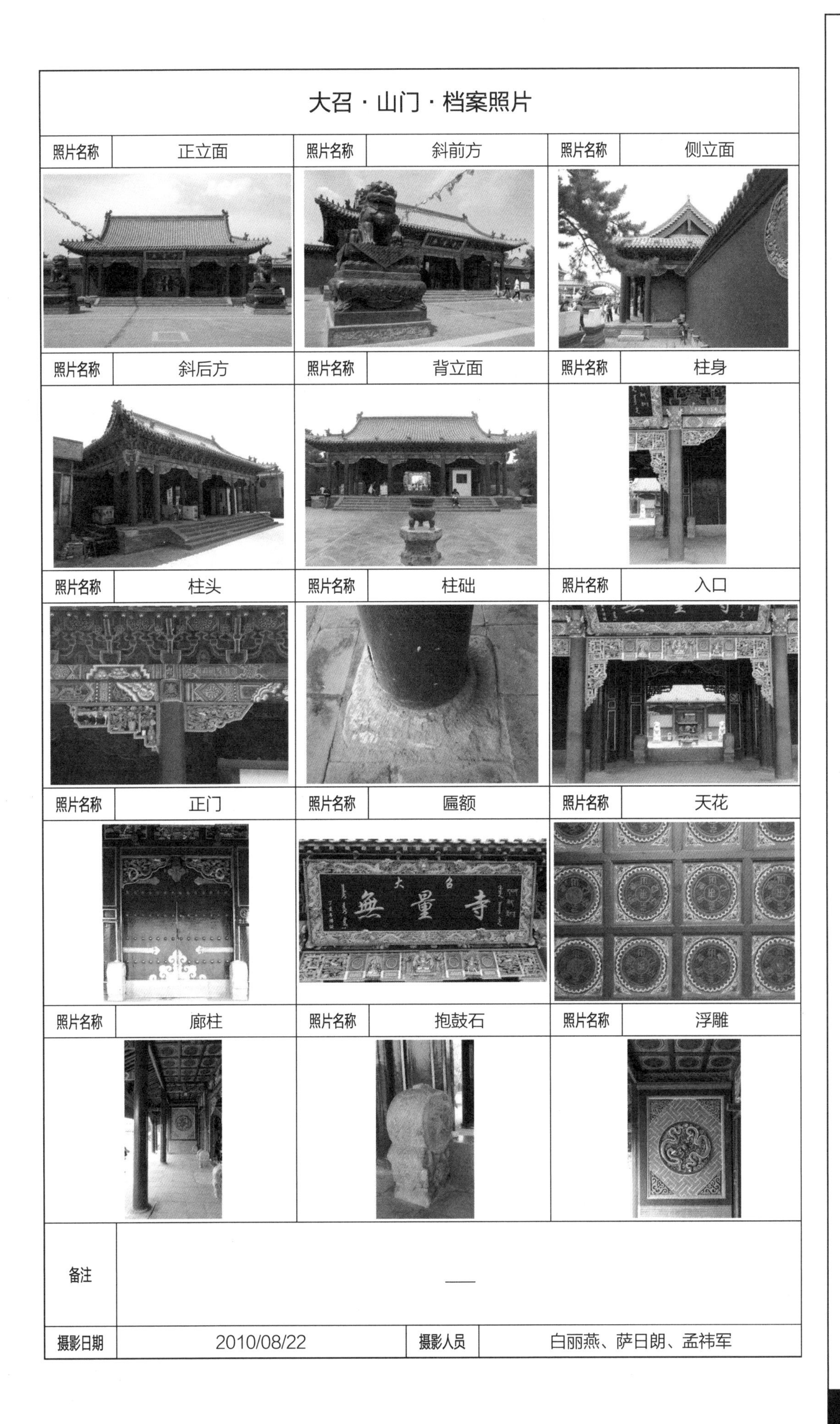

大召·山门·档案照片

照片名称	正立面	照片名称	斜前方	照片名称	侧立面
照片名称	斜后方	照片名称	背立面	照片名称	柱身
照片名称	柱头	照片名称	柱础	照片名称	入口
照片名称	正门	照片名称	匾额	照片名称	天花
照片名称	廊柱	照片名称	抱鼓石	照片名称	浮雕
备注	—				
摄影日期	2010/08/22	摄影人员	白丽燕、萨日朗、孟祎军		

山门正立面图

山门斜前方

山门背立面（左图）
山门斜后方（右图）

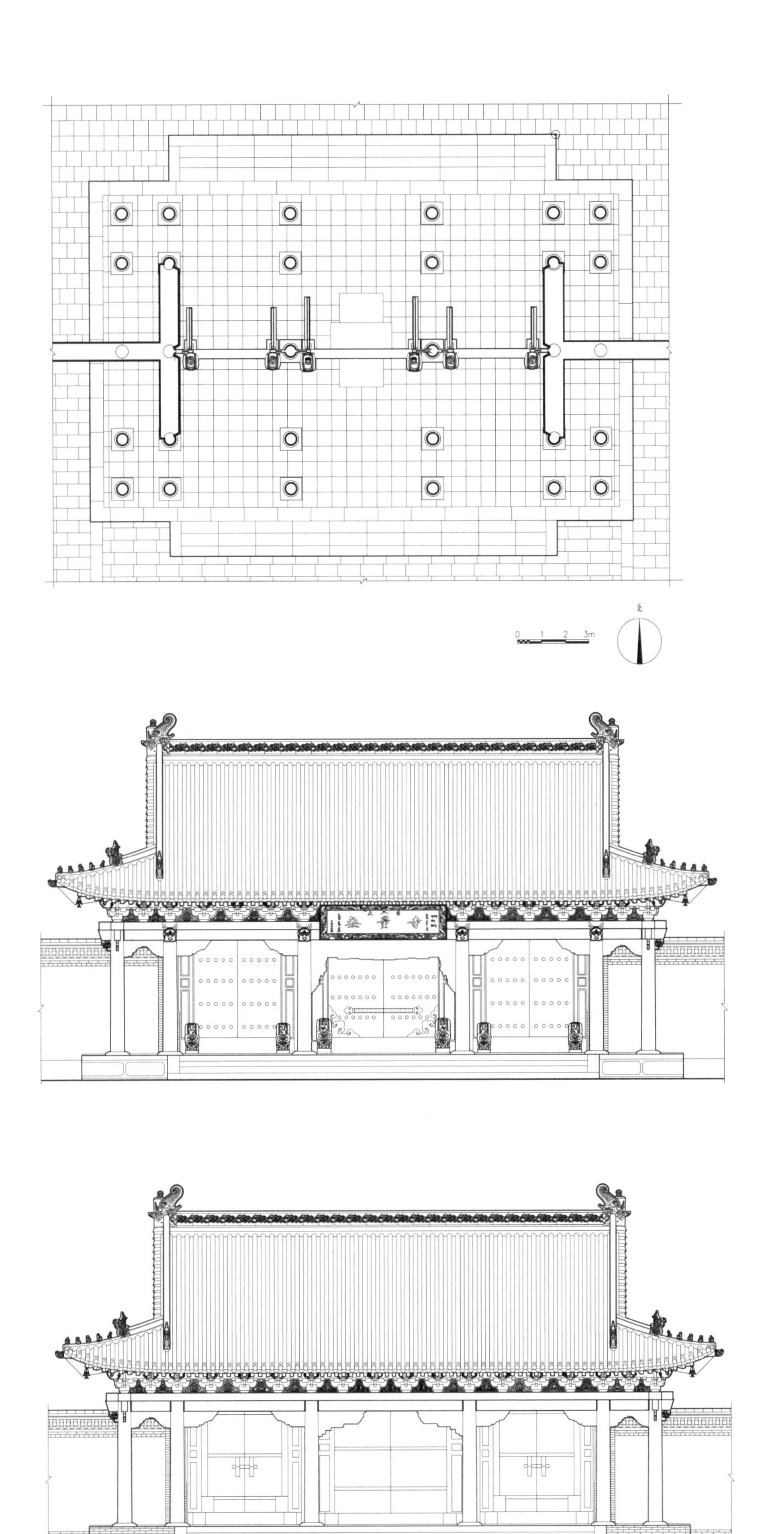

山门一层平面图

山门正立面图

山门背立面图

山门侧立面图

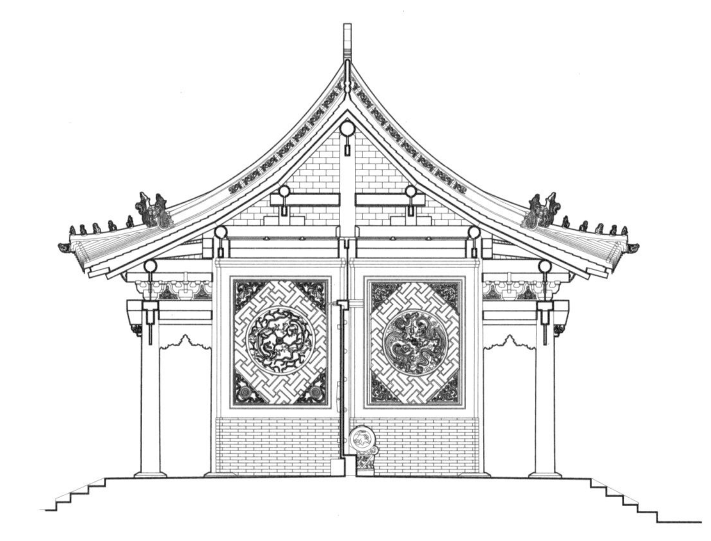

山门横剖面图

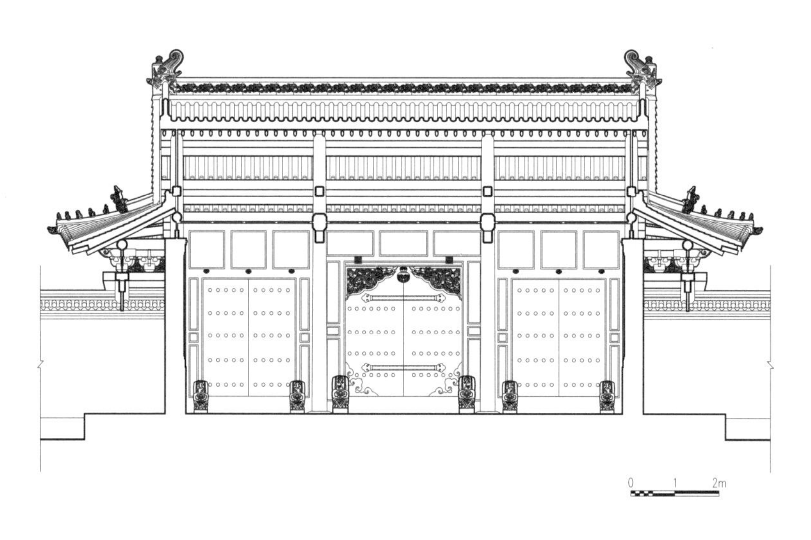

山门纵剖面图

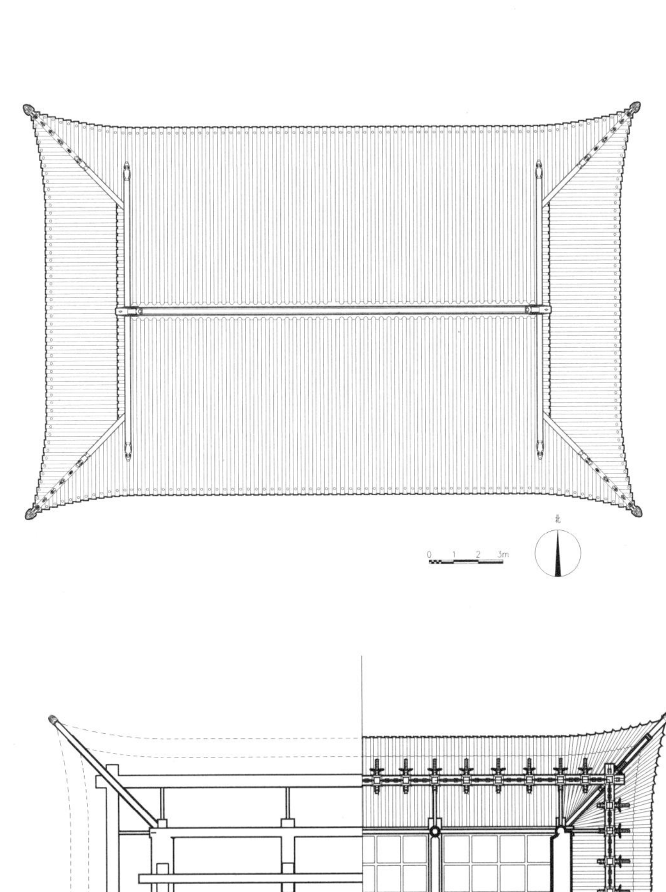

山门屋顶平面图

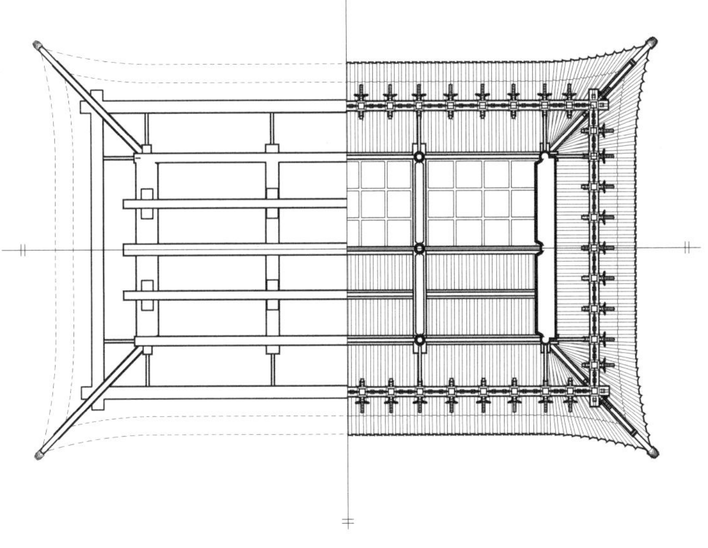

山门梁架分布图

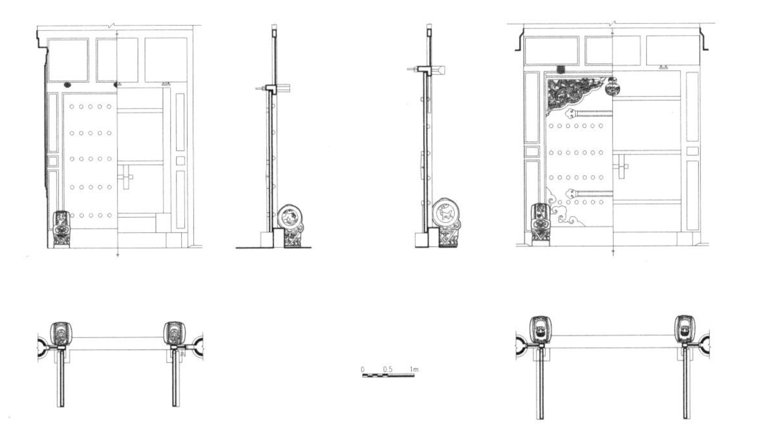

山门正门平、立、剖面图

1.2 大召·天王殿

单位:毫米

建筑名称	汉语正式名称	天王殿	俗称	— —													
概述	初建年	万历十五年（1587年)始建	建筑朝向	南	建筑层数	—											
	建筑简要描述	汉式建筑，中间开门两边实墙															
	重建重修记载	——															
		信息来源	——														
结构规模	结构形式	砖木	相连的建筑	无	室内天井	无											
	建筑平面形式	方形	外廊形式	回廊													
	通面阔	9340	开间数	3	明间	3540	次间	2900	梢间	——	次梢间	——	尽间	——			
	通进深	5000	进深数	1	进深尺寸（前→后）	5000											
	柱子数量	——	柱子间距	横向尺寸	——	（藏式建筑结构体系填写此栏，不含廊柱）											
				纵向尺寸													
	其他	——															
建筑主体（大木作）（石作）（瓦作）	屋顶	屋顶形式	歇山	瓦作	琉璃瓦												
	外墙	主体材料	青砖	材料规格	280×135×50	饰面颜色	暗红										
		墙体收分	无	边玛檐墙	无	边玛材料	——										
	斗栱、梁架	斗栱	无	平身科斗口尺寸	——	梁架关系	不详（有吊顶）										
	柱、柱式（前廊柱）	形式	汉	柱身断面形状	圆	断面尺寸	直径 D=240	（在没有前廊柱的情况下，填写室内柱及其特征）									
		柱身材料	木	柱身收分	有	栌斗、托木	无	雀替	有								
		柱础	不可见	柱础形状	不详	柱础尺寸	不详										
	台基	台基类型	普通台基	台基高度	610	台基地面铺设材料	石材										
	其他	——															
装修（小木作）（彩画）	门(正面)	隔扇门	门楣	无	堆经	无	门帘	有									
	窗（正面）	无	窗楣	无	窗套	无	窗帘	无									
	室内隔扇	隔扇	无	隔扇位置	——												
	室内地面、楼面	地面材料及规格	砖石	楼面材料及规格	无												
	室内楼梯	楼梯	无	楼梯位置	——	楼梯材料	——	梯段宽度	——								
	天花、藻井	天花	有	天花类型	井口	藻井	无	藻井类型	——								
	彩画	柱头	有	柱身	无	梁架	有	走马板	有								
		门、窗	无	天花	有	藻井	——	其他彩画	——								
	其他	悬塑	无	佛龛	无	匾额	无										
装饰	室内	帷幔	无	幕帘彩绘	无	壁画	有	唐卡	无								
		经幡	无	经幢	无	柱毯	无	其他	——								
	室外	玛尼轮	无	苏勒德	无	宝顶	有	祥麟法轮	无								
		四角经幢	无	经幡	无	铜饰	无	石刻、砖雕	无								
		仙人走兽	1+4	壁画	无	其他	——										
陈设	室内	主佛像	四大金刚佛像	佛像基座	其他												
		法座	无	藏经橱	无	经床	无	诵经桌	无	法鼓	无	玛尼轮	无	坛城	无	其他	——
	室外	旗杆	无	苏勒德	无	狮子	有	经幡	无	玛尼轮	无	香炉	有	五供	无	其他	——
	其他	——															
备注	——																
调查日期	2010/05/01	调查人员	白丽燕、萨日朗	整理日期	2010/05/01	整理人员	萨日朗										

天王殿基本概况表1

大召·天王殿·档案照片					
照片名称	正立面	照片名称	斜前方	照片名称	斜后方
照片名称	背立面	照片名称	柱身	照片名称	柱头
照片名称	台基	照片名称	门	照片名称	外墙匾
照片名称	室内正面	照片名称	室内侧面	照片名称	佛像
照片名称	天花	照片名称	地面	照片名称	室内局部
备注	——				
摄影日期	2010/05/01	摄影人员	白丽燕、萨日朗、孟祎军		

天王殿基本概况表2

天王殿正立面

天王殿斜前方

天王殿斜后方（左图）
天王殿背立面（右图）

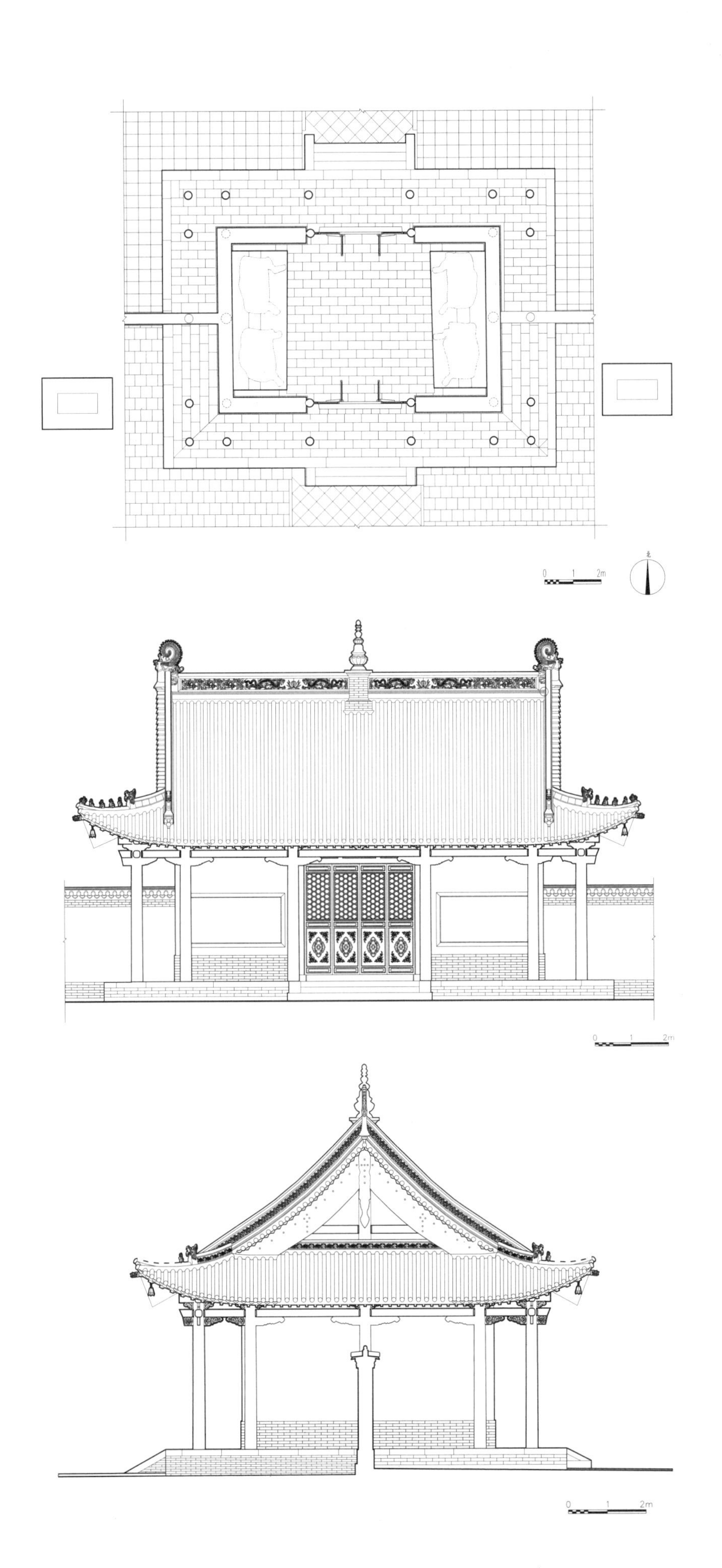

天王殿一层平面图

天王殿立面图

天王殿侧立面图

天王殿纵剖面图

0 1 2m

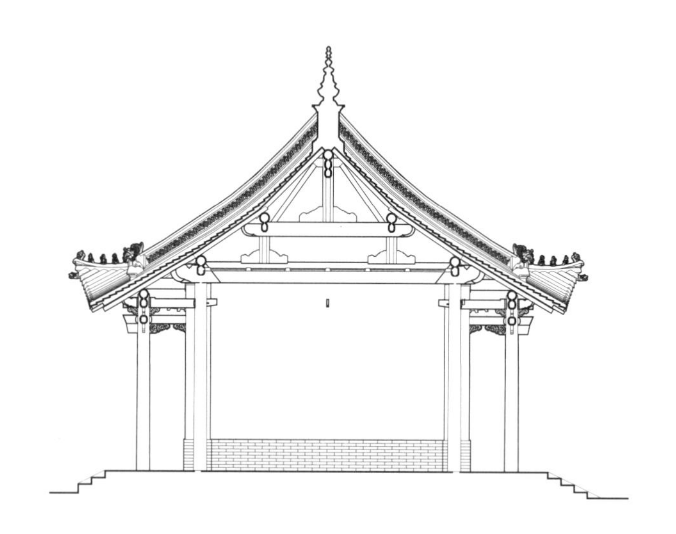
天王殿横剖面图

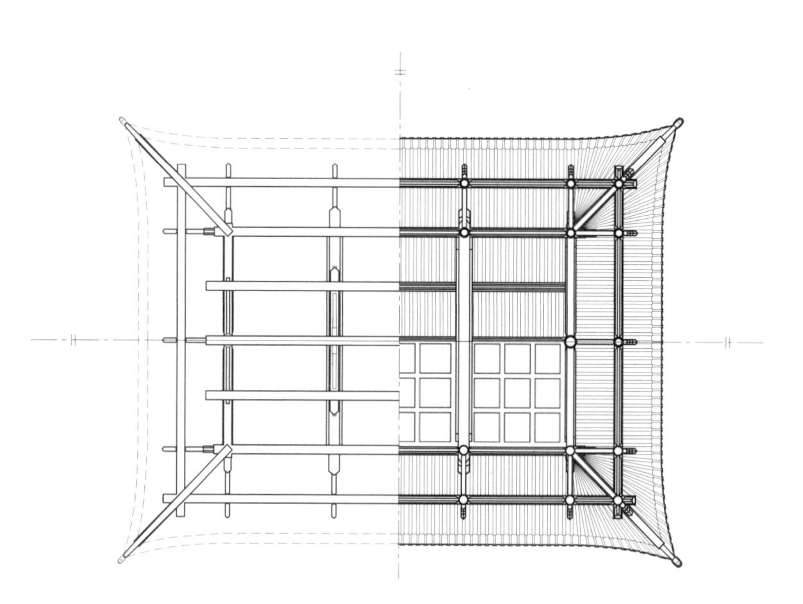
天王殿梁架平面图

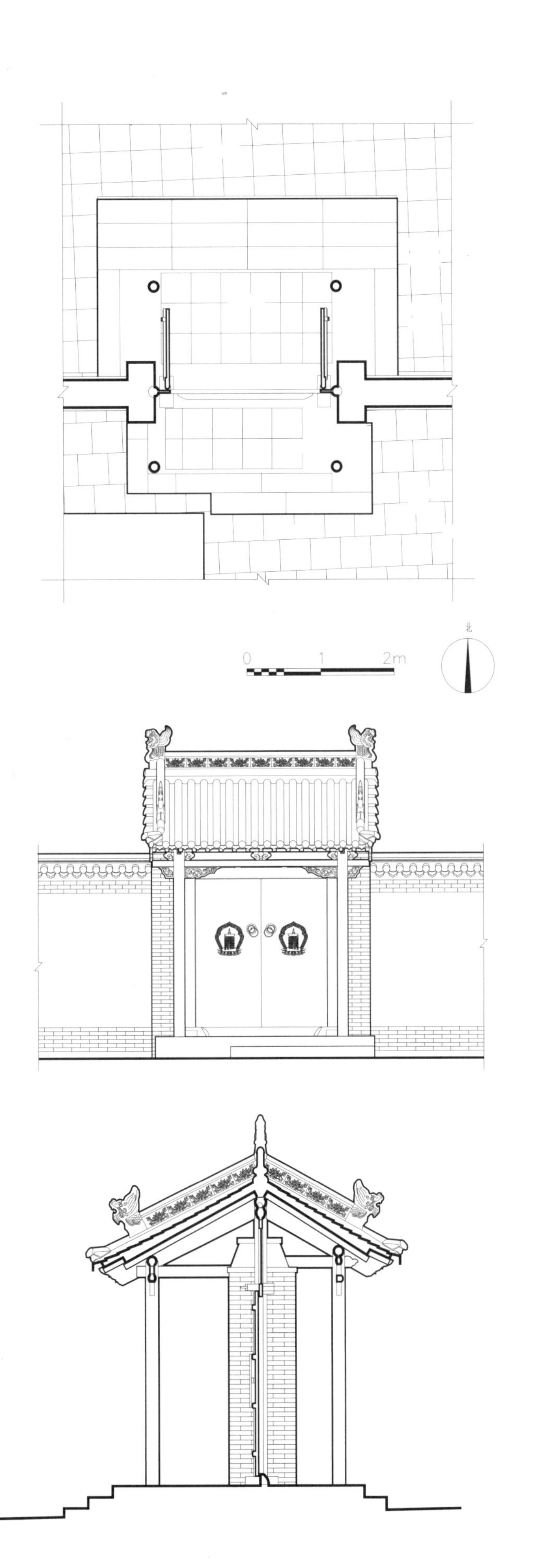

天王殿侧门一层平面图

天王殿侧门立面图

天王殿侧门纵剖面图

1.3 大召·普明佛殿

单位:毫米

建筑名称	汉语正式名称	普明佛殿		俗称	——												
概述	初建年	万历十五年（1587年）始建		建筑朝向	东	建筑层数	一										
	建筑简要描述	汉式建筑，歇山屋顶，汉式结构体系															
	重建重修记载	——															
		信息来源	——														
结构规模	结构形式	砖木混合	相连的建筑	无	室内天井	无											
	建筑平面形式	方形平面	外廊形式	前廊													
	通面阔	12610	开间数	5	明间	3350	次间	3200	梢间	1430	次梢间	——	尽间	——			
	通进深	8800	进深数	4	进深尺寸（前→后）	1500→2900→2900→1500											
	柱子数量	——	柱子间距	横向尺寸	——	（藏式建筑结构体系填写此栏，不含廊柱）											
				纵向尺寸													
	其他	——															
建筑主体（大木作）（石作）（瓦作）	屋顶	屋顶形式	歇山	瓦作	琉璃瓦												
	外墙	主体材料	青砖	材料规格	280×135×50	饰面颜色	红										
		墙体收分	无	边玛檐墙	无	边玛材料	——										
	斗栱、梁架	斗栱	有	平身科斗口尺寸	不详	梁架关系	不详（有吊顶）										
	柱、柱式（前廊柱）	形式	汉	柱身断面形状	圆	断面尺寸	直径 $D=200$		（在没有前廊柱的情况下，填写室内柱及其特征）								
		柱身材料	木	柱身收分	有	栌斗、托木	无	雀替	有								
		柱础	有	柱础形状	方形	柱础尺寸	420×420										
	台基	台基类型	石材	台基高度	600	台基地面铺设材料	石砖										
	其他	——															
装修（小木作）（彩画）	门(正面)	隔扇门	门楣	无	堆经	无	门帘	无									
	窗（正面）	槛窗	窗楣	无	窗套	无	窗帘	无									
	室内隔扇	隔扇	无	隔扇位置	——												
	室内地面、楼面	地面材料及规格	方砖270×270	楼面材料及规格	无												
	室内楼梯	楼梯	无	楼梯位置	——	楼梯材料	——	梯段宽度	——								
	天花、藻井	天花	有	天花类型	井口	藻井	无	藻井类型	——								
	彩画	柱头	有	柱身	无	梁架	有	走马板	无								
		门、窗	无	天花	有	藻井	——	其他彩画	——								
	其他	悬塑	无	佛龛	无	匾额	无										
装饰	室内	帷幔	无	幕帘彩绘	无	壁画	无	唐卡	无								
		经幡	无	经幢	无	柱毯	无	其他	——								
	室外	玛尼轮	无	苏勒德	无	宝顶	无	祥麟法轮	无								
		四角经幢	无	经幡	无	铜饰	无	石刻、砖雕	无								
		仙人走兽	1+4	壁画	无	其他	——										
陈设	室内	主佛像	普明大日如来佛像	佛像基座	莲花座												
		法座	无	藏经橱	无	经床	无	诵经桌	有	法鼓	无	玛尼轮	无	坛城	无	其他	——
	室外	旗杆	无	苏勒德	无	狮子	无	经幡	无	玛尼轮	无	香炉	无	五供	无	其他	——
	其他	——															
备注	——																
调查日期	2010/05/01	调查人员	白丽燕、萨日朗	整理日期	2010/05/01	整理人员	萨日朗										

普明佛殿基本概况表1

大召·普明佛殿·档案照片

照片名称	正立面	照片名称	斜前方	照片名称	侧立面
照片名称	柱身	照片名称	柱头	照片名称	柱础
照片名称	柱廊	照片名称	室外壁画	照片名称	明间门前台阶
照片名称	室内侧面	照片名称	室内正面	照片名称	室内天花
照片名称	室内局部	照片名称	室内地面	照片名称	佛像

备注	—

摄影日期	2010/08/22	摄影人员	张 宇

普明佛殿正立面

普明佛殿侧立面（左图）
普明佛殿柱廊（右图）

普明佛殿室内正面（左图）
普明佛殿室外柱头（右图）

1.4 大召 · 长寿佛殿

单位:毫米

建筑名称	汉语正式名称	长寿佛殿	俗称	——													
概述	初建年	万历十五年（1587年）始建	建筑朝向	西	建筑层数	一											
	建筑简要描述	汉式建筑，歇山屋顶，汉式结构体系															
	重建重修记载	2003年进行维修															
		信息来源	大召大事记														
结构规模	结构形式	砖木混合	相连的建筑	无	室内天井	无											
	建筑平面形式	方形平面	外廊形式	前廊													
	通面阔	12610	开间数	5	明间	3350	次间	3200	梢间	1430	次梢间	——	尽间	——			
	通进深	8800	进深数	4	进深尺寸（前→后）	1500→2900→2900→1500											
	柱子数量	——	柱子间距	横向尺寸	——	（藏式建筑结构体系填写此栏，不含廊柱）											
				纵向尺寸													
	其他	——															
建筑主体（大木作）（石作）（瓦作）	屋顶	屋顶形式	歇山	瓦作	琉璃瓦												
	外墙	主体材料	青砖	材料规格	280×140×50	饰面颜色	红										
		墙体收分	无	边玛檐墙	无	边玛材料	——										
	斗栱、梁架	斗栱	有	平身科斗口尺寸	不详	梁架关系	不详（吊顶）										
	柱、柱式（前廊柱）	形式	汉	柱身断面形状	圆	断面尺寸	直径 $D=200$	（在没有前廊柱的情况下，填写室内柱及其特征）									
		柱身材料	木	柱身收分	有	栌斗、托木	无	雀替	有								
		柱础	有	柱础形状	方形	柱础尺寸	420×420										
	台基	台基类型	石材	台基高度	600	台基地面铺设材料	石砖										
	其他	——															
装修（小木作）（彩画）	门(正面)	隔扇门	门楣	无	堆经	无	门帘	无									
	窗（正面）	槛窗	窗楣	无	窗套	无	窗帘	无									
	室内隔扇	隔扇	无	隔扇位置	——												
	室内地面、楼面	地面材料及规格	方砖，270×270	楼面材料及规格	——												
	室内楼梯	楼梯	无	楼梯位置	——	楼梯材料	无	梯段宽度	——								
	天花、藻井	天花	有	天花类型	井口	藻井	无	藻井类型	——								
	彩画	柱头	有	柱身	无	梁架	有	走马板	有								
		门、窗	无	天花	有	藻井	无	其他彩画	——								
	其他	悬塑	无	佛龛	无	匾额	无										
装饰	室内	帷幔	无	幕帘彩绘	无	壁画	无	唐卡	无								
		经幡	无	经幢	无	柱毯	无	其他	——								
	室外	玛尼轮	无	苏勒德	无	宝顶	无	祥麟法轮	无								
		四角经幢	无	经幡	无	铜饰	无	石刻、砖雕	无								
		仙人走兽	1+4	壁画	无	其他	——										
陈设	室内	主佛像	长寿佛佛像	佛像基座	莲花座												
		法座	无	藏经橱	无	经床	无	诵经桌	有	法鼓	无	玛尼轮	无	坛城	无	其他	——
	室外	旗杆	无	苏勒德	无	狮子	无	经幡	无	玛尼轮	无	香炉	无	五供	无	其他	——
	其他	——															
备注	——																
调查日期	2010/05/01	调查人员	白丽燕、萨日朗	整理日期	2010/05/02	整理人员	萨日朗										

长寿佛殿基本概况表1

<table>
<tr><td colspan="6">大召·长寿佛殿·档案照片</td></tr>
<tr><td>照片名称</td><td>正立面</td><td>照片名称</td><td>斜前方</td><td>照片名称</td><td>侧立面</td></tr>
<tr><td colspan="2"></td><td colspan="2"></td><td colspan="2"></td></tr>
<tr><td>照片名称</td><td>斜后方</td><td>照片名称</td><td>北立面</td><td>照片名称</td><td>柱身</td></tr>
<tr><td colspan="2"></td><td colspan="2"></td><td colspan="2"></td></tr>
<tr><td>照片名称</td><td>柱头</td><td>照片名称</td><td>柱础</td><td>照片名称</td><td>门</td></tr>
<tr><td colspan="2"></td><td colspan="2"></td><td colspan="2"></td></tr>
<tr><td>照片名称</td><td>室内正面</td><td>照片名称</td><td>室内侧面</td><td>照片名称</td><td>室内天花</td></tr>
<tr><td colspan="2"></td><td colspan="2"></td><td colspan="2"></td></tr>
<tr><td>照片名称</td><td>室内地面</td><td>照片名称</td><td>佛像</td><td>照片名称</td><td>室内装饰</td></tr>
<tr><td colspan="2"></td><td colspan="2"></td><td colspan="2"></td></tr>
<tr><td>备注</td><td colspan="5">——</td></tr>
<tr><td>摄影日期</td><td colspan="2">2010/08/22</td><td>摄影人员</td><td colspan="2">张 宇</td></tr>
</table>

长寿佛殿基本概况表2

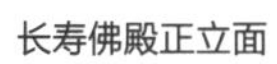

长寿佛殿正立面

长寿佛殿门（左图）

长寿佛殿侧立面（右图）

长寿佛殿室内正面

1.5 大召 · 菩提过殿

单位:毫米

建筑名称	汉语正式名称	菩提过殿	俗称	——													
概述	初建年	万历十五年（1587年）始建	建筑朝向	南	建筑层数	一											
	建筑简要描述	汉式建筑，前殿为卷棚屋顶，后殿为歇山顶															
	重建重修记载	1990年进行了装修、彩画															
		信息来源	大召大事记														
结构规模	结构形式	砖木混合	相连的建筑	无	室内天井	无											
	建筑平面形式	整体凸字形	外廊形式	无													
	通面阔	18930	开间数	7	明间	3250	次间	3200	梢间	2900	次梢间	——	尽间	1740			
	通进深	6530	进深数	3	进深尺寸（前→后）	1730→2400→2400											
	柱子数量	——	柱子间距	横向尺寸	——	（藏式建筑结构体系填写此栏，不含廊柱）											
				纵向尺寸													
	其他	——															
建筑主体（大木作）（石作）（瓦作）	屋顶	屋顶形式	卷棚顶	瓦作	布瓦												
	外墙	主体材料	青砖	材料规格	300×145×60	饰面颜色	红										
		墙体收分	无	边玛檐墙	无	边玛材料	——										
	斗栱、梁架	斗栱	有	平身科斗口尺寸	不详	梁架关系	不详（有吊顶）										
	柱、柱式（前廊柱）	形式	汉	柱身断面形状	圆	断面尺寸	直径 $D=260$	（在没有前廊柱的情况下，填写室内柱及其特征）									
		柱身材料	木	柱身收分	有	栌斗、托木	无	雀替	有								
		柱础	不可见	柱础形状	不详	柱础尺寸	不详										
	台基	台基类型	普通台基	台基高度	600	台基地面铺设材料	石材										
	其他	——															
装修（小木作）（彩画）	门(正面)	隔扇门	门楣	无	堆经	无	门帘	无									
	窗（正面）	槛窗	窗楣	无	窗套	无	窗帘	无									
	室内隔扇	隔扇	有	隔扇位置	位于菩提过殿与药师殿与护法殿之间												
	室内地面、楼面	地面材料及规格	石材600×300	楼面材料及规格	——												
	室内楼梯	楼梯	无	楼梯位置	——	楼梯材料	——	梯段宽度	——								
	天花、藻井	天花	有	天花类型	井口	藻井	无	藻井类型	——								
	彩画	柱头	有	柱身	无	梁架	有	走马板	有								
		门、窗	无	天花	有	藻井	——	其他彩画	——								
	其他	悬塑	无	佛龛	有	匾额	“佛泽万物”										
装饰	室内	帷幔	无	幕帘彩绘	无	壁画	有	唐卡	无								
		经幡	无	经幢	无	柱毯	无	其他	——								
	室外	玛尼轮	无	苏勒德	有	宝顶	有	祥麟法轮	有								
		四角经幢	无	经幡	有	铜饰	无	石刻、砖雕	无								
		仙人走兽	1+4	壁画	有	其他	——										
陈设	室内	主佛像	无	佛像基座	无												
		法座	无	藏经橱	无	经床	无	诵经桌	无	法鼓	无	玛尼轮	有	坛城	无	其他	——
	室外	旗杆	有	苏勒德	有	狮子	有	经幡	有	玛尼轮	有	香炉	有	五供	无	其他	——
	其他	——															
备注	菩提过殿由南北两个空间组成，南侧为经堂，北侧为佛殿。佛殿分为东西两个佛殿，东侧为药师佛殿，西侧为密宗护法殿，中间以过道分隔																
调查日期	2010/05/01	调查人员	白丽燕、萨日朗	整理日期	2010/05/12	整理人员	萨日朗										

菩提过殿基本概况表1

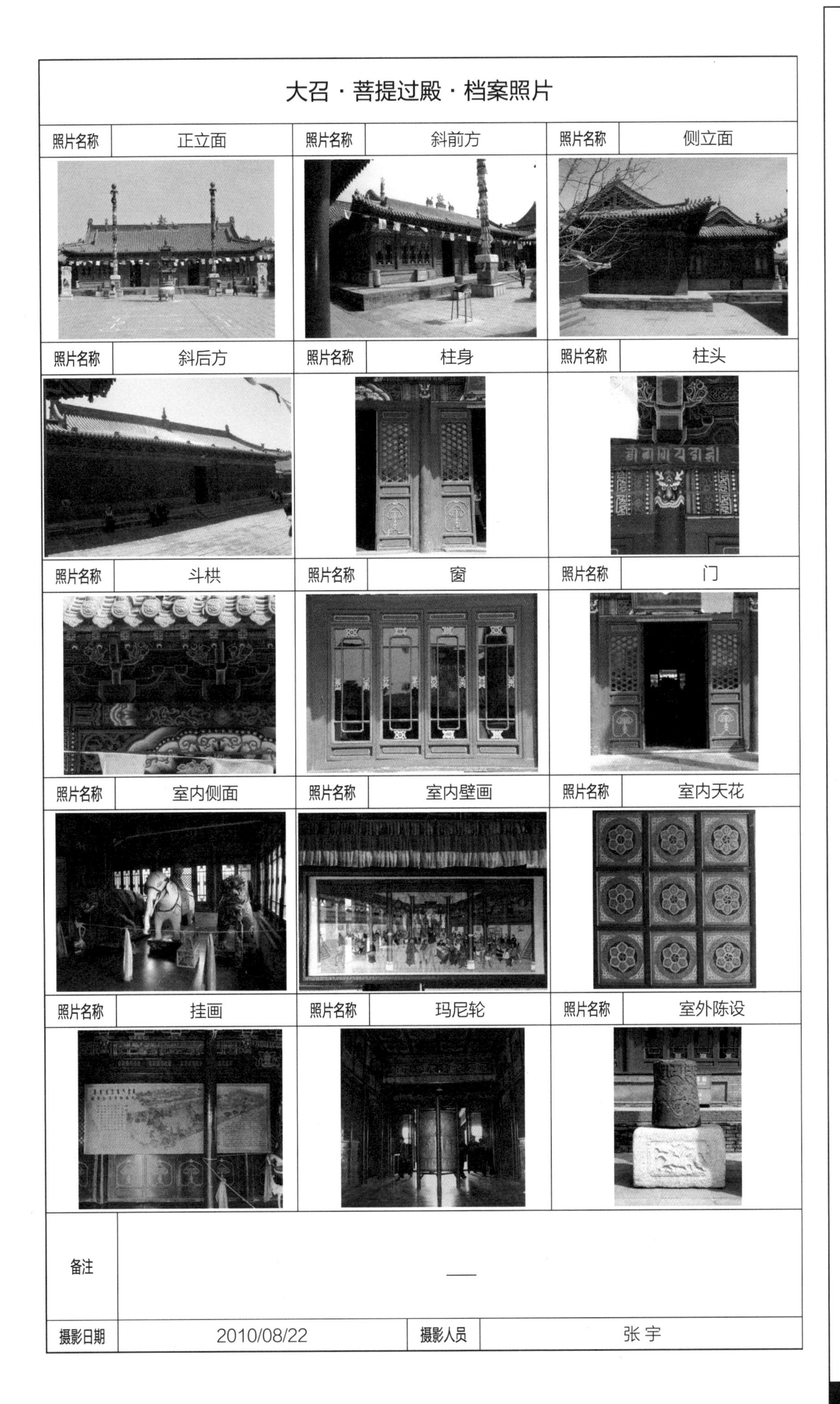

大召・菩提过殿・档案照片

照片名称	正立面	照片名称	斜前方	照片名称	侧立面
照片名称	斜后方	照片名称	柱身	照片名称	柱头
照片名称	斗栱	照片名称	窗	照片名称	门
照片名称	室内侧面	照片名称	室内壁画	照片名称	室内天花
照片名称	挂画	照片名称	玛尼轮	照片名称	室外陈设
备注	—				
摄影日期	2010/08/22	摄影人员	张 宇		

菩提过殿正立面

菩提过殿侧立面

菩提过殿玛尼轮
（左图）
菩提过殿室内天花
（右图）

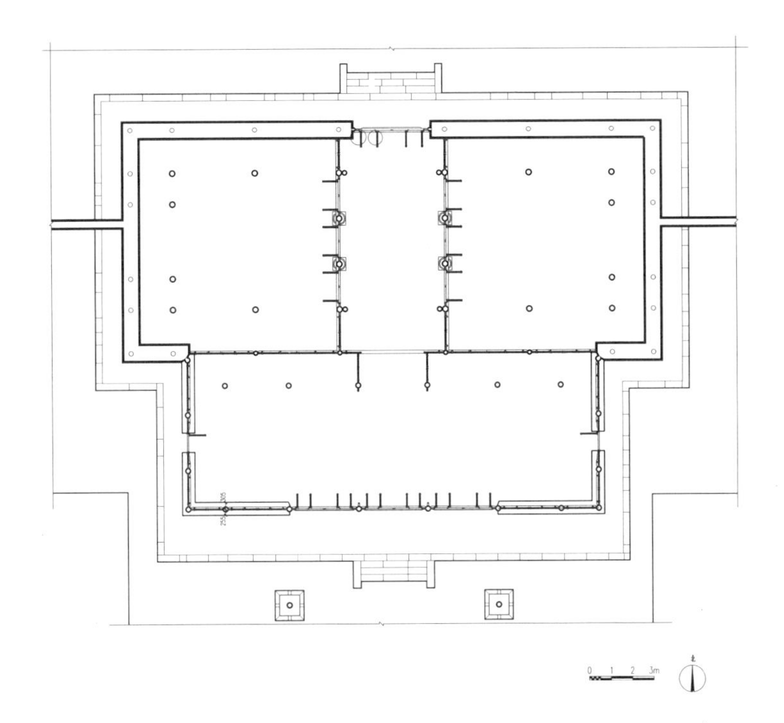

菩提过殿一层平面图

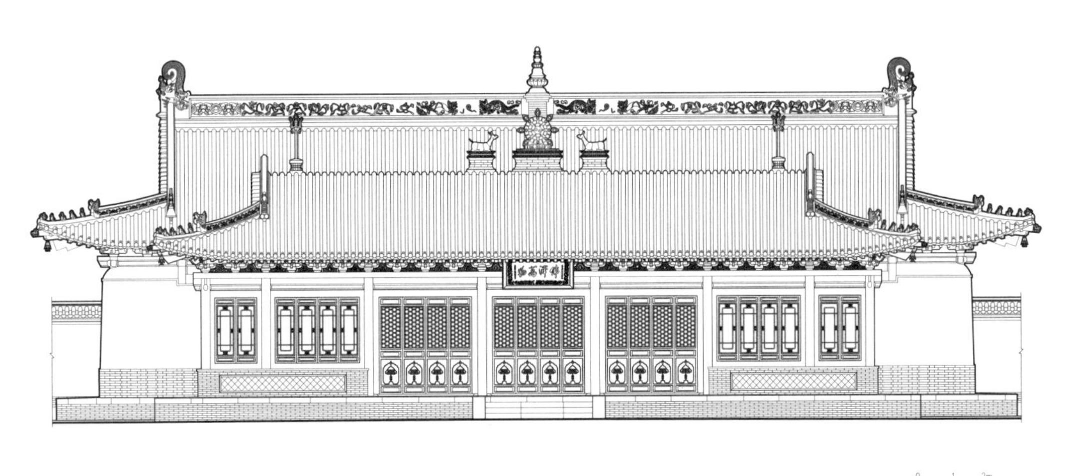

菩提过殿立面图

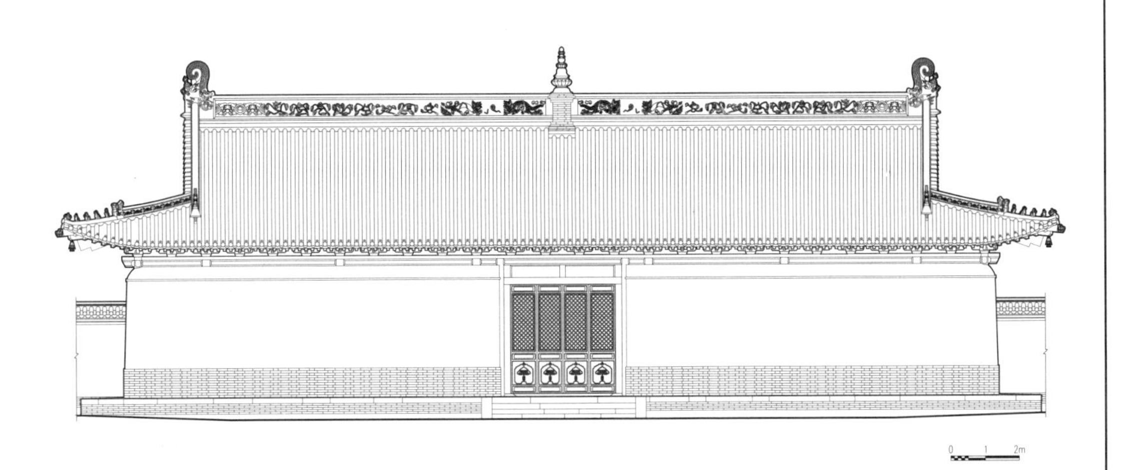

菩提过殿背立面图

菩提过殿侧立面图

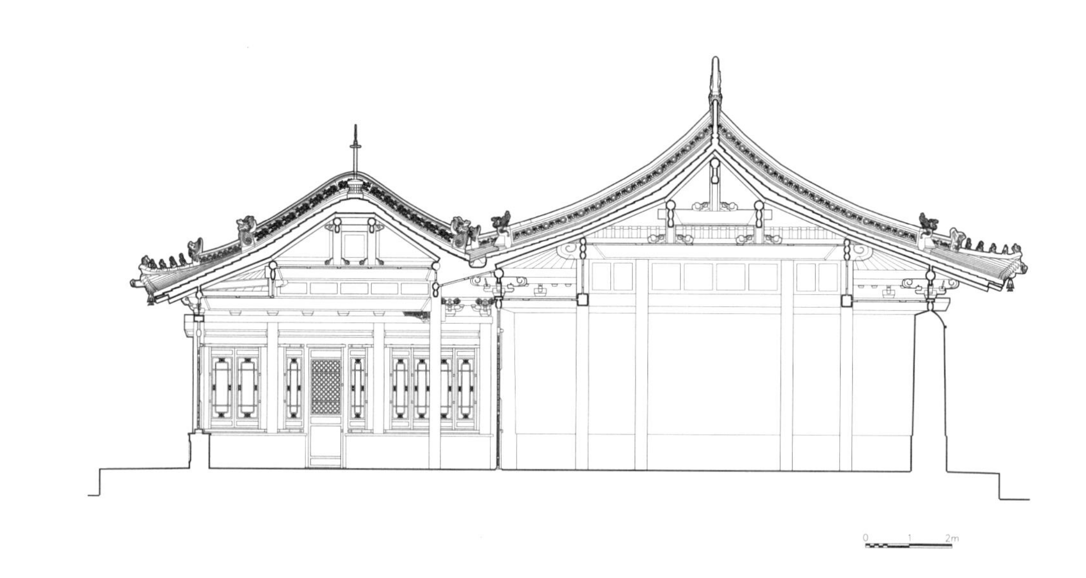

菩提过殿横剖面图

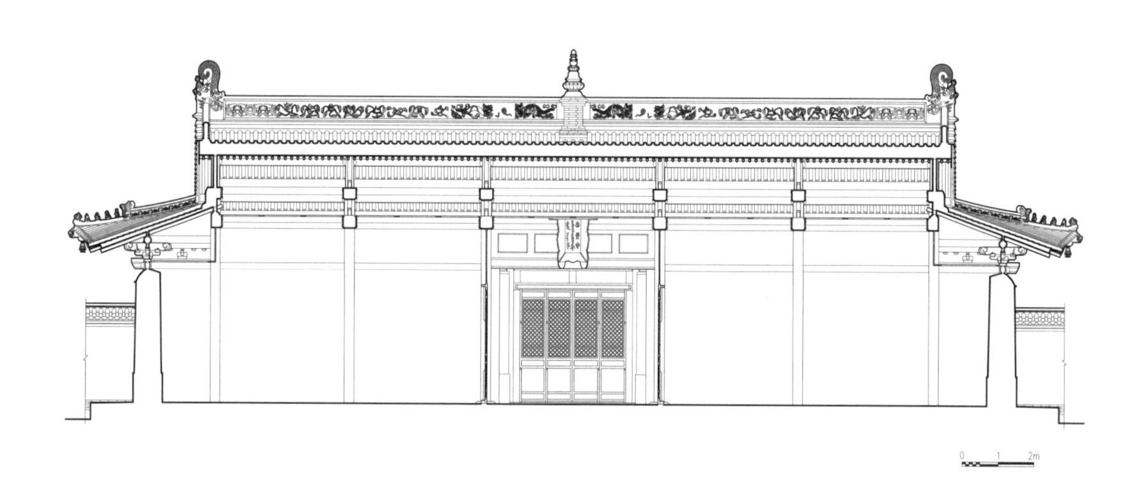

菩提过殿纵剖面图

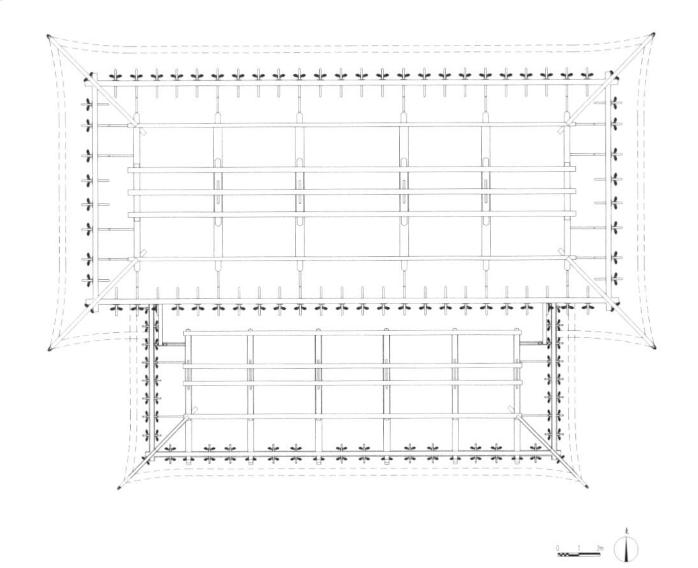

菩提过殿梁架分布图1

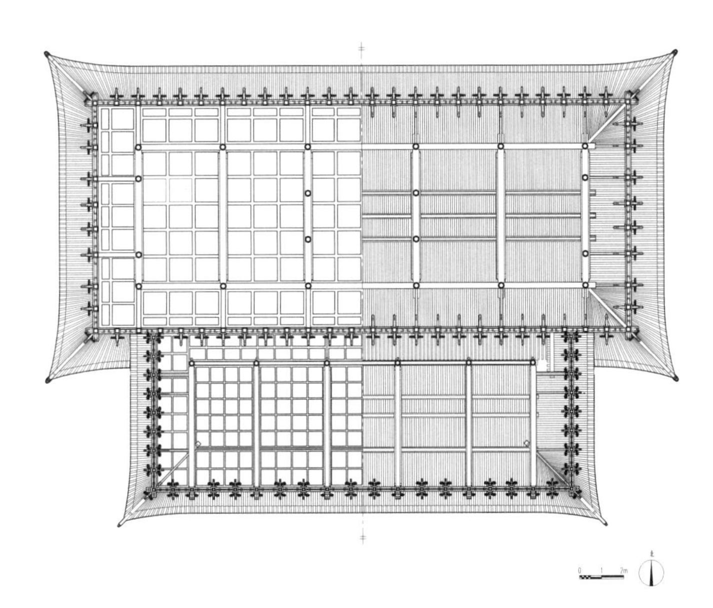

菩提过殿梁架分布图2

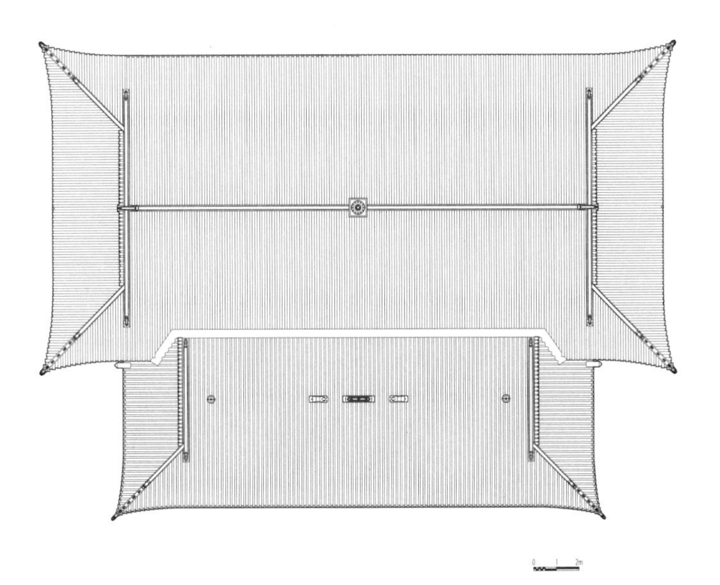

菩提过殿屋顶平面图

1.6 大召·密集佛殿

单位:毫米

建筑名称	汉语正式名称	密集佛殿	俗称	——													
概述	初建年	万历十五年（1587年）始建	建筑朝向	东	建筑层数	一											
	建筑简要描述	汉式建筑，硬山屋顶，汉式结构体系															
	重建重修记载	1982年重修															
		信息来源	调研访谈记录														
结构规模	结构形式	砖木混合	相连的建筑	无	室内天井	无											
	建筑平面形式	长方形	外廊形式	前廊式													
	通面阔	12390	开间数	5	明间	3530	次间	3000	梢间	1430	次梢间	——	尽间	——			
	通进深	8540	进深数	4	进深尺寸（前→后）	1390—2840—2900—1410											
	柱子数量	——	柱子间距	横向尺寸	——	（藏式建筑结构体系填写此栏，不含廊柱）											
				纵向尺寸													
	其他	——															
建筑主体（大木作）（石作）（瓦作）	屋顶	屋顶形式	硬山	瓦作	琉璃瓦												
	外墙	主体材料	青砖	材料规格	500×200×70	饰面颜色	红										
		墙体收分	无	边玛檐墙	无	边玛材料	——										
	斗栱、梁架	斗栱	无	平身科斗口尺寸	——	梁架关系	不详（有吊顶）										
	柱、柱式（前廊柱）	形式	汉式	柱身断面形状	圆	断面尺寸	直径D=220	（在没有前廊柱的情况下，填写室内柱及其特征）									
		柱身材料	木材	柱身收分	无	栌斗、托木	无	雀替	无								
		柱础	不可见	柱础形状	不详	柱础尺寸	不详										
	台基	台基类型	普通台基	台基高度	460	台基地面铺设材料	方砖										
	其他	——															
装修（小木作）（彩画）	门(正面)	隔扇门	门楣	无	堆经	无	门帘	无									
	窗（正面）	隔扇窗	窗楣	无	窗套	无	窗帘	无									
	室内隔扇	隔扇	无	隔扇位置	——												
	室内地面、楼面	地面材料及规格	青砖320×320	楼面材料及规格	无												
	室内楼梯	楼梯	无	楼梯位置	——	楼梯材料	——	梯段宽度	——								
	天花、藻井	天花	有	天花类型	井口天花	藻井	无	藻井类型	——								
	彩画	柱头	无	柱身	无	梁架	有	走马板	有								
		门、窗	无	天花	有	藻井	——	其他彩画	——								
	其他	悬塑	无	佛龛	有	匾额	无										
装饰	室内	帷幔	无	幕帘彩绘	无	壁画	无	唐卡	无								
		经幡	无	经幢	有	柱毯	无	其他	——								
	室外	玛尼轮	无	苏勒德	无	宝顶	无	祥麟法轮	无								
		四角经幢	无	经幡	无	铜饰	无	石刻、砖雕	无								
		仙人走兽	无	壁画	有	其他	——										
陈设	室内	主佛像	宗喀巴大师	佛像基座	莲花座												
		法座	无	藏经橱	无	经床	无	诵经桌	有	法鼓	无	玛尼轮	无	坛城	无	其他	——
	室外	旗杆	无	苏勒德	无	狮子	无	经幡	无	玛尼轮	无	香炉	无	五供	无	其他	——
	其他	——															
备注	——																
调查日期	2010/05/01	调查人员	白丽燕、萨日朗	整理日期	2010/05/12	整理人员	萨日朗										

密集佛殿基本概况表1

大召·密集佛殿·档案照片					
照片名称 	正立面	照片名称 	斜前方	照片名称 	侧立面
照片名称 	鸟瞰图	照片名称 	前廊	照片名称 	柱头
照片名称 	柱础	照片名称 	门	照片名称 	室外彩画
照片名称 	室内侧面	照片名称 	室内正面	照片名称 	佛龛
照片名称 	供桌	照片名称 	柱饰	照片名称 	天花
备注	——				
摄影日期	2010/08/22	摄影人员	白丽燕、萨日朗、孟祎军		

密集佛殿正立面

密集佛殿鸟瞰图

密集佛殿斜前方（左图）
密集佛殿侧立面（右图）

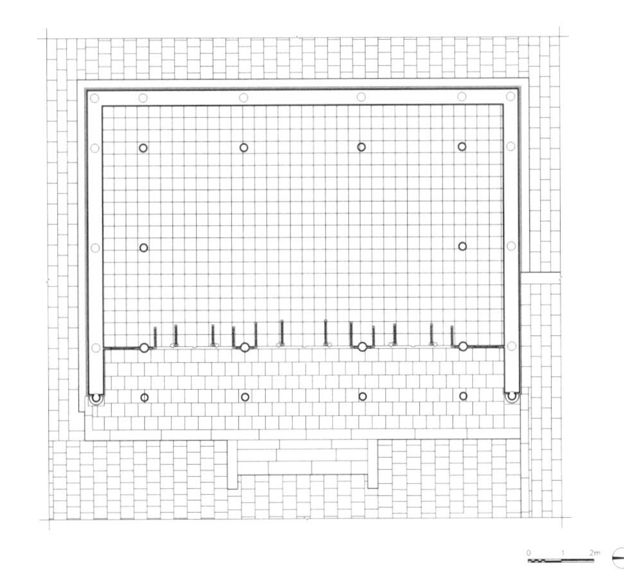

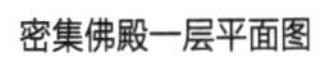

密集佛殿一层平面图

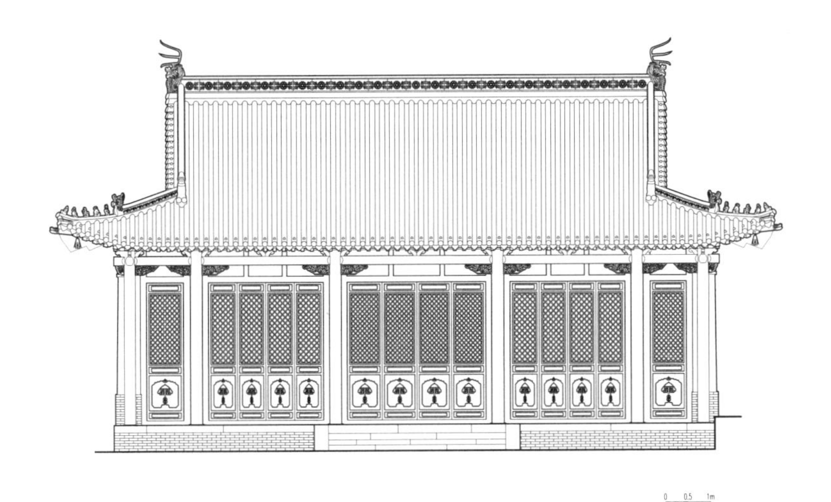

密集佛殿正立面图

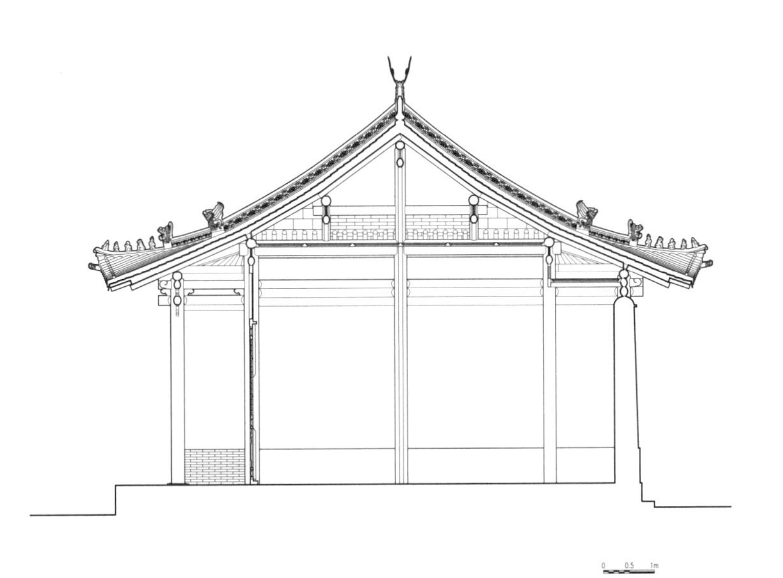

密集佛殿剖面图

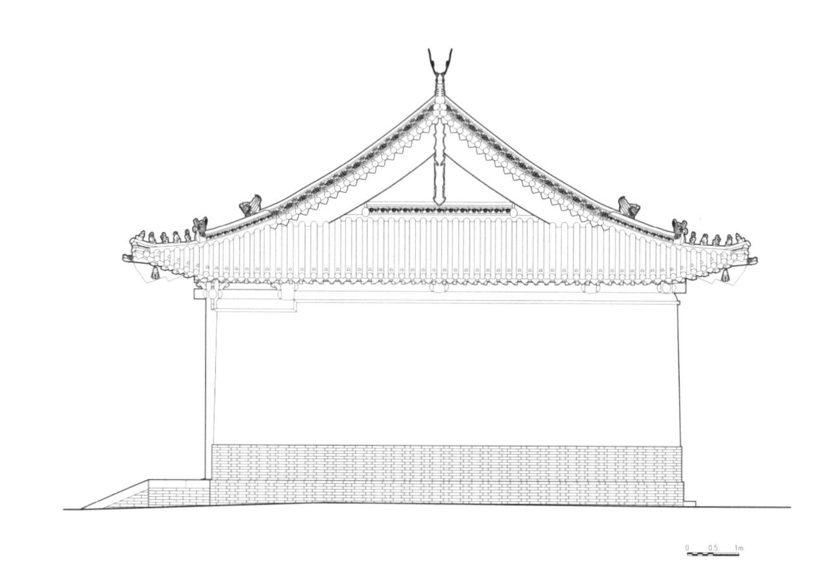

密集佛殿侧立面图

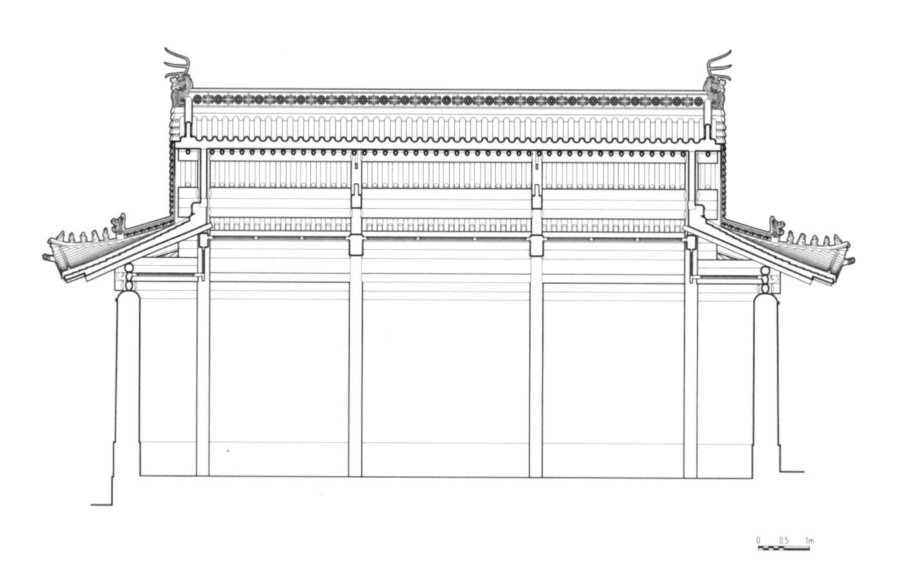

密集佛殿纵剖面图

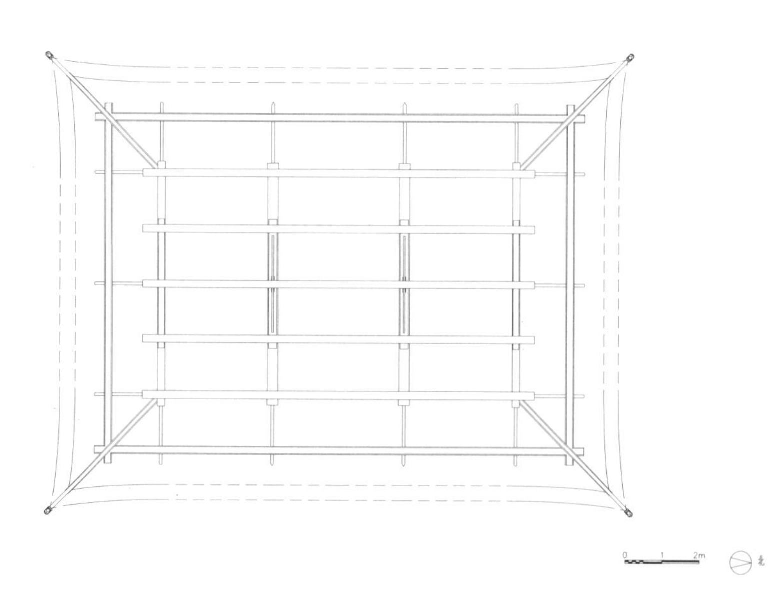

密集佛殿梁架分布图

1.7 大召·胜乐佛殿

单位:毫米

建筑名称	汉语正式名称	胜乐佛殿					俗称	——							
概述	初建年	万历十五年(1587年)始建					建筑朝向	西		建筑层数		一			
	建筑简要描述	汉式建筑,歇山屋顶,汉式结构体系													
	重建重修记载	——													
		信息来源	——												
结构规模	结构形式	砖木混合		相连的建筑	无					室内天井		无			
	建筑平面形式	长方形		外廊形式	前廊										
	通面阔	12900	开间数	5	明间	3500	次间	3300	梢间	1400	次梢间	——	尽间	——	
	通进深	8540	进深数	4	进深尺寸(前→后)			1390—2840—2900—1410							
	柱子数量	——	柱子间距	横向尺寸	——			(藏式建筑结构体系填写此栏,不含廊柱)							
				纵向尺寸											
	其他	——													
建筑主体(大木作)(石作)(瓦作)	屋顶	屋顶形式	歇山					瓦作		琉璃瓦					
	外墙	主体材料	青砖	材料规格	270×130×50			饰面颜色		红					
		墙体收分	无	边玛檐墙	无			边玛材料		——					
	斗栱、梁架	斗栱	无	平身科斗口尺寸		——		梁架关系		不详(有吊顶)					
	柱、柱式(前廊柱)	形式	汉式	柱身断面形状	圆	断面尺寸		直径 $D=200$			(在没有前廊柱的情况下,填写室内柱及其特征)				
		柱身材料	木材	柱身收分	无	栌斗、托木		无	雀替	有					
		柱础	有	柱础形状	方	柱础尺寸		400×400							
	台基	台基类型	砖砌台明	台基高度	530	台基地面铺设材料				青砖					
	其他	——													
装修(小木作)(彩画)	门(正面)	隔扇门		门楣	无	堆经		无		门帘		无			
	窗(正面)	槛窗		窗楣	无	窗套		无		窗帘		无			
	室内隔扇	隔扇	无	隔扇位置	——										
	室内地面、楼面	地面材料及规格		青砖 360×360		楼面材料及规格				——					
	室内楼梯	楼梯	无	楼梯位置	——	楼梯材料		——		楼段宽度		——			
	天花、藻井	天花	有	天花类型	井口天花	藻井		无		藻井类型		——			
	彩画	柱头	有	柱身	无	梁架		有		走马板		有			
		门、窗	无	天花	有	藻井		——		其他彩画		——			
	其他	悬塑	无	佛龛	有	匾额		无							
装饰	室内	帷幔	有	幕帘彩绘	无	壁画		无		唐卡		无			
		经幡	无	经幢	无	柱毯		无		其他		——			
	室外	玛尼轮	无	苏勒德	无	宝顶		无		祥麟法轮		无			
		四角经幢	无	经幡	无	铜饰		无		石刻、砖雕		无			
		仙人走兽	无	壁画	无	其他		——							
陈设	室内	主佛像	胜乐佛像			佛像基座		莲花座							
		法座	无	藏经橱	无	经床	无	诵经桌	无	法鼓	无	玛尼轮	无	坛城	无	其他	——
	室外	旗杆	无	苏勒德	无	狮子	无	经幡	无	玛尼轮	无	香炉	无	五供	无	其他	——
	其他	——													
备注	——														
调查日期	2010/05/01	调查人员	白丽燕、萨日朗	整理日期	2010/05/12	整理人员	萨日朗								

胜乐佛殿基本概况表1

大召·胜乐佛殿·档案照片

照片名称	正立面	照片名称	斜前方	照片名称	侧立面
照片名称	斜后方	照片名称	柱廊、柱身	照片名称	柱头
照片名称	柱础	照片名称	正门	照片名称	彩画
照片名称	室内正面	照片名称	室内侧面	照片名称	天花
照片名称	佛龛	照片名称	室内柱子	照片名称	佛像
备注	—				
摄影日期	2010/08/22	摄影人员	白丽燕、萨日朗、孟祎军		

胜乐佛殿基本概况表2

胜乐佛殿正立面

胜乐佛殿斜前方（左图）
胜乐佛殿侧立面（右图）

胜乐佛殿鸟瞰图（左图）
胜乐佛殿柱廊（右图）

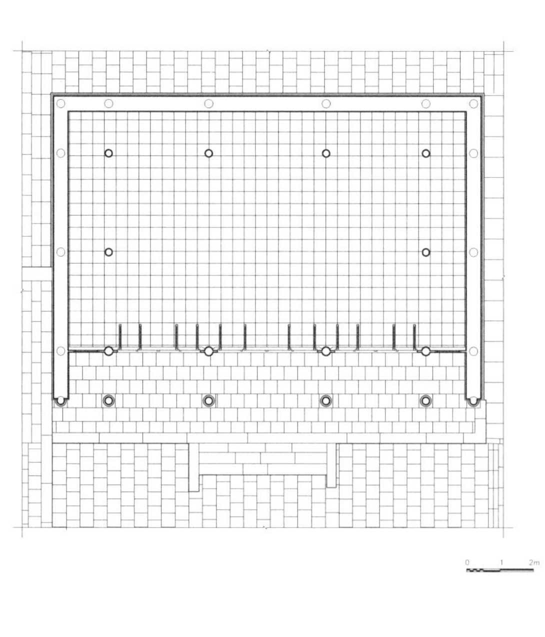

胜乐佛殿一层平面图

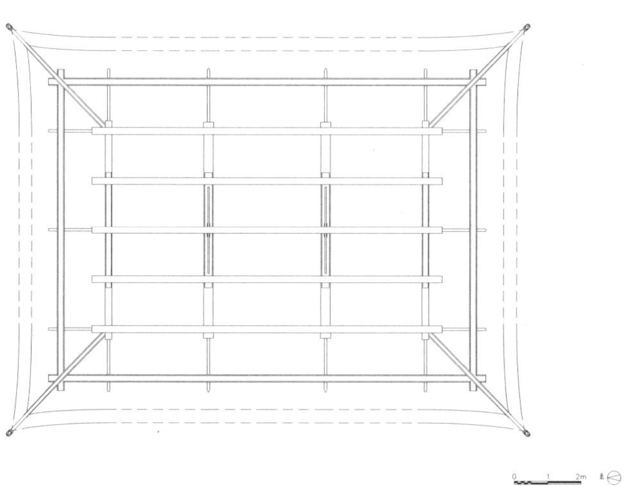

胜乐佛殿梁架平面图1

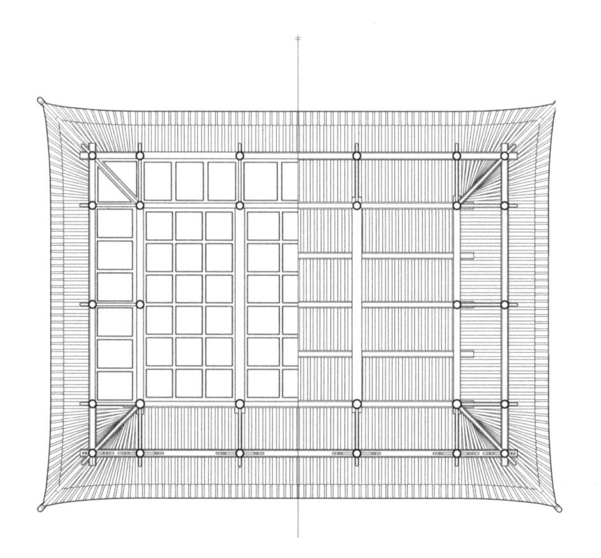

胜乐佛殿梁架平面图2

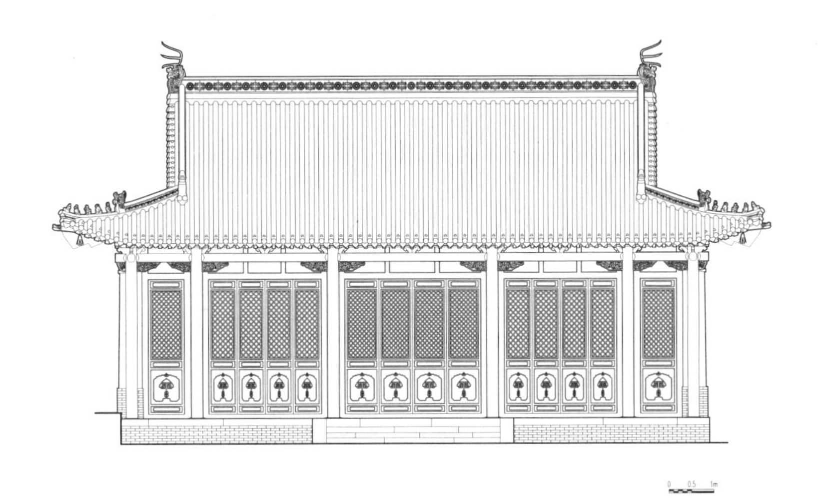

胜乐佛殿正立面图

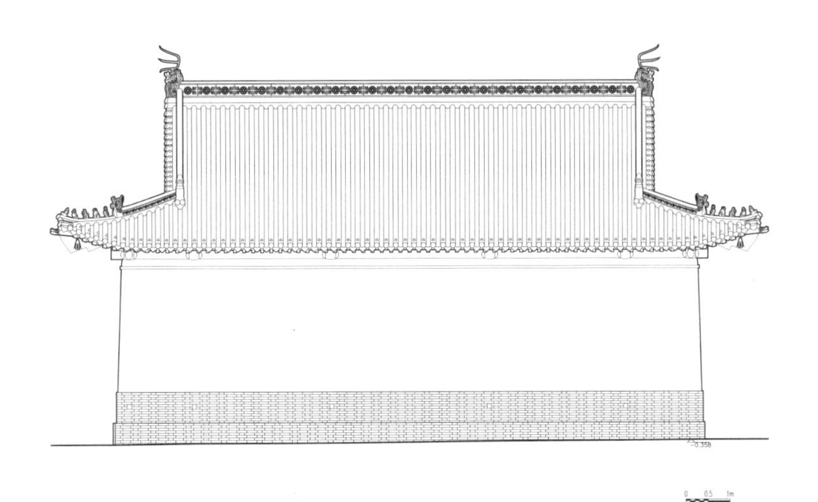

胜乐佛殿背立面图

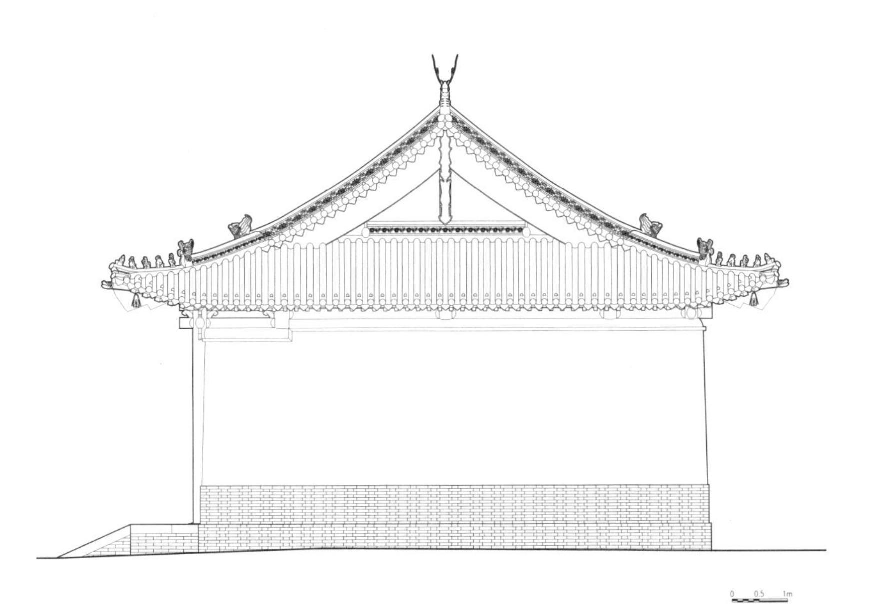

胜乐佛殿侧立面图

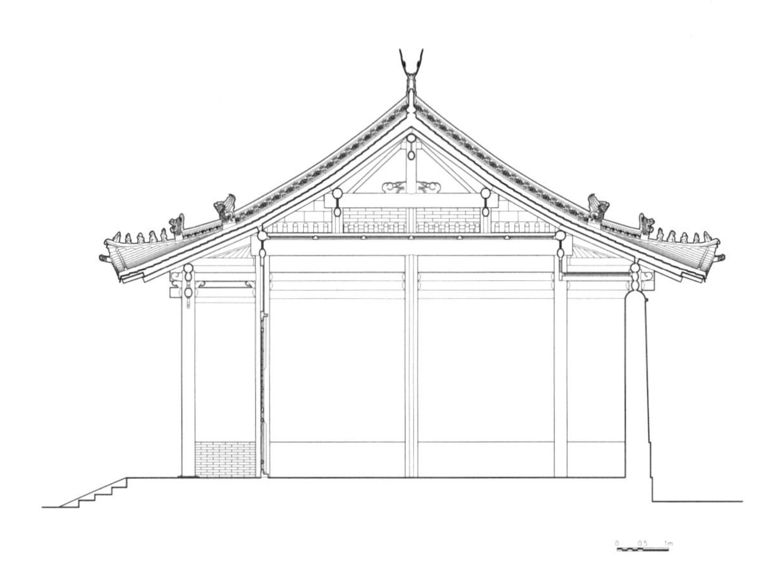

胜乐佛殿横剖面图1

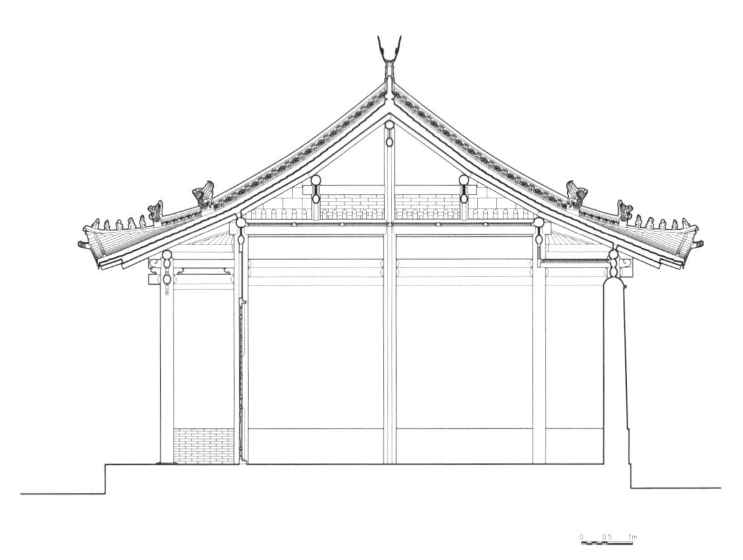

胜乐佛殿横剖面图2

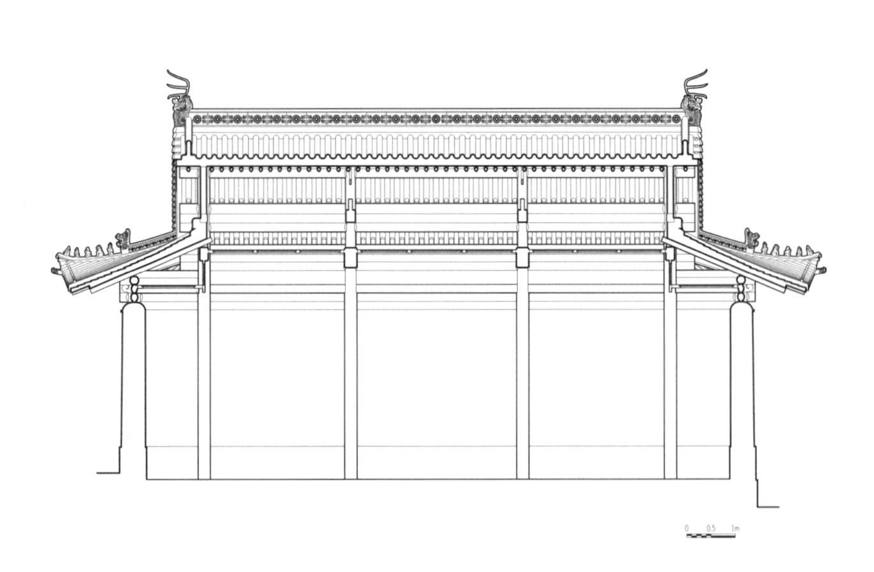

胜乐佛殿纵剖面图

1.8 大召·大雄宝殿

单位:毫米

建筑名称	汉语正式名称	大雄宝殿	俗称	大雄宝殿		
概述	初建年	万历十五年（1587年）始建	建筑朝向	南	建筑层数	二
	建筑简要描述	汉藏结合式，藏式分格装饰，歇山屋顶，汉式结构体系				
	重建重修记载	1985—1986年进行了抢救性维修、2003年进行了全面维修				
		信息来源	大召大记事			

结构规模												
	结构形式	砖木混合	相连的建筑	无			室内天井	都纲法式				
	建筑平面形式	凸字形	外廊形式	前廊、半回廊								
	通面阔	20560	开间数	7	明间	3700	次间	3600	梢间	1950	次梢间	——
											尽间	2880
	通进深	37580	进深数	12	进深尺寸（前→后）	80→3400→3750→3550→3750→3670→1230→3200→3300→3250→3300→2600						
	柱子数量	——	柱子间距	横向尺寸 / 纵向尺寸	——	（藏式建筑结构体系填写此栏，不含廊柱）						
	其他	——										

建筑主体（大木作）（石作）（瓦作）									
	屋顶	屋顶形式	歇山			瓦作	黄色琉璃瓦		
	外墙	主体材料	青砖	材料规格	500×200×70	饰面颜色	青砖色		
		墙体收分	有	边玛檐墙	有	边玛材料	边玛草		
	斗栱、梁架	斗栱	有	平身科斗口尺寸	不详	梁架关系	不详（有吊顶）		
	柱、柱式（前廊柱）	形式	汉式	柱身断面形状	圆	断面尺寸	直径 D=360		（在没有前廊柱的情况下，填写室内柱及其特征）
		柱身材料	木材	柱身收分	无	栌斗、托木	有	雀替 有	
		柱础	有	柱础形状	方上圆	柱础尺寸	580×580、直径 D=500		
	台基	台基类型	普通台基	台基高度	700	台基地面铺设材料	方砖		
	其他	——							

装修（小木作）（彩画）									
	门(正面)	板门		门楣	有	堆经	有	门帘	无
	窗（正面窗）	一层殿门两侧藏式盲窗、二层隔扇窗		窗楣	有	窗套	无	窗帘	无
	室内隔扇	隔扇	有	隔扇位置	经堂佛殿之间以隔扇相隔				
	室内地面、楼面	地面材料及规格		木地板、100宽		楼面材料及规格		木地板、100宽	
	室内楼梯	楼梯	有	楼梯位置	殿门左侧	楼梯材料	木	梯段宽度	900
	天花、藻井	天花	无	天花类型	井口天花	藻井	有	藻井类型	八方形
	彩画	柱头	有	柱身	无	梁架	有	走马板	无
		门、窗	有	天花	有	藻井	有	其他彩画	——
	其他	悬塑	无	佛龛	无	匾额	“普渡慈航”，壬午年		

装饰									
	室内	帷幔	无	幕帘彩绘	有	壁画	有	唐卡	有
		经幡	有	经幢	有	柱毯	有	其他	——
	室外	玛尼轮	有	苏勒德	有	宝顶	有	祥麟法轮	无
		四角经幢	有	经幡	有	铜饰	有	石刻、砖雕	无
		仙人走兽	1+4	壁画	有	其他	——		

陈设																	
	室内	主佛像	释迦牟尼佛像					佛像基座	莲花座								
		法座	有	藏经橱	无	经床	有	诵经桌	有	法鼓	有	玛尼轮	无	坛城	有	其他	——
	室外	旗杆	无	苏勒德	有	狮子	有	经幡	有	玛尼轮	有	香炉	有	五供	有	其他	——
	其他	——															

备注	——						
调查日期	2010/05/01	调查人员	白丽燕、萨日朗	整理日期	2010/05/12	整理人员	萨日朗

大雄宝殿基本概况表1

大召·大雄宝殿·档案照片

照片名称	正立面	照片名称	斜前方	照片名称	侧立面
照片名称	斗栱	照片名称	柱身	照片名称	柱头
照片名称	柱础	照片名称	门	照片名称	窗
照片名称	诵经区	照片名称	转经道	照片名称	天花
照片名称	佛殿正面	照片名称	佛像	照片名称	室内坛城
备注	—				
摄影日期	2010/08/22	摄影人员	白丽燕、萨日朗、孟祎军		

大雄宝殿基本概况表2

大雄宝殿正立面

大雄宝殿西立面（左图）
大雄宝殿斜前方（右图）

大雄宝殿诵经区（左图）
大雄宝殿转经道（右图）

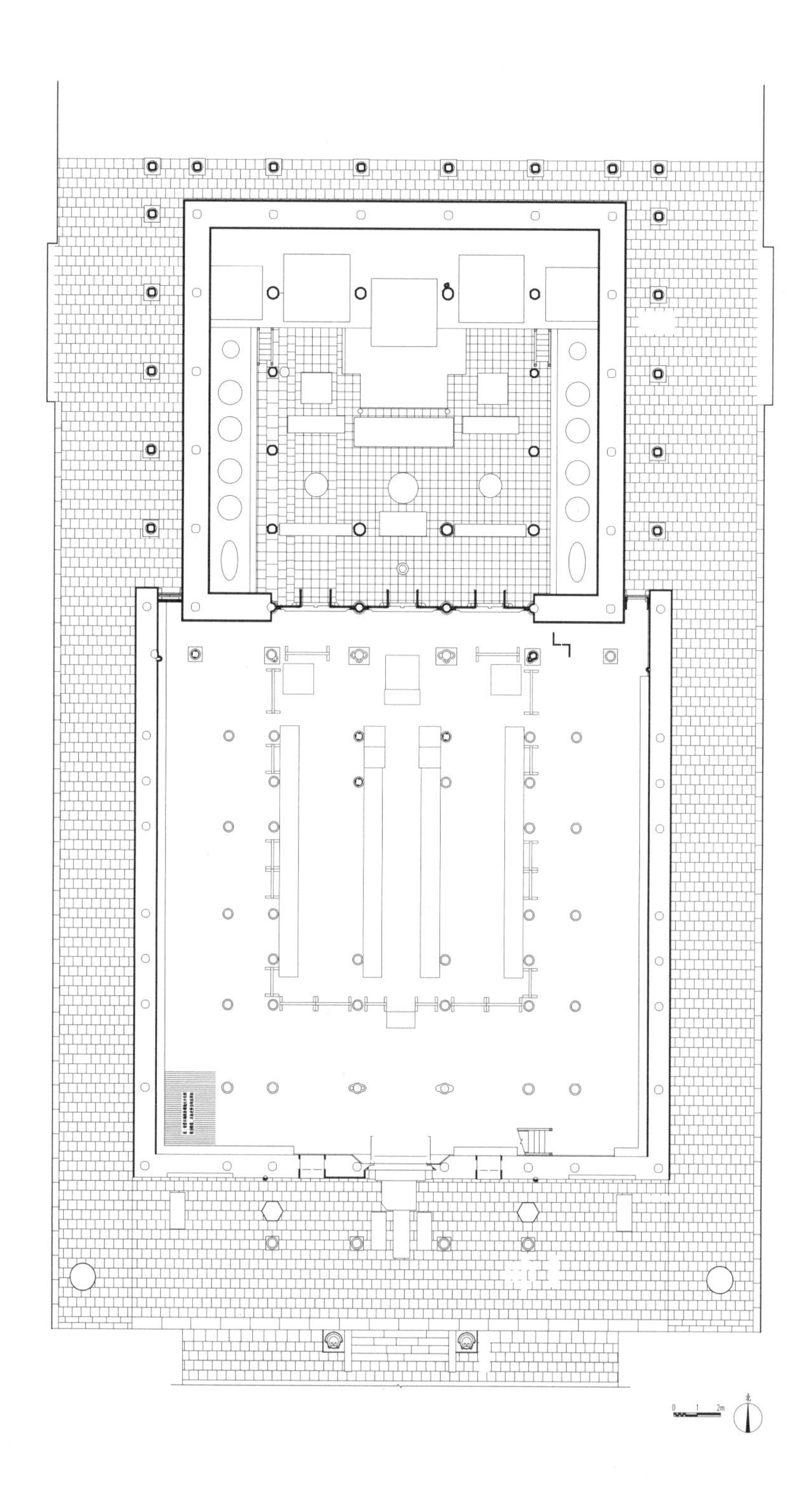

大雄宝殿一层平面图

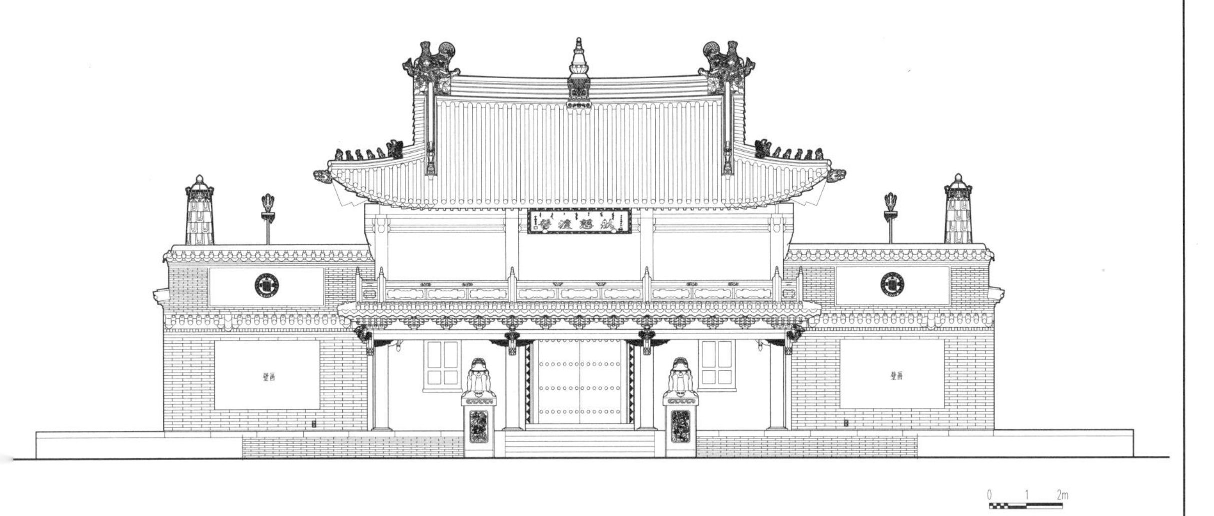

大雄宝殿正立面图

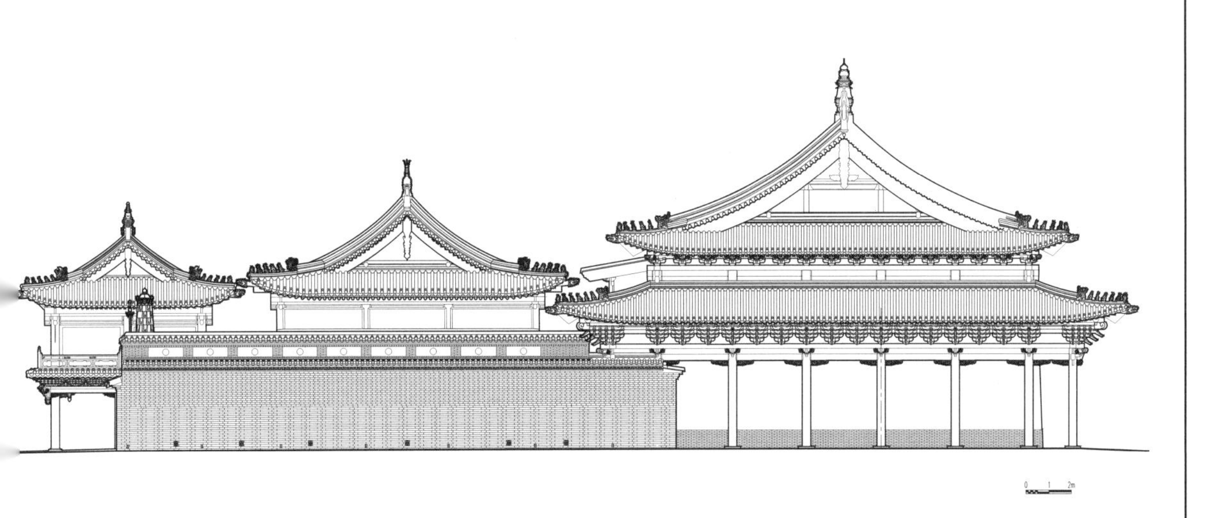

大雄宝殿侧立面图

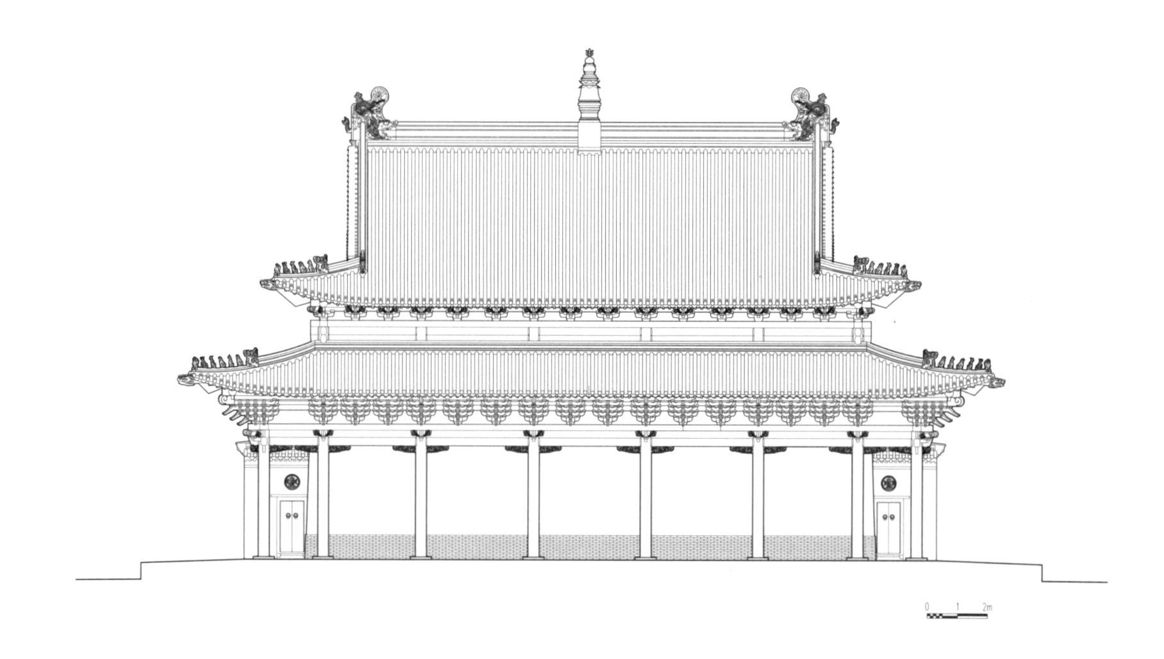

大雄宝殿背立面图

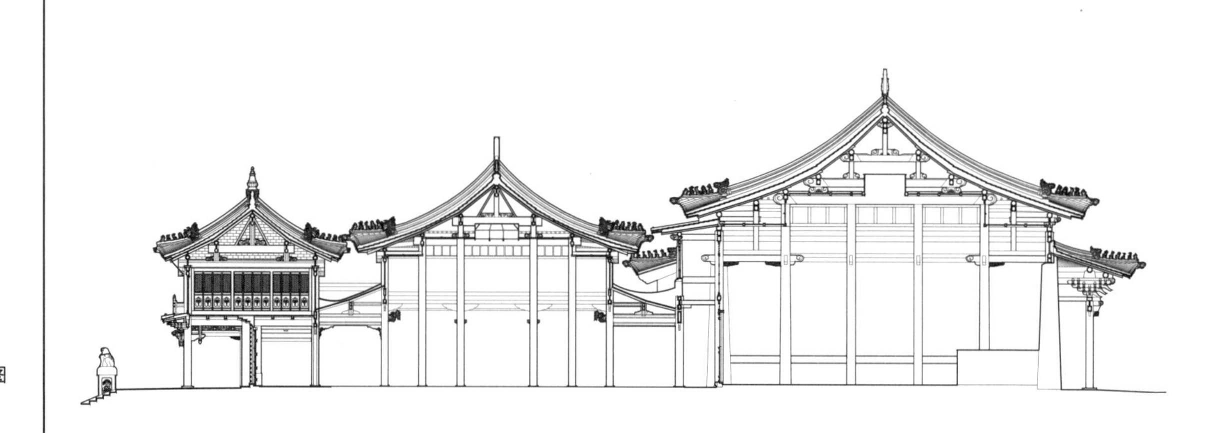

大雄宝殿横剖面图

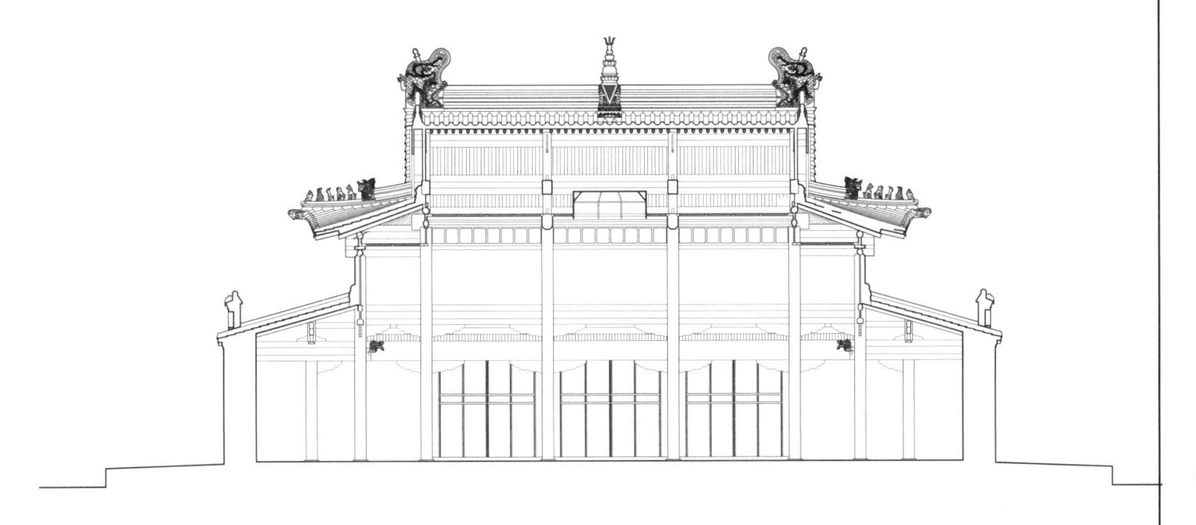

大雄宝殿纵剖面图

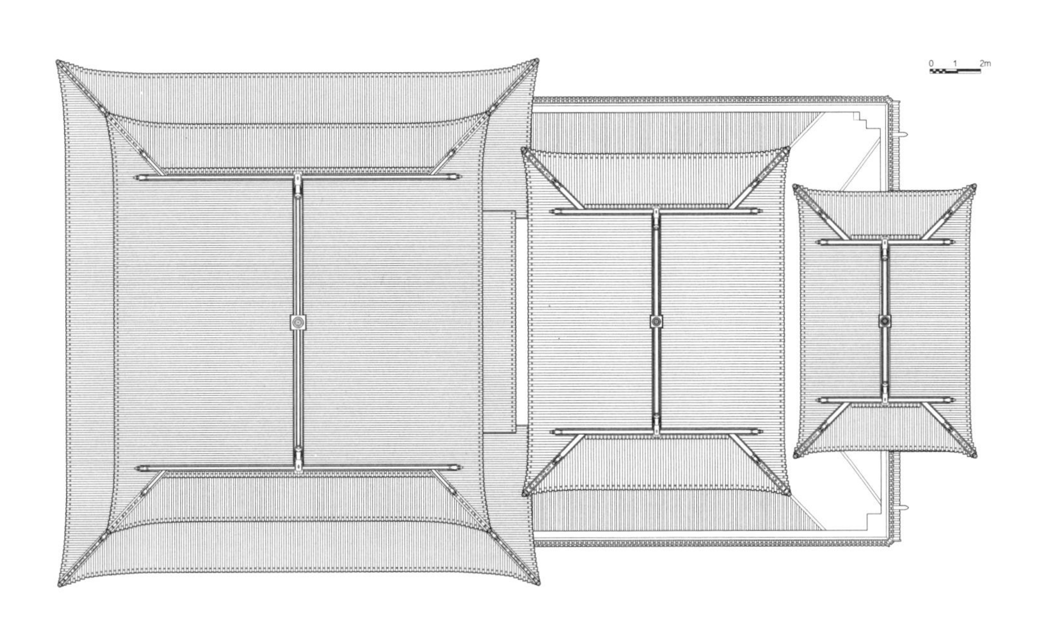

大雄宝殿屋顶平面图

1.9 大召·九间楼

单位:毫米

建筑名称	汉语正式名称	——	俗称	九间楼	
概述	初建年	万历十五年（1587年）始建	建筑朝向	南	建筑层数 二
	建筑简要描述	汉式建筑，硬山屋顶，屋顶为灰色布瓦，是一座二层的配有东西两个耳房的古建筑			
	重建重修记载	——			
		信息来源	——		

结构规模														
	结构形式	砖木混合	相连的建筑	无			室内天井	无						
	建筑平面形式	长方形	外廊形式	前廊										
	通面阔	26900	开间数	9	明间	3160	次间	3100	梢间	3100	次梢间	2600	尽间	2600
	通进深	6090	进深数	2	进深尺寸（前→后）		4900→1190							
	柱子数量	——	柱子间距	横向尺寸	——		（藏式建筑结构体系填写此栏，不含廊柱）							
				纵向尺寸										
	其他	——												

建筑主体（大木作）（石作）（瓦作）									
屋顶	屋顶形式	硬山				瓦作	布瓦		
外墙	主体材料	青砖	材料规格	300×130×50		饰面颜色	青砖色		
	墙体收分	无	边玛檐墙	无		边玛材料	无		
斗栱、梁架	斗栱	无	平身科斗口尺寸		——	梁架关系	不详（吊顶）		
柱、柱式（前廊柱）	形式	汉式	柱身断面形状	圆	断面尺寸	直径 D=240			（在没有前廊柱的情况下，填写室内柱及其特征）
	柱身材料	木材	柱身收分	无	栌斗、托木	有	雀替	有	
	柱础	有	柱础形状	方	柱础尺寸	570×570			
台基	台基类型	砖砌台明	台基高度	480	台基地面铺设材料		方砖		
其他	——								

装修（小木作）（彩画）								
门(正面)	隔扇门		门楣	无	堆经	无	门帘	无
窗（正面）	槛窗		窗楣	无	窗套	无	窗帘	无
室内隔扇	隔扇	有	隔扇位置	东西两面阔的位置				
室内地面、楼面	地面材料及规格		方砖280×280		楼面材料及规格		木板,宽80	
室内楼梯	楼梯	有	楼梯位置	西侧靠墙	楼梯材料	木材	梯段宽度	1100
天花、藻井	天花	有	天花类型	海墁天花	藻井	无	藻井类型	——
彩画	柱头	无	柱身	无	梁架	无	走马板	有
	门、窗	无	天花	无	藻井	无	其他彩画	——
其他	悬塑	无	佛龛	无	匾额	无		

装饰								
室内	帷幔	无	幕帘彩绘	无	壁画	无	唐卡	无
	经幡	无	经幢	无	柱毯	无	其他	——
室外	玛尼轮	无	苏勒德	无	宝顶	无	祥麟法轮	无
	四角经幢	无	经幡	无	铜饰	无	石刻、砖雕	无
	仙人走兽	无	壁画	无	其他	——		

陈设																
室内	主佛像		无						佛像基座		——					
	法座	——	藏经橱	——	经床	——	诵经桌	——	法鼓	——	玛尼轮	——	坛城	——	其他	——
室外	旗杆	无	苏勒德	无	狮子	有	经幡	无	玛尼轮	无	香炉	有	五供	无	其他	——
其他	——															

备注	一层室内现已用为库房，二层室内在装修						
调查日期	2010/05/01	调查人员	白丽燕、萨日朗	整理日期	2010/05/12	整理人员	萨日朗

九间楼基本概况表1

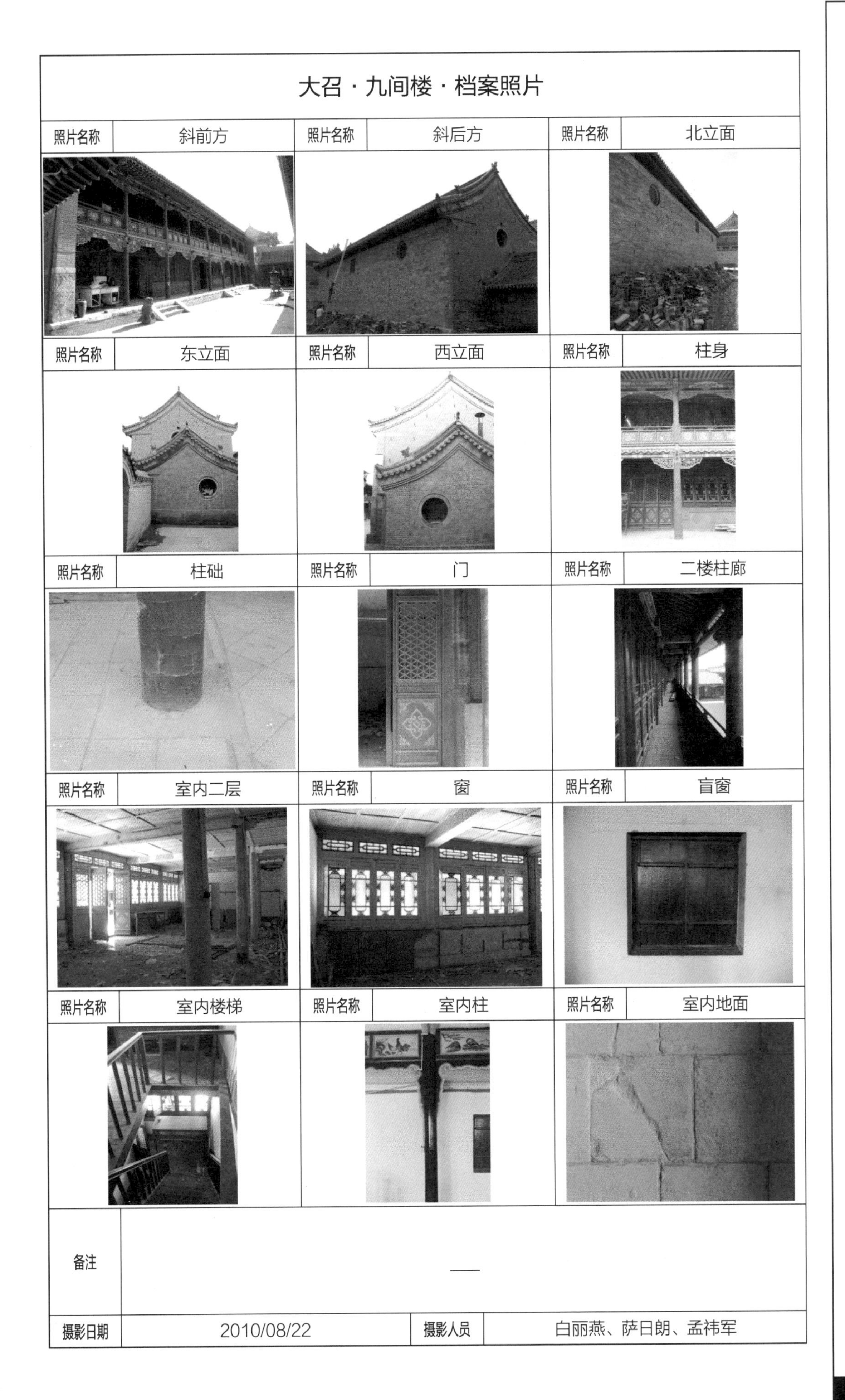

大召·九间楼·档案照片

照片名称	斜前方	照片名称	斜后方	照片名称	北立面
照片名称	东立面	照片名称	西立面	照片名称	柱身
照片名称	柱础	照片名称	门	照片名称	二楼柱廊
照片名称	室内二层	照片名称	窗	照片名称	盲窗
照片名称	室内楼梯	照片名称	室内柱	照片名称	室内地面

备注	—		
摄影日期	2010/08/22	摄影人员	白丽燕、萨日朗、孟祎军

九间楼斜前方

九间楼斜后方（左图）
九间楼东立面（右图）

九间楼室内（左图）
九间楼二楼柱廊（右图）

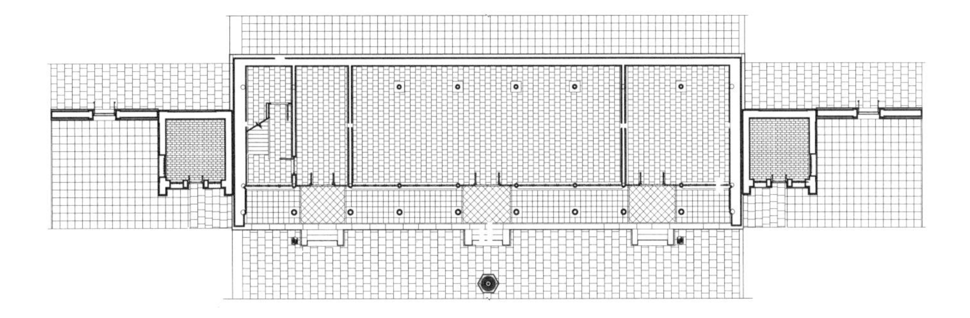

九间楼一层平面图

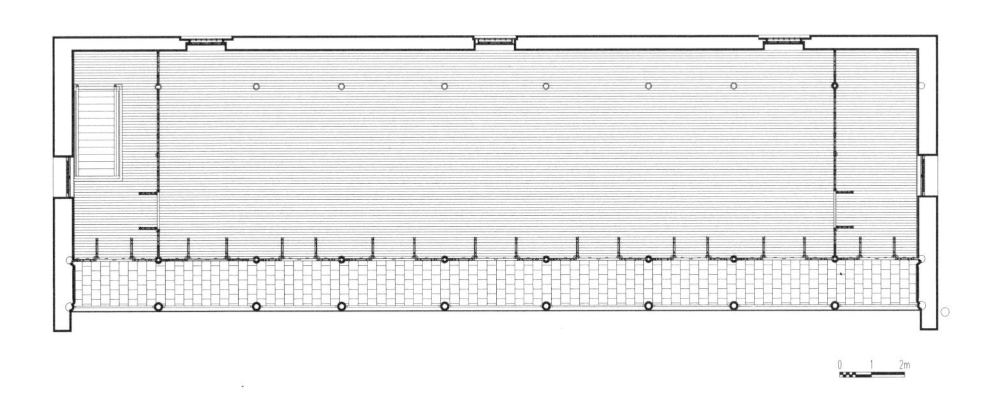

九间楼二层平面图

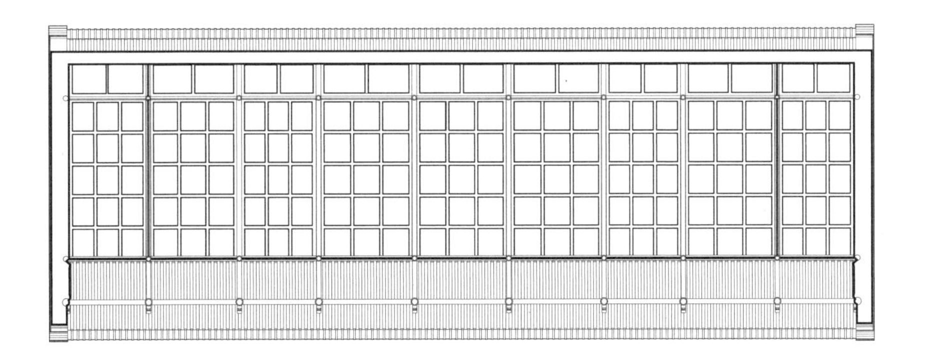

九间楼顶棚平面图

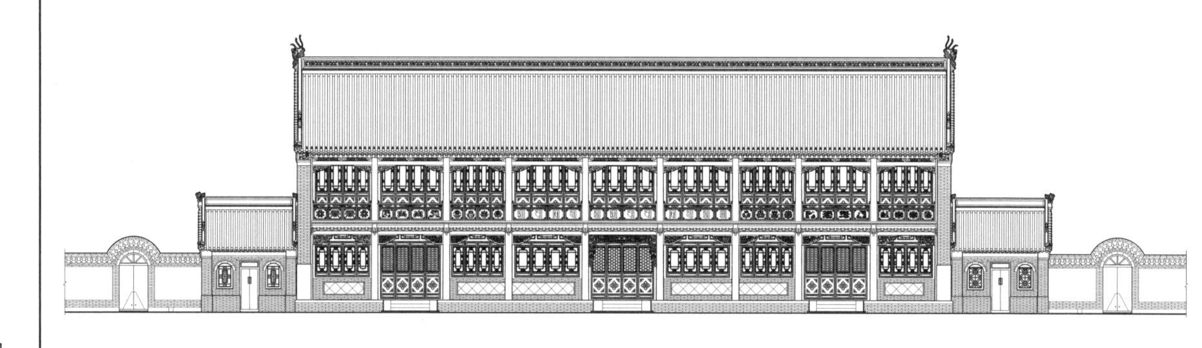

九间楼正立面图

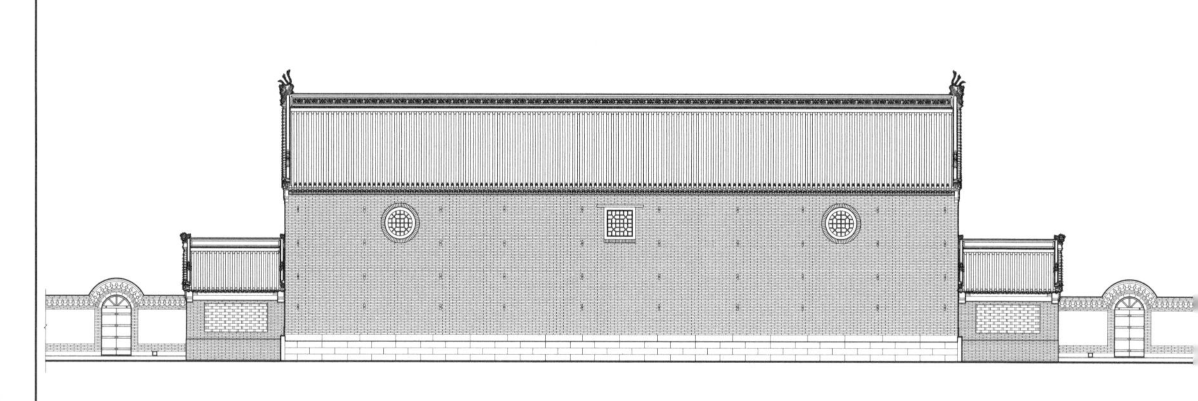

九间楼背立面图

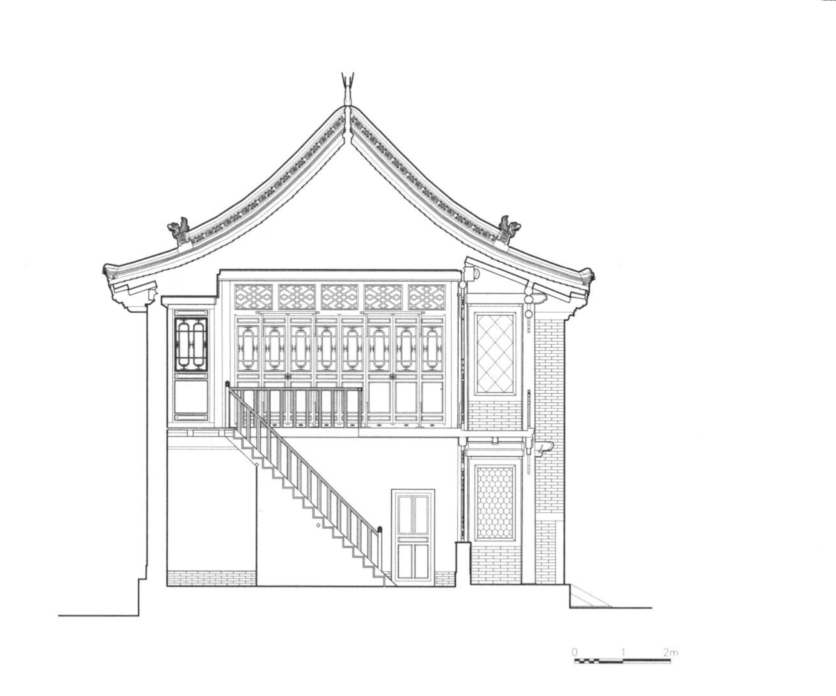

九间楼剖面图

1.10 大召·乃春庙

单位:毫米

建筑名称	汉语正式名称	护法殿	俗称	乃春庙		
概述	初建年	万历十五年（1587年）始建	建筑朝向	南	建筑层数	二
	建筑简要描述	汉式梁架结构，外墙藏式装饰,经堂以藏式为主，佛殿以汉式为主				
	重建重修记载	2004—2005年进行修缮				
		信息来源	大召大事记			

结构规模														
结构形式	砖木混合	相连的建筑	无	室内天井	都纲法式									
建筑平面形式	凸字形	外廊形式	前廊、半回廊											
通面阔	10040	开间数	5	明间	3220	次间	1930	梢间	1480	次梢间	——	尽间	——	
通进深	20160	进深数	8	进深尺寸（前→后）	2530→2800→2900→2900→1940→1300→2880→2910									
柱子数量	——	柱子间距	横向尺寸 / 纵向尺寸	——	（藏式建筑结构体系填写此栏，不含廊柱）									
其他	——													

建筑主体（大木作）（石作）（瓦作）										
屋顶	屋顶形式	歇山	瓦作	布瓦						
外墙	主体材料	土坯	材料规格	140×290×60	饰面颜色	青砖色				
	墙体收分	无	边玛檐墙	有	边玛材料	边玛草				
斗栱、梁架	斗栱	有	平身科斗口尺寸	50	梁架关系	抬梁式，五檩前廊				
柱、柱式（前廊柱）	形式	汉式	柱身断面形状	圆	断面尺寸	直径 $D=360$			（在没有前廊柱的情况下，填写室内柱及其特征）	
	柱身材料	木材	柱身收分	无	栌斗、托木	无	雀替	有		
	柱础	有	柱础形状	方形	柱础尺寸	500×500				
台基	台基类型	普通式台基	台基高度	150	台基地面铺设材料	条石				
其他	经堂：单檐歇山，佛殿：重檐歇山									

装修（小木作）（彩画）								
门(正面)	板门		门楣	无	堆经	无	门帘	无
窗（正面）	无窗（二层有隔扇）		窗楣	无	窗套	无	窗帘	无
室内隔扇	隔扇	有	隔扇位置	经堂与佛殿之间				
室内地面、楼面	地面材料及规格		木板，宽80,长不规则		楼面材料及规格		不规则木板	
室内楼梯	楼梯	有	楼梯位置	进门左侧	楼梯材料	木	梯段宽度	890
天花、藻井	天花	有	天花类型	井口	藻井	无	藻井类型	——
彩画	柱头	有	柱身	无	梁架	有	走马板	有
	门、窗	无	天花	有	藻井	——	其他彩画	斗栱有
其他	悬塑	无	佛龛	无	匾额	“乃春庙”，己卯年立		

装饰								
室内	帷幔	无	幕帘彩绘	有	壁画	有	唐卡	有
	经幡	无	经幢	有	柱毯	无	其他	——
室外	玛尼轮	无	苏勒德	无	宝顶	有	祥麟法轮	无
	四角经幢	有	经幡	有	铜饰	有	石刻、砖雕	有
	仙人走兽	1+3	壁画	有	其他	——		

陈设																
室内	主佛像	三世佛像							佛像基座		莲花座					
	法座	有	藏经橱	无	经床	有	诵经桌	有	法鼓	有	玛尼轮	无	坛城	无	其他	——
室外	旗杆	无	苏勒德	无	狮子	有	经幡	有	玛尼轮	无	香炉	有	五供	无	其他	——
其他	——															

备注	乃春是护法的一类，此建筑以前是护法殿，现在改为三世佛殿，但名称还叫乃春庙						
调查日期	2010/05/01	调查人员	白丽燕、萨日朗	整理日期	2010/05/12	整理人员	萨日朗

乃春庙基本概况表1

大召·乃春庙·档案照片

照片名称 	正立面	照片名称 	斜前方	照片名称 	侧立面
照片名称 	斜后方	照片名称 	北立面	照片名称 	局部
照片名称 	柱身	照片名称 	柱头	照片名称 	柱础
照片名称 	经堂室内正面	照片名称 	经堂室内侧面	照片名称 	经堂室内局部
照片名称 	经堂装饰	照片名称 	佛殿室内佛像	照片名称 	佛殿室内侧面
备注	—				
摄影日期	2010/08/22	摄影人员	白丽燕、萨日朗、孟祎军		

乃春庙基本概况表2

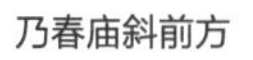

乃春庙斜前方

乃春庙经堂室内局部（左图）
乃春庙室外香炉（右图）

乃春庙佛殿室内正面（左图）
乃春庙经堂室内天花（右图）

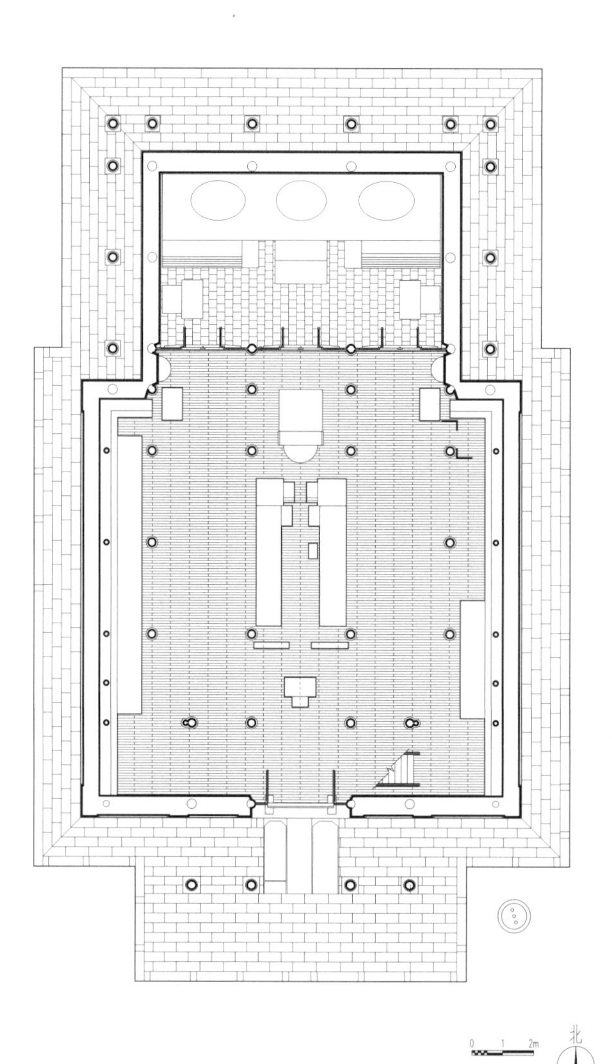

乃春庙一层平面图

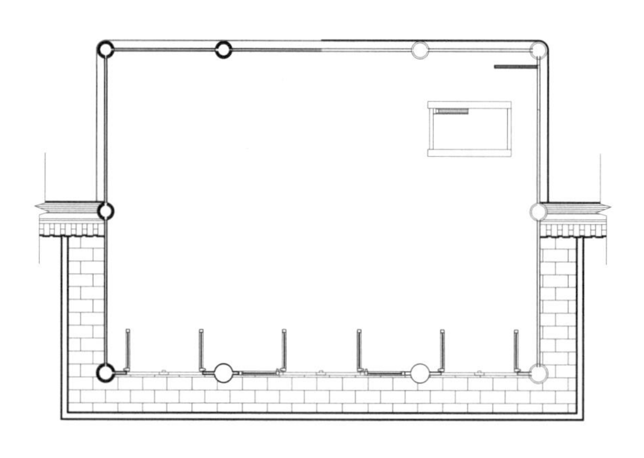

乃春庙二层平面图

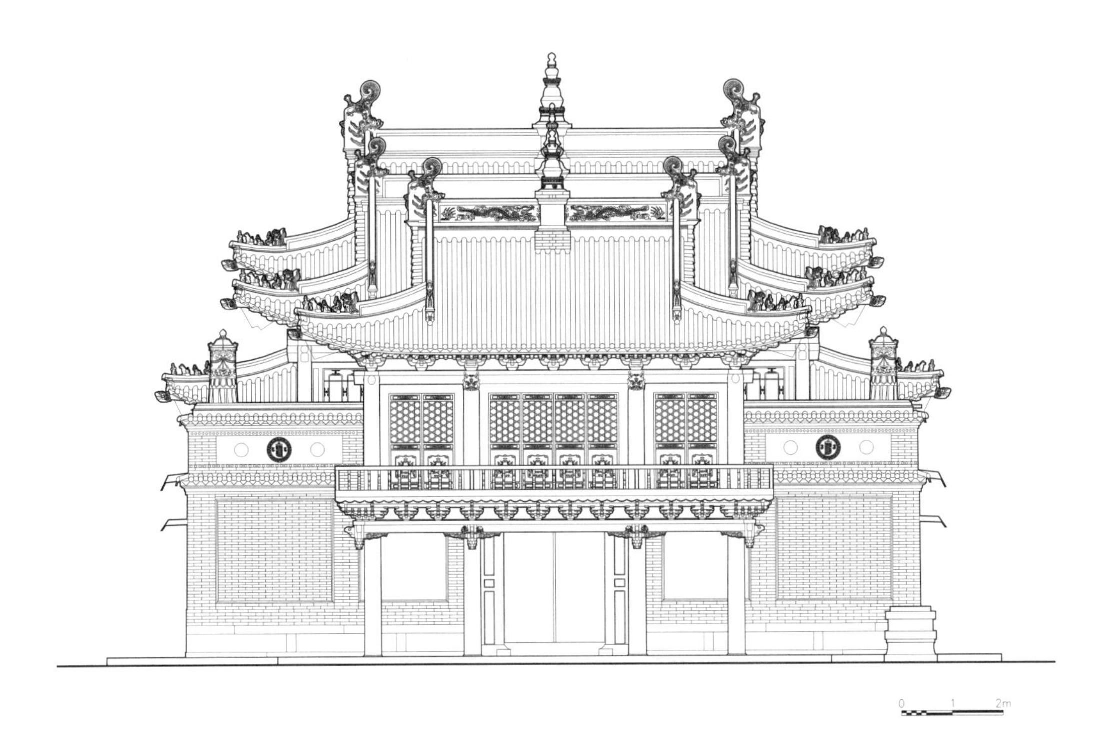

乃春庙经堂立面图

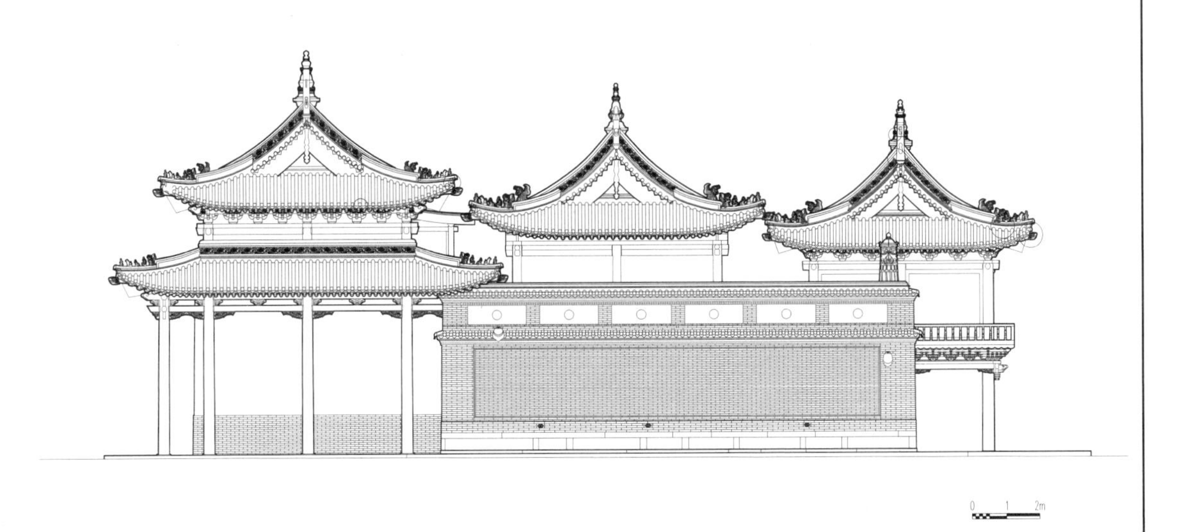

乃春庙经堂侧立面图

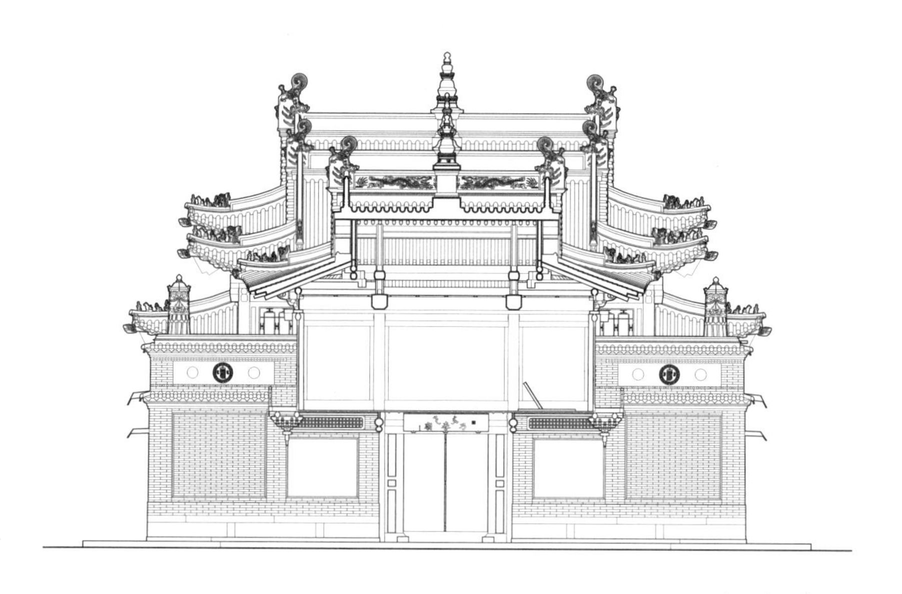

乃春庙纵剖面图1

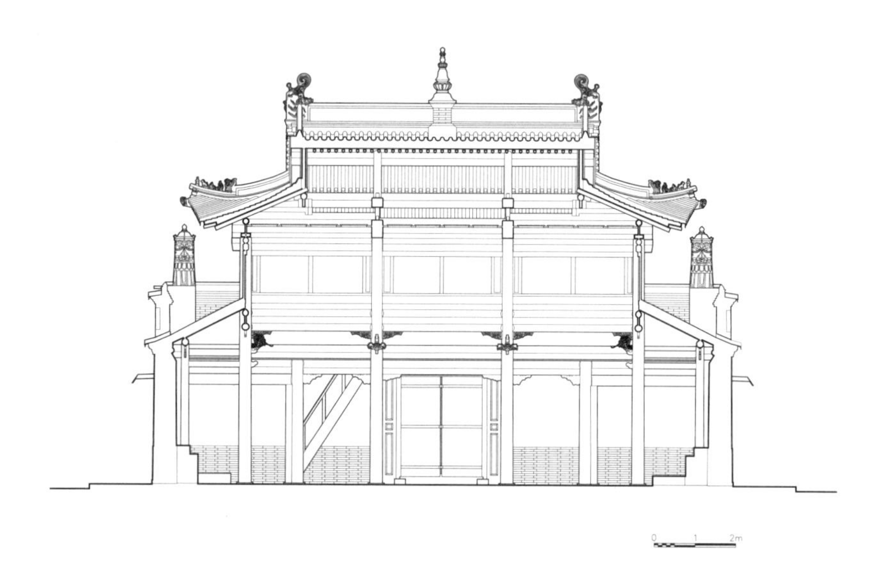

乃春庙纵剖面图2

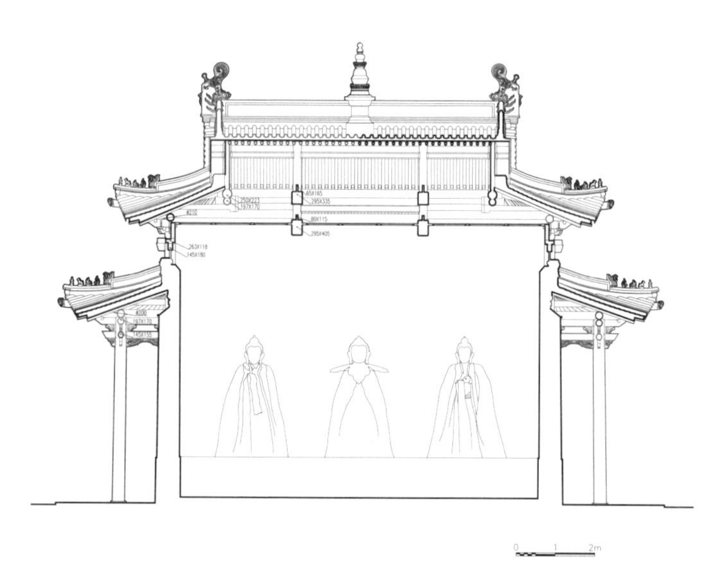

乃春庙纵剖面图3

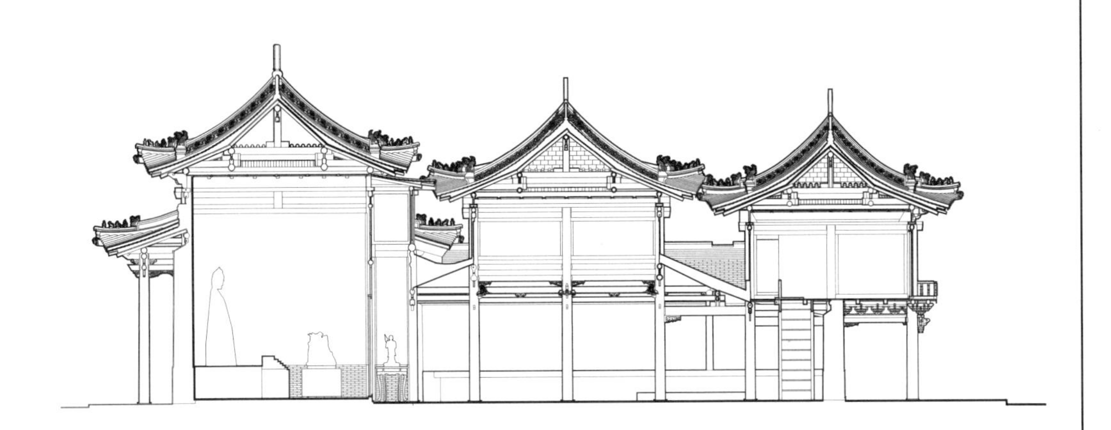

乃春庙经堂横剖面图

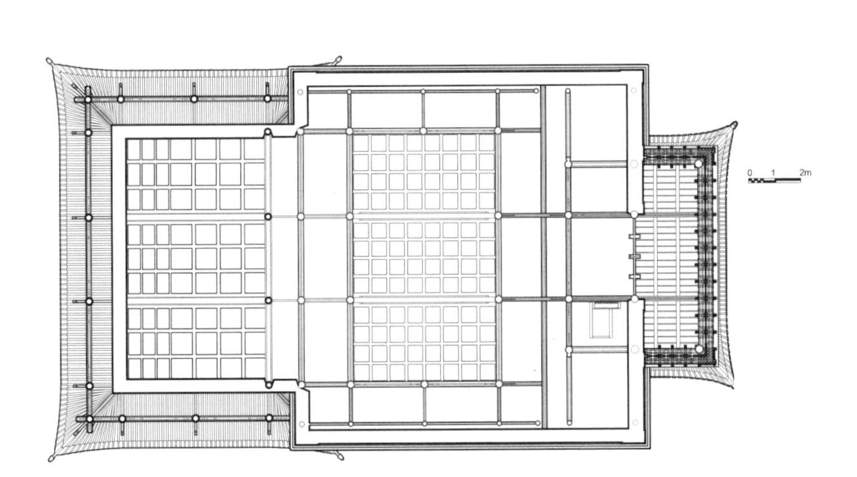

乃春庙经堂梁架平面图

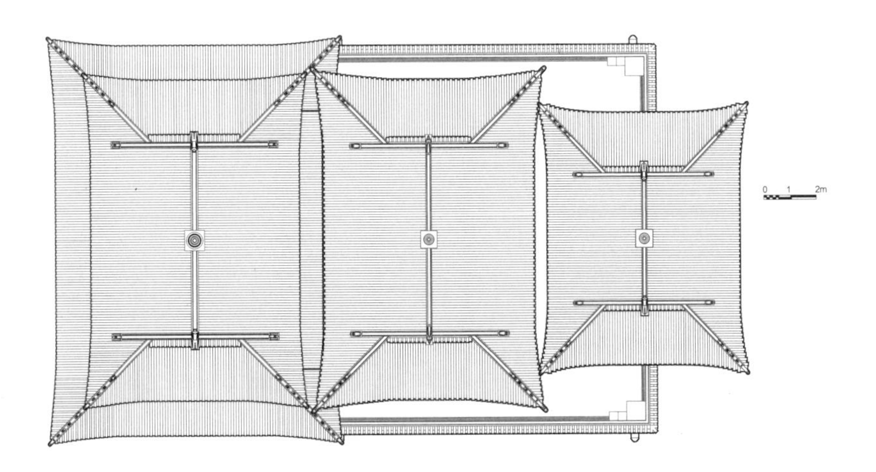

乃春庙屋顶平面图

1.11 大召·藏经阁

单位:毫米

建筑名称	汉语正式名称	藏经阁	俗称	——													
概述	初建年	万历十五年（1587年）始建	建筑朝向	南	建筑层数	二											
	建筑简要描述	汉式建筑，硬山屋顶，汉式结构体系															
	重建重修记载	2004—2005年进行修缮															
		信息来源	大召大事记														
结构规模	结构形式	钢筋混凝土仿木结构	相连的建筑	无	室内天井	无											
	建筑平面形式	长方形	外廊形式	前廊													
	通面阔	195500	开间数	7	明间	3350	次间	3250	梢间	3300	次梢间	——	尽间	3200			
	通进深	85400	进深数	2	进深尺寸（前→后）	1430→7110											
	柱子数量	——	柱子间距	横向尺寸	——	（藏式建筑结构体系填写此栏，不含廊柱）											
				纵向尺寸													
	其他	——															
建筑主体（大木作）（石作）（瓦作）	屋顶	屋顶形式	硬山	瓦作	布瓦												
	外墙	主体材料	青砖	材料规格	140×280×55	饰面颜色	无饰面										
		墙体收分	无	边玛檐墙	无	边玛材料	——										
	斗栱、梁架	斗栱	无	平身科斗口尺寸	不详	梁架关系	五檩前廊										
	柱、柱式（前廊柱）	形式	汉式	柱身断面形状	圆形	断面尺寸	直径$D=280$	（在没有前廊柱的情况下，填写室内柱及其特征）									
		柱身材料	混凝土	柱身收分	无	栌斗、托木	无	雀替	有								
		柱础	无	柱础形状	——	柱础尺寸	——										
	台基	台基类型	普通台基	台基高度	430	台基地面铺设材料	砖										
	其他	——															
装修（小木作）（彩画）	门(正面)	隔扇门	门楣	无	堆经	无	门帘	有									
	窗（正面）	槛窗	窗楣	无	窗套	无	窗帘	无									
	室内隔扇	隔扇	有	隔扇位置	室内左右侧												
	室内地面、楼面	地面材料及规格	方砖750×750	楼面材料及规格	不规则木板												
	室内楼梯	楼梯	有	楼梯位置	进门左右侧	楼梯材料	木	梯段宽度	880								
	天花、藻井	天花	有	天花类型	井口天花	藻井	无	藻井类型	——								
	彩画	柱头	无	柱身	无	梁架	有	走马板	无								
		门、窗	无	天花	有	藻井	——	其他彩画	——								
	其他	悬塑	无	佛龛	无	匾额	“藏经阁”2004年六月初八，何佳嵘敬献										
装饰	室内	帷幔	无	幕帘彩绘	有	壁画	有	唐卡	无								
		经幡	有	经幢	无	柱毯	无	其他	——								
	室外	玛尼轮	无	苏勒德	有	宝顶	无	祥麟法轮	无								
		四角经幢	无	经幡	无	铜饰	无	石刻、砖雕	无								
		仙人走兽	1+3	壁画	无	其他	二龙戏珠										
陈设	室内	主佛像	宗喀巴佛像	佛像基座	莲花座												
		法座	无	藏经橱	有	经床	无	诵经桌	无	法鼓	无	玛尼轮	无	坛城	无	其他	——
	室外	旗杆	无	苏勒德	无	狮子	有	经幡	有	玛尼轮	无	香炉	有	五供	无	其他	——
	其他	——															
备注	——																
调查日期	2010/05/01	调查人员	白丽燕、萨日朗	整理日期	2010/05/12	整理人员	萨日朗										

藏经阁基本概况表1

大召·藏经阁·档案照片

照片名称	斜前方	照片名称	正立面	照片名称	前廊
照片名称	室外柱廊	照片名称	柱身	照片名称	柱头
照片名称	柱础	照片名称	台基	照片名称	门
照片名称	室内正面	照片名称	室内侧面	照片名称	室内地面
照片名称	室内天花	照片名称	室内局部	照片名称	室内佛像

备注	—		
摄影日期	2010/08/22	摄影人员	白丽燕、萨日朗、孟祎军

藏经阁正立面

藏经阁主入口（左图）
藏经阁斜前方（右图）

藏经阁室内正面（左图）
藏经阁室内佛像（右图）

1.12 大召·庇佑殿

单位:毫米

建筑名称	汉语正式名称	庇佑殿	俗称	——													
概述	初建年	万历十五年（1587年）始建	建筑朝向	南	建筑层数	二											
	建筑简要描述	汉藏结合式，汉式结构体系，两个歇山屋顶组合，前廊及正面俩侧为藏式风格															
	重建重修记载	原建筑已毁，2009年恢复重建															
		信息来源	大召大事记														
结构规模	结构形式	砖木混合	相连的建筑	无	室内天井	都纲法式											
	建筑平面形式	长方形	外廊形式	前廊													
	通面阔	13010	开间数	5	明间	4450	次间	1830	梢间	2450	次梢间	——	尽间	——			
	通进深	19400	进深数	5	进深尺寸（前→后）	3550→3360→3570→4460→4460											
	柱子数量	——	柱子间距	横向尺寸	——	（藏式建筑结构体系填写此栏，不含廊柱）											
				纵向尺寸													
	其他	——															
建筑主体（大木作）（石作）（瓦作）	屋顶	屋顶形式	歇山	瓦作	琉璃瓦												
	外墙	主体材料	青砖	材料规格	240×100×50	饰面颜色	蓝色琉璃										
		墙体收分	有	边玛檐墙	有	边玛材料	青砖，局部蓝色琉璃										
	斗栱、梁架	斗栱	有	平身科斗口尺寸	110	梁架关系	抬梁式，7檩前廊										
	柱、柱式（前廊柱）	形式	藏式	柱身断面形状	12楞柱	断面尺寸	440×440	（在没有前廊柱的情况下，填写室内柱及其特征）									
		柱身材料	木材	柱身收分	有	栌斗、托木	无	雀替	有								
		柱础	有	柱础形状	方形	柱础尺寸	550×550										
	台基	台基类型	普通台基	台基高度	830	台基地面铺设材料	石材										
	其他	——															
装修（小木作）（彩画）	门(正面)	——	门楣	——	堆经	——	门帘	——									
	窗（正面）	藏式明窗	窗楣	无	窗套	无	窗帘	无									
	室内隔扇	隔扇	有	隔扇位置	二层												
	室内地面、楼面	地面材料及规格	青砖 120×240×60	楼面材料及规格	木材，规格不均												
	室内楼梯	楼梯	有	楼梯位置	进正门左侧	楼梯材料	木	梯段宽度	820								
	天花、藻井	天花	有	天花类型	井口天花	藻井	有	藻井类型	四方变圆								
	彩画	柱头	有	柱身	无	梁架	有	走马板	无								
		门、窗	——	天花	有	藻井	有	其他彩画	——								
	其他	悬塑	无	佛龛	无	匾额	无										
装饰	室内	帷幔	——	幕帘彩绘	——	壁画	——	唐卡	——								
		经幡	——	经幢	——	柱毯	——	其他	——								
	室外	玛尼轮	无	苏勒德	无	宝顶	有	祥麟法轮	有								
		四角经幢	无	经幡	无	铜饰	有	石刻、砖雕	无								
		仙人走兽	1+4	壁画	无	其他	歇山屋顶装饰										
陈设	室内	主佛像	大白伞盖佛	佛像基座	须弥座												
		法座	——	藏经橱	——	经床	——	诵经桌	——	法鼓	——	玛尼轮	——	坛城	——	其他	——
	室外	旗杆	无	苏勒德	无	狮子	无	经幡	无	玛尼轮	无	香炉	无	五供	无	其他	——
	其他	——															
备注	此建筑在测绘的时候处于施工状态																
调查日期	2010/05/01	调查人员	白丽燕、萨日朗	整理日期	2010/05/01	整理人员	萨日朗										

庇佑殿基本概况表1

大召·庇佑殿·档案照片					
照片名称 	正立面	照片名称 	斜前方	照片名称 	二层侧面
照片名称 	北立面	照片名称 	墙体材料	照片名称 	边玛墙
照片名称 	柱身	照片名称 	柱头	照片名称 	柱础
照片名称 	室内正面	照片名称 	室内二层正面	照片名称 	室内局部
照片名称 	室内柱身	照片名称 	室内二层回廊	照片名称 	藻井
备注	—				
摄影日期	2010/08/22、2012/08/28	摄影人员	白丽燕、萨日朗、孟祎军		

庇佑殿基本概况表2

庇佑殿正立面

庇佑殿二层正面（左图）
庇佑殿柱身（右图）

庇佑殿室内局部（左图）
庇佑殿室内柱头（右图）

1.13 大召·公中仓佛殿

单位:毫米

建筑名称	汉语正式名称	公中仓佛殿	俗称	——										
概述	初建年	万历十五年（1587年）始建	建筑朝向	南	建筑层数	二								
	建筑简要描述	汉藏结合式，汉式屋顶，正殿前廊两侧有藏式窗												
	重建重修记载	原建筑已毁，2009年恢复重建												
		信息来源	大召大事记											
结构规模	结构形式	砖木混合	相连的建筑	无	室内天井	无								
	建筑平面形式	长方形	外廊形式	前廊										
	通面阔	28910	开间数	9	明间	3970	次间	3950	梢间	1850	次梢间	4690	尽间	1980
	通进深	19840	进深数	5	进深尺寸（前→后）	3970→3980→3970→3970→3950								
	柱子数量	——	柱子间距	横向尺寸	——	（藏式建筑结构体系填写此栏，不含廊柱）								
				纵向尺寸										
	其他	——												
建筑主体（大木作）（石作）（瓦作）	屋顶	屋顶形式	歇山屋顶，局部平屋顶	瓦作	琉璃瓦									
	外墙	主体材料	青砖	材料规格	青砖 220×30×60	饰面颜色	青色							
		墙体收分	有	边玛檐墙	有	边玛材料	青砖							
	斗栱、梁架	斗栱	有	平身科斗口尺寸	170	梁架关系	抬梁式，九檩前廊							
	柱、柱式（前廊柱）	形式	藏式	柱身断面形状	12楞柱	断面尺寸	440×440	（在没有前廊柱的情况下，填写室内柱及其特征）						
		柱身材料	木材	柱身收分	有	栌斗、托木	无	雀替	有					
		柱础	有	柱础形状	方形	柱础尺寸	550×550							
	台基	台基类型	普通台基	台基高度	870	台基地面铺设材料	石材							
	其他	——												
装修（小木作）（彩画）	门(正面)	——	门楣	——	堆经	——	门帘	——						
	窗（正面）	藏式明窗	窗楣	有	窗套	有	窗帘	无						
	室内隔扇	隔扇	有	隔扇位置	大殿东西两侧，位置对称									
	室内地面、楼面	地面材料及规格	——	楼面材料及规格	木板，规格不均									
	室内楼梯	楼梯	有	楼梯位置	进大殿正前方	楼梯材料	木	梯段宽度	1100					
	天花、藻井	天花	无	天花类型	——	藻井	无	藻井类型	——					
	彩画	柱头	无	柱身	无	梁架	——	走马板	无					
		门、窗	无	天花	——	藻井	——	其他彩画	——					
	其他	悬塑	无	佛龛	无	匾额	——							
装饰	室内	帷幔	无	幕帘彩绘	无	壁画	无	唐卡	无					
		经幡	无	经幢	无	柱毯	无	其他	——					
	室外	玛尼轮	无	苏勒德	无	宝顶	有	祥麟法轮	有					
		四角经幢	无	经幡	无	铜饰	有	石刻、砖雕	——					
		仙人走兽	无	壁画	无	其他	——							
陈设	室内	主佛像	——	佛像基座	——									
		法座	——	藏经橱	——	经床	——	诵经桌	——	法鼓	——	玛尼轮	——	坛城	——	其他	——
	室外	旗杆	——	苏勒德	——	狮子	——	经幡	——	玛尼轮	——	香炉	——	五供	——	其他	——
	其他	——												
备注	此建筑在测绘过程中处于未完工状态													
调查日期	2010/05/01	调查人员	白丽燕、萨日朗	整理日期	2010/05/12	整理人员	萨日朗							

公中仓佛殿基本概况表1

大召·公中仓佛殿·档案照片					
照片名称	正立面	照片名称	侧立面	照片名称	斜前方
照片名称	窗	照片名称	檐墙	照片名称	二层窗
照片名称	斗栱	照片名称	屋顶装饰	照片名称	台阶
照片名称	边玛檐部	照片名称	室外柱子	照片名称	室外柱头
照片名称	室外柱础	照片名称	室内柱子	照片名称	室内隔断
备注	现为寺庙办公室				
摄影日期	2010/08/22、2012/08/29	摄影人员	白丽燕、萨日朗、张宇		

公中仓佛殿正立面

公中仓佛殿侧立面

公中仓佛殿屋顶装饰
（左图）
公中仓佛殿柱头
（右图）

1.14 大召·菩萨殿经堂

单位:毫米

建筑名称	汉语正式名称	菩萨殿经堂	俗称	——										
概述	初建年	万历十五年（1587年）始建	建筑朝向	南	建筑层数	二								
	建筑简要描述	以汉式结构为主，局部有藏式装饰												
	重建重修记载	原建筑已毁，2007年恢复重建												
		信息来源	《大召大记事》											
结构规模	结构形式	砖木混合	相连的建筑	与佛殿咬合相连	室内天井	都纲法式								
	建筑平面形式	方形	外廊形式	前廊										
	通面阔	17900	开间数	5	明间	3900	次间	3300	梢间	3700	次梢间	——	尽间	——
	通进深	22150	进深数	7	进深尺寸（前→后）	2700→3500→3500→3500→3500→3500→1950								
	柱子数量	24	柱子间距	横向尺寸	——	（藏式建筑结构体系填写此栏，不含廊柱）								
				纵向尺寸	——									
	其他	——												
建筑主体（大木作）（石作）（瓦作）	屋顶	屋顶形式	入口上部重檐歇山，经堂上部为单檐歇山	瓦作	黄琉璃									
	外墙	主体材料	青砖	材料规格	青砖，270×125×55	饰面颜色	青砖色							
		墙体收分	有	边玛檐墙	有	边玛材料	边玛草							
	斗栱、梁架	斗栱	有	平身科斗口尺寸	不详	梁架关系	不详（有吊顶）							
	柱、柱式（前廊柱）	形式	汉式	柱身断面形状	圆	断面尺寸	直径 D = 450	（在没有前廊柱的情况下，填写室内柱及其特征）						
		柱身材料	木	柱身收分	无	栌斗、托木	无	雀替	有					
		柱础	有	柱础形状	上圆下方	柱础尺寸	650×650、直径D = 650							
	台基	台基类型	带勾栏式台基	台基高度	710	台基地面铺设材料	方砖290×290，条砖145×275							
	其他	——												
装修（小木作）（彩画）	门(正面)	板门	门楣	有	堆经	有	门帘	无						
	窗（正面）	一层为牖窗，二层为隔扇	窗楣	无	窗套	无	窗帘	无						
	室内隔扇	隔扇	无	隔扇位置	——									
	室内地面、楼面	地面材料及规格	方砖290×290，条砖145×275	楼面材料及规格	木板，长宽不定									
	室内楼梯	楼梯	有	楼梯位置	进南门右侧	楼梯材料	木	梯段宽度	900					
	天花、藻井	天花	有	天花类型	井口天花	藻井	有	藻井类型	八方变圆					
	彩画	柱头	有	柱身	无	梁架	有	走马板	有					
		门、窗	无	天花	有	藻井	有	其他彩画	——					
	其他	悬塑	无	佛龛	无	匾额	“菩萨殿”，2007年书，高延青							
装饰	室内	帷幔	有	幕帘彩绘	无	壁画	无	唐卡	有					
		经幡	无	经幢	有	柱毯	无	其他	——					
	室外	玛尼轮	无	苏勒德	有	宝顶	有	祥麟法轮	无					
		四角经幢	有	经幡	无	铜饰	有	石刻、砖雕	有					
		仙人走兽	1+4	壁画	无	其他	——							
陈设	室内	主佛像	无	佛像基座	——									
		法座	有	藏经橱	无	经床	有	诵经桌	有	法鼓	无	玛尼轮	有	坛城	无	其他	——
	室外	旗杆	无	苏勒德	无	狮子	有	经幡	无	玛尼轮	无	香炉	有	五供	无	其他	——
	其他	——												
备注	——													
调查日期	2010/05/01	调查人员	白丽燕、萨日朗	整理日期	2010/05/09	整理人员	萨日朗							

菩萨殿经堂基本概况表1

大召·菩萨殿经堂·档案照片

照片名称 	正立面	照片名称 	斜前方	照片名称 	东立面
照片名称 	西立面	照片名称 	局部	照片名称 	柱头
照片名称 	门	照片名称 	窗	照片名称 	带勾栏式台基
照片名称 	室内正面	照片名称 	室内藻井	照片名称 	室内壁画
照片名称 	二层侧面	照片名称 	二层栏杆	照片名称 	室内装饰
备注	—				
摄影日期	2010/08/22	摄影人员	白丽燕、萨日朗、孟祎军		

菩萨殿经堂基本概况表2

菩萨殿经堂正立面

菩萨殿经堂柱廊（左图）
菩萨殿经堂柱头（右图）

1.15 大召·菩萨殿佛殿

单位:毫米

建筑名称	汉语正式名称	菩萨殿佛殿	俗称	——													
概述	初建年	万历十五年(1587年)始建	建筑朝向	南	建筑层数	二											
	建筑简要描述	汉式结构体系															
	重建重修记载	原建筑已毁，2007年恢复重建															
		信息来源	大召大事记														
结构规模	结构形式	砖木混合	相连的建筑	南侧与经堂咬合相连	室内天井	有											
	建筑平面形式	方形	外廊形式	半回廊													
	通面阔	12700	开间数	3	明间	4500	次间	4100	梢间	——	次梢间	——	尽间	——			
	通进深	13200	进深数	3	进深尺寸(前→后)	4400→4400→4400											
	柱子数量	——	柱子间距	横向尺寸	——	(藏式建筑结构体系填写此栏，不含廊柱)											
				纵向尺寸	——												
	其他	——															
建筑主体(大木作)(石作)(瓦作)	屋顶	屋顶形式	重檐歇山	瓦作	黄琉璃												
	外墙	主体材料	青砖	材料规格	270×125×55	饰面颜色	红										
		墙体收分	无	边玛檐墙	无	边玛材料	——										
	斗栱、梁架	斗栱	有	平身科斗口尺寸	不详	梁架关系	不详(有吊顶)										
	柱、柱式(前廊柱)	形式	汉式	柱身断面形状	圆	断面尺寸	直径 D = 450	(在没有前廊柱的情况下，填写室内柱及其特征)									
		柱身材料	木	柱身收分	不详	栌斗、托木	无	雀替	有								
		柱础	有	柱础形状	上圆下方	柱础尺寸	650×650、直径 D = 650										
	台基	台基类型	带勾栏式台基	台基高度	710	台基地面铺设材料	方砖290×290，条砖145×275										
	其他	——															
装修(小木作)(彩画)	门(正面)	板门	门楣	有	堆经	有	门帘	无									
	窗(正面)	无	窗楣	无	窗套	无	窗帘	无									
	室内隔扇	隔扇	无	隔扇位置	——												
	室内地面、楼面	地面材料及规格	方砖290×290，条砖145×275	楼面材料及规格	——												
	室内楼梯	楼梯	佛殿与经堂共用楼梯	楼梯位置	——	楼梯材料	——	梯段宽度	——								
	天花、藻井	天花	有	天花类型	井口天花	藻井	无	藻井类型	——								
	彩画	柱头	有	柱身	无	梁架	有	走马板	无								
		门、窗	无	天花	有	藻井	无	其他彩画	——								
	其他	悬塑	无	佛龛	无	匾额	无										
装饰	室内	帷幔	有	幕帘彩绘	无	壁画	无	唐卡	无								
		经幡	无	经幢	有	柱毯	无	其他	——								
	室外	玛尼轮	无	苏勒德	无	宝顶	有	祥麟法轮	无								
		四角经幢	无	经幡	无	铜饰	无	石刻、砖雕	无								
		仙人走兽	1+4	壁画	无	其他	——										
陈设	室内	主佛像	面南为观音菩萨，面北为绿渡母	佛像基座	须弥座												
		法座	无	藏经橱	无	经床	无	诵经桌	无	法鼓	无	玛尼轮	无	坛城	无	其他	——
	室外	旗杆	无	苏勒德	无	狮子	无	经幡	无	玛尼轮	无	香炉	无	五供	无	其他	——
	其他	——															
备注	——																
调查日期	2010/05/01	调查人员	白丽燕、萨日朗	整理日期	2010/05/09	整理人员	萨日朗										

菩萨殿佛殿基本概况表1

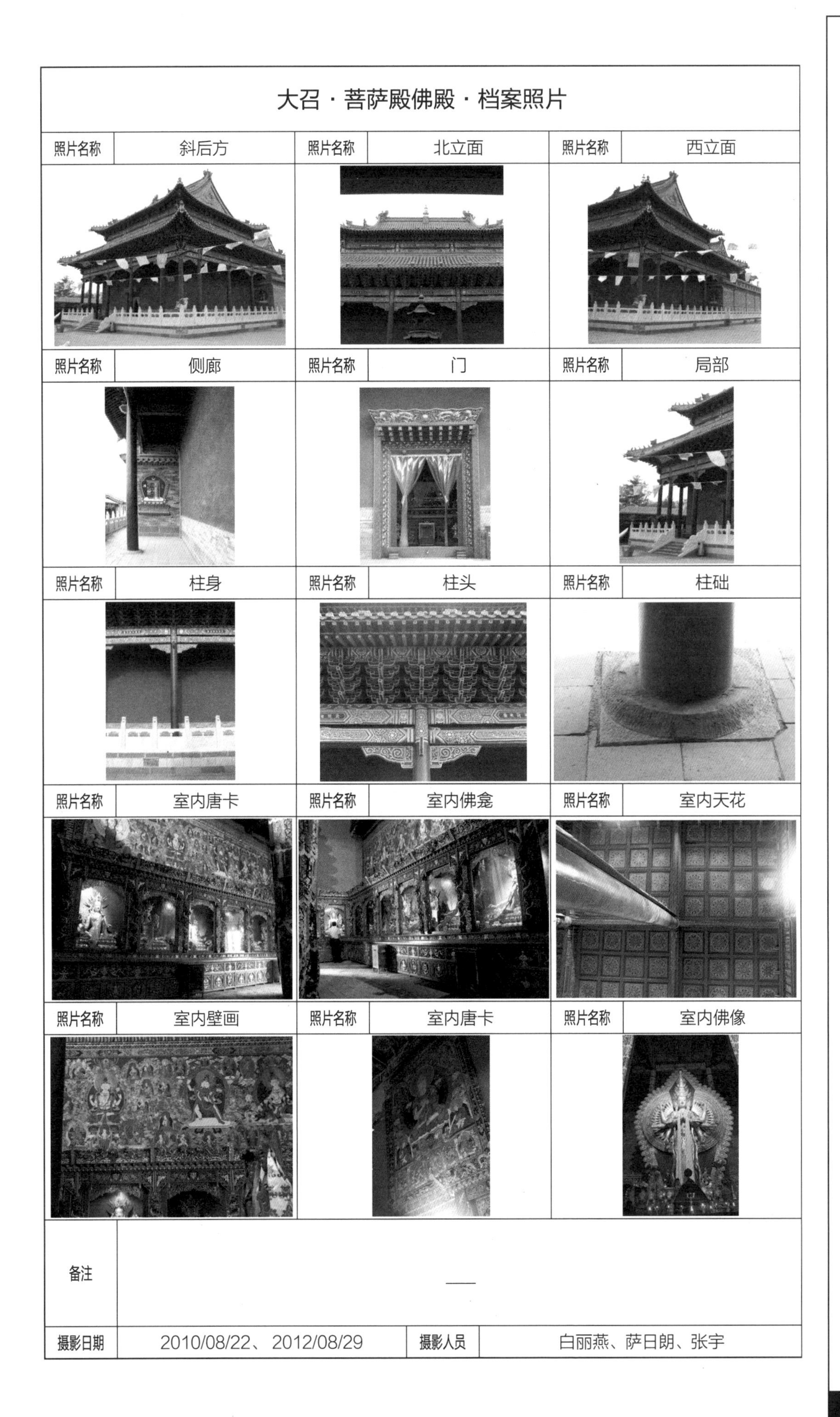

大召・菩萨殿佛殿・档案照片

照片名称	斜后方	照片名称	北立面	照片名称	西立面
照片名称	侧廊	照片名称	门	照片名称	局部
照片名称	柱身	照片名称	柱头	照片名称	柱础
照片名称	室内唐卡	照片名称	室内佛龛	照片名称	室内天花
照片名称	室内壁画	照片名称	室内唐卡	照片名称	室内佛像
备注	—				
摄影日期	2010/08/22、2012/08/29	摄影人员	白丽燕、萨日朗、张宇		

菩萨殿佛殿斜后方

菩萨殿佛殿室内佛龛（左图）

菩萨殿佛殿唐卡（左图）
菩萨殿佛殿佛像（右图）

1.16 大召·玉佛殿

单位:毫米

建筑名称	汉语正式名称	释迦牟尼佛殿	俗称	玉佛殿													
概述	初建年	万历十五年（1587年）始建	建筑朝向	南	建筑层数	二
	建筑简要描述	主体外墙为藏式，屋顶为汉式
	重建重修记载	原建筑已毁，2007年恢复重建
		信息来源	大召大事记
结构规模	结构形式	砖 混凝土柱	相连的建筑	无	室内天井	有
	建筑平面形式	凸字形	外廊形式	前廊 后半回廊		
	通面阔	14820	开间数	5	明间	4020	次间	3300	梢间	2100	次梢间	——	尽间	——
	通进深	18710	进深数	6	进深尺寸（前→后）	4350→2150→2120→3600→4350→2200
	柱子数量	——	柱子间距	横向尺寸	——	（藏式建筑结构体系填写此栏，不含廊柱）
				纵向尺寸		
	其他	——
建筑主体（大木作）（石作）（瓦作）	屋顶	屋顶形式	前歇山 后重檐歇山	瓦作	金色琉璃瓦
	外墙	主体材料	青砖	材料规格	270×130×58	饰面颜色	青砖色
		墙体收分	有	边玛檐墙	有	边玛材料	红柳
	斗栱、梁架	斗栱	有	平身科斗口尺寸	80	梁架关系	不详（有吊顶）
	柱、柱式（前廊柱）	形式	汉式	柱身断面形状	圆	断面尺寸	直径 D=2600	（在没有前廊柱的情况下，填写室内柱及其特征）
		柱身材料	混凝土	柱身收分	有	栌斗、托木	无	雀替	有
		柱础	有	柱础形状	覆盆	柱础尺寸	610×610
	台基	台基类型	有	台基高度	450	台基地面铺设材料	石材
	其他	——
装修（小木作）（彩画）	门(正面)	板门	门楣	有	堆经	有	门帘	无
	窗（正面）	隔扇	窗楣	无	窗套	无	窗帘	无
	室内隔扇	隔扇	有	隔扇位置	二层外廊口
	室内地面、楼面	地面材料及规格	方砖+青砖 290×290	楼面材料及规格	木材 295×795
	室内楼梯	楼梯	有	楼梯位置	第四开间第五进	楼梯材料	木	梯段宽度	890
	天花、藻井	天花	有	天花类型	井口	藻井	无	藻井类型	——
	彩画	柱头	有	柱身	无	梁架	有	走马板	有
		门、窗	无	天花	有	藻井	无	其他彩画	——
	其他	悬塑	无	佛龛	有	匾额	“玉佛殿”
装饰	室内	帷幔	无	幕帘彩绘	无	壁画	无	唐卡	有
		经幡	有	经幢	有	柱毯	无	其他	——
	室外	玛尼轮	无	苏勒德	无	宝顶	有	祥麟法轮	无
		四角经幢	有	经幡	有	铜饰	有	石刻、砖雕	有
		仙人走兽	1+4	壁画	有	其他	——
陈设	室内	主佛像	释迦牟尼佛像	佛像基座	莲花座
		法座	无	藏经橱	有	经床	无	诵经桌	有	法鼓	有	玛尼轮	无	坛城	有	其他	——
	室外	旗杆	无	苏勒德	有	狮子	无	经幡	有	玛尼轮	无	香炉	无	五供	有	其他	——
	其他	——
备注	——
调查日期	2010/05/01	调查人员	白丽燕、萨日朗	整理日期	2010/05/09	整理人员	萨日朗

玉佛殿基本概况表1

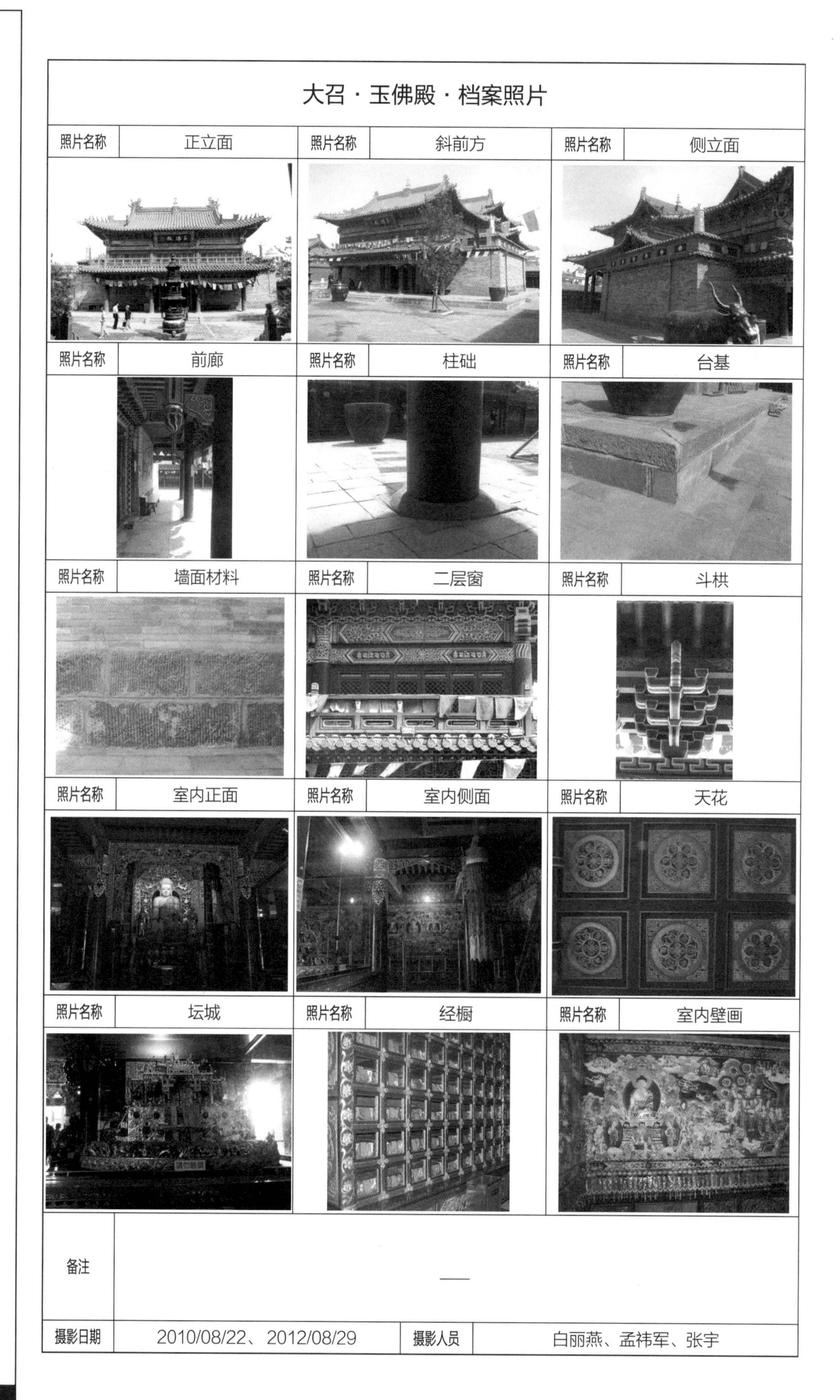

大召·玉佛殿·档案照片					
照片名称	正立面	照片名称	斜前方	照片名称	侧立面
照片名称	前廊	照片名称	柱础	照片名称	台基
照片名称	墙面材料	照片名称	二层窗	照片名称	斗栱
照片名称	室内正面	照片名称	室内侧面	照片名称	天花
照片名称	坛城	照片名称	经橱	照片名称	室内壁画
备注	—				
摄影日期	2010/08/22、2012/08/29	摄影人员	白丽燕、孟祎军、张宇		

玉佛殿基本概况表2

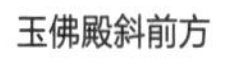
玉佛殿斜前方

玉佛殿侧立面（左图）
玉佛殿斗栱（右图）

玉佛殿前廊柱（左图）
玉佛殿室内正面（右图）

1.17 大召·大乐殿

单位:毫米

建筑名称	汉语正式名称	大乐殿、弥勒佛殿	俗称	——													
概述	初建年	万历十五年（1587年）始建	建筑朝向	南	建筑层数	二											
	建筑简要描述	汉藏结合式，汉式结构体系，2个歇山屋顶组合，前廊及正面两侧为藏式风格															
	重建重修记载	原建筑已毁，2008年重建															
		信息来源	大召大事记														
结构规模	结构形式	砖木混合	相连的建筑	无	室内天井	有											
	建筑平面形式	长方形	外廊形式	前廊													
	通面阔	13010	开间数	5	明间	4450	次间	1830	梢间	2450	次梢间	——	尽间	——			
	通进深	19400	进深数	5	进深尺寸（前→后）	3550→3360→3570→4460→4460											
	柱子数量	——	柱子间距	横向尺寸	——	（藏式建筑结构体系填写此栏，不含廊柱）											
				纵向尺寸													
	其他	——															
建筑主体（大木作）（石作）（瓦作）	屋顶	屋顶形式	歇山	瓦作	琉璃瓦												
	外墙	主体材料	青砖	材料规格	240×105×53	饰面颜色	蓝色琉璃										
		墙体收分	有	边玛檐墙	有	边玛材料	青砖，局部蓝色琉璃										
	斗栱、梁架	斗栱	有	平身科斗口尺寸	110	梁架关系	抬梁式，7檩前廊										
	柱、柱式（前廊柱）	形式	藏式	柱身断面形状	12楞柱	断面尺寸	440×440	（在没有前廊柱的情况下，填写室内柱及其特征）									
		柱身材料	木材	柱身收分	有	栌斗、托木	无	雀替	有								
		柱础	有	柱础形状	方形	柱础尺寸	550×550										
	台基	台基类型	普通台基	台基高度	830	台基地面铺设材料	石材										
	其他	——															
装修（小木作）（彩画）	门(正面)	板门	门楣	有	堆经	有	门帘	无									
	窗（正面）	藏式明窗	窗楣	无	窗套	无	窗帘	无									
	室内隔扇	隔扇	有	隔扇位置	二层												
	室内地面、楼面	地面材料及规格	青砖 300×300	楼面材料及规格	木材，规格不均												
	室内楼梯	楼梯	有	楼梯位置	进正门左侧	楼梯材料	木	梯段宽度	820								
	天花、藻井	天花	有	天花类型	井口天花	藻井	有	藻井类型	四方变圆								
	彩画	柱头	有	柱身	无	梁架	有	走马板	无								
		门、窗	——	天花	有	藻井	有	其他彩画	——								
	其他	悬塑	无	佛龛	无	匾额	无										
装饰	室内	帷幔	——	幕帘彩绘	——	壁画	——	唐卡	——								
		经幡	——	经幢	——	柱毯	——	其他	——								
	室外	玛尼轮	无	苏勒德	无	宝顶	有	祥麟法轮	有								
		四角经幢	无	经幡	无	铜饰	有	石刻、砖雕	无								
		仙人走兽	1+4	壁画	无	其他	——										
陈设	室内	主佛像	弥勒佛像	佛像基座	须弥座												
		法座	——	藏经橱	——	经床	——	诵经桌	——	法鼓	——	玛尼轮	——	坛城	——	其他	——
	室外	旗杆	无	苏勒德	无	狮子	无	经幡	无	玛尼轮	无	香炉	无	五供	无	其他	——
	其他	——															
备注	——																
调查日期	2010/05/01	调查人员	白丽燕、萨日朗	整理日期	2010/05/12	整理人员	萨日朗										

大乐殿基本概况表1

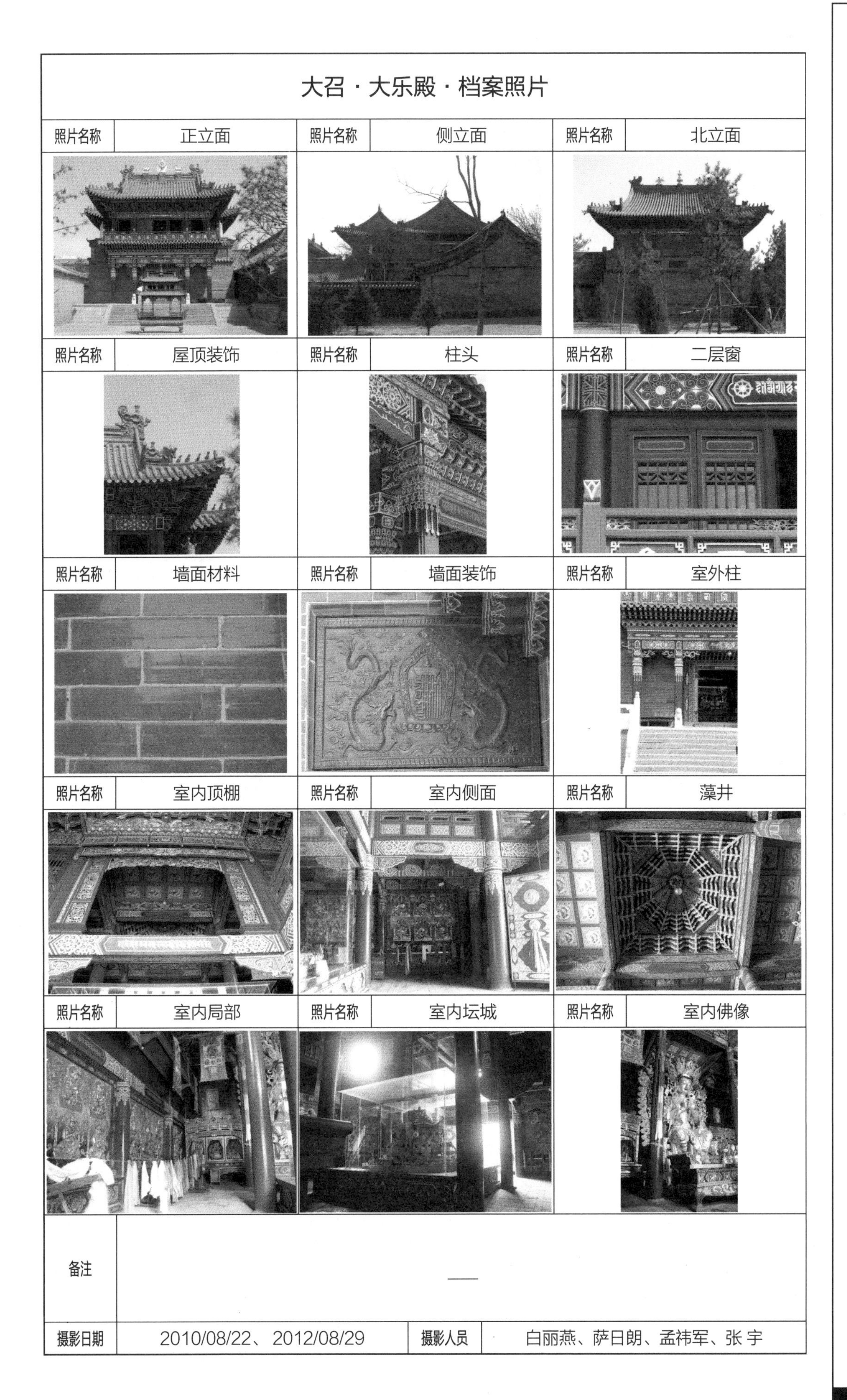

大召·大乐殿·档案照片

照片名称	正立面	照片名称	侧立面	照片名称	北立面
照片名称	屋顶装饰	照片名称	柱头	照片名称	二层窗
照片名称	墙面材料	照片名称	墙面装饰	照片名称	室外柱
照片名称	室内顶棚	照片名称	室内侧面	照片名称	藻井
照片名称	室内局部	照片名称	室内坛城	照片名称	室内佛像
备注	—				
摄影日期	2010/08/22、2012/08/29	摄影人员	白丽燕、萨日朗、孟祎军、张宇		

大乐殿基本概况表2

大乐殿正立面

大乐殿柱头（左图）
大乐殿侧立面（右图）

大乐殿二层窗（左图）
大乐殿背立面（右图）

1.18 大召·喇嘛印务处

单位:毫米

建筑名称	汉语正式名称	喇嘛印务处				俗称	——							
概述	初建年	康熙二十四年(1685年)建成				建筑朝向	南		建筑层数	一				
	建筑简要描述	汉式结构体系，用一墙体与大召寺内建筑相隔，室内已重新装修为现代风格，供园林管理办公使用												
	重建重修记载	——												
		信息来源	——											
结构规模	结构形式	砖木混合		相连的建筑	无				室内天井	无				
	建筑平面形式	长方形		外廊形式	无									
	通面阔	15500	开间数	5	明间	3100	次间	3100	梢间	3100	次梢间	——	尽间	——
	通进深	8350	进深数	3	进深尺寸(前→后)		1500→5650→1200							
	柱子数量	——	柱子间距	横向尺寸	——		(藏式建筑结构体系填写此栏，不含廊柱)							
				纵向尺寸	——									
	其他	——												
建筑主体(大木作)(石作)(瓦作)	屋顶	屋顶形式	歇山				瓦作	布瓦						
	外墙	主体材料	青砖	材料规格	300×120×50		饰面颜色	青砖色						
		墙体收分	无	边玛檐墙	无		边玛材料	——						
	斗栱、梁架	斗栱	无	平身科斗口尺寸		——	梁架关系	不详(有吊顶)						
	柱、柱式(前廊柱)	形式	汉式	柱身断面形状	圆	断面尺寸	直径 D=250		(在没有前廊柱的情况下，填写室内柱及其特征)					
		柱身材料	木材	柱身收分	无	栌斗、托木	无	雀替	无					
		柱础	无	柱础形状	——	柱础尺寸	——							
	台基	台基类型	普通式台基	台基高度	320	台基地面铺设材料		长方形灰砖						
	其他	——												
装修(小木作)(彩画)	门(正面)	板门		门楣	无	堆经	无	门帘	无					
	窗(正面)	槛窗		窗楣	无	窗套	有	窗帘	无					
	室内隔扇	隔扇	无	隔扇位置	——									
	室内地面、楼面	地面材料及规格		方灰砖，规格不详		楼面材料及规格		——						
	室内楼梯	楼梯	无	楼梯位置	——	楼梯材料	——	梯段宽度	——					
	天花、藻井	天花	有	天花类型	天花板	藻井	无	藻井类型	——					
	彩画	柱头	无	柱身	无	梁架	无	走马板	无					
		门、窗	无	天花	无	藻井	无	其他彩画	——					
	其他	悬塑	无	佛龛	无	匾额	无							
装饰	室内	帷幔	无	幕帘彩绘	无	壁画	无	唐卡	无					
		经幡	无	经幢	无	柱毯	无	其他	无					
	室外	玛尼轮	无	苏勒德	无	宝顶	无	祥麟法轮	无					
		四角经幢	无	经幡	无	铜饰	无	石刻、砖雕	无					
		仙人走兽	1+4	壁画	无	其他	——							
陈设	室内	主佛像	无			佛像基座	无							
		法座	无	藏经橱	无	经床	无	诵经桌	无	法鼓	无	玛尼轮	无	坛城	无	其他	——
	室外	旗杆	无	苏勒德	无	狮子	无	经幡	无	玛尼轮	无	香炉	无	五供	无	其他	——
	其他	——												
备注	——													
调查日期	2010/05/01	调查人员	白丽燕、萨日朗	整理日期	2010/05/12	整理人员	萨日朗							

喇嘛印务处基本概况表

喇嘛印务处正立面

喇嘛印务处门（左图）
喇嘛印务处柱身（右图）

喇嘛印务处台阶（左图）
喇嘛印务处大门（右图）

席力图召

Xiretzhao Temple

2 席力图召 Xiretzhao Temple

席力图召鸟瞰图

席力图召为呼和浩特七大召之一，系内蒙古地区最早建立的藏传佛教格鲁派寺庙之一。蒙古古籍载，大召与该寺为土默特阿拉坦汗与三世达赖喇嘛在世时创建的两座召庙。康熙三十三年（1694年），清廷御赐满、蒙古、汉、藏四体“延寿寺”匾额。寺庙管辖巧尔齐召（延禧寺，1819年摆脱席力图召管辖，独立成寺，成为呼和浩特八小召之一）、东乌素图召（广寿寺）、查干哈达召（永安寺）、希拉木仁庙（普会寺）等4座属庙。寺内珍藏康熙亲征噶尔丹的汉白玉纪功碑等历史文物。

明万历十三年（1585年），土默特阿拉坦汗之子僧格杜棱汗于三世达赖喇嘛索南嘉措来呼和浩特时修建一座小庙。后由第一世席力图活佛扩建。其寺院被称为席力图召。该寺历经雍正、咸丰、光绪等年间的数次大规模扩建后成为呼和浩特规模最大的寺庙。

寺庙建筑风格为汉藏结合式建筑。寺庙中有3间天王殿、5间菩提过殿、81间大雄宝殿、大佛殿、18间二层九间楼等5座主殿及牌楼、钟楼、鼓楼、东西廊房、东西碑亭等殿宇楼阁。大院西侧有乃春庙、前殿、丹珠尔殿3座殿宇，东侧有古佛殿、前殿、甘珠尔殿3座殿宇。大院西侧有席力图呼图克图拉布隆一座。大院东侧有汉白玉双耳白塔1座。

“文化大革命”中寺庙严重受损，随着城市的发展，该寺大面积房屋被占用或改建。1981年起，开始进行修缮，成为呼和浩特保留较完整的寺庙之一。

参考文献：

[1] 金峰整理注释.呼和浩特召庙（蒙古文）.呼和浩特：内蒙古人民出版社,1982,11.
[2] 金启孮.呼和浩特召庙、清真寺历史概述.中国蒙古史学会年会论文,1981.

席力图召大雄宝殿斜前方

<table>
<tr><th colspan="8">席力图召 · 基本概况 1</th></tr>
<tr><td rowspan="3">寺院蒙古语藏语名称</td><td>蒙古语</td><td>ᠰᠢᠷᠡᠭᠡᠲᠦ ᠵᠤᠤ</td><td rowspan="2">寺院汉语名称</td><td>汉语正式名称</td><td colspan="3">延寿寺</td></tr>
<tr><td>藏语</td><td>བྱང་ཆུབ་གླིང་།</td><td>俗称</td><td colspan="3">席力图召</td></tr>
<tr><td>汉语语义</td><td>首席、法座</td><td colspan="2">寺院汉语名称的由来</td><td colspan="3">清廷赐名</td></tr>
<tr><td>所在地</td><td colspan="4">呼和浩特旧城玉泉区石头巷北端</td><td>东经 ——</td><td colspan="2">北纬 ——</td></tr>
<tr><td>初建年</td><td colspan="2">万历十三年（1585年）</td><td>保护等级</td><td colspan="4">自治区级文物保护单位</td></tr>
<tr><td>盛期时间</td><td colspan="2">——</td><td colspan="2">盛期喇嘛僧/现有喇嘛僧数</td><td colspan="3">不详</td></tr>
<tr><td rowspan="2">历史沿革</td><td colspan="7">1581年，新建寺院西侧的古佛殿。
1603年，席力图一世全面扩建寺庙。
1694年，席力图四世扩修寺院。
1703年，立满、汉、蒙古、藏四种文字镌刻的征噶尔丹纪功碑。
1859年，席力图九世重修寺庙。
1859年，重修殿基，增高数尺。
1891年，重修寺庙。
1887年，该寺全部建筑因失火而烧毁，席力图九世仅用三年完成了修复工程。
1943年，殿内起火，佛殿与九间楼全部被焚毁。
1953年、1956年、1959年，三次重修寺庙。
1966年，庙内陈设大量丢失，建筑严重受损。
1970年，乃春庙遭火灾。
1981年，修缮天王殿、大雄宝殿、菩提过殿。
1993年，席力图十一世修缮塑像。
2004年，修缮彩画。
2006年，开始重建大佛殿、弥勒佛殿、殿前院落及侧门</td></tr>
<tr><td>资料来源</td><td colspan="6">[1] 金峰整理注释.呼和浩特召庙（蒙古文）.呼和浩特：内蒙古人民出版社,1982,11.
[2] 金启孮.呼和浩特召庙、清真寺历史概述.中国蒙古史学会年会论文,1981.
[3] 实地调研、调研访谈、据召内口碑所传</td></tr>
<tr><td rowspan="2">现状描述</td><td colspan="4" rowspan="2">席力图召内的大部分原有建筑都重新修葺过，现中轴线上的系列建筑及白塔保存较好，东侧轴线上的喇嘛宿舍院落被破坏，现在只有部分屋顶尚能辨认，屋身已被居民改造居住，中轴线上的九间楼和佛殿2006年开始重建，2007年完工</td><td>描述时间</td><td colspan="2">2010/05/09</td></tr>
<tr><td>描述人</td><td colspan="2">萨日朗</td></tr>
<tr><td>调查日期</td><td colspan="2">2010/05/01</td><td>调查人员</td><td colspan="4">白丽燕、萨日朗</td></tr>
</table>

席力图召基本概况表1

席力图召·基本概况 2

现存建筑				已毁建筑		
现存建筑	牌楼	天王殿	过殿	已毁建筑	美岱庙	——
	大雄宝殿	九间楼	古佛殿		弥勒殿	——
	护法殿	观音殿	度母殿		——	——
	活佛府	喇嘛住所	长寿塔		信息来源	《内蒙古寺庙》

区位图

总平面图

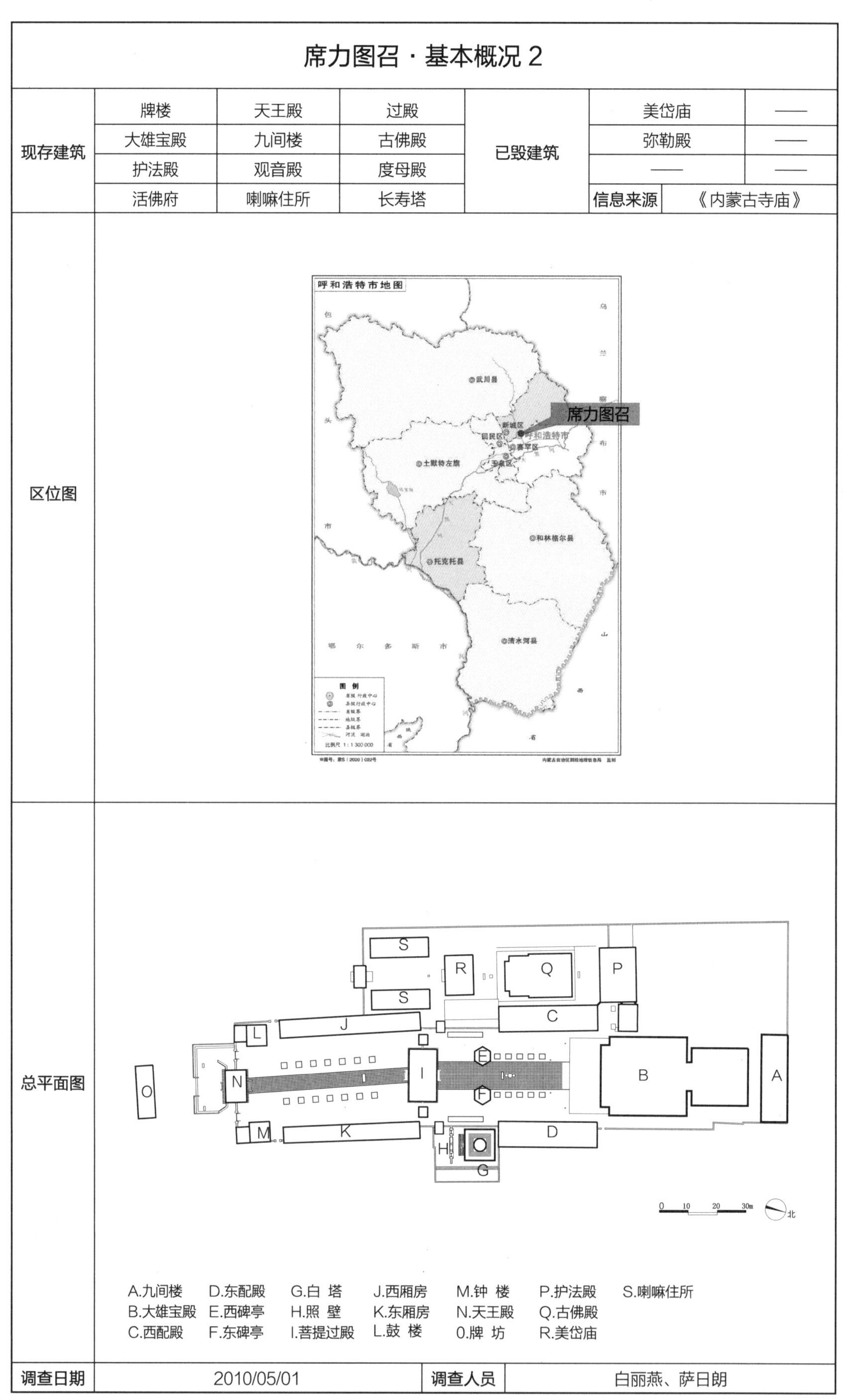

A.九间楼 D.东配殿 G.白 塔 J.西厢房 M.钟 楼 P.护法殿 S.喇嘛住所
B.大雄宝殿 E.西碑亭 H.照 壁 K.东厢房 N.天王殿 Q.古佛殿
C.西配殿 F.东碑亭 I.菩提过殿 L.鼓 楼 0.牌 坊 R.美岱庙

调查日期	2010/05/01	调查人员	白丽燕、萨日朗

席力图召基本概况表2

注：本召庙测绘图由天津大学王其亨教授为首的国家精品课程《古建筑测绘》教研组合作测绘并提供。

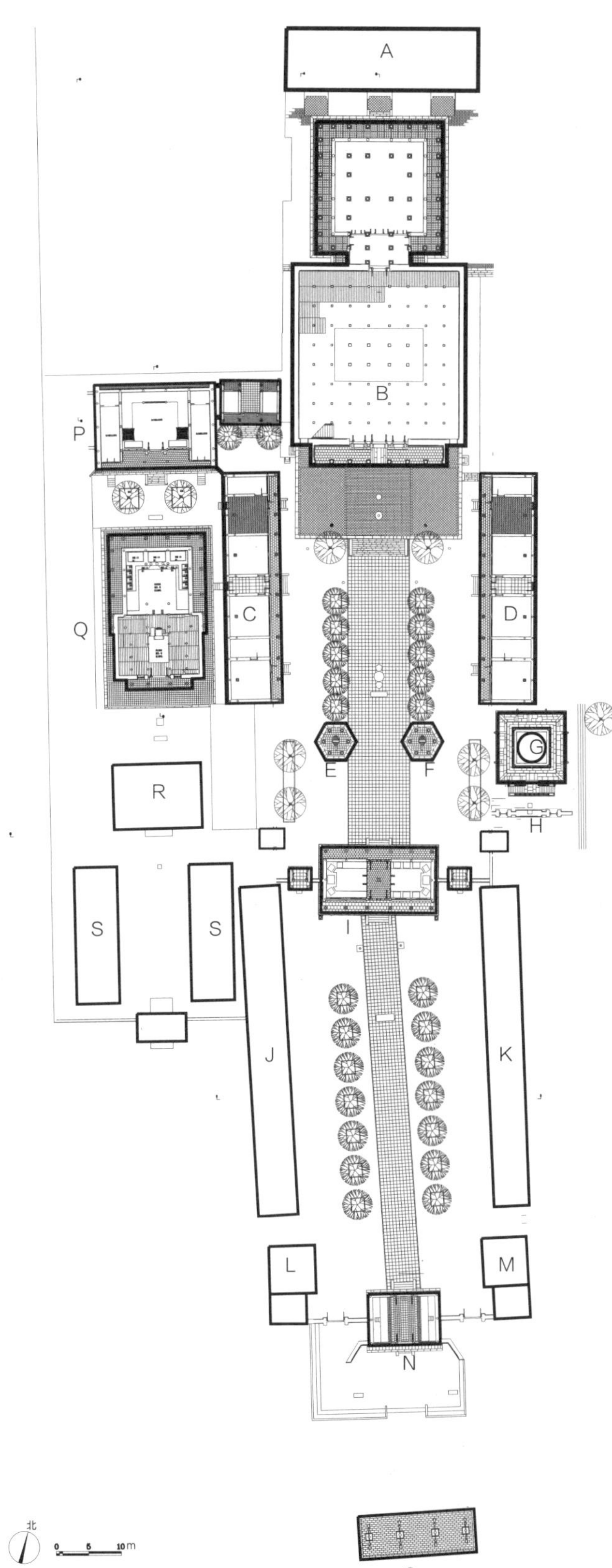

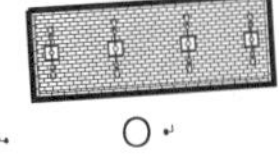

A.九间楼
B.大雄宝殿
C.西配殿
D.东配殿
E.西碑亭
F.东碑亭
G.白 塔
H.照 壁
I.菩提过殿
J.西厢房
K.东厢房
L.鼓 楼
M.钟 楼
N.天王殿
O.牌 坊
P.护法殿
Q.古佛殿
R.美岱庙
S.喇嘛住所

席力图召总平面图

2.1 席力图召·牌楼

牌楼正前方

牌楼斜前方

牌楼局部1（左图）
牌楼局部2（右图）

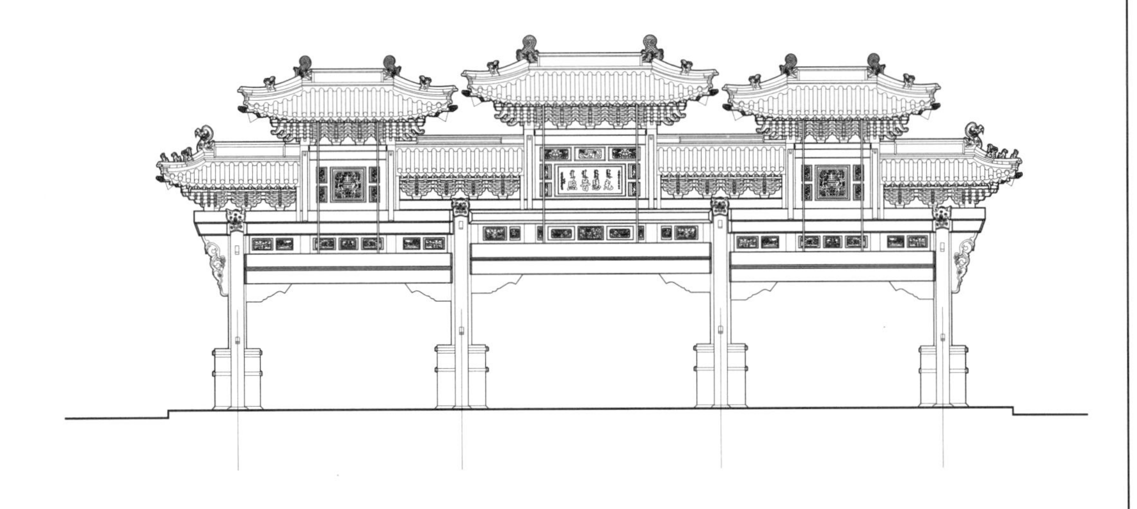

牌楼正立面

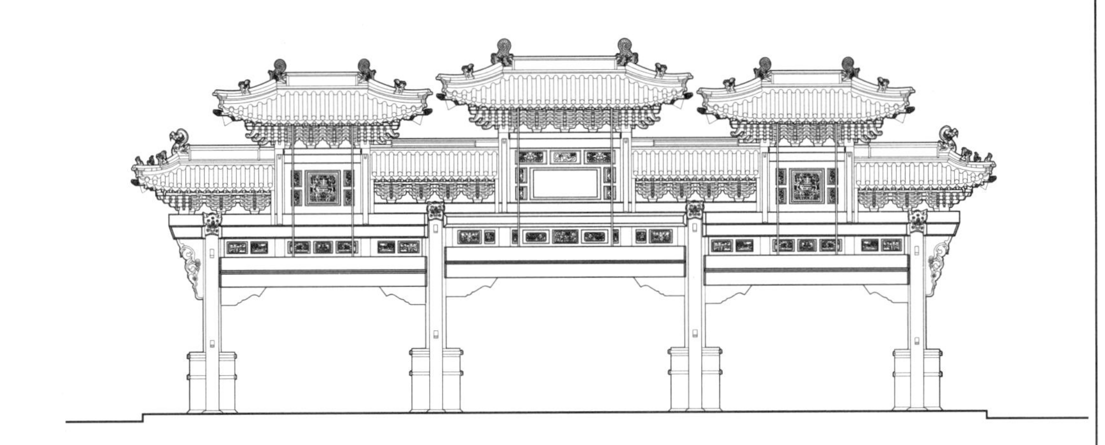

牌楼背立面

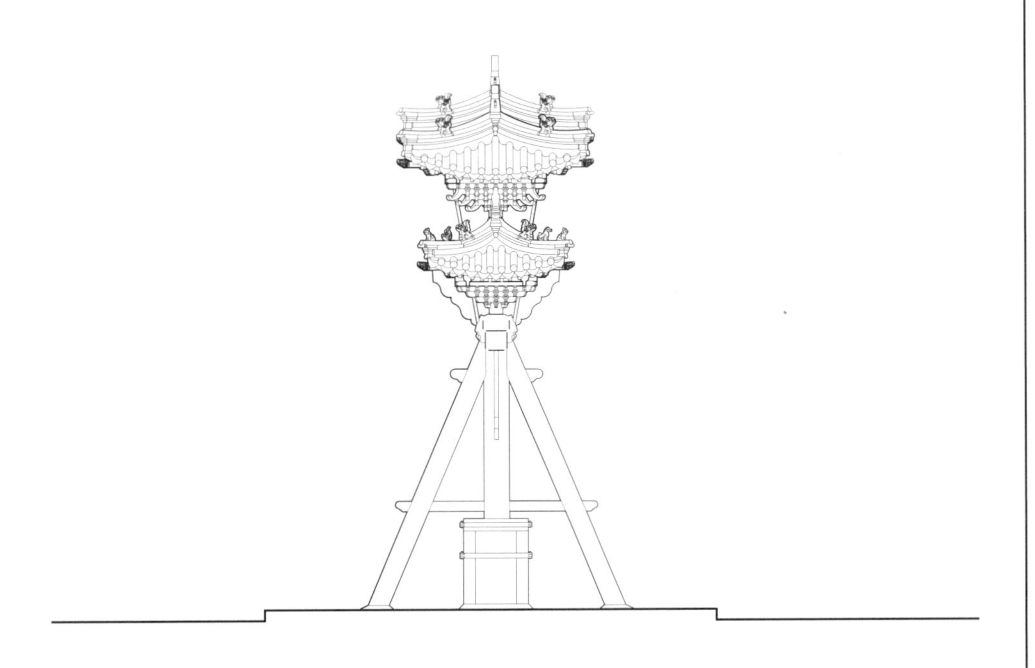

牌楼侧立面

2.2 席力图召·天王殿

单位:毫米

建筑名称	汉语正式名称	天王殿	俗称	——													
概述	初建年	万历十三年（1585年）	建筑朝向	南	建筑层数	一											
概述	建筑简要描述	汉式建筑、歇山式屋顶、建筑结构为五架抬梁式砖木结构															
概述	重建重修记载	——															
概述	重建重修记载	信息来源	——														
结构规模	结构形式	砖木结构	相连的建筑	无	室内天井	无											
结构规模	建筑平面形式	长方形	外廊形式	无	室内天井	无											
结构规模	通面阔	10940	开间数	3	明间	3920	次间	——	梢间	3520	次梢间	——	尽间	——			
结构规模	通进深	6780	进深数	2	进深尺寸（前→后）	3390→3390											
结构规模	柱子数量	——	柱子间距	横向尺寸	——	（藏式建筑结构体系填写此栏，不含廊柱）											
结构规模	柱子数量	——	柱子间距	纵向尺寸	——	（藏式建筑结构体系填写此栏，不含廊柱）											
结构规模	其他	——															
建筑主体（大木作）（石作）（瓦作）	屋顶	屋顶形式	歇山	瓦作	绿琉璃黄剪边												
建筑主体（大木作）（石作）（瓦作）	外墙	主体材料	砖	材料规格	260×50×130	饰面颜色	灰色、红色										
建筑主体（大木作）（石作）（瓦作）	外墙	墙体收分	无	边玛檐墙	无	边玛材料	——										
建筑主体（大木作）（石作）（瓦作）	斗栱、梁架	斗栱	有	平身科斗口尺寸	不详	梁架关系	七架梁										
建筑主体（大木作）（石作）（瓦作）	柱、柱式（前廊柱）	形式	汉式	柱身断面形状	方形	断面尺寸	260×260	（在没有前廊柱的情况下，填写室内柱及其特征）									
建筑主体（大木作）（石作）（瓦作）	柱、柱式（前廊柱）	柱身材料	木材	柱身收分	无	栌斗、托木	无	雀替	无								
建筑主体（大木作）（石作）（瓦作）	柱、柱式（前廊柱）	柱础	有	柱础形状	方形	柱础尺寸	600×600										
建筑主体（大木作）（石作）（瓦作）	台基	台基类型	普通	台基高度	500	台基地面铺设材料	石材										
建筑主体（大木作）（石作）（瓦作）	其他	——															
装修（小木作）（彩画）	门(正面)	撒带门	门楣	有	堆经	有	门帘	无									
装修（小木作）（彩画）	窗（正面）	槛窗	窗楣	无	窗套	无	窗帘	无									
装修（小木作）（彩画）	室内隔扇	隔扇	无	隔扇位置	——												
装修（小木作）（彩画）	室内地面、楼面	地面材料及规格	石材 350×350/270×270	楼面材料及规格	——												
装修（小木作）（彩画）	室内楼梯	楼梯	无	楼梯位置	——	楼梯材料	——	梯段宽度	——								
装修（小木作）（彩画）	天花、藻井	天花	无	天花类型	——	藻井	无	藻井类型	——								
装修（小木作）（彩画）	彩画	柱头	有	柱身	无	梁架	有	走马板	有								
装修（小木作）（彩画）	彩画	门、窗	有	天花	无	藻井	无	其他彩画	无								
装修（小木作）（彩画）	其他	悬塑	无	佛龛	无	匾额	灵光四澈（延寿寺）										
装饰	室内	帷幔	无	幕帘彩绘	无	壁画	有	唐卡	无								
装饰	室内	经幡	无	经幢	无	柱毯	无	其他	——								
装饰	室外	玛尼轮	无	苏勒德	无	宝顶	无	祥麟法轮	无								
装饰	室外	四角经幢	无	经幡	无	铜饰	无	石刻、砖雕	无								
装饰	室外	仙人走兽	1+3	壁画	有	其他	——										
陈设	室内	主佛像	四大天王	佛像基座	无												
陈设	室内	法座	无	藏经橱	无	经床	无	诵经桌	无	法鼓	无	玛尼轮	无	坛城	无	其他	——
陈设	室外	旗杆	无	苏勒德	无	狮子	无	经幡	无	玛尼轮	无	香炉	无	五供	无	其他	——
陈设	其他	——															
备注	——																
调查日期	2010/05/01	调查人员	白丽燕、萨日朗	整理日期	2010/05/09	整理人员	萨日朗										

天王殿基本概况表1

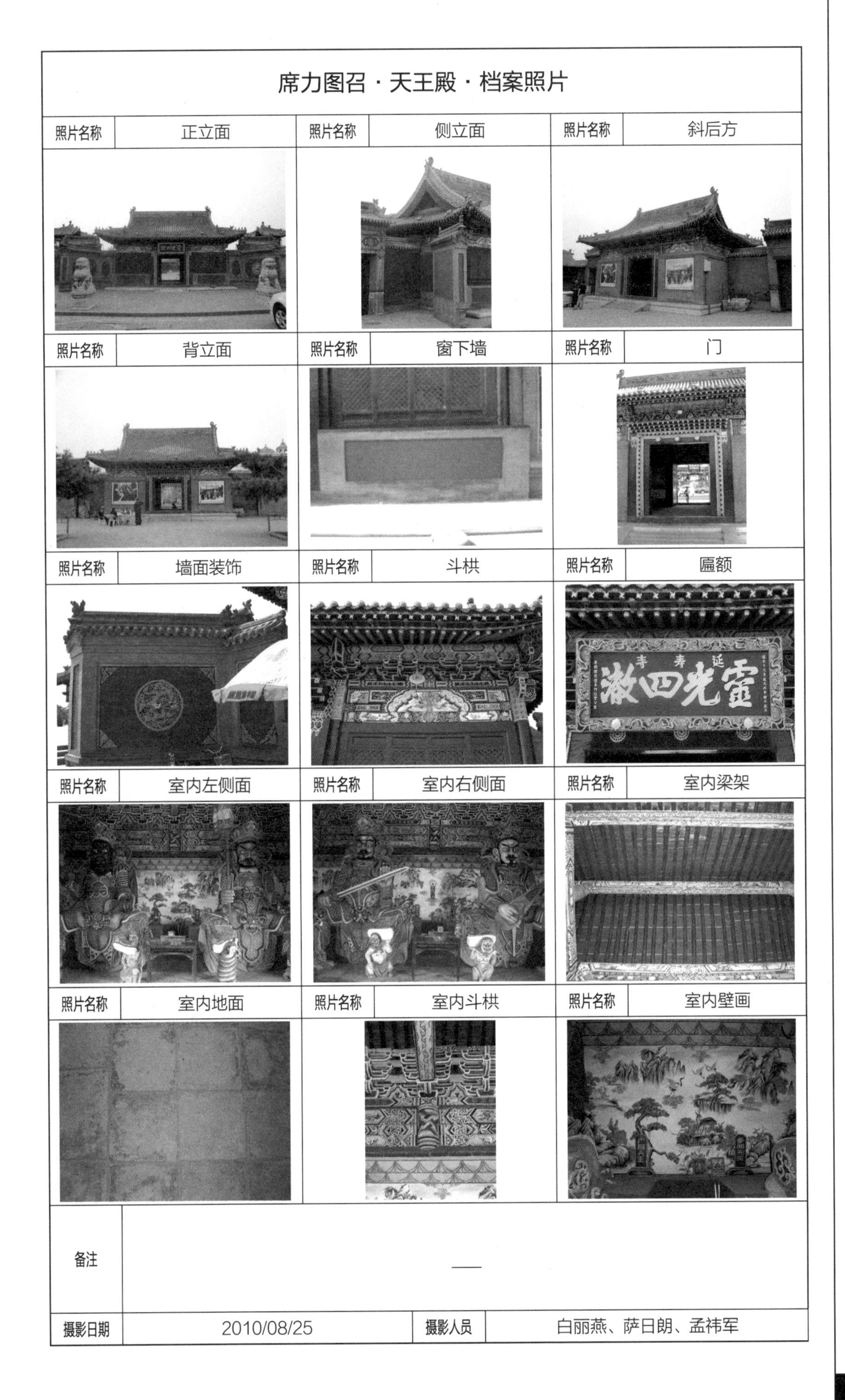

席力图召·天王殿·档案照片

照片名称	正立面	照片名称	侧立面	照片名称	斜后方
照片名称	背立面	照片名称	窗下墙	照片名称	门
照片名称	墙面装饰	照片名称	斗栱	照片名称	匾额
照片名称	室内左侧面	照片名称	室内右侧面	照片名称	室内梁架
照片名称	室内地面	照片名称	室内斗栱	照片名称	室内壁画
备注	—				
摄影日期	2010/08/25	摄影人员	白丽燕、萨日朗、孟祎军		

天王殿正立面

天王殿斜后方（左图）
天王殿佛像 （右图）

天王殿斜前方（左图）
天王殿佛像（右图）

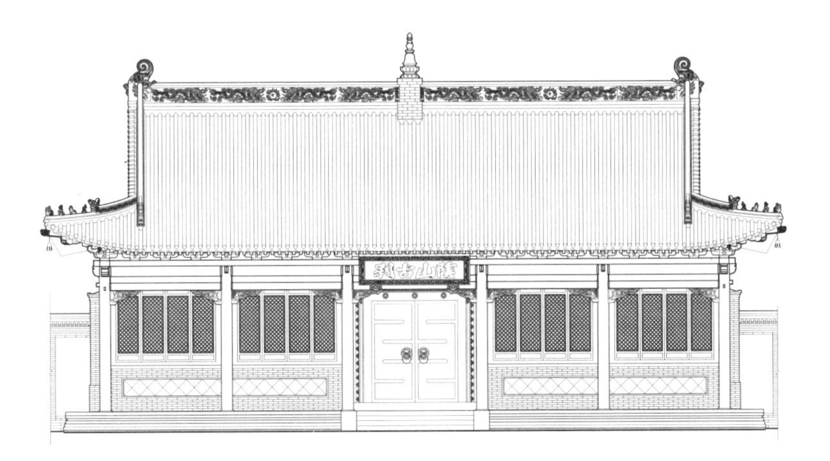

天王殿正立面图

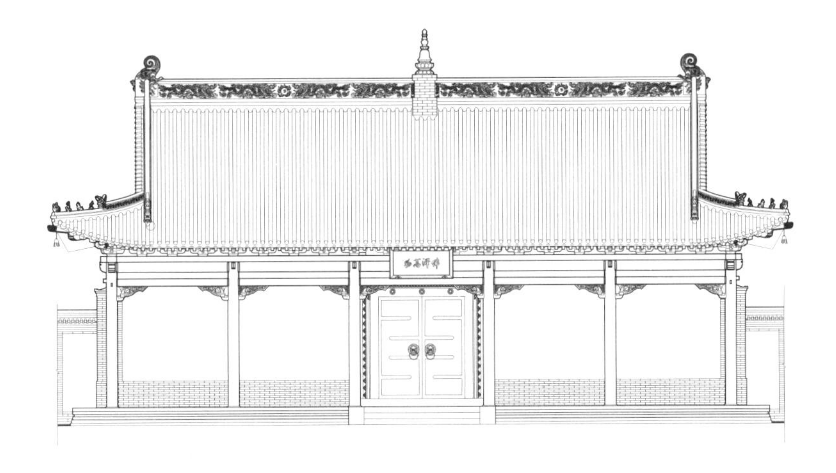

天王殿背立面图

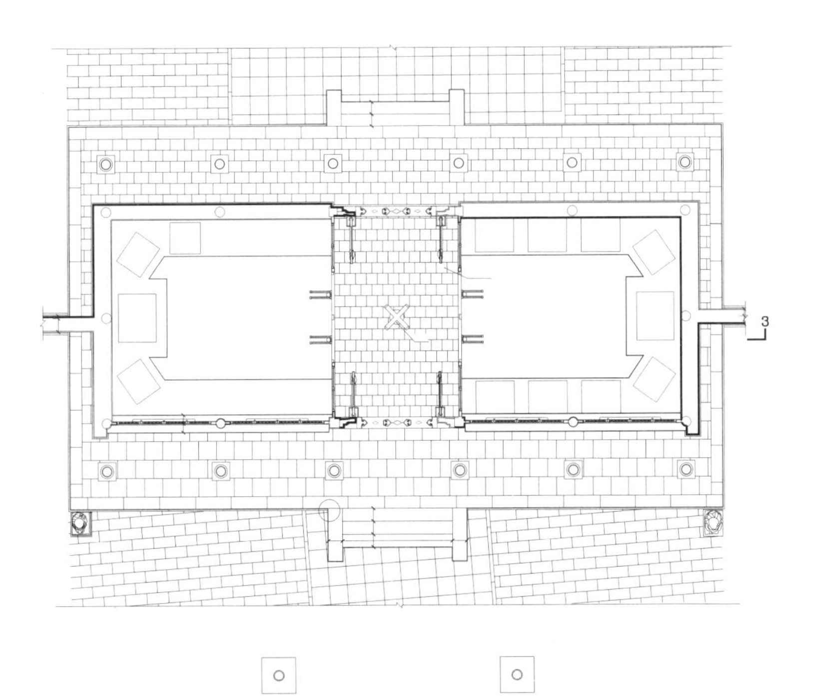

天王殿平面图

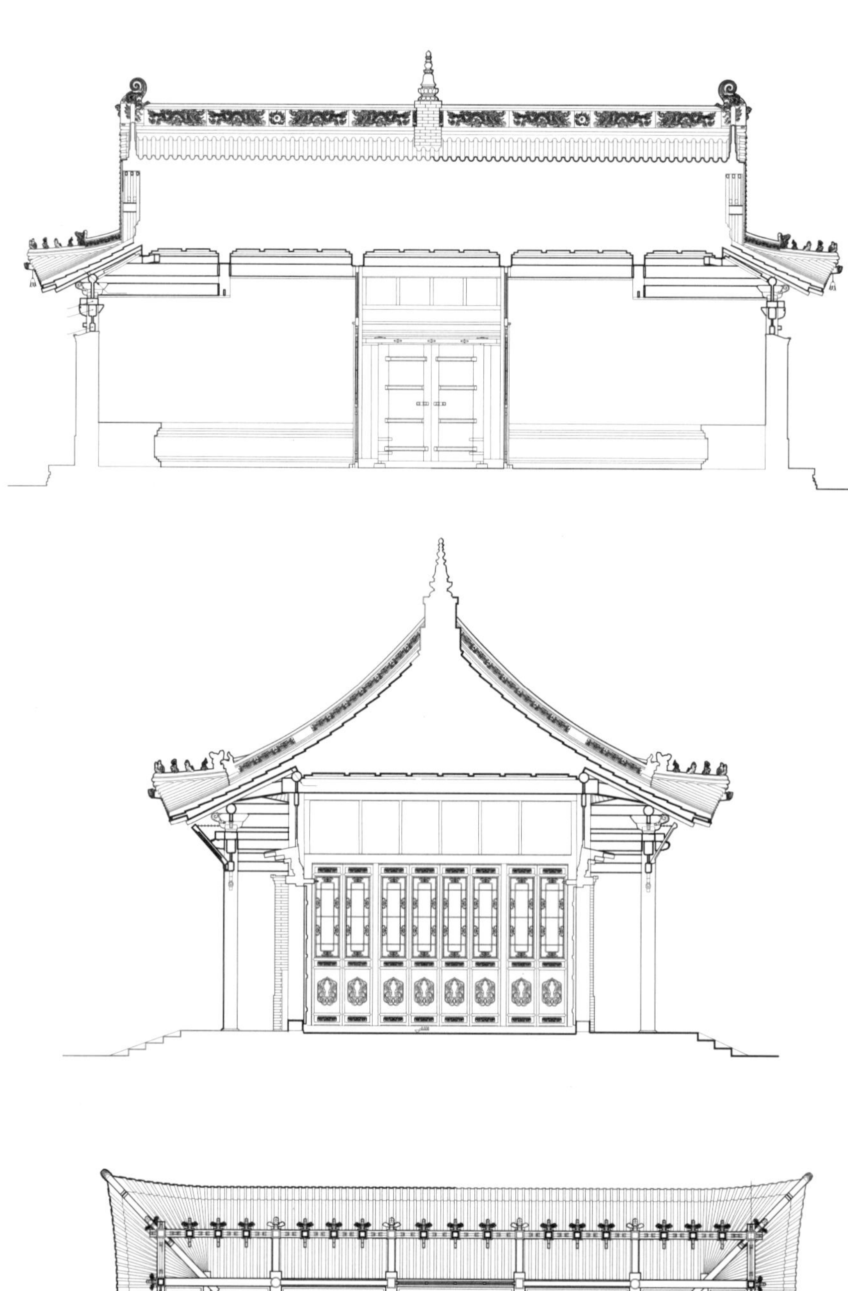

天王殿剖面图1

天王殿剖面图2

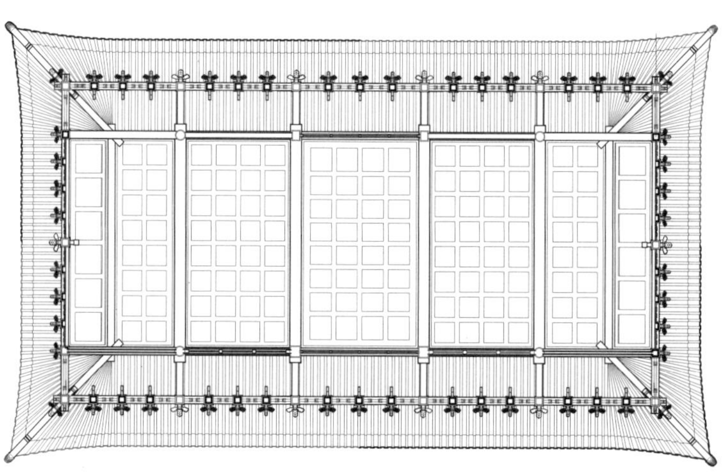

天王殿梁架图

2.3 席力图召 · 菩提过殿

单位:毫米

建筑名称	汉语正式名称	菩提过殿					俗称	过殿					
概述	初建年	——					建筑朝向	南	建筑层数	一			
	建筑简要描述	汉式建筑											
	重建重修记载	——											
		信息来源	——										
结构规模	结构形式	砖木结构		相连的建筑	无			室内天井	无				
	建筑平面形式	长方形		外廊形式	前后廊								
	通面阔	16180	开间数	5	明间	3500	次间	3140	梢间	3200	次梢间	——	尽间 ——
	通进深	8400	进深数	4	进深尺寸（前→后）		1370→2830→1370→2830						
	柱子数量	——	柱子间距	横向尺寸	——		（藏式建筑结构体系填写此栏，不含廊柱）						
				纵向尺寸	——								
	其他	——											
建筑主体（大木作）（石作）（瓦作）	屋顶	屋顶形式	歇山				瓦作	绿琉璃黄剪边					
	外墙	主体材料	砖	材料规格	258×132×50		饰面颜色	灰色					
		墙体收分	无	边玛檐墙	无		边玛材料	——					
	斗栱、梁架	斗栱	有	平身科斗口尺寸		不详	梁架关系	五架梁					
	柱、柱式（前廊柱）	形式	汉式	柱身断面形状	圆形	断面尺寸	直径 $D=300$		（在没有前廊柱的情况下，填写室内柱及其特征）				
		柱身材料	木材	柱身收分	无	栌斗、托木	无	雀替	有				
		柱础	有	柱础形状	圆形	柱础尺寸	直径 $D=370$						
	台基	台基类型	普通	台基高度	500	台基地面铺设材料		石材					
	其他	——											
装修（小木作）（彩画）	门(正面)	撒带门		门楣	有	堆经	有	门帘	有				
	窗（正面）	槛窗		窗楣	无	窗套	无	窗帘	有				
	室内隔扇	隔扇	无	隔扇位置	——								
	室内地面、楼面	地面材料及规格		方砖 300×300		楼面材料及规格		——					
	室内楼梯	楼梯	无	楼梯位置	——	楼梯材料	——	梯段宽度	——				
	天花、藻井	天花	有	天花类型	井口天花	藻井	无	藻井类型	——				
	彩画	柱头	有	柱身	无	梁架	有	走马板	有				
		门、窗	无	天花	有	藻井	无	其他彩画	无				
	其他	悬塑	无	佛龛	无	匾额	阴山古刹						
装饰	室内	帷幔	无	幕帘彩绘	无	壁画	有	唐卡	无				
		经幡	无	经幢	无	柱毯	无	其他	——				
	室外	玛尼轮	无	苏勒德	无	宝顶	有	祥麟法轮	无				
		四角经幢	无	经幡	无	铜饰	无	石刻、砖雕	无				
		仙人走兽	1+3	壁画	无	其他	——						
陈设	室内	主佛像	东：药师佛　西：释迦牟尼佛			佛像基座	莲花座						
		法座	无	藏经橱	无	经床	有	诵经桌	无	法鼓	无	玛尼轮 有	坛城 无 其他 ——
	室外	旗杆	有	苏勒德	无	狮子	有	经幡	无	玛尼轮	无	香炉 有	五供 无 其他 ——
	其他	——											
备注	——												
调查日期	2010/05/01	调查人员	白丽燕、萨日朗	整理日期	2010/05/09	整理人员	萨日朗						

菩提过殿基本概况表1

席力图召·菩提过殿·档案照片					
照片名称 	斜前方	照片名称 	正立面	照片名称 	侧立面
照片名称 	背立面	照片名称 	柱身	照片名称 	柱头
照片名称 	柱础	照片名称 	台基	照片名称 	正门
照片名称 	室内正面	照片名称 	佛像	照片名称 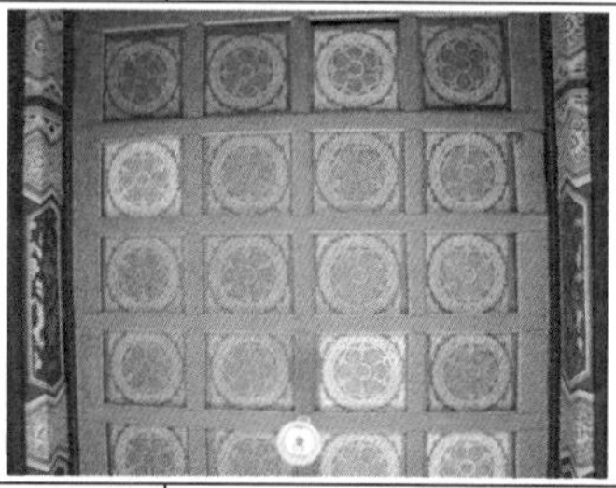	室内天花
照片名称 	壁画	照片名称 	玛尼轮	照片名称 	彩画
备注	—				
摄影日期	2010/08/25	摄影人员	白丽燕、萨日朗、孟祎军		

菩提过殿基本概况表2

菩提过殿正立面

菩提过殿背立面（左图）
菩提过殿侧立面（右图）

菩提过殿室内正面（左图）
菩提过殿外廊（右图）

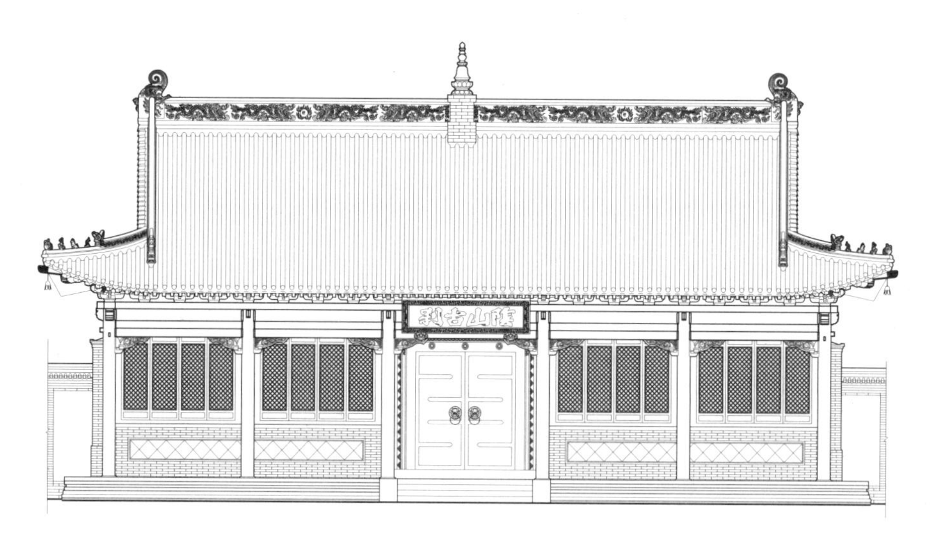

菩提过殿正立面图

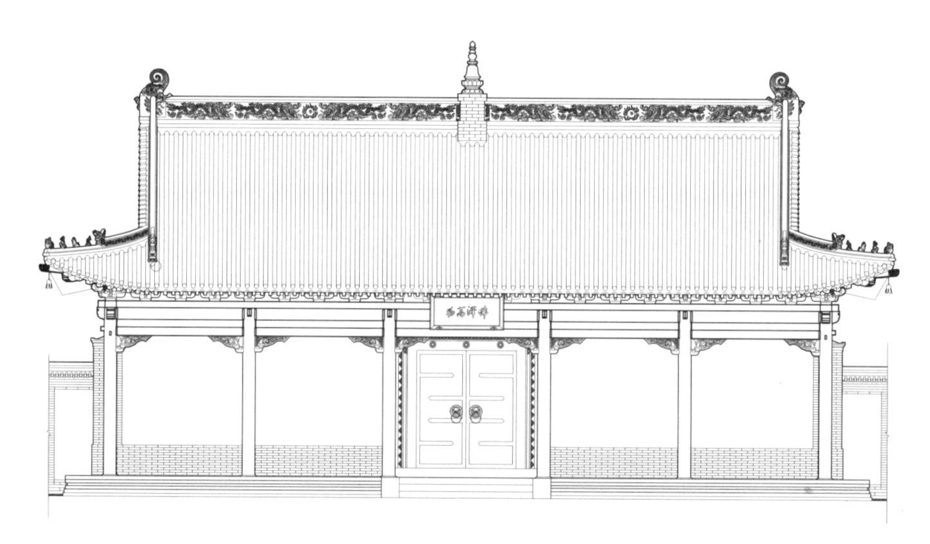

菩提过殿背立面图

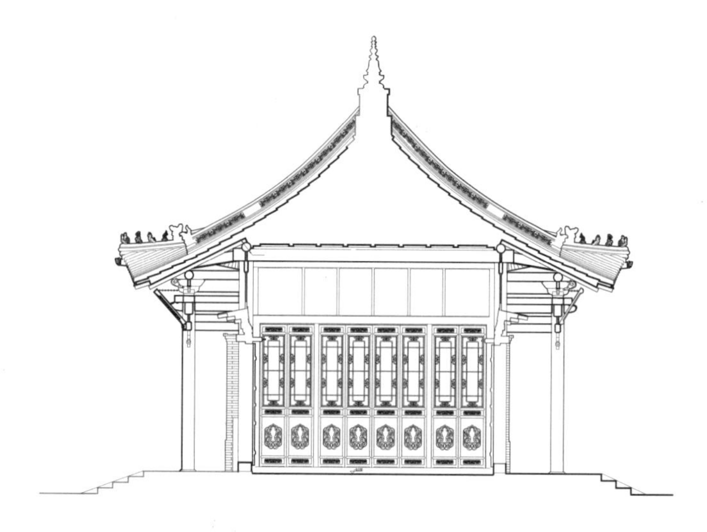

菩提过殿剖面图

2.4 席力图召 · 碑亭

碑亭正视图

碑亭石碑基座（左图）
碑亭石碑顶端（右图）

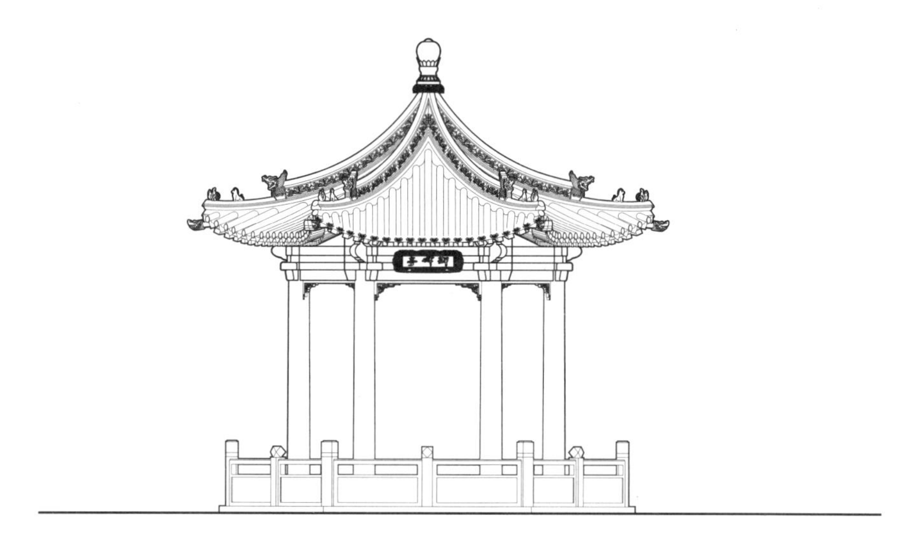

碑亭正立面图

碑亭剖面图

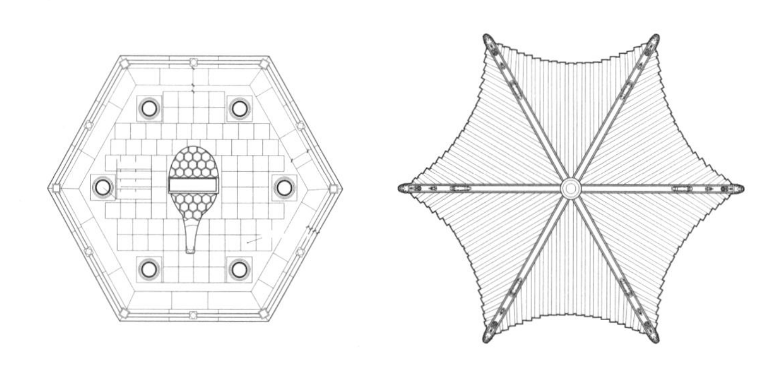

碑亭平面图（左图）
碑亭屋顶平面图（右图）

2.5 席力图召·配殿

东配殿斜前方

西配殿斜前方

配殿风门（左图）
配殿风门（左图）

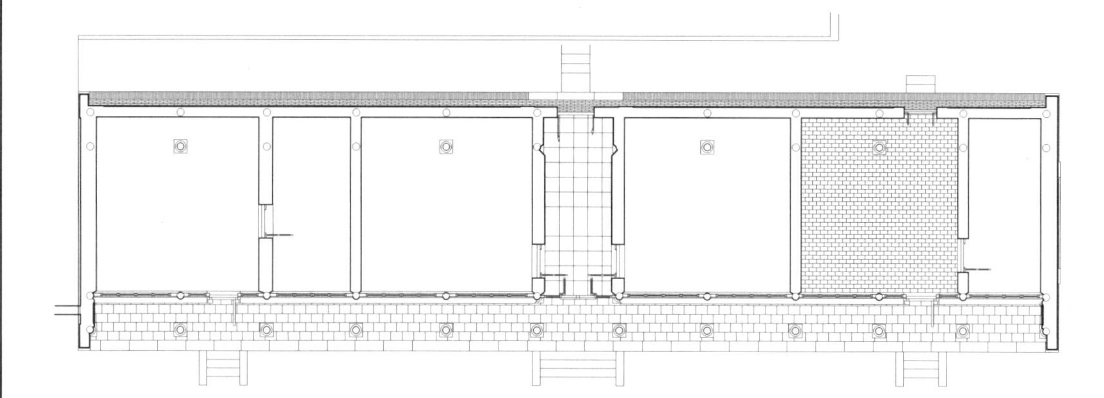

西配殿平面图

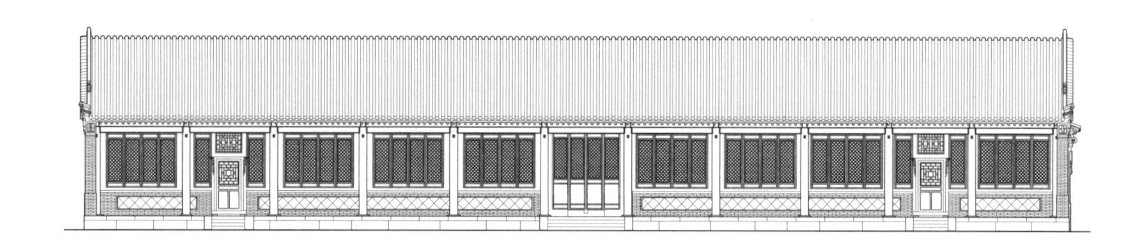

西配殿正立面图

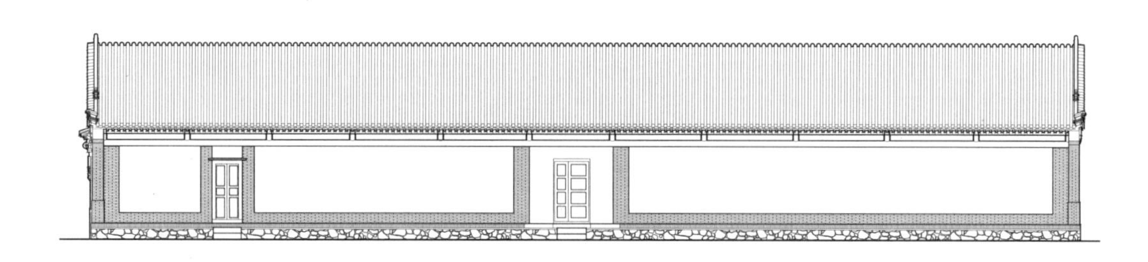

西配殿背立面图

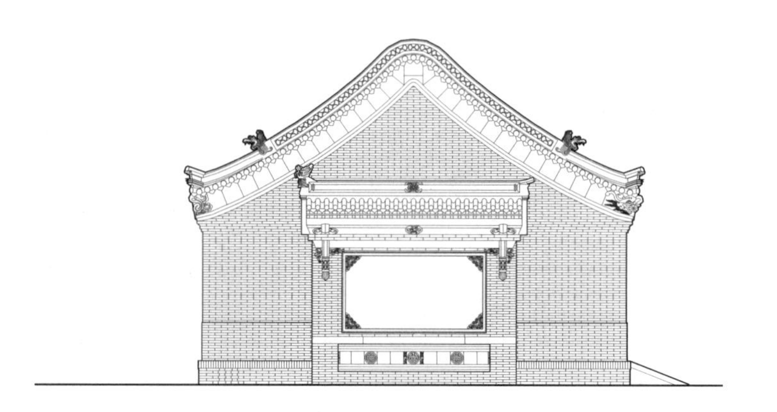

西配殿侧立面图

2.6 席力图召 · 大雄宝殿

单位:毫米

建筑名称	汉语正式名称	大雄宝殿	俗称	——													
概述	初建年	——	建筑朝向	南	建筑层数	二											
	建筑简要描述	汉藏结合式，经堂为藏式，佛殿为汉式															
	重建重修记载	佛殿为重建建筑，2006年开工，2009年完工															
		信息来源	据本寺喇嘛口述														
结构规模	结构形式	砖木混合	相连的建筑	经堂与佛殿以过亭相连	室内天井	无											
	建筑平面形式	凸字形	外廊形式	前廊式													
	通面阔	34240	开间数	7	明间	3130	次间	2800	梢间	2830	次梢间	——	尽间	2790			
	通进深	38920	进深数	14	进深尺寸（前→后）	2490→2870→2930→2910→2860→2890→2890→2900→2490→2400→2780→3500→2730→2280											
	柱子数量	64	柱子间距	横向尺寸	——	（藏式建筑结构体系填写此栏，不含廊柱）											
				纵向尺寸													
	其他	——															
建筑主体（大木作）（石作）（瓦作）	屋顶	屋顶形式	重檐歇山	瓦作	绿琉璃金剪边												
	外墙	主体材料	砖	材料规格	270×130×50	饰面颜色	灰色										
		墙体收分	有	边玛檐墙	有	边玛材料	边玛草										
	斗栱、梁架	斗栱	无	平身科斗口尺寸	——	梁架关系	——										
	柱、柱式（前廊柱）	形式	藏式	柱身断面形状	十二楞柱	断面尺寸	340×340	（在没有前廊柱的情况下，填写室内柱及其特征）									
		柱身材料	木材	柱身收分	有	栌斗、托木	有	雀替	无								
		柱础	有	柱础形状	方形	柱础尺寸	440×440										
	台基	台基类型	普通台基	台基高度	1350	台基地面铺设材料	方砖、红砖										
	其他	——															
装修（小木作）（彩画）	门(正面)	板门	门楣	有	堆经	有	门帘	无									
	窗（正面）	殿门两侧藏式盲窗、二层为隔扇窗	窗楣	有	窗套	有	窗帘	无									
	室内隔扇	隔扇	无	隔扇位置	——												
	室内地面、楼面	地面材料及规格	木地板、长210宽150	楼面材料及规格	木地板、长210宽150												
	室内楼梯	楼梯	有	楼梯位置	殿内右侧靠墙	楼梯材料	木	梯段宽度	800								
	天花、藻井	天花	有	天花类型	井口天花	藻井	有	藻井类型	八边形								
	彩画	柱头	有	柱身	有	梁架	有	走马板	——								
		门、窗	有	天花	有	藻井	有	其他彩画	——								
	其他	悬塑	无	佛龛	有	匾额	能仁显化										
装饰	室内	帷幔	无	幕帘彩绘	无	壁画	有	唐卡	有								
		经幡	无	经幢	有	柱毯	有	其他	——								
	室外	玛尼轮	有	苏勒德	有	宝顶	有	祥麟法轮	有								
		四角经幢	有	经幡	无	铜饰	有	石刻、砖雕	有								
		仙人走兽	无	壁画	有	其他	——										
陈设	室内	主佛像	释迦牟尼佛	佛像基座	须弥座												
		法座	有	藏经橱	有	经床	有	诵经桌	有	法鼓	有	玛尼轮	有	坛城	无	其他	——
	室外	旗杆	无	苏勒德	无	狮子	无	经幡	无	玛尼轮	有	香炉	有	五供	无	其他	——
	其他	——															
备注	过亭尺寸：横向——2830→3570→2750　纵向——2300→2000																
调查日期	2010/05/01	调查人员	白丽燕、萨日朗	整理日期	2010/05/09	整理人员	萨日朗										

大雄宝殿基本概况表1

席力图召·大雄宝殿·档案照片					
照片名称 	斜前方	照片名称 	正立面	照片名称 	东立面
照片名称 	西立面	照片名称 	柱身	照片名称 	柱头
照片名称 	台基	照片名称 	门	照片名称 	盲窗
照片名称 	室内正面	照片名称 	室内侧面	照片名称 	室内柱身
照片名称 	室内柱头	照片名称 	室内楼梯	照片名称 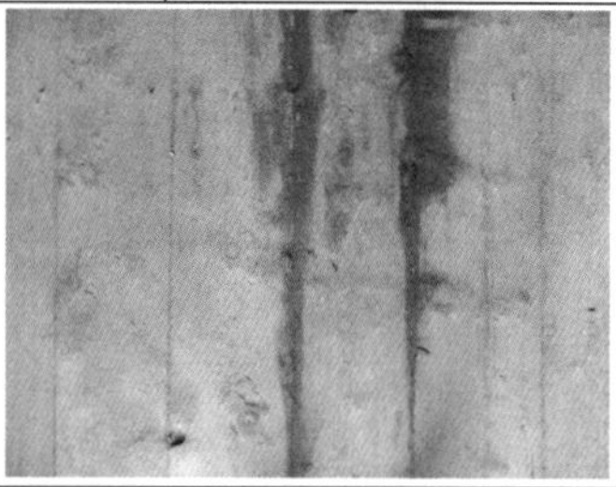	室内地面
备注	—				
摄影日期	2010/08/25	摄影人员	白丽燕、萨日朗、孟祎军		

大雄宝殿基本概况表2

大雄宝殿正立面

大雄宝殿经堂斜前方

大雄宝殿经堂柱廊
（左图）
大雄宝殿佛殿斜后方
（右图）

大雄宝殿一层平面图

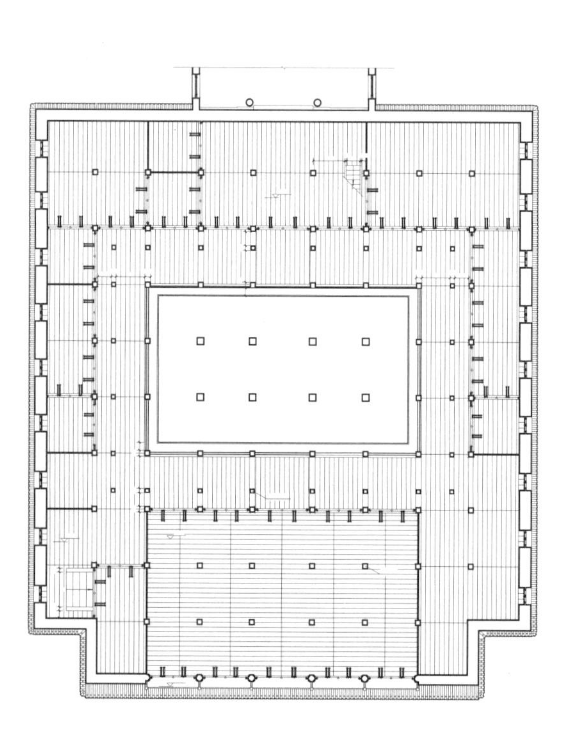

大雄宝殿二层平面图

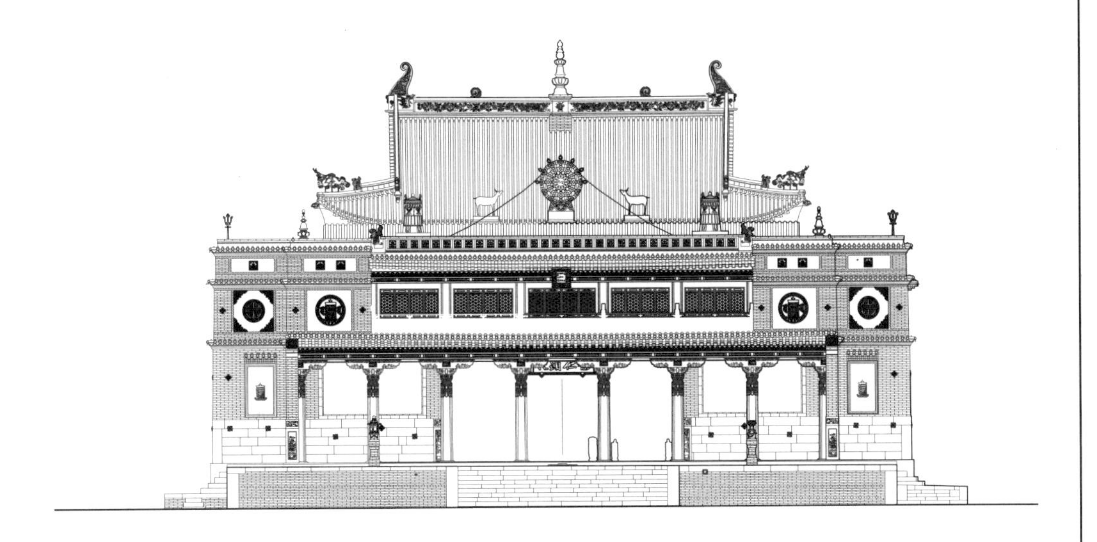

大雄宝殿正立面图

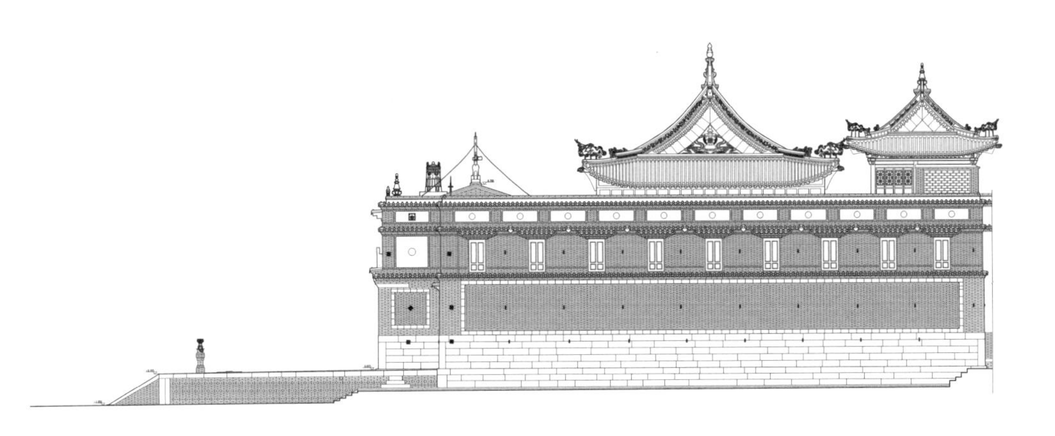

大雄宝殿侧立面图

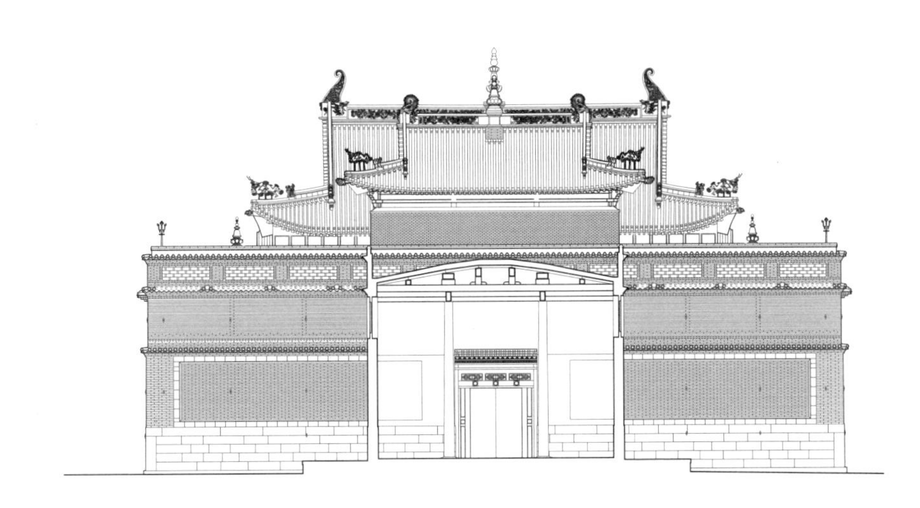

大雄宝殿背立面图

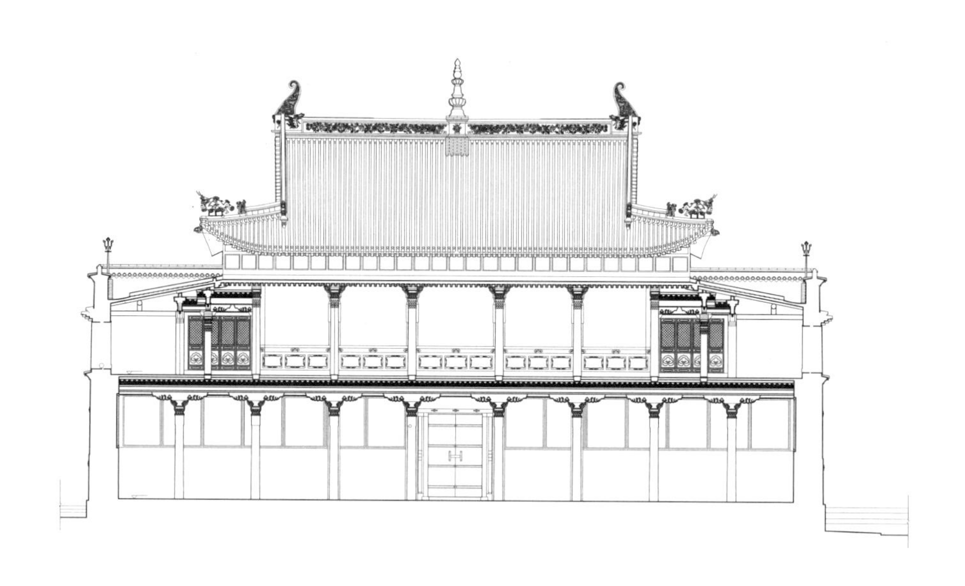

大雄宝殿剖面图1

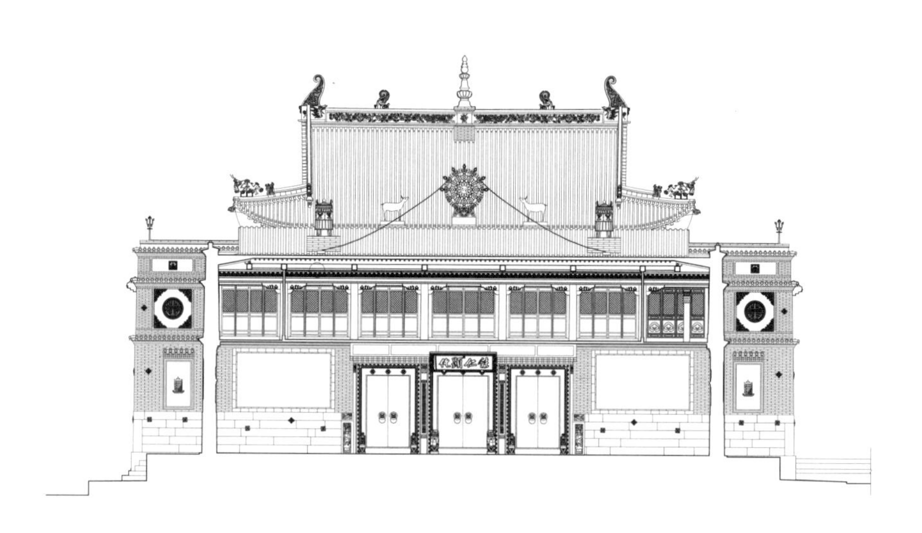

大雄宝殿剖面图2

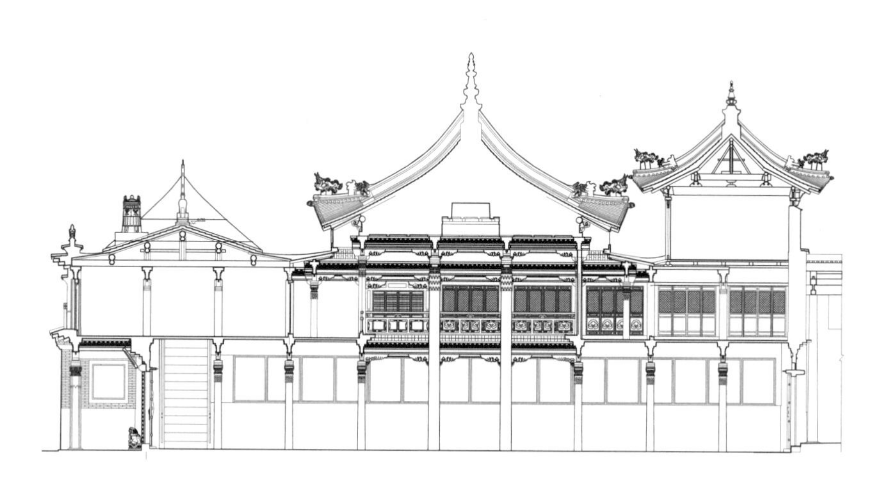

大雄宝殿剖面图3

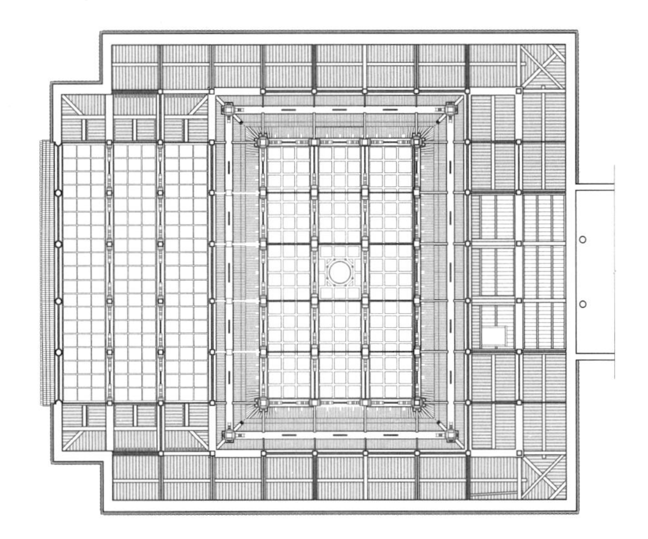

大雄宝殿梁架平面图

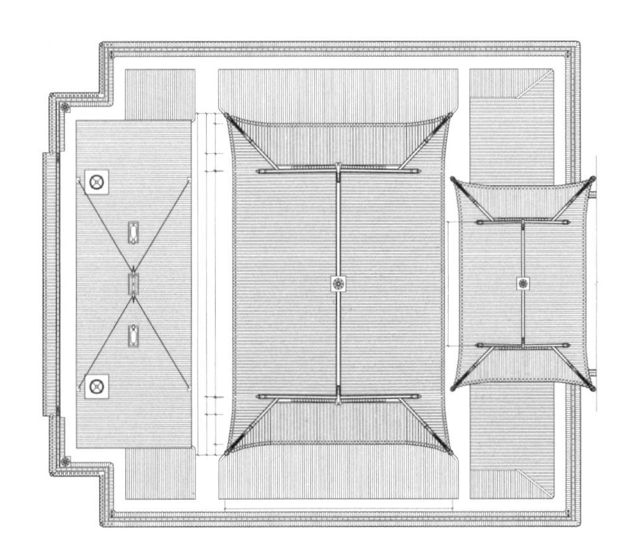

大雄宝殿屋顶平面图

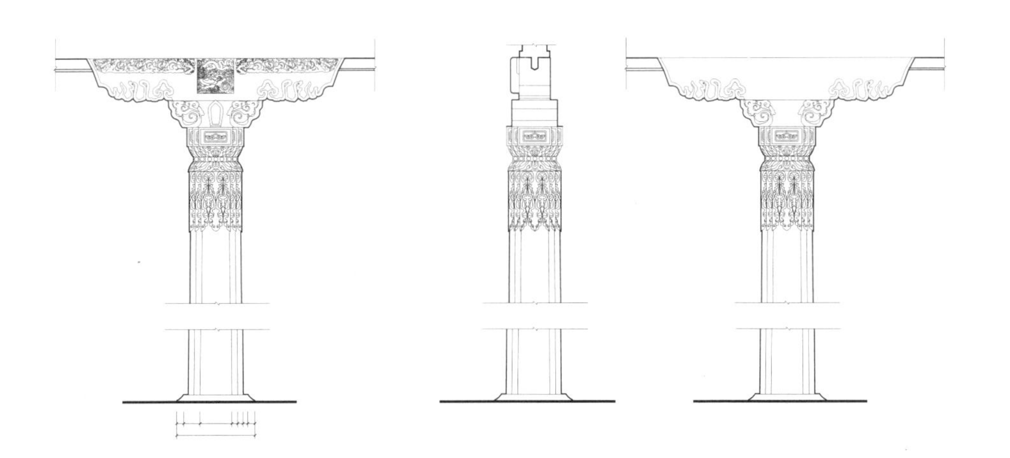

大雄宝殿棱柱立面图

2.7 席力图召 · 照壁

照壁斜前方

照壁东侧门（左图）
照壁顶部（右图）

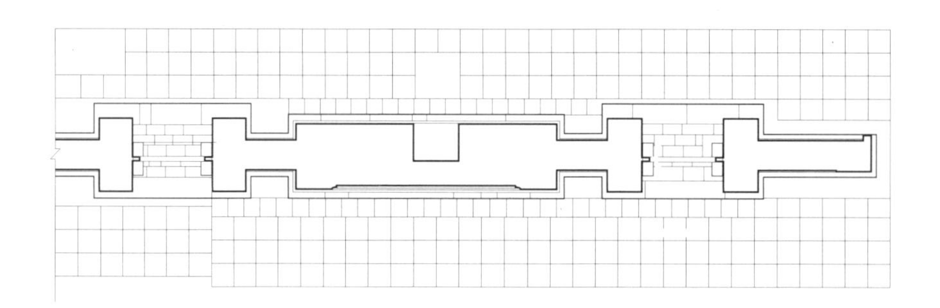

照壁平面图

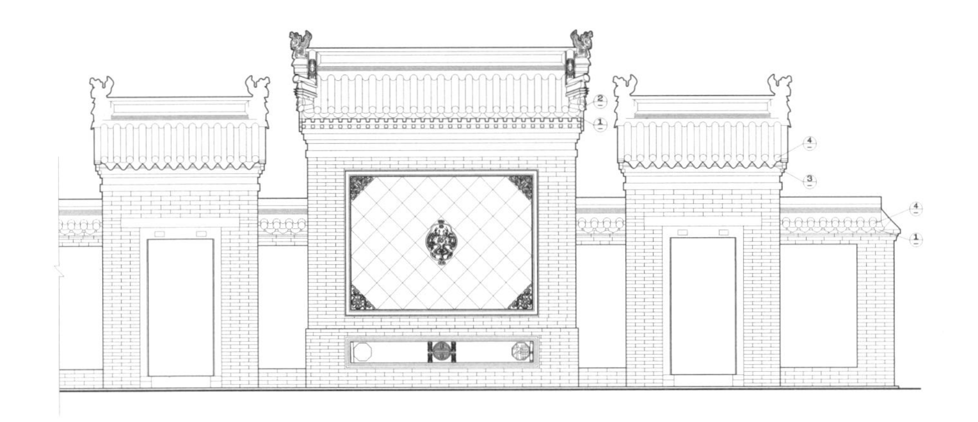

照壁正立面图

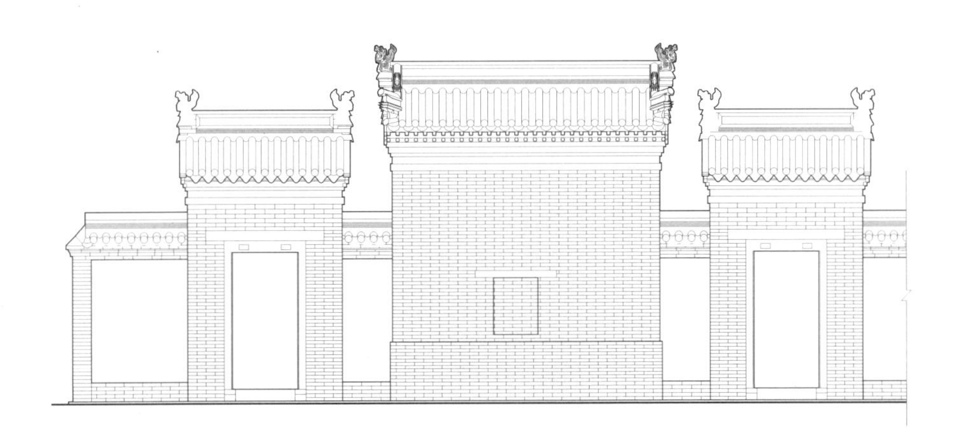

照壁背立面图

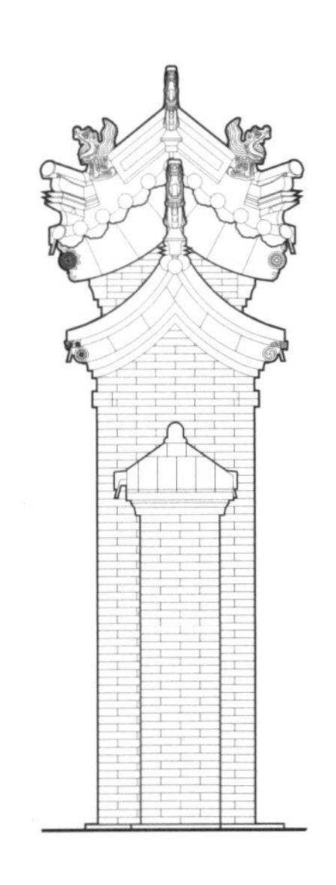

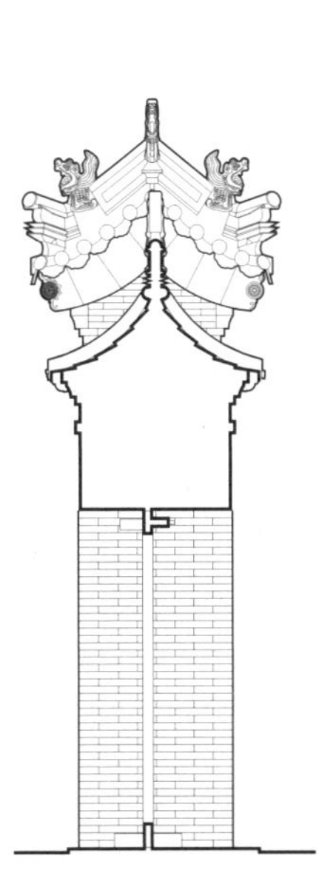

照壁侧立面图（左图）
照壁侧剖面图（右图）

2.8 席力图召·白塔

白塔斜前方

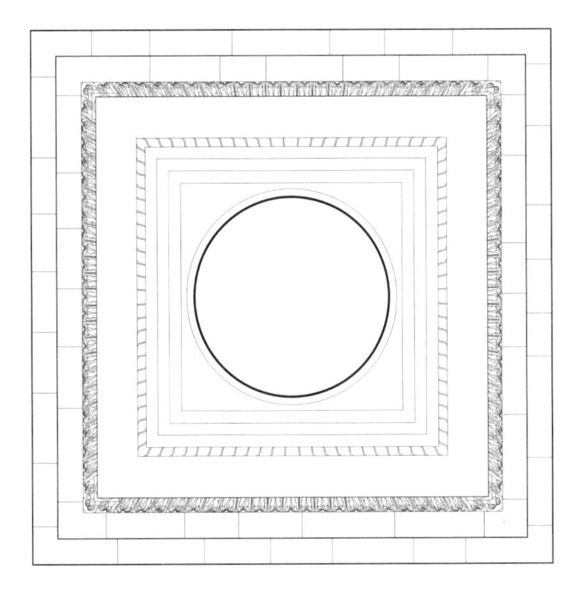

白塔平面图 1

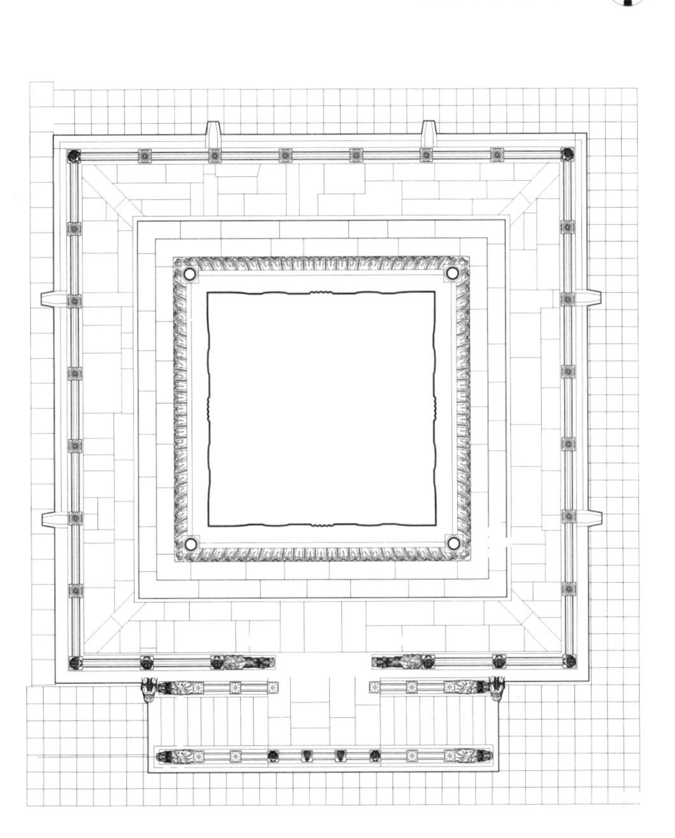

白塔平面图 2

白塔正立面图

白塔背立面图

2.9 席力图召・九间楼

九间楼斜前方

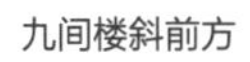

九间楼柱子（左图）
九间楼柱头（右图）

九间楼隔扇门（左图）
九间楼窗（右图）

2.10 席力图召 · 古佛殿

单位:毫米

建筑名称	汉语正式名称	古佛殿	俗称	——													
概述	初建年	——	建筑朝向	南	建筑层数	二											
	建筑简要描述	汉藏结合，汉式结构体系，正面两侧为藏式风格															
	重建重修记载	——															
		信息来源	——														
结构规模	结构形式	砖木混合	相连的建筑	无	室内天井	无											
	建筑平面形式	凸字形	外廊形式	前廊，半回廊													
	通面阔	13550	开间数	5	明间	3330	次间	2550	梢间	2560	次梢间	——	尽间	——			
	通进深	20250	进深数	9	进深尺寸（前→后）	1830→1850→1890→1630→930→1830→2580-3850→2930→1820											
	柱子数量	——	柱子间距	横向尺寸	——	（藏式建筑结构体系填写此栏，不含廊柱）											
				纵向尺寸	——												
	其他	——															
建筑主体（大木作）（石作）（瓦作）	屋顶	屋顶形式	经堂是歇山（局部平屋顶），佛殿是二重歇山	瓦作	琉璃瓦												
	外墙	主体材料	青砖,土坯	材料规格	青砖 290×180×50	饰面颜色	青、红										
		墙体收分	有	边玛檐墙	有	边玛材料	边玛草										
	斗栱、梁架	斗栱	有	平身科斗口尺寸	60	梁架关系	抬梁式										
	柱、柱式（前廊柱）	形式	藏式	柱身断面形状	12楞柱	断面尺寸	290×290	（在没有前廊柱的情况下，填写室内柱及其特征）									
		柱身材料	木材	柱身收分	有	栌斗、托木	无	雀替	有								
		柱础	有	柱础形状	方形	柱础尺寸	450×450										
	台基	台基类型	普通台基	台基高度	110	台基地面铺设材料	石材										
	其他	——															
装修（小木作）（彩画）	门(正面)	板门	门楣	有	堆经	有	门帘	有									
	窗（正面）	藏式明窗	窗楣	有	窗套	有	窗帘	无									
	室内隔扇	隔扇	有	隔扇位置	经堂与佛殿之间												
	室内地面、楼面	地面材料及规格	木板2120×270	楼面材料及规格	木板，规格不均												
	室内楼梯	楼梯	有	楼梯位置	进正门左侧	楼梯材料	木	梯段宽度	600								
	天花、藻井	天花	有	天花类型	井口天花	藻井	有	藻井类型	八角形								
	彩画	柱头	有	柱身	无	梁架	有	走马板	无								
		门、窗	有	天花	有	藻井	有	其他彩画	——								
	其他	悬塑	无	佛龛	有	匾额	古佛殿										
装饰	室内	帷幔	无	幕帘彩绘	无	壁画	有	唐卡	有								
		经幡	有	经幢	有	柱毯	无	其他	——								
	室外	玛尼轮	无	苏勒德	无	宝顶	有	祥麟法轮	有								
		四角经幢	无	经幡	有	铜饰	有	石刻、砖雕	无								
		仙人走兽	1+2	壁画	有	其他	——										
陈设	室内	主佛像	释迦牟尼佛	佛像基座	须弥座												
		法座	有	藏经橱	有	经床	有	诵经桌	有	法鼓	有	玛尼轮	无	坛城	无	其他	——
	室外	旗杆	有	苏勒德	无	狮子	无	经幡	有	玛尼轮	有	香炉	有	五供	无	其他	——
	其他	——															
备注	——																
调查日期	2010/05/01	调查人员	白丽燕、萨日朗	整理日期	2010/05/09	整理人员	萨日朗										

古佛殿基本概况表1

席力图召·古佛殿·档案照片

照片名称	斜前方	照片名称	正前方	照片名称	背立面
照片名称	前廊	照片名称	周围廊	照片名称	室外柱子
照片名称	室外局部	照片名称	正门	照片名称	室外窗
照片名称	室内正面	照片名称	室内佛像	照片名称	室内天花
照片名称	室内局部	照片名称	室内壁画	照片名称	室内唐卡
备注	——				
摄影日期	2010/08/26	摄影人员	白丽燕、萨日朗、孟祎军		

古佛殿基本概况表2

古佛殿斜前方

古佛殿正立面（左图）
古佛殿柱廊（右图）

古佛殿背立面（左图）
古佛殿室外陈设（右图）

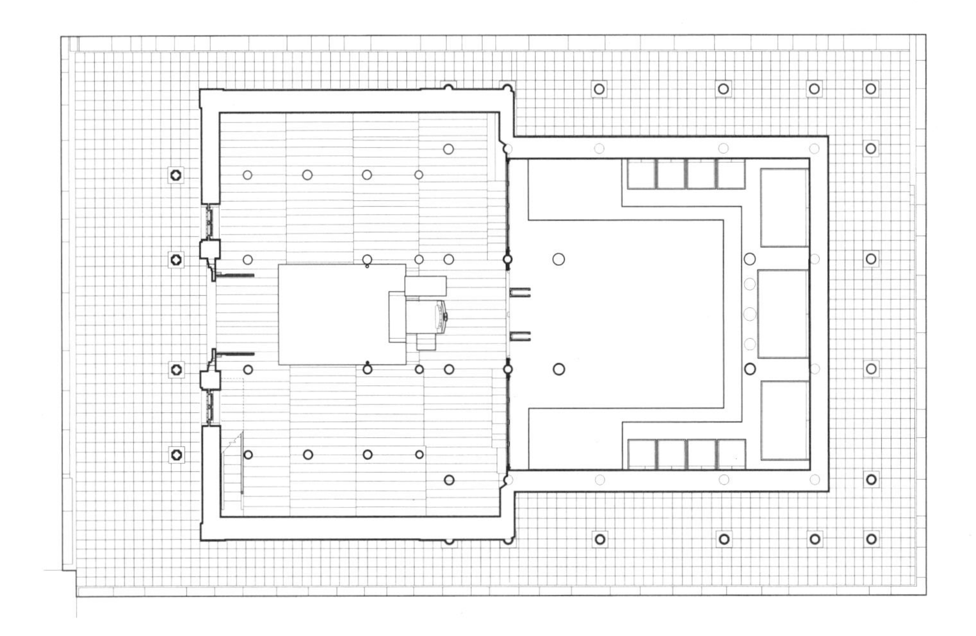

古佛殿一层平面图

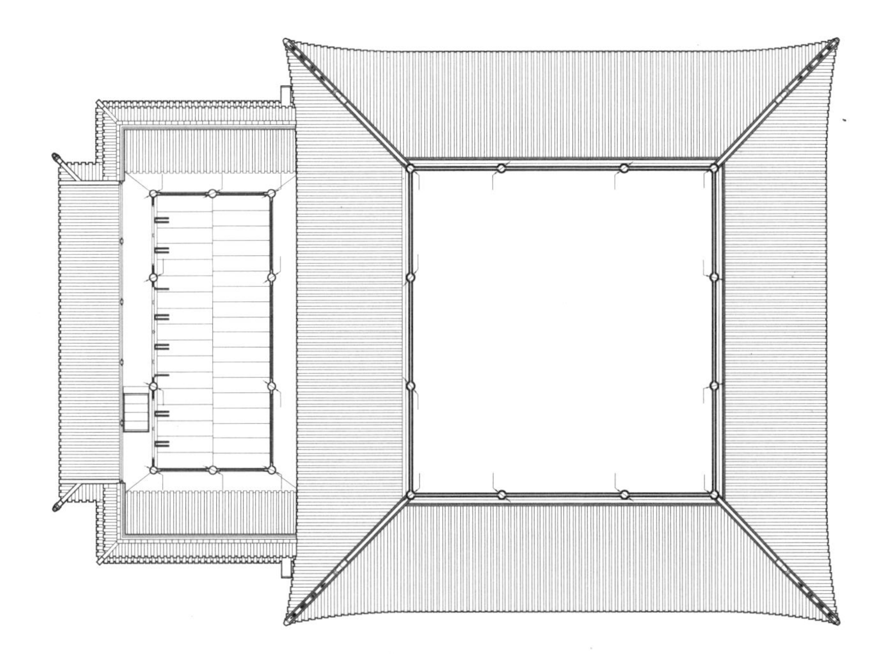

北

古佛殿二层平面图

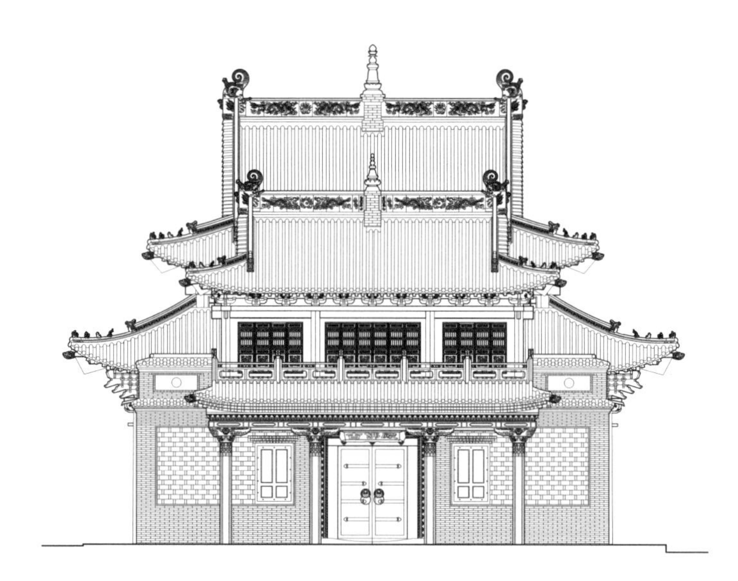

古佛殿正立面图

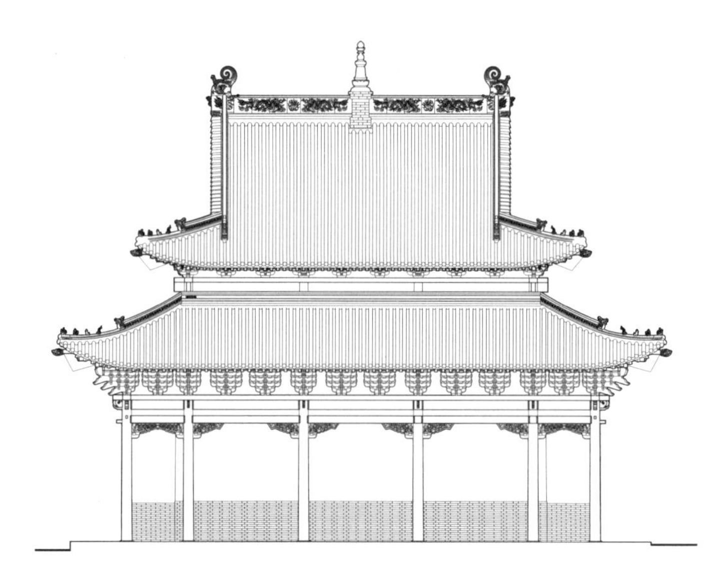

古佛殿背立面图

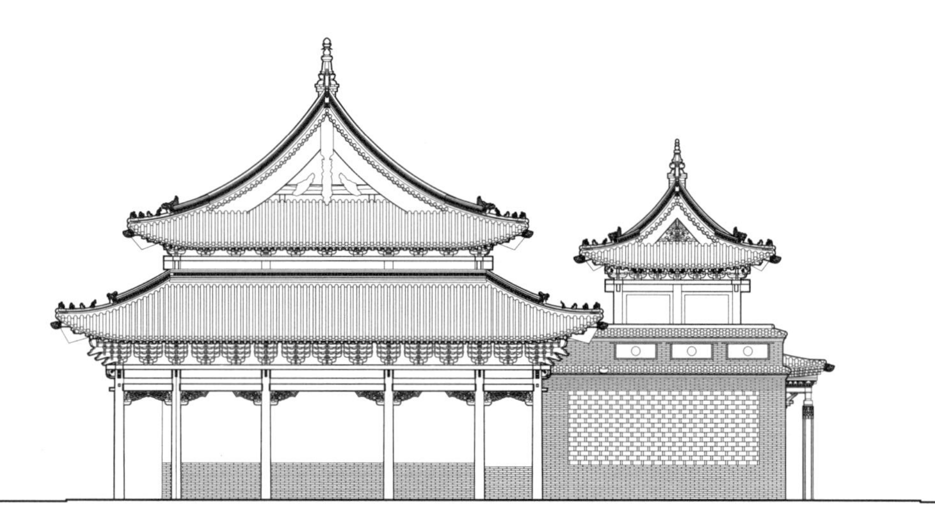

古佛殿侧立面图

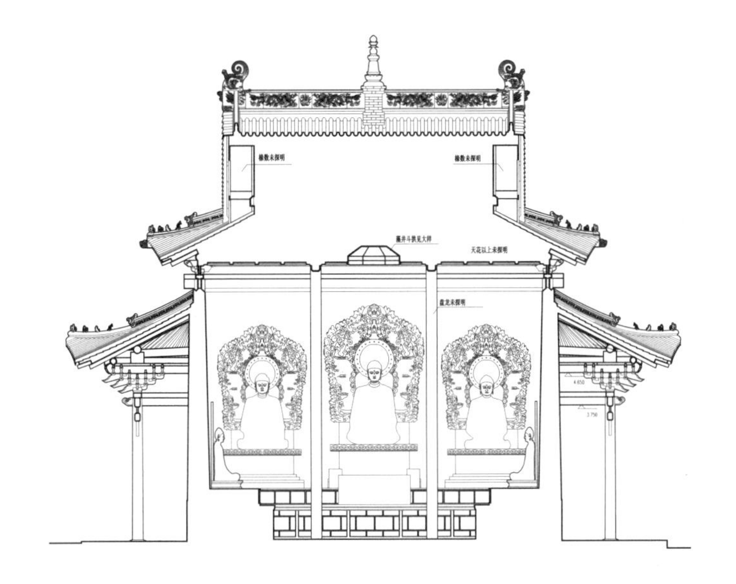

古佛殿剖面图1

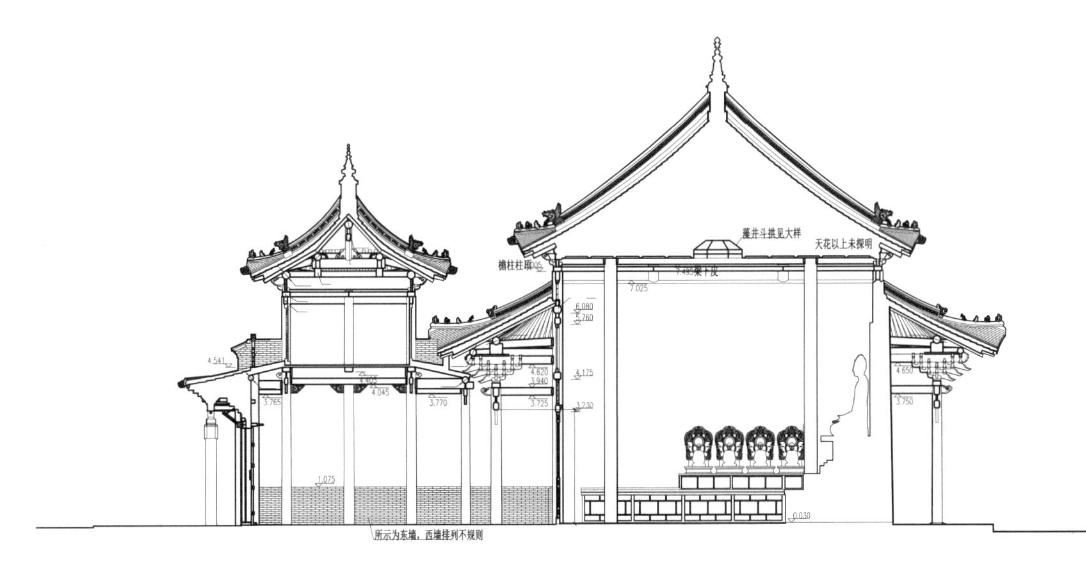

古佛殿剖面图2

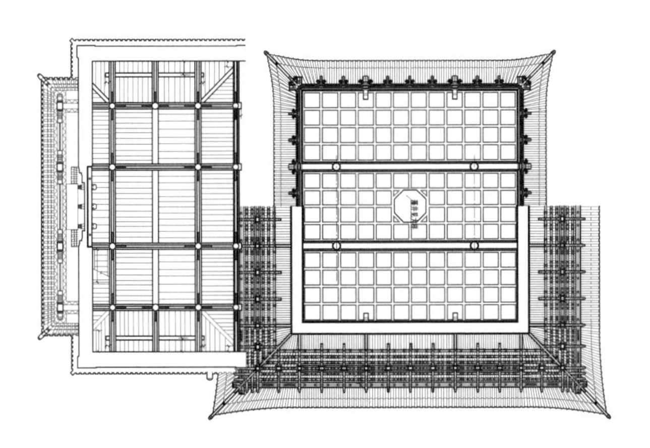

古佛殿梁架平面图

2.11 席力图召·护法殿

单位:毫米

建筑名称	汉语正式名称	护法殿	俗称	——
概述	初建年	——	建筑朝向	南	建筑层数	一
	建筑简要描述	该建筑为一个五开间硬山分隔为三殿，其中护法殿居中为三开间
	重建重修记载	1993年修缮塑像，改为密宗护法殿
		信息来源	——
结构规模	结构形式	砖木	相连的建筑	无	室内天井	无
	建筑平面形式	方形	外廊形式	前廊		
	通面阔	26920	开间数	5	明间	3530	次间	3500	梢间	3230	次梢间	——	尽间	——
	通进深	室内9000	进深数	1	进深尺寸（前→后）	9000
	柱子数量	——	柱子间距	横向尺寸	——	（藏式建筑结构体系填写此栏，不含廊柱）
				纵向尺寸	——	
	其他	——
建筑主体（大木作）（石作）（瓦作）	屋顶	屋顶形式	硬山	瓦作	布瓦
	外墙	主体材料	青砖	材料规格	270×130×50	饰面颜色	青
		墙体收分	有	边玛檐墙	无	边玛材料	无
	斗栱、梁架	斗栱	无	平身科斗口尺寸	——	梁架关系	抬梁
	柱、柱式（前廊柱）	形式	汉	柱身断面形状	圆	断面尺寸	直径$D=210$	（在没有前廊柱的情况下，填写室内柱及其特征）
		柱身材料	木	柱身收分	有	栌斗、托木	无	雀替	有	
		柱础	不可见	柱础形状	不可见	柱础尺寸	不详	
	台基	台基类型	普通台基	台基高度	550	台基地面铺设材料	条石1150×300×130
	其他	——
装修（小木作）（彩画）	门(正面)	隔扇门	门楣	无	堆经	无	门帘	无
	窗（正面）	槛窗	窗楣	无	窗套	无	窗帘	无
	室内隔扇	隔扇	无	隔扇位置	无
	室内地面、楼面	地面材料及规格	有地毯，材料规格不详	楼面材料及规格	无
	室内楼梯	楼梯	无	楼梯位置	无	楼梯材料	无	梯段宽度	无
	天花、藻井	天花	有	天花类型	井口天花	藻井	无	藻井类型	无
	彩画	柱头	无	柱身	无	梁架	无	走马板	无
		门、窗	无	天花	有	藻井	无	其他彩画	——
	其他	悬塑	无	佛龛	无	匾额	观音殿、护法殿、度母殿
装饰	室内	帷幔	无	幕帘彩绘	无	壁画	有	唐卡	无
		经幡	有	经幢	无	柱毯	无	其他	——
	室外	玛尼轮	无	苏勒德	无	宝顶	无	祥麟法轮	无
		四角经幢	无	经幡	无	铜饰	无	石刻、砖雕	有
		仙人走兽	1+2	壁画	无	其他	——
陈设	室内	主佛像	观音、大威德金刚、度母佛	佛像基座	不可见
		法座	无	藏经橱	无	经床	无	诵经桌	无	法鼓	无	玛尼轮	无	坛城	无	其他	——
	室外	旗杆	无	苏勒德	无	狮子	无	经幡	无	玛尼轮	无	香炉	有	五供	无	其他	——
	其他	——
备注	——
调查日期	2010/05/01	调查人员	白丽燕、萨日朗	整理日期	2010/05/09	整理人员	萨日朗

护法殿基本概况表1

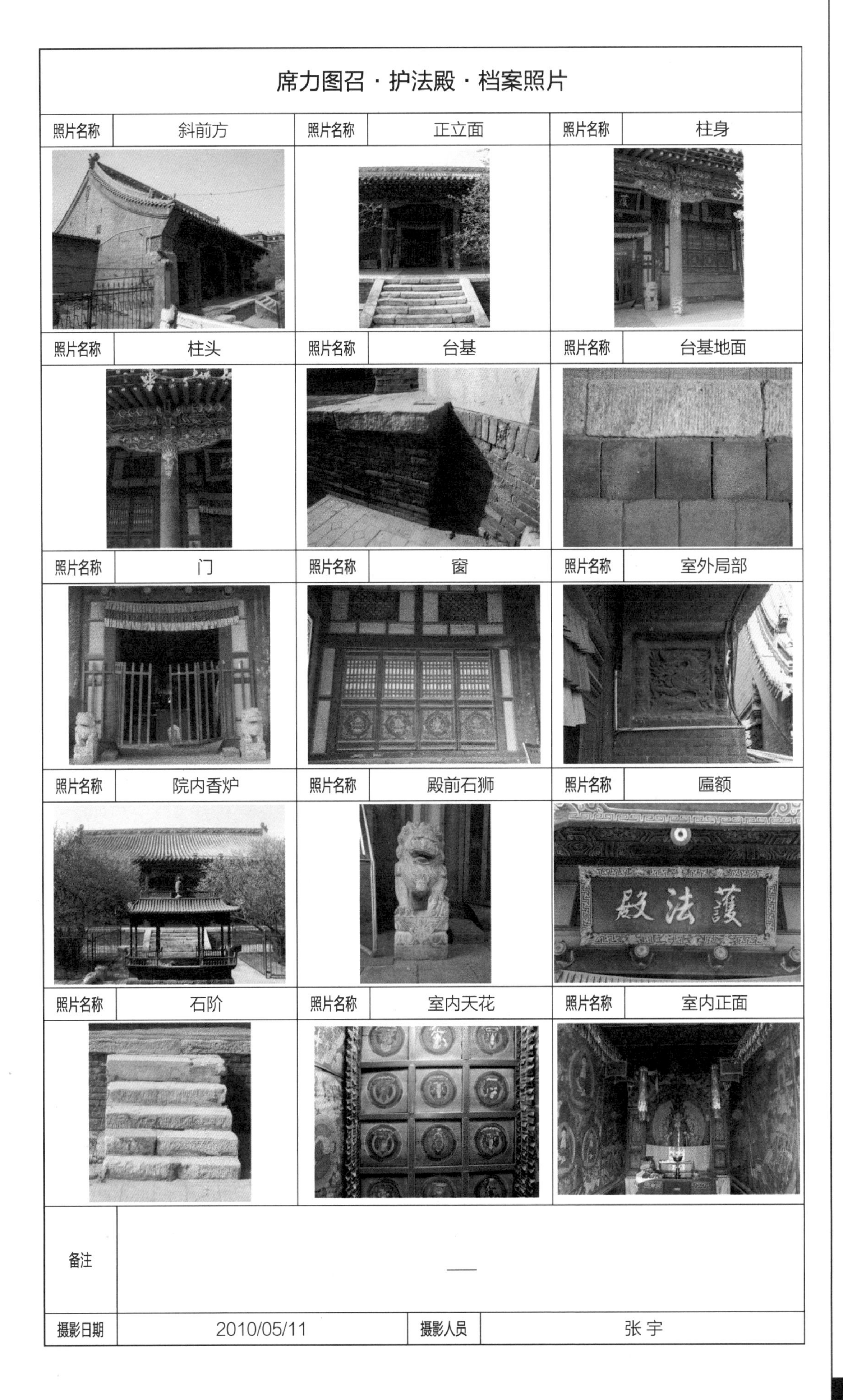

席力图召·护法殿·档案照片

照片名称	斜前方	照片名称	正立面	照片名称	柱身
照片名称	柱头	照片名称	台基	照片名称	台基地面
照片名称	门	照片名称	窗	照片名称	室外局部
照片名称	院内香炉	照片名称	殿前石狮	照片名称	匾额
照片名称	石阶	照片名称	室内天花	照片名称	室内正面
备注	—				
摄影日期	2010/05/11	摄影人员	张 宇		

护法殿基本概况表2

护法殿斜前方

护法殿西侧一间为度母殿，图为度母殿斜前方（左图）

护法殿东侧一间为观音殿，图为观音殿正立面（右图）

护法殿院内香炉（左图）

护法殿正门（右图）

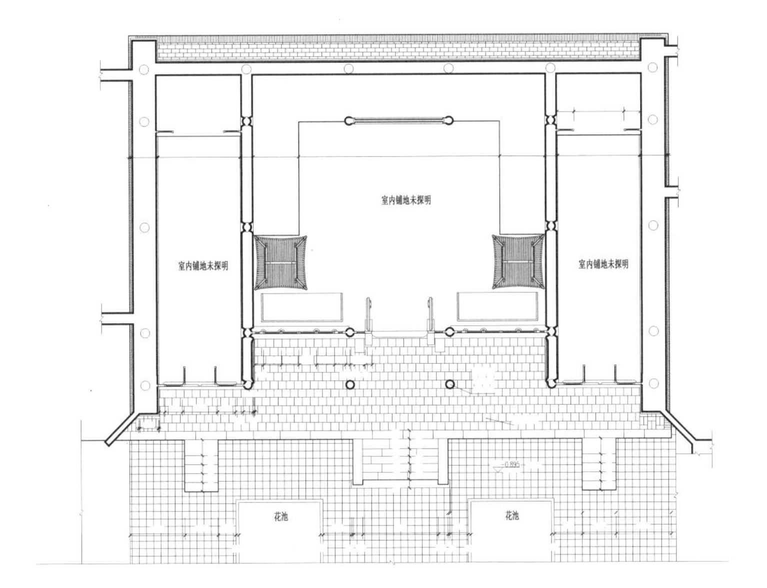

护法殿一层平面图

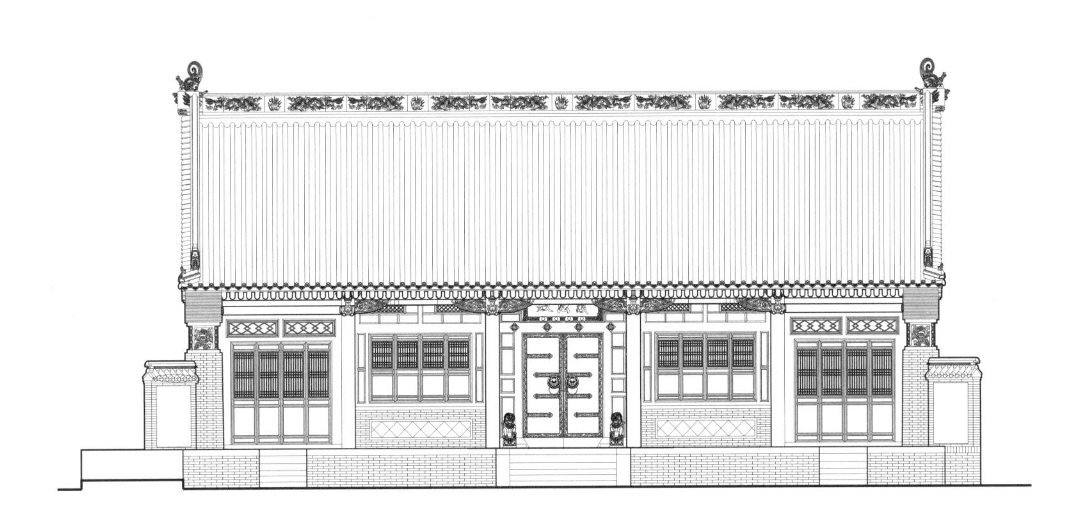

护法殿立面图

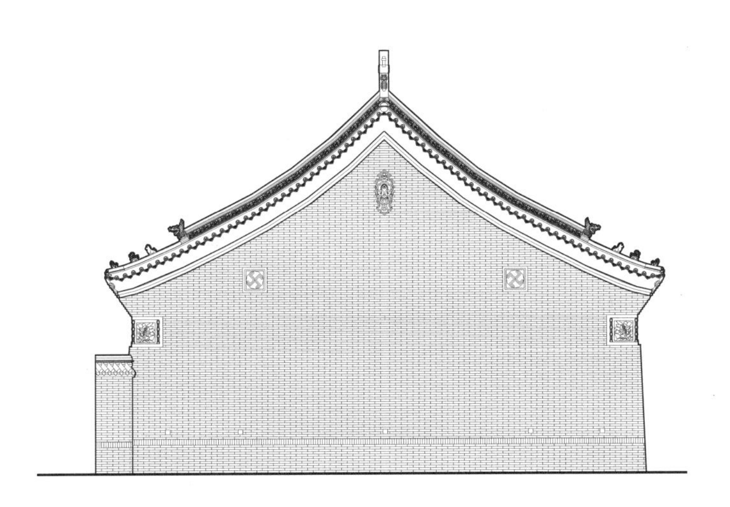

护法殿侧立面图

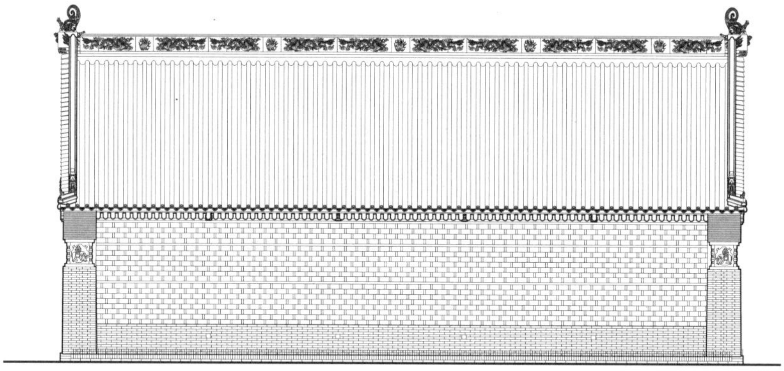

护法殿背立面图

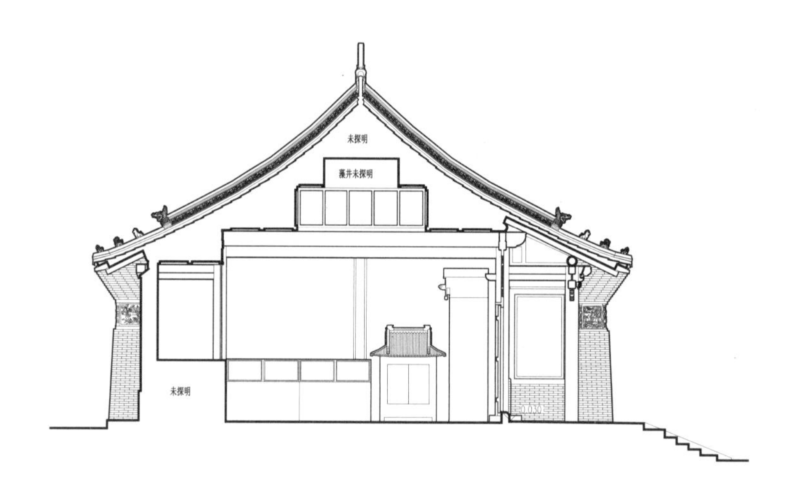

护法殿剖面图

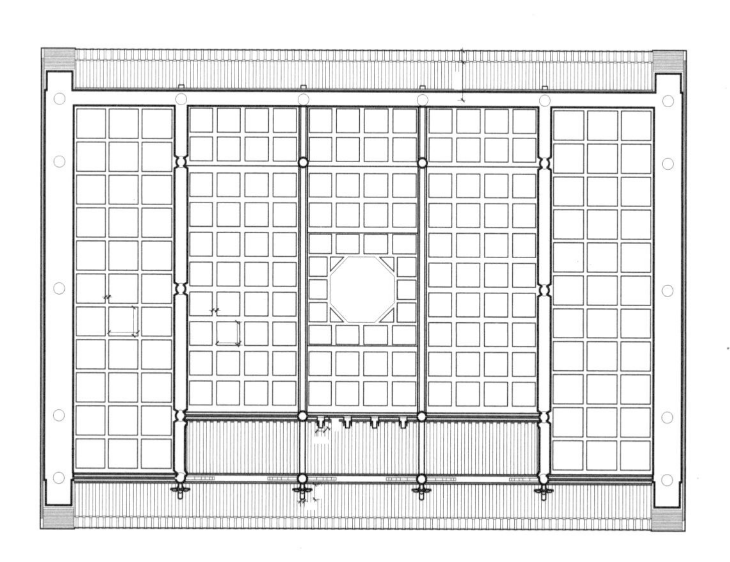

护法殿梁架平面图

2.12 席力图召·活佛府

活佛府鸟瞰图

活佛府山门背立面（左图）

活佛府山门入口（右图）

活佛府耳房（左图）

活佛府柱廊（右图）

2.13 席力图召·其它建筑

钟楼斜前方（左图）
钟楼正立面（右图）

东配殿柱廊（左图）
西配殿工艺品店（右图）

喇嘛住所

小召

Xiaozhao Temple

3 小召 Xiaozhao Temple

小召牌楼斜前方

小召为呼和浩特七大召之一，康熙三十五年（1696年），清廷御赐满、蒙古、汉三种文字的“崇福寺”匾额。寺庙管辖五塔寺（慈灯寺）、岱海寺（荟安寺）、登努斯台召（善缘寺）3座属庙。康熙年间，小召受到清廷许多优遇，独许该寺喇嘛诵读蒙古文经典。寺内珍藏康熙亲征噶尔丹的汉白玉纪功碑以及现已展存内蒙古博物院的康熙甲胄、弓箭等珍贵文物。

明天启末年或崇祯年间，土默特阿拉坦汗之孙俄木布·洪台吉将其父不他失里建于大召东侧的三世佛殿扩建成一座寺庙。因修建者辈分使然，阿拉坦汗兴建的“弘慈寺”被称作大召，而其孙所建的寺庙被称作小召。也因汉文典籍中将阿拉坦汗之后的明代蒙古部族首领称作“小王子”，故此取小召之称。至清顺治初年已残破失修，第一世内齐托音在呼和浩特小召南住修期间劝土默特都统古禄格修葺小召，古禄格派遣拉布太扎兰章京承修，并利用剩余建材在呼和浩特另建一小寺，俗称拉布太召。

寺庙建筑风格为汉式建筑。据“内齐托音二世传”记载，康熙三十五年（1696年），寺庙有双层佛殿、双层大雄宝殿、释迦牟尼殿、度母殿、三怙主殿、大威德金刚殿、罗汉殿、双层神殿、释迦牟尼殿（内齐托音之弟子毕力衮达赖所建）、三长寿佛殿等殿堂，并沿中轴线对称分布。寺庙正殿两侧各有一座碑亭，立有康熙亲征噶尔丹的汉白玉纪功碑。

内齐托音七世圆寂后寺庙逐渐衰落，至光绪十九年（1893年）时已破败不堪，僧人四散。中华人民共和国成立初期小召已基本颓废，“文化大革命”期间原建筑大部分被拆除，现留有山门前的牌楼一座。现小召原址上建有一座小学。

参考文献:

［1］金峰整理注释.呼和浩特召庙（蒙古文）.呼和浩特：内蒙古人民出版社，1982,11.

［2］金启孮.呼和浩特召庙、清真寺历史概述.中国蒙古史学会年会论文,1981.

［3］金峰整理.漠南大活佛传（蒙古文）.海拉尔：内蒙古文化出版社,2009,12.

<table>
<tr><td colspan="8">小召 · 基本概况 1</td></tr>
<tr><td rowspan="3">寺院蒙古语藏语名称</td><td>蒙古语</td><td>ᠪᠠᠭ᠎ᠠ ᠵᠤᠤ</td><td rowspan="2">寺院汉语名称</td><td>汉语正式名称</td><td colspan="3">崇福寺</td></tr>
<tr><td>藏语</td><td>——</td><td>俗称</td><td colspan="3">小召</td></tr>
<tr><td>汉语语义</td><td>——</td><td colspan="2">寺院汉语名称的由来</td><td colspan="3">清廷赐名</td></tr>
<tr><td>所在地</td><td colspan="4">呼和浩特市玉泉区小召前街</td><td>东经 ——</td><td colspan="2">北纬 ——</td></tr>
<tr><td>初建年</td><td colspan="2">始建于1621年</td><td>保护等级</td><td colspan="4">——</td></tr>
<tr><td>盛期时间</td><td colspan="2">1698年</td><td colspan="2">盛期喇嘛僧/现有喇嘛僧数</td><td colspan="3">150/0人</td></tr>
<tr><td rowspan="2">历史沿革</td><td colspan="7">1696年，修缮扩建寺庙。
1696年，康熙出征噶尔丹，途径归化，下榻小召，将御用戎装武器赐予该寺，并敕准建造刻有平定噶尔丹事迹的青色琉璃瓦石碑。
1703年，立满、汉、蒙古、藏四种文字镌刻的征噶尔丹纪功碑</td></tr>
<tr><td>资料来源</td><td colspan="6">［1］金峰整理注释.呼和浩特召庙（蒙古文）.呼和浩特：内蒙古人民出版社，1982.11.
［2］金启倧.呼和浩特召庙、清真寺历史概述.中国蒙古史学会年会论文,1981.
［3］金峰整理.漠南大活佛传（蒙古文）.海拉尔：内蒙古文化出版社，2009,12.</td></tr>
<tr><td rowspan="2">现状描述</td><td colspan="4" rowspan="2">“文化大革命”期间原建筑大部分被拆除，现留有山门前的牌楼一座。近年来，人民政府拨款重新修葺了小召牌楼，加筑了围墙，保护了这座三百年前的古代建筑，现在也成为名胜古迹之一</td><td>描述时间</td><td colspan="2">2010/07/15</td></tr>
<tr><td>描述人</td><td colspan="2">萨日朗</td></tr>
<tr><td>调查日期</td><td colspan="2">2010/07/12</td><td>调查人员</td><td colspan="4">白丽燕、萨日朗</td></tr>
</table>

小召基本概况表1

<table>
<tr><td colspan="7">小召 · 基本概况 2</td></tr>
<tr><td rowspan="4">现存建筑</td><td>牌楼</td><td rowspan="4">已毁建筑</td><td>大雄宝殿</td><td>释迦牟尼殿</td><td>度母殿</td><td>三怙主殿</td></tr>
<tr><td>——</td><td>大威德金刚殿</td><td>罗汉殿</td><td>双层神殿</td><td>三长寿佛殿</td></tr>
<tr><td>——</td><td>庙仓</td><td>僧仓</td><td>——</td><td>——</td></tr>
<tr><td>——</td><td>信息来源</td><td colspan="3">1.调研访谈记录.
2.乔吉，马永真. 内蒙古古城[M]. 呼和浩特：内蒙古人民出版社,2003,3.</td></tr>
<tr><td>区位图</td><td colspan="6"> </td></tr>
<tr><td>总平面图</td><td colspan="6">A.牌 楼</td></tr>
</table>

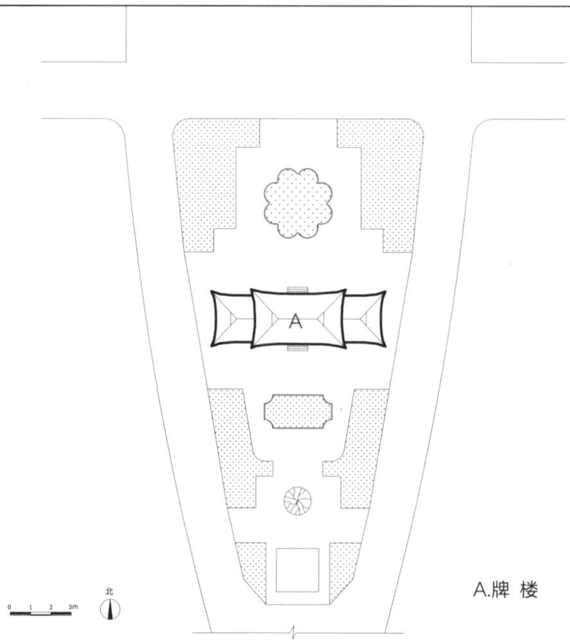

小召基本概况表2

牌楼正前方图

牌楼背立面（左图）
牌楼侧立面（右图）

牌楼柱础（左图）
牌楼斗栱（右图）

五塔寺

Tavensobrega Temple

4 五塔寺 Tavensobrega Temple

五塔寺金刚舍利宝塔斜前方

五塔寺为小召3座属庙之一。系雍正与乾隆时期在归化城新建的呼和浩特掌印扎萨克达喇嘛印务处所辖5座属庙之一。雍正十年（1732年），清廷御赐满、蒙古、汉三体“慈灯寺”匾额。该寺以一座“金刚座舍利宝塔”著称，该塔紧靠原寺院北墙，塔内珍藏三幅石刻图：蒙古文天文图、须弥山图及六道轮回图，其中蒙古文天文图最为珍贵，是迄今为止世界上唯一用蒙古文标注的一幅天文图。

雍正五年（1727年），时任呼和浩特副扎萨克达喇嘛的小召喇嘛阳察尔济忽必勒汗一世因洞礼年班居住于京城期间，呈请清廷在归化城东南康乐街南初建该寺。据传，该寺建筑动工在新城（绥远城）以前，完工在新城以后，故又称新召。

寺庙建筑风格为汉式建筑，寺庙原有殿宇分三层，每层都有三座佛殿，第一院内的正殿为三世佛殿，两座配殿分别为大慈寺及度母殿。第二院内的正殿为瓦齐尔-达拉寺，两座配殿分别为无量光佛殿及阿摩果西德丁殿。第三院内的正殿为大日如来佛殿，两座配殿分别为阿克硕毕佛殿及剌特纳-萨姆巴瓦佛殿。金刚座舍利宝塔是五塔寺留存至今的惟一的古建筑，成为第三批全国重点文物保护单位。

该寺至1893年时，已完全荒废。1977年起，开始修葺寺庙。至2007年已基本恢复寺庙原规模。

参考文献：

[1] 金峰整理注释.呼和浩特召庙.呼和浩特：内蒙古人民出版社，1982,11.
[2]（俄）阿马波慈德涅夫.蒙古及蒙古人.刘汉明等译.呼和浩特：内蒙古人民出版社，1983,5.
[3] 荣祥,荣赓麟.土默特沿革（征求意见稿）.土默特左旗文化局编辑,1981,6.
[4]（德）W · 海西希.宝鬘.（伊西班丹著“蒙古喇嘛寺及蒙古编年史”1985年成书）哥本哈根影印版,1961.

五塔寺金刚舍利宝塔正面图

五塔寺·基本概况 1

寺院蒙古语藏语名称	蒙古语	[illegible]	寺院汉语名称	汉语正式名称	慈灯寺		
	藏语	——		俗称	五塔寺		
	汉语语义	五塔寺	寺院汉语名称的由来		清廷赐名		
所在地	呼和浩特市玉泉区			东经	——	北纬	——
初建年	雍正五年（1727年）		保护等级	——			
盛期时间	——		盛期喇嘛僧/现有喇嘛僧数		——		
历史沿革	1727年，始建寺庙。 1945—1949年间，金刚座舍利宝塔曾作为炮台，拆去门窗佛像。 “文化大革命”前后，寺院曾一度改造成为五塔寺小学，佛教殿堂荡然无存。 1977年，修葺五塔寺，成立了呼和浩特文物管理所。 1982年，成立五塔文物保管所。 1988年，成为全国重点文物保护单位。 2007年，恢复原貌。主要发掘复原的是从五塔寺的山门进入后第一进院的三世佛殿和两侧的两座佛塔。在一期工程修复了大日如来殿、金刚萨埵殿等8座殿的基础上，二期工程集中在第二、三进院，扩建一座三世佛殿和两座配殿						
	资料来源	[1] 金峰整理注释.呼和浩特召庙.呼和浩特：内蒙古人民出版社，1982,11. [2]（俄）阿马波慈德涅夫.蒙古及蒙古人.刘汉明等译.呼和浩特：内蒙古人民出版社，1983,5. [3] 荣祥，荣赓麟.土默特沿革（征求意见稿）.土默特左旗文化局编辑,1981,6. [4] 调研访谈记录.					
现状描述	五塔寺除金刚宝塔之外其他建筑皆为新建，如今五塔寺已恢复原貌，现占地面积约10000平方米，殿宇规模宏大，雄伟壮观，属明清时代汉藏结合的建筑风格，殿内塑有贴金铜佛像、壁画、唐卡、堆秀和法器等珍贵的佛教艺术品，以金刚座舍利宝塔为主轴的五塔寺又完整地呈现在游人和信众面前			描述时间	2010/11/07		
				描述人	萨日朗		
调查日期	2010/11/07		调查人员	白丽燕、萨日朗			

五塔寺基本概况表1

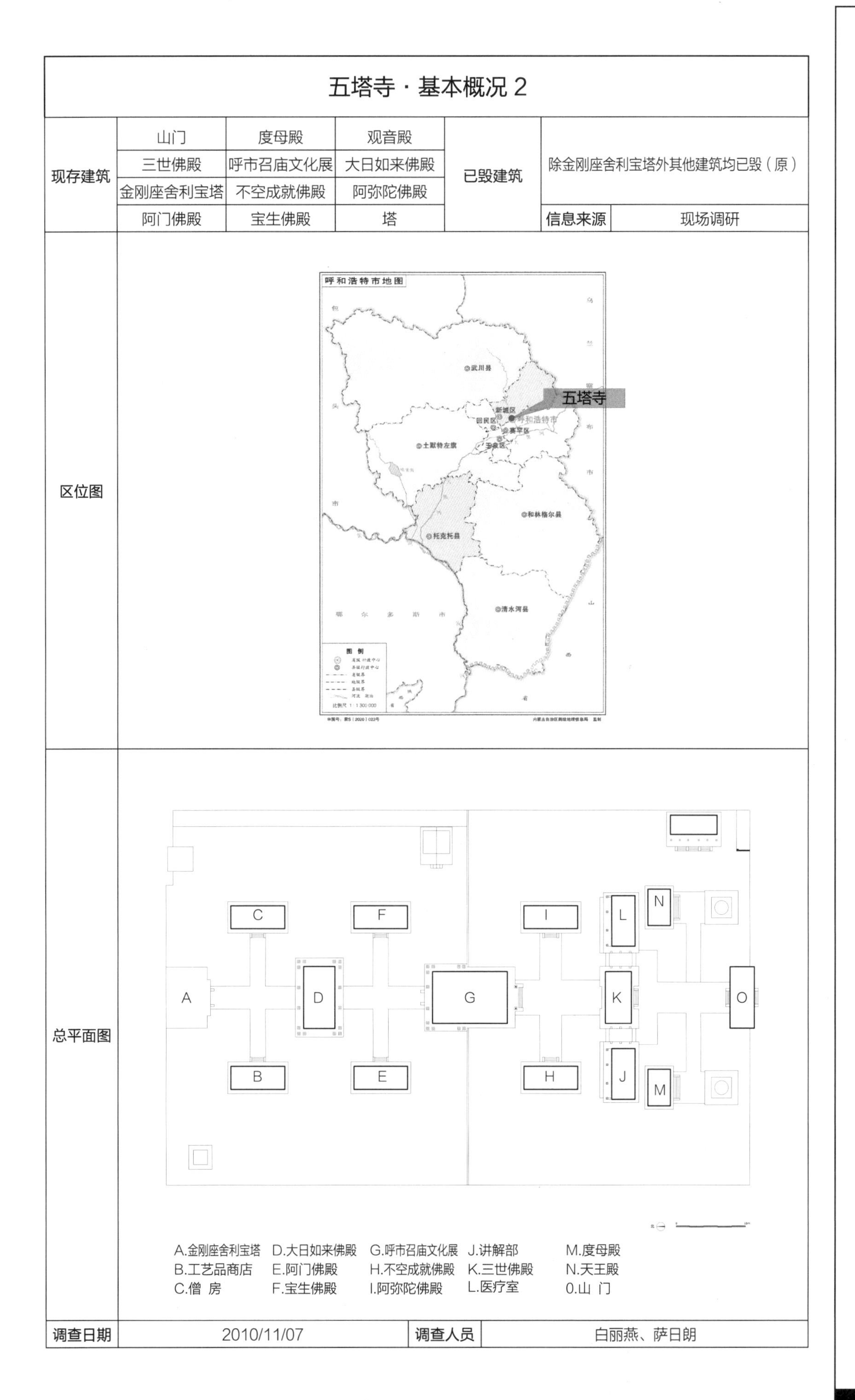

五塔寺 · 基本概况 2

现存建筑	山门	度母殿	观音殿	已毁建筑	除金刚座舍利宝塔外其他建筑均已毁（原）	
	三世佛殿	呼市召庙文化展	大日如来佛殿			
	金刚座舍利宝塔	不空成就佛殿	阿弥陀佛殿			
	阿门佛殿	宝生佛殿	塔		信息来源	现场调研

区位图

总平面图

A.金刚座舍利宝塔　D.大日如来佛殿　G.呼市召庙文化展　J.讲解部　M.度母殿
B.工艺品商店　E.阿门佛殿　H.不空成就佛殿　K.三世佛殿　N.天王殿
C.僧 房　F.宝生佛殿　I.阿弥陀佛殿　L.医疗室　0.山 门

调查日期	2010/11/07	调查人员	白丽燕、萨日朗

五塔寺基本概况表2

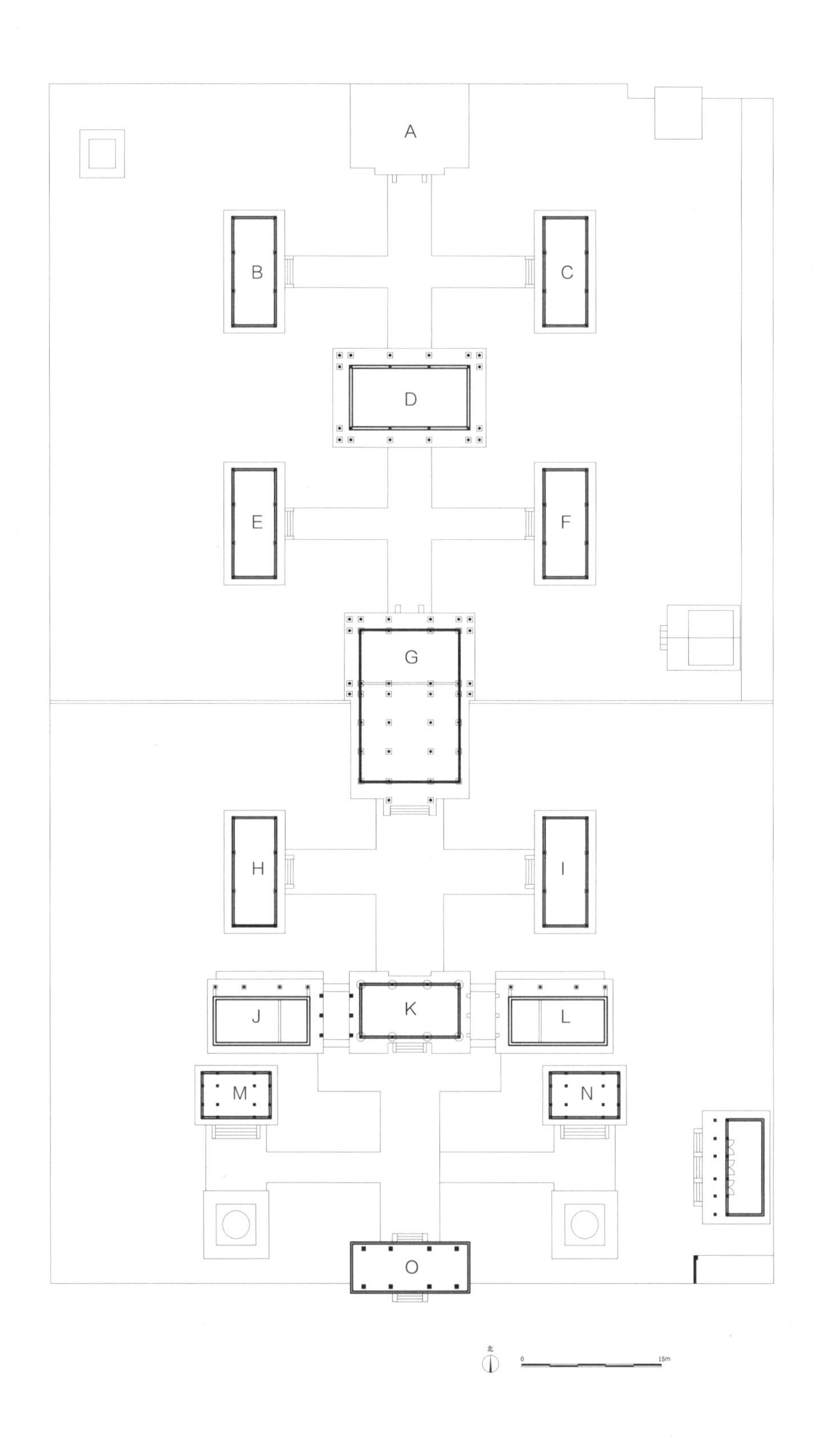

A.金刚座舍利宝塔　D.大日如来佛殿　G.呼市召庙文化展　J.讲解部　M.度母殿
B.工艺品商店　E.阿门佛殿　H.不空成就佛殿　K.三世佛殿　N.天王殿
C.僧 房　F.宝生佛殿　I.阿弥陀佛殿　L.医疗室　0.山 门

五塔寺总平面图

4.1 五塔寺·三世佛殿

单位:毫米

建筑名称	汉语正式名称	三世佛殿		俗称	——														
概述	初建年	雍正五年(1727年)		建筑朝向	南	建筑层数	一												
	建筑简要描述	汉式建筑，歇山屋顶，汉式结构体系																	
	重建重修记载	原建筑已毁，2007年重建																	
		信息来源	——																
结构规模	结构形式	砖木混合	相连的建筑	无	室内天井	无													
	建筑平面形式	长方形	外廊形式	前廊式															
	通面阔	10800	开间数	3	明间	3800	次间	3500	梢间	——	次梢间	——	尽间	——					
	通进深	5400	进深数	2	进深尺寸(前→后)	2700→2700													
	柱子数量	——	柱子间距	横向尺寸	——	(藏式建筑结构体系填写此栏，不含廊柱)													
				纵向尺寸	——														
	其他	——																	
建筑主体(大木作)(石作)(瓦作)	屋顶	屋顶形式	歇山	瓦作	布瓦														
	外墙	主体材料	青砖	材料规格	270×130×50	饰面颜色	深灰色												
		墙体收分	无	边玛檐墙	无	边玛材料	——												
	斗栱、梁架	斗栱	有	平身科斗口尺寸	不详	梁架关系	台梁式，五檩无廊												
	柱、柱式(前廊柱)	形式	汉式	柱身断面形状	圆形	断面尺寸	直径 D=270	(在没有前廊柱的情况下，填写室内柱及其特征)											
		柱身材料	木材	柱身收分	无	栌斗、托木	无	雀替	无										
		柱础	有	柱础形状	圆形	柱础尺寸	直径 D=470												
	台基	台基类型	普通台基	台基高度	760	台基地面铺设材料	青砖、石条												
	其他	——																	
装修(小木作)(彩画)	门(正面)	隔扇门	门楣	无	堆经	无	门帘	无											
	窗(正面)	槛窗	窗楣	无	窗套	有	窗帘	无											
	室内隔扇	隔扇	无	隔扇位置	——														
	室内地面、楼面	地面材料及规格	地砖800×800	楼面材料及规格	——														
	室内楼梯	楼梯	无	楼梯位置	——	楼梯材料	——	楼段宽度	——										
	天花、藻井	天花	无	天花类型	——	藻井	无	藻井类型	——										
	彩画	柱头	有	柱身	无	梁架	有	走马板	——										
		门、窗	有	天花	——	藻井	——	其他彩画	——										
	其他	悬塑	无	佛龛	无	匾额	无												
装饰	室内	帷幔	无	幕帘彩绘	无	壁画	有	唐卡	无										
		经幡	无	经幢	有	柱毯	无	其他	——										
	室外	玛尼轮	无	苏勒德	无	宝顶	有	祥麟法轮	无										
		四角经幢	无	经幡	无	铜饰	无	石刻、砖雕	无										
		仙人走兽	1+3	壁画	无	其他	——												
陈设	室内	主佛像	三世佛	佛像基座	莲花座														
		法座	无	藏经橱	无	经床	无	诵经桌	无	法鼓	无	玛尼轮	无	坛城	无	其他	——		
	室外	旗杆	无	苏勒德	无	狮子	无	经幡	无	玛尼轮	无	香炉	有	五供	无	其他	——		
	其他	——																	
备注	——																		
调查日期	2010/11/07	调查人员	白丽燕、萨日朗	整理日期	2010/11/10	整理人员	萨日朗												

三世佛殿基本概况表1

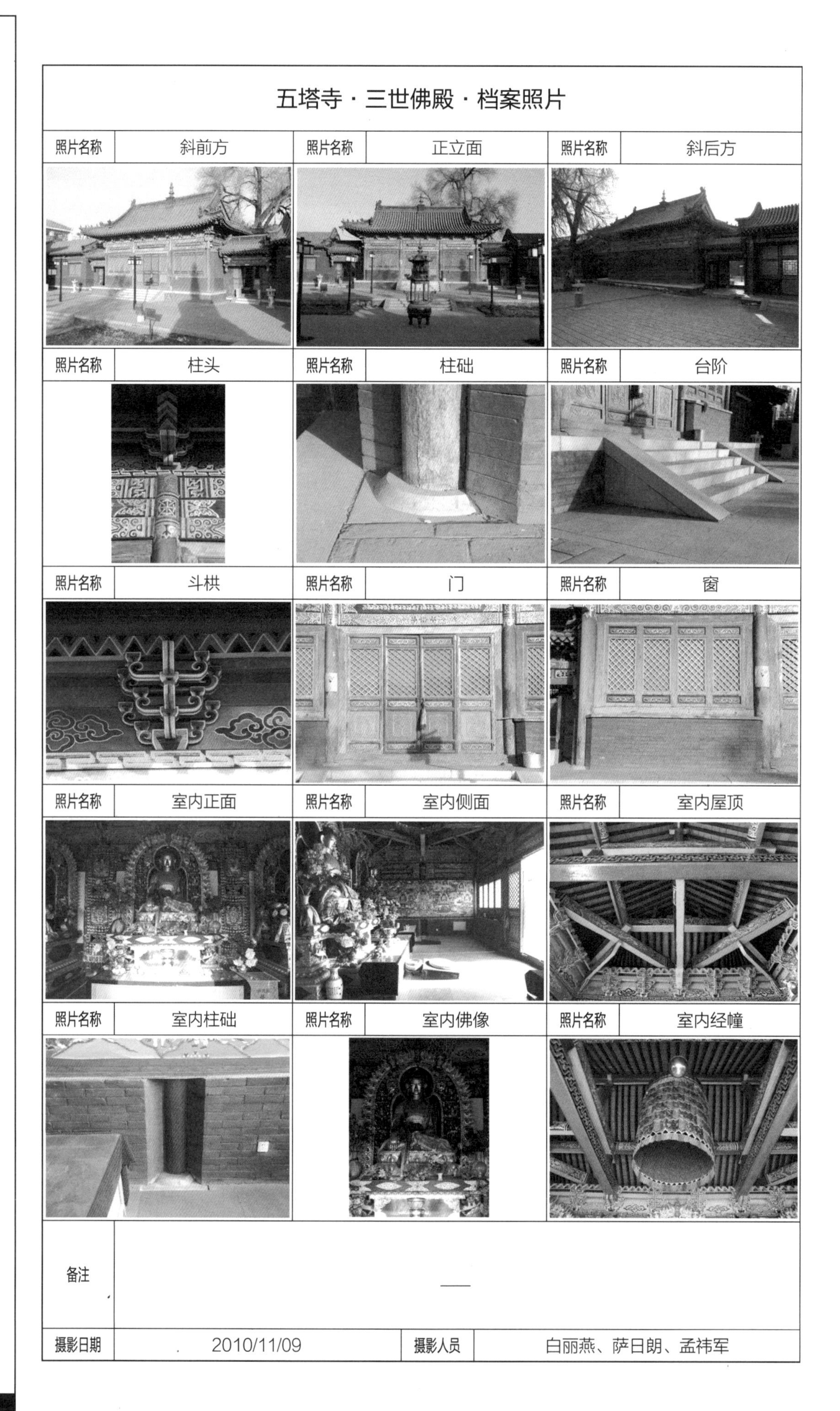

五塔寺·三世佛殿·档案照片

照片名称	斜前方	照片名称	正立面	照片名称	斜后方
照片名称	柱头	照片名称	柱础	照片名称	台阶
照片名称	斗栱	照片名称	门	照片名称	窗
照片名称	室内正面	照片名称	室内侧面	照片名称	室内屋顶
照片名称	室内柱础	照片名称	室内佛像	照片名称	室内经幢
备注	—				
摄影日期	2010/11/09	摄影人员	白丽燕、萨日朗、孟祎军		

三世佛殿基本概况表2

三世佛殿正立面

三世佛殿斜后方

三世佛殿宝顶（左图）
三世佛殿西侧门（右图）

4.2 五塔寺 · 度母殿

度母殿斜前方图
（左上图）
度母殿室外局部
（左下图）
度母殿室内正面
（右图）

4.3 五塔寺 · 观音殿

观音殿斜前方图（左图）
观音殿侧立面（右图）

观音殿正门（左图）
观音殿室内正面
（右图）

4.4 五塔寺·经堂

单位:毫米

建筑名称	汉语正式名称	经堂	俗称	——													
概述	初建年	雍正五年(1727年)始建	建筑朝向	南	建筑层数	一											
	建筑简要描述	藏汉式建筑，歇山屋顶，藏式梁架关系															
	重建重修记载	2007重建															
		信息来源	——														
结构规模	结构形式	砖木混合	相连的建筑	无	室内天井	无											
	建筑平面形式	长方形	外廊形式	半回廊													
	通面阔	10500	开间数	3	明间	4500	次间	3000	梢间	——	次梢间	——	尽间	——			
	通进深	15700	进深数	5	进深尺寸(前→后)	3100→3000→3000→1100→5500											
	柱子数量	——	柱子间距	横向尺寸	——	(藏式建筑结构体系填写此栏，不含廊柱)											
				纵向尺寸	——												
	其他	——															
建筑主体(大木作)(石作)(瓦作)	屋顶	屋顶形式	(前)歇山屋顶、(后)重檐歇山屋顶	瓦作	布瓦												
	外墙	主体材料	青砖	材料规格	270×130×50	饰面颜色	深灰色、朱红色										
		墙体收分	有	边玛檐墙	有	边玛材料	边玛草										
	斗栱、梁架	斗栱	有	平身科斗口尺寸	不详	梁架关系	梁纵排架										
	柱、柱式(前廊柱)	形式	藏式	柱身断面形状	十二楞柱	断面尺寸	300×300	(在没有前廊柱的情况下，填写室内柱及其特征)									
		柱身材料	木材	柱身收分	有	栌斗、托木	有	雀替	无								
		柱础	有	柱础形状	方+圆形	柱础尺寸	670×670、直径 D=400										
	台基	台基类型	普通台基	台基高度	750	台基地面铺设材料	青砖、石条										
	其他	——															
装修(小木作)(彩画)	门(正面)	藏式门	门楣	有	堆经	有	门帘	无									
	窗(正面)	藏式暗窗	窗楣	有	窗套	有	窗帘	无									
	室内隔扇	隔扇	有	隔扇位置	进深为第五排柱的位置												
	室内地面、楼面	地面材料及规格	地砖800×800	楼面材料及规格	无												
	室内楼梯	楼梯	无	楼梯位置	——	楼梯材料	——	楼段宽度	——								
	天花、藻井	天花	有	天花类型	井口天花	藻井	无	藻井类型	——								
	彩画	柱头	有	柱身	无	梁架	有	走马板	——								
		门、窗	有	天花	无	藻井	——	其他彩画	——								
	其他	悬塑	无	佛龛	无	匾额	无										
装饰	室内	帷幔	有	幕帘彩绘	无	壁画	无	唐卡	有								
		经幡	无	经幢	有	柱毯	无	其他	——								
	室外	玛尼轮	无	苏勒德	无	宝顶	有	祥麟法轮	有								
		四角经幢	无	经幡	无	铜饰	有	石刻、砖雕	无								
		仙人走兽	1+2	壁画	无	其他	——										
陈设	室内	主佛像	18个佛像陈设于佛龛内	佛像基座	无												
		法座	无	藏经橱	无	经床	无	诵经桌	无	法鼓	有	玛尼轮	有	坛城	有	其他	——
	室外	旗杆	无	苏勒德	无	狮子	无	经幡	无	玛尼轮	无	香炉	有	五供	无	其他	——
	其他	——															
备注	——																
调查日期	2010/11/07	调查人员	白丽燕、萨日朗	整理日期	2010/11/10	整理人员	萨日朗										

经堂基本概况表1

五塔寺·经堂·档案照片					
照片名称 	斜前方	照片名称 	正立面	照片名称 	侧立面
照片名称 	斜后方	照片名称 	背立面	照片名称 	侧立面
照片名称 	柱身	照片名称 	柱头	照片名称 	柱础
照片名称 	室内正面	照片名称 	室内侧面	照片名称 	佛龛
照片名称 	室内幢	照片名称 	廊部梁架	照片名称 	坛城
备注	—				
摄影日期	2010/11/09	摄影人员	白丽燕、萨日朗、孟祎军		

经堂基本概况表2

经堂斜前方

经堂正立面

经堂斜后方（左图）
经堂室内局部（右图）

4.5 五塔寺·大日如来佛殿

单位:毫米

建筑名称	汉语正式名称	大日如来佛殿		俗称	——												
概述	初建年	——		建筑朝向	南	建筑层数	一										
	建筑简要描述	汉式建筑，重檐歇山屋顶，汉式结构体系															
	重建重修记载	——															
		信息来源	——														
结构规模	结构形式	砖木混合	相连的建筑	无	室内天井	无											
	建筑平面形式	长方形	外廊形式	回廊													
	通面阔	12600	开间数	3	明间	4200	次间	4200	梢间	——	次梢间	——	尽间	——			
	通进深	6400	进深数	1	进深尺寸（前→后）	6400											
	柱子数量	——	柱子间距	横向尺寸	——	（藏式建筑结构体系填写此栏，不含廊柱）											
				纵向尺寸	——												
	其他	——															
建筑主体（大木作）（石作）（瓦作）	屋顶	屋顶形式	重檐歇山	瓦作	布瓦												
	外墙	主体材料	青砖	材料规格	270×130×50	饰面颜色	深灰色										
		墙体收分	无	边玛檐墙	无	边玛材料	——										
	斗栱、梁架	斗栱	有	平身科斗口尺寸	不详	梁架关系	不详（有吊顶）										
	柱、柱式（前廊柱）	形式	汉式	柱身断面形状	圆	断面尺寸	直径 D=300		（在没有前廊柱的情况下，填写室内柱及其特征）								
		柱身材料	木材	柱身收分	无	栌斗、托木	无	雀替	有								
		柱础	有	柱础形状	方形＋圆形	柱础尺寸	（下）600×600、（上）直径 D=400										
	台基	台基类型	普通台基	台基高度	450	台基地面铺设材料	青砖、石条										
	其他	——															
装修（小木作）（彩画）	门(正面)	隔扇门		门楣	无	堆经	无	门帘	无								
	窗（正面）	槛窗		窗楣	无	窗套	有	窗帘	无								
	室内隔扇	隔扇	无	隔扇位置	——												
	室内地面、楼面	地面材料及规格		地砖800×800		楼面材料及规格	——										
	室内楼梯	楼梯	无	楼梯位置	——	楼梯材料	——	楼段宽度	——								
	天花、藻井	天花	有	天花类型	井口	藻井	无	藻井类型	——								
	彩画	柱头	有	柱身	无	梁架	有	走马板	——								
		门、窗	有	天花	有	藻井	——	其他彩画	——								
	其他	悬塑	无	佛龛	无	匾额	慈灯寺										
装饰	室内	帷幔	无	幕帘彩绘	无	壁画	有	唐卡	无								
		经幡	无	经幢	无	柱毯	无	其他	——								
	室外	玛尼轮	无	苏勒德	无	宝顶	有	祥麟法轮	无								
		四角经幢	无	经幡	无	铜饰	无	石刻、砖雕	无								
		仙人走兽	1＋4	壁画	无	其他	殿前两侧有千佛灯										
陈设	室内	主佛像	如来佛			佛像基座	莲花座										
		法座	无	藏经橱	无	经床	无	诵经桌	无	法鼓	无	玛尼轮	无	坛城	无	其他	——
	室外	旗杆	无	苏勒德	无	狮子	无	经幡	无	玛尼轮	无	香炉	有	五供	无	其他	——
	其他	——															
备注	——																
调查日期	2010/11/07	调查人员	白丽燕、萨日朗	整理日期	2010/11/10	整理人员	萨日朗										

大日如来佛殿基本概况表1

五塔寺·大日如来佛殿·档案照片

照片名称	斜前方	照片名称	正立面	照片名称	侧立面
照片名称	背立面	照片名称	斜后方	照片名称	斗栱
照片名称	柱身	照片名称	柱头	照片名称	柱础
照片名称	室内佛像	照片名称	室内侧面	照片名称	室内彩画
照片名称	室内天花	照片名称	室内顶棚	照片名称	曼荼罗彩画
备注	—				
摄影日期	2010/11/09	摄影人员	白丽燕、萨日朗、孟祎军		

大日如来佛殿基本概况
表2

大日如来佛殿正立面

大日如来佛殿斜后方（左图）
大日如来佛殿侧立面（右图）

大日如来佛殿斜前方

4.6 五塔寺·金刚座舍利宝塔

金刚座舍利宝塔正面图

金刚座舍利宝塔侧立面（左图）
金刚座舍利宝塔局部（右图）

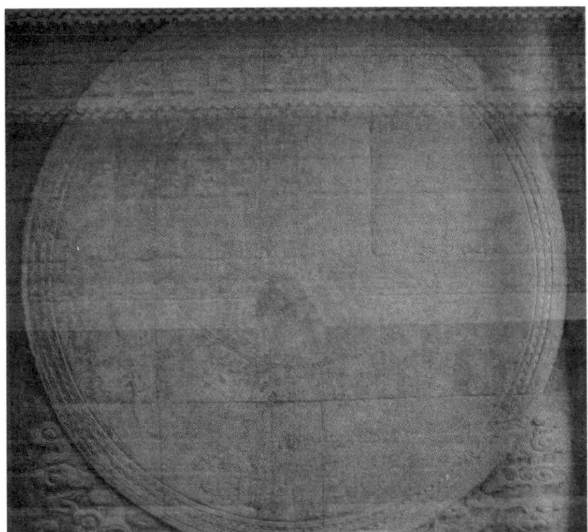

石刻蒙文天文图（左图）
金刚座舍利宝塔局部（右图）

0 1 2 3 4 5m 北

金刚座舍利宝塔
一层平面图

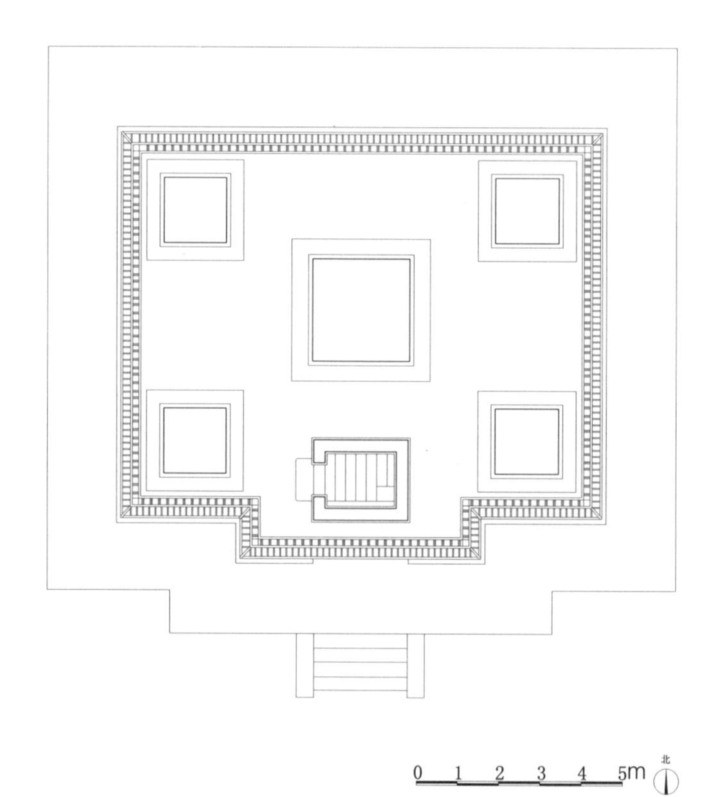

金刚座舍利宝塔
二层平面图

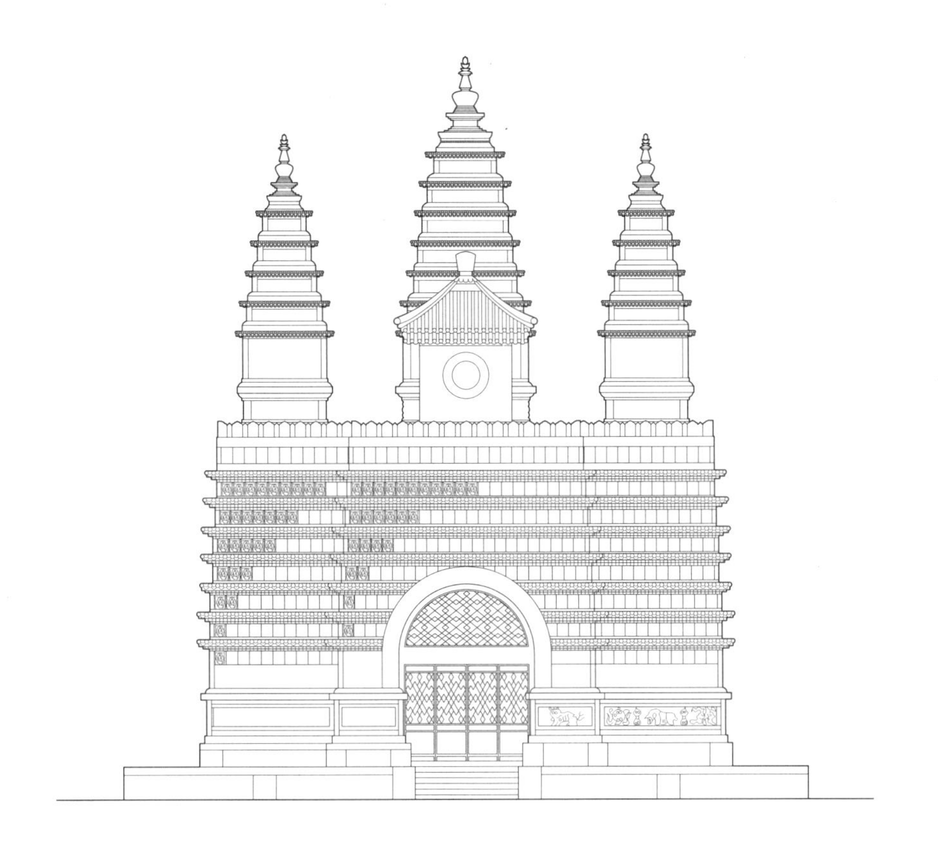

金刚座舍利宝塔南立面图

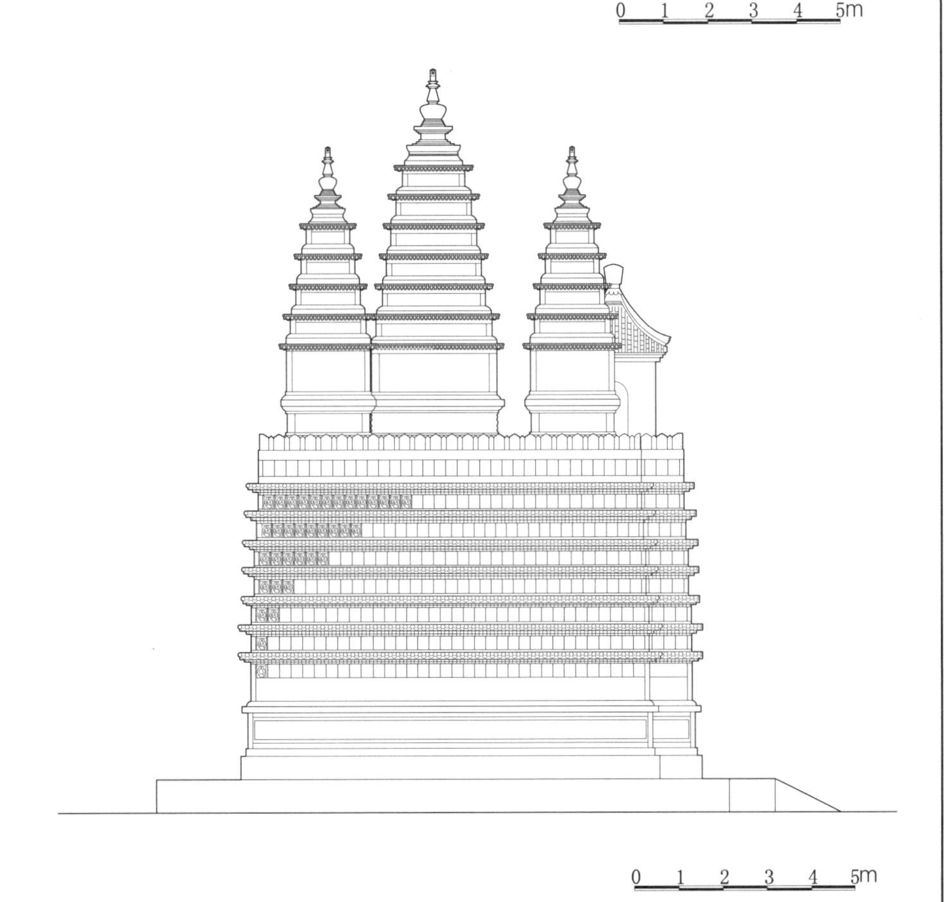

金刚座舍利宝塔西立面图

金刚座舍利宝塔
剖面图

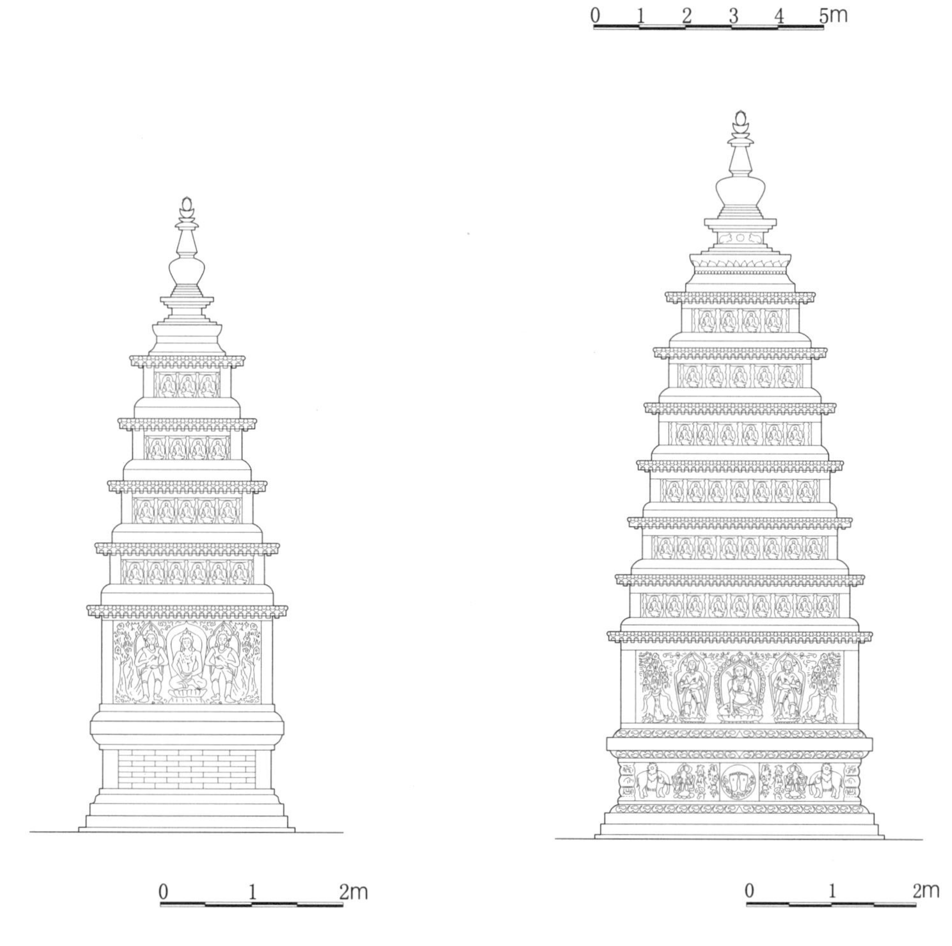

金刚座舍利宝塔
西北塔南立面（左图）
金刚座舍利宝塔
中塔南立面（右图）

4.7 五塔寺·其他建筑

不空成就佛殿斜前方图

办公室图

山门斜后方图

5

乌素图召

Osutozhao Temple

5 乌素图召 Osutozhao Temple

乌素图召鸟瞰图

乌素图召，俗称西乌素图召，为呼和浩特八小召之一。呼和浩特曾有两座乌素图召，均位于城西北大青山之阳。东乌素图召（广寿寺）为席力图召属庙。西乌素图召由毗邻相连的7座寺庙组成，即以庆缘寺为中心，东有长寿寺，西有东茶坊，东北有法禧寺，西北有药王寺，正北为罗汉寺，其北为法成广寿寺。乾隆四十八年（1783年），清廷御赐满、蒙古、汉、藏四体“庆缘寺”匾额。庆缘寺管辖里素召（增福寺）、乔尔吉拉然巴召（法禧寺）两座属庙。寺中曾珍藏松巴堪布·益西班觉所著《如意宝树史》木刻板。

明万历三十四年（1606年），由第一世察哈尔迪彦其呼图克图初建，并由希古尔、贝拉等为首的蒙古匠人建造了第一座寺庙。庆缘寺为乌素图召庙宇群的主庙，其他各寺均由该寺呼图克图及其僧俗弟子所修建。而关于寺庙具体数量及营建次序上后人观点不一。仅就始建年代与最初寺庙有两种说法：一说为明万历年间建造，最初寺庙为庆缘寺，另一说为明隆庆年间建造，最初寺庙为法成广寿寺。依后一说法，法成广寿寺与东茶坊为明隆庆年间所建寺院，俗称察哈尔庙，此后依次建造了明万历年间的庆缘寺、康熙年间的长寿寺、雍正年间的法禧寺、乾隆年间的罗汉寺及最后建的药王寺。

寺院建筑风格为汉藏结合式风格。庆缘寺由两进院落组成，前院有天王殿、大雄宝殿、东西配殿、5间过殿，后院即佛爷府，内有5间双层大厅、东西厢房各5间、东北西北角房各3间。法成广寿寺由山门、佛殿、东西厢房各3间、八角楼构成。东茶坊因内设僧众的膳房与水房，故称茶坊。长寿寺为三进院落，有天王殿、汉式佛殿、东西厢房各3间、钟楼、鼓楼、3间大厅、东西配房、东西僧舍各3间、东北角小楼等建筑以及蒙汉汉白玉石碑各1座，内记寺庙维修事迹。法禧寺有天王殿、汉藏风格的佛殿、东西配房、东北角高台小楼等建筑及茶坊院。罗汉寺有佛殿、东西厢房各3间。药王寺为一处四合小院。寺庙有2座白塔。

“文化大革命”中寺庙严重受损，但多数建筑留存至今，成为内蒙古地区保存较为完好的古寺庙群。

参考文献：

[1] 金峰整理注释.呼和浩特召庙（蒙古文）.呼和浩特：内蒙古人民出版社，1982,11.
[2] 政协呼和浩特市委员会文史资料组.呼和浩特文史资料（第一辑）（内部资料），1982,12.

乌素图召·基本概况 1

寺院蒙古语藏语名称	蒙古语	[illegible]	寺院汉语名称	汉语正式名称	庆缘寺
	藏语	——		俗称	乌素图召
	汉语语义	有水的地方	寺院汉语名称的由来		清廷赐名
所在地	呼和浩特市回民区攸攸板镇西乌素图村		东经 111° 33′	北纬 40° 49′	
初建年	隆庆元年（1567年）		保护等级	区级重点文物保护单位	
盛期时间	——		盛期喇嘛僧/现有喇嘛僧数	200人/8人	

历史沿革

1567—1572年，萨木腾阿斯尔创建法成广寿寺。

1582—1583年，萨木腾阿斯尔创建东茶坊。

1606年，新建庆缘寺，萨木腾阿斯尔创建。

1694—1739年，新建罗汉寺。

1697年，达赖嘉木苏达喇嘛改建长寿寺。该寺在雍正至嘉庆年间共修葺多次。

1724年，由罗桑旺吉勒喇嘛始建法禧寺，1785年清廷赐名“法禧寺”。是年，达喇嘛达赖嘉木苏重修长寿寺。

1777年，第五次修葺长寿寺。

1782年，全面维修和扩建寺庙。

1799年，第六次修葺长寿寺。

1802年，第七次修葺长寿寺。

20世纪30年代，拆除东茶坊。

1956年，修葺乌素图召。

1959年，修葺乌素图召。

“文化大革命”时期，寺庙严重受损，长寿寺等庙宇被拆除。

2006年，粉刷维修乌素图召

资料来源

［1］金峰整理注释.呼和浩特召庙（蒙古文）.呼和浩特：内蒙古人民出版社，1982,11.

［2］政协呼和浩特市委员会文史资料组.呼和浩特文史资料（第一辑）（内部资料）,1982,12.

现状描述	乌素图召依山就势，现由庆缘寺、长寿寺、法禧寺、罗汉寺、药王寺五座寺庙组成，保存比较完好。主要遗存建筑有4座主殿、1座活佛府及各寺的山门、罗汉殿、厢房、僧舍等。法成广寿寺、东茶坊被毁	描述时间	2010/05
		描述人	萨日朗
调查日期	2010/04/16	调查人员	白丽燕、萨日朗

乌素图召基本概况表1

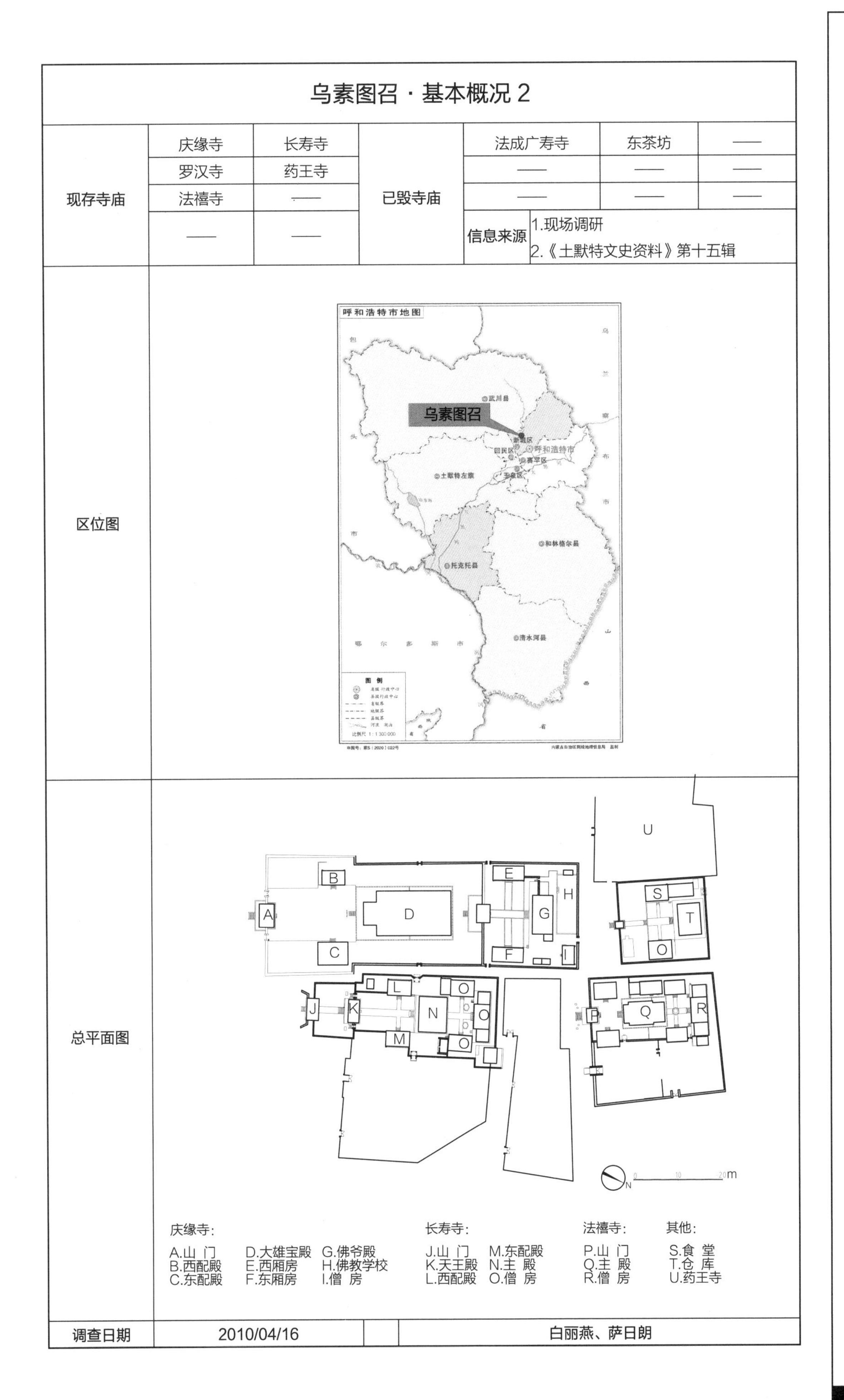

乌素图召·基本概况 2

现存寺庙	庆缘寺	长寿寺	已毁寺庙	法成广寿寺	东茶坊	——
	罗汉寺	药王寺		——	——	——
	法禧寺	——		——	——	——
	——	——	信息来源	1.现场调研 2.《土默特文史资料》第十五辑		

区位图	

总平面图	

庆缘寺：
A.山 门　D.大雄宝殿　G.佛爷殿
B.西配殿　E.西厢房　H.佛教学校
C.东配殿　F.东厢房　I.僧 房

长寿寺：
J.山 门　M.东配殿
K.天王殿　N.主 殿
L.西配殿　O.僧 房

法禧寺：
P.山 门
Q.主 殿
R.僧 房

其他：
S.食 堂
T.仓 库
U.药王寺

调查日期	2010/04/16		白丽燕、萨日朗

U
T
S
R
Q
P
H
I
G
E
F
O
N
M
L
K
J
D
B
C
A
0 10 20 m
庆缘寺:
A.山 门
B.西配殿
C.东配殿
D.大雄宝殿
E.西厢房
F.东厢房
G.佛爷殿
H.佛教学校
I.僧 房
长寿寺:
J.山 门
K.天王殿
L.西配殿
M.东配殿
N.主 殿
O.僧 房
法禧寺:
P.山 门
Q.主 殿
R.僧 房
其他:
S.食 堂
T.仓 库
U.药王寺

乌素图召总平面图

5.1 乌素图召·庆缘寺大雄宝殿

单位:毫米

建筑名称	汉语正式名称	庆缘寺大雄宝殿	俗称	经堂													
概述	初建年	万历三十四年（1606年）	建筑朝向	南	建筑层数	二											
	建筑简要描述	汉藏结合式，汉式结构体系，藏式风格装饰，前廊为汉式柱廊															
	重建重修记载	乾隆四十七年（公元1782年）第五世察哈尔迪彦齐对乌素图召进行了大的维修和扩建，现在的庆缘寺基本上保持了当时的规模															
		信息来源	《绥远厅简志》第九卷记载														
结构规模	结构形式	砖木混合	相连的建筑	无	室内天井	——											
	建筑平面形式	凸字形	外廊形式	前廊、半回廊													
	通面阔	18850	开间数	7	明间	3280	次间	2910	梢间	1985	次梢间	2890	尽间	——			
	通进深	28660	进深数	11	进深尺寸（前→后）	2980→3570→2590→2590→2600→3110→2850→2580→2630→2580→2900											
	柱子数量	——	柱子间距	横向尺寸	——												
				纵向尺寸	——												
	其他	——															
建筑主体（大木作）（石作）（瓦作）	屋顶	屋顶形式	重檐歇山	瓦作	布瓦												
	外墙	主体材料	青砖	材料规格	270×130×50	饰面颜色	白										
		墙体收分	有	边玛檐墙	有	边玛材料	砖										
	斗栱、梁架	斗栱	有	平身科斗口尺寸	不详	梁架关系	不详（有吊顶）										
	柱、柱式（前廊柱）	形式	汉式	柱身断面形状	圆	断面尺寸	直径D=390	（在没有前廊柱的情况下，填写室内柱及其特征）									
		柱身材料	木	柱身收分	无	栌斗、托木	无	雀替	有								
		柱础	有	柱础形状	方	柱础尺寸	620×620										
	台基	台基类型	普通台基	台基高度	1850	台基地面铺设材料	青方砖										
	其他	——															
装修（小木作）（彩画）	门(正面)	藏式门	门楣	有	堆经	有	门帘	无									
	窗（正面）	一层藏式盲窗、二层隔扇窗	窗楣	无	窗套	一层无，二层有	窗帘	无									
	室内隔扇	隔扇	有	隔扇位置	经堂与佛殿以隔扇相隔												
	室内地面、楼面	地面材料及规格	青方砖270×270	楼面材料及规格	木板，规格不均												
	室内楼梯	楼梯	有	楼梯位置	正门左侧	楼梯材料	木材	梯段宽度	800								
	天花、藻井	天花	有	天花类型	井口	藻井	有	藻井类型	八边形								
	彩画	柱头	有	柱身	无	梁架	有	走马板	有								
		门、窗	无	天花	有	藻井	有	其他彩画	——								
	其他	悬塑	无	佛龛	无	匾额	“庆缘寺”乾隆四十八年										
装饰	室内	帷幔	有	幕帘彩绘	无	壁画	有	唐卡	有								
		经幡	有	经幢	有	柱毯	无	其他	——								
	室外	玛尼轮	有	苏勒德	无	宝顶	有	祥麟法轮	无								
		四角经幢	无	经幡	有	铜饰	无	石刻、砖雕	无								
		仙人走兽	1+4	壁画	无	其他	——										
陈设	室内	主佛像	五方佛、宗喀巴及两个弟子	佛像基座	莲花座												
		法座	有	藏经橱	无	经床	有	诵经桌	有	法鼓	有	玛尼轮	有	坛城	无	其他	——
	室外	旗杆	无	苏勒德	无	狮子	无	经幡	有	玛尼轮	有	香炉	有	五供	无	其他	——
	其他	——															
备注	——																
调查日期	2010/04/16	调查人员	白丽燕、萨日朗	整理日期	2010/04/22	整理人员	萨日朗										

庆缘寺大雄宝殿基本概况
表1

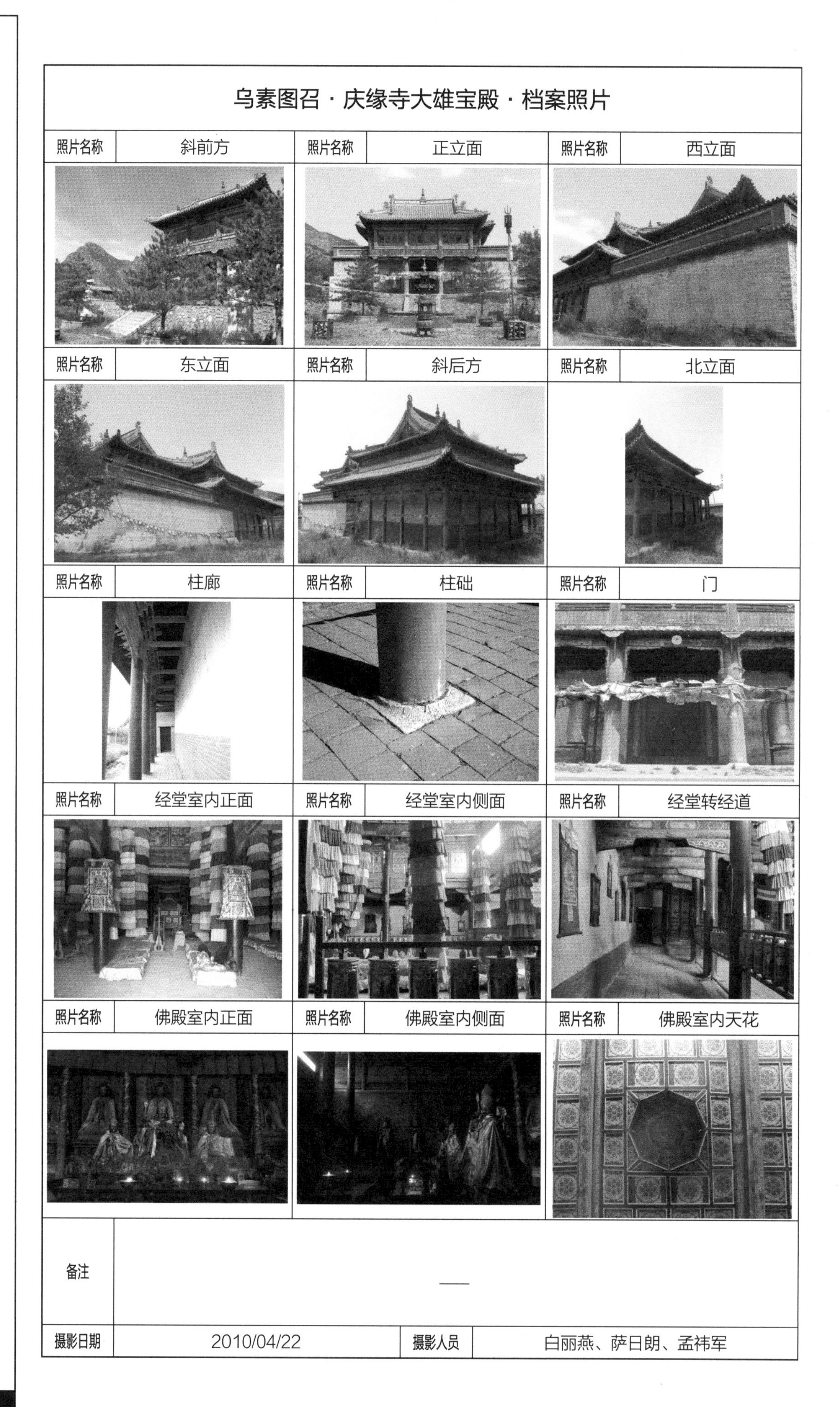

乌素图召·庆缘寺大雄宝殿·档案照片

照片名称	斜前方	照片名称	正立面	照片名称	西立面
照片名称	东立面	照片名称	斜后方	照片名称	北立面
照片名称	柱廊	照片名称	柱础	照片名称	门
照片名称	经堂室内正面	照片名称	经堂室内侧面	照片名称	经堂转经道
照片名称	佛殿室内正面	照片名称	佛殿室内侧面	照片名称	佛殿室内天花
备注	——				
摄影日期	2010/04/22	摄影人员	白丽燕、萨日朗、孟祎军		

庆缘寺大雄宝殿基本概况
表2

庆缘寺大雄宝殿正立面

庆缘寺大雄宝殿斜前方

庆缘寺大雄宝殿东立面（左图）

庆缘寺大雄宝殿斜后方（右图）

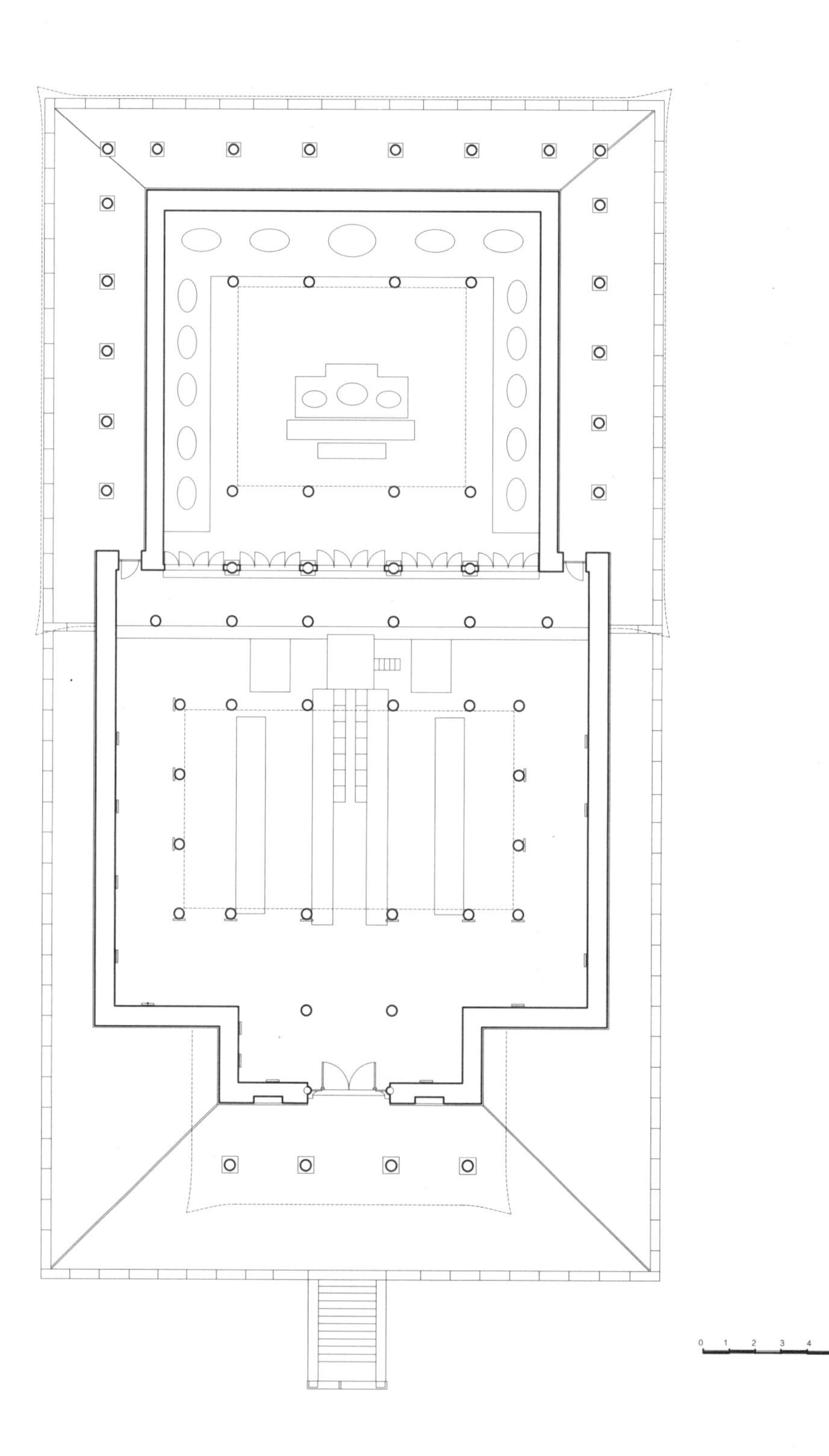

庆缘寺大雄宝殿
一层平面图

5.2 乌素图召·庆缘寺东、西厢房

庆缘寺大雄宝殿东厢房正立面

东厢房侧立面（左图）
东厢房前廊柱身（右图）

西厢房侧立面（左图）
西厢房正立面（右图）

5.3 乌素图召·庆缘寺活佛府

庆缘寺活佛府佛爷殿正立面

活佛府佛爷殿背立面（左图）

活佛府佛爷殿室外柱（右图）

佛爷殿东厢房正立面

5.4 乌素图召·庆缘寺山门

庆缘寺山门正立面

庆缘寺山门背立面

5.5 乌素图召·长寿寺大雄宝殿

单位:毫米

建筑名称	汉语正式名称	长寿寺		俗称	大雄宝殿		
概述	初建年	康熙三十六年（1697年）		建筑朝向	南	建筑层数	二
	建筑简要描述	汉式建筑，重檐歇山屋顶，汉式结构体系					
	重建重修记载	清朝时共修葺六次，最后一次完成于嘉庆七年(1802年)，原大殿在“文化大革命”时遭到破坏，残存无几，现存大殿为20世纪80年代按大殿原样在原址上修复					
		信息来源	据寺碑记载				

结构规模												
结构形式	砖木混合		相连的建筑	无				室内天井		无		
建筑平面形式	方形		外廊形式	回廊								
通面阔	12480	开间数	5	明间	3220	次间	2870	梢间	1710	次梢间	——	尽间 ——
通进深	9460	进深数	4	进深尺寸（前→后）				1750→3000→2990→1720				
柱子数量	——	柱子间距	横向尺寸	——				（藏式建筑结构体系填写此栏，不含廊柱）				
			纵向尺寸	——								
其他	——											

建筑主体（大木作）（石作）（瓦作）									
屋顶	屋顶形式	重檐歇山			瓦作	布瓦			
外墙	主体材料	青砖	材料规格	270×130×50	饰面颜色	红			
	墙体收分	无	边玛檐墙	无	边玛材料	——			
斗栱、梁架	斗栱	有	平身科斗口尺寸	不详	梁架关系	不详（有吊顶）			
柱、柱式（前廊柱）	形式	汉式	柱身断面形状	圆	断面尺寸	直径 $D=270$			（在没有前廊柱的情况下，填写室内柱及其特征）
	柱身材料	木材	柱身收分	无	栌斗、托木	无	雀替	有	
	柱础	有	柱础形状	圆形	柱础尺寸	直径 $D=550$			
台基	台基类型	普通台基	台基高度	1450	台基地面铺设材料		青方砖		
其他	——								

装修（小木作）（彩画）								
门(正面)	隔扇门		门楣	无	堆经	无	门帘	无
窗（正面）	隔扇窗		窗楣	无	窗套	无	窗帘	无
室内隔扇	隔扇	无	隔扇位置	——				
室内地面、楼面	地面材料及规格		瓷砖 60×60		楼面材料及规格		木板 规格不均	
室内楼梯	楼梯	有	楼梯位置	进正门左侧	楼梯材料	木材	梯段宽度	780
天花、藻井	天花	有	天花类型	井口天花	藻井	无	藻井类型	——
彩画	柱头	有	柱身	无	梁架	有	走马板	有
	门、窗	无	天花	有	藻井	——	其他彩画	——
其他	悬塑	无	佛龛	无	匾额	无		

装饰								
室内	帷幔	无	幕帘彩绘	无	壁画	有	唐卡	有
	经幡	无	经幢	有	柱毯	无	其他	——
室外	玛尼轮	无	苏勒德	无	宝顶	有	祥麟法轮	无
	四角经幢	无	经幡	有	铜饰	无	石刻、砖雕	无
	仙人走兽	1+4	壁画	有	其他	——		

陈设																
室内	主佛像	释迦牟尼三世佛、度母像							佛像基座		莲花座					
	法座	有	藏经橱	无	经床	有	诵经桌	有	法鼓	有	玛尼轮	无	坛城	无	其他	——
室外	旗杆	无	苏勒德	无	狮子	无	经幡	有	玛尼轮	无	香炉	无	五供	无	其他	——
其他	——															

备注	——						
调查日期	2010/04/16	调查人员	白丽燕、萨日朗	整理日期	2010/04/22	整理人员	萨日朗

长寿寺大雄宝殿基本概况表1

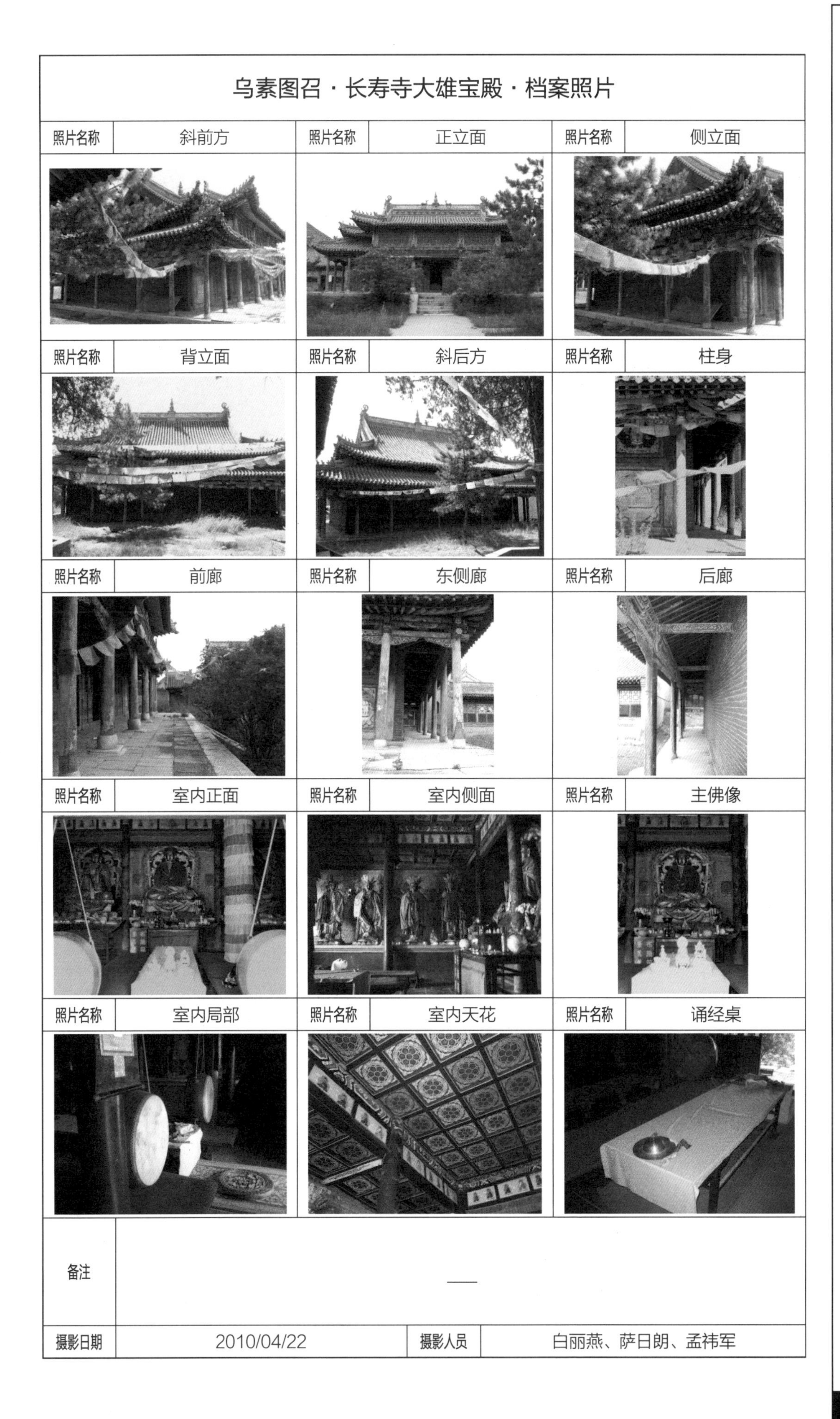

乌素图召·长寿寺大雄宝殿·档案照片

照片名称	斜前方	照片名称	正立面	照片名称	侧立面
照片名称	背立面	照片名称	斜后方	照片名称	柱身
照片名称	前廊	照片名称	东侧廊	照片名称	后廊
照片名称	室内正面	照片名称	室内侧面	照片名称	主佛像
照片名称	室内局部	照片名称	室内天花	照片名称	诵经桌
备注	—				
摄影日期	2010/04/22	摄影人员	白丽燕、萨日朗、孟祎军		

长寿寺大雄宝殿基本概况表2

长寿寺大雄宝殿正立面

长寿寺大雄宝殿柱廊
（左图）

长寿寺大雄宝殿斜前方
（右图）

长寿寺大雄宝殿斜后方
（左图）

长寿寺大雄宝殿前廊
（右图）

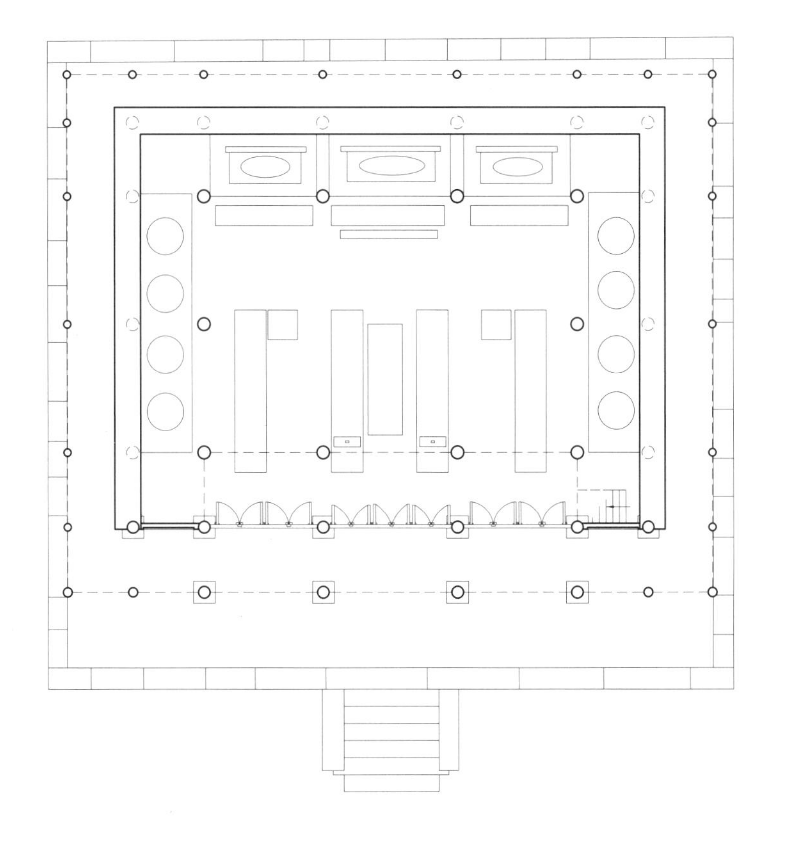

长寿寺大雄宝殿
一层平面图

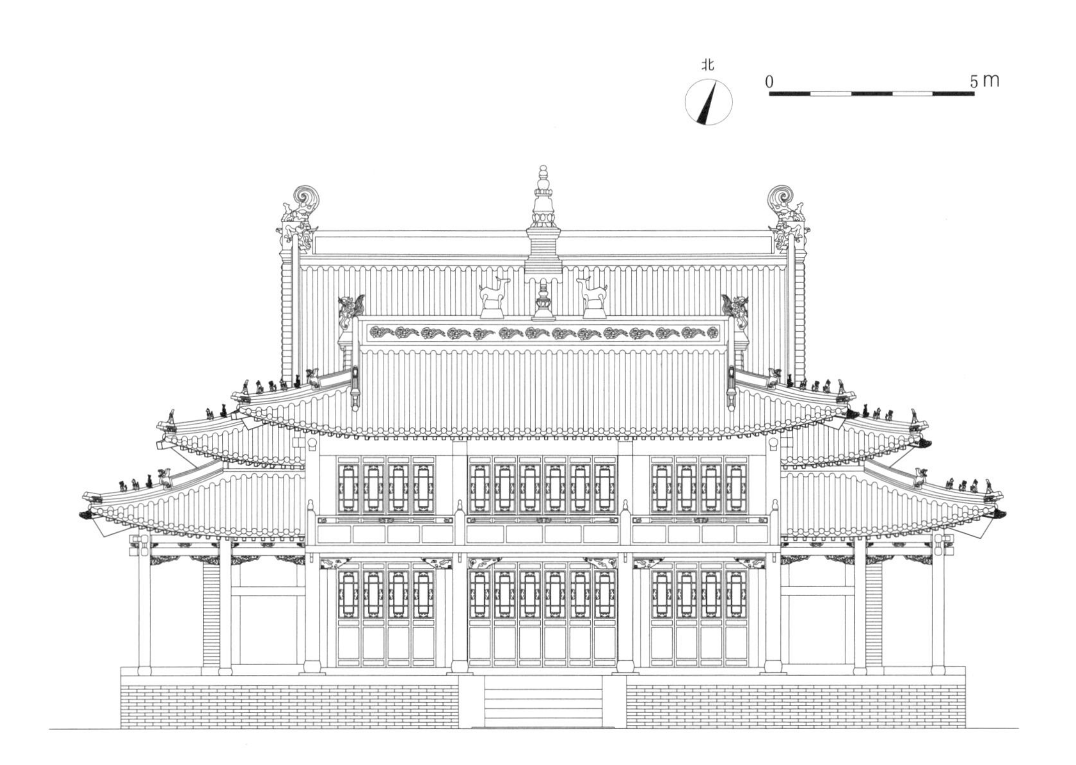

长寿寺大雄宝殿
正立面图

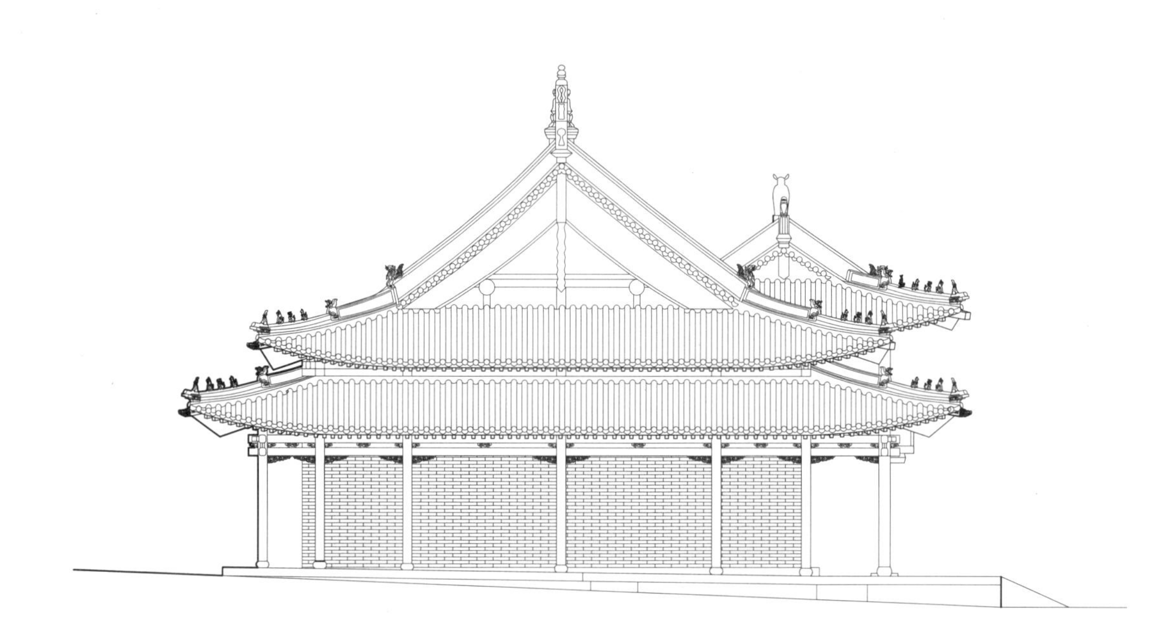

长寿寺大雄宝殿侧立面图

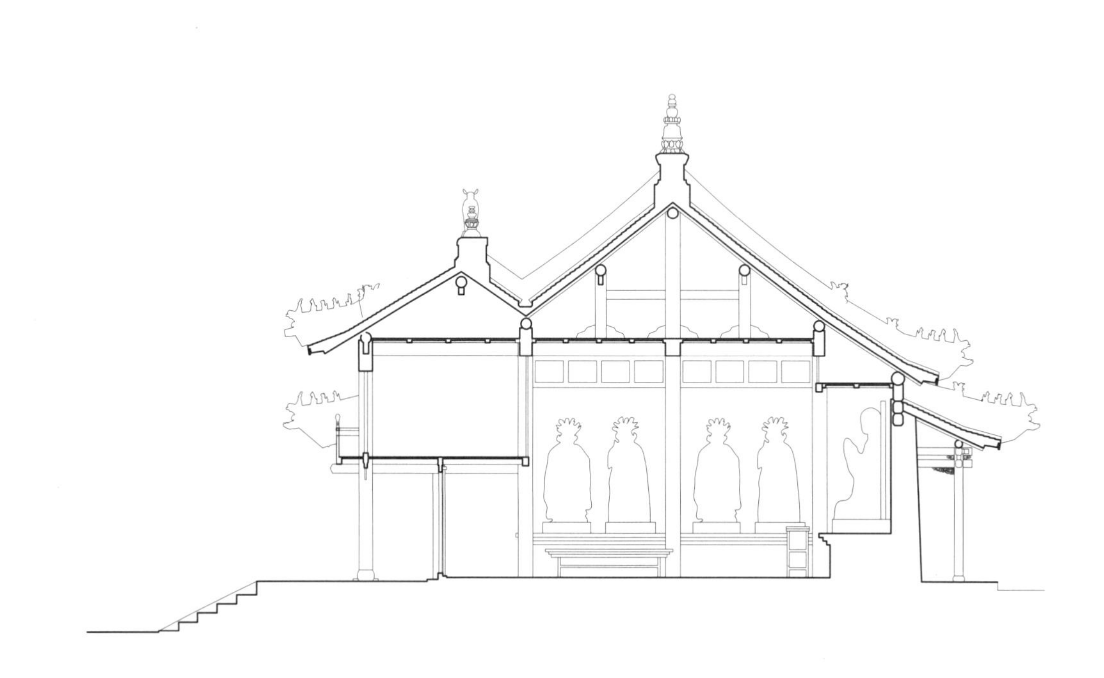

长寿寺大雄宝殿剖面图

5.6 乌素图召 · 长寿寺其他建筑

长寿寺天王殿背立面

长寿寺西配殿正立面（左图）
长寿寺东配殿斜前方（右图）

长寿寺后院局部图

5.7 乌素图召 · 法禧寺大雄宝殿

单位:毫米

建筑名称	汉语正式名称	法禧寺				俗称	大雄宝殿					
概述	初建年	雍正三年（1725年）				建筑朝向	南		建筑层数	二		
	建筑简要描述	藏式建筑，藏式结构体系，前廊为藏式柱，经堂内部为都刚法式空间格局										
	重建重修记载	三庆纳尔布喇嘛执事后重修了本寺院的围墙，粉刷全寺殿顶。1956年，进行了一次土木维修，2006年，法禧寺再次进行了粉刷维修										
		信息来源	据寺内喇嘛口述									
结构规模	结构形式	砖木混合		相连的建筑	无				室内天井	——		
	建筑平面形式	凸字形		外廊形式	前廊							
	通面阔	9730	开间数	5	明间	2610	次间	1580	梢间	1980	次梢间	——	尽间	——
	通进深	14890	进深数	6	进深尺寸（前→后）				2680→2680→2550→1600→2550→2830			
	柱子数量	14（4圆10方）	柱子间距	横向尺寸	1580→2610→1580				（藏式建筑结构体系填写此栏，不含廊柱）			
				纵向尺寸	2680→2550							
	其他	——										
建筑主体（大木作）（石作）（瓦作）	屋顶	屋顶形式	密肋平顶				瓦作	绿琉璃瓦				
	外墙	主体材料	青砖	材料规格	270×130×50		饰面颜色	白色				
		墙体收分	有	边玛檐墙	有		边玛材料	边玛草				
	斗栱、梁架	斗栱	无	平身科斗口尺寸		——	梁架关系	不详（有吊顶）				
	柱、柱式（前廊柱）	形式	藏式	柱身断面形状	十二楞柱	断面尺寸	300×300		（在没有前廊柱的情况下，填写室内柱及其特征）			
		柱身材料	木材	柱身收分	有	栌斗、托木	有	雀替	有			
		柱础	有	柱础形状	方形	柱础尺寸	470×470					
	台基	台基类型	普通台基	台基高度	1480	台基地面铺设材料		方青砖				
	其他	——										
装修（小木作）（彩画）	门(正面)	藏式门		门楣	有	堆经	有	门帘	无			
	窗（正面）	藏式盲窗		窗楣	有	窗套	有	窗帘	无			
	室内隔扇	隔扇	有	隔扇位置	经堂与佛殿以隔扇相隔							
	室内地面、楼面	地面材料及规格		水泥地面		楼面材料及规格		木板 规格不均				
	室内楼梯	楼梯	有	楼梯位置	进正门左侧	楼梯材料	木材	梯段宽度	800			
	天花、藻井	天花	有	天花类型	井口天花	藻井	有	藻井类型	四方变八方			
	彩画	柱头	有	柱身	有	梁架	有	走马板	有			
		门、窗	有	天花	有	藻井	有	其他彩画	——			
	其他	悬塑	无	佛龛	无	匾额	“法禧寺”，1994年重立					
装饰	室内	帷幔	有	幕帘彩绘	无	壁画	无	唐卡	有			
		经幡	有	经幢	有	柱毯	无	其他	——			
	室外	玛尼轮	无	苏勒德	无	宝顶	无	祥麟法轮	无			
		四角经幢	无	经幡	无	铜饰	无	石刻、砖雕	无			
		仙人走兽	无	壁画	无	其他	——					
陈设	室内	主佛像	释迦牟尼三世佛、宗喀巴、度母像			佛像基座	须弥座					
		法座	有	藏经橱	无	经床	有	诵经桌	有	法鼓	有	玛尼轮	无	坛城	无	其他	——
	室外	旗杆	无	苏勒德	无	狮子	无	经幡	无	玛尼轮	无	香炉	有	五供	无	其他	——
	其他	——										
备注	——											
调查日期	2010/04/16	调查人员	白丽燕、萨日朗	整理日期	2010/04/22	整理人员	萨日朗					

法禧寺大雄宝殿基本概况
表1

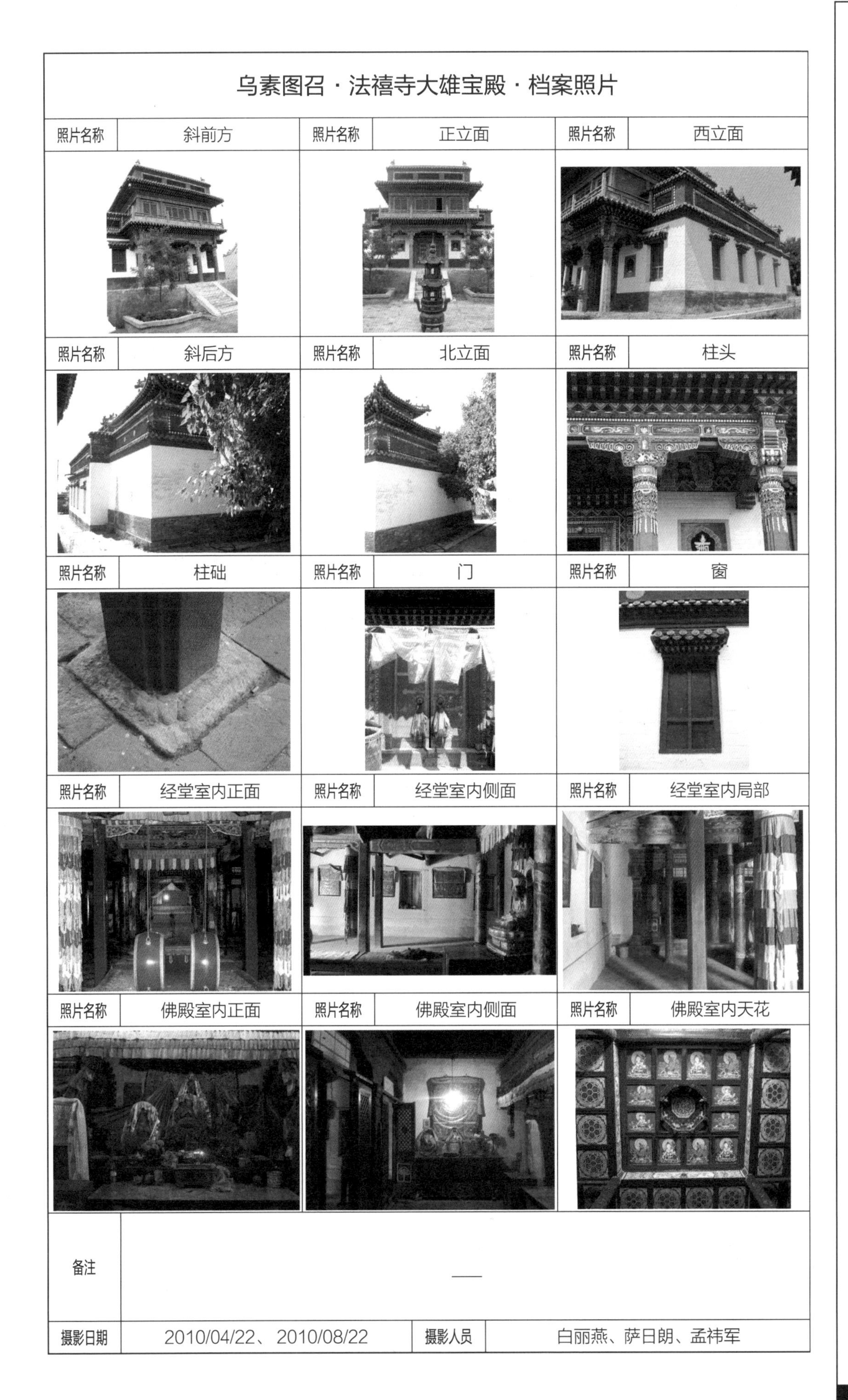

乌素图召·法禧寺大雄宝殿·档案照片

照片名称	斜前方	照片名称	正立面	照片名称	西立面
照片名称	斜后方	照片名称	北立面	照片名称	柱头
照片名称	柱础	照片名称	门	照片名称	窗
照片名称	经堂室内正面	照片名称	经堂室内侧面	照片名称	经堂室内局部
照片名称	佛殿室内正面	照片名称	佛殿室内侧面	照片名称	佛殿室内天花
备注	—				
摄影日期	2010/04/22、2010/08/22	摄影人员	白丽燕、萨日朗、孟祎军		

法喜寺大雄宝殿基本概况表2

法禧寺大雄宝殿斜前方图

法禧寺大雄宝殿室外柱（左图）
法禧寺大雄宝殿室外窗（右图）

法禧寺大雄宝殿室内正面

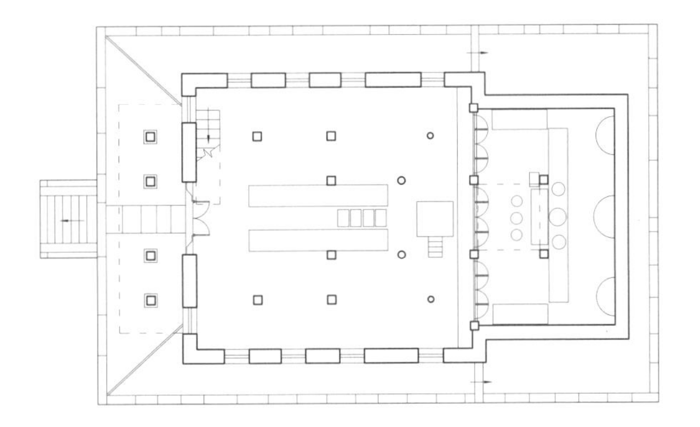

法禧寺大雄宝殿平面图

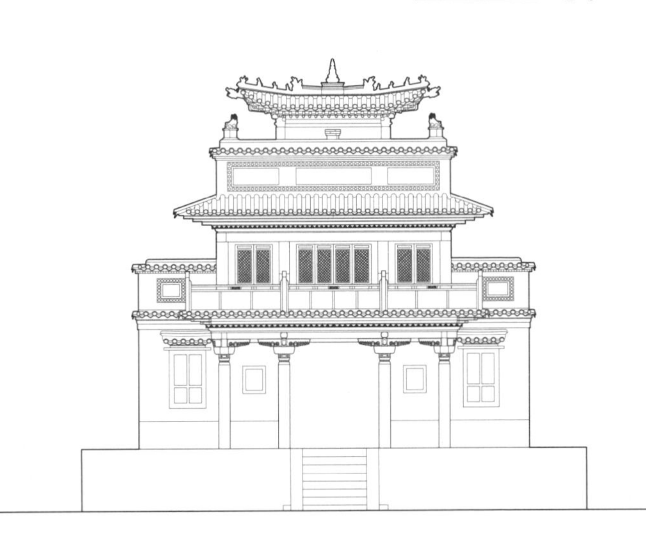

法禧寺大雄宝殿正立面图

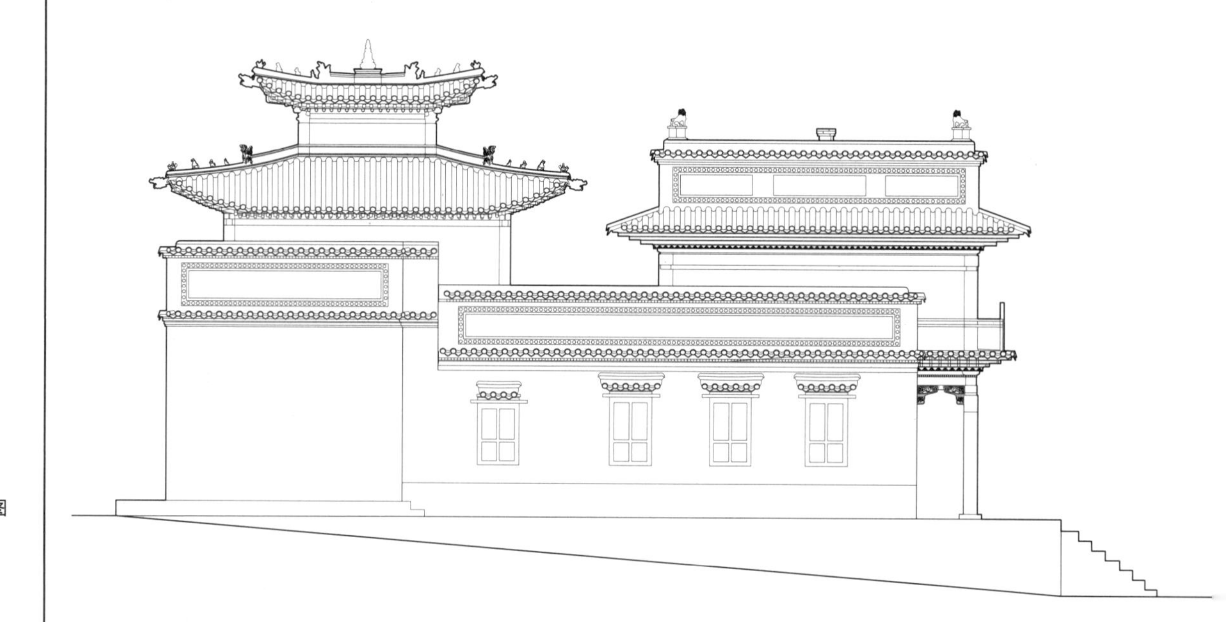

法禧寺大雄宝殿西立面图

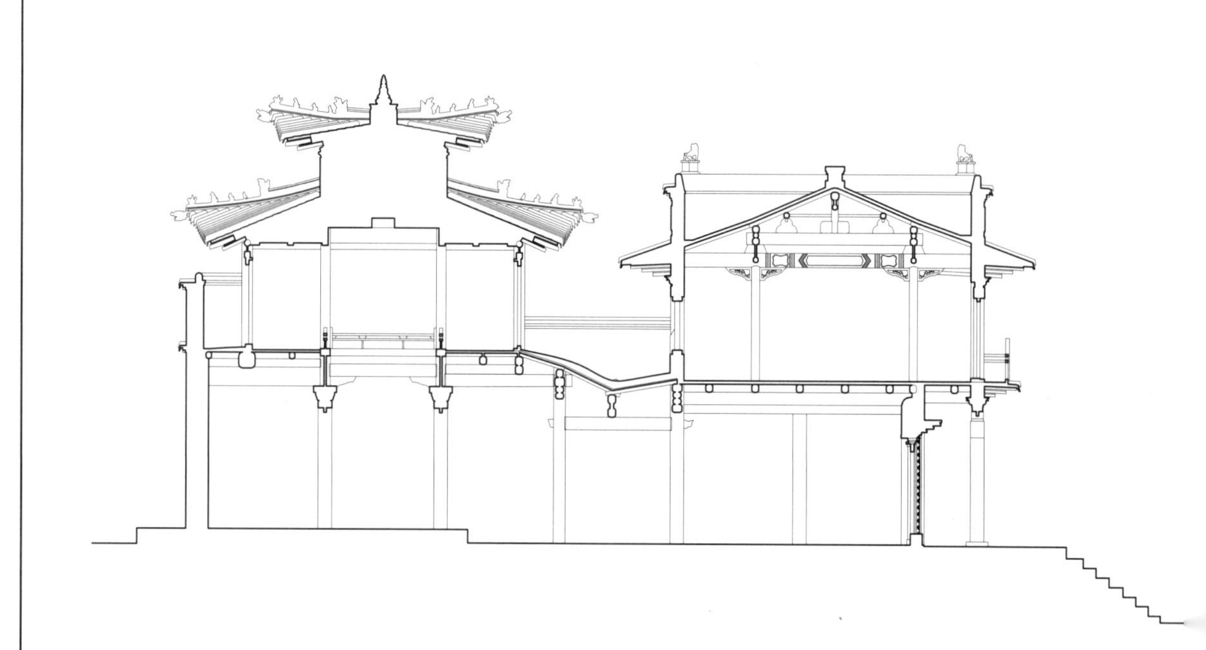

法禧寺大雄宝殿剖面图

5.8 乌素图召·罗汉寺大雄宝殿

单位:毫米

建筑名称	汉语正式名称	罗汉寺	俗称	——													
概述	初建年	康熙三十三年（1694年）	建筑朝向	南	建筑层数	一											
	建筑简要描述	汉式建筑															
	重建重修记载	此寺一直没有改建及扩建，寺院基本保持原样，有过多次修补															
		信息来源	据寺内喇嘛口述														
结构规模	结构形式	砖木混合	相连的建筑	无	室内天井	无											
	建筑平面形式	长方形	外廊形式	前廊													
	通面阔	12600	开间数	5	明间	3000	次间	3000	梢间	1800	次梢间	——	尽间	——			
	通进深	7800	进深数	3	进深尺寸（前→后）	3000→3000→1800											
	柱子数量	——	柱子间距	横向尺寸	——	（藏式建筑结构体系填写此栏，不含廊柱）											
				纵向尺寸	——												
	其他	——															
建筑主体（大木作）（石作）（瓦作）	屋顶	屋顶形式	单檐九脊歇山	瓦作	布瓦												
	外墙	主体材料	青砖	材料规格	270×130×50	饰面颜色	红										
		墙体收分	无	边玛檐墙	无	边玛材料	——										
	斗栱、梁架	斗栱	无	平身科斗口尺寸	——	梁架关系	不详（有吊顶）										
	柱、柱式（前廊柱）	形式	汉式	柱身断面形状	圆	断面尺寸	直径 $D=220$	（在没有前廊柱的情况下，填写室内柱及其特征）									
		柱身材料	木材	柱身收分	无	栌斗、托木	无	雀替	有								
		柱础	有	柱础形状	方形	柱础尺寸	400×400										
	台基	台基类型	普通台基	台基高度	1350	台基地面铺设材料	青方砖										
	其他	——															
装修（小木作）（彩画）	门(正面)	隔扇门	门楣	无	堆经	无	门帘	无									
	窗（正面）	隔扇窗	窗楣	无	窗套	无	窗帘	无									
	室内隔扇	隔扇	无	隔扇位置	——												
	室内地面、楼面	地面材料及规格	青方砖400×400	楼面材料及规格	——												
	室内楼梯	楼梯	无	楼梯位置	——	楼梯材料	——	楼段宽度	——								
	天花、藻井	天花	有	天花类型	井口	藻井	无	藻井类型	——								
	彩画	柱头	有	柱身	无	梁架	有	走马板	有								
		门、窗	无	天花	无	藻井	——	其他彩画	——								
	其他	悬塑	无	佛龛	无	匾额	无										
装饰	室内	帷幔	无	幕帘彩绘	无	壁画	无	唐卡	无								
		经幡	无	经幢	无	柱毯	无	其他	——								
	室外	玛尼轮	无	苏勒德	无	宝顶	有	祥麟法轮	无								
		四角经幢	无	经幡	无	铜饰	无	石刻、砖雕	无								
		仙人走兽	无	壁画	无	其他	——										
陈设	室内	主佛像	无	佛像基座	无												
		法座	无	藏经橱	无	经床	无	诵经桌	无	法鼓	无	玛尼轮	无	坛城	无	其他	——
	室外	旗杆	无	苏勒德	无	狮子	无	经幡	无	玛尼轮	无	香炉	无	五供	无	其他	——
	其他	——															
备注	室内现已用为库房																
调查日期	2010/04/16	调查人员	白丽燕、萨日朗	整理日期	2010/04/22	整理人员	萨日朗										

罗汉寺大雄宝殿基本概况表1

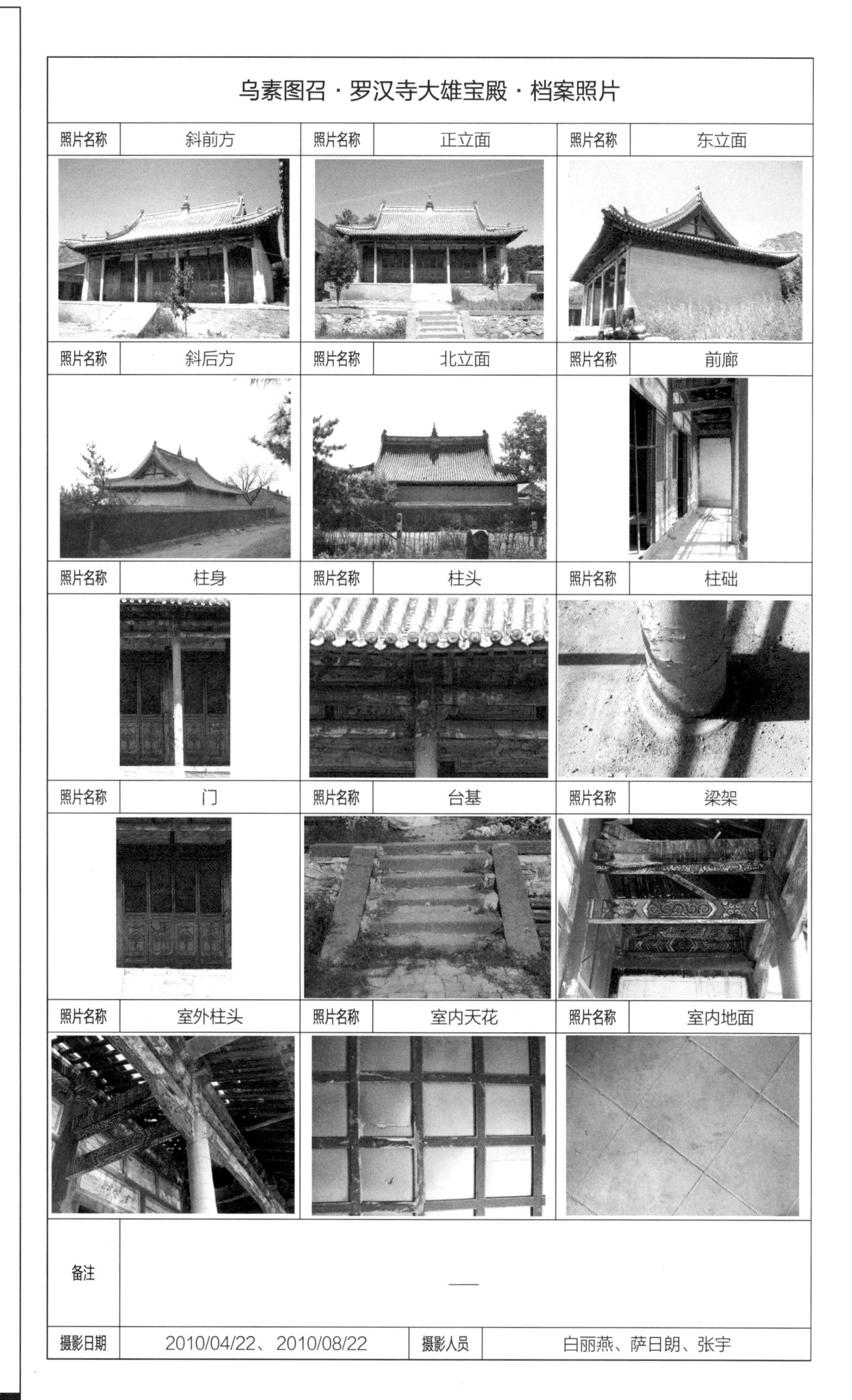

乌素图召·罗汉寺大雄宝殿·档案照片

照片名称	斜前方	照片名称	正立面	照片名称	东立面
照片名称	斜后方	照片名称	北立面	照片名称	前廊
照片名称	柱身	照片名称	柱头	照片名称	柱础
照片名称	门	照片名称	台基	照片名称	梁架
照片名称	室外柱头	照片名称	室内天花	照片名称	室内地面
备注	—				
摄影日期	2010/04/22、2010/08/22	摄影人员	白丽燕、萨日朗、张宇		

罗汉寺大雄宝殿基本概况
表2

罗汉寺大雄宝殿正立面

罗汉寺大雄宝殿斜后方

5.9 乌素图召·其他建筑

白塔石碑

6 喇嘛洞召

Lama-in-agui Temple

6 喇嘛洞召 Lama-in-agui Temple

喇嘛召洞鸟瞰图

呼和浩特城北大青山有两座喇嘛洞，一座位于城西八十里，乾隆四十九年（1784年），清廷御赐满、蒙古、汉、藏四体“广化寺”匾额，俗称西喇嘛洞；一座位于城东北六十里处额奇特沟内，清廷赐名“崇禧寺”，俗称东喇嘛洞。两座喇嘛洞同属呼和浩特八小召之列。现存西喇嘛洞一座，简称喇嘛洞。寺庙管辖沙尔沁召、珠尔沟召、明安召、黑格林召、祝乐庆召等属庙。

明万历年间，博格多察罕喇嘛云游呼和浩特西北山洞，修行坐禅。该僧圆寂后，其弟子道宝迪彦其·赤列嘉木苏继法座，于崇祯年间（1628-1644年），在其师修业道场——乌素图河源得力格尔阿贵处始建该寺。赤列扎木苏之弟子吹斯嘎巴继法座时增建一座大佛殿。四世呼图克图将旧寺庙从乌素图河源迁移至河源下方，并加以扩建。

寺庙建筑风格为汉藏结合式建筑。喇嘛洞又称银洞，寺庙位于银洞南坡，分前后两院。前院有天王殿、大经堂、大雄宝殿、欢喜佛殿等殿宇，后部建有佛爷府，西北建有三座白色覆钵式喇嘛塔。后寺建在山腰的洞前，凿山为洞，与洞连成一体。

寺庙在“文化大革命”中严重受损，至1982年仅存拉布隆院。1985年起重修寺庙。

参考文献:

[1] 金峰整理注释.呼和浩特召庙（蒙古文）.呼和浩特：内蒙古人民出版社,1982,11.

[2] 乔吉.内蒙古寺庙.呼和浩特:内蒙古人民出版社,1994,8.

[3] 唐吉思.蒙古族佛教文化调查研究.沈阳:辽宁民族出版社,2010,12.

喇嘛召洞前视图

<table>
<tr><th colspan="8">喇嘛洞召 · 基本概况 1</th></tr>
<tr><td rowspan="3">寺院蒙古语藏语名称</td><td>蒙古语</td><td>[illegible]</td><td rowspan="2">寺院汉语名称</td><td>汉语正式名称</td><td colspan="3">广化寺</td></tr>
<tr><td>藏语</td><td>ཀུན་འདུལ་གླིང་།</td><td>俗称</td><td colspan="3">喇嘛洞召</td></tr>
<tr><td>汉语语义</td><td>喇嘛的山洞</td><td colspan="2">寺院汉语名称的由来</td><td colspan="3">清廷赐名</td></tr>
<tr><td>所在地</td><td colspan="3">呼和浩特土默特左旗</td><td>东经</td><td>111° 17′</td><td>北纬</td><td>40° 47′</td></tr>
<tr><td>初建年</td><td colspan="2">万历四十三（1615年）</td><td>保护等级</td><td colspan="4">自治区级重点保护单位</td></tr>
<tr><td>盛期时间</td><td colspan="2">——</td><td colspan="2">盛期喇嘛僧/现有喇嘛僧数</td><td colspan="3">400/9位</td></tr>
<tr><td rowspan="2">历史沿革</td><td colspan="7">1627—1655年间，新建5间寺庙。
1658年，另建一座9间大佛殿，班禅赐名为“广安寺”。
1719年，将寺庙由乌素图河源迁移至河源下方，加以扩建。
1783年，扩建寺庙，增修佛殿、佛像。
1966年，拆毁拉布隆，各殿内佛像被砸毁。
1967年、1968年、1970年，先后拆毁东、西八角楼与主殿，木材运往他处。
1985年，重建寺庙</td></tr>
<tr><td>资料来源</td><td colspan="6">[1] 金峰整理注释.呼和浩特召庙（蒙古文）.呼和浩特:内蒙古人民出版社,1982,11.
[2] 乔吉.内蒙古寺庙.呼和浩特:内蒙古人民出版社,1994,8.
[3] 唐吉思.蒙古族佛教文化调查研究.沈阳:辽宁民族出版社,2010,12.
[4] 调研访谈记录、土默特志载.</td></tr>
<tr><td rowspan="2">现状描述</td><td colspan="4" rowspan="2">喇嘛洞召1985年重建，现有建筑均为新建建筑。由银洞、佛爷府、经堂等组成。银洞两侧的悬崖峭壁上，布满了大大小小的岩刻造像。银洞建在山腰上，凿山为洞，建楼3层，与洞连为一体。寺下石阶140级，直达佛爷府。寺院掩映于古松丛林中，环境幽静，景色优美，现仍不失为大青山胜景之首</td><td>描述时间</td><td colspan="2">2010/11/22</td></tr>
<tr><td>描述人</td><td colspan="2">萨日朗</td></tr>
<tr><td>调查日期</td><td colspan="2">2010/11/15</td><td>调查人员</td><td colspan="4">白丽燕、萨日朗</td></tr>
</table>

喇嘛洞召基本概况表1

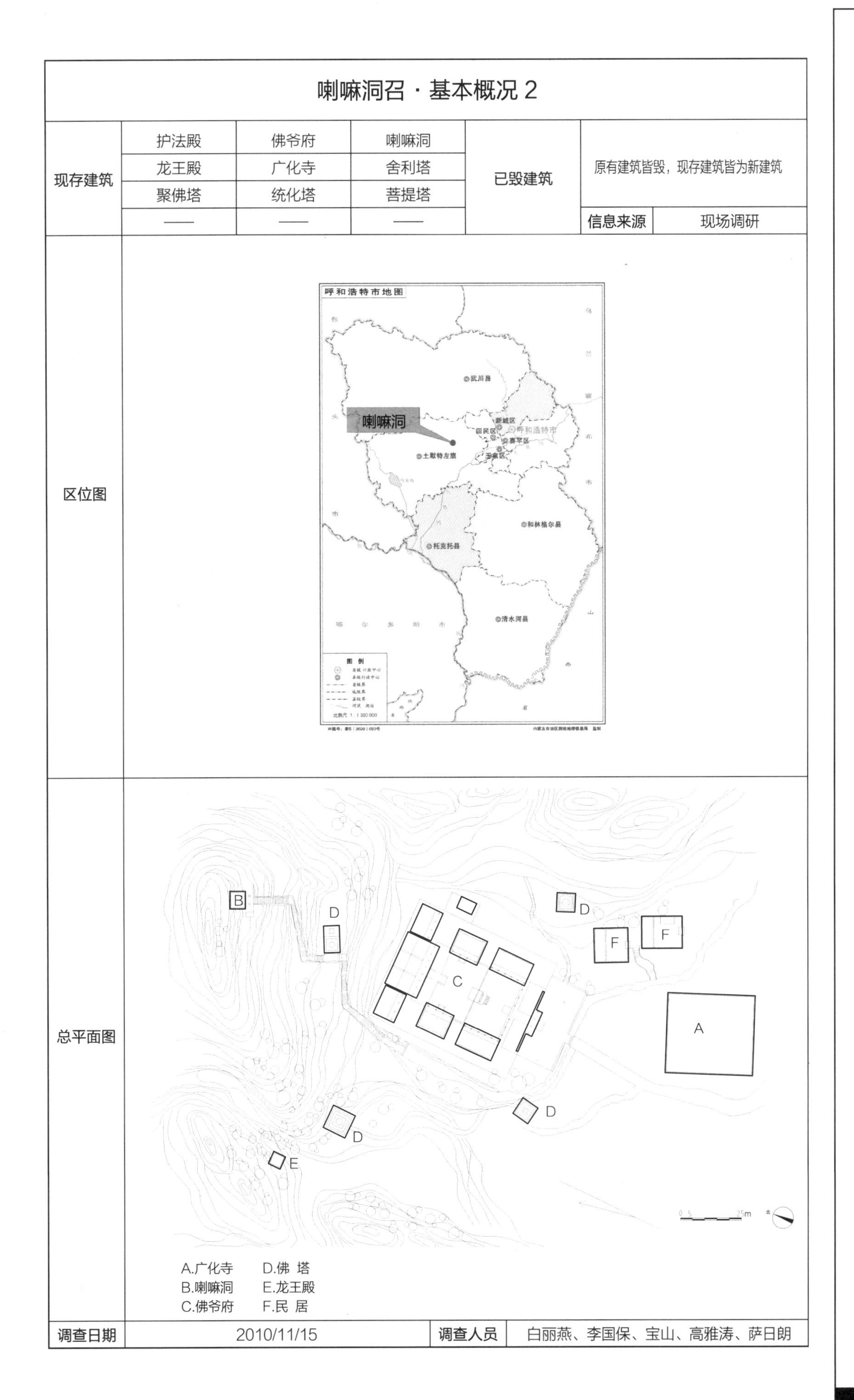

喇嘛洞召·基本概况 2

现存建筑	护法殿	佛爷府	喇嘛洞	已毁建筑	原有建筑皆毁，现存建筑皆为新建筑	
	龙王殿	广化寺	舍利塔			
	聚佛塔	统化塔	菩提塔			
	——	——	——		信息来源	现场调研
区位图						
总平面图	A.广化寺 D.佛 塔 B.喇嘛洞 E.龙王殿 C.佛爷府 F.民 居					
调查日期	2010/11/15		调查人员	白丽燕、李国保、宝山、高雅涛、萨日朗		

A.广化寺　D.佛 塔
B.喇嘛洞　E.龙王殿
C.佛爷府　F.民 居

喇嘛洞召总平面图

6.1 喇嘛洞召 · 喇嘛洞

单位:毫米

建筑名称	汉语正式名称	喇嘛洞					俗称	银洞									
概述	初建年	万历初年（1573年）					建筑朝向	南			建筑层数	三					
	建筑简要描述	藏式															
	重建重修记载	1985年重建															
		信息来源	据召内喇嘛口述														
结构规模	结构形式	砖木混合	相连的建筑	无	室内天井	无											
	建筑平面形式	凸字形	外廊形式	廊柱													
	通面阔	6600	开间数	3	明间	2400	次间	2150	次间	——	梢间	——	尽间	——			
	通进深	4380	进深数	2	进深尺寸（前→后）	1900→2480											
	柱子数量	2	柱子间距	横向尺寸	2400	（藏式建筑结构体系填写此栏，不含廊柱）											
				纵向尺寸	——												
	其他	——															
建筑主体（大木作）（石作）（瓦作）	屋顶	屋顶形式	藏式平顶	瓦作	琉璃瓦												
	外墙	主体材料	青砖	材料规格	270×130×50	饰面颜色	白色										
		墙体收分	有	边玛檐墙	有	边玛材料	砖										
	斗栱、梁架	斗栱	无	平身科斗口尺寸	——	梁架关系	——										
	柱、柱式（前廊柱）	形式	藏式柱	柱身断面形状	小八角柱	断面尺寸	直径D=165	（在没有前廊柱的情况下，填写室内柱及其特征）									
		柱身材料	木材	柱身收分	无	栌斗、托木	有	雀替	无								
		柱础	有	柱础形状	圆	柱础尺寸	直径D=270										
	台基	台基类型	普通台基	台基高度	750	台基地面铺设材料	石条、砖										
	其他	——															
装修（小木作）（彩画）	门(正面)	板门	门楣	有	堆经	有	门帘	有									
	窗（正面）	藏式明窗	窗楣	有	窗套	有	窗帘	无									
	室内隔扇	隔扇	无	隔扇位置	——												
	室内地面、楼面	地面材料及规格	方砖	楼面材料及规格	不规格木条、六边形地砖												
	室内楼梯	楼梯	有	楼梯位置	入口左侧	楼梯材料	木	梯段宽度	700								
	天花、藻井	天花	有	天花类型	井口天花	藻井	无	藻井类型	——								
	彩画	柱头	有	柱身	无	梁架	有	走马板	无								
		门、窗	无	天花	有	藻井	——	其他彩画	——								
	其他	悬塑	无	佛龛	有	匾额	喇嘛洞										
装饰	室内	帷幔	有	幕帘彩绘	无	壁画	无	唐卡	有								
		经幡	无	经幢	无	柱毯	无	其他	——								
	室外	玛尼轮	无	苏勒德	无	宝顶	有	祥麟法轮	有								
		四角经幢	无	经幡	有	铜饰	无	石刻、砖雕	有								
		仙人走兽	无	壁画	无	其他	——										
陈设	室内	主佛像	释迦牟尼佛像	佛像基座	莲花座												
		法座	无	藏经橱	无	经床	无	诵经桌	无	法鼓	无	玛尼轮	无	坛城	无	其他	——
	室外	旗杆	无	苏勒德	无	狮子	无	经幡	有	玛尼轮	无	香炉	有	五供	无	其他	——
	其他	——															
备注	——																
调查日期	2010/11/15	调查人员	白丽燕、萨日朗	整理日期	2010/11/22	整理人员	萨日朗										

喇嘛洞基本概况表1

喇嘛洞召 · 喇嘛洞 · 基本概况

单位:毫米

建筑名称	汉语正式名称	喇嘛洞	俗称	银洞			
概述	初建年	明万历四十三年（1615年）	建筑朝向	坐北朝南	建筑层数	三（第二层）	
	建筑简要描述	藏式					
	重建重修记载	1985年重建					
		信息来源	据召内喇嘛口述				

结构规模															
	结构形式	砖木混合	相连的建筑	无	室内天井	无									
	建筑平面形式	凸字形	外廊形式	廊柱											
	通面阔	6600	开间数	3	明间	2400	次间	2150	梢间	——	次梢间	——	尽间	——	
	通进深	6180	进深数	3	进深尺寸（前→后）	1900→2480→1800									
	柱子数量	4	柱子间距	横向尺寸	2150—2400—2250	（藏式建筑结构体系填写此栏，不含廊柱）									
				纵向尺寸	1900—2480—1800										
	其他	——													

建筑主体（大木作）（石作）（瓦作）											
	屋顶	屋顶形式	藏式平顶	瓦作	琉璃瓦						
	外墙	主体材料	青砖	材料规格	270×130×50	饰面颜色	白色				
		墙体收分	有	边玛檐墙	有	边玛材料	砖				
	斗栱、梁架	斗栱	无	平身科斗口尺寸	——	梁架关系	——				
	柱、柱式（前廊柱）	形式	藏式柱	柱身断面形状	小八角柱	断面尺寸	直径 $D=165$	（在没有前廊柱的情况下，填写室内柱及其特征）			
		柱身材料	木材	柱身收分	无	栌斗、托木	有	雀替	无		
		柱础	无	柱础形状	——	柱础尺寸	——				
	台基	台基类型	普通台基	台基高度	750	台基地面铺设材料	石条、砖				
	其他	——									

装修（小木作）（彩画）									
	门(正面)	隔扇门	门楣	有	堆经	有	门帘	有	
	窗（正面）	槛窗	窗楣	有	窗套	有	窗帘	无	
	室内隔扇	隔扇	无	隔扇位置	——				
	室内地面、楼面	地面材料及规格	——	楼面材料及规格	不规格木条、六边形地砖				
	室内楼梯	楼梯	有	楼梯位置	左侧	楼梯材料	木	梯段宽度	700
	天花、藻井	天花	有	天花类型	井口天花	藻井	无	藻井类型	——
	彩画	柱头	有	柱身	无	梁架	有	走马板	无
		门、窗	无	天花	有	藻井	——	其他彩画	——
	其他	悬塑	无	佛龛	无	匾额	喇嘛洞		

装饰									
	室内	帷幔	有	幕帘彩绘	无	壁画	有	唐卡	无
		经幡	无	经幢	有	柱毯	无	其他	——
	室外	玛尼轮	无	苏勒德	无	宝顶	有	祥麟法轮	有
		四角经幢	无	经幡	有	铜饰	无	石刻、砖雕	有
		仙人走兽	无	壁画	无	其他	——		

陈设																	
	室内	主佛像	十八罗汉像、释迦牟尼像、阿南迦叶像	佛像基座	莲花座												
		法座	无	藏经橱	无	经床	无	诵经桌	无	法鼓	无	玛尼轮	无	坛城	无	其他	——
	室外	旗杆	无	苏勒德	无	狮子	无	经幡	有	玛尼轮	无	香炉	有	五供	无	其他	——
	其他	——															

备注	——						
调查日期	2010/11/15	调查人员	白丽燕、萨日朗	整理日期	2010/11/22	整理人员	萨日朗

喇嘛洞基本概况表2

喇嘛洞召 · 喇嘛洞 · 基本概况

单位:毫米

建筑名称	汉语正式名称	喇嘛洞	俗称	银洞													
概述	初建年	万历四十三年（1615年）	建筑朝向	南	建筑层数	三（第三层）											
	建筑简要描述	藏式															
	重建重修记载	1985年重建															
		信息来源	据召内喇嘛口述														
结构规模	结构形式	砖木混合	相连的建筑	无	室内天井	无											
	建筑平面形式	凸字形	外廊形式	廊柱													
	通面阔	6600	开间数	3	明间	2400	次间	2150	次间	——	梢间	——	尽间	——			
	通进深	8380	进深数	4	进深尺寸（前→后）	1900→2480→1800→2200											
	柱子数量	8	柱子间距	横向尺寸	2150—2400—2250	（藏式建筑结构体系填写此栏，不含廊柱）											
				纵向尺寸	1900—2480—1800—2200												
	其他	——															
建筑主体（大木作）（石作）（瓦作）	屋顶	屋顶形式	藏式平顶	瓦作	琉璃瓦												
	外墙	主体材料	青砖	材料规格	270×130×50	饰面颜色	白色										
		墙体收分	有	边玛檐墙	有	边玛材料	砖										
	斗栱、梁架	斗栱	无	平身科斗口尺寸	——	梁架关系	——										
	柱、柱式（前廊柱）	形式	藏式柱	柱身断面形状	小八角柱	断面尺寸	直径 $D=165$	（在没有前廊柱的情况下，填写室内柱及其特征）									
		柱身材料	木材	柱身收分	无	栌斗、托木	有	雀替	无								
		柱础	无	柱础形状	——	柱础尺寸	——										
	台基	台基类型	普通台基	台基高度	750	台基地面铺设材料	石条、砖										
	其他	——															
装修（小木作）（彩画）	门(正面)	隔扇门	门楣	有	堆经	有	门帘	有									
	窗（正面）	槛窗	窗楣	有	窗套	有	窗帘	无									
	室内隔扇	隔扇	无	隔扇位置	——												
	室内地面、楼面	地面材料及规格	——	楼面材料及规格	不规格木条、水泥地面												
	室内楼梯	楼梯	有	楼梯位置	室外左侧	楼梯材料	铁	梯段宽度	800								
	天花、藻井	天花	有	天花类型	井口天花	藻井	无	藻井类型	——								
	彩画	柱头	有	柱身	无	梁架	有	走马板	无								
		门、窗	无	天花	有	藻井	——	其他彩画	——								
	其他	悬塑	无	佛龛	无	匾额	喇嘛洞										
装饰	室内	帷幔	无	幕帘彩绘	无	壁画	有	唐卡	无								
		经幡	无	经幢	有	柱毯	无	其他	佛帘、哈达								
	室外	玛尼轮	无	苏勒德	无	宝顶	有	祥麟法轮	有								
		四角经幢	无	经幡	有	铜饰	无	石刻、砖雕	有								
		仙人走兽	无	壁画	无	其他	佛帘										
陈设	室内	主佛像	胜乐金刚佛像	佛像基座	莲花座												
		法座	无	藏经橱	无	经床	无	诵经桌	无	法鼓	无	玛尼轮	无	坛城	无	其他	——
	室外	旗杆	无	苏勒德	无	狮子	无	经幡	有	玛尼轮	无	香炉	有	五供	无	其他	——
	其他	——															
备注	——																
调查日期	2010/11/15	调查人员	白丽燕、萨日朗	整理日期	2010/11/22	整理人员	萨日朗										

喇嘛洞基本概况表3

喇嘛洞召 · 喇嘛洞 · 档案照片

照片名称	正前方	照片名称	斜前方	照片名称	门
照片名称	窗	照片名称	柱础	照片名称	匾
照片名称	一层室内正面	照片名称	一层室内侧面	照片名称	一层室内天花
照片名称	二层室内正面	照片名称	二层室内侧面	照片名称	二层室内天花
照片名称	三层室内正面	照片名称	三层室内侧面	照片名称	三层室内局部
备注	——				
摄影日期	2010/11/22	摄影人员	高亚涛		

喇嘛洞基本概况表4

喇嘛洞召正前方

喇嘛洞前廊（左图）
喇嘛洞窗（中图）
喇嘛洞室外楼梯（右图）

喇嘛洞岩画（左图）
喇嘛洞三层室内正面（右图）

6.2 喇嘛洞召·佛爷府

单位:毫米

建筑名称	汉语正式名称	佛爷府	俗称	——													
概述	初建年	——	建筑朝向	南	建筑层数	二											
	建筑简要描述	汉式															
	重建重修记载	1985年重建															
		信息来源	据召内喇嘛口述														
结构规模	结构形式	砖木混合	相连的建筑	无	室内天井	无											
	建筑平面形式	长方形	外廊形式	前廊式													
	通面阔	22330	开间数	7	明间	3370	次间	3150	梢间	3200	次梢间	——	尽间	3150			
	通进深	8350	进深数	3	进深尺寸（前→后）	1200→6100→1050											
	柱子数量	——	柱子间距	横向尺寸	——	（藏式建筑结构体系填写此栏，不含廊柱）											
				纵向尺寸	——												
	其他	——															
建筑主体（大木作）（石作）（瓦作）	屋顶	屋顶形式	硬山屋顶	瓦作	布瓦												
	外墙	主体材料	青砖	材料规格	270×130×50	饰面颜色	灰色										
		墙体收分	无	边玛檐墙	无	边玛材料	——										
	斗栱、梁架	斗栱	无	平身科斗口尺寸	——	梁架关系	——										
	柱、柱式（前廊柱）	形式	汉式	柱身断面形状	圆形	断面尺寸	直径D=220	（在没有前廊柱的情况下，填写室内柱及其特征）									
		柱身材料	木材	柱身收分	无	栌斗、托木	无	雀替	有								
		柱础	有	柱础形状	圆形	柱础尺寸	直径D=350										
	台基	台基类型	复合台基	台基高度	310	台基地面铺设材料	青砖										
	其他	——															
装修（小木作）（彩画）	门(正面)	隔扇门	门楣	无	堆经	无	门帘	无									
	窗（正面）	槛窗	窗楣	无	窗套	有	窗帘	无									
	室内隔扇	隔扇	无	隔扇位置	——												
	室内地面、楼面	地面材料及规格	地砖160×160	楼面材料及规格	不规格木条												
	室内楼梯	楼梯	有	楼梯位置	殿内右北角	楼梯材料	木	梯段宽度	700								
	天花、藻井	天花	有	天花类型	海墁天花	藻井	无	藻井类型	——								
	彩画	柱头	无	柱身	无	梁架	无	走马板	——								
		门、窗	无	天花	无	藻井	——	其他彩画	——								
	其他	悬塑	无	佛龛	有	匾额	无										
装饰	室内	帷幔	有	幕帘彩绘	无	壁画	无	唐卡	有								
		经幡	无	经幢	有	柱毯	无	其他	——								
	室外	玛尼轮	无	苏勒德	无	宝顶	有	祥麟法轮	无								
		四角经幢	无	经幡	有	铜饰	无	石刻、砖雕	无								
		仙人走兽	无	壁画	有	其他	——										
陈设	室内	主佛像	宗喀巴大师	佛像基座	莲花座												
		法座	有	藏经橱	无	经床	有	诵经桌	有	法鼓	有	玛尼轮	无	坛城	无	其他	——
	室外	旗杆	无	苏勒德	无	狮子	无	经幡	有	玛尼轮	无	香炉	有	五供	无	其他	——
	其他	——															
备注	——																
调查日期	2010/11/15	调查人员	白丽燕、萨日朗	整理日期	2010/11/21	整理人员	萨日朗										

佛爷府基本概况表1

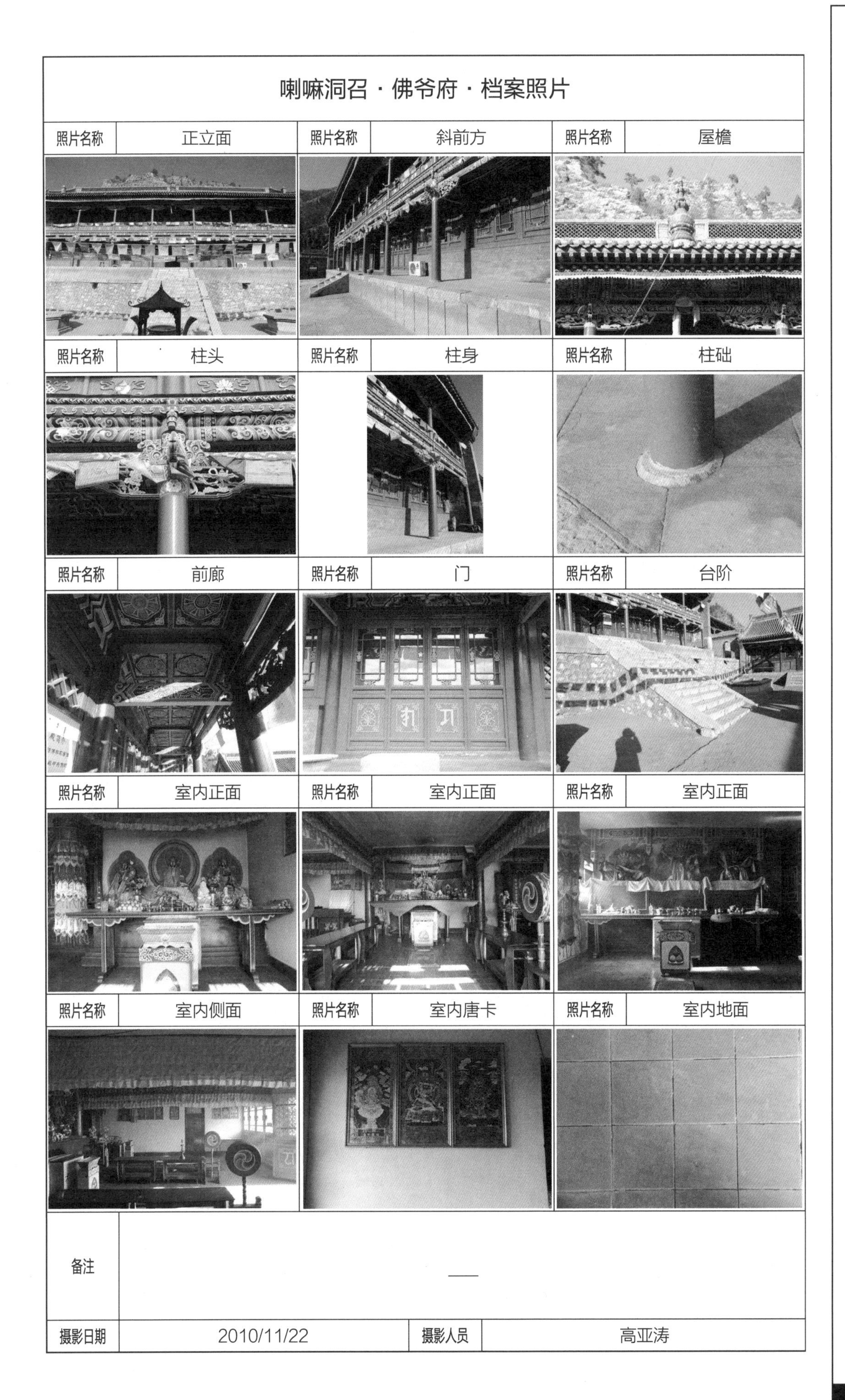

喇嘛洞召・佛爷府・档案照片

照片名称	正立面	照片名称	斜前方	照片名称	屋檐
照片名称	柱头	照片名称	柱身	照片名称	柱础
照片名称	前廊	照片名称	门	照片名称	台阶
照片名称	室内正面	照片名称	室内正面	照片名称	室内正面
照片名称	室内侧面	照片名称	室内唐卡	照片名称	室内地面
备注	—				
摄影日期	2010/11/22	摄影人员	高亚涛		

佛爷府正立面

佛爷府斜前方（左图）
佛爷府柱身（右图）

佛爷府室内侧面

6.3 喇嘛洞召・护法殿

护法殿正立面

6.4 喇嘛洞召・广化寺

广化寺配殿图（左图）
广化寺四臂观音殿（右图）

广化寺天王殿（左图）
广化寺大经堂斜前方（右图）

6.5 喇嘛洞召 · 白塔

聚佛塔正立面

聚佛塔侧立面
（左图）

菩提塔斜前方
（左图）

聚佛塔侧立面
（右图）

图书在版编目（CIP）数据

内蒙古召庙建筑．上册 = INNER MONGOLIA TEMPLE ARCHITECTURE（VOL.1）/ 张鹏举著．-- 北京 ：中国建筑工业出版社，2020.12
ISBN 978-7-112-20818-0

Ⅰ．①内… Ⅱ．①张… Ⅲ．①喇嘛宗一宗教建筑一研究一内蒙古 Ⅳ．①TU-098.3

中国版本图书馆 CIP 数据核字（2019）第 284278 号

本书作为多项国家自然科学基金资助项目的研究成果之一，是对内蒙古地区召庙建筑全面调研和测绘基础上的系统归档以及发生、发展和特征的综述，将成为此类课题后续研究的基础素材。本书适用于建筑学等相关科研人员、高校教师和学生阅读参考。

责任编辑：唐　旭　吴　绫　张　华
文字编辑：李东禧
版式设计：李国保　张　宇
责任校对：王　烨

内蒙古召庙建筑　上册
INNER MONGOLIA TEMPLE ARCHITECTURE（VOL.1）
张鹏举 著
ZHANG PENGJU
*
中国建筑工业出版社出版、发行（北京海淀三里河路 9 号）
各地新华书店、建筑书店经销
北京富诚彩色印刷有限公司印刷
*
开本：880 毫米 ×1230 毫米　1/16　印张：39　字数：1151 千字
2020 年 12 月第一版　2020 年 12 月第一次印刷
定价：358.00 元
ISBN 978-7-112-20818-0
(35145)